FREQUENCY DOMAIN AND STATE SPACE METHODS
FOR LINEAR SYSTEMS

FREQUENCY DOMAIN AND STATE SPACE METHODS FOR LINEAR SYSTEMS

edited by

Christopher I. BYRNES
Arizona State University
Tempe, AZ 85281, U.S.A.

and

Anders LINDQUIST
Royal Institute of Technology
100 44 Stockholm, Sweden

1986

NORTH-HOLLAND
AMSTERDAM · NEW YORK · OXFORD · TOKYO

ISBN: 0 444 70038 2

Publishers:

ELSEVIER SCIENCE PUBLISHERS B.V.
P.O. Box 1991
1000 BZ Amsterdam
The Netherlands

Sole distributors for the U.S.A. and Canada:

ELSEVIER SCIENCE PUBLISHERS COMPANY, INC.
52 Vanderbilt Avenue
New York, N.Y. 10017
U.S.A.

Library of Congress Cataloging-in-Publication Data

Frequency domain and state space methods for linear systems.

Selected papers originally presented at the 7th International Symposium on the Mathematical Theory of Networks and Systems, held at the Royal Institute of Technology in Stockholm, June 10-14, 1985.
Includes index.
1. System analysis--Congresses. 2. State space methods--Congresses. 3. Linear systems--Congresses. I. Byrnes, Christopher I., 1949- . II. Lindquist, Anders. III. MTNS International Symposium (7th : 1985 : Kungl. Tekniska högskolan)
QA402.F74 1986 003 86-9029
ISBN 0-444-70038-2 (U.S.)

PREFACE

With its origins in classical frequency domain control, in the past twenty-five years analytic state space methods have extended the applicability of linear systems theory to a wide variety of control applications for such tasks as decoupling, stabilization and pole placement of multivariable systems. On the other hand, in the past decade the algebraic approach to the state space theory has inspired the application of methods from algebra, algebraic geometry and complex analysis to the parallel development of a truly multivariable frequency domain theory for linear systems, leading to a virtual renaissance of frequency domain methods. Inevitably, these two approaches (which use a common algebraic, analytic and geometric framework) have lead to significant new advances in the solution of many problems in classical and modern linear control with their methodology and insights permeating many aspects of deterministic and stochastic systems theory.

The papers in this volume, which is devoted to such important developments in linear systems theory and its applications, were selected from the invited and contributed papers presented at the 7th International Symposium on the Mathematical Theory of Networks and Systems held at the Royal Institute of Technology in Stockholm on June 10-14, 1985.

We would like to take this opportunity to thank the following research agencies for their generous support of MTNS-85: the National Swedish Board for Technical Development (STU), the Office of Navel Research Branch Office, London (ONRL), the Research Institute of National Defence (FOA), the Swedish Institute for Applied Mathematics (ITM), the Swedish Natural Science Research Council (NFR) and the USAF European Office of Aerospace Research and Development (EOARD).

Christopher I. Byrnes and Anders Lindquist
Stockholm, January 29, 1986

CONTENTS

1. ALGEBRAIC SYSTEMS

Frequency Domain and State Space Methods for Linear Systems
C.I. Byrnes and A. Lindquist (Editors)

On discrete n-dimensional linear constant systems

L. Baratchart

INRIA
Route des Lucioles
06560 Valbonne FRANCE

ABSTRACT

This paper deals with realization theory of discrete, multi-input multi-output, linear, constant n-d systems over a field. Global and local state, rationality, and relations between them are discussed. A link with partial differential equations is briefly mentioned.

1. Introduction

For over ten years, discrete linear system theory has been intensively studied and is still an active field of research. A seemingly natural generalization of this theory to signals with n indices has been called call n-dimensional (in short: n-d) system theory. Comparatively, it has raised moderate interest. Some work has concerned n=2, but almost none with the general case. Tentative explanations include the following.

First, the connection between 1-d sytems and linear differential equations, especially their identification and control, has been a strong motivation for the development of the theory, while the relevance of a potential n-d theory to physical problems is less clear. Image processing apparently motivated much of the 2-d work (e. g. [1,2,6,16,17,18,19]). However, the significance of linearity is not so clear in this context, except to the extent that it makes computations easier. In fact, it seems [22] that the simplest 2d-model, recognized as the easy case of the theory, has also proved the most useful in approximation and synthesis [1,2,19]. A link between n-d systems and partial differential equations is suggested in [8,12].

Second, since most results in the 1-d theory over a field K are related to the Euclidean character of $K[Z]$, these tools are no longer available in the n-d case which really belongs to commutative algebra. For $n=2$, work has been done by considering $K[Z_1,Z_2]$ as $K[Z_1][Z_2]$, and exploiting the principality of the base ring. We may cite [12,17] for the factorization of polynomial matrices, while [6] uses realization theory over rings which is especially powerful in the principal case [21]. Some more direct approaches, like [1,2,9,10,14,16,18], have also been made to examine realization theories of rational 2-d systems that are suggested by the 1-d case. They introduce several "local" state space models and minimality notions for them, as opposed to a "global" state space which is always infinite dimensional. Though this is not our concern here, we may note that stability has also been studied ([9,15]) in connection with the location of poles.

The object of the paper is to discuss some aspects of the general situation when dealing with an n-d system over a field with m inputs and p outputs. We parallel more or less the development of most papers (e.g.[11,16]) dealing with 2-d realization theory: we speak of global state, local state, and rationality of the transfer function. However, our definition of the global state allows us to explain the special role played by Attasi's model [1,2]. We also give a formal definition of a local state with respect to the input-ouput behaviour. In the next section, we introduce the formal setting and discuss rationality in some details. At the end of the paper, we describe briefly a link with a partial differential analytic problem in the small.

Our main results are:

Rational fractions have invertible denominators.

Global state is finite dimensional if and only if the transfer function is rational and its denominator splits.

Local state is finite dimensional if and only if the transfer function is rational.

2. Notations and preliminaries

We let K be a field, $Z=(Z_1,...,Z_n)$ be variables. We let further $K[Z]=K[Z_1,...,Z_n]$ be the polynomial ring, $K(Z)$ the rational field, $A((Z))=K[[Z_1^{-1},...,Z_n^{-1}]]$ the power series ring, $M=(Z_1^{-1},\ldots,Z_n^{-1})$ its maximal ideal, and $B((Z))$ the quotient ring of $A((Z))$ with respect to the multiplicative system $U=\{Z_1^{-i_1}\cdots Z_n^{-i_n}, (i_1,\ldots,i_n)\in\mathbb{N}^n\}$. Elements of $B((Z))$ are thus power series: $S=\sum_{(e_1,\ldots,e_n)\in\mathbb{Z}^n} S_{e_1,\ldots,e_n} Z_1^{-e_1}\cdots Z_n^{-e_n}$ satisfying the following hypothesis:

there exists $k\in\mathbb{Z}$, *such that* : $S_{e_1,\ldots,e_n}\neq 0$ *implies* $e_i>k, \forall i\in\{1,...,n\}$

Elements of M will be called strictly proper. If S has *no* strictly proper monomials, we will say it is antiproper. The terminology extends to matrices in the obvious way. If $n>1$, $B((Z))$ is no longer a field. Units are of the form : $Z_1^{i_1}\cdots Z_n^{i_n}S$, where $S\in A((Z))$ has nonzero constant term, the i_j being in $\mathbb{Z}$. $B((Z))$ contains $K[Z]$, and a polynomial will be said to split if it can be written as a product : $Q(Z)=Q_1(Z_1)Q_2(Z_2)\cdots Q_n(Z_n)$ with $Q_i\in K[Z_i]$.

We define an n-d system to be a linear map: $B((Z))^m \rightarrow B((Z))^p$ whose matrix T is strictly proper. We call T the transfer function of the system.

Now, in the previously cited work, concerned with n=2, rational fractions are naturally written with a denominator which is a unit in $B((Z))$, that is a polynomial of the form: $Z_1^{k_1}\cdots Z_n^{k_n}+\sum_{i_j\le k_j,\, i_r<k_r \text{ for some } r} a_{i_1,\ldots,i_n}Z_1^{i_1}\cdots Z_n^{i_n}$. Strict properness means then that the numerator N is subject to $deg_{Z_i}N<k_i$. Strangely enough, it is never asked wether the denominator *has* to be a unit for a reduced fraction to be strictly proper. This is a true fact which is not so trivial when $n>1$:

Proposition 1 : If P and Q are coprime polynomials such that $\frac{P}{Q}$, which belongs to the quotient field of $B((Z))$, actually lies in $B((Z))$, then Q is a unit in $B((Z))$.

proof: We denote $K[Z_1^{-1},\ldots,Z_n^{-1}]$ by $K[Z^{-1}]$ for simplicity. Multiplying respectively P and Q by suitable members of U, We obtain a new fraction $\frac{N}{D}$, where N and D are in $K[Z^{-1}]$, and are coprime in that ring. Let J be the ideal generated by the Z_i^{-1}'s in $K[Z^{-1}]$. If f is an irreducible factor of D in $K[Z^{-1}]$ lying in J, the principal ideal (f) is closed for the J-adic topology ([23] vol 1). Therefore, the completion of $\frac{K[Z^{-1}]}{(f)}$ is naturally isomorphic to $\frac{A((Z))}{fA((Z))}$ ([23] vol 2). Since N is not a zero divisor in the former ring, it is not a zero divisor in the latter ([23] vol 2). Since a polynomial not in J is a unit in $A((Z))$, this means (recall that $A((Z))$ is factorial) that N and D are coprime in $A((Z))$; since $\frac{N}{D}\in A((Z))$, D is a unit in that ring and Q is a unit in $B((Z))$. Q.E.D.

3. Global state-space and Attasi's model.

For any $S\in B((Z))$, we define $\Theta(S)$ to be its strictly proper part. By a state module, or a global state-space, we mean a factorization: $K[Z]^m \rightarrow X \rightarrow M^p$ of the map $\Theta(T.)$ into a surjective $K[Z]$-module homomorphism followed by an injective one. Abstract nonsense shows that this is well-defined up to an isomorphism, so that we sometimes speak of "the" state-module X of the system.

We let now $W = Ker\,\Theta(T.) = \{P \in K[Z]^m, TP \text{ is antiproper } \}$, and define X to be $\frac{K[Z]^m}{W}$. We denote by π the canonical map: $K[Z]^m \to X$. It is obvious that X, together with π and the induced map, is a state-module for T. If $u=\sum u_{j_1,\ldots,j_n} Z_1^{-j_1} \cdots Z_n^{-j_n}$ is an input in $B((Z))$, and $y=Tu$ the associated output, the state $x_{i_1,\ldots,i_n}$ at $(i_1,\ldots,i_n)$ will be the canonical image by π of the "past input": $\sum_{j_r \le i_r} u_{j_1,\ldots,j_n} Z_1^{i_1-j_1} \cdots Z_n^{i_n-j_n}$. We define then:

$$F_i : X \to X \quad by \quad F_i(x)=Z_i x$$
$$H \;: X \to K^p \quad by \quad H(\pi(P))=\left[TP\right]_{1,1,\ldots,1}$$

The F_i 's are pairwise commuting $K[Z]$-morphisms, while H is K-linear. It is easily seen that:

$$x_{i_1,\ldots,i_{r-1},i_r+1,i_{r+1},\ldots,i_n} = F_{i_r}(x_{i_1,\ldots,i_n})$$
$$+\ \pi\left[\sum_{j_k \le i_k \text{ if } k \ne r,\, j_r=i_r+1} u_{j_1,\ldots,j_n} Z_1^{i_1-j_1} \cdots Z_{r-1}^{i_{r-1}-j_{r-1}} Z_{r+1}^{i_{r+1}-j_{r+1}} \cdots Z_n^{i_n-j_n}\right]$$

$$y_{i_1,\ldots,i_n}=H(x_{i_1-1,\ldots,i_n-1})$$

The state-module we define here is not the same than the global state-space introduced in the 2-d theory by the previously mentioned authors: instead of allowing to compute the output in the absence of new inputs in the future, it only allows it when all nonzero inputs are in the past. The difference comes from the fact that the future and the past of a given point fail to exhaust the set of indices. Our X is finitely generated over $K[Z]$, and finite-dimensional over K in a special case. This is the object of the next result:

Theorem 1: X is finite dimensional over K if and only if $T=\frac{N}{Q}$ with $N \in K[Z]^{m \times p}$, and Q splits in $K[Z]$. If $Q=\prod_{i=1}^{n} Q_i$, with $Q_i \in K[Z_i]$, and $I=(Q_1,\ldots,Q_n)$ is the ideal generated by the Q_i's in $K[Z]$, then $W=(IK[Z]^m : N)$. In this case, we may identify the F_i's and H with matrices, and denoting by G the matrix associated with the restriction of π to K^m, we have:

$$T=H\left[Z_1 Id-F_1\right]^{-1}\left[Z_2 Id-F_2\right]^{-1},\ldots,\left[Z_n Id-F_n\right]^{-1}G \tag{1}$$

proof: Since a full proof of the theorem can be found in [3], we will only sketch the proof of the first assertion. Suppose with the above notations that Q splits. X is a quotient module of $\frac{K[Z]^m}{IK[Z]^m}$, and this latter module is clearly finite-dimensional over K. Conversely, if X is finite-dimensional over K, it is also *a fortiori* finitely generated over $K[Z_i]$, and by a well-known theorem [5] is a torsion module over $K[Z_i]$. This gives for each i a polynomial $Q_i \in K[Z_i]$ such that $TQ_iK[Z]^m$ contains only antiproper series. Necessity is then a consequence of the following lemma:

lemma: If $S \in M$, and for each $i \in \{1,\ldots,n\}$, $Q_i \in K[Z_i]$ is such that $Q_i S$ is antiproper, then $S=\frac{R}{\prod_i Q_i}$ with $R \in K[Z]$.

The proof of the lemma is carried out by induction. Q.E.D.

We want to point out that the assertion just proved is not new, and that the splitting condition is very familiar in the bilinear realization theory of polynomial systems [20]. In fact, if $dim_K X<\infty$ is interpreted as the finiteness of a Hankel rank, the result is a consequence of [7], already mentioned in [11]. It can also be obtained by a more elementary and combinatorial approach as in [20]. These papers treat the scalar case only, but as long as finiteness is concerned, the passage to the general case is straightforward. The proof we just sketched, however, is shorter, and applies for any

m and p. It avoids reference to recognizable power series and remains notationaly simple. From (1), one can deduce, by an appropriate factorization of the Hankel matrix (see [2]), that $dim_K X$, which is obviously bounded from below by the Hankel rank is in fact equal to it.

The 2-d (scalar) state-space model introduced in [1,2] corresponds precisely to the case when X is finite dimensional. The above results allows one to develop the same theory from a polynomial point of view for arbitrary m, p, n. Note that the density theorem becomes trivial in this context, and that the hypothesis that the F_i's commute pairwise is superfluous since,by the theorem, this can always be achieved (for $m=p=1$, this is also a consequence of [7], mentioned in [11]). The computation of a minimal realization from the transfer function becomes effective *via* standard-bases algorithms [13]. Note also that since X is of finite length, it breaks up into a direct sum of its localized at associated primes [4]. This is an analogue of the cyclic decomposition of the 1-d case.

4. Local state-space.

We adopt the following convention: if $\varepsilon = (\varepsilon_i)$ belongs to $D = \{0,1\}^n$, we denote by $l(\varepsilon)$ the integer $\sum_i \varepsilon_i$, by $1-\varepsilon$ the element $(1-\varepsilon_i)$ of D, and by Z^ε the monomial $Z_1^{\varepsilon_1} \cdots Z_n^{\varepsilon_n}$. By a local state space for the system associated with T, we mean a *K−vector space* Y together with a linear map $f : K[Z]^m \to Y$ and a section g of f (i.e. a linear map $: Y \to K[Z]^m$ such that $fg = id$) satisfying the following conditions:

$Ker f$ is contained in $Ker H\pi$, and if we let $h = gf$, we have for any $v \in K[Z]^m$:

$$\sum_{\varepsilon \in D} (-1)^{l(\varepsilon)} Z^\varepsilon h(Z^{1-\varepsilon} v) \in Ker f \tag{2}$$

Note that X is an obvious local state-space with $f = \pi$, and g any section. The local state $w_{i_1,\ldots,i_n}$ at $(i_1,\ldots,i_n)$ will be the image by f of the "past input": $\sum_{j_r \le i_r} u_{j_1,\ldots,j_n} Z_1^{i_1-j_1} \cdots Z_n^{i_n-j_n}$. The reason why introducing this local space lies in the following theorem.

Theorem 2: A finite K-dimensional local state space exists if and only if T is rational. Choosing some basis in it, one gets then matrices $B, C, A_\varepsilon, \varepsilon \in D - \{(0,\ldots,0)\}$, such that

$$w_{i_1,\ldots,i_n} = \sum_\varepsilon A_\varepsilon w_{i_1-\varepsilon_1,\ldots,i_n-\varepsilon_n} + B u_{i_1,\ldots,i_n} \tag{3}$$

$$y_{i_1,\ldots,i_n} = C w_{i_1-1,\ldots,i_n-1}$$

Moreover, we have:

$$T = C\left[Z_1 \cdots Z_n Id - \sum_\varepsilon Z^{1-\varepsilon} A_\varepsilon\right]^{-1} B \tag{4}$$

indications on the proof: If T is rational, and q is a common denominator for its entries, Euclidean division by q with respect to each Z_i yields a remainder whose polynomial part is well-defined, and defines a map f which satisfies all our requirements with $Y = \{v \in K[Z]^m, deg_{Z_i} v < deg_{Z_i} q\}$. Conversely, (3) can be computed from (2), and (4) follows then from (3), implying obviously that T is rational. Q.E.D.

When a denominator of T splits, X is already a finite-dimensional local state-space, and the above construction yields $A_\varepsilon = (-1)^{1+l(\varepsilon)} \prod_i F_i^{\varepsilon_i}$, so that (4) is consistent with (1). According to [11] and the example presented therein, the minimality of a (2-d) state satisfying updating equations like (3) is apparently not related to the concept of controllability, nor of observability. This is perhaps not too surprising, since the local state space is *not* a $K[Z]$-module, but only a K-vector space. We will not discuss this matter here, but let us point out that the local space constructed in the above proof is generally not minimal.

5. A Link with an analytic differential problem

This is an attempt to motivate the study of n-d systems by something else than image processing. We first adopt some notations. If $e = (e_i)$ is in $\mathbb{Z}^n$, we abbreviate by $e!$ the integer $e_1!...e_n!$, by Z^e the monomial $Z_1^{e_1} \cdots Z_n^{e_n}$, and by Z^{-e} its inverse. We denote by O_p the space of functions that are analytic in a neighborhood of 0 in $\mathbb{R}^n$ with values in $\mathbb{R}^p$. To every member of O_p defined in a neighborhood of 0 by $f(Z) = \sum_{e_i \geq 0} \frac{f_e}{e!} Z^e$, we associate a member of $A((Z))$ (here K=$\mathbb{R}$) defined by $S(f) = \sum_e f_e Z_{-e}$ (this is Fourier transform really). If S belongs to $B((Z))$, we let $PP(S)$ be its proper part (that is $\Theta(S)$ plus its constant term), and $E(S) = S - PP(S)$. Finally, if $M(\partial)$ is a matrix whose entries are polynomials in the $\frac{\partial}{\partial Z_i}$'s, we let M be the polynomial matrix obtained by replacing $\frac{\partial}{\partial Z_i}$ by Z_i. Let now $M(\partial)$ be regular, of size $p \times p$.

Theorem 3 : The partial differential equation $M(\partial)f = u$, where $f \in O_p$ is the unknown subject to the condition $E(S(f)) = E(S(f_0))$, admits a solution for every u and f_0 in O_p if and only if M^{-1} is proper (i.e. a constant matrix plus a transfer function in the sense of our definition). The solution is then given by $S(f) = M^{-1}\left[PP(S(u)) + E(M\ S(f_0))\right]$

We will not prove the theorem here. Let us mention that if det$M$ splits, the equation breaks up into a cascade of ordinary differential equations. In this context, to control the system means to control behaviour of the derivatives of f at 0.

6. References

[1] Attasi S. "Systèmes linéaires homogènes à deux indices", Rapport Laboria n0 31, 1973.

[2] Attasi S. "Modélisation et traitement des suites à deux indices", Rapport Laboria 1975.

[3] Baratchart L. "Sur la représentation de Nerode des systèmes multiindiciels", submitted to C. R. Acad. Sc., Paris.

[4] Bourbaki N. "Algèbre commutative", Masson, Paris 1985.

[5] Bourbaki N. "Algèbre" chap. 7, Masson, Paris 1981.

[6] Eising R. "2-d Systems, an algebraic approach", Ph.D. Thesis, Mathematish Centrum, Amsterdam, 1979.

[7] Fliess M. "Matrices de Hankel" J. Math. pures et appl., 53,1974, p.197 -224.

[8] Fliess M. "Une théorie fonctionnelle de la réalisation en filtrage multidimensionnel, échantillonné, récurrent", Inf. and Control, 43, 1979, p. 338-355.

[9] Fornasini E. "Research topics in linear systems on partially ordered time sets" in "Analyse des systèmes" Asterisque 75-76, 1980.

[10] Fornasini E. & Marchesini G. "Algebraic realization theory of 2-d filters" in "Variable structure Systems", A. Ruberti & R. Mohler Eds, Springer 1975.

[11] Fornasini E. & Marchesini G. "State space realization theory of 2-d filters', IEEE Trans. Aut. Cont., vol. AC-21 Aug.1976.

[12] Frost M. G. & Storey C. "Equivalence of a matrix over $R[s,z]$ with its Smith form", Int. J. Cont., vol 28 n0 5.

[13] Galligo A. "Algorithmes de construction de bases standards", Preprint Univ. Nice 1983.

[14] Givone D.D. & Roesser R.P. "Minimization of multidimensional linear iterative circuits", IEEE Trans. Comput., vol C-22, July 1973.

[15] Huang T.S. "Stability of 2-d recursive filters", IEEE Trans. Audio Elect. , vol. AU-20, June 1972.

[16] Kung S.Y., Lévy B.C., Morf M., Kailath T. "New results in 2-d systems theory, part II : 2-d state space models", Proc. IEEE, vol. 65, n0 6, June 1977.

[17] Morf M., Lévy B.C., Kung S.Y. "New results in 2-d systems theory, part I : 2-d polynomial matrices", Proc. IEEE, vol. 65, n0 6, June 1977.

[18] Roesser R.P. "A discrete state space model for linear image proccessing", IEEE Trans. Aut. Cont., vol. AC-20, Feb. 1975.

[19] Rosencher E. "Approximation rationnelledes filtres à 1 ou 2 indices: une approche Hilbertienne", Thèse de docteur ingénieur, Paris IX, 1978.

[20] Rugh W.J. "Nonlinear System Theory: The Volterra/Wiener Approach", Johns Hopkins Univ. Press, Baltimore, MD, 1981.

[21] Rouchaleau Y. "Linear discrete time finite dimensional dynamical systems over some classes of commutative rings", Ph.D. Thesis, Stanford, 1972.

[22] Willsky A.S. "Digital Signal Processing and Control and Estimation Theory", MIT Press, Cambridge Mass., 1979.

[23] Zariski O. & Samuel P. "Commutative algebra", 2 Vol., Springer 1979.

Frequency Domain and State Space Methods for Linear Systems
C.I. Byrnes and A. Lindquist (Editors)

FACTORIZATIONS OVER THE RING OF PROPER RATIONAL STABLE FUNCTIONS APPLICATION TO DECOUPLING

C. COMMAULT and J.M. DION

Laboratoire d'Automatique de Grenoble
(Laboratoire associé au C.N.R.S. n° 228)
E.N.S.I.E.G. - I.N.P.G.
B.P. 46 - 38402 Saint-Martin-d'Hères
FRANCE

In this paper the use of factorizations over the ring of proper rational functions for solving some decoupling problems is presented. We consider linear multivariable systems whose transfer matrix T(s) is a pxm proper rational matrix. The main tools for this study are Smith and Hermite forms over the P.I.D. of proper rational stable functions. The use of these tools allows to exhibit quite easily fundamental invariants for the problem under consideration, and to make proofs clear and consise.

1. INTRODUCTION

In the last few years, there has been an increasing interest in factorizations of proper rational transfer functions over useful rings as the ring of proper rational functions or proper stable rational functions. Three types of factorizations are now classical namely :

- The Matrix Fraction Description of transfer matrices over the ring of proper rational stable functions, which allowed to solve interesting problems as the parametrization of all stabilizing compensators. See [1], [2].

- The Hermite form, which is the simplest form achievable by right multiplication with unimodular matrices. See [3].

- The Smith form, which is the simplest form achievable by right and left multiplication with unimodular matrices. See [4], [5].

The use of these factorizations allows to exhibit quite easily fundamental invariants of the transfer matrix for solving classical control problems.

In this paper we will use these factorizations to solve in a unified way various decoupling problems. For authoritive references representing important steps in the development of decoupling theory. See [6], [7], [8].

We will consider several decoupling problems and various types of proper compensators. All the considered compensators must satisfy the following properties:

i) The transfer matrix of the compensated system has the same rank as T(s).

ii) The compensated system is input-output stable (external stability) and no unstable pole-zero concellation occurs (internal stability).

Roughly speaking, the situation may be summarized as follows :

i) Block triangular decoupling

The solvability condition is given in terms of independance among the rows of T(s). The important point is that this condition is the same for any type of admissible compensator i.e. pure precompensators, precompensators implementable by dynamic (or static) state feedback on a minimal realization of T(s).

ii) Block diagonal decoupling

By pure precompensation the solvability condition is also expressed in terms of independance among the transfer matrix rows. When dealing with feedback implementations of these precompensators the situation becomes considerably more complex. This problem is not solved for all cases. We consider two important situations.

First we consider compensators which are implementable by feedback with non singular input transformation. The decoupling condition is given in terms of the "infinite" and "unstable" independance of the transfer matrix rows. This condition turns out to be the same for static and dynamic state feedback.

Then we consider the row by row decoupling problem with possibly singular input transformation. For stable dynamic state feedback implementable compensators the decoupling condition is roughly speaking the following : "the number of inputs must be large enough in order to compensate some rank defection for infinite and unstable frequencies".

The aim of the paper is to exhibit the important underlying structural properties for each problem, in particular the role of infinite and unstable zeros as well as some decoupling invariants will be emphasized.

The paper is organized as follows :

In Section 2 we will briefly recall some mathematical tools concerning factorizations of transfer matrices over the P.I.D. of proper rational stable functions. In Section 3, we will study the block triangular decoupling. Section 4 is devoted to the analysis of block diagonal decoupling by state feedback with non singular input transformation, in this section we consider also the row by row decoupling problem with possibly singular input transformation.

2. PRELIMINARIES

Let C_g be an arbitrary symetric subset of the complex plane intersecting the real axis. A rational function is stable if it has all its poles in C_g. Let $R_{ps}(s)$ be the ring of proper stable rational functions.

As in the case of polynomials we can build on $R_{ps}(s)$, Smith forms, Hermite forms, left and right coprime factorizations. This results from the existence on this ring of a "degree". $R_{ps}(s)$ is an Euclidian domain, defining the degree δ of $f(s) \in R_{ps}(s)$ as

$\delta(f(s)) = d_1(f(s)) + d_2(f(s))$ where $d_1(f(s))$ is the infinite zero order of $f(s)$ and $d_2(f(s))$ is the number of unstable zeros of $f(s)$ counted with their multiplicity.

We briefly recall now what are the Smith forms and Hermite forms over $R_{ps}(s)$.

Theorem 1 [9]

Let T(s) be a pxm proper rational stable matrix. T(s) can be factorized in $T(s) = B_1(s)\, \Lambda(s)\, B_2(s)$ where $B_1(s)$ and $B_2(s)$ are bicausal and bistable and

$$\Lambda(s) = \begin{bmatrix} \Delta(s) & 0 \\ 0 & 0 \end{bmatrix}$$

where $\Delta(s) = \text{diag}\,(f_1(s),\ldots, f_r(s))$.

r is equal to the rank of T(s) over the rational functions.

Moreover $f_i(s)$ divides $f_{i+1}(s)$ for $i = 1,\ldots, r-1$
$\Lambda(s)$ is called a <u>Smith form</u> of T(s) over $R_{ps}(s)$.

Notice that $\Delta(s)$ contains all the system's infinite and unstable zeros. The $f_i(s)$'s are defined up to units of $R_{ps}(s)$, in order to get the uniqueness of $\Lambda(s)$ it is possible to impose that $f_i(s)$ does not contain any unit except 1.

Recall now the Hermite form over $R_{ps}(s)$.

<u>Theorem 2</u> [3], [10]

Let T(s) be a pxm surjective proper stable rational matrix. T(s) can be factorized in $T(s) = H(s)\,B(s)$ where B(s) is a bicausal and bistable matrix (unit of $R_{ps}(s)$) and $H(s) = [\vec{H}(s), 0]$

$$\text{where } \bar{H}(s) = \begin{bmatrix} h_{11}(s) & & 0 \\ & \ddots & \\ h_{ij}(s) & & h_{pp}(s) \end{bmatrix}$$

Furthermore the degree of $h_{ij}(s)$ is lower than the degree of $h_{ii}(s)$ for $j < i$. H(s) is called a Hermite form of T(s) over $R_{ps}(s)$.

H(s) is non unique, it is defined up to units of $R_{ps}(s)$. Uniqueness of H(s) may be obtained by adding some conditions see [3].

Consider the following example [3].

$$T(s) = \begin{bmatrix} \dfrac{1}{s+1} & \dfrac{1}{s+2} \\[2ex] \dfrac{1}{s+3} & \dfrac{1}{s+5} \end{bmatrix}$$

Choose the stability domain as the open left half plane

The Hermite form of T(s) over $R_{ps}(s)$ is

$$H(s) = \begin{bmatrix} \dfrac{1}{\pi} & 0 \\[2ex] \dfrac{s+(a-1)/2}{\pi^2} & \dfrac{s-1}{\pi^3} \end{bmatrix} \quad \text{where } \pi = s+a \text{ is any stable polynomial}$$

The Smith form of T(s) over $R_{ps}(s)$ is :

$$S(s) = \begin{bmatrix} \frac{1}{\pi} & 0 \\ 0 & \frac{s-1}{\pi^3} \end{bmatrix}$$

Notice that $S(s)$ exhibits the two infinite zeros (of respective order 1 and 2) and the unstable zero (located at $s = 1$) of $T(s)$.

In the sequel we will use a generalization of the interactor matrix defined in [11].

Definition 1

Let $T(s)$ be a (pxm) proper rational stable matrix. Consider $H(s) = [\bar{H}(s), 0]$ a Hermite form of $T(s)$ over $R_{ps}(s)$. The rational matrix $I_g(s) = \bar{H}^{-1}(s)$ is called a generalized interactor of $T(s)$ over $R_{ps}(s)$.

If one considers $H(s)$ the Hermite form of $T(s)$ over $R_{ps}(s)$ we will speak of the generalized interactor of $T(s)$.

In this paper we will focus our attention on decoupling compensators which are feedback implementable. This type of solution is highly desirable from a practical point of view. In order to do this we need some complements about feedback compensators.

Let $T(s)$ be a pxm strictly proper rational transfer matrix and (A, B, C) be a realization of $T(s)$ i.e.

$$\dot{x} = Ax + Bu \qquad x \in R^n,\ u \in R^m$$
$$y = Cx \qquad y \in R^p$$
with $$T(s) = C(sI - A)^{-1} B$$

In the following we will restrict our attention to pxm strictly proper transfer matrices $T(s)$ with null static kernel i.e. $\mathrm{Ker}(T(s)) \cap R^m = 0$. This in order to give a unified presentation, although some results hold in more general cases.

There is no serious restriction in considering only transfer matrices with zero static kernel, since this can always be achieved by eliminating the non independant inputs.

In the state space framework and for a strictly proper transfer matrix $T(s)$ it can be proved that :

$T(s)$ possesses a null static kernel if and only if B is injective in any realization (A, B, C) of $T(s)$.

We consider the effect of a dynamic state feedback defined by $u = F(s)\,x + Gv$ where $F(s)$ is a mxn proper rational transfer matrix and G is a square full rank constant matrix.

The closed-loop transfer matrix is then.

$$\begin{aligned} T_{FG}(s) &= C(sI - A - BF(s))^{-1} BG \\ &= T(s)\,(I - F(s)\,(sI - A)^{-1} B)^{-1} G \\ &= T(s)\,B(s) \end{aligned}$$

where $B(s)$ is easily seen to be a bicausal matrix.

The converse problem was studied in [12] and [13]. More precisely we will use the following theorem from [13], which is below reformulated using a more traditional terminology.

<u>Theorem 3</u>

Let T(s) be a pxm strictly proper rational matrix with null static kernel and (A, B, C) be a realization of T(s). Let C(s) be a mxn proper rational compensator, there exists a mxn proper rational matrix F(s) and a square full rank constant matrix G such that

$$T(s)\,C(s) = C(sI - A - B\,F(s))^{-1}\,BG$$

if and only if C(s) is bicausal.

The effect of a bicausal precompensator is then equivalent to the effect of a dynamic state feedback with non singular input transformation G. This is true in any realization and in particular in a minimal one. It is shown in [12] that the effect of a bicausal precompensator is also equivalent to the effect of a static state feedback, but acting on a possibly non minimal realization of T(s).

When dealing with stability the above theorem can be stated as follows.

<u>Theorem 4</u>

Let T(s) be a pxm strictly proper stable rational matrix with null static kernel and (A, B, C) be a stable realization of T(s). Let C(s) be a mxn proper rational compensator. C(s) is dynamic state feedback implementable (u = F(s) x + Gv with G regular) with stability (internal and external) on the realization (A, B, C) if and only if C(s) is bicausal and stable.

The proof is made by using a right coprime factorization of T(s) and some results of [1] see [14].

We will consider in the following several decoupling problems. In order to avoid trivialities we will require the compensated system to be as "output controllable" as the initial system.
We will say that the compensator C(s) is <u>admissible</u> if

$$\text{rank } T(s)\,C(s) = \text{rank } T(s)$$

This admissibility condition is equivalent to the preservations of the C^∞ controlled output trajectories.

When dealing with stability we will require the compensated system to be input-output stable (external stability) and that no unstable cancellation occurs (internal stability).

3. BLOCK TRIANGULAR DECOUPLING

First we will define the problem under consideration.
Let T(s) be a (pxm) proper rational matrix and let $(p_1, \ldots, p_k)$ be a given set of non zero positive integers satisfying

$$\sum_{i=1}^{k} p_i = p.$$ The system with transfer matrix T(s) is said to be stable <u>block triangular decoupled relatively to the partition $\{p_i\}$</u> if there exist non zero positive integers $m_1 \ldots m_k$ satisfying

$$\sum_{i=1}^{k} m_i = m$$ such that T(s) has the block triangular form :

$$T(s) = \begin{bmatrix} T_{11}(s) & & 0 \\ \vdots & \ddots & \\ T_{k1} & \cdots & T_{kk}(s) \end{bmatrix}$$

with $T_{ij}(s) \in R_{ps}^{p_i \times m_j}(s)$ for $i \geq j$

$T_{ij}(s) = 0$ for $i < j$

$T_{ii} \neq 0$ for any i

An admissible stable block triangular decoupling compensator C(s) must be such that :

- C(s) is admissible and proper
- T(s) C(s) is stable block triangular decoupled
- C(s) is stable.

We will consider admissible stable block triangular decoupling precompensators implementable by static state feedback u = F x + Gv on a minimal realization of T(s). This problem can be solved easily using factorizations over $R_{ps}(s)$. One has the following result.

Theorem 5 [15]

Let T(s) be a (pxm) proper rational transfer matrix decomposed according to the partition $(p_1, \ldots, p_k)$ as

$$T(s) = \begin{bmatrix} T_1(s) \\ \vdots \\ T_k(s) \end{bmatrix}$$

with $T_i(s)$ are $p_i$xm proper rational matrices.

Consider the block triangular decoupling problem relatively to $(p_1, \ldots, p_k)$. The following statements are then equivalent :

i) There exists an admissible decoupling proper precompensator.

ii) There exists an admissible decoupling static state feedback compensator implementable on a minimal realization of T(s).

iii) There exists a stable admissible decoupling static state feedback compensator implementable on a minimal realization of T(s).

iv) $$\dim \left(\bigcap_{i=1}^{j-1} \ker T_i(s) \right) > \dim \left(\bigcap_{i=1}^{j} \ker T_i(s) \right) \text{ for } j = 2 \ldots k$$

and $\quad \dim (\ker T_1(s)) < m.$

Roughly speaking this theorem gives rank conditions (iv) which are fairly weak for T(s) to be block triangular decouplable with internal and external stability by a static state feedback compensator.

The solvability condition is given in terms of independance among the rows of T(s). The important point is that this condition is the same for any type of admissible compensators i.e. pure precompensators, precompensators implementable by dynamic or static state feedback with stability or not on a minimal realization of T(s).

The proof is given in [15]. The necessity proof is direct. We will give the main idea of the sufficiency. First make the system stable by static state feedback. Suppose that condition (iv) is satisfied, it is possible to construct a right coprime polynomial factorization of T(s).

$$T(s) = N(s)\, D^{-1}(s)$$

where N(s) has the prespecified block triangular structure. Then factorize $D^{-1}(s) = H(s)\, B(s)$ where H(s) is the Hermite form of $D^{-1}(s)$ over $R_{ps}(s)$ and B(s) is bicausal and bistable.

When using the precompensator $C(s) = B^{-1}(s)$ one obtains

$T(s)\, C(s) = N(s)\, H(s)$ which is triangular block decoupled with stability then using a result of [12] the static state feedback implementation follows.

Consider now the block diagonal decoupling problem.

4. BLOCK DIAGONAL DECOUPLING

The problem formulation is similar to the preceding one, the difference is that we impose now the compensated system T(s) C(s) to be diagonal bloc decoupled.

More precisely, there exist non zero positive integers $m_1, \ldots, m_k$ satisfying

$\sum_{i=1}^{k} m_i = m$ such that T(s) C(s) has the block diagonal form :

$$T(s)\, C(s) = \bar{T}(s) = \begin{bmatrix} \bar{T}_{11}(s) & & 0 \\ & \ddots & \\ 0 & & \bar{T}_{kk}(s) \end{bmatrix}$$

with $\quad \bar{T}_{ii}(s)$ are non null $(p_i \times m_i)$ proper rational stable matrices

$$\bar{T}_{ij}(s) = 0 \qquad \text{for } i = j$$

As precedently the precompensator C(s) must be admissible and proper. When dealing with stability C(s) and T(s) C(s) have to be stable.

By pure precompensation the solvability condition of the block decoupling problem is given in terms of independance among the transfer matrix rows. When dealing with feedback implementations of these precompensators the situation becomes considerably more complex, contrasting with the block triangular decoupling problem.

This problem is not solved for all cases. We will consider two important situations :

* First we will consider compensators which are implementable by feedback with non singular input transformation

$$u = F x + Gv \qquad G \text{ regular}$$

* Then we will consider the row by row decoupling problem with possibly non regular input transformations. In this case the dynamic state feedback decoupling is achievable if the number of independant inputs is sufficiently large. Notice that the static state feedback decoupling problem is not solved when the number of outputs is larger or equal to 3.

4. 1. Block diagonal decoupling problem with G non singular

For the row by row decoupling (diagonal decoupling) problem it is shown in [16] that a linear square invertible system is decouplable if and only if its infinite zero orders are equal to the orders of the infinite zeros of each row. We will consider now the block decoupling problem (with G non singular), this problem was solved in [17] by transfer matrix arguments and in [18] via a geometric approach.

Let T(s) be a (pxm) proper rational matrix decomposed as previously according to the partition $(p_1, \ldots, p_k)$,

$$T(s) = \begin{bmatrix} T_1(s) \\ \vdots \\ T_k(s) \end{bmatrix}$$

One can state the following theorem [17].

Theorem 6

Let T(s) be a pxm proper rational matrix. Let $(p_1, \ldots, p_k)$ be a given set of non zero positive integers satisfying

$$\sum_{i=1}^{k} p_i = p.$$

The system whose transfer matrix is T(s) can be block decoupled relatively to the partition $(p_1, \ldots, p_k)$ by a proper admissible compensator implementable by static state feedback, u = F x + Gv, G regular, on a minimal realization of T(s) if and only if the infinite structure of T(s) is equal to the union of the infinite structures of the $T_i(s)$'s.

Furthermore if the unstable transmission zeros of T(s) are given by the unstable transmission zeros of the $T_i(s)$'s decoupling is achievable with stability.

The infinite structure of T(s) as well as the infinite structure of the $T_i(s)$'s are invariants under right multiplication by bicausal matrices (or feedback compensators).

The condition of theorem 6 is clearly satisfied for the decoupled system, then it is a necessary condition to be satisfied on T(s) for decoupling.

Without stability the proof is made by using Smith forms over the P.I.D. of proper rational functions. When dealing with stability the proof is made by using Smith forms over the P.I.D. of proper rational stable functions. In the proof is also used a previous result of [19] where the decoupling condition is given in terms of the independance "at infinity" and "for unstable frequencies" of the transfer matrix row blocks $T_i(s)$.

This condition turns out to be the same for static and dynamic state feedback decoupling (with G regular). When G is singular the problem becomes considerably more complex, in this case the set of all precompensators does not form a transformation group as in the regular case. the way the infinite zero orders are changed via such transformations is rather intricate. For these reasons we will consider the row by row decoupling problem restricting our attention to the precompensators which are implementable by dynamic state feedback $u = F(s)\,x + Gv$, G possibly non regular, on a minimal realization of the process.

4. 2. Dynamic state feedback decoupling

Then we consider the row by row decoupling problem with possibly singular input transformation. We will see that for dynamic state feedback implementable precompensators the decoupling condition is, roughly speaking, the following : "The number of process inputs must be large enough in order to compensate some rank defection for infinite frequencies". Remark that in the dynamic state feedback decoupling case we have the possibility to create new unstable zeros via the unstable poles of F(s) so the condition is expressed uniquely in terms of infinite frequencies.

In order to present clearly the necessary and sufficient condition we need some other invariants which are the following : Consider the Hermite form of T(s) over the ring of proper rational functions (and not proper rational stable functions) $H(s) = [H_1(s),\ 0]$.

The inverse I(s) of $H_1(s)$ is polynomial and is usually called the interactor.

Definition 2

Let T(s) be a (pxm) proper rational matrix and I(s) its interactor. Denote $I_i(s)$ the ith column of I(s).

Let r_i be the positive integer such that

$$\lim_{s\to\infty} I_i(s)\, s^{-r_i} = t_i, \qquad \text{and } t_i \text{ is a non null constant vector when}$$

$I_i(s)$ is non null.

The column rank at infinity k of I(s) is defined as the rank over the real field of the matrix $[t_1, \ldots, t_m]$.

One has the following result [14].

Theorem 7

The system whose transfer matrix T(s) is a (pxm) strictly proper stable rational surjective transfer matrix with null static kernel, is decouplable with stability by an admissible dynamic state feedback implementable on a minimal realization of T(s) if and only if

$$m \geq 2p-k$$

where k is the infinite column rank of I(s) the interactor of T(s) over $R_p(s)$.

It turns out that when the system is decouplable it can also be decoupled with stability. The proof is made by using appropriate factorizations of T(s) via two preliminary theorems see [14].

Roughly speaking, the idea of the proof is the following.

Let T(s) be factorized in $T(s) = [I^{-1}(s),\ 0]\ B(s)$

We have to find an mxp compensator C(s) which is a piece of a bicausal matrix such that T(s) C(s) be a decoupled transfer matrix D(s).

Denote

$$V(s) = \begin{bmatrix} V_1(s) \\ \\ V_2(s) \end{bmatrix} = B(s)\ C(s) \text{ where } V_1(s) \text{ is a pxp matrix}$$

Since $V_1(s) = I(s)\ D(s)$, the infinite column rank of $V_1(s)$ is equal to k. V(s) must have an infinite column rank equal to p, this implies that $V_2(s)$ has at least p-k rows.

It follows that $m - p \geq p-k$ (necessity). The sufficiency part is constructive.

In other words the number of additional inputs should be large enough in order to compensate the intrinsic non decouplability of the Hermite form.

We will now present some direct consequences of this theorem.

* Remark 1 :

When $m \geq 2p-1$, for any surjective system, decoupling is always achievable, because k is at least 1. Furthermore this bound (2p-1) can be reached, as shown on the following example.

$$T(s) = \begin{bmatrix} \pi^{-1} & 0 & 0 & \pi^{-2} & 0 \\ 0 & \pi^{-1} & 0 & \pi^{-3} & 0 \\ \pi^{-1} & \pi^{-1} & \pi^{-2} & -\pi^{-2}-\pi^{-3} & -\pi^{-3} \end{bmatrix} \text{. The Hermite form is :}$$

$$H(s) = \begin{bmatrix} \pi^{-1} & 0 & 0 & 0 & 0 \\ 0 & \pi^{-1} & 0 & 0 & 0 \\ -\pi^{-1} & -\pi^{-1} & -\pi^{-2} & 0 & 0 \end{bmatrix}$$

where $\pi = s+a$ is a stable polynomial with the usual stability domain (open left half plane).

$$T(s) = [I^{-1}(s), 0]\, B(s)$$

where $$I(s) = \begin{bmatrix} \pi & 0 & 0 \\ 0 & \pi & 0 \\ \pi^2 & \pi^2 & \pi^2 \end{bmatrix}$$ and $B(s)$ is bicausal

In this case the column rank at infinity k of I(s) is one, then the five inputs are necessary for decoupling.

At this point we note that dynamic decoupling is actually related with the column rank at infinity of I(s), and not with the row rank at infinity of the transfer matrix, which could have been reasonably conjectured from existing literature [8]. In the previous example the row rank at infinity of T(s) is two while k is one.

* Remark 2 :

When decoupling is not achievable this theorem gives the minimum number of inputs to be added i.e. 2p-k-m. This number is related with the complexity of the precompensator which is needed for decoupling.

The Hermite form of T(s) over $R_{ps}(s)$ provides us other useful invariants related with the minimum Mac Millan degree achievable by feedback.

Define first these invariants :

Definition 3

Let T(s) be a (pxm) surjective proper stable rational matrix and $I_g(s)$ its generalized interactor.

Denote q_i the number of infinite and unstable poles of the ith column of $I_g(s)$ counted with their multiplicity. The integers $q_1 \ldots q_p$ are called the stable decoupling invariants of T(s).

$$\delta = \sum_{i=1}^{p} q_i$$ is called the stable decoupling degree of T(s)

Notice that the stable decoupling invariants of T(s) are invariant under the right multiplication of T(s) by bicausal and bistable matrices.

As shown in the following theorem the q_i's are closely related with the "simplest" decoupled system achievable by stable admissible compensation.

Theorem 8 [14]

Consider a system whose transfer matrix T(s) is a (pxm) strictly proper stable rational surjective transfer matrix with null static kernel

i) The system is decouplable by an admissible proper precompensator with stability, the minimal Mac Millan degree achievable for the decoupled system is δ the stable decoupling degree of T(s), in this case the ith diagonal entry of the decoupled system has Mac Millan degree q_i.

ii) If the system is decouplable by an admissible dynamic state feedback compensator with stability on a minimal realization of T(s), the minimal Mac Millan degree achievable for the decoupled system is δ the stable decoupling degree of T(s), in this case the i-th diagonal entry of the decoupled system has Mac Millan degree q_i.

The proof is not given here for space limitation. Other results are given in [14] for example the stable decoupling with stable dynamic state feedback is solved. A frequent use of the Smith and Hermite forms is made in order to prove such results.

Remark

We have considered dynamic state feedback decoupling compensators. For static state feedback compensators the decoupling problem with G possibly non regular is solved in the two output case only [20].

Roughly speaking the two output system of transfer T(s) is decouplable if the degree ν of a minimal basis of Ker T(s) is large enough (in geometrical terms this degree is equal to the dimension of the largest controllability subspace of Ker C).

Consider the following transfer matrix

$$T(s) = \begin{bmatrix} s^{-2} & 0 & s^{-3} \\ -s^{-1} & s^{-3} & -s^{-2} \end{bmatrix}$$

Since the two rows are dependant at infinity this system is not decoupable by static state feedback with regular G, but this system is decouplable by dynamic state feedback, see theorem 7.

This system is no more decouplable by static state feedback with singular G. In fact ν = 1 which is not sufficient to meet the condition of [20].

REFERENCES

[1] F.M. CALLIER, C.A. DESOER, "Multivariable feedback systems", New-York, Springer Verlag, 1982.

[2] M. VIDYASAGAR, N., VISWANADHAN, "Algebraic design techniques for reliable stabilization", I.E.E.E. Trans. Auto. Cont., AC-27, pp. 1085-1096, 1982.

[3] A.S. MORSE, "Systems invariants under feedback and cascade control", Proceedings Int. Symp., Udine, Springer Verlag, New-York, 1975.

[4] L. PERNEBO, "An algebraic theory for the design of controllers for linear multivariable systems - Parts I and II", I.E.E.E. Trans. Auto. Cont., AC-26, pp. 171-194, 1981.

[5] J.M. DION, C. COMMAULT, "Smith Mac Millan factorizations at infinity of rational matrix functions and their control interpretation". Syst. and Cont. lett., vol. 1, pp. 312-320, 1982

[6] P.L. FALB, W.A. WOLOVICH, "Decoupling in the design and synthesis of multivariable control systems", I.E.E.E. trans. Auto. Cont., AC-12, n° 6, pp. 651-669, 1967.

[7] A.S. MORSE, W.M. WONHAM, "Status of noninteracting control", I.E.E.E. Trans. on Auto. Cont., AC-16, n° 6, pp. 568-581, 1973.

[8] M.L.J. HAUTUS, M. HEYMANN, "Linear feedback decoupling, transfer function analysis", I.E.E.E. Trans. on Auto. Cont., AC-28, pp. 823-832, 1983.

[9] M. MARCUS, "Introduction to modern algebra", Marcel Dekker, New-York, 1978.

[10] N.D. HUNG, B.D.O. ANDERSON, "Triangularization technique for the design of multivariable control system", I.E.E.E. Trans. on Auto. Cont., AC-24, n° 3, pp. 455-460.

[11] W.A. WOLOVICH, P.L. FALB, "Invariants and canonical forms under dynamic compensation" SIAM J. on Cont. and Opt., n° 14, pp. 996-1008, 1976.

[12] M.L.J. HAUTUS, M. HEYMANN, "Linear feedback - An algebraic approach", SIAM J. on Cont. and Opt., n° 16, pp. 83-105, 1978.

[13] J. HAMMER, M. HEYMANN, "Causal factorization and linear feedback", SIAM J. on Cont. and Opt., vol. 19, n° 4, pp. 445-468.

[14] J.M. DION, C. COMMAULT, "On linear dynamic state feedback decoupling", Submitted for publication 1985.

[15] C. COMMAULT, J.M. DION, "Transfer matrix approach to the triangular block decoupling problem", Automatica, vol. 19, n° 5, pp. 533-542, 1983.

[16] J. DESCUSSE, J.M. DION, "On the structure at infinity of linear decouplable systems", I.E.E.E. Trans. on Auto. Cont., AC-27, n° 4, pp. 971-974, 1982.

[17] J.M. DION, "Feedback block decoupling and infinite structure of linear systems", Int. Journal of Control, March 1983.

[18] J. DESCUSSE, J.F. LAFAY, M. MALABRE, "On the structure at infinity of linear block decouplable systems : the general case", I.E.E.E. Trans. on Auto. Cont., AC-28, Octobre 1983.

[19] M.L.J. HAUTUS, M. HEYMANN, "New results in linear decoupling", Proceedings of the INRIA Conference Analysis and Optimization of Systems Versailles, Springer Verlag, pp. 562-577, 1980.

[20] J. DESCUSSE, J.F. LAFAY, M. MALABRE, "Solution of the static state feedback decoupling problem for linear systems with two inputs", I.E.E.E. Trans. on Auto. Cont., to appear.

Frequency Domain and State Space Methods for Linear Systems
C.I. Byrnes and A. Lindquist (Editors)

ZEROS OF CASCADE COMPOSITIONS

G. Conte A.M. Perdon
Ist. Matematica Univ. di Genova
Genova, Italy

The zero structure of the cascade composition $H(z) = F(z)\ G(z)$ of two transfer functions $F(z)$ and $G(z)$ is analysed using the notion of local zero and pole module. A precise description, in algebraic terms, of the cancellation and generation of zeros in the cascade composition is given.

1. INTRODUCTION

In this paper we study the zeros of the cascade composition $H(z) = F(z)G(z)$ of two transfer functions $F(z)$ and $G(z)$. It is known, and can be shown by easy examples, that, together with the expected pole/zero cancellation between the factors $F(z)$ and $G(z)$ (dynamical cancellation), a crucial role in determining the zeros of $H(z)$ is played by the interaction of the zero structures with the null spaces of $F(z)$ and $G(z)$ and by the interaction between the null spaces themselves. In particular, it may happen that some zeros of $F(z)$ and $G(z)$ disappear without a corresponding cancellation of poles (we speak in this case of trivial cancellation or cancellation due to rank reasons) and, also, that new zeros, which were not present in $F(z)$ or $G(z)$, appear in in $H(z)$ (see [4]).

Clearly a better insight into these phenomena may be useful in solving factorization problems, such as exact model matching, minimal design and minimal inverse problem (see [2] and the references therein), with additional constraints. Our aim is therefore to give a precise description, in algebraic terms, of the cancellation and generation of zeros in the cascade composition.

In our investigation we focus on the local zero structure at α, where α is any point of the base field k or $\alpha=\infty$, and we make use of the notion of local zero module at α, Z_α, introduced in [5]. Essentially, what we want is a description of $Z_\alpha(H)$ in terms of $Z_\alpha(F)$, $Z_\alpha(G)$, Ker F, Ker G, Im F and Im G. This is obtained constructing a short exact sequence whose central term is $Z_\alpha(H)$ and whose other terms are modules with a structure strictly related to the phenomena of dynamical and trivial cancellation and of generation of zeros previously mentioned. These modules, in particular, are defined by using the zero modules, the kernels and the images of F(z) and G(z) through successive reductions of the general case to the case in which F(z) is injective and to those in which F(z) has neither poles nor zeros at the chosen α (compare with [7], where other results are derived from a slightly different point of view). It is still worthwhile to remark that in the representations we use, the absence of any of the phenomena of cancellation or generation of zeros in a specific case means the annihilation of a certain module suitably defined or the isomorphism of two others (see sect. 3 for details).

2. PRELIMINARIES AND NOTATIONS

Let k be an algebraically closed field and let k[z] and k(z) denote respectively the ring of polynomials and the field of rational functions in the variable z over k.

Given a point $\alpha \in k$, any $g(z) \in k(z)$ can be represented as $g(z) = (z-\alpha)^r(p(z)/q(z))$ with $r \in \mathbf{Z}$, p(z) and q(z) polynomials not divisible by $(z-\alpha)$. The <u>valuation of g(z) at α</u> is then defined by $v_\alpha(g) = r$. If $g(z) \in k(z)$ is written as the quotient of polynomials $g(z) = p(z)/q(z)$, the <u>valuation of g(z) at infinity</u> is defined by $v_\infty(g) = \deg q - \deg p$. For $\alpha \in k \cup \{\infty\}$, we denote by O_α the subring of k(z) given by

$$O_\alpha = \{ g(z) \in k(z) \text{ such that } v_\alpha(g) \geq 0 \}$$

Any O_α is a local ring whose maximal ideal $\mathbf{m}_\alpha$ consists of all the elements g(z) such that $v_\alpha(g) \geq 1$.

We denote by η_α a fixed generator of $\mathbf{m}_\alpha$.

Given a k-vector space V, we denote by V(z) the k(z)-vector space $V \otimes_k k(z)$ and by $\Omega_\alpha V$ the O_α-module $V \otimes_k O_\alpha$. $\Omega_\alpha V$ is, in obvious way, an O_α-submodule of V(z).

Given two k-vector spaces $U \equiv k^m$ and $Y \equiv k^p$, by a transfer function T(z) we mean a k(z)-linear map $T(z): U(z) \rightarrow Y(z)$, or, equivalently, the $p \times m$ matrix of rational functions which represents T(z) with respect to the canonical k(z)-bases of U(z) and Y(). For any $\alpha \in k \cup \{\infty\}$, any transfer function can be represented as a right or left fraction $T(z) = N_\alpha(z)D_\alpha^{-1}(z) = \tilde{D}_\alpha^{-1}(z)\tilde{N}_\alpha(z)$ of matrices with elements in O_α, with $D_\alpha(z)$, respectively $\tilde{D}_\alpha(z)$, non singular. We say that a right fractional representation is coprime when the matrices $N_\alpha(z)$, $D_\alpha(z)$ are coprime over O_α. This means that any common right factor of $N_\alpha(z)$ and $D_\alpha(z)$ is α-unimodular (i.e. it has an inverse with elements in O_α) or, equivalently, that there exist matrices $R_\alpha(z)$, $S_\alpha(z)$ with elements in O_α such that $R_\alpha(z)N_\alpha(z) + S_\alpha(z)D_\alpha(z) = I$. Dually for left fractional representations. Coprime fractional representations are essentially unique.

The poles and zeros at α of a transfer function T(z) can be obtained in the following way (compare with [3]). Let s be the minimum integer such that all the elements of $\eta_\alpha{}^S\, T(z)$ are in O_α, denote by $\bar{M}_\alpha(z)$ the Smith form over O_α of $\eta_\alpha{}^S\, T(z)$ and let

$$M_\alpha(z) = \eta_\alpha{}^{-S}\, \bar{M}_\alpha(z) = \text{block diag}\{\, \text{diag}\{\, \eta_\alpha{}^{\nu 1},..,\eta_\alpha{}^{\nu r}\},\ 0\,\}.$$

Then T(z) is said to have a pole at α of order ν_i for any $\nu_i < 0$ and a zero at α of order ν_j for any $\nu_j > 0$. The sequence $\{\ \nu_1,.., \nu_r\ \}$ is called pole/zero structure of T(z) at α. More abstractly, we can associate to the transfer function $T(z) : U(z) \to Y(z)$ two O_α-modules, for any α, called respectively the module of poles at α, $X_\alpha(T)$, and the module of zeros at α, $Z_\alpha(T)$. The first one, which coincides essentially with the state module of the local realization at α described in [5], is defined by:

$$X_\alpha(T) = \frac{T(\Omega_\alpha U) + \Omega_\alpha Y}{\Omega_\alpha Y}$$

and the second one, which represents the local analogous of the (finite) zero module introduced by B.Wyman and M.Sain in [15], is defined by:

$$Z_\alpha(T) = \frac{T^{-1}(\Omega_\alpha Y) + \Omega_\alpha U}{\text{Ker}\, T + \Omega_\alpha U}$$

Both these modules and their properties are described in detail in the case $\alpha=\infty$ in [1].

The connections between $X_\alpha(T)$, $Z_\alpha(T)$ and the classical notions of pole and zero at α can be established using a coprime fractional representation of T(z) over O_α of the form $T(z) = D_\alpha{}^{-1}(z)N_\alpha(z)$. Assuming that $\{\nu_1,..,\nu_r\}$ is the pole/zero structure of T(z) at α and that $\nu_1 \le \nu_2 \le \cdots < 0 < \nu_j \le \cdots \le \nu_r$, one has, in fact, that the non trivial elements in the Smith form over O_α of $D_\alpha(z)$ and $N_\alpha(z)$ are given, respectively, by $\eta_\alpha{}^{-\nu 1}, \ldots, \eta_\alpha{}^{-\nu i}$ and by $\eta_\alpha{}^{\nu j}, \cdots, \eta_\alpha{}^{\nu r}$.

On the other hand, one can show that $X_\alpha(T)$ is isomorphic to the O_α-module $\Omega_\alpha Y/D_\alpha(\Omega_\alpha Y)$ and that $Z_\alpha(T)$ is isomorphic to the torsion part of $\Omega_\alpha Y/N_\alpha(\Omega_\alpha U)$ (see, for $\alpha=\infty$, [1]). Hence, the invariant factors of $X_\alpha(T)$ and those of $Z_\alpha(T)$ over O_α determine the pole/zero structure at α of $T(z)$. Moreover, $X_\alpha(T)$ and $Z_\alpha(T)$ are shown in this way to be finite dimensional k-vector spaces such that $\dim_k X_\alpha(T)$ = (total number of poles at α) and $\dim_k Z_\alpha(T)$ = (total number of zeros at α).

<u>2.1 Remark</u> It is useful to note that:

i) $T(z)$ has no poles at α iff $T(\Omega_\alpha U) \subseteq \Omega_\alpha Y$ iff $\Omega_\alpha U + \mathrm{Ker}\, T \subseteq T^{-1}(\Omega_\alpha Y)$

ii) $T(z)$ has no zeros at α iff $T^{-1}(\Omega_\alpha Y) \subseteq \Omega_\alpha U + \mathrm{Ker}\, T$

iii) $T(z)$ has neither poles nor zeros at α iff $T^{-1}(\Omega_\alpha Y) = \Omega_\alpha U + \mathrm{Ker}\, T$.

3. MAIN RESULT

In this section we consider the problem of describing the zero structure of the cascade composition $H(z) = F(z)G(z)$, where $G(z)$: $U(z) \to Y(z)$ and $F(z)$: $Y(z) \to W(z)$ are two given transfer functions of dimension pxm and pxq, in terms of zeros and poles of $F(z)$ and $G(z)$. We focus our attention at the local structure at α, where α belongs to $k \cup \{\infty\}$, and we start by considering two special cases, namely:

(A) $F(z)$ is injective;

(B) $F(z)$ has neither poles nor zeros at α.

Then, we will investigate the general case by successive reduction to (A) and (B).

To begin with, assuming that $H(z)$, $F(z)$, $G(z)$ and α are as above, let us introduce the following O_α-module:

$$(3.1)\quad Z_\alpha^F(G) = \frac{G^{-1}(\Omega_\alpha Y \cap (F^{-1}(\Omega_\alpha W) + G(\Omega_\alpha U))) + \Omega_\alpha U + \mathrm{Ker}\, FG}{\mathrm{Ker}\, FG + \Omega_\alpha U}$$

<u>3.2 Remark</u> In our framework, $Z_\alpha^F(G)$ represents "the zeros at α of $G(z)$ which are zeros at α of $H(z)$". The precise (algebraic) sense of this assertion will be established below (compare with [1], [2], [6]).

CASE (A)

In this subsection we assume that F(z) is injective. In this case (3.1) becomes:

$$(3.3)\qquad Z_\alpha^F(G) = \frac{G^{-1}(\Omega_\alpha Y \cap (F^{-1}(\Omega_\alpha W) + G(\Omega_\alpha U))) + \Omega_\alpha U}{\operatorname{Ker} G + \Omega_\alpha U}$$

and we can state the following

3.4 Proposition i) There exists a natural inclusion $i_1 : Z_\alpha{}^F(G) \to Z_\alpha(G)$.
ii) There exists a natural inclusion $i_2 : Z_\alpha{}^F(G) \to Z_\alpha(H)$.

Proof. i) Obvious.
ii) By Ker G = Ker H and by $G^{-1}(\Omega_\alpha Y \cap (F^{-1}(\Omega_\alpha W) + G(\Omega_\alpha U))) \subseteq G^{-1}F^{-1}(\Omega_\alpha W) + \Omega_\alpha U = H^{-1}(\Omega_\alpha W) + \Omega_\alpha U$.

Let us introduce now the following O_α-module

$$(3.5)\qquad Z_\alpha^G(F) := \frac{F^{-1}(\Omega_\alpha W \cap \operatorname{Im} G + G(\Omega_\alpha U) + \Omega_\alpha Y}{G(\Omega_\alpha U) + \Omega_\alpha Y}$$

3.6 Remark Analogously to 3.2, we can say that $Z_\alpha{}^G(F)$ represents " the zeros at α of F(z) which are zeros at α of H(z)".

3.7 Proposition i) $Z_\alpha{}^G(F)$ can be viewed, in a natural way, as a subquotient (i.e. quotient of a submodule) of $Z_\alpha(F)$.
ii) G(z) induces an O_α-isomorphism $g : Z_\alpha(H)/Z_\alpha{}^F(G) \to Z_\alpha{}^G(F)$ given by g([u]) = [G(u)].

Proof . i) Let us consider the two submodules of $Z_\alpha(F)$ given by

$$\frac{F^{-1}(\Omega_\alpha W) \cap \operatorname{Im} G + \Omega_\alpha Y}{\Omega_\alpha Y} \qquad \text{and by}$$

$$\frac{(G(\Omega_\alpha U) + \Omega_\alpha Y) \cap (F^{-1}(\Omega_\alpha W) \cap \operatorname{Im} G) + \Omega_\alpha Y}{\Omega_\alpha Y}$$

Clearly, the first contains the second and, by a natural isomorphism, their quotient can be identified with $Z_\alpha{}^G(F)$.

ii) Let us consider the following diagram

$$\begin{array}{ccc}
G^{-1}(\Omega_\alpha Y \cap (F^{-1}(\Omega_\alpha W) + G(\Omega_\alpha U))) + \Omega_\alpha U & \xrightarrow{G(z)} & G(\Omega_\alpha U) + \Omega_\alpha Y \\
\downarrow & & \downarrow \\
G^{-1}F^{-1}(\Omega_\alpha W) + \Omega_\alpha U & \xrightarrow{G(z)} & F^{-1}(\Omega_\alpha W) \cap \operatorname{Im} G + G(\Omega_\alpha U) + \Omega_\alpha Y \\
\downarrow & & \downarrow \\
Z_\alpha(H)/Z_\alpha^F(G) & \dashrightarrow{g} & Z_\alpha^G(F)
\end{array}$$

where the quotient modules are identified up to isomorphism.
Now, if, for some $u \in G^{-1}(F^{-1}(\Omega_\alpha W))$, $g([u]) = 0$, then $G(u) \in G(\Omega_\alpha U) + \Omega_\alpha Y$ and this implies that, for some $u' \in \Omega_\alpha U$,
$(u-u') \in G^{-1}(\Omega_\alpha Y \cap (F^{-1}(\Omega_\alpha W) + G(\Omega_\alpha U)))$. Hence $[u] = 0$ and g is injective. On the other hand, if $[v] \in Z_\alpha{}^G(F)$, then, for some $v' \in \Omega_\alpha Y$, $(v-v')$ belongs to $F^{-1}(\Omega_\alpha W) \cap \operatorname{Im} G + G(\Omega_\alpha U)$, that is, $(v-v') = G(v'')$, with v'' in $G^{-1}F^{-1}(\Omega_\alpha W) + \Omega_\alpha U$. Hence $[v] = g([v''])$ and g is surjective.

By 3.6 and 3.7, we can summarize the situation in Case (A) as follows:

<u>3.8 Proposition</u> The sequence of O_α-modules and morphisms

$$0 \to Z_\alpha{}^F(G) \xrightarrow{i_2} Z_\alpha(H) \xrightarrow{p} Z_\alpha{}^G(F) \to 0,$$

where p denotes the canonical projection composed with g, is exact.

Essentially, this says, in a precise algebraic sense, that, in the case we are considering, the zeros at any $\alpha \in k \cup \{\infty\}$ of the cascade composition $H(z) = F(z)G(z)$ can be decomposed into two parts which correspond to two subsets, in a suitable sense, of the zeros at α of $G(z)$ and $F(z)$.

Since 3.7 implies, in particular, that $Z_\alpha(H) = Z_\alpha{}^F(G) \oplus_K Z_\alpha{}^G(F)$ as k-vector spaces, the total number of zeros at α of $H(z)$ cannot exceed the sum of the total number of zeros at α of G(z) and F(z). Therefore, if G(z) and F(z) have no zeros at α, no zeros can appear in H(z).

<u>3.9 Remark</u> It is useful to note that, when F(z) has no poles at α, i.e. $\Omega_\alpha Y \subseteq F^{-1}(\Omega_\alpha W)$, $Z_\alpha{}^F(G)$ coincides with $Z_\alpha(G)$. More generally, we can say, to describe 3.2, that $Z_\alpha{}^F(G)$ represents the zeros of G(z) which do not cancel out together with some pole of F(z).
On the other hand, looking at (3.5), we can say that the possible cancellation of zeros of F(z), which in our representation is shown by the formal difference between $Z_\alpha(F)$ and $Z_\alpha{}^G(F)$, may be caused by the interaction between $F^{-1}(\Omega_\alpha W)$ and the image of G(z), described by their intersection, and by the presence of poles at α in G(z), described by $G(\Omega_\alpha U)+\Omega_\alpha Y$. If we call respectively trivial and dynamical the zero cancellation in the two cases, we have clearly that, when G(z) is surjective (and also transposing G(z) and F(z) we are still in Case (A)), $Z_\alpha{}^G(F)$ is a quotient of $Z_\alpha(F)$ and only dynamical cancellations take place.

<u>3.10 Proposition</u> i) We have $Z_\alpha{}^F(G) = Z_\alpha(G)$ iff

$$G^{-1}(\Omega_\alpha Y) \subset G^{-1}(\Omega_\alpha Y \cap (F^{-1}(\Omega_\alpha W)+G(\Omega_\alpha U)))+\Omega_\alpha U.$$

ii) We have $Z_\alpha{}^G(F) = Z_\alpha(F)$ iff

$$F^{-1}(\Omega_\alpha W) \subset F^{-1}(\Omega_\alpha W) \cap \operatorname{Im} G + \Omega_\alpha Y \quad \text{and}$$

$$(G(\Omega_\alpha U)+\Omega_\alpha Y) \cap (F^{-1}(\Omega_\alpha W)+\Omega_\alpha Y) \subset \Omega_\alpha Y.$$

Proof. i) ii) It is easily seen by (3.3) and (3.5).

CASE (B)

<u>In this subsection we assume that F(z) has neither poles nor zeros at α.</u>
In this case (3.1) becomes

$$Z_\alpha^F(G) = \frac{G^{-1}(\Omega_\alpha Y) + \Omega_\alpha U + \operatorname{Ker} FG}{\operatorname{Ker} FG + \Omega_\alpha U} \tag{3.11}$$

<u>3.12 Proposition</u> i) $Z_\alpha{}^F(G)$ can be viewed, in a natural way, as a quotient of

$Z_\alpha(G)$.

ii) There exists a natural inclusion $i_3 : Z_\alpha{}^F(G) \to Z_\alpha(H)$.

Proof. i) $Z_\alpha{}^F(G)$ is naturally isomorphic to the quotient of $Z_\alpha(G)$ by the submodule

$$\frac{(\mathrm{Ker}\, FG + \Omega_\alpha U) \cap (G^{-1}(\Omega_\alpha Y) + \Omega_\alpha U)}{\mathrm{Ker}\, G + \Omega_\alpha U}.$$

ii) By 2.1, $F^{-1}(\Omega_\alpha W) = \Omega_\alpha Y + \mathrm{Ker}\, F$. Now, we have $G^{-1}(\Omega_\alpha Y) + \mathrm{Ker}\, FG =$ $G^{-1}(\Omega_\alpha Y) + G^{-1}(\mathrm{Ker}\, F) \subseteq G^{-1}(\Omega_\alpha Y + \mathrm{Ker}\, F) = G^{-1}(F^{-1}(\Omega_\alpha W)) = H^{-1}(\Omega_\alpha W)$ and this induces the inclusion.

In this case there is no contribution to $Z_\alpha(H)$ given by $Z_\alpha(F)$, since we assumed the latter to be zero.
However, it is known that, also when $Z_\alpha(G) = 0$, some zeros at α may appear in $H(z)$ (compare with [4]. Therefore, we consider the O_α-module:

$$M_\alpha(H) = Z_\alpha(H)/Z_\alpha^F(G) \cong \frac{G^{-1}(\Omega_\alpha Y + \mathrm{Ker}\, F) + \Omega_\alpha U}{G^{-1}(\Omega_\alpha Y) + G^{-1}(\mathrm{Ker}\, F) + \Omega_\alpha U} \tag{3.13}$$

The following sequence, which is exact by definition, can be used to summarize the situation in Case (B):

$$0 \to Z_\alpha{}^F(G) \xrightarrow{i_2} Z_\alpha(H) \to M_\alpha(H) \to 0 \tag{3.14}$$

This is to say, in a precise algebraic sense, that in the case we are considering, the zeros at $\alpha \in k\cup\{\infty\}$ of the cascade composition $H(z) = F(z)G(z)$ can be decomposed into two parts, which correspond, respectively, to a subset, in a suitable sense, of the zeros at α of $G(z)$ and to the zeros generated by the interaction of $F(z)$ and $G(z)$. We will come back to this point later on.

<u>3.15 Proposition</u> i) $M_\alpha(H) = 0$ iff
$G^{-1}(\Omega_\alpha Y + \mathrm{Ker}\, F) \subseteq G^{-1}(\Omega_\alpha Y) + G^{-1}(\mathrm{Ker}\, F) + \Omega_\alpha U$.

ii) Ker F $\subseteq$ Im G implies $M_\alpha(H) = 0$.

Proof. i) Obvious by (3.13).

ii) Let $u \in G^{-1}(\Omega_\alpha Y + \text{Ker } F)$, then $G(u-u') \in \Omega_\alpha Y$ for some $u' \in G^{-1}(\text{Ker } F)$. Hence, $u = (u-u') + u' \in G^{-1}(\Omega_\alpha Y) + G^{-1}(\text{Ker } F)$ and the conclusion follows from i).

<u>3.16 Remark</u> It is obvious that, in the case we are considering, only trivial cancellations of the zeros of G(z) are possible. (3.11)shows that they are due to the difference between Ker G = $G^{-1}(0)$ and Ker FG = $G^{-1}(\text{Ker } F)$. On the other hand, the **interaction** between F(z) and G(z) which is responsible for the presence of the component $M_\alpha(H)$, is described by the difference, up to $\Omega_\alpha U$, between $G^{-1}(\Omega_\alpha Y + \text{Ker } F)$ and $G^{-1}(\Omega_\alpha Y) + G^{-1}(\text{Ker } F)$. This clarifies the role of the right and left null spaces in connection with the behaviour of zeros in cascade compositions.

<u>3.17 Examples</u>

i) Let us consider the transfer functions

$$G(z) = [\,(3z-1)/z^2 \quad -2/z^2\,]' \quad \text{and} \qquad F(z) = [\,1/z^2 \quad 1/z\,].$$

They have no zeros at any finite α (and one zero each at $\alpha=\infty$), but their cascade composition $H(z) = F(z)G(z) = (z-1)/z^4$ has, in particular, one zero at $\alpha=1$. Actually, we can see that $1/(z-1)\Omega_1 U = 1/(z-1)\mathcal{Q} \subseteq G^{-1}(\Omega_1 Y + \text{Ker } F)$. In fact

$$G(u/(z-1)) = ((3z-1)u/z^2(z-1) \quad -2u/z^2(z-1))' =$$
$$(\,u/z^2 \quad 0)' + (2u/(z(z-1)) \quad -2u/z^2(z-1))',$$

but $1/(z-1)\Omega_1 U \subseteq G^{-1}(\Omega_1 Y) + G^{-1}(\text{Ker } F) + \Omega_1 U$.

More precisely, we have $M_1(H) = 1/(z-1)\mathcal{Q}/O_1 = Z_1(H)$ and $\dim_K Z_1(H) = 1$.

ii) Let us take

$$F(z) = \left(\frac{1}{z^2} \quad \frac{1}{z} \quad 0\,\right) \quad \text{and} \qquad G(z) = \begin{bmatrix} z & 0 & z \\ 1 & 0 & 1 \\ 0 & 1 & 1 \end{bmatrix}$$

G(z) is not surjective, but G(U(z)) is easily seen to contain Ker F. We have, therefore, that H(z) has no zeros except possibly at ∞, where F(z) has one zero. In fact H(z) = [2/z 0 2/z]. Before investigating the situation at ∞, we need to describe the general case.

GENERAL CASE

In this subsection we remove any assumption on $F(z)$

Let $\alpha \in k \cup \{\infty\}$ be given and factor $F(z)$ as $F(z) = \tilde{N}_\alpha(z)\tilde{D}_\alpha^{-1}(z)$, where $\tilde{N}_\alpha(z)$, $\tilde{D}_\alpha(z)$ have elements in O_α and are coprime. If rank $\tilde{N}_\alpha(z) = r < m$, choosing a suitable α-unimodular matrix $P(z)$ we can write

$$(3.18)\ F(z) = \tilde{N}_\alpha(z)P(z)P^{-1}(z)\tilde{D}_\alpha^{-1}(z) = N_\alpha(z)\,(I_r \quad 0)D_\alpha^{-1}(z) = N_\alpha(z)\,J\,D_\alpha^{-1}(z)$$

where $N_\alpha(z)$ is injective of dimension $p \times r$ and $J = (\, I_r \quad 0\,)$ has dimension $r \times m$. Now we have

$$(3.19)\ H(z) = F(z)G(z) = N_\alpha(z) \cdot J \cdot D_\alpha^{-1}(z)G(z)$$

and, writing $G_1(z) = D_\alpha^{-1}(z)G(z)$, $G_2(z) = J\,G_1(z)$, we can describe $Z_\alpha(H)$ by means of the following diagram where the short sequences are exact:

$$\begin{array}{ccccccccc}
 & Z_\alpha(G) & & 0 & & & & & \\
 & \cup| & & \downarrow & & & & & \\
 & Z_\alpha(G_1) & \longrightarrow & Z_\alpha^J(G_1) & \longrightarrow & 0 & & & \\
 & & & \downarrow & & & & & \\
(3.20) & 0 \longrightarrow Z_\alpha(G_2) & = & Z_\alpha^{N_\alpha}(G_2) & \xrightarrow{i_2} & Z_\alpha(H) & \xrightarrow{p} & Z_\alpha^{G_2}(N_\alpha) & \longrightarrow 0 \\
 & & & \downarrow & & & & & \\
 & & & M_\alpha(G_2) & & & & & \\
 & & & \downarrow & & & & & \\
 & & & 0 & & & & &
\end{array}$$

The modules appearing in (3.20) can be characterized in the following way. Since $D_\alpha^{-1}(z)$ is injective and has no zeros at α, by (A), we have

$$Z_\alpha(G_1) = Z_\alpha^{D_\alpha^{-1}}(G) = \frac{G^{-1}(\Omega_\alpha Y \cap (D_\alpha(\Omega_\alpha Y) + G(\Omega_\alpha U))) + \Omega_\alpha U}{\operatorname{Ker} G + \Omega_\alpha U} \subseteq Z_\alpha(G)$$

and, by (3.11) and (3.13),

$$Z_\alpha^J(G_1) = \frac{G^{-1}(D_\alpha(\Omega_\alpha Y)) + \Omega_\alpha U + G^{-1}D_\alpha \operatorname{Ker} J}{G^{-1}D_\alpha(\operatorname{Ker} J) + \Omega_\alpha U}$$

$$M_\alpha(G_2) = \frac{G^{-1}D_\alpha(\Omega_\alpha Y + \operatorname{Ker} J) + \Omega_\alpha Y}{G^{-1}D_\alpha(\Omega_\alpha Y) + G^{-1}D_\alpha(\operatorname{Ker} J) + \Omega_\alpha Y}$$

By (B), using $Z_\alpha{}^J(G_1)$ and $M_\alpha(G_2)$, we obtain the vertical exact sequence which describes $Z_\alpha(G_2)$. Since $N_\alpha(z)$ is injective and has no poles at α, we have, by (3.3) and 3.9, the equality $Z_\alpha(G_2) = Z_\alpha{}^{N\alpha}(G_2)$ and, by (3.5),

$$Z_\alpha^{G_2}(N_\alpha) = \frac{(N_\alpha^{-1}(\Omega_\alpha Y) \cap JD_\alpha^{-1}G(U(z)) + JD_\alpha^{-1}G(\Omega_\alpha U) + \Omega_\alpha W}{JD_\alpha^{-1}G(\Omega_\alpha U) + \Omega_\alpha W}.$$

Using $Z_\alpha(G_2)$ and $Z_\alpha{}^{G2}(N_\alpha)$, by (A), we obtain the horizontal exact sequence which describes $Z_\alpha(H)$.

3.21 Remark The above diagram shows that we can think of the zeros at α of $H(z)$ as generated in the following way. First, the zeros at α of $G(z)$ are reduced by cancellation with the poles of $D_\alpha{}^{-1}(z)$ and by the trivial cancellation due to the interaction with the kernel of J. The same phenomenon adds new zeros, so that we obtain $Z_\alpha(G_2.)$ Then the zeros of $N_\alpha(z)$, i.e. of $F(z)$, are reduced by cancellation to $Z_\alpha{}^{G2}(N_\alpha)$ and added to $Z_\alpha(G_2)$. We note that also a completely equivalent dual description could be given.

3.22 Examples

i) Let us consider again the transfer function

$$F(z) = \left(\frac{1}{z^2} \quad \frac{1}{z} \quad 0\right) \quad \text{and} \quad G(z) = \begin{bmatrix} z & 0 & z \\ 1 & 0 & 1 \\ 0 & 1 & 1 \end{bmatrix}$$

introduced in 3.17 and let us choose $\alpha=\infty$. Then, (3.19) becomes

$$H(z) = F(z)G(z) = (\ 1/z\ \ 0\ \ 0\ )\, D_\infty{}^{-1}(z)\, G(z) =$$

$$= 1/z(1\ \ 0\ \ 0)\, D_\infty{}^{-1}(z)G(z) =$$

$$= (1/z^2 \ \ 1/z \ \ 0)\begin{bmatrix} 0 & 1 & 0 \\ 1 & -1/z & 0 \\ 0 & 0 & 1 \end{bmatrix}\begin{bmatrix} 0 & 1 & 0 \\ 1 & -1/z & 0 \\ 0 & 0 & 1 \end{bmatrix}^{-1} G(z) =$$

$$= (1/z \ \ 0 \ \ 0)\, D_\infty^{-1}(z)\, G(z) = 1/z(1 \ \ 0 \ \ 0)\, D_\infty^{-1}(z)\, G(z) =$$

$$= 1/z J\, D_\infty^{-1}(z) G(z)$$

with $D_\infty^{-1}(z)$ ∞-unimodular. Since $G(z)$ has no zeros at ∞, by (3.20), $Z_\infty(JD_\infty^{-1}G) = M_\infty(JD_\infty^{-1}G)$ and, moreover, $M_\infty(JD_\infty^{-1}G) = 0$. In fact, an easy computation shows that if $D_\infty^{-1}G(\, u_1\, u_2\, u_3\,)' \in \Omega_\infty Y + \mathrm{Ker}J$, then $u_1 + u_3 = u \in O_\infty$ and $D_\infty^{-1}G(\, u\, 0\, 0\,)' \in \Omega_\infty Y$.
Hence $(\, u_1\, u_2\, u_3\,)' = (\, u\, 0\, 0\,)' + (\, -u_3\, u_2\, u_3\,)' \in G^{-1}D_\infty(\Omega_\infty Y) + G^{-1}D_\infty(\mathrm{Ker}\, J)$. As a consequence $Z_\infty(G_2) = 0$ and $Z_\infty(H) = Z_\infty^{G2}(1/z)$, moreover, since $JD_\infty^{-1}G(U(z)) = k(z)$ and $JD_\infty^{-1}G(\Omega_\infty U) = O_\infty$, $Z_\infty^{G2}(1/z) = 1/z \cdot O_\infty / O_\infty$ and $\dim_k Z_\infty^{G2}(1/z) = 1$.

REFERENCES

[1] Conte,G. and Perdon,A.M., Infinite zero module and infinite pole module, Proc. VII Int. Conf. on Analysis and Opt. of Systems, Nice, Lecture Notes in Control and Info. Science, Springer-Verlag 62 (1984) 302-315

[2] Conte, G. and Perdon,A.M., Zero module and factorization problems, to appear on A.M.S. Series in Contemporary Mathematics (1985)

[3] Kailath , T., Linear Systems, (Prentice Hall , 1980)

[4] Kung ,S. and Kailath, T., Some notes on valuation theory in linear systems, Proc. IEEE Conf. on Decision and Control, San Diego (1978)

[5] Wyman, B., Conte ,G. and Perdon, A.M., Local and global linear system theory, to appear in Proc. MTNS 85, Stokholm (1985)

[6] Wyman,B. and Sain, M., The zero module and essential inverse systems, IEEE Trans. on Circuit and Systems, CAS-28 (1981) 112-126

[7] Wyman, B. and Sain, M., Exact sequences for pole zero cancellation, Proc. MTNS 1981, Santa Monica (1981)

Frequency Domain and State Space Methods for Linear Systems
C.I. Byrnes and A. Lindquist (Editors)

MULTIVARIABLE POLE-ZERO CANCELLATION: NEW DEFINITION AND APPLICATION

Takao Fujii and Nobuhide Suda

Department of Control Engineering
Faculty of Engineering Science
Osaka University
Toyonaka, Osaka 560
JAPAN

We propose a new definition of multivariable pole-zero cancellation in the series connection of two linear multivariable systems or in the product of two rational transfer matrices of the systems. Our definition is based on certain output and input direction vectors related closely to the system pole and zero, respectively. The notions of pole-zero cancellation or the two directions are then applied to derive conditions for controllability, observability and so on of various composite systems such as series, parallel and feedback composite systems. They are also applied to characterize the minimality of a linear multivariable stochastic system.

1. INTRODUCTION

Consider the series connection S_{12} of two linear time-invariant continuous- or discrete-time systems S_1, S_2 (Fig.1) that have minimal state space system descriptions and rational transfer function matrices G_1, G_2. It is well known in the scalar case [1] that this cascade system S_{12} with the natural state variable is observable (controllable) iff no pole of G_1 (G_2) is a zero of G_2 (G_1), and hence S_{12} is minimal iff G_1 and G_2 have no pole-zero cancellations.

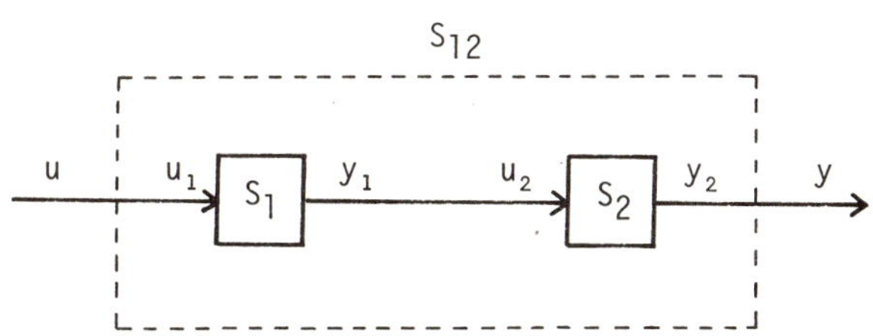

$$S_i \ (i=1,2) \left\{ \begin{array}{ll} \dot{x}_i(t) = A_i x(t)+B_i u(t), & y_i(t) = C_i x(t)+E_i u(t) \\ G_i(s) = C_i(sI-A_i)^{-1}B_i+E_i & \\ \text{or} & \\ x_i(k+1) = A_i x_i(k)+B_i u_i(k), & y_i(k) = C_i x(k)+E_i u(k) \\ G_i(z) = C_i(zI-A_i)^{-1}B_i+E_i & \end{array} \right.$$

Figure 1 The series connection of S_1 followed by S_2

Similar results on the minimality of composite systems are also known for the feedback connection (Fig.2) and the parallel connection (Fig.3) of S_1, S_2 in the scalar case [1].

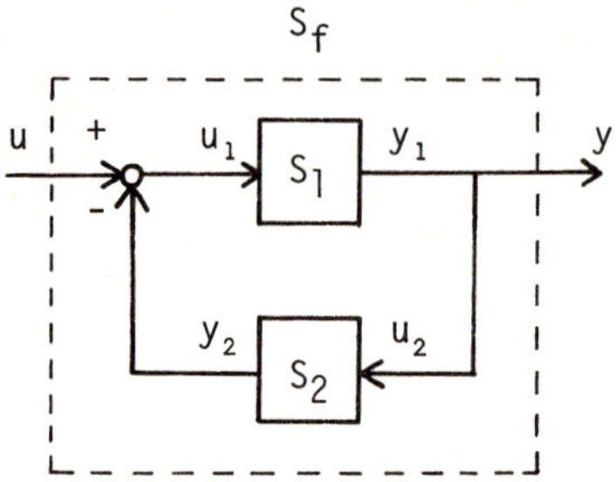

Figure 2 Feedback connection

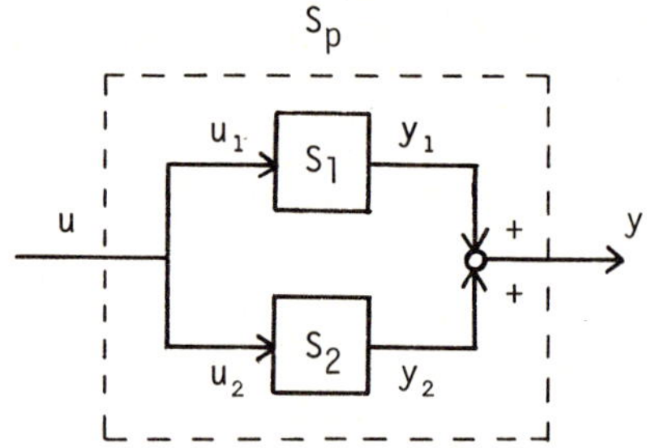

Figure 3 Parallel connection

Multivariable extensions of these results are carried out in [2], [3] and [4] based on the matrix fraction descriptions of G_1, G_2, and in [5] based on the Laurent expansions of G_1, G_2. Particulaly in [4] they propose a precise definition of multivariable pole-zero cancellation in G_2G_1 using McMillan degree ideas, and relate it with the coprimeness of polynomial matrices associated with the matrix fraction descriptions of G_1, G_2. The authors in [5] also used a similar notion of pole-zero cancellation based on the Laurent expansion of the square rational matrix in order to characterize the minimality of the square factorization $G(s) = G_1(s)G_2(s)$ with $\det G(s) \neq 0$.

In this paper we shall propose a new definition of multivariable pole-zero cancellation in the state-space setting as the most natural extension of the scalar one. The key role in our definition is played by two direction vectors in the output and input spaces as described in [6] and termed an *output pole-direction* and an *input zero-direction*, which are closely related to the poles and zeros of a system, respectively. Analogously as in the scalar case, we define multivariable pole-zero cancellation in G_2G_1 roughly as cancellation between these two direction vectors associated with any common pole of G_1 (G_2) and zero of G_2 (G_1). This notion of pole-zero cancellation and the preceding direction vectors are then used to express conditions for controllability, observability, stabilizability and detectability of the composite systems as shown in Figures 1 to 3. Finally they are also applied to characterize the minimality of a linear multivariable system with noise inputs in a stochastic sense.

2. POLE- AND ZERO- DIRECTIONS

Consider the linear time-invariant system S with the minimal state space description

$$\dot{x} = Ax + Bu, \quad x(0) = x_0 \tag{1a}$$

$$y = Cx + Eu \tag{1b}$$

and the rational transfer function matrix G(s) in left-coprime fractional form:

$$G(s) = D_\ell(s)^{-1}N_\ell(s) \tag{2}$$

We denote its state and output responses by $x(t; x_0, u)$ and $y(t; x_0, u)$, respectively. To begin with, we propose a precise and more general definition of the so-called *pole direction* and *zero direction* as described in [6]. We note that if λ is a pole of S, then there exists an initial state $x_0 \neq 0$ such that for all $t \geq 0$

$$x(t;\ x_0,\ 0) = x_0 e^{\lambda t} \tag{3a}$$

$$y(t;\ x_0,\ 0) = pe^{\lambda t} \quad \text{for some nonzero vector } p \tag{3b}$$

and vice versa. This fact motivates the following definition of such an output direction vector p as above.

Definition 1. A nonzero vector p is an *output pole-direction* of S associated with a number λ iff the *zero-input* responses at the state and output of S are of the form (3) for some $x_0 \neq 0$ (see Fig.4 for the illustration).

Dual to this notion is an *input zero-direction*, which is defined in view of the well-known transmission-blocking property of the system zero [7] as follows:

Definition 2. A nonzero vector g is an *input zero-direction* of S associated with a number λ iff there exists an initial state x_0 such that for all $t \geq 0$

$$x(t;\ x_0,\ ge^{\lambda t}) = x_0 e^{\lambda t} \tag{4a}$$

$$y(t;\ x_0,\ ge^{\lambda t}) = 0 \tag{4b}$$

(see Fig.5 for the illustration and note the symmetry of Figures 4 and 5).

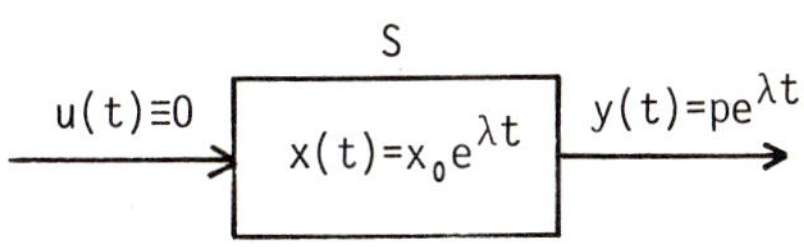

Figure 4 Output pole-direction

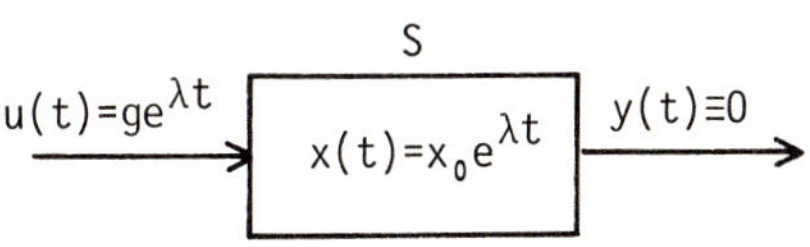

Figure 5 Input zero-direction

Remark 1. The discrete-time versions of the above definitions can be obtained by replacing t and $e^{\lambda t}$ in Definitions 1 and 2 with k and λ^k respectively, and regarding S as the linear time-invariant discrete-time system with the minimal state space description

$$x(k+1) = Ax(k) + Bu(k), \qquad x(0) = x_0 \tag{5a}$$

$$y(k) = Cx(k) + Eu(k) \tag{5b}$$

and the rational transfer matrix G(z) in left-coprime fractional form:

$$G(z) = D_\ell(z)^{-1} N_\ell(z)$$

Remark 2. Let G(s) as in (2) be an $m \times r$ rational function matrix. In the case of $m \geq r$, the above state-space definitions of the two direction vectors can be rephrased without introducing the state space (i.e., only in terms of the input and output responses) based on the following characterizations of poles and zeros given by [7].

Theorem III [7]
A number λ is a pole of G(s) iff

C1) there is an input of the form

$$u(t) = \sum_\alpha u_\alpha \delta^{(\alpha)}(t)$$

(here u_α is an r-dimensional vector and $\delta^{(\alpha)}(t)$ denotes the α th generalized derivative of the unit Dirac impulse function $\delta(t)$) such that the corresponding zero-state response has the property that

$$y(t) = pe^{\lambda t} \qquad \text{for all } t > 0$$

where p is some nonzero vector.

Theorem I [7]
Let $m \geq r$ and normal rank $G(s) = r$. If a number λ is a zero of G(s), then

C2) there exists a nonzero vector g and a polynomial $\Sigma_\alpha m_\alpha s^\alpha = m(s)$ such that for the input

$$u(t) = 1(t)e^{\lambda t}g + \sum_\alpha m_\alpha \delta^{(\alpha)}(t)$$

the zero-state response has the property

$$y(t) = 0 \qquad \text{for all } t > 0$$

where 1(t) denotes the Heaviside unit step function.

The output pole-direction and input zero-direction of S as defined in Definitions 1 and 2 turn out to coincide with those vectors p and g as stated in Theorems III and I, respectively (see [6] for details).

Remark 3. In [5] they also defined similar notions of the output pole-direction and the input zero-direction as stated above, which they call a *pole-vector* and a *root-vector* (or eigenvector), respectively. They consider only the case of $m = r$ and $\det G(s) \not\equiv 0$, and defined these two vectors based on the Laurent expansion of G(s) at λ

$$G(s) = \sum_{j=-q}^{\infty} (s - \lambda)^j A_j \qquad (q \geq 0)$$

as follows:

(i) An m-dimensional vector $p \neq 0$ is called a *pole-vector* of G at λ if

C1') there exist vectors $x_1, \cdots, x_q$ such that

$$A_{-q} x_q + A_{-q+1} x_{q-1} + \cdots + A_{-1} x_1 = p$$

$$A_{-q} x_i + A_{-q+1} x_{i-1} + \cdots + A_{-q+i-1} x_1 = 0 \qquad (1 \leq i \leq q)$$

(ii) An m-dimensional vector g is called a *root-vector* of G at λ if

C2') there exist vectors $x_1, \cdots, x_q$ such that

$$A_{-q} x_i + A_{-q+1} x_{i-1} + \cdots + A_{-q+i-1} x_1 + A_{-q+i} g = 0 \qquad (0 \leq i \leq q)$$

It is of interest to study the relationship between the time-domain characterizations of poles and zeros as stated in Remark 2 and the frequency-domain definitions of pole-vector and root-vector as stated just above, thereby clarify the relationship between the two notions of output pole-direction and pole-vector as well as those of input zero-direction and root-vector. A simple caluculation shows that the preceding two conditions C1 and C1' are equivalent, and so are conditions C2 and C2' as well. We therefore conclude from the observation stated in Remark 2 that the pole-vector and the root-vector definied in [5] are respectively the same as the output pole-direction and the input zero-direction defined previously in the particular case of $m = r$ and $\det G(s) \not\equiv 0$.

We denote the sets of all output pole-directions and all input zero-directions of S associated with λ by $P_\lambda(S)$ and $Z_\lambda(S)$ respectively. Then the following characterizations of these two sets are obtained.

Proposition 1. (Characterization of output pole-direction)
Let p be a nonzero vector. Then the following conditions are equivalent.

(i) $p \in P_\lambda(S)$

(ii) $p = Cx_0, \quad Ax_0 = \lambda x_0 \quad$ for some $x_0 \neq 0$

(iii) $D_\ell(\lambda)p = 0$

(Proof)

(i) $\Longleftrightarrow$ (ii) : Substituting (3) into (1) yields

$$\begin{aligned} \lambda x_0 e^{\lambda t} &= A x_0 e^{\lambda t} \\ p e^{\lambda t} &= C x_0 e^{\lambda t} \end{aligned} \tag{6}$$

This is equivalent to (3) by the uniqueness property of solutions to (1) for x_0 and u, and is also equivalent to (ii) since $e^{\lambda t} \neq 0$ for all t. Hence (i) $\Longleftrightarrow$ (ii) is obvious by Definition 1.

(ii) $\Longleftrightarrow$ (iii) : By Lemma A (i) in the appendix, there exists a minimal realization $(\tilde{A}, \tilde{B}, \tilde{C}, E)$ of G(s) and a polynomial matrix L(s) such that

$$F(s)\cdot H(s) \triangleq [\overbrace{D_\ell(s)}^{m} \quad \overbrace{L(s)}^{n}]\cdot \overbrace{\begin{bmatrix} \tilde{C} \\ \tilde{A} - sI \end{bmatrix}}^{n} \begin{matrix} \}m \\ \}n \end{matrix} = 0 \tag{7}$$

where $D_\ell(s)$ and L(s) are left-coprime. By definition (A, B, C, E) and $(\tilde{A}, \tilde{B}, \tilde{C}, E)$ are both minimal realizations of G(s), so that they are strictly system equivalent. Hence (ii) is equivalent to saying that

$$\text{(ii')} \qquad \begin{bmatrix} p \\ 0 \end{bmatrix} = H(\lambda)\,\tilde{x}_0 \qquad \text{for some } \tilde{x}_0 \neq 0. \tag{8}$$

The implication (ii') $\Longrightarrow$ (iii) is an immediate consequence of (7), and (iii) $\Longrightarrow$ (ii') follows by noting $[p^T\ 0]^T \in \operatorname{Ker} F(\lambda)$ by (iii), and $\operatorname{Ker} F(\lambda) = \operatorname{Im} H(\lambda)$ by (7) since $D_\ell(s)$ and L(s) are left-coprime, $\tilde{C}$ and $\tilde{A} - sI$ are right-coprime, and so $\dim \operatorname{Ker} F(\lambda) = n = \dim \operatorname{Im} H(\lambda)$. QED.

Proposition 2. (Characterization of input zero-direction)
Let g be a nonzero vector. Then the following conditions are equivalent.

(i) $g \in Z_\lambda(S)$

(ii) $\begin{bmatrix} A - \lambda I & B \\ C & E \end{bmatrix} \begin{bmatrix} x_0 \\ g \end{bmatrix} = 0 \quad$ for some x_0

(iii) $N_\ell(\lambda)g = 0$

(Proof)

(i) $\Longleftrightarrow$ (ii) : Since (4) $\Longleftrightarrow$ (ii) as shown in Theorem 3 of [6], this is obvious by Definition 2.

(ii) $\Longleftrightarrow$ (iii) : By Lemma A (ii) in the appendix, there exist a minimal realization $(\tilde{A}, \tilde{B}, \tilde{C}, E)$ of G(s) and polynomial matrices X(s), Y(s) and L(s) such that

$$M(s)\begin{bmatrix} s\tilde{I}-\tilde{A} & \tilde{B} \\ -\tilde{C} & E \end{bmatrix} = \begin{bmatrix} I & X(s)\tilde{B}+Y(s)E \\ 0 & N_\ell(s) \end{bmatrix} \tag{9}$$

where M(s) is a unimodular matrix defined by

$$M(s) \triangleq \begin{bmatrix} X(s) & Y(s) \\ L(s) & D_\ell(s) \end{bmatrix}$$

Since (A, B, C, E) and $(\tilde{A}, \tilde{B}, \tilde{C}, E)$ are strictly system equivalent (see the proof of (ii) $\Longleftrightarrow$ (iii) of Prop.1 for the reason), (ii) is equivalent to saying that

(ii') $$\begin{bmatrix} \tilde{A}-\lambda I & \tilde{B} \\ \tilde{C} & E \end{bmatrix}\begin{bmatrix} \tilde{x}_0 \\ g \end{bmatrix} = 0 \qquad \text{for some } \tilde{x}_0$$

Multiplying $[\ \tilde{x}_0^T \ \ g^T\ ]^T$ from the right side on (9) and setting $s=\lambda$ yields

$$M(\lambda)\begin{bmatrix} \lambda I-\tilde{A} & \tilde{B} \\ -\tilde{C} & E \end{bmatrix}\begin{bmatrix} \tilde{x}_0 \\ g \end{bmatrix} = \begin{bmatrix} I & X(\lambda)\tilde{B}+Y(\lambda)E \\ 0 & N_\ell(\lambda) \end{bmatrix}\begin{bmatrix} \tilde{x}_0 \\ g \end{bmatrix} \tag{10}$$

Using this relation toger with (ii) $\Longleftrightarrow$ (ii') and det $M(\lambda) \neq 0$ enables us to show (ii) $\Longleftrightarrow$ (iii) immediately. In fact, (ii) $\Rightarrow$ (iii) follows by substituting (ii') into (10), and (iii) $\Rightarrow$ (ii) follows by substituting (iii) into (10) and setting

$$\tilde{x}_0 = -[\,X(\lambda)\tilde{B}+Y(\lambda)E\,]\,g$$

Q E D.

From these characterizations, we observe that $P_\lambda(S) \neq \phi$ iff λ is a pole of S, i.e., λ is an eigenvector of A, and that $Z_\lambda(S) \neq \phi$ if λ is a zero of S, i.e.,

$$\text{rank}\begin{bmatrix} A-\lambda I & B \\ C & E \end{bmatrix} < \text{normal rank}\begin{bmatrix} A-sI & B \\ C & E \end{bmatrix}$$

and vice versa if S is left invertible (see [6] for the definition of zeros). Obviously, these two observations indicate a close relationship between poles and output pole-directions as well as zeros and input zero-directions. Particularly, the first observation allows us to call a $p \in P_\lambda(S)$, from now on, an output pole-direction of S associated with a *pole* of S. Note, however, that not all input zero-directions correspond to some zeros, unlike the zero directions in [6], where they consider only those zero-directions associated with *zeros*.

We recall the definitions of poles and zeros of G(s) based on the coprime factorization (2) due to [7] that λ is a pole of G(s) $\Longleftrightarrow$ det $D_\ell(\lambda)=0$, and λ is a zero of G(s) $\Longleftrightarrow$ rank $N_\ell(\lambda)<$ normal rank $N_\ell(s)$. It is then obvious that the two characterizations (iii) given in Propositions 1 and 2 are compatible with these definitions of pole and zeros. In addition, these characterizations enable us to obtain the numbers of independent output pole-directions and input zero-directions of S associated with λ simply as m-rank $D_\ell(\lambda)$ and r-rank $N_\ell(\lambda)$, respectively.

Remark 4. It is easy to show that the above characterizations of output pole-directions and input zero-directions as well as the observations just following are also valid without any changes for the discrete-time versions as described in Remark 1.

3. MULTIVARIABLE POLE-ZERO CANCELLATION (DEFINITION AND APPLICATION)

Using the two directions defined above enables us to obtain very easily necessary and sufficient conditions for controllability, observability and so on of the composite systems as stated in Section 1.

(I) Series connection

Theorem 1. Let S_{12} be the series connection of S_1 followed by S_2 (Fig.1) with the natural state variable.

(i) S_{12} is observable (detectable) if and only if

C1 (C1') : $P_\lambda(S_1) \cap Z_\lambda(S_2) = \phi$ for all (unstable) poles λ of S_1

(ii) S_{12} is controllable (stabilizable) if and only if

C2 (C2') : $P_\lambda(S_2{}^d) \cap Z_\lambda(S_1{}^d) = \phi$ for all (unstable) poles λ of S_2

Here, λ being *unstable* means Re $\lambda \geq 0$, and $S_1{}^d$ ($S_2{}^d$) is the dual system of S_1 (S_2).

(Proof)

(i) To prove this, we use the known fact that S is unobservable iff there exist a number λ and an initial state $x_0 \neq 0$ such that $Ax_0 = \lambda x_0$ and $Cx_0 = 0$, or equivalently, for all $t \geq 0$

$$x(t; x_0, 0) = x_0 e^{\lambda t}$$

$$y(t; x_0, 0) = 0$$

From this fact we observe that the cascade system S_{12} with the state variable $\hat{x} = [x_1^T \ x_2^T]^T$ is unobservable iff there exist a number λ and an initial sate $\hat{x}_0 = [x_{10}^T \ x_{20}^T]^T \neq 0$ such that the zero-input responses at the state and output of S_{12} become

$$x_1(t; x_{10}, 0) = x_{10} e^{\lambda t} \tag{11a}$$

$$x_2(t; x_{20}, u_2) = x_{20} e^{\lambda t} \tag{11b}$$

$$y(t; \hat{x}_0, 0) = y_2(t; x_{20}, u_2) = 0 \tag{12}$$

for all $t \geq 0$. We also observe in this case that the corresponding zero-input response at the output of S_1, which is equal to the corresponding input function to S_2, becomes

$$y_1(t; x_{10}, 0) = u_2(t) = p_1 e^{\lambda t} \tag{13}$$

where $p_1 \triangleq C_1 x_{10} \neq 0$, since otherwise $x_{10} = 0$ [$x_{20} = 0$] by (11a), (13) [(11b) - (13)] and the observability assumption of S_1 [S_2], a contradiction to $\hat{x}_0 \neq 0$, so that $x_{10} \neq 0$. We finally observe from Definitions 1 and 2 that $p_1 \in P_\lambda(S_1) \cap Z_\lambda(S_2)$ iff (11) - (13) hold for some $x_{10} \neq 0$ and x_{20}. We therefore conclude that S_{12} is unobservable iff

$$P_\lambda(S_1) \cap Z_\lambda(S_2) \neq \phi$$

This completes the proof for the *observability* case. The detectability case can be verified analogously by using the fact that S is *undetectable* iff $Ax_0 = \lambda x_0$ and $Cx_0 = 0$ for some λ with Re $\lambda \geq 0$ and some $x_0 \neq 0$.

(ii) This is a direct consequence of (i) and the fact that S_{12} is controllable (stabilizable) iff its dual system $S_{12}{}^d$, or equivalently, a series connection of $S_2{}^d$ followed by $S_1{}^d$ is observable (detectable). Q E D.

Remark 5. The discrete-time versions of Theorem 1 can be obtained if in Theorem 1 we use the discrete-time versions of the sets $P_\lambda(\cdot)$ and $Z_\lambda(\cdot)$, and we mean $|\lambda| \geq 1$ by λ being unstable.

Remark 6. It is a direct consequence of Propositions 1 and 2 to rewrite the four conditions in Theorem 1 in terms of the coprime fraction descriptions of G_1 and G_2:

$$G_i(s) = D_{\ell i}(s)^{-1} N_{\ell i}(s) = N_{ri}(s)\, D_{ri}(s)^{-1} \qquad i = 1,\ 2$$

as follows

(i) C1 (C1') $\Longleftrightarrow$ $\begin{bmatrix} D_{\ell 1}(\lambda) \\ N_{\ell 2}(\lambda) \end{bmatrix}$ has full column rank for all (unstable) poles λ of S_1

(ii) C2 (C2') $\Longleftrightarrow$ $[\, D_{r2}(\lambda) \quad N_{r1}(\lambda)\,]$ has full row rank for all (unstable) poles λ of S_2

from which part of the results in [4] follows.

Remark 7. Conditions C1 and C2 are quite similar to those conditions used by Bart et al. [5, Theorem 4.6] for characterizing the minimality of a square factorization $G(s) = G_1(s)\, G_2(s)$ with $m = n$ and $\det G(s) \not\equiv 0$. In their conditions the sets of pole- and root-vectors as stated in Remark 3 and denoted by Pol $(G\,;\lambda)$ and Ker $(G\,;\lambda)$, are used in place of $P_\lambda(S)$ and $Z_\lambda(S)$, respectively. Note, however, that their result [5, Theorem 4.6] may not be generalized to the non-square case of G(s), or even to the square and singular case of G(s), since in the proof of this result they use the following property

$$\mathrm{Ker}\,(\,G_i\ ;\ \lambda\,) = \mathrm{Pol}\,(\,G_i{}^{-1}\ ;\ \lambda\,), \qquad i = 1,\ 2$$

which requires $\det G_i(s) \not\equiv 0$ $(i = 1,\ 2)$ implicitly. In addition to that, their proof is based on the minimal realizations of G_1 and G_2, which is not so consistent with the definitions of pole- and root-vectors based on the Laurent expansion of a rational function matrix. From these observations together with our proof we believe that our result (i.e. Theorem 1) is more general and easier to understand intuitively than the corresponding result by Bart et al. [5, Theorem 4.6].

Noting the similarity of Theorem 1 to the corresponding results for the scalar case as described in §1, it may be natural to define multivariable pole-zero cancellation as follows:

Definition 3. In the cascade system S_{12} of S_1 followed by S_2 as shown in Fig.1, G_2G_1 has no (unstable) pole-zero cancellations if and only if Conditions C1 and C2 (C1' and C2') are both satisfied.

With this notion of pole-zero cancellation, a concise characterization of minimality of the cascade system follows directly from Theorem 1.

Corollary 1. The cascade system S_{12} of S_1 followed by S_2 is controllable and observable (stabilizable and detectable) if and only if G_2G_1 has no (unstable) pole-zero cancellations.

Comment : The same expression as this is already known [4] with a different notion of pole-zero cancellation : *G_2G_1 has no pole-zero cancellations iff the McMillan degree of G_2G_1 equals to the sum of those of G_2 and G_1*. Since our definition is based on two direction vectors related directly to the system pole and zero, and moreover it includes the cancellation of these two vectors in addition to the usual pole-zero cancellation in the scalar case, we believe that our definition of multivariable pole-zero cancellation is more natural than the one in [4] as a multivariable extension of the scalar definition. It is important to note that the inclusion of cancellation of *vectors* in our definition explains intuitively the intrinsic difference between pole-zero cancellations in the scalar and multivariable cases.

(II) Feedback connection

We know [1] that the feedback connection S_f of S_1, S_2 shown in Fig.2 is controllable (observable) iff so is the series connection of S_1 (S_2) followed by S_2 (S_1), under the assumption of $\det(I + G_1(\infty)G_2(\infty)) \neq 0$. The same is true for stabilizability and detectability. Hence the corresponding results for the feedback connection to Theorem 1 and Corollary 1 are obvious from these two results, and hence they are omitted.

(III) Parallel connection

Theorem 2. Let S_p be the parallel connection of S_1, S_2 (Fig.3) with the natural state variable, and $S_1{}^d$ ($S_2{}^d$) be the dual system of S_1 (S_2).

(i) S_p is observable (detectable) if and only if

B1 (B1'): $P_\lambda(S_1) \cap P_\lambda(S_2) = \phi$ for all (unstable) poles λ of S_1, S_2

(ii) S_p is controllable (stabilizable) if and only if

B2 (B2'): $P_\lambda(S_1{}^d) \cap P_\lambda(S_2{}^d) = \phi$ for all (unstable) poles λ of S_1, S_2

(Proof)

(i) (If) Assume that the parallel system S_p with the state variable $\hat{x}=[x_1{}^T\ x_2{}^T]^T$ is unobservable. Then there exist a number λ and an initial state $\hat{x}_0=[x_{10}{}^T\ x_{20}{}^T] \neq 0$ such that the zero-input responses at the state and output of S_p become

$$x_1(t;\ x_{10},\ 0) = x_{10}\,e^{\lambda t} \tag{14a}$$

$$x_2(t;\ x_{20},\ 0) = x_{20}\,e^{\lambda t} \tag{14b}$$

$$y(t;\ \hat{x}_0,\ 0) = y_1(t;\ x_{10},\ 0) + y_2(t;\ x_{20},\ 0) = 0 \tag{15}$$

for all $t \geq 0$. Here

$$y_1(t;\ x_{10},\ 0) = p_1 e^{\lambda t}, \qquad p_1 \triangleq C_1 x_{10} \tag{16a}$$

$$y_2(t;\ x_{20},\ 0) = p_2 e^{\lambda t}, \qquad p_2 \triangleq C_2 x_{20} \tag{16b}$$

and hence by (15) we have

$$p_1 = -p_2 \tag{17}$$

Moreover it follows easily from (14) - (16) and the observability assumption of S_1, S_2 that

$$p_1 \neq 0 \quad \text{and} \quad p_2 \neq 0 \tag{18}$$

so that

$$x_{10} \neq 0 \quad \text{and} \quad x_{20} \neq 0 \tag{19}$$

Combining (14a), (16a), (18a) and (19a) yields by Definition 1

$$p_1 \in P_\lambda(S_1)$$

Analogously by (14b), (16b), (18b) and (19b) we can obtain

$$-p_2 \in P_\lambda(S_2)$$

We therefore conclude from (17) that $P_\lambda(S_1) \cap P_\lambda(S_2) \neq \phi$.

(Only if) Let $p \in P_\lambda(S_1) \cap P_\lambda(S_2)$. Then by definition there exist some nonzero initial states x_{10} and x_{20} such that

$$x_1(t; x_{10}, 0) = x_{10} e^{\lambda t}, \quad y_1(t; x_{10}, 0) = p e^{\lambda t} \tag{20}$$

$$x_2(t; x_{20}, 0) = x_{20} e^{\lambda t}, \quad y_2(t; x_{20}, 0) = p e^{\lambda t} \tag{21}$$

Since $p = C_2 x_{20}$ in (21), by choosing $\tilde{x}_{20} \overset{\Delta}{=} -x_{20}$ in place of x_{20} in (21), we have

$$x_2(t; \tilde{x}_{20}, 0) = \tilde{x}_{20} e^{\lambda t}, \quad y_2(t; \tilde{x}_{20}, 0) = -p e^{\lambda t} \tag{22}$$

It then follows from (20b) and (22b) that the zero-input response at the output of Sp for the initial state $\hat{x}_0 = [\, x_{10}^T \;\; \tilde{x}_{20}^T \,]^T \neq 0$ satisfies

$$y(t; \hat{x}_0, 0) = y_1(t; x_{10}, 0) + y_2(t; \tilde{x}_{20}, 0) = 0 \qquad \text{for all } t \geq 0 \tag{23}$$

From (20a), (22a) and (23) we conclude that Sp is unobservable.

The proof stated above is for the observability case. The proof for the detectability case follows the same line except that λ is restricted to be unstable.

(ii) This is a direct consequence of (i) and the fact that Sp is controllable (stabilizable) iff its dual system Sp^d, or equivalently, a parallel connection of S_1^d and S_2^d is observable (detectable). Q E D.

Remark 8. It is also an obvious matter to rewrite the four conditions in Theorem 2 in terms of the coprime fraction descriptions of G_1, G_2, like we did in Remark 6; e.g.,

$$\text{B1 (B1')} \Longleftrightarrow \begin{bmatrix} D_{\ell 1}(\lambda) \\ D_{\ell 2}(\lambda) \end{bmatrix} \text{ has full rank for all (unstable) poles } \lambda \text{ of } S_1 \text{ and } S_2$$

from which Theorem 2 of [3] follows as a collary of Theorem 2.

Remark 9. Noting that in the scalar case, B1 and B2 as in Theorem 2 never hold if S_1 and S_2 have a common pole, we easily obtain the well known scalar result [1] stated as : *Sp is controllable and observable iff S_1 and S_2 have no common poles.*

4. APPLICATION TO MINIMALITY CHARACTERIZATION OF STOCHASTIC SYSTEMS

As the final application, we characterize the minimality of a stochastic system in terms of the new notion of pole-zero cancellation defined above.

Let us consider the stochastic system S with the minimal state space description (1), where x is the n-dimensional state process, y is the m-dimensional output process u is the r-dimensional zero mean white noise process, and A is stable, i.e.,

A has all its eigenvalues in the open left-half complex plane. We write $\mathcal{S}$ for the class of such a system with the fixed number of outputs and constant matrices A, B, C and E as described above. Then the output spectral density matrix of S is given by

$$\Phi(s) = G(s)\ G(-s)^{T} \tag{24a}$$

where

$$G(s) = C(sI - A)^{-1}B + E \tag{24b}$$

This relation plays an important role in the subsequent development.

The problem of interest is concerned with the minimality of $S \in \mathcal{S}$ in the following sense.

Definition 4. We say that $S \in \mathcal{S}$ is minimal, if it has the smallest dimension among all those systems in $\mathcal{S}$ which have the same output covariance as S in the steady state.

Note that this sense of minimality of the stochastic system S is the same as that of a (stable) spectral factor G(s) in the minimal spectral factorization problem of $\Phi(s)$ posed by Anderson [8] (see also [5]). This problem is stated as: *Given a spectral density matrix $\Phi(s)$, find (A, B, C, E) with A stable such that (i) (24) holds and (ii) A has the smallest possible dimension.* If A, B, C and E satisfy (i), then G(s) given by (24b) is called a (stable) spectral factor of $\Phi(s)$, and it is, in particular, called a *minimal* spectral factor if it also satisfies (ii). It is shown in [8] that the spectral factor G(s) is minimal in this sense iff

$$\delta[\,G(s)\,] = \frac{1}{2}\,\delta[\,\Phi(s)\,]$$

or equivalently

$$\delta[\,G(s)\ G(-s)^{T}\,] = 2n \tag{25}$$

Moreover, (25) is equivalent to saying that a series connection of S with the transfer function matrix G(s) followed by its adjoint system with the transfer function matrix $G(-s)^{T}$ is controllable and observable.

Combining these observations with Corollary 1 yields our desired result as follows.

Theorem 3. A system $S \in \mathcal{S}$ is minimal if and only if $G(s)G(-s)^{T}$ has no pole-zero cancellations, in other words, no output pole-direction of S associated with any pole λ of S is not an input zero-direction of the adjoint system of S associated with λ.

5. CONCLUSIONS

We have proposed a new simple definition of multivariable pole-zero cancellation including unstable pole-zero cancellation as the most natural extension of the scalar one. This has enabled us to give some new compact characterizations of the minimality of various composite systems as well as stochastic systems. This notion of pole-zero cancellation provides us with an intuitive understanding not only of how pole-zero cancellations are related to the minimality of composite systems, but also of what is the intrinsic difference between pole-zero cancellations in the scalar and multivariable cases. Moreover it is highly consistent with the interesting dynamical interpretation of poles and zeros as shown in [6], [7]. From these points of view we believe that our definition give some new insights into the solutions of certain problems connected with pole-zero cancellation such as stability of feedback systems (see e.g. [4]) etc.

APPENDIX

Lemma A

For the coprime factorization (2), there exists a minimal realization ($\tilde{A}$, $\tilde{B}$, $\tilde{C}$, E) of G(s) together with polynomial matrices L(s), X(s) and Y(s) such that

(i) $$D_\ell(s)\tilde{C} = L(s)(sI - \tilde{A})$$

(ii) $$\begin{bmatrix} X(s) & Y(s) \\ L(s) & D_\ell(s) \end{bmatrix} \begin{bmatrix} sI - \tilde{A} & \tilde{B} \\ -\tilde{C} & E \end{bmatrix} = \begin{bmatrix} I & X(s)\tilde{B} + Y(s)E \\ 0 & N_\ell(s) \end{bmatrix}$$

where $D_\ell(s)$ and L(s) are left-coprime.

(Proof) This can be verified by following the same proof as that of Theorem H-6 [1] except that the factorization of G(s) corresponding to (H-19) in the proof [1], i.e.,

$$G(s) = E + D_\ell^{-1}(s)\bar{N}(s) = D_\ell^{-1}(s)\,[\,D_\ell(s)E + \bar{N}(s)\,] \triangleq D_\ell^{-1}(s)\,N_\ell(s)$$

is replaced by

$$G(s) = D_\ell^{-1}(s)N_\ell(s) = E + D_\ell^{-1}(s)\,[\,N_\ell(s) - D_\ell(s)E\,] \triangleq E + D_\ell^{-1}\bar{N}(s)$$

where $D_\ell(s)$ and $\bar{N}(s) \triangleq N_\ell(s) - D_\ell(s)E$ are left-coprime since so are $D_\ell(s)$ and $N_\ell(s)$ by assumption. Q.E.D.

ACKNOWLEDGEMENT

The authors are deeply grateful for generous invaluable comments by Professor Jan C. Willems on the idea of multivariable pole-zero cancellation proposed here. They are also grateful to Prof. M.A. Kashoek for mentioning the reference [5].

REFERENCES

[1] C. T. Chen, *Linear System Theory and Design*. New York: Holt Rinehart and Winston, 1984

[2] F. M. Callier and C. D. Nahum, "Necessary and sufficient conditions for the complete controllability and observability of systems in series using the coprime factorization of a rational matrix," *IEEE Trans. Circuits Syst.*, Vol. CAS-22, pp.90 - 95, Feb., 1975.

[3] W. A. Wolovich and H. L. Hwang, "Composite System Controllability and Observability," *Automatica*, Vol.10, pp.209 - 212, 1974.

[4] B. D. O. Anderson and M. R. Gevers, "On multivariable pole-zero cancellations and the stability of feedback systems," *IEEE Trans. Circuits Syst.*, Vol. CAS-28, pp.830 - 833, 1981.

[5] H. Bart, I. Gohberg and M. A. Kaashoek, *Minimal Factorization of Matrix and Operator Functions*. Basel: Birkhauser Verlag, 1979.

[6] A. G. J. MacFarlane and N. Karcanias, "Poles and zeros of linear multivariable systems: a survey of the algebraic, geometric and complex-variable theory," *Int. J. Control*, Vol.24, pp.33 - 74, 1976.

[7] C. A. Desoer and J. D. Schulman, "Zeros and poles of matrix transfer functions and their dynamical interpretation," *IEEE Trans. Circuits Syst.*, Vol. CAS - 21, pp.3 - 8, 1974.

[8] B. D. O. Anderson, "The inverse problem of stationary covariance generation," *J. Stat. Phys.*, pp.133 - 147, 1969.

Frequency Domain and State Space Methods for Linear Systems
C.I. Byrnes and A. Lindquist (Editors)

STRUCTURAL MODIFICATIONS OF LINEAR SYSTEMS A MATRIX PENCIL APPROACH

Jean Jacques LOISEAU

Laboratoire d'Automatique, E.N.S.M.
Unité Associée au C.N.R.S.
1 rue de la Noë 44072 NANTES Cédex France

The Morse's structural invariants of a linear system are modified by using non regular static state feedback control laws. This modification is translated in matrix pencil terms, using the relationships which exist between these invariants and the Kronecker's invariants of the singular pencil of the system. The study is particularized to the placement by static state feedback of the infinite zero orders.

1. INTRODUCTION

The Morse's structural invariants [1], and particularly the list I_4, which is the list of the infinite zero orders of the system [2] play a key-role in the solution of many control problems.

The decoupling problem, for instance, is solvable using regular static state feedback laws $u = Fx + Gv$ (regular stands for G invertible) if and only if the Morse's list I_4 of (C,A,B) equals the union of the Morse's list I_4 of each row or block-subsystem [3],[4].

When regular solution do not exist, one has to understand how some structural properties of (C,A,B), and more particularly, in this context, the infinite zero orders, can be modified through static state feedback transformations where G is only monic [5].

The aim of this paper is to present this "structural modification problem" within a matrix pencil approach, using the relationships which exist between the geometric Morse's decomposition of systems and the Kronecker's theory of singular pencils [6].

The paper is organized as follows. In § 2, we summarize the Morse's results, the Kronecker's theory and we recall the link between both approaches. In § 3, we translate in matrix pencil terms the problem consisting in the placement of the four Morse's lists. This study is particularized in § 4 to the placement of the Morse's lists I_4 and I_2. Finally § 5 is devoted to some concluding remarks : we make a link with the work of Heymann [7] and we introduce a recent result [8] which characterizes all the lists I_4 that can be obtained through static state feedback transformations.

2. PRELIMINARIES

2.1. Morse's lists

We shall consider linear time-invariant (C,A,B) systems described by the following equations :

$$\dot{x} = Ax + Bu$$

$$y = Cx$$

where $x \in \mathcal{X} \simeq \mathcal{R}^n$, $y \in \mathcal{Y} \simeq \mathcal{R}^p$, $u \in \mathcal{U} \simeq \mathcal{R}^m$, B is monic and C is epic. Let $\mathcal{B}$ denote the image of B and $\mathcal{K}$ the kernel of C.

Morse (1973) exhibited four lists which form a complete set of invariants for (C,A,B) under the following transformation group :

$$(C,A,B) \rightarrow (SCT^{-1},\ T(A + BF + RC)T^{-1},\ TBG)$$

where S,T,G are isomorphisms and R,F are homomorphisms.

The four Morse's lists can be deduced from the different steps of the following algorithms [9] :

- The Invariant Subspace Algorithm (I.S.A.) :

$$\mathcal{V}_K^o = X,\ \mathcal{V}_K = K \cap A^{-1}(\mathcal{B} + \mathcal{V}_K^{\mu-1}) \text{ for } \mu \geqslant 1,$$

which is non increasing and converges to $\mathcal{V}_K^*$, the supremal (A,B) invariants subspace included in K.

- The Conditionnaly Invariant Subspace Algorithm (C.I.S.A.) :

$$\mathcal{S}_B^o = 0,\ \mathcal{S}_B^\mu = \mathcal{S} + A(K \cap \mathcal{S}_B^{\mu-1}) \text{ for } \mu \geqslant 1,$$

which is non decreasing and converges to $\mathcal{S}_B^*$, the infimal (C,A) invariant subspace containing $\mathcal{B}$.

Indeed, we have :

- The Morse's list I_1 is the list of invariant polynomials $p_i(s)$ of the map $(\mathcal{V}_K^*/\mathcal{V}_K^* \cap \mathcal{S}_B^* | A + BF | \mathcal{V}_K^*/\mathcal{V}_K^* \cap \mathcal{S}_B^*)$ induced in the quotient space $\mathcal{V}_K^*/\mathcal{V}_K^* \cap \mathcal{S}_B^*$ by any map $(A + BF)$ such that $(A + BF)\ \mathcal{V}_K^* \subset \mathcal{V}_K^*$.

- <u>The Morse's list I_2</u> is the list of positive integers σ_i :

$$\sigma_i = \text{card}\{\ j \mid \alpha_j \geqslant i\} \quad \text{for } i \geqslant 1, \tag{2.1}$$

where card $\{.\}$ stands for the number of elements of $\{\ .\}$, and where :

$$\sigma_j = \dim\ (\mathcal{S}_B^j \cap \mathcal{V}_K^*/\mathcal{S}_B^{j-1} \cap \mathcal{V}_K^*) \text{ for } j \geqslant 1 \tag{2.2}$$

- <u>The Morse's list I_3</u> is the list of positive integers τ_i :

$$\tau_i = \text{card}\{\ j \mid \beta_j \geqslant i\} \quad \text{for } i \geqslant 1, \tag{2.3}$$

where :

$$\beta_j = \dim\ (\mathcal{V}_K^{j-1} + \mathcal{S}_B^*/\mathcal{V}_K^j + \mathcal{S}_B^*) \text{ for } j \geqslant 1. \tag{2.4}$$

- <u>The Morse's list I_4</u> is the list of positive integers n_i :

$$n_i = \text{card}\{\ j \mid p_j \geqslant i\} \quad \text{for } i \geqslant 1, \tag{2.5}$$

with :

$$p_i = \dim\ (\mathcal{S}_B^j + \mathcal{V}_K^*/\mathcal{S}_B^{j-1} + \mathcal{V}_K^*) \text{ for } j \geqslant 1. \tag{2.6}$$

Note that the above definitions of I_1, I_2, I_3 are those given by Morse [1], whereas the definition of I_4 is due to Dion and Commault [2].

2.2. Kronecker's theory of singular pencils

Kronecker characterized at the end of the 19th century the orbit of a pencil (sE-H) under left and right changes of bases :

$$(sE - H) \rightarrow P\ (sE - H)\ Q,$$

where P and Q are isomorphisms.

The Kronecker's normal form of (sE - H) associated with this transformation group is a quasi diagonal form where each diagonal block corresponds to one of the Kronecker's invariants (see Gantmacher [6] for a complete information) :

$$P(sE - H)\,Q = \mathrm{diag}\{L_{\varepsilon_1},\dots,L_{\varepsilon_{\alpha_1}};L_{\nu_1},\dots,L_{\nu_{p_2}};L_{\eta_1},\dots,L_{\eta_{\beta_1}};L_{z_1},\dots,L_{z_r}\}$$

with :

$$L_{\varepsilon_i} = \underbrace{\begin{bmatrix} 1 & s & 0 & & (0) \\ 0 & 1 & s & & \\ & (0) & \ddots & \ddots & \\ & & & 1 & s \end{bmatrix}}_{\varepsilon_i + 1} \Big\updownarrow \varepsilon_i \qquad (2.8) \qquad L_{\eta_i} = \underbrace{\begin{bmatrix} 1 & 0 & & \\ s & 1 & (0) & \\ 0 & s & \ddots & \\ & (0) & \ddots & 1 \\ & & & s \end{bmatrix}}_{\eta_i} \Big\updownarrow \eta_i + 1 \qquad (2.9)$$

$$L_{\nu_i} = \underbrace{\begin{bmatrix} 1 & s & & (0) \\ 0 & 1 & \ddots & \\ & & \ddots & s \\ (0) & & & 1 \end{bmatrix}}_{\nu_i} \Big\updownarrow \nu_i \qquad (2.10) \qquad L_{\lambda_i} = \underbrace{\begin{bmatrix} s-z_i & 1 & & (0) \\ 0 & s-z_i & \ddots & \\ & & \ddots & 1 \\ (0) & & & s-z_i \end{bmatrix}}_{\lambda_i} \Big\updownarrow \lambda_i \qquad (2.11)$$

The non negative integers ε_i (resp. η_i) are the column (resp. row) minimal indices, the positive integers ν_i are the infinite elementary divisors, and the polynomials $(s - z_i)^{\lambda_i}$ are the finite elementary divisors of the pencil (sE - H).

Note that a zero column (resp. row) index do not correspond to a diagonal block but to a zero column (resp. row).

2.3. The restricted matrix pencil

The Morse's decomposition of systems is closely related to the Kronecker's theory of matrix pencils. Indeed one can associate to a (C,A,B) system the so-called restricted matrix pencil (sE - H) :

$$(sE - H) := N(s1 - A)K$$

where $K : \mathcal{K} \to \mathcal{X}$ is a basis matrix for $\mathcal{K}$, and $N : \mathcal{X} \to \mathcal{X}/\mathcal{B}$ is a left annihilator for B.

This pencil has been studied in detail by Jaffe and Karcanias [10], who obtained a complete classification of the geometric concepts in matrix pencil terms. A well known result (see for instance [11]) is that the following relationships hold between the four Morse's lists of (C,A,B) and the Kronecker's invariants of its restricted pencil : we have with the previous notations :

$$\varepsilon_i + 1 = \sigma_i \quad , \quad \text{for } i = 1 \text{ to } \alpha_1 \qquad (2.12)$$

$$\eta_i + 1 = \tau_i \quad , \quad \text{for } i = 1 \text{ to } \beta_1 \qquad (2.13)$$

$$\nu_i + 1 = n_i \quad , \quad \text{for } i = 1 \text{ to } p_2 \qquad (2.14)$$

The list $\{(s - z_i)^{\lambda_i}\}$ is the list of elementary divisors associated (by a well-known bijection [9]) with the list of invariant factors I_1. (2.15)

The infinite zeros of the system of which the order is 1 do not appear in the restricted pencil. This is due to the fact that these zeros correspond to a part of $\mathcal{B}$ which disappears with N.

Nethertheless, this loss of structural information can be compensated by the knowledge of either m, p or n, since :

$$p_1 = m - \alpha_1 ,$$

or $$p_1 = p - \beta_1$$

or $$\sum_{i=1}^{p_1} n_i = n - \sum_{i=1}^{\alpha_1} \sigma_i - \sum_{i=1}^{\beta_1} \tau_i - \sum_{i=1}^{r} \lambda_i$$

The following property is thus clear.

THEOREM 1.

Let (C,A,B) and (C',A',B') be two systems, and let (sE - H) and (sE' - H') be their corresponding restricted matrix pencils.

There then exist isomorphisms T,G and S, and homomorphisms F and R which verify :

$$A' = T\,(A + BF + RC)\,T^{-1}$$
$$B' = TBG$$
$$C' = SCT^{-1} ,$$

if and only if the systems have the same dimension and there exist isomorphisms P and Q which verify :

$$P\,(sE' - H') = (sE - H)\,Q$$

We shall now use this link to translate in matrix pencil terms the problem of the modification of the structural invariants of a system under non regular feedback transformations.

3. THE PLACEMENT OF THE FOUR MORSE'S LISTS

Let (sE - H) be the restricted matrix pencil associated with a (C,A,B) system to which the Morse's lists are $\{p_i(s)\}$, $\{\sigma_i\}$, $\{\tau_i\}$ and $\{n_i\}$.

Let $\{p_i'(s)\}$ be a list of invariant factors (i.e. a list of polynomials with adequate division properties) and $\{\sigma_i'\}$, $\{\tau_i'\}$ and $\{n_i'\}$ three non increasing lists of positive integers, and let (sE' - H') be any matrix pencil corresponding to these lists in the sense of (2.12), (2.13), (2.14) and (2.15).

We shall prove the following :

THEOREM 2.

There exist an isomorphism T and homomorphisms F and G (G monic) such that the Morse's lists of the closed-loop system $(C',A',B') = (CT^{-1}, T(A + BF)T^{-1}, TBG)$ are $\{p_i'(s)\}$, $\{\sigma_i'\}$, $\{\tau_i'\}$ and $\{n_i'\}$ if and only if the two following conditions hold:

(i) $\mathrm{card}\{n_i'\} + \mathrm{card}\{\tau_i'\} = \mathrm{card}\{n_i\} + \mathrm{card}\{\tau_i\}$

(ii) there exist an isomorphism Q and an epimorphism P which satisfy :

$$P\,(sE' - H') = (sE - H)\,Q$$

PROOF

NECESSITY : Since $(C',A',B') = (CT^{-1}, T(A + BF)T^{-1}, TBG)$, the restricted pencil $N'(sI - A')K'$ of (C',A',B') can be described by :

$$K' = TK$$
$$PN'T = N ,$$

where P is the canonical epimorphism from X/Im TBG to X/ImB. We now have the following :

$$
\begin{aligned}
N(s1 - A)K &= N(s1 - (A + BF))\ K \\
&= NT^{-1}(s1 - T(A + BF)\ T^{-1})\ TK \\
&= PN'(s1 - A')\ K',
\end{aligned}
$$

and the necessity of (ii) follows from the fact that $(sE' - H')$ and $N'(s1 - A')K'$ have the same Kronecker's invariants.

Condition (i) means that both systems have the same dimension , which is obviously necessary.

SUFFICIENCY : In order to prove the sufficiency, let us consider a triple (C',A',B') of which the Morse's lists are $\{p_i'(s)\}$, $\{\sigma_i'\}$, $\{\tau_i'\}$ and $\{n_i'\}$. The restricted pencil associated with (C',A',B') is a pencil $N'(s1 - A')K'$ which is equivalent to $(sE' - H')$. With a judicious choice of bases we can assume that :

$$sE' - H' = N'\ (s1 - A')\ K'$$

From (i), one can see that the output space of (C',A',B') is isomorphic to the output space $\mathcal{Y}$ of (C',A',B'). From this, and since Q is an isomorphism, one can see that the state space of (C',A',B') is also isomorphic to the state space $\mathcal{X}$ of (C,A,B). Indeed we will consider that C' acts from $\mathcal{X}$ to $\mathcal{Y}$, A' acts from $\mathcal{X}$ to $\mathcal{X}$ and B' acts to $\mathcal{X}$ from an input space $\mathcal{V}$.

Define now N'' like follows :

$$N'' = PN'$$

N'' is epic, and acts from $\mathcal{X}$ to $\mathcal{X}/\mathcal{B}$. Let B'' be a basis matrix of ker N''. There then exists a monic matrix $G'' : \mathcal{V} \to \mathcal{U}$, verifying the following :

$$B' = B''G''$$

The pencil associated with (C',A',B'') is $P(sE' - H')$ which is equivalent to $(sE-H)$. We then have (see Theorem 1) :

$$(C',A',B'') = (SCT^{-1},\ T(A + BF + RC)T^{-1},\ TBG)$$

and consequently :

$$(C',A',B') = (SCT^{-1},\ T(A + BF + RC)T^{-1},\ TBGG'')$$

The system $(C,\ A + BF,\ BGG'')$ is then equivalent to (C',A',B'), and its Morse's lists are thus $\{p_i'(s)\}$, $\{\sigma_i'\}$, $\{\tau_i'\}$ and $\{n_i'\}$, which ends the proof.

4. THE PLACEMENT OF THE MORSE'S LISTS I_2 AND I_4

Consider now a system (C,A,B) to which the Morse's lists I_2 and I_4 are $\{\sigma_i\}$ and $\{n_i\}$. Let $\{\sigma_i'\}$ and $\{n_i'\}$ be two non increasing lists of positive integers.

Let $(sE_1 - H_1)$ be a matrix pencil associated with the lists $\{\sigma_i\}$ and $\{n_i\}$, and $(sE_1' - H_1')$ be a matrix pencil corresponding to the lists $\{\sigma_i'\}$ and $\{n_i'\}$ in the sense of (2.12) and (2.14).

THEOREM 3.

There exist an isomorphism T, and homomorphisms F and G (G monic) such that the Morse's lists I_2 and I_4 of the closed loop system $(C',A',B') = (CT^{-1},T(A+BF)T^{-1},TBG)$ are respectively $\{\sigma_i'\}$ and $\{n_i'\}$ if and only if the two following conditions are verified :

(i) card $\{j | n_j' \leq i\} \leq$ card $\{j | n_j \leq i\}$ for $i \geq 1$.

(ii) there exist an homomorphism P_1 and a monomorphism Q_1 such that :

$$P_1(sE_1' - H_1') = (sE_1 - H_1)\ Q_1$$

PROOF

NECESSITY : The necessity of (ii) comes directly from Theorem 2. The pencils $(sE-H)$ and $(sE'-H')$ of Theorem 2 can be chosen in their Kronecker normal form. This quasi diagonal form induces a block-partitionned form for P and Q.

An easy calculus, based on the form of the diagonal blocks of (sE - H) and (sE'- H') (see (2.7), (2.8), (2.9), (2.10)) shows that some blocks of P and Q must be 0. We have in fact the following :

$$P = \begin{bmatrix} P_{11} & 0 & P_{13} & 0 \\ P_{21} & P_{22} & P_{23} & P_{24} \\ 0 & 0 & P_{33} & 0 \\ 0 & 0 & P_{43} & P_{44} \end{bmatrix} \qquad Q = \begin{bmatrix} Q_{11} & 0 & Q_{13} & 0 \\ Q_{21} & Q_{22} & Q_{23} & Q_{24} \\ 0 & 0 & Q_{33} & 0 \\ 0 & 0 & Q_{43} & Q_{44} \end{bmatrix}$$

Consequently, one can choose P_1 and Q_1 like follows :

$$P_1 = \begin{bmatrix} P_{22} & P_{24} \\ 0 & P_{44} \end{bmatrix} \qquad Q_1 = \begin{bmatrix} Q_{22} & Q_{24} \\ 0 & Q_{44} \end{bmatrix}$$

Q_1 is monic because Q is an isomorphism (but nothing can be said about P_1).

The necessity of (i) can be deduced from the following geometric definition of the Morse's list I_4 [12] :

$$\dim \left(\frac{S_B^i + K}{K}\right) = p_1 - p_{i+1} \text{ , for } i \geq 1, \tag{4.1}$$

and from the fact that (see (2.5)) :

$$p_1 - p_{i+1} = \text{card } \{j \mid n_j \leq i\}, \text{ for } i \geq 1. \tag{4.2}$$

Indeed, for the closed loop system the steps of the algorithm (CISA) :

$$S_{B'}^0 = 0 \text{ ; } S_{B'}^i = B' + (A + BF)\ (K \cap S_{B'}^{i-1}) \text{ , for } i \geq 1,$$

where $B' := \text{Im } BG$,

obviously satisfy : $S_{B'}^i \subset S_B^i$ for $i \geq 0$. This gives :

$$\dim \left(\frac{S_{B'}^i + K}{K}\right) \leq \dim \left(\frac{S_B^i + K}{K}\right) \text{ , for } i \geq 0,$$

which, combined with (4.1) and (4.2), shows the necessity of (i).

The sufficiency is much more intricate. We will use the following Lemma :

LEMMA 4.

$\{\sigma_i\}_{\alpha_1}$ and $\{n_i\}_{p_1}$ are the Morse's lists I_2 and I_4 of (C,A,B) if and only if it exists a subspace S of the state space X, and an $F : X \to U$, having the following properties :

$$S = \bigoplus_{i=1}^{\alpha_1} R_i \bigoplus_{i=1}^{p_1} S_i \quad , \tag{4.3}$$

where :

- For i = 1 to α_1, R_i is a controllability subspace [9] (generated by a single vector b_i of B) and included in K :

$$R_i = \text{span } \{b_i, A_F\, b_i, \dots, A_F^{\sigma_i - 1}\, b_i\}$$

with $b_i \in B$, $A_F^{j-1} b_i \in K$ for $1 \leq j \leq \sigma_i$, $A_F^{\sigma_i} b_i = 0$

and $A_F = A + BF$, $F : X \to U$.

- For i = 1 to p_1, S_i is an almost controllability prime subspace [2] :

$$S_i = \text{span}\ \{b_i, A_F\, b_i, \ldots, A_F^{n_i-1}\ b_i\},$$

with $b_i \in B$, $A_F^{j-1}\ b_i \in K$ for $1 \leq j \leq n_i-1$, $A_F^{n_i-1} b_i \notin K$.

- B is spanned by the vectors b_i and b_i :

$$B = \text{span}\{\, b_1, \ldots, b_{\alpha_1}\ ;\ b_1, \ldots, b_{p_1}\} \tag{4.4}$$

Otherwise said :

$$\dim B = \alpha_1 + p_1$$

- S is a (C,A) invariant subspace ; that is :

$$S \supset A(K \cap S) \tag{4.5}$$

Otherwise said :

$$\dim C\, S = p_1$$

PROOF

The necessity comes by choosing $S = S_B^*$. (4.3) is then the well-known decomposition of S_B^* into elementary controllability and almost controllability subspaces (see [1], [2],[9],[13] and [14]).

The sufficiency comes from Willems [13]. The subspace S is both a (C,A) invariant subspace containing B such that

$$S = B + A(K \cap S)$$

and an almost controllability subspace.

It thus follows from Theorem 7, p. 241 of [13] that :

$$S = S_B^*$$

and, using the definitions (2.1), (2.2), (2.5), (2.6), one can verify that the Morse's lists I_2 and I_4 of (C,A,B) are $\{\sigma_i\}_{\alpha_1}$ and $\{n_i\}_{\alpha_1}$, which ends the proof.

REMARK 5.

One can see that this decomposition of S_B^* is closely related to the Kronecker's decomposition of $(sE_1 - H_1)$. With the previous notation, we have :

$$(sE_1 - H_1) = \text{diag}\ \{L_{\varepsilon_1}, \ldots, L_{\varepsilon_{\alpha_1}}\ ;\ L_{\nu_1}, \ldots, L_{\nu_{p_1}}\}$$

Each block L_{ε_i} can be considered as acting from R_i to $\frac{R_i + B}{B}$ whereas L_{ν_i} from $K \quad S_i$ to $\frac{S_i + B}{B}$.

Consequently we shall assume in the following that the diagonal blocks of $(sE_1 - H_1)$ are written with respect to the bases described in Lemma 4.

We shall now complete the proof of Theorem 3 :

SUFFICIENCY : We previously have noted that P_1 and Q_1 are block-partitionned following the diagonal form of $(sE_1 - H_1)$ and $(sE_1' - H_1')$. In fact each of the blocks of P_1 and Q_1 satisfies an equation of the form :

$$P_{ij}\ L'_j = L_i\ Q_{ij}\ ,$$

where L_i (resp. L'_j) is one of the elementary diagonal blocks of $(sE_1 - H_1)$ (resp. $(sE'_1 - H'_1)$).

Indeed, to each pair (σ'_i, σ'_j) corresponds a block with the following form :

$$Q_{\sigma_i\sigma'_j} = \underbrace{\begin{bmatrix} a^1_{ij} & a^2_{ij} & \cdots\cdots & a^{\sigma'_j-\sigma_i+1}_{ij} & & (0) \\ \ddots & \ddots & & \ddots & & \\ (0) & & a^1_{ij} & a^2_{ij} & \cdots\cdots & a^{\sigma'_j-\sigma_i+1}_{ij} \end{bmatrix}}_{\sigma'_j} \Bigg\}\ \sigma_i \tag{4.6}$$

To each pair (σ_i, n'_j) corresponds a block with the form :

$$Q_{\sigma_i n'_j} = \underbrace{\begin{bmatrix} d^1_{ij} & d^2_{ij} & \cdots & d^{n'_j-1}_{ij} \\ & d^1_{ij} & & \vdots \\ & & \ddots & d^{n'_j-\sigma_i}_{ij} \\ (0) & & d^1_{ij} \cdots & d^{n'_j-\sigma_i}_{ij} \end{bmatrix}}_{n'_j - 1} \Bigg\}\ \sigma_i \tag{4.7}$$

Finally, to each pair (n_i, n'_j) corresponds a block with the form :

$$Q_{n_i n'_j} = \underbrace{\begin{bmatrix} 0\ldots0 & c^{n'_j-n_i+1}_{ij} & \cdots & c^{n'_j-2}_{ij} & c^{n'_j-1}_{ij} \\ & & & & c^{n'_j-2}_{ij} \\ & & \ddots & & \vdots \\ & (0) & & & c^{n'_j-n_i+1}_{ij} \end{bmatrix}}_{n'_j - 1} \Bigg\}\ n_i - 1 \tag{4.8}$$

The monic matrix Q_{22}, partitionned according to the lists $\{\sigma_i\}$ and $\{\sigma'_i\}$ is the basis matrix of a controllability subspace $\mathcal{R}'$ included in $\mathcal{R}^{*\prime}_{K}$ [15].

We have

$$\mathcal{R}' = \mathcal{R}'_1 \oplus \ldots \oplus \mathcal{R}'_{\alpha_1} \tag{4.9}$$

where :

$$\mathcal{R}'_i = \operatorname{span}\{\, b'_i,\ A_{F'}\, b'_i,\ \ldots,\ A_{F'}^{\sigma'_i-1}\, b'_i \}$$

where b'_i and F' are characterized by $Q_{\sigma_i\sigma'_j}$:

$$b'_i = \sum_{j=1}^{\alpha_1} a^1_{ji}\, b_j\ , \text{ for } i = 1 \text{ to } \alpha_1$$

$$BF'(A + BF')^{k-1}\, b_i' = \sum_{j=1}^{\alpha_1} a_{ji}^{k+1}\, b_j, \text{ for } k = 1 \text{ to } \sigma_i' - 1$$

Because of the form of $Q_{\sigma_i \sigma_j'}$, we have the following :

$$A_{F'}^{\sigma_i'}\, b_i' = \sum_{j=1}^{\alpha_1} a_{ji}^{\sigma_i' - \sigma_j' + 1}\quad A_F^{\sigma_j}\, b_j$$

$$= 0$$

In the same way, the matrix $\begin{bmatrix} Q_{24} \\ \hline Q_{44} \end{bmatrix}$, partitionned according to the lists $\{\sigma_i\}$, $\{\sigma_i'\}$ and $\{n_i'\}$ is the basis matrix of an almost controllability subspace R_a' included in K :

$$R_a' = R_{a_1}' \oplus R_{a_2}' \oplus \ldots \oplus \quad R_{a_{p_2}}'$$

where :

$$R_{ai}' = \text{span}\{\, b_i',\ A_{F'}\, b_i',\ \ldots,\ A_{F'}^{n_i'-2}\ b_i'\}$$

$$b_i' = \sum_{j=1}^{p_2} c_{ji}^1\, b_j + \sum_{j=1}^{\alpha_1} d_{ji}^1\, b_j \text{ , for } i = 1 \text{ to } p_2'$$

$$BF'(A + BF')^{k-1}\, b_i' = \sum_{j=1}^{p_2} c_{ji}^{k+1}\quad b_j + \sum_{j=1}^{\alpha_1} d_{ji}^{k+1}\, b_j$$

$$\text{for } k = 1 \text{ to } n_i' - 2.$$

In order to prove using Lemma 4 that $\{\sigma_i'\}$ and $\{n_i'\}$ are the Morse's lists I_2 and I_4 of a closed-loop system, we have to choose F' and the vectors b_i' such that the vectors $A_{F'}^{n_i'-1}\ b_i'$ are independent and do not belong to K.
According to the form of $Q_{n_i n_j'}$, this choice is possible if and only if the following matrix Q_2 can be choosen to be monic :

$$Q_2 = \underbrace{\boxed{c_{ij}^{n_j' - n_i + 1}}}_{p_1'} \updownarrow p_1$$

Since the only condition that the elements of Q_2 have to satisfy (see (4.8)) is :

$$c_{ij}^{n_j' - n_i + 1} = 0 \text{ if } n_i > n_j' \text{ ,}$$

an obvious necessary and sufficient condition for the existence of such a monic matrix Q_2 is :

$$\text{card}\{\, j \,|\, n_j \leqslant i\} \geqslant \text{card}\ \{j \,|\, n_j' \leqslant i\} \quad \text{for } i \geqslant 1.$$

Thus, when conditions (i) and (ii) of Theorem 3 are verified, there exists a subspace S' such that :

$$S' = \bigoplus_{i=1}^{\alpha_1} R_i' \bigoplus_{i=1}^{p_1'} S_i'$$

where :

$$S_i' = \text{span}\{\, b_i',\ A_{F'}\, b_i',\ \ldots,\ A_{F'}^{n_i'-1}\ b_i'\} \text{ for } i = 1 \text{ to } p_1' \text{ ,}$$

and such that :

$$\dim C\,S' = \dim \operatorname{span}\{ A_{F'}^{n_1'-1}\, b_1', \ldots, A_{F'}^{n_{p_1'}'-1}\, b_{p_1'}' \}$$

$$= p_1' .$$

Define now G like follows :

$$\operatorname{Im} BG = \operatorname{span}\{ b_1', \ldots, b_{\alpha_1'}' \,;\, b_1' \,, \ldots, b_{p_1'}' \} \,;$$

Lemma 4 then shows that $\{\sigma_i'\}$ and $\{n_i'\}$ are the Morse's lists I_2 and I_4 of $(C, A + BF', BG)$.

4. CONCLUDING REMARKS

The problem of the placement of the Morse's lists has been investigated via a matrix pencil approach. This can be seen as a simplification. Indeed for Theorem 2, the research of the matrix Q is easier than the research of T, F and G.

The main contribution of this matrix pencil approach is to give an intuitive understanding of the problem. Unfortunately the use of matrix concepts often leads to intricate proofs (like for Theorem 3). The solution of the problem of Morse's lists placement finally involves the introduction of geometric concepts which clarify the matrix formulation.

Particular cases of Theorem 3 are the placement of the Morse's list I_2 and the placement of the list I_4 (infinite zero orders). For the list I_2, the existence of a monic matrix Q_{22} corresponds to the existence of a controllability subspace included in Ker C for which Q_{22} is a basis matrix (see Warren and Eckberg [15]). Following Heymann [7], such a matrix exists if and only if :

$$\sum_{j \in J(i)} \sigma_i \geqslant \sum_{j \in J'(i)} \sigma_i' \,, \ \forall i \quad 1 \tag{5.1}$$

where $J(i) = \{ k \mid \sigma_k \leqslant i \}$

In the same way, a similar characterization for the placement of the infinite zero orders has been recently obtained, starting from the present matrix pencil approach [8].

Let $\{\Delta_i\}$ denote the non-increasing list obtained from the list of differencies $\{p_i' - p_i\}$ after reordering. It then exist $F : X \to U$ and $G : V \to U$ (G monic) such that $\{n_i'\}$ is the list of the infinite zero orders of $(C, A + BF, BG)$ if and only if the following conditions hold :

$$p_1 - p_i \geqslant p_1' - p_i' \,, \text{ for } i \geqslant 1 \tag{5.2}$$

$$\sum_{j=1}^{i} \alpha_j \geqslant \sum_{j=1}^{i} \Delta_j \quad, \text{ for } i \geqslant 1. \tag{5.3}$$

(5.2) is clearly equivalent to the first condition of Theorem 3. In fact one can see that this first condition expresses that the infinite zero orders can only be increased. Indeed, condition (5.2) is equivalent to the following :

$$p_1 \geqslant p_1'$$

and :

$$n_{p_1'-j}' \geqslant n_{p_1-j} \,, \text{ for } j = 0 \text{ to } p_1' - 1.$$

In the same way, (5.3) traduces the second condition of Theorem 3. Otherwise said, (5.3) imply the existence of a monic matrix $Q = \frac{Q_{24}}{Q_{44}}$ corresponding to the lists $\{\sigma_i\}$, $\{n_i\}$ and $\{n_i'\}$ in the sense of Theorem 3.

ACKNOWLEDGMENT

I would like to thank M. Malabre for his suggestions, and to have presented for me this work at Stockholm.

REFERENCES

[1] Morse, A.S., Structural invariants of linear multivariable systems, SIAM J. Control 11 (1973) 446-465.

[2] Commault, C. and Dion, J.M., Structure at infinity of linear multivariable systems : a geometric approach, 20th IEEE Conference on Decision and Control (San Diego, 1981).

[3] Dion, J.M., Feedback block decoupling and infinite structure of linear systems, Internat. J. Control. 37 (1983) 521-533.

[4] Descusse, J., Lafay, J.F. and Malabre, M., On the structure of linear block-decouplable systems : the general case, IEEE Trans. Automat. Control 28 (1983) 1115-1118.

[5] Descusse, J., Lafay, J.F. and Malabre, M., Further results on Morgan's problem, Systems & Control Letters 4 (1984) pp 203-208.

[6] Gantmacher, F.R., Theorie des matrices (Dunod, Paris, 1966).

[7] Heymann, M., Controllability subspaces and feedback simulation, SIAM J.Control 14 (1976) 769-789.

[8] Loiseau, J.J., Sur la modification de la structure à l'infini par un retour d'état statique, Submitted to SIAM J. Control (1985).

[9] Wonham, W.M., Linear multivariable control, a geometric approach (Springer-Verlag, New York, 1979).

[10] Jaffe, S. and Karcanias, N., Matrix pencil characterization of almost (A,B)-invariant subspaces : A classification of geometric concepts, Internat. J. Control 33 (1981) 51-93.

[11] Loiseau, J.J., Some geometric considerations about the Kronecker normal form, to appear in Internat. J. Control (1985).

[12] Malabre, M., Structure à l'infini des triplets invariants. Application à la poursuite parfaite de modèle, in : Analysis and Optimization of Systems, Lecture Notes in Control and Information Sciences, vol. 44 (Springer-Verlag, New-York, 1982).

[13] Willems, J.C., Almost A(mod B)-invariant subspaces, Asterisque 75-76 (1980) 239-248.

[14] Willems, J.C., Almost invariant subspaces : an approach to high gain feedback design - Part 1 : almost controlled invariant subspaces, IEEE Trans. Automat. Control 26 (1981) 235-252.

[15] Warren, M.E. and Eckberg, A.E., On the dimensions of controllability subspaces : a characterization via polynomial matrices and Kronecker invariants, SIAM J. Control 13 (1975) 434-445.

Frequency Domain and State Space Methods for Linear Systems
C.I. Byrnes and A. Lindquist (Editors)

POLYNOMIAL AND STABLE-PROPER FACTORIZATIONS OF TRANSFER-FUNCTION MATRICES FOR LINEAR TIME-VARYING SYSTEMS*

Kameshwar R. Poolla
Dept. of Electrical and Computer Engineering
University of Illinois
Urbana, IL 61801

Edward W. Kamen
Dept. of Electrical Engineering
University of Florida
Gainesville, FL 32611

Pramod P. Khargonekar
Dept. of Electrical Engineering
University of Minnesota
Minneapolis, MN 55455

This paper is a summary of results obtained by the authors on linear time-varying discrete-time systems over the past two years. Much of this work is based on a new transfer-function type theory for linear time-varying systems based on skew (noncommutative) rings. Using this framework we have been able to generalize much of the existing polynomial model theory. Specifically, we have treated polynomial realization theory (FUHRMANN) and polynomial factorization theory. We have also examined in detail the problems of stabilization and stable-proper factorizations for linear time-varying systems. One of our most significant results is the equivalence of dynamic and memoryless state feedback in the context of stabilization.

1. INTRODUCTION

This paper is a summary of results on linear time-varying discrete-time systems obtained by the authors over the last two years. All proofs are omitted; they may be found in Kamen, Khargonekar, and Poolla (1985), Khargonekar and Poolla (1985), and Poolla and Khargonekar (1985).

Many frequency-domain based synthesis methods used to design controllers for linear time-invariant plants begin with left- and/or right-polynomial and/or stable-proper coprime factorizations of the plant transfer-function matrix, see for example, Zames and Francis (1983). Given the power and utility of these representations, it seems natural to develop a corresponding factorization theory for time-varying systems. In order to realize this objective we need to incorporate time-variance in a natural way into a transform type description of the system behavior. Attempts have been made in the past to develop such a theory (e.g., the system function of Zadeh (1950), but until recently, there has been no theory that has met with much success.

In Section 2 of this paper, we establish some notation, and we outline the basic elements of a transfer-function approach to linear time-varying systems based on skew (noncommutative) rings developed by Kamen, Khargonekar, and Poolla (1985). Following this in Section 3, we generalize much of the existing polynomial realization theory of Fuhrmann to linear time-varying systems. This generalization is entirely nontrivial because of the noncommutative nature of our framework. Then, in Section 4, we treat polynomial factorization theory. In glaring contrast to

*This work was supported in part by the National Science Foundation under Grant Nos. ECS-8200607 and ECS-8400832.

time-invariant systems, not all time-varying input/output operators admit polynomial coprime factorizations (see Example (4.3)). Thus, before we use these factorizations for solving feedback control problems we first characterize those linear time-varying input/output operators that admit factorizations (see Theorem (4.2)).

In Section 5 we briefly review some well-known concepts related to the stability of linear time-varying systems. Then, in Section 6 we introduce the key notion of asycontrollability of a linear time-varying system Σ which is equivalent to being able to stabilize Σ using dynamic state feedback. Anderson and Moore (1981) have defined a notion of stabilizability which is equivalent to being able to stabilize Σ via memoryless state feedback. One of our deepest results (Theorem (6.4)) is the equivalence of stabilizability and asycontrollability. This result in particular implies that dynamics in state feedback buy nothing extra as far as the problem of stabilization is concerned.

Finally, in Section 7 we introduce stable-proper factorizations for linear-time-varying systems. Again, in contrast to time-invariant systems, not all time-varying input/output maps admit stable-proper factorizations (see Example (7.4)). Hence we first characterize the existence of stable-proper factorizations (see Theorem (7.2)). Following this one can systematically employ the axiomatic theory of Desoer, et al. (1980) to study feedback control problems, for instance by obtaining a complete parameterization of all controllers that internally stabilize a linear-time-varying plant.

For a more leisurely exposition of the results in this paper, the interested reader may consult the dissertation of Poolla (1984).

2. THE BASIC FRAMEWORK

With $\mathbb{Z}$ = set of integers and $\mathbb{R}$ = field of real numbers, let A denote $\mathbb{R}$ - linear space of all functions from $\mathbb{Z}$ into $\mathbb{R}$. With the operations of pointwise addition and pointwise multiplication, it is easy to see that A is a commutative ring with identity 1, where $1(k) = 1$ for all $k \in \mathbb{Z}$. Let σ denote the right-shift operator on A defined by

$$(\sigma\alpha)(k) = \alpha(k-1) \qquad \text{all} \quad k \in \mathbb{Z}.$$

Let A_+ denote the subring of A consisting of all functions with support bounded on the left; that is, for any α A_+ there is an integer k_α (depending on α in general) such that $\alpha(k) = 0$ for all $k \leq k_\alpha$. Let $\ell^\infty(\mathbb{Z})$ denote the set of all bounded time-functions. Clearly, $\ell^\infty(\mathbb{Z})$ is a difference subring of A.

(2.1) DEFINITION. Let m and p be positive integers. An m-input p-output linear time-varying causal input/output map f is an $\mathbb{R}$ linear map

$$f : A_+^m \to A_+^p$$

such that if $u(k) = 0$, $k \leq k_u$, for some $u \in A_+^m$, then $f(u)(k) = 0$ for all $k \leq k_u$.

It is well-known that for any input/output map f as defined above, there exists a p×m matrix function $W_f(i,j)$ such that for any $u \in A_+^m$,

$$f(u)(i) = \sum_{j=-\infty}^{i} W_f(i,j)u(j).$$

The matrix function W_f is the unit-pulse response function associated with the input/output map f. Note that by causality, $W_f(i,j)$ is not defined for $i < j$.

Our next concept is the notion of a system.

(2.2) DEFINITION. An m-input p-output n-dimensional linear time-varying system Σ over A is a quadruple $\Sigma = (F,G,H,J)$ of matrices over A where F is n×n, G is n×m, H is p×n, and J is p×m.

With a system $\Sigma = (F,G,H,J)$, we shall associate the dynamical equations

$$x(j+1) = F(j)x(j) + G(j)u(j)$$

$$y(j) = H(j)x(j) + J(j)u(j),$$

where u(j), x(j), y(j) have the usual interpretation.

A system $\Sigma = (F,G,H,J)$ or the pair (F,G) over A is said to be reachable in N steps if there is a positive integer N such that for any $j \in \mathbb{Z}$ and any $x \in \mathbf{R}^n$ there exists an input sequence u(j-N),u(j-N+1),...,u(j-1) which drives Σ from the zero state at time j-N to the state x at time j. The dual notion of observability in N steps has the obvious system-theoretic interpretation, and a system is called canonical if and only if it is both reachable and observable in N steps.

Let R_N denote the N-step reachability matrix

$$R_N: = [G \mid F(\sigma G) \mid \cdots \mid F(\sigma F) \cdots (\sigma^{N-2}F)(\sigma^{N-1}G)].$$

Weiss (1972) has shown that $\Sigma = (F,G,H,J)$ is reachable in N steps if and only if rank $R_N(j) = n$, for all j in $\mathbb{Z}$. Notice that this condition is equivalent to the right invertibility of R_N over the ring A. Dual criteria for observability also exist.

As is well-known, the unit-pulse response function W_Σ associated with the system $\Sigma = (F,G,H,J)$ is given by

$$W_\Sigma(i,j) = \begin{cases} H(i)\, F(i-1)\, F(i-2) \ldots F(j+1)G(j), & i > j \\ J(j), & i = j \\ \text{not defined}, & i < j. \end{cases}$$

The input/output behavior of a system Σ is described by its input/output map f_Σ where

$$f_\Sigma(u)(i) = \sum_{k=-\infty}^{i} W_\Sigma(i,j)u(j), \qquad u \in A_+^m.$$

Given an input/output map f: $A_+^m \to A_+^p$, a realization of f is a system $\Sigma = (F,G,H,J)$ over A such that $f_\Sigma = f$. For results on realizability, we refer the reader to Weiss (1972), and Ferrer and Kamen (1984).

The commutative rings of polynomials, power series and formal Laurent series all with coefficients in the reals $\mathbf{R}$, play a central role in the transfer-function theory of linear time-invariant systems. For time-varying systems the analogous objects are skew (noncommutative) rings with coefficients in the ring of time functions.

More precisely, with z equal to an indeterminate, let $A((z^{-1}))$ denote the set of all formal Laurent series of the form

$$\sum_{r=-N}^{\infty} a^{-r}\alpha_r, \qquad \alpha_r \text{ in } A.$$

With the usual addition and multiplication defined by

$$z^r z^t = z^{r+t},$$

$$\alpha z = z(\sigma\alpha), \qquad \alpha \text{ in } A$$

where $(\sigma\alpha)(k) = \alpha(k-1)$, $A((z^{-1}))$ is a noncommutative ring with identity, called the skew ring of formal Laurent series over A. There are two important subrings of $A((z^{-1}))$: The skew ring of polynomial $A[z]$ and the skew ring of formal power series $A[[z^{-1}]]$. These have the obvious definitions.

Define a projection map

$$\pi: A((z^{-1})) \to A((z^{-1})) : \sum_{r=-N}^{\infty} z^{-r}\alpha_r \to \sum_{r=1}^{\infty} z^{-r}\alpha_r.$$

For any α in $A((z^{-1}))$, let $(\alpha)_+ := \alpha - \pi(\alpha)$. By $(\alpha)_0$, we shall mean the constant coefficient of α, and α is said to be strictly proper if $\pi(\alpha) = \alpha$.

Given any $r\times r$ skew polynomial matrix

$$Q = \sum_{i=0}^{q} z^i Q_i,$$

the degree of Q is the largest integer q such that $Q_q \neq 0$ and Q is said to be monic if and only if $Q_q = I$. Also, Q is said to be invertible if and only if there exists an $r\times r$ matrix Q^{-1} with entries in $A((z^{-1}))$ such that $QQ^{-1} = Q^{-1}Q = I$. In general left-invertibility of Q does not imply right-invertibility of Q, the pathology being due to the skew nature of our rings.

We now describe our transfer function approach to linear time-varying systems. All proofs are omitted; they can be found in Kamen, Khargonekar, and Poolla (1985).

Again, let A_+ denote the subring of A consisting of all functions $\alpha: \mathbb{Z} \to \mathbb{R}$ with support bounded on the left. Let Δ denote the unit-impulse at the origin, i.e., $\Delta(k) = 1$ if $k = 0$, and $\Delta(k) = 0$ otherwise. Given any b in A_+, the (generalized) z-transform of b written B(z) is defined to be the skew Laurent series

$$B(z) = \sum_r z^{-r} b(r)\Delta.$$

Let f be an input/output map, and let W_f denote the unit-pulse response function associated with f. For each integer $r \geq 0$, define a $p\times m$ matrix W_r over A by

$$W_r(j) = W_f(r+j, j), \qquad j \text{ in } \mathbb{Z}.$$

(2.5) DEFINITION. The (formal) transfer function matrix $\hat{W}_f(z)$ associated with the input/output map f is the $p\times m$ matrix over $A[[z^{-1}]]$ defined by

$$\hat{W}_f(z) = \sum_{r=0}^{\infty} z^{-r} W_r.$$

The input/output difference equation $y = f(u)$ can now be characterized in terms of the transfer-function matrix $\hat{W}_f(z)$ as follows.

(2.6) PROPOSITION. Let f be an input/output map. Let y be the output resulting from the input u in A_+^m. Let Y(z) and U(z) denote the (generalized) z-transforms of y and u. Then,

$$Y(z) = \hat{W}_f(z)U(z). \quad (2.7) \qquad \square$$

Note the close resemblance of (2.7) to the time-invariant transfer-function theory. This analogy to the time-invariant theory is further illustrated by the

(2.8) PROPOSITION. Let $\Sigma = (F,G,H,J)$ be a linear time-varying system with input/output map f_Σ. Then the transfer-function matrix $\hat{W}_\Sigma$ is associated with f_Σ is given by

$$\hat{W}(z) = H(zI-F)^{-1}G + J. \quad (2.9) \qquad \square$$

Despite the close resemalance these two results bear to the time-invariant theory, it must be emphasized that (2.7) and (2.9) are computed via the skew (noncommutative) multiplication defined earlier. For instance, (zI-F) has a unique inverse given by

$$(zI-F)^{-1} = z^{-1}(I + Fz^{-1} + F(\sigma F)z^{-2} + \cdots).$$

Given an input/output map f, polynomial factorizations of $\hat{W}_f(z)$ as $\hat{W}_f(z) = P(z)Q^{-1}(z)R(z)$ correspond in a natural way to input/output difference equations. For linear time-invariant systems, this was first observed by Rosenbrock (1970). The detailed study of such factorizations is the subject of the next two sections.

3. POLYNOMIAL REALIZATION THEORY

We will now consider matrix fraction descriptions of a given transfer function. Let P, Q, R, and S be polynomial matrices such that Q is invertible. Let f be an input/output map with the associated transfer matrix

$$\hat{W}_f(z) = PQ^{-1}R + S.$$

For time-invariant systems over fields, Fuhrmann (1976) gave a natural realization for f in terms of the polynomial matrices P, Q, R, and S. For time-invariant systems over arbitrary commutative rings, Khargonekar (1982) has obtained corresponding results. We now proceed to derive results for time-varying systems analogous to those of Fuhrmann and of Khargonekar. We first develop the machinery with which we can obtain these natural realizations for time-varying systems.

Let Q be an $r \times r$ (skew) polynomial matrix, invertible over $A((z^{-1}))$. Define a right A-module

$$X_Q := \{x \text{ in } A^r[z] : Q^{-1}x \text{ is strictly proper}\}.$$

Define a right A-linear projection map

$$\pi_Q : A^r[z] \to X_Q : x \to Q\pi(Q^{-1}x),$$

where $\pi(Q^{-1}x)$ is the strictly proper part of $Q^{-1}x$. Clearly, π_Q is surjective.

(3.1) PROPOSITION. X_Q is a finitely generated, free right A-module. $\square$

We can view X_Q as a right A[z]-module by defining right multiplication by z in X_Q as

$$x \,.\, z := \pi_Q(xz), \qquad x \text{ in } X_Q.$$

This right A[z]-module structure of X_Q is fundamental for obtaining the natural realizations alluded to in the beginning of this section (see Theorem (3.3)).

(3.2) REMARK. The right A-module X_Q will serve as the state space for our natural realization. The above proposition tells us that for any polynomial matrix Q, X_Q admits a basis, i.e., X_Q is isomorphic as a right A-module to A^n for some integer n. This result is very important in ensuring that our realizations are free, i.e., have constant state-space dimension. □

We are now in a position to state the following result.

(3.3) THEOREM. *Let* P, Q, R, *and* S *be* p×r, r×r, r×m, *and* p×m *polynomial matrices such that* Q *is invertible over* $A((z^{-1}))$. *Let* f *be an input/output map with the associated transfer matrix*

$$W_f(z) = PQ^{-1}R + S.$$

Define the maps, ϕ, Γ, *and* ψ *by*

$$\phi : X_Q \to X_Q : x \to \pi_Q(xz)$$

$$\Gamma : A^m \to X_Q : u \to \pi_Q(Ru)$$

$$\psi : X_Q \to A^p : x \to (PQ^{-1}xz)_0.$$

Let F, G, *and* H *be the matrix representations of* ϕ, Γ, *and* ψ *respectively, with reference to some (fixed) basis of* X_Q, *and let* $J := (PQ^{-1}R + S)_0$. *Then* $\Sigma(P,Q,R,S) := (F,G,H,J)$ *is a realization of* f.

(3.4) REMARK. We could have chosen a different basis $\{\hat{b}_1,\hat{b}_2,\ldots,\hat{b}_n\}$ for X_Q and obtained a different realization $\hat{\Sigma} = (\hat{F},\hat{G},\hat{H},\hat{J})$ for the input/output map f. In this event $\hat{\Sigma}$ and $\Sigma = (F,G,H,J)$ are A-isomorphic. We shall call any realization of f obtained as in Theorem (3.3) the *Fuhrmann realization* associated with the polynomial matrix fraction representation $W_f = PQ^{-1}R + S$. □

The Fuhrmann realization, say for systems over fields, is known to be useful in the time-invariant case for tackling many control theoretic problems such as stabilization, regulation, tracking, etc. We now continue to develop the theory by examining the relationship between control theoretic properties of the Fuhrmann realization and algebraic properties of the associated polynomial matrix representation. We begin by characterizing reachability in N stpes.

(3.5) THEOREM. *Let* P, Q, *and* R *be* p×r, r×r, *and* r×m *polynomial matrices over* A[z] *such that* Q *is invertible over* $A((z^{-1}))$. *Let* f *be an input/output map with the associated transfer matrix*

$$W_f(z) = PQ^{-1}R.$$

Then, the Fuhrmann realization $\Sigma(P,Q,R) = (F,G,H)$ *is reachable in* N *steps at all times, for some positive integer* N *if and only if there exist polynomial matrices* Y_1 *and* Y_2 *such that*

$$QY_1 + RY_2 = I.$$

It is possible to continue this development and examine issues of strict-system-equivalence for time-varying systems. The interested reader may consult Khargonekar and Poolla (1985).

4. POLYNOMIAL FACTORIZATIONS OF TRANSFER-FUNCTION MATRICES

In this sections, we shall consider the issues of existence and computation of polynomial matrix fraction representations. We would like to emphasize that all of our results are constructive.

Many frequency-domain methods used to design controllers for time-invariant systems begin with polynomial factorizations of the plant transfer-function matrix G(x) of the form

(4.1)
$$G(s) = Q^{-1}(s)R(s), \quad \text{with} \quad Q(s)Y_1(s) + R(s)Y_2(s) = I.$$

See for example the books of Rosenbrock (1970) and Wolovich (1974). Here, Q, R, Y_1, and Y_2 are polynomial matrices, and (4.1) is referred to as a left-Bezout polynomial factorization. For time-invariant systems, the existence of these factorizations is not an issue: any plant transfer function matrix admits left- or right-Bezout polynomial factorizations. This, however, is not the case with time-varying systems (see Example (4.5)). We therefore need to understand what class of time-varying systems admit such factorizations before we use them to design control systems. Theorem (4.2) answers just this issue and essentially states that a transfer-function matrix admits Bezout factorizations if and only if the associated input/output map admits a canonical realization.

(4.2) THEOREM. Let f be an input/output map, and let $\hat{W}_f(z)$ be the transfer function matrix associated with f. Then, the following statements are equivalent.

(a) There exist polynomial matrices Q, R, Y_1, and Y_2 over A[z] with Q invertible and such that

$$\hat{W}_f(z) = Q^{-1}R, \qquad QY_1 + RY_2 = I,$$

i.e., $\hat{W}_f(z)$ admits a right-Bezout polynomial factorization.

(b) There exist polynomial matrices Q, Q, Y_3, and Y_4 with Q_1 invertible and such that

$$\hat{W}_f(z) = PQ_1^{-1}, \qquad Y_3P + Y_4Q_1 = I,$$

i.e., $\hat{W}_f(z)$ admits a left-Bezout polynomial factorization.

(c) f admits a canonical realization.

We would like to emphasize that the proof of the above theorem is constructive enabling one to explicitly determine polynomial factorizations from canonical realizations.

Not all time-varying systems admit canonical realizations. This is demonstrated by the following example.

(4.3) EXAMPLE. Consider the linear time varying system $\Sigma = (F,g,h)$ where $F(k) = h(k) = 1$ for all k in Z, and $g = \Delta =$ the unit pulse concentrated at the origin. Let f_Σ be the input/output map associated with Σ. We will show that f_Σ does not admit a canonical realization, and therefore, by Theorem (4.2) $\hat{W}_f(Z)$ does not admit left- or right-polynomial factorizations.

Suppose f_Σ did admit a canonical realization. Then, by Theorem (4.2), there exist polynomial matrices P and Q with Q invertible and such that

(4.4) $$h(zI - F)^{-1}g = (z-1)^{-1}\Delta = PQ^{-1}.$$

Let $Q = \sum_{i=0}^{n} z^i Q_i$. From (4.4) it follows that $(z-1)^{-1}\Delta Q = P$ is a polynomial. The coefficient of z^{-1} above must be zero:

$$0 = \Delta Q_0 + (\sigma\Delta)Q_1 + \cdots + (\sigma^n\Delta)Q_n.$$

Multiplying this equation on the left by $(\sigma^i\Delta)$ we conclude that $(\sigma^i\Delta)Q_i = 0$ for $i = 0,1,\ldots,n$. Thus $\Delta Q(z) = 0$ which is impossible since Q is assumed to be invertible. □

The central theorem of this paper, Theorem (4.2), is stated for canonical linear time-varying systems. Under the weaker hypothesis of reachability we can obtain the following theorem (which is actually an intermediate result used in proving Theorem (4.2)).

(4.5) THEOREM. Let $\Sigma = (F,G,H)$ be a linear time-varying system that is reachable in N steps. Then, there exist polynomial matrices P and Q with Q such that

$$W(z) = H(zI-F)^{-1}G = PQ^{-1}$$

and the Fuhrmann realization $\Sigma(P,Q,I)$ is A-isomorphic to $\Sigma = (F,G,H)$. □

(4.6) REMARK. Recently, the use of stable-proper factorizations in control problems has become more popular, one advantage being that the properness of compensators is automatic. See for example Desoer et al (1980), Saeks and Murray (1981). The detailed study of the existence of such factorizations is treated in the remainder of this paper. □

We would like to remark in closing that the systematic procedure digen in this section for computing Bezout polynomial factorizations for time-varying systems carried through with minor modifications for systems over a principal ideal domain. The interested reader may refer to Poolla and Khargonekar (1983) for more details.

5. STABILITY

In this section we briefly review some well-known concepts dealing with the stability of linear time-varying systems, and we relate these concepts to the transfer-function theory based on skew-rings described in Section 2. There is extensive literature available on the stability of linear time-varying systems. For example, the interested reader may consult Anderson and Moore (1981), Willems (1970), Cesari (1963), Freedman and Zames (1968), etc.

For any vector x in $\mathbb{R}^n$ let $\|x\|$ denote the Euclidean norm of x. For an m×n matrix M over $\mathbb{R}$, let $\|M\| := \sup_x \|Mx\|/\|x\|$. For an n×m matrix N over A, define the norm of N to be $\|N\| := \sup_t \|N(t)\|$.

Let $f : A_+^m \to A_+^p$ be an input/output map. We shall say that f is bounded input-bounded output (BIBO) stable if and only if for any bounded input sequence, the resulting output is bounded. Let $\Sigma = (F,G,H,J)$ be any (fixed) realization of f.

Consider the free behavior of the system Σ described by the vector difference equation

(5.1) $$x(k+1) = F(k)x(k).$$

The system Σ is said to be <u>internally uniformly asymptotically stable (u.a.s.)</u> if and only if for every number $\varepsilon > 0$, there exists a positive integer N_ε such that for any initial time t_0 in $\mathbb{Z}$ and any initial state $x(t_0)$ with $\|x(t_0)\| \leq 1$, we have that $\|x(t_0+i)\| \leq \varepsilon$ for all $i \geq N_\varepsilon$. Here, $x(t_0+i)$ is the solution of (5.1) at time t_0+i starting from initial state $x(t_0)$.

Let $P(z)$ be a $n \times m$ matrix over the skew ring $A((z^{-1}))$, i.e., $P(z)$ is a matrix Laurent series of the form

$$P(z) = \sum_{i=-N}^{\infty} z^{-i} P_i .$$

(5.2) DEFINITION. <u>The matrix Laurent series</u> $P(z)$ <u>is said to be stable if and only if</u>

$$\|P_i\| \to 0 \quad \text{as} \quad i \to \infty .$$

In terms of this notion, we can characterize internal stability as follows (see Green and Kamen (1983) and Kamen, Khargonekar, and Poolla (1984, Prop. (4.4))).

(5.3) PROPOSITION. <u>Let $\Sigma = (F,G,H)$ be a linear time-varying system over A. Then, Σ is internally u.a.s. if and only if $(zI-F)^{-1}$ is a stable matrix power series.</u> □

Note the similarity this result bears to the time-invariant theory. Let f be an input/output map, and let $\Sigma = (F,G,H,J)$ over A be any (fixed) realization of f. In contrast with the time-invariant case, internal stability of Σ does <u>not</u> in general imply BIBO stability of f. However, for the class of <u>bounded</u> linear time-varying systems, we have the following result.

(5.4) PROPOSITION. <u>Let f be an input/output map and let $\Sigma = (F,G,H,J)$ over $\ell^\infty(\mathbb{Z})$ be any (fixed) realization of f. Suppose Σ is internally u.a.s. Then f is BIBO stable.</u> □

The following example illustrates a phenomena peculiar to time-varying systems with <u>unbounded</u> coefficients.

(5.5) EXAMPLE. Consider the time-varying system $\Sigma = (F,G,H)$ over A where $F = 1/2$, $G = 1$, and

$$H(k) = \begin{cases} 0 , & k \leq 0 \\ 2^k , & k > 0 . \end{cases}$$

From Proposition (5.3) it is evident that Σ is internally u.a.s. This system, however, is <u>not</u> BIBO stable because the output $y(k)$ resulting from the input $u = \Delta$ = unit impulse at the origin grows without bound. Notice that Proposition (5.4) does not apply here because H is <u>not</u> over $\ell^\infty(\mathbb{Z})$. □

In subsequent sections, we shall deal only with linear time-varying systems with <u>bounded</u> coefficients, i.e., defined over the difference subring $\ell^\infty(\mathbb{Z}) \subseteq A$. We shall require that any controller we design be over $\ell^\infty(\mathbb{Z})$. These are physically reasonable constraints since most time-varying plants that arise in practice have bounded time variation, and, implementation of controllers with unbounded coefficients would be numerically ill-conditioned. (Technically, these constraints substantially complicate proofs and make results harder to obtain.)

6. STABILIZABILITY AND ASYCONTROLLABILITY

In this section, we introduce the key notion of asycontrollability, which is closely related to being able to stabilize a linear time-varying system by dynamic state feedback. Anderson and Moore (1981) have defined the notion of stabilizability which is equivalent to being able to stabilize a linear time-varying system by nondynamic (i.e., memoryless) state feedback. The central result in this section, Theorem (6.4), shows the equivalence of stabilizability and asycontrollability. A striking conclusion of this theorem is that dynamics in state feedback buy nothing extra as far as the problem of stabilization is concerned.

We being with the following key definition:

(6.1) DEFINITION. Let $\varepsilon > 0$ be any (fixed) real number. A system $\Sigma = (F,G,H)$ over $\ell^\infty(Z)$ is said to be asycontrollable if and only if there exists a real number $\hat{M}$ and an integer N such that for any initial time t_0 in Z and any initial state ξ in R^n with $\|\xi\| = 1$, there exists an input sequence $u_{t_{0,\xi}}(t_0)$, $u_{t_{0,\xi}}(t_0+1), \ldots,$ $u_{t_{0,\xi}}(t_0+N-1)$ which results in the state trajectory $x_{t_{0,\xi}}(t_0) = , x_{t_{0,\xi}}(t_0+1), \ldots,$ $x_{t_{0,\xi}}(t_0+N)$ and such that

$$\|x_{t_{0,\xi}}(t_0+N)\| < \varepsilon$$

$$\|x_{t_{0,\xi}}(t_0+k)\|, \quad \|u_{t_{0,\xi}}(t_0+k)\| < \hat{M}.$$ □

This technical definition has a rather complex precise system-theoretic interpretation in terms of stabilizing Σ via an open-loop control law. Roughly speaking, a system is asycontrollable if and only if it can be driven to zero "final" state asymptotically, using uniformly bounded input sequences along uniformly bounded state trajectories. The phrase asycontrollability (from asymptotically controllable) is borrowed from Khargonekar and Sontag (1982) for time-invariant systems over rings.

(6.2) REMARK. It can be shown (see Poolla (1985)) that asycontrollability is equivalent to the existence of stable Laurent series Y_1 and Y_2 over $\ell^\infty(\mathbb{Z})((z^{-1}))$ such that $(zI-F)Y_1 + GY_2 = I$. In particular, this means that if Σ can be stabilized using a dynamic state-feedback controller, then Σ is asycontrollable. □

Let $\Sigma = (F,G,H)$ be a linear time-varying system over $\ell^\infty(\mathbb{Z})$. Anderson and Moore (1981) have defined a notion of stabilizability which intuitively corresponds to requiring that unstable modes be controllable. The authors then show that this notion is equivalent to the existence of a stabilizing memoryless state feedback law $u(k) = -L(k)x(k)$. We shall take this to be the definition of stabilizability. In terms of our skew-ring framework, we phrase this as follows:

(6.3) DEFINITION. Let $\Sigma = (F,G,H)$ over $\ell^\infty(\mathbb{Z})$ be a linear time-varying system. Then, Σ is said to be stabilizable if and only if there exists an $m\times n$ matrix L over $\ell^\infty(\mathbb{Z})$ such that $(zI-F+GL)^{-1}$ is a stable matrix power series. □

One can similarly define the dual notion of detectability.

Let $\Sigma = (F,G,H)$ over $\ell^\infty(\mathbb{Z})$ be a stabilizable linear time-varying system. Let L be an $m\times n$ feedback matrix as in Definition (6.3). Notice that we can write

$$(zI-F)Y_1 + GY_2 = I,$$

where $Y_1 = (zI-F+GL)^{-1}$ and $Y_2 = LY_1$. Since Y_1 and Y_2 are stable Laurent series, it follows from Remark (6.2) that Σ is asycontrollable. Thus, stabilizability *implies* asycontrollability. The much more difficult converse is also true:

(6.4) THEOREM. *Let $\Sigma = (F,G,H,J)$ be a linear time-varying system over $\ell^\infty(\mathbb{Z})$. Then, Σ is asycontrollable if and only if Σ is stabilizable.* □

Recall (see Remark (6.2)) that if a linear time-varying system Σ can be stabilized by a dynamic state-feedback controller, then Σ is asycontrollable. This observation together with the Definition (6.3) of stabilizability immediately offers the following surprising conclusion.

(6.5) COROLLARY. *If a linear time-varying system can be stabilized using dynamic state-feedback, then Σ can also be stabilized using memoryless state-feedback.* □

(6.6) REMARK. Indeed the above corollary is not (necessarily) *expected*, because there are classes of systems (for example delay systems, see Kamen (1982, Ex. 3, p. 371)) for which it is *not* true. For time-invariant systems over a field, it is well-known that dynamic state feedback is equivalent to memoryless state feedback as far as the problem of stabilization is concerned. The proof of this fact (see any modern text on control theory) relies heavily on the Kalman canonical decomposition. No such decomposition exists for time-varying systems because the "dimension" of the reachable space could depend on time. □

We shall say that an input/output map f can be *stabilized by dynamic output feedback* if and only if there exists a realization $\Sigma = (F,G,H,J)$ of f and a controller $\Sigma_c = (F_c, G_c, G_c)$ both over $\ell^\infty(\mathbb{Z})$ such that Σ_c internally stabilizes Σ. Dualizing Corollary (6.5) we immediately obtain the following result:

(6.7) THEOREM. *An input/output map f can be stabilized by dynamic output feedback if and only if f admits a stabilizable and detectable realization.*

Since stabilizability and asycontrollability are equivalent notions, we shall henceforth only speak of stabilizability.

It is important to find "nice" necessary and sufficient tests for stabilizability. This problem appears to be quite formidable unless one specializes to particular classes (e.g., periodic) of time-varying systems. We do have, however, the following sufficient condition.

(6.8) THEOREM. *Let $\Sigma = (F,G,H,J)$ over $\ell^\infty(\mathbb{Z})$ be a lienar time-varying system. Suppose that one can find integers N and K and a real number $\varepsilon > 0$ with the following property: For any integer t_0, there exists some t_1 with $t_0 \le T \le t_0 + K$ such that*

$$\det[R_N R_N'](t_1) \ge \varepsilon > 0, \tag{6.9}$$

where R_N is the N step reachability matrix

$$R_N := [G \quad F\sigma G \; \ldots \; F\sigma F \ldots \sigma^{N-2} F \sigma^{N-1} G].$$

Then, Σ is stabilizable.

We are now in a position to study stable-proper factorizations for linear time-varying systems.

7. STABLE-PROPER FACTORIZATIONS

Of late, the use of stable-proper factorizations for control system design has become increasingly popular. See for example Vidyasagar (1978), Desoer, et al. (1980), Saeks and Murray (1981), Khargonekar and Sontag (1982), etc. In this section we investigate in detail stable-proper factorizations for time-varying systems.

Define the set RP (for rational and proper) to be

$$RP := \{\phi \text{ in } \ell^\infty(\mathbb{Z})[[z^{-1}]] : \phi = \alpha(zI-\beta)^{-1}\gamma + \delta\}$$

where α, β, γ, and δ are matrices of compatible dimensions over $\ell^\infty(\mathbb{Z})$. It is easy to verify that RP forms a ring with the usual skew-multiplication and addition defined in $\ell^\infty(\mathbb{Z})[[z^{-1}]]$. Also, define the subring RP_s by

$$RP_s := \{\phi = \alpha(zI-\beta)^{-1}\gamma + \delta \text{ in } RP : (zI-\beta)^{-1} \text{ is stable}\}.$$

We shall call RP_s the <u>ring of stable, proper, rational functions</u> (the subscript s denotes <u>s</u>table). We shall abandon rigorous nomenclature and loosely refer to elements in RP_s or matrices over RP_s as being <u>stable-proper</u>.

Let D(z) be an n×n matrix over RP. We shall say that D(z) is bicausal if and only if D has an inverse $D^{-1}(z)$ also over RP. Let f be an input/output map and let $\hat{W}_f(z)$ be its associated transfer-function matrix. Then, $\hat{W}_f(z)$ is said to admit a <u>right-Bezout stable-proper factorization</u> if and only if there exist stable-proper matrices (i.e., over RP_s) N, D, X, and Y with D bicausal and such that

$$\hat{W}_f(z) = ND^{-1}, \qquad XN + YD = I. \tag{7.1}$$

One can also similarly define <u>left-Bezout stable-proper factorization</u>. Such factorizations have been found to be extremely useful in tackling many control-theoretic problems (for example, H^∞-optimal controller design, see Zames and Francis (1983), Francis and Zames (1984), Francis, Helton, and Zames (1984)). We first characterize input/output maps whose transfer-functions admit left- and/or right-Bezout stable-proper factorizations. We have the following central result.

(7.2) THEOREM. <u>Let f be an input/output map and let $\hat{W}_f(z)$ be its p×m associated transfer function matrix over $\ell^\infty(\mathbb{Z})[[z^{-1}]]$. Then, the following are equivalent:</u>

(a) <u>$\hat{W}_f(z)$ admits a right-Bezout stable-proper factorization</u>.

(b) <u>$\hat{W}_f(z)$ admits a left-Bezout stable-proper factorization</u>.

(c) <u>f admits a stabilizable and detectable realization</u>.

(d) <u>f can be stabilized by dynamic output feedback</u>.

(7.3) REMARK. In proving the above result, we have made critical use of Theorem (6.4); the fundamental result of the previous section. Recall that Theorem (6.4) states that asycontrollability is equivalent to stabilizability and the rather difficult proof of this result may be found in Poolla and Khargonekar (1983). There appears to be no way to circumvent Theorem (6.4) in studying the existence of stable-proper factorizations. □

Theorem (7.2) essentially states that a time-varying system admits stable-proper factorizations if and only if it admits a stabilizable and detectable realization. This characterization is non-trivial because (in glaring contrast with time-<u>invariant</u> systems), <u>not all time-varying systems admit stabilizable realizations</u> (and therefore by Theorem (7.2), stable-proper factorizations). We illustrate this with the following example.

(7.4) EXAMPLE. Consider the linear time-varying system $\Sigma = (F,G,H)$ over $\ell^\infty(\mathbb{Z})$ where $F(k) = G(k) = 1$ for all k in $\mathbb{Z}$ and $G = \Delta =$ the unit pulse concentrated at the origin. Let f_Σ be the input/output map associated with Σ. Suppose f_Σ admits a stabilizable realization $\bar{\Sigma} = (\bar{F},\bar{G},\bar{H})$. This would imply that for any initial state ξ at time t=1, there exists an open loop control law that drives $\bar{\Sigma}$ from $\bar{x}(1) = \xi$ to zero final state (and therefore zero final output) asymptotically. However, from the unit-pulse response function of f_Σ, it is evident that application of inputs after t=1 has no effect on the output. Thus, it must be that

(7.5)
$$\lim_{k\to\infty} \|\bar{H}(k)\bar{F}(k-1)\bar{F}(k-1) \ldots \bar{F}(1)\| = 0.$$

However,
$$W_{f_\Sigma}(z) := H(zI-F)^{-1}G = z^{-1}\Delta + z^{-2}\Delta + \cdots$$
$$= \bar{H}(zI-\bar{F})^{-1}\bar{G}.$$

Consequently, for all integers $k \geq 1$

$$\bar{H}(k)\bar{F}(k-1) \ldots \bar{F}(1)\bar{G}(0) = 1.$$

This, together with the fact that $\bar{G}$ is over $\ell^\infty(\mathbb{Z})$, renders (7.5) impossible. Therefore, f_Σ does not admit a stabilizable realization which implies (by Theorem (7.2)) that $\hat{W}_{f_\Sigma}(z)$ does not admit a stable-proper factorization. □

Armed with the powerful tool of stable-proper factorizations, we can systematically apply the axiomatic theory of Desoer et al. (1980) to study feedback control problems for linear time-varying plants. For instance we can readily obtain a complete parameterization of all stabilizing controllers:

(7.6) THEOREM. Let $\Sigma = (F,G,H,J)$ be a stabilizable and detectable plant. Let

$$\hat{W}_\Sigma(z) = ND^{-1} = D_1^{-1}N_1,$$
$$XN + YD = I = N_1X_1 + D_1Y_1,$$

be any left- and right-stable-proper factorizations of W_Σ. Then, a controller Σ_c internally stabilizes Σ if and only if the controller transfer function $W_{\Sigma_c}(z)$ is of the form

(7.7)
$$\hat{W}_{\Sigma_c}(z) = (MN_1 + Y)^{-1}(-MD_1 + X)$$

for some matrix M over RP_s. □

REFERENCES

Anderson, B. D. P., and J. B. Moore (1981). "Detectability and stabilizability of time-varying discrete-time linear systems," *SIAM J. Control and Optimization, 19:* 20-32.

Cesari, L. (1963). *Asymptotic Behavior and Stability Problems in Ordinary Differential Equations,* Academic Press, New York.

Desoer, C. A., R. W. Liu, J. Murray, and R. Saeks (1980). "Feedback system design: the fractional representation approach to analysis and synthesis," *IEEE Transactions on Automatic Control, AC-25:* 399-412.

Ferrer, J. J. and E. W. Kamen (1985). "Realization of linear time-varying discrete-time systems," in preparation.

Freedman, M. and G. Zames (1968). "Logarithmic variation criteria for the stability of systems with time-varying gains," *SIAM J. of Control, 6:* 487-507.

Fuhrmann, P. A. (1976). "Algebraic system theory: An analyst's point of view," *J. of the Franklin Institute, 301:* 521-540.

Green, W. L. and E. W. Kamen (1984). "On stability of linear difference equations with time-varying coefficients," in preparation.

Kamen, E. W., P. P. Khargonekar, and K. Poolla (1985). "A transfer function approach to linear time-varying discrete-time systems," *SIAM J. of Control and Optimization, 23:* 550-565.

Khargonekar, P. P. (1982). "On matrix fraction representations for linear systems over commutative rings," *SIAM J. of Control and Optimization, 20:* 172-197.

Khargonekar, P. P. and K. Poolla (1985). "On polynomial fraction representations for linear time-varying systems," *Linear Algebra and its Applications,* to appear.

Khargonekar, P. P. and E. D. Sontag (1982). "On the relation between stable matrix fraction factorization and regulable realizations of linear systems over rings," *IEEE Trans. on Automatic Control, AC-27:* 627-638.

Poolla, K. (1984). "Linear-time-varying systems: Representation and control via transfer function matrices," University of Florida, Gainesville.

Poolla, K. and P. P. Khargonekar (1983). "Fractional representations for systems over a P.I.D.: A constructive technique," *Systems and Control Letters, 3:* 145-150.

Poolla, K. and P. P. Khargonekar (1985). "Stabilizability and stable-proper factorizations for linear time-varying systems," submitted for publication.

Rosenbrock, H. H. (1970). *State Space and Multivariable Theory,* Wiley, New York.

Saeks, R. and J. Murray (1981). "Feedback system design: The tracking and disturbance rejection problems," *IEEE Trans. on Automatic Control, AC-26:* 203-217.

Vidyasagar, M. (1978). "On the use of right-coprime factorizations in distributed feedback systems containing unstable subsystems," *IEEE Trans. on Circuits and Systems, CAS-25:* 916-921.

Weiss, L. (1972). "Controllability, realization, and stability of discrete-time systems," *SIAM J. of Control, 10:* 230-251.

Willems, J. L. (1970). *Stability Theory of Dynamic Systems,* Nelson, London.

Zadeh, L. A. (1950). "Frequency analysis of variable networks," *Proc. IRE, 38:* 291-299.

Zames, G. and B. Francis (1983). "Feedback, minimax sensitivity, and optimal robustness," *IEEE Trans. on Automatic Control, AC-28:* 585-601.

Frequency Domain and State Space Methods for Linear Systems
C.I. Byrnes and A. Lindquist (Editors)

INVARIANT FACTORS OF ALGEBRAIC COMBINATIONS OF MATRICES

Robert C. Thompson

Mathematics Department,
University of California,
Santa Barbara, Ca., 93106,
U.S.A.

For matrices over a principal ideal domain, this paper studies the Smith invariants of elementary algebraic combinations of matrices with given Smith invariants, including a sum, a product, a submatrix, a greatest common divisor or least common multiple, and a rank one perturbation. Proofs are omitted. The heaviest emphasis is on the problem of the invariant factors of a product, including a conjectured resolution based on links to the algebra of symmetric polynomials and to the analytical problem of the eigenvalues of a sum of Hermitian matrices.

INTRODUCTION.

We work with matrices over a principal ideal domain R. If A is such a matrix, it is well known that unimodular matrices U and V exist such that UAV is diagonal with the diagonal entries forming a divisibility chain. The important Smith invariants (or invariant factors) are the elements in this divisibility chain. We shall discuss certain facts concerning the Smith invariants of various algebraic combinations of matrices.

This paper is based on a portion of a lecture the author gave at the 1984 conference on linear algebra and control theory at Bowdoin, Maine, and on his talk at the MTNS-85 conference at Stockholm, Sweden. He wishes to thank B. N. Datta for inviting him to participate in both conferences. At the Bowdoin meeting, a lecture on a similar theme was given by E. Marques de Sa, and the author wishes to acknowledge the stimulus afforded by hearing Sa's presentation. As will be seen below, the stimulus afforded by the MTNS-85 meeting also was considerable.

At several places in the following discussion a *general principle* will appear: *If a matrix theorem involves three matrices and has both a necessary and a sufficient aspect, if the sufficiency part is not substantially harder to prove than the necessity then only part of the truth has been revealed.* This principle will established by examples. In one example it will occur as a result involving two matrices that in a sharper focus is a corollary of a more difficult theorem involving three.

SUMS.

The following theorem provides a typical and not too complicated illustration of the sort of result often found in the subject under review. In this theorem, it is convenient to take the invariant factors in *increasing* order, as follows: Let the invariant factors of n×n matrices A, B, and C be $a_1|a_2|\ldots$, $b_1|b_2|\ldots$, and $c_1|c_2|\ldots$, respectively, where | signifies divisibility.

THEOREM. *If* $(\cdot,\cdot)$ *denotes greatest common divisor and* $\equiv$ *congruence, then the relations*

$$(a_i,b_j)|c_{i+j-1}, \qquad (a_i,c_j)|b_{i+j-1}, \qquad (b_i,c_j)|a_{i+j-1},$$

$$\det A \equiv (-1)^n \det B \pmod{c_1}, \qquad \det B \equiv \det C \pmod{a_1},$$

$$\det A \equiv \det C \pmod{b_1},$$

hold for the invariant factors of a sum C = A + B.

These conditions hold for all indices i, j for which the subscripts do not exceed the size of the matrices. The subscript on the congruence moduli is intended to be 1, not i, and it is also assumed that any element of *R* dividing all elements of A, B, and C was cancelled in a preliminary simplification. These divisibility conditions constitute necessary conditions for the existence of matrices A, B, C = A + B with prescribed invariant factors. The conditions are not sufficient for 1×1 matrices, but are sufficient for 2×2 and 3×3. Their sufficiency in dimensions higher than 3×3 is unknown. The proof of the necessity of these conditions (in every dimension) is not too hard, but the sufficiency argument is already a considerable challange in dimensions 2 and 3, see [45]. (Here we have the first example illustrating our general principle.) There perhaps are other conditions to be discovered in dimensions higher than 3, but every reasonable candidate for an additional constraint on the invariant factors of a sum has been shown to be invalid or unnecessary. These facts were described in some detail in a widely read article in the American Mathematical Monthly [43]. It is probably true that a successful assault on the 4×4 case will reveal whatever pattern and shape the problem has.

Apart from its challange as a hard diophantine problem, it seems difficult to explain why the study of the invariant factors of a matrix sum is important. There appears to be no known structural framework within which the problem sits as a natural candidate for consideration. The best justification, undoubtedly, is this: In number theory, it is well recognized that additive problems are intrinsically more difficult than multiplicative ones (Fermat's last theorem, the Goldbach problem, etc.) and especially so when the additive problem involves quantities that naturally arise multiplicatively (e.g., primes.) The invariant factor problem for matrix sums seems to fit this mold.

PRODUCTS.

For convenience of exposition, take now the invariant factors of the n×n matrices A, B, C in *decreasing* order: $a_n|\ldots|a_1$, and similarly for B and C. Suppose that C = AB, instead of C = A + B.

THEOREM. *There is a large family of divisibility relations linking the invariant factors of* A, B, *and* C = AB, *having the form*

$$c_{k_1}\cdots c_{k_m} \,|\, a_{i_1}\cdots a_{i_m} b_{j_1}\cdots b_{j_m} \tag{1}$$

for certain index sets $i_1 < \dots < i_m$, $j_1 < \dots < j_m$, $k_1 < \dots < k_m$. *(When* $m = n$, *(1) is be equality to within a unit factor.)*

The somewhat technical description of these index sets is given in the language of Young tableaux. Specifically, one forms a tableau with row lengths (top row first) $i_m - m, \dots, i_1 - 1$, then augments the rows with cells to obtain a tableau with row lengths $k_m - m, \dots, k_1 - 1$. The augmenting cells are filled with, first, $j_m - m$ symbols named x_m, next, $j_{m-1} - (m-1)$ symbols named $x_{m-1}, \dots$, finally, $j_1 - 1$ symbols named x_1, so that various combinatorial conditions hold: after each set of symbols is added, the configuration is still a Young tableau, also, when read down a column no symbol occurs more than once, and further, the so called *lattice rule* holds: when the augmenting symbols are read right to left row by row, beginning at the top row and ending at the bottom, at every stage in the reading process the number of occurrences of symbol x_t is at least the number of occurrences of x_{t-1}, for all t. (These rules may be found in [44], or see the attractive book [24] or article [21] for similar descriptions.) The simplest but significantly nontrivial special case of a family of indices meeting these intricate rules is the divisibility relation (1) displayed above in which the k's are the following functions of the i's and j's:

$$k_t = i_t + j_t - t \qquad \text{for } t = 1, \dots, m, \tag{2}$$

an index family first found in quite another context [36]. Note that this special case, when $m = 1$, reduces to

$$c_{i+j-1} \mid a_i b_j,$$

a condition much resembling the additive dress conditions in the previous section. However, further additive dress conditions mimicking (2) are known to be inessential, a somewhat mysterious fact.

The divisibility relations (1) with these very combinatorial subscripts defined via tableaux is proved in this way: First, one proves that the existence of matrices A, B, $C = AB$ with prescribed invariant factors is equivalent to the existence of R modules $\mathcal{A}$, $\mathcal{B}$, $\mathcal{C}$ with $\mathcal{A} \subseteq \mathcal{C}$ and $\mathcal{B} = \mathcal{C}/\mathcal{A}$. (This seems well known in control theory.) Then a theorem of T. Klein [18] (that isn't well known but should be) is invoked in which she established necessary and sufficient conditions for the existence of a triple of modules as just described with the individual terms having prescribed invariant factors. However, Klein's necessary and sufficient conditions for the existence of a triple of modules are not divisibility conditions, even though they are presented in the language of Young tableaux. They differ from the tableau conditons described above in that Klein specifies *column* rather than *row* lengths in her tableau. It requires, apparently, an intricate combinatorial argument [44] involving the construction of lattice paths in the two dimensional coordinate plane to show that her conditions imply the divisibility formulas (1) cited above. This argument is a kind of duality theorem connecting partitions with conjugate partitions, and seems significant in its own right as a purely combinatorial result. The degree of intricacy of the proof, however, makes one wonder whether a more conceptual presentation of the same ideas isn't possible. But it is worth noting that essentially the same kind of lattice path reasoning has appeared elsewhere in the combinatorial literature [30]. Be that as it may, the divisibility relations (1) for the invariant factors of a matrix product are surely established, albeit by a round-about route.

Having established inequalities (1), the author perceives that two (at least) shortcomings remain in the understanding of the invariant factors of a product (others may have differing perceptions):

(a) It has not yet been shown that the divisibility relations (1) constitute sufficient conditions for the existence of matrices A, B, C = AB having prescribed invariant factors;

(b) The description of the index sets (subscripts in (1)) isn't explicit, the test for the allowability of a proposed set of subscripts being that a tableau satisfying very combinatorial constraints be constructible.

Work is actively under way on a direct attack to remove these shortcomings, by attempting to show that Klein's sufficient conditions are met, but the combinatorics involved is extremely intransigent. Example: In the 4x4 case there are 42 conditions (1). There is a recursive formulation, originally due to A. Horn [12], that one conjectures exactly describes our index sets. This conjecture is reached by studying a rather similar situation involving the classification of the allowable eigenvalues of a sum of Hermitian matrices, and for which B. V. Lidskii has announced a solution [23]. Following the lead suggested by Lidskii's announcement, the author has indeed shown [47] that the tableau-defined index sets allowable in (1) do satisfy this recursion, but he has not yet been shown that every sequence produced by the recursion is in fact given by tableau, and therefore allowable. However, it is fair to state that author's attempted direct approach, using combinatorics to pass between the conjectured necessary and sufficient conditions (1) and Klein's theorem, has so far dealt successfully only with the necessary conditions satisfied by the invariant factors of a product, and unsuccessfully with their sufficiency. To experts in these kinds of problems the sufficiency is always recognizable as the tougher half of the challange. (This is the second example supporting the principle announced earlier.)

For an approach to the invariant factors of a product that establishes divisibility relations like (1) by directly working inside the modules, rather than attempting to use combinatorics to reach out to Klein's result, see [48]. And, for an independent and self contained proof of the special case (2) of (1), see [42]. It is worth observing that special cases of (2) have recently appeared in the literature. (Carelessly, the author has lost these references.)

The question of the invariant factors of a matrix product has depth extending much beyond the facts so far explained. It will, therefore, be discussed again in the last section of this paper, where another approach will be described. But, before moving to the next topic, it is worthwhile to present some historical background behind Klein's theorem. In group theory there is the well known structure theorem decomposing finite abelian groups into cyclic constituents, a theorem every serious student learns at the start of his career. A most natural question, which the group theorist Philip Hall perhaps first asked, is (for a finite abelian group) *how do the structural invariants of a group, a subgroup, and the quotient by the subgroup, relate?* According to a remark in [18], Hall lectured on this topic in 1955 at a St. Andrews colloquium, and there he appears to have observed the intimate connection between this problem and the algebra of symmetric polynomials (see below.) The Schubert calculus of algebraic geometry (defined below) seems also to be related to the issue of the invariant factors of a module, a submodule, and the quotient by the submodule. Hall appears to have been first with this observation, too, for in a lecture he gave at a 1959 Canadian Mathematical Congress conference in Banff on the algebra of symmetric polynomials he noted that the Schubert calculus has combinatorics similar to that of the symmetric polynomials [9].

Apparently influenced by Hall's observations, the problem of the invariant factors of a module, a submodule, and the quotient by the submodule was further examined by the algebraist J. A. Green, concluding with the two definitive papers [18,19] by Green's student T. Klein. The time has surely come to exploit the Hall, Green, Klein results to finish the invariant factor problem for products, and since Klein did all the module theory that seems needed, the work remaining appears to be pure combinatorics (or some substitute for the combinatorics.)

The connection between the invariant factor problem and the algebra of symmetric polynomials is this: If an algebra (now called the *Hall algebra*) is constructed having as structure constants (relative to a suitable basis) the number of distinct ways of constructing a module containing a submodule, with module, submodule, and quotient all having specified Smith invariants, then this algebra is isomorphic to the symmetric polynomial algebra. (See [24] for details.) There also is a natural isomorphism between a quotient algebra of the symmetric polynomial algebra and the intersection ring of the Grassmann manifold, this latter ring having multiplication of Schubert cycles as its multiplication, explaining (perhaps) the remark about the Schubert calculus. (This is described, with references, in [14].) These observations should convincingly reveal that the invariant factor problem for products involves a significant breadth of ideas.

The Schubert calculus is this: First, a nested tower of subspaces in a vector space is called a *flag*. Given three flags comprising subspaces in general position with fixed dimensions, a *Schubert condition* is that *one* subspace of specified dimension should exist meeting *all* the subspaces of the flags in specified ways. The Schubert calculus may be viewed as a study of the allowable dimensions of the spaces in the flags for which a Schubert condition is satisfied. The books [10,11] explain the rather intricate subject in detail. A point worth observing (for later comment) is that the vector spaces are over an algebraically closed field.

SUBMATRICES.

Let the invariant factors of A, B, C again be in increasing order, and now suppose that A and B are complementary submatrices in C, so that

$$C = \begin{bmatrix} A & * \\ * & B \end{bmatrix}.$$

Let C be $n \times n$, A be $p \times p$, and B be $q \times q$.

THEOREM. *The following inequalities connect the invariant factors of the full matrix C with those of its submatrix A:*

$$c_i \mid a_i \mid c_{i+2q}, \qquad i = 1, 2, \ldots, p, \tag{3}$$

(where $c_t = 0$ *if* $t > n$*,). These inequalities are necessary and*

sufficient for the existence of a matrix C *containing a submatrix* A *with both having prescribed invariant factors.*

This theorem was discovered independently by the author and E. Marques de Sa [38,32], but it was later found that an equivalent fact in the language of modules appears in Bourbaki [2], and perhaps is originally due to A. Chatalet [reference unknown]. A further contribution is by P. Joyce [15], who explained the connection with the Ulm invariants [16] of abelian group theory. It turns out for this theorem that the sufficiency part of the proof is not noticeably more difficult than the necessity part, perhaps (in view of the general principle) suggesting that further facts remain to be found. Obviously, there is more to be found, since the theorem does not mention the submatrix B complementary to A. Of course,a similar necessary and sufficient statement connects C with B, but then A is not mentioned.

Surely, though, when A, B, and C are considered simultaneously, further conditions may be expected. However, and this is surprising, there are no further conditions:

THEOREM. *The divisibility relations (3) connecting the invariant factors of* A *and* C, *together with the analogous divisibility relations connecting the invariant factors of* B *and* C, *comprise necessary and sufficient conditions for the existence of a matrix* C *having complementary blocks* A *and* B, *with* C, A, B *all having prescribed invariant factors.*

This result appears in [39]. And now it is true that sufficiency is much harder to establish than necessity, conforming to the general principle. (But there is still another fact, see the section below on similarity invariant factors.)

COMMON DIVISORS AND COMMON MULTIPLES.

Matrices A and B over the domain *R* are well known to possess a least common left multiple, call it M, and a greatest common right divisor, to be called D. Applying the facts described above, certain conclusions may be obtained about the interaction of the invariant factors of M and D with those of A and B. Specifically [46], if invariant factors are taken in increasing order, with those of M being $m_1|m_2|\ldots$, and those of D being $d_1|d_2|\ldots$, then (for nxn matrices)

THEOREM. $d_i\,|\,(a_i,\ b_i)\,|\,d_{2i-1},\ m_{2i-n}\,|\,\langle a_i,\ b_i\rangle\,|\,m_i,\ i = 1,\ldots,\ n.$

Here, $(\cdot,\cdot)$ is the greatest common divisor of ring elements as before and $\langle\cdot,\cdot\rangle$ the least common multiple. (Take $m_t = 1$ if $t \leq 0$, and $d_t = 0$ if $t > n$.) Moreover, these constraints are sharp in this sense:

THEOREM. *If elements* m_i *are given for all* i, $1 \leq i \leq n$, *and elements* a_i *and* b_i *are given for a fixed* i, *then matrices* A, B *exist for which the* M *has all these* m_i, *and* A, B *have (for the fixed* i) *their* a_i *and* b_i *equal to the given values. A similar statement holds if* A, B, *and* D *are considered together, with all* d_i *and one pair* a_i, b_i *prescribed.*

Stronger statements should be expected if A, B, M, D are considered simultaneously, since it can be shown that M and D may be chosen so that $M = AD^{-1}B$. So the proper vehicle for investigating the interaction of A, B, M, and D appears to be the material above on products, in a generalized form involving three factors (instead of two.) However, the author got nothing significant by pursuing this idea, and another approach may be needed. These results are in the unpublished manuscript [46], but, since greatest common divisors and least common multiples have been heavily investigated (in the framework of polynomial matrices) by I. C. Gohberg and coworkers (see the references in [8]), [46] may not contain substantially new facts.

MATRICES OF ALGEBRAIC INTEGERS

Sometimes one can go beyond the realm of principal ideal domains, and this is the case when the ground ring R containing the matrix elements is a ring of algebraic integers, a ring intensely interesting to any number theorist. So let R be a ring of algebraic integers. Then R is usually not a principal ideal ring, but it is well known from class field theory [31] that a finite algebraic extension exists such that in the integer ring R' of the algebraic extension every ideal generated by elements of R becomes principal (even though R' is not usually a principal ideal ring.) Let A be a matrix over R. Then it no longer has invariant factors in R, at least not usually. But not all is lost, since we have this fact:

THEOREM. *A matrix with algebraic integer entries from a ring R has invariant factors in the ring R' making every R ideal principal.*

An extensive discussion of this theorem and many related facts appears in [28]. Its ideas draw heavily on earlier work, going back to the historically significant mathematicians E. Steinitz [34] and W. Krull [20], with recent work by L. Levy [22]. *For these invariant factors, that usually lie in the extended ring R', we assert that all the previous theorems hold.* Thus we succeed, to some extent, in going beyond the realm of principal ideal domains, though not very far beyond. An unexpected aspect is that the proof of the existence of the invariant factors (in the extended ring R') is *heavily categorical,* and at the present time no really good computational procedure has been worked out for calculating these seemingly remote algebraic integers. Perhaps there is no need to calculate them since they only reflect the behaviour of ideals within R, with all theorems provable by working directly with the ideals. (There is a calculational, procedure, though [28], but its defect is that it seems to work in a ring much larger than R'.)

Sometimes, however, when a matrix has elements in a ring R of algebraic integers which is not a principal ideal domain, the invariant factors of A will still be in R. A nontrivial example of this situation is the Schur matrix

$$A = [a^{ij}]_{0\leqslant i,j<n},$$

where a is an algebraic integer in R.

THEOREM. *It can be proved that the invariant factors of the Schur matrix are already in the ground ring R of algebraic integers* containing a, *and explicit formulas can be given for them.*

This result is in [27] and does not require that the ground ring R be a principal ideal ring. The formulas are particularly nice when $n = p$ is a prime integer and $a = 1 - \zeta$ is the well known prime in the integer ring of the field of the primitive pth root of unity ζ.

Before leaving this topic, an interesting point is that Steinitz appears to have anticipated Hall in the discovery of the Hall algebra, see [13].

INVARIANT FACTORS AND EIGENVALUES.

Again, let A be a matrix with entries in a ring R of algebraic integers. Then A has eigenvalues, which are algebraic integers in a larger ring, and it has invariant factors, which also are algebraic integers in a larger ring. What connection exists between the invariant factors and the eigenvalues?

THEOREM. *If the invariant factors are* $a_1|\dots|a_n$, *and the eigenvalues are* $e_1,\dots, e_n$, *the latter in any convenient order, then a necessary and sufficient condition for the existence of a matrix of algebraic integers having these eigenvalues and invariant factors is*

$$a_1\dots a_k | e_{i_1}\dots e_{i_k}$$

for all index sets $i_1 < \dots < i_k$, $k = 1,\dots, n$, *with equality (to within a unit) when* $k = n$.

The sufficiency part of the proof constructs the required matrix over the smallest ring containing all a_i and e_i. A very explicit, one line, construction is due to R. Guralnick, and is so clever that it almost is a test of intelligence to find it. It appears in [28]. (So sufficiency proofs are not always hard, however, this one involves only one matrix, not contradicting the general principle.)

RANK ONE PERTURBATIONS.

Although the general situation concerning the behaviour of invariant factors when matrices add is open, there is a case in which decisive results are available.

Let two sets of invariant factors, in increasing order, be given, $a_1|a_2|\dots$ and $b_1|b_2|\dots$, and ask: when are there matrices A and B having these invariant factors and differing by at most rank one? The answer is:

THEOREM. *There exists matrices* A, B, *with prescribed invariant factors and with* rank$(A - B)$ *at most one, if and only if*

$$a_1|b_2|a_3|b_4|\dots, \quad \text{and} \quad b_1|a_2|b_3|a_4|\dots. \tag{4}$$

This intriguing theorem [40] is closely related to the submatrix embedding discussed above, and is not too hard. The reason it isn't hard, of course, is that it involves only two matrices, the third matrix $X = A - B$ being unrestricted (apart from rank.)

SIMILARITY INVARIANT FACTORS.

The invariant factors for similarity are the invariant factors of the characteristic matrix $xI-A$, where A has entries in a field and x is a polynomial indeterminate. The special form of this polynomial matrix imposes additional constraints on the similarity invariants of algebraic combinations of matrices. There appears to be much information in an unpublished Portuguese language dissertation of E. Marques de Sa [33]. However, let us discuss a case for which most of the pertinent facts are published.

Let A be a principal submatrix of a larger matrix C, and ask:

What connection exists between the similarity invariants of A and C? This question is completely answered:

THEOREM. *Over a field, there exists a matrix* C *containing a given matrix* A *as a principal submatrix, with both having prescribed similarity invariants, if and only if the invariant factors satisfy the divisibility conditions (3) and the degree condition.*

The *degree condition* is the statement that *the sum of the degrees of a proposed set of similarity invariant factors of an* n×n *matrix* C *must sum to* n (and a corresponding statement for the principal submatrix A of C.) This degree condition makes the sufficiency proofs very much harder, when the submatrix has deficiency strictly more than one, since these proofs go by induction on the dimension, and the degree constraint has be carried along as the induction runs. In fact, in the original papers of Sa [32] and Thompson [38] in which this matrix-submatrix similarity invariant factor theorem was first established, coping with the degree condition added greatly to the length and difficulty of the proofs, and it is this part of these papers that renders them significant. Thompson used a lengthy, rather brute force, technique to overcome the difficulty with the degrees, whereas Sa invented a clever argument involving (unbelievedly!) the approximation property of a Lebesgue integral by integrals of step functions. Sa's proof established more than Thompson's, namely an attractive *convexity property possessed by the polynomial degree*. Later, another rather combinatorial proof covering the same difficulty with the degrees was obtained by P. Y. Cheng [6] (but not including the convexity.) These three proofs have little in common, and all are hard to follow. A reasonable judgment is that a really conceptual proof covering this awkward point is still to be found. And it's probably also true that no one of the three authors understands the other two proofs.

In any event, here we have another aspect of the original submatrix embedding theorem for which sufficiency is much harder to prove than necessity. This version involves only two matrices (C, and its submatrix A), apparently contradicting the general principle, but the problem has the additional degree constraint, a restriction that adequately substitutes for the absence of a third matrix.

However, a complete theorem should involve the similarity invariants of a full matrix C and of complementary principal blocks A and B, i.e., not just C and A alone. For the available information on this topic, Sa's thesis and the work of other members of the Portuguese linear algebra school should be consulted. By virtue of the general principal, the complete discussion cannot be easy.

SIMILARITY INVARIANTS OF A SUM OR PRODUCT.

The Portuguese group under the leadership of G. N. de Oliveira has worked on these questions, and their papers should be consulted. The study of similarity invariant factors in some ways is different from the corresponding study over rings. It is surprising how strong the degree condition is. For similarity invariants, one often is reduced to elaborate computations with matrix elements. A typical example of this technique may be seen in the paper [29] of da Silva, de Oliveira, and Sa. Some work imitating [29] on the similarity invariants of products was done by Thompson, but not written up for publication. In general, divisibility conditions like those in the in the sections above on sums and products are to be expected, but in the presence of the degree constraint they often say nothing, except when the matrices are very close to being scalar, and then they powerfully reflect the special properties of scalar matrices.

However, the rank one perturbation theorem discussed above, see the section containing (4), is very little harder for similarity invariants than it is for invariant factors over rings. Here, given a proposed set of similarity invariants, one asks when are there matrices A and B over a field having these invariants and differing by rank at most one? *The necessary and sufficient conditions for this are the divisibility relations (4) and the degree condition.* The proof [40] is only slightly harder than the corresponding proof over rings, and, of course, since it only involves only two matrices cannot have nearly the complexity of those involving three.

The Portuguese group has been extremely active on a whole range of related questions, involving invariant factors together with analogous theorems about singular values. See, for example, [5], and related work by Carlson [4].

FURTHER DISCUSSION OF THE INVARIANT FACTORS OF A PRODUCT.

We intend to return to working with a product of matrices over a principal ideal domain R, but to explain our theme we must first make certain remarks about the deep and interesting problem of the eigenvalues of a sum of Hermitian matrices. Let, momentarily, A, B, and C be n×n Hermitian matrices with eigenvalues

$$\alpha_1 \geqslant \dots \geqslant \alpha_n, \quad \beta_1 \geqslant \dots \geqslant \beta_n, \quad \gamma_1 \geqslant \dots \geqslant \gamma_n,$$

respectively. A conjecture, due to A. Horn [12], is that Hermitian matrices A, B, C = A + B exist with the prescribed numbers just listed as eigenvalues, if and only if a set of linear inequalities of the form

$$\gamma_{k_1} + \dots + \gamma_{k_m} \leqslant \alpha_{i_1} + \dots + \alpha_{i_m} + \beta_{j_1} + \dots + \beta_{j_m} \tag{5}$$

holds, where the index sets are three strictly increasing sequences, as in (1). (Equality is required in this inequality when m = n.) Moreover, Horn suggested a recursion formula for these sequences, as follows: Subscripts $i_1 < \dots < i_m$, $j_1 < \dots < j_m$, $k_1 < \dots < k_m$ are admissible if, and only if,

$$i_1 + \dots + i_m + j_1 + \dots + j_m = k_1 \dots + k_m + m(m+1)/2,$$

and

$$i_{p_1} + \dots + i_{p_s} + j_{q_1} + \dots + j_{q_s} \leqslant k_{r_1} + \dots + k_{r_s} + s(s+1)/2$$

for all previously defined admissible sequences $p_1 < \dots < p_s$, $q_1 < \dots < q_s$, $r_1 < \dots < r_s$. ("Previously defined" means s < m.)

As already stated, B. V. Lidskii [23] has announced a proof of this conjecture, but the proof appears still to be unpublished. In [47] the author has shown that the tableau-defined sequences in the discussion earlier of products do satisfy this recursion, and he views this as almost compelling evidence that the invariant factors of a product of matrices over a domain are described in precisely the same way as the eigenvalues of a sum of Hermitian matrices. But there is even more compelling evidence, as we shall now explain.

It is known, and indeed this is an idea appearing in an old paper of Berezin and Gel'fand [1], that the eigenvalue problem for a sum of Hermitian matrices is closely tied to to the question of how the tensor product of irreducible representations of the unitary group U(n) decomposes into irreducible representations. These irreducible representations are known from Lie theory to be described by Young tableaux, see for example [26]. But, beyond this, B. V. Lidskii's announcement contains the following, precise, statement:

(Lidskii): *There exist representations of the unitary group* U(n) *with highest weights* $a_1 \geqslant \ldots \geqslant a_n$ *and* $b_1 \geqslant \ldots \geqslant b_n$, *such that their tensor product contains the representation with highest weight* $c_1 \geqslant \ldots \geqslant c_n$, *if and only if the* a's, b's, c's *satisfy the class of inequalities (5)* (the subscripts running over all admissible sequences, and including equality in (5) when m = n.)

One perspective on this result is that it is about integer vectors, a fortiori about integer matrices, by extension about matrices over principal ideal domains, and so is tantalizingly close to solving the invariant factor problem for products. This, though, supplies no details to justify the optimism, and a better perspective is afforded by the following sequence of facts.

(i) By Lie theory the compact unitary group corresponds in a natural way to the noncompact general linear group, both groups having representations described by tableaux.

(ii) The representation theory for the general linear group is naturally dual to the representation theory for the symmetric group.

(iii) There is a famous combinatorial rule, the *Littlewood-Richardson rule*, that describes how the tensor product of irreducible representations of the symmetric group decomposes into irreducible representations [21, or 35]. Symmetric group representations are described by tableaux, and the rule is that the tensor product of two representations contains a third if, and only if, a tableau filled in with entries is constructible exactly as described below (1).

(iv) As shown in [24], the Littlewood-Richardson rule also describes the nonzero coefficients in the Hall algebra (the structure constants of the algebra), with a coefficient (= structure constant) being nonzero precisely when a module exists containing a submodule with module, submodule, and quotient all having prescribed Smith invariants.

It seems that this line of thought is very close to solving the invariant factor problem for products, and may indeed be a solution. If it is, the reason is that the representation theory finds the missing combinatorics by transferring the logic to a case where it already has been done, thanks to Lidskii's theorem.

ADDITIONAL REMARKS.

(a) It is not hard, using certain material on relations matrices in [44], for example, to see that the invariant factor problem for products is essentially equivalent to the invariant factor problem for a block triangular matrix C with diagonal blocks A, B, the off-diagonal corner block being variable. For some early work on the latter problem (for similarity) see [3]. There seems considerable current interest in this block triangular version of the invariant factor problem for products.

(b) The depth and breadth of ideas before us clearly is remarkable, but here is an unexpected example. Physicists have learned that elementary particles belong to irreducible representations of Lie groups, and particle interactions producing offspring to the decomposition of the tensor product of irreducibles. This point appears in the elementary and well written book [7] (unfortunately, without proofs). Our realm of ideas, therefore, includes some modern particle physics. Going farther, particle physics has been revolutionized by the quark concept, particles with numbers like 2/3 and -1/3 as their charges. But these numbers were already to be found (as coming from representations) in the Berezin-Gel'fand paper [2], apparently written before quarks were invented. This paper should be famous.

(c) It is known that the Schubert flags furnish a *cell decomposition* of the Grassmann manifold [10] (see [25] for cell decompositions in general.) The author had known this fact for some time, but had not paid attention to it. However, at the MTNS-85 meeting, M. Shayman gave a remarkable lecture on the Riccati equation in which the Schubert flags appeared prominently. With great lucidity, Shayman described the *recursive definition* of a cell decomposition Now, it is known, from a (difficult) 1966 paper by B. P. Zwahlen [49], from the author's own insight, and from the work of his student S. Johnson [14], that the Schubert flags more or less directly correspond to inequalities satisfied by the eigenvalues of a sum of Hermitian matrices. The new idea that impinged itself on the author's mind, during Shayman's lecture, is that the recursive description of the cell decomposition resembles the recursive description of the inequalities (5) satisfied by the eigenvalues of a sum of Hermitians. *This is not an accident,* and is an idea very much worth investigating.

As a side remark, Zwahlen's 1966 paper is difficult because it's author did not use the Schubert calculus, did not require algebraic closure (usually essential for the validity of the Schubert calculus), see [37], and did not possess the "general position" concept of algebraic geometry, resulting in a consideration of a plethora of special geometric configurations.

(d) How the Schubert flags directly relate to the invariant factor problem isn't presently clear to the author, but two hints are: nested towers of submodules in prime characteristic are already to be seen in Hall's Banff paper [9] and in Macdonald's book [24].

(e) The study of the Schubert flags is Hilbert's fifteenth problem, see the very fine article by Kleiman [17]. Among the many insights to be found in this paper is the point that a focus of thought about an algebraic object is its cohomology ring (for the problem of the invariant factors of products the cohomology ring appears to be the Hall algebra), and that different objects may have different cohomology. The extent to which the invariant factor problem for products differs (if it does differ) from the eigenvalue problem for a sum of Hermitians will be expressible in terms of the differences in the two cohomologies (but the expection is that the cohomologies are the same.) Kleiman also indicates how the level of abstract insight extends even further, mentioning K theory, Chern classes, singularities of mappings, etc., etc. The wealth of opportunity for further understanding of the invariant factors of a matrix product is plainly visible.

For some mathematicians, though, the presence of too much

abstract structure decreases interest, rather than increasing it. The author is sympathetic to this viewpoint, and for this reason is as much attracted to the problem of the invariant factors of a matrix sum as he is to the problem for products. Each mathematician to his own taste.

ACKNOWLEDGMENT

The preparation of this paper was supported in part by grants from the National Science Foundation and the Flight Dynamics Laboratory at Wright Patterson Air Force Base.

REFERENCES

[1] Berezin, F. A., and Gel'fand, I., Some remarks on spherical functions on symmetric Riemannian manifolds, Tr. Mosk. Mat. Obshch. 5 (1956) 311-351.

[2] Bourbaki, N., Elements de Mathematique, Algebra, Chapter 7, Modules sur les anneaux principaux, (Hermann, Paris, 1952) 104-105.

[3] Carlson, D., Inequalities for the degrees of elementary divisors of modules, Lin. Alg. Appl. 5 (1972) 293-298.

[4] Carlson, D., Minimax and interlacing theorems for matrices, Lin. Alg. Appl. 54 (1983) 153-172.

[5] Carlson, D., and Marques de Sa, E., Generalized minimax and interlacing theorems, Lin. Mult. Alg. 15 (1984) 77-103.

[6] Cheng, P. Y., Interlacing divisibility of similarity invariant factors and its application, University of California at Davis (1978).

[7] Dodd, J. E., The ideas of particle physics (an introduction for scientists) (Cambridge University Press, Cambridge, 1984) 57.

[8] Gohberg, I., Lancaster, P., and Rodman, L., Matrix polynomials, (Academic Press, New York, 1982).

[9] Hall, P., The algebra of partitions, Canadian Mathematical Congress Procedings, Banff (1959).

[10] Harris, P. and Griffiths, J., Principles of algebraic geometry, (Wiley-Interscience, New York, 1978), 193.

[11] Hodge, W. V. D., and Pedoe, D., Methods of algebraic geometry, (Cambridge University Press, Cambridge, 1952), vol. 2.

[12] Horn, A., Eigenvalues of a sum of Hermitian matrices, Pacific J. Math. 12 (1962) 225-241.

[13] Johnson, K., On a forgotten note by Ernst Steinitz on the theory of abelian groups, Bull. London Math Soc. 14 (1982) 353-355.

[14] Johnson, S., The Schubert calculus and eigenvalue inequalities for sums of Hermitian matrices, University of California at Santa Barbara (1979).

[15] Joyce, Peter M., Quotients of finitely generated modules by cy-

clic submodules, University of Kentucky (1983).

[16] Kaplansky, I., Infinite abelian groups, (University of Michigan Press, Ann Arbor, 1956.)

[17] Kleiman, S. L., Rigorous foundations of Schubert's enumerative calculus, in Mathematical developments arising from Hilbert's problems, Proceedings of symposia in pure mathematics 28 (American Mathematical Society, Providence, 1976) 445-482.

[18] Klein, T., The multiplication of Schur functions and extensions of p modules, J. London Math. Soc. 43 (1968) 280-284.

[19] Klein, T., The Hall polynomial, J. Algebra 12 (1969) 61-68.

[20] Krull, W., Matrizen, Moduln, und verallgemeinerte Abelschen Gruppen in Bereich der ganzen Zahlen, Heidelburger Akademie der Wissenschaften 2 (1932) 13-38.

[21] Lam, T. Y., Young diagrams, Schur functions, the Gale Ryser theorem, and a conjecture of Snapper, J. Pure Appl. Alg., 10 (1977) 81-94.

[22] Levy, L., Almost diagonal matrices over Dedekind domains, Math. Z. 124 (1972) 89-99.

[23] Lidskii, B. V., Spectral polyhedron of a sum of two Hermitian matrices, Functional Anal. Appl. 10 (1982) 76-77 (Russian), 139-140 (English).

[24] Macdonald, I., Symmetric functions and Hall polynomials, Oxford mathematical monographs (Oxford University Press, Oxford, 1979.)

[25] Massey, W. S., Homology and cohomology theory (Dekker, New York, 1979) Chapter 3.

[26] Naimark, M. A., and Stern, I., Theory of group representations, Grundlehren der Math. Wissenschaften 246 (Springer Verlag, New York, 1982).

[27] Newman, M., and Thompson, Robert C., Invariant factors of Vandermonde and generalized Schur matrices, manuscript.

[28] Newman, M., and Thompson, Robert C., Matrices over rings of algebraic integers, manuscript.

[29] de Oliveira, G. N., Marques de Sa, E., and Dias da Silva, J. A., On the eigenvalues of $A + XBX^{-1}$, Lin. Mult. Alg. 5, (1977) 119-128.

[30] Remmel, J., Bijective proofs for formulae for the number of standard Young tableaux, Lin. Mult. Alg., 11 (1982) 45-100.

[31] Ribenboim, P., Algebraic numbers (Wiley, New York, 1972).

[32] Marques de Sa, E., Embedding conditions for λ matrices, Lin. Alg. Appl. 24 (1979) 32-50.

[33] Marques de Sa, E., Imersao de matrizes e entrelacamento de factores invariantes, Coimbra (Portugal), 1979.

[34] Steinitz, E., Rechteckige Systeme und Moduln in algebraischen Zahlkorper, I, II, Math. Ann., 71 (1912) 328-354, 72 (1912) 297-345.

[35] Thomas, G. P., On Schensted's construction and the multiplication of Schur functions, in Kung, J. S., ed., Young tableaux in combinatorics, invariant theory, and algebra (Academic Press, New York, 1982), 81-105.

[36] Thompson, Robert C., (with Freede, L. J.), On the eigenvalues of sums of Hermitian matrices, Lin. Alg. Appl. 4 (1971) 369-376.

[37] Thompson, Robert C., and Therianos, S., On a construction of B. P. Zwahlen, Lin. Mult. Alg. 1 (1974) 309-325.

[38] Thompson, Robert C., Interlacing inequalities for invariant factors, Lin. Alg. Appl. 24 (1979) 1-31.

[39] Thompson, Robert C., Invariant factors of complementary minors of integral matrices, Houston J. Math. 5 (1979) 421-426.

[40] Thompson, Robert C., Invariant factors under rank one perturbations, Canad. J. Math. 22 (1980) 240-245.

[41] Thompson, Robert C., The Smith invariants of a matrix sum, Proc. Amer. Math. Soc. 78 (1980) 162-165.

[42] Thompson, Robert C., An inequality for invariant factors, Proc. Amer. Math. Soc. 86 (1982) 9-11.

[43] Thompson, Robert C., Author vs. referee: A case history for middle level mathematicians, Amer. Math. Monthly 90 (1983) 661-668.

[44] Thompson, Robert C., Smith invariants of a product of integral matrices, in Contemporary mathematics (Amer. Math. Soc., Providence, in press.)

[45] Thompson, Robert C., Sums of integral matrices, manuscript.

[46] Thompson, Robert C., Left multiples and right divisors of integral matrices, manuscript.

[47] Thompson, Robert C., Divisibility relations satisfied by the invariant factors of a matrix product, manuscript.

[48] Zabella, I., Inequalities for the Weyr characteristic of modules, Algebra lineal y applicaciones, Universidad del Pais Vasco (1984) 432-442.

[49] Zwahlen, B. P., Uber die Eigenwerte der Summe zweier selbstadjungierter Operatoren, Comment. Math. Helv 40 (1966) 81-116.

Frequency Domain and State Space Methods for Linear Systems
C.I. Byrnes and A. Lindquist (Editors)

STATE GROUP REALIZATIONS OF MATRICES OF RATIONAL NUMBERS

Harald K. Wimmer

Mathematisches Institut
Universität Würzburg
D-8700 Würzburg, W.Germany

Matrices of rational numbers are studied with methods from linear systems theory.

The concept of state space realization for matrices of rational functions has an analogue with matrices of rational numbers, which we call state group realization. In this note we report on results on such realizations. For proofs we refer to [2],[3],[4].

The starting point of our investigation is a summation formula for rational numbers. Let p be a prime. If $w \in \mathbb{Q}$ has the p-adic expansion

$$w = \alpha_{-N}p^{-N} + \dots + \alpha_{-1}p^{-1} + \alpha_0 + \alpha_1 p + \dots , \quad 0 \leqq \alpha_i < p,$$

we call

$$(w)_{-1,p} := \alpha_{-1} \tag{1}$$

the <u>residue</u> of w at p. The <u>principal part</u> of w at p [1] is

$$h_p(w) := \sum_{\nu=1}^{N} \alpha_{-\nu}p^{-\nu} \tag{2}$$

The following summation formula holds (see e.g. [1, p. 314]).

LEMMA. Any rational number w has a unique decomposition as

$$w = \sum_q h_q(w) + m, \quad m \in \mathbb{Z}, \tag{3}$$

where the summation is over all prime divisors of the denominator of w.

For $w \in \mathbb{Q}$ given by (3) we define

$$\pi w := \sum_q h_q(w). \tag{4}$$

The preceding definitions (1), (2) and (4) will be extended component-wise to vectors and matrices of rational numbers. - The set $\pi\mathbb{Q}$ is a group under the addition

$$a + b := \pi(a+b), \quad a,b \in \pi\mathbb{Q},$$

and $\pi\mathbb{Q} \cong \mathbb{Q}/Z$.

For an integer matrix $M = (m_{ij})$ we shall write

$$0 \leqq M < p$$

if all entries m_{ij} satisfy $0 \leqq m_{ij} < p$.

To motivate our subsequent definition let us consider first a matrix $W \in F^{s\times t}(z)$ over a field F. In a state space realization

$$W(z) = C(zI-A)^{-1}B + D(z) \tag{5}$$

of W the matrices A,B and C are constant and $D \in F^{s\times t}[z]$. If F is algebraically closed one can obtain a decomposition (5) such that zI-A = P-N with

$$P = \operatorname{diag}\left((z-\lambda_1)I, \ldots (z-\lambda_k)I\right) \text{ and } N = \operatorname{diag}(N_1,\ldots,N_k)$$

where $\lambda_1\ldots,\lambda_k$ are the distinct eigenvalues of A and N is nilpotent and partitioned according to P. We note that the nonzero entries of P are prime elements of F[z]. That B and C are constant matrices is equivalent to the fact that the elements of $(zI-A)^{-1}$ and $C(zI-A)^{-1}$ are strictly proper rational.

DEFINITION. Let $W \in \mathbb{Q}^{s\times t}$ be given and let P-N,C,B and D be integer matrices of size $m\times m$, $s\times m$, $m\times t$ and $s\times t$ respectively. We call

$$W = C(P-N)^{-1}B + D \tag{6}$$

a <u>state group realization</u> of W if

$$P-N = \operatorname{diag}(p_1I-N_1,\ldots,p_kI-N_k), \tag{7a}$$

$$C = (C_1,\ldots,C_k),\; B^T = (B_1^T,\ldots,B_k^T), \tag{7b}$$

such that the matrices in (7) are partitioned correspondingly, the numbers $p_1,\ldots,p_k$ are distinct primes, the matrices $N_1,\ldots,N_k$ are nilpotent, and furthermore $(P-N)^{-1} \in \pi\mathbb{Q}^{m\times m}$, $(P-N)^{-1} \in \pi\mathbb{Q}^{m\times t}$ and $C(P-N)^{-1} \in \pi\mathbb{Q}^{s\times t}$. The size m of P-N is called the <u>straight dimension</u> of (6). A <u>minimal realization</u> has minimal straight dimension.

As in the case of matrices of rational functions one can construct a minimal realization using the Smith-MacMillan form of W. - In linear systems theory minimality can be characterized by rank and coprimeness conditions. The same is possible for matrices over $\mathbb{Q}$.

DEFINITION. Let $H \in \mathbb{Z}^{m\times m}$ be given and let X and Y be in $\mathbb{Z}^{m\times q}$ and $\mathbb{Z}^{r\times m}$ respectively. We write

$$\text{column rank } X = m (\text{mod } H)$$

if for each $v \in \mathbb{Z}^m$ there exists a $y \in \mathbb{Z}^q$ such that $v \equiv Xy \pmod{H}$, i.e. $v = Xy + Hz$ for some $z \in \mathbb{Z}^m$. We define

$$\text{row rank } Y(\text{mod } H) := \text{column rank } Y^T (\text{mod } H^T).$$

THEOREM. Let $W = C(P-N)^{-1}B + D$ be a state group realization of $W \in \mathbb{Q}^{s\times t}$ of straight dimension m. The following statements are equivalent.

(a) The realization is minimal.
(b) The pair (P-N,B) is left coprime and the pair $\binom{P-N}{C}$ is right coprime.

(c) column rank $(N,B) = m(\text{mod } P)$

and

$$\text{row rank } \binom{N}{C} = m(\text{mod } P).$$

(d) column rank $(B, NB, \ldots, N^{m-1}B) = m(\text{mod } P)$ and

$$\text{row rank} \begin{pmatrix} C \\ CN \\ \vdots \\ CN^{m-1} \end{pmatrix} = m(\text{mod } P) .$$

We will construct the counterpart to a state space model. Let $W \in \mathbb{Q}^{s \times t}$ have the realization

$$W = VG^{-1}U + D \ , \ G \in \mathbb{Z}^{n \times n} \ , \ \det G \neq 0 \tag{8}$$

where the integer matrices U,V,D are of size $n \times t$, $s \times n$, $s \times t$, respectively. We call the factor group

$$\mathbb{Z}^n/G\mathbb{Z}^n \tag{9}$$

a <u>state group</u> associated to (8). To obtain a concrete representation of (9) we define a map $\pi_G : \mathbb{Z}^n \to \mathbb{Z}^n$ by

$$\pi_G h := G\pi G^{-1}h, \ h \in \mathbb{Z}^n,$$

and put

$$V_G := \text{Im } \pi_G .$$

Then $V_G \subset \mathbb{Z}^n$ and V_G is a group under the addition $v_1 + v_2 := \pi_G(v_1 + v_2)$, $v_i \in V_G$, and

$$V_G = \mathbb{Z}^n/G\mathbb{Z}^n .$$

Occasionally we will write

$$v_1 \equiv v_2(\text{mod } G)$$

if $\pi_G v_1 = \pi_G v_2$ or equivalently $v_2 = v_1 + Gh$, for some $h \in \mathbb{Z}^n$.

We require an endomorphism of V_G which should play the role of a shift operator. Let A be an abelian group of order $p_1^{m_1} \cdots p_k^{m_k}$ and let

$$A = A(p_1) \oplus \ldots \oplus A(p_k)$$

be its primary decomposition such that the $A(p_i)$'s are the p_i-components of A. For an element $a \in A$ given by $a = \Sigma a_i$, $a_i \in A(p_i)$, put

$$Sa := \Sigma \, p_i \, a_i .$$

Then S is a nilpotent endomorphism of A.

The notion of straight basis allows us to pass from mappings to matrices.

DEFINITION. An m_i-tupel $\Sigma_i = (\sigma_{i1},\ldots,\sigma_{i,m_i})$ of elements of $A(p_i)$ is called a <u>straight basis</u> of $A(p_i)$, if each $a_i \in A(p_i)$ can be expressed uniquely as

$$a_i = \sum_{\mu=1}^{m_i} c_\mu \sigma_{i_\mu}, c_\mu \in \{0,1,\ldots,p_i-1\}. \tag{10}$$

We call $\Sigma = (\Sigma_1,\ldots,\Sigma_k)$ a straight basis of A, if each Σ_i is a straight basis of $A(p_i)$, $i=1,\ldots,k$. We define

$$\text{str.dim } A := \sum_{i=1}^{k} m_i.$$

The <u>straight dimension</u> of A is the cardinality of any straight basis of A.

If we regard A as a right $\mathbb{Z}$-module then we can write (10) as $a_i = \Sigma_i c_i$ with $c_i = (c_1,\ldots,c_{m_i})^T$, $0 \leq c_i < p_i$. The action of S on Σ_i, can be described by the matrix $p_i I$ and also by a nilpotent matrix $N_i \in \mathbb{Z}^{m_i \times m_i}$, i.e.

$$S\Sigma_i = \Sigma_i p_i = \Sigma_i N_i.$$

The matrix N_i is unique if we require $(p_i I - N_i)^{-1} \in \pi\mathbb{Q}^{m_i \times m_i}$. For a straight basis $\Sigma = (\Sigma_1,\ldots,\Sigma_k)$ of A there exist matrices

$$P = \text{diag}(p_1 I,\ldots,p_k I) \quad \text{and} \quad N = \text{diag}(N_1,\ldots,N_k) \tag{11}$$

such that

$$S\Sigma = \Sigma P = \Sigma N.$$

We call P and N the shift matrices associated to Σ. - If the matrix N in (11) is in (lower triangular) Jordan form we call Σ a <u>Jordan basis</u> of A.

From a realization (8) we can obtain a state group realization. It is no essential restriction to assume $W = h_p(W)$.

THEOREM. Let $W \in \mathbb{Q}^{s\times t}$ be given by

$$W = \sum_{\nu \geq 0} W_\nu p^{-(\nu+1)}, \quad 0 \leq W_\nu < p, \quad \nu \geq 0,$$

and let W have the realization $W = VG^{-1}U + D$ such that $|\det G| = p^m$. For a fixed straight basis Σ of V_G let N be the associated shift matrix, i.e. $S\Sigma = \Sigma N$, and $(pI-N)^{-1} \in \pi\mathbb{Q}^{m\times m}$. Define the homomorphism

$$\tilde{B} : Z^t \to V_{pI-N}, \quad \tilde{B}\xi := \pi_G U\xi,$$

and the map

$$\tilde{C} : V_{pI-N} \to \mathbb{Z}^s \ , \ \tilde{C}h := (VG^{-1}h)_{-1,p}.$$

Then

$$\tilde{C}S^{\nu}\tilde{B}\, e_i = W_{\nu} e_i \quad , \quad i=1,\dots,t \ , \ \nu \geqq 0.$$

Let the matrix $B \in \mathbb{Z}^{m\times t}$ be given by $B = (\tilde{B}\, e_1,\dots,\tilde{B}\, e_t)$ and set $\tilde{C}(pI-N)^{-1}h := \sum_{\nu\geqq 0} p^{-(\nu+1)}\tilde{C}N^{\nu}h$. Then there exists a matrix $C \in \mathbb{Z}^{s\times m}$ such that

$$\pi C(pI-N)^{-1}h = \tilde{C}(pI-N)^{-1}h, \quad h \in V_{pI-N}$$

and there is a $\hat{D} \in \mathbb{Z}^{s\times t}$ such that $W = C(pI-N)^{-1}B + \hat{D}$ is a state group realization.

The relation between two minimal realizations is described in the following result.

THEOREM. Let $W = C(P-N)^{-1}B + D$ and $W = \tilde{C}(\tilde{P}-\tilde{N})^{-1}\tilde{B} + \tilde{D}$ be two irreducible state group realizations of straight dimension m. Then there exist matrices $K, \tilde{K}, R, \tilde{R} \in \mathbb{Z}^{m\times m}$ which have the intertwinging properties

$$\tilde{K}(P-N) = (\tilde{P}-\tilde{N})R \ , \ K(\tilde{P}-\tilde{N}) = (P-N)\tilde{R}$$

and which satisfy

$$K\tilde{K} \equiv I(\text{mod } P-N) \ , \ \tilde{K}K \equiv I(\text{mod } \tilde{P}-\tilde{N}),$$

$$(\tilde{R}R)^T \equiv I(\text{mod}(P-N)^T) \ , \ (R\tilde{R})^T \equiv I(\text{mod } (\tilde{P}-\tilde{N})^T) \ .$$

The matrices of the two minimal realizations are linked by

$$\tilde{P} \equiv \tilde{K}PK \equiv \tilde{N} \equiv \tilde{K}NK \ (\text{mod } \tilde{P}-\tilde{N}),$$

$$\tilde{B} \equiv \tilde{K}B \ (\text{mod } \tilde{P}-\tilde{N}).$$

$$\tilde{C}^T \equiv (C\tilde{R})^T \ (\text{mod } (\tilde{P}-\tilde{N})^T).$$

Furthermore

$$\pi \ (\tilde{P}-\tilde{N})^{-1} = \pi \ R(P-N)^{-1}K.$$

We will next discuss duality. Let us remark already here that the matrix $\tilde{K}$ yields an isomorphism $f : V_{P-N} \to V_{\tilde{P}-\tilde{N}}$ given by

$$fv = \pi_{\tilde{P}-\tilde{N}}\ \tilde{K}v \ , \ v \in V_{P-N},$$

and the matrix $\tilde{R}$ corresponds to an isomorphism $g : V_{(P-N)^T} \to V_{(\tilde{P}-\tilde{N})^T}$ which is dual to f^{-1}.

Recall that a character of an abelian group A (in additive notation) is a homomorphism of A into the group $\pi\mathbb{Q}$. Let A* denote the character group of A. According to the following theorem $(V_G)^*$ can be identified with V_{G^T}.

THEOREM. For $x \in V_{G^T}$ and $v \in V_G$ put

$$\varphi(x,v) := \pi x^T G^{-1} v$$

and

$$\varphi_x(v) := \varphi(x,v) \ .$$

Then (a) the map $\varphi : V_{G^T} \times V_G \to \pi\mathbb{Q}$ is bilinear and non-degenerate and

(b) the map $\Phi : x \mapsto \varphi_x$ is an isomorphism from V_{G^T} onto $(V_G)^*$.

Let A be a finite abelian group with a straight basis $\Sigma = (\sigma_1,\ldots,\sigma_m)$ and let $X = (\chi_1,\ldots,\chi_m)^T$ be a straight basis of A*. Put $\chi_i\sigma_j := \chi_i(\sigma_j)$ and

$$X\Sigma = \begin{pmatrix} \chi_1 \\ \cdot \\ \cdot \\ \cdot \\ \chi_m \end{pmatrix} (\sigma_1,\ldots,\sigma_m) \ .$$

We regard A* as a left Z-module and will arrange the elements of a straight basis always in a column.

DEFINITION. Let Σ and X as above be straight bases of A and A* respectively, such that $S\Sigma = \Sigma P = \Sigma N$. We call Σ and X dual to each other if

$$X\Sigma = \pi(P-N)^{-1} \tag{12}$$

holds.

In the case of a Jordan basis we can replace (12) by "orthogonality" conditions. For $w \in \mathbb{Q}$ put $(w)_{-1} = \sum_q (w)_{-1,q}$.

REMARK. For a Jordan basis Σ the condition (12) is equivalent to

$$\Big(\chi_i(\sigma_j)\Big)_{-1} = \delta_{ij}, \ i,j=1,\ldots,m.$$

There is a relation between minimal realizations of G^{-1} and pairs of dual straight bases of V_G and V_{G^T}.

THEOREM. Let the columns of M, $H^T \in \mathbb{Z}^{n\times m}$ be a pair of dual straight bases of V_G, V_{G^T} such that

$$\pi H G^{-1} M = \pi(P-N)^{-1}.$$

Then the matrices $C := G^{-1}M(P-N)$ and $B := (P-N)HG^{-1}$ are integer matrices, and $G^{-1} = C(P-N)^{-1}B + D$ is a minimal realization for a suitable $D \in \mathbb{Z}^{n\times n}$.

For an integer matrix G with $G^{-1} \in \pi\mathbb{Q}^{n\times n}$ we have a factorization which corresponds to a transformation into a Jordan normal form.

THEOREM. Let $G \in \mathbb{Z}^{n\times n}$ be non-singular such that $G^{-1} \in {}_{\pi}\mathbb{Q}^{n\times n}$ holds. If the columns of $M, H^T \in \mathbb{Z}^{n\times m}$ form a pair of dual Jordan bases of V_G, V_{G^T} with associated shift matrices P and N, i.e. $\pi HG^{-1}M = (P-N)^{-1}$ then G can be factored as

$$G = M(P-N)H. \tag{13}$$

The invariant factors of G and of P-N are the same.

The matrix M in (13) is made up of blocks of vectors which can be regarded as analogues of Jordan chains (consisting of an eigenvector and generalized eigenvectors). To each elementary divisor p^r of G corresponds a chain $\{m_1,\ldots,m_r\}$ of columns of M which have the following property. There are vectors $c_i \in \mathbb{Z}^n$, $i=1,\ldots,r$, with $0 \leq c_i < p$ such that

$$Gc_1 = pm_1,\ Gc_2 = pm_2-m_1,\ldots,Gc_r = pm_r-m_{r-1}.$$

REFERENCES

1. H. Hasse, Number Theory, Grundlehren d. mathematischen Wissenschaften 229, Springer-Verlag, Berlin 1980.

2. H.K. Wimmer, Realizations of matrices of rational numbers, submitted to J. Number Theory.

3. H.K. Wimmer, Matrices of rational numbers, state group realizations and straight bases of abelian groups, submitted to Systems and Control Letters.

4. H.K. Wimmer, Duality of finite abelian groups and a Jordan normal form of integer matrices, submitted to SIAM J. Discr. Algebr. Methods.

2. ALGEBRAIC GEOMETRIC CONTROL THEORY

Frequency Domain and State Space Methods for Linear Systems
C.I. Byrnes and A. Lindquist (Editors)
Elsevier Science Publishers B.V. (North-Holland), 1986

CONTROLLER DESIGN FOR 2D SYSTEMS

Mauro Bisiacco, Ettore Fornasini, Giovanni Marchesini

Istituto di Elettrotecnica e di Elettronica
Via Gradenigo, 6/A 35131 Padova, Italy

In this paper the problem of constructing output feedback compensators for 2D systems is considered. The bounds on the algebraic variety associated to the closed loop characteristic polynomial are evidentiated and an explicit technique for synthesizing the compensator is given.

INTRODUCTION

Very recently some papers dealing with 2D systems have been concerned with the problem of constructing dynamic compensators having observer-controller structure [1,2].

The principal tool considered for solving this problem is constituted by 2D analogs of the polynomial matrices involved in the PBH tests for controllability and reconstructibility. In particular, the rank analysis of these matrices provides the bounds on the performances of asymptotic observers and state feedback controllers.

The aim of this paper is to analyze the 2D output feedback scheme using the polynomial matrix approach and to investigate what is the structure of the characteristic polynomial as the output feedback compensator is varied. Moreover we shall introduce a general procedure for synthesizing the compensator.

In this framework the 2D analogs of the polynomial matrices involved in the PBH tests play a fundamental role, in the sense that the unique constraint on the variety of the characteristic polynomial is that it has to include the set of points where at least one of these matrices does not have full rank.

Obviously, this set depends on the system that realizes the transfer matrix. However, it contains the so called "rank singularities", that are characteristic of the transfer matrix only, and are invariant both with respect to the realization and to the output feedback compensation policy.

SOME PROPERTIES OF 2D TRANSFER MATRICES

Many papers appeared in the recent literature, concerning with the theory of matrix fractions in two indeterminates [see for instance 3]. In the following we shall briefly review some well known fundamental properties of this class of matrices with the purpose of stating notations and terminology. We shall rather dwell upon the analysis of the so called "rank singularities" that provide the extension to the matrix case of the non essential singularities of the second kind for rational functions in two variables.

As we shall see, rank singularities constitute the fundamental constraint we are faced by, when we use output feedback techniques for modifying the dynamics of 2D systems.

Consider a strictly proper 2D transfer matrix $W(z_1,z_2) \in \mathbb{R}(z_1,z_2)^{p\times m}$ and let

$$N_R(z_1,z_2)D_R^{-1}(z_1,z_2) = W(z_1,z_2) \tag{1}$$

be a right matrix fraction description (MFD) of W.

The polynomial matrices N_R and D_R(*) are said <u>right factor coprime</u> (and, consequently, $N_R D_R^{-1}$ is a right factor coprime MFD) if for any polynomial matrix X such that

$$N_R = \tilde{N}_R X, \qquad D_R = \tilde{D}_R X$$

with $\tilde{N}_R$ and $\tilde{D}_R$ polynomial matrices, we have $\det X = \text{const}$.

The following theorem provides an extension of coprimeness condition to polynomial matrices in two indeterminates.

<u>Theorem 1</u> [3] Let $N_R \in R[z_1,z_2]^{p\times m}$ and $D_R \in R[z_1,z_2]^{m\times m}$ and consider the matrix fraction $N_R D_R^{-1}$. Then the following facts are equivalent

(i) N_R and D_R are right factor coprime;

(ii) the Bézout equation

$$X_R D_R + Y_R N_R = I \tag{2}$$

is solvable both with X_R and Y_R having elements in $R(z_1)[z_2]$ and in $R(z_2)[z_1]$;

(iii) N_R and D_R are 1D right coprime both on $R(z_1)[z_2]$ and $R(z_2)[z_1]$;

(iv)

$$[N_R' \quad D_R'] \tag{3}$$

is full rank for any generic point of $\mathcal{V}$, $\mathcal{V}$ being the algebraic curve associated with the equation $\det D_R = 0$ over the complex field $\mathbb{C}$.

Assuming $W(z_1,z_2)$ is strictly proper is equivalent to require that every right coprime MFD $N_R D_R^{-1} = W$ satisfies

(i) $N_R(0,0) = 0$

(ii) $D_R(0,0)$ is invertible, so we can assume $D_R(0,0) = I$.

Clearly, all definitions and statements have analogs for left MFDs.

We shall say that N_R and D_R are <u>zero right coprime</u> if (3) has full rank for any $(z_1,z_2) \in \mathbb{C}\times\mathbb{C}$. It can be shown that N_R and D_R are zero right coprime if and only if there exist polynomial matrices X_R and Y_R such that the Bézout identity (2) holds.

(*) In the sequel the arguments z_1 and z_2 of polynomial and rational matrices in two variables will be omitted, when unnecessary.

In general, factor coprimeness assumption does not imply zero coprimeness. In fact, even when N_R and D_R are factor coprime, the rank of (3) may be less than m on a finite subset of $\mathbb{C} \times \mathbb{C}$. The elements of this set, called "rank singularities", are the intersections of the algebraic curves associated with the minors of maximal order of (3).

As a consequence of the following Theorem, the set of rank singularities does not depend on the factor coprime MFD of W and for this reason it will be denoted by $\mathcal{V}(W)$.

<u>Theorem 2.</u> Let $N_R D_R^{-1}$ and $D_L^{-1} N_L$ be right and left coprime MFDs of W. If $\mathcal{V}_R(W)$ and $\mathcal{V}_L(W)$ are the (finite) subsets of $\mathbb{C} \times \mathbb{C}$ where $[N_R' \quad D_D']$ and $[N_L \quad D_L]$ are not full rank, then $\mathcal{V}_R(W) = \mathcal{V}_L(W)$.

proof. By contradiction. Assume $(z_{10}, z_{20}) \in \mathcal{V}_L(W)$ and $(z_{10}, z_{20}) \notin \mathcal{V}_R(W)$. Consider the line

$$\begin{aligned} z_1 &= z_{10} + \alpha t \\ z_2 &= z_{20} + \beta t \end{aligned} \tag{4}$$

with α and β such that (4) does not intersect $\mathcal{V}_R(W)$ and let $N_R(t) D_R^{-1}(t)$ and $D_L^{-1}(t) N_L(t)$ denote the restrictions of $N_R D_R^{-1}$ and $D_L^{-1} N_L$ to the line (4).

$N_R(t) D_R^{-1}(t)$ is a 1D coprime MFD since $[N_R'(t) \quad D_R'(t)]$ is full rank for every complex t. On the contrary, $[N_L(0) \quad D_L(0)]$ is not full rank, so $D_L^{-1}(t) N_L(t)$ is not a 1D coprime MFD and

$$\begin{aligned} N_L(t) &= \tilde{N}_L(t) R(t) \\ D_L(t) &= \tilde{D}_L(t) R(t) \end{aligned}$$

where $\tilde{N}_L$ and $\tilde{D}_L$ are polynomial matrices, $R(t)$ is not unimodular and $\tilde{D}_L^{-1}(t) \tilde{N}_L(t)$ is a left coprime MFD.

Then

$$\det \tilde{D}_L(t) = \det D_R(t) \tag{5}$$

On the other side, because of 2D coprimeness, we have

$$\det D_L(z_1, z_2) = \det D_R(z_1, z_2)$$

hence

$$\det D_L(t) = \det D_R(t) \tag{6}$$

Since

$$\det D_L(t) = \det \tilde{D}_L(t) \det R(t)$$

(5) and (6) lead to a contradiction.

As a consequence of Theorem 2 if $N_R D_R^{-1}$ and $\bar{N}_R \bar{D}_R^{-1}$ are two right coprime MFDs and $D_L^{-1} N_L$ is a left coprime MFD, denoting by $\mathcal{V}_R(W)$, $\bar{\mathcal{V}}_R(W)$ and $\mathcal{V}_L(W)$ their rank singularity sets, we have

$$\mathscr{V}_R(W) = \mathscr{V}_L(W) = \bar{\mathscr{V}}_R(W)$$

that proves the invariance of the rank singularities with respect to coprime MFDs.

CONNECTIONS BETWEEN $\mathscr{V}(W)$ AND THE STRUCTURE OF REALIZATIONS OF $N_R D_R^{-1}$

Consider a 2D system $\Sigma = (A_1, A_2, B_1, B_2, C, D)$ given by

$$\begin{aligned} x(h+1,k+1) &= A_1 x(h,k+1) + A_2 x(h+1,k) + B_1 u(h,k+1) + B_2 u(h+1,k) \\ y(h,k) &= C\, x(h,k) + D\, u(h,k) \end{aligned} \tag{7}$$

where the local state x is an n-dimensional vector over the real field $\mathbb{R}$, input and output functions take values in $\mathbb{R}^m$ and $\mathbb{R}^p$, A_1, A_2, B_1, B_2, C and D are matrices of suitable dimensions with entries in $\mathbb{R}$ [4].

Denote by

$$\mathscr{X}_0 = \sum_{-\infty}^{+\infty}{}_i\; x(i,-i) z_1^i z_2^{-i}$$

the global state on the separation set

$$\mathscr{C}_0 = \{(i,j):\ i+j = 0\}$$

and by

$$X(z_1,z_2) = \sum_{i+j\geq 0} x(i,j) z_1^i z_2^j, \qquad U(z_1,z_2) = \sum_{i+j\geq 0} u(i,j) z_1^i z_2^j$$

$$Y(z_1,z_2) = \sum_{i+j\geq 0} y(i,j) z_1^i z_2^j$$

the state, input and output functions respectively.

Then from (7) one gets

$$(I - A_1 z_1 - A_2 z_2) X(z_1,z_2) - (B_1 z_1 + B_2 z_2) U(z_1,z_2) = \mathscr{X}_0 \tag{8}$$

and

$$Y(z_1,z_2) = C\, X(z_1,z_2) + D\, U(z_1,z_2). \tag{9}$$

Assuming zero initial conditions $\mathscr{X}_0 = 0$, the input-output map is given by

$$Y(z_1,z_2) = W_\Sigma(z_1,z_2) U(z_1,z_2)$$

where

$$C(I - A_1 z_1 - A_2 z_2)^{-1}(B_1 z_1 + B_2 z_2) + D \triangleq W_\Sigma(z_1,z_2) \tag{10}$$

is the transfer matrix of Σ.

A system Σ is strictly causal when D=0 and is finite-memory if for any set of initial conditions $\mathscr{X}_0$, the free state evolution goes to zero in a finite number of steps.

The following matrices

$$[I-A_1z_1-A_2z_2 \quad B_1z_1+B_2z_2] \tag{11}$$

and

$$\begin{bmatrix} I-A_1z_1-A_2z_2 \\ C \end{bmatrix} \tag{12}$$

are of paramount importance in analyzing the internal structure of Σ and in designing feedback control policies.

Matrices (11) and (12) are related to the notions of controllability and causal reconstructibility [1,2], in the sense that Σ is controllable if and only if (11) is full rank for every (z_1,z_2) and is causally reconstructible if and only if (12) is full rank for every (z_1,z_2).

The role played by the corresponding 1D matrices

$$[I-Az \quad Bz] \quad \text{and} \quad \begin{bmatrix} I-Az \\ C \end{bmatrix} \tag{13}$$

in defining how the dynamics of output feedback connected systems can be modified, is well known. Also, we can always realize a proper transfer matrix W(z) so that (13) have both full rank for preassigned values of z, and in particular for every z. In the latter case, the realization is controllable and reconstructible and dynamic output feedback allows to obtain arbitrary characteristic polynomials.

The situation is quite different in the 2D case, since there exist transfer matrices $W(z_1,z_2)$ that cannot be realized by 2D systems having (11) and (12) of full rank for every (z_1,z_2).

This is a direct consequence of the following Theorem.

Theorem 3. Consider a strictly proper transfer matrix W and let $N_RD_R^{-1}$ be a right coprime MFD of W. If $\Sigma=(A_1,A_2,B_1,B_2,C)$ is a realization of $N_RD_R^{-1}$ and $(z_{10},z_{20})\in\mathscr{V}(W)$, then one of the matrices (11) and (12) has not full rank for (z_{10},z_{20}).

proof. Given $(z_{10},z_{20})\in\mathscr{V}(W)$, consider a straight line

$$z_1 = z_{10}+\alpha t \ , \quad z_2 = z_{20}+\beta t \tag{14}$$

where α and β are chosen so that $\det(I-A_1z_1-A_2z_2)$ is not identically zero on this line.

Substituting (14) in the expression of the transfer matrix of Σ, we obtain

$$
\begin{aligned}
&C(I-(A_1z_{10}+A_2z_{20})-t(\alpha A_1+\beta A_2))^{-1}(B_1z_{10}+B_2z_{20}+t(\alpha B_1+\beta B_2)) \\
&= C\, V(t)^{-1}U(t) \overset{\Delta}{=} W_{\Sigma}(t)
\end{aligned}
\tag{15}
$$

where

$$V(t) = I-(A_1z_{10}+A_2z_{20}) - t(\alpha A_1+\beta A_2)$$

$$U(t) = B_1z_{10} + B_2z_{20} + t(\alpha B_1+\beta B_2)$$

Now, letting $R(t)$ be the g c r d of C and $V(t)$, we have

$$
\begin{aligned}
C &= \hat{C}(t)R(t) \\
V(t) &= \tilde{V}(t)R(t)
\end{aligned}
\tag{16}
$$

and letting $L(t)$ be the g c l d of $\tilde{V}(t)$ and $U(t)$, we have

$$
\begin{aligned}
\tilde{V}(t) &= L(t)\hat{V}(t) \\
U(t) &= L(t)\hat{U}(t)
\end{aligned}
\tag{17}
$$

Using (15), (16) and (17) we obtain

$$W_{\Sigma}(t) = \hat{C}(t)\hat{V}^{-1}(t)\hat{U}(t)$$

where $\hat{C}(t)$ and $\hat{V}(t)$ are right coprime, $\hat{V}(t)$ and $\hat{U}(t)$ are left coprime and

$$V(t) = L(t)\hat{V}(t)R(t) \tag{18}$$

Considering $N_R D_R^{-1}$ on (14) we obtain

$$N_R(z_{10}+\alpha t,\ z_{20}+\beta t)D_R^{-1}(z_{10}+\alpha t,\ z_{20}+\beta t) = N_R(t)D_R^{-1}(t)$$

where

$$N_R(t) \overset{\Delta}{=} N_R(z_{10}+\alpha t,\ z_{20}+\beta t)$$

$$D_R(t) \overset{\Delta}{=} D_R(z_{10}+\alpha t,\ z_{20}+\beta t)$$

If $S(t)$ is the g c r d of $N_R(t)$ and $D_R(t)$, then

$$
\begin{aligned}
N_R(t) &= \hat{N}_R(t)S(t) \\
D_R(t) &= \hat{D}_R(t)S(t)
\end{aligned}
\tag{19}
$$

Using the identity

$$\hat{N}_R(t)\hat{D}_R^{-1}(t) = \hat{C}(t)\hat{V}^{-1}(t)\hat{U}(t)$$

and recalling the coprimeness properties involved, we have

$$\det \hat{D}_R(t) = \det \hat{V}(t) \tag{20}$$

Recalling that

$$N_R D_R^{-1} = C(I-A_1z_1-A_2z_2)^{-1}(B_1z_1+B_2z_2)$$

and that N_R and D_R are factor coprime, there exists a polynomial $h(z_1,z_2)$ such that:

$$\det(I-A_1z_1-A_2z_2) = h(z_1,z_2)\det D_R$$

which gives

$$\det V(t) = h(t)\det D_R(t) \tag{21}$$

Because of (18) and (20) we obtain

$$\begin{aligned}\det V(t) &= \det L(t)\ \det \hat{V}(t)\ \det R(t) = \\ &= \det L(t)\ \det \hat{D}_R(t)\ \det R(t)\end{aligned}$$

Now, recalling (19) and (21) we have

$$h(t)\ \det S(t)\ \det \hat{D}(t) = \det L(t)\ \det \hat{D}_R(t)\ \det R(t)$$

hence

$$h(t)\ \det S(t) = \det L(t)\ \det R(t) \tag{22}$$

Assume now, by contradiction, that both (11) and (12) have full rank for (z_{10},z_{20}). Then $[V'(0)\ \ C']$ and $[V(0)\ \ U(0)]$ have full rank. Consequently, since the product matrix

$$\begin{bmatrix}\tilde{V}(0)\\ \hat{C}(0)\end{bmatrix} R(0) = \begin{bmatrix}V(0)\\ C\end{bmatrix}$$

has full rank, R(0) has full rank too. Similarly, since

$$\begin{aligned}L(0)[\hat{V}(0)R(0)\ \ \hat{U}(0)] &= [\tilde{V}(0)R(0)\ \ U(0)] = \\ &= [V(0)\ \ U(0)]\end{aligned}$$

has full rank, so L(0) has.

On the other side, since the matrix on the right hand side of the following identity

$$\begin{bmatrix}\hat{N}_R(0)\\ \hat{D}_R(0)\end{bmatrix} S(0) = \begin{bmatrix}N_R(0)\\ D_R(0)\end{bmatrix}$$

has not full rank and $[\hat{N}_R(0)\ \ \hat{D}_R(0)]$ has full rank because of the coprimeness of $\hat{N}_R(t)$ and $\hat{D}_R(t)$, it follows that S(0) is singular.

Now, letting t=0 in (22) we have

$$0 = h(0)\ \det S(0) = \det L(0)\ \det R(0)$$

that is a contradiction since L(0) and R(0) are non singular.

We shall now characterize the realizations of the matrix fraction $N_R D_R^{-1}$ having the property that the subset of $\mathbb{C}\times\mathbb{C}$ where at least one of the matrices (11) and (12) has not full rank, coincides with $\mathscr{V}(W)$.

The following Theorem provides the answer to this problem.

Theorem 4. Let $N_R D_R^{-1}$ be a right coprime, proper matrix fraction and let $\Sigma = (A_1, A_2, B_1, B_2, C)$ be a realization of it. Then

$$\det D_R \mid \det(I-A_1z_1-A_2z_2) \tag{23}$$

and the following facts are equivalent:

(i) $\det(I-A_1z_1-A_2z_2) = \det D_R$;

(ii) $C(I-A_1z_1-A_2z_2)^{-1}$ and $(I-A_1z_1-A_2z_2)^{-1}(B_1z_1+B_2z_2)$ are respectively right coprime and left coprime matrix fractions;

(iii) the subset of $\mathbb{C}\times\mathbb{C}$ where at least one of the matrices (11) and (12) has not full rank, coincides with $\mathscr{V}(W)$.

proof. (ii) $\rightarrow$ (i) recalling Theorem 5.5 in [3], if in the following equality

$$C(I-A_1z_1-A_2z_2)^{-1}(B_1z_1+B_2z_2) = N_R D_R^{-1}$$

$(C,(I-A_1z_1-A_2z_2))$ are right coprime and $((I-A_1z_1-A_2z_2),(B_1z_1+B_2z_2))$ are left coprime, then $\det(I-A_1z_1-A_2z_2) = \det D_R$.

In general, if only (N_R, D_R) are right coprime, the matrix product $C(I-A_1z_1-A_2z_2)^{-1}(B_1z_1+B_2z_2)$ can be put in the form $T\,V^{-1}U$, where (T,V) are right coprime, (V,U) are left coprime and $\det V \mid \det(I-A_1z_1-A_2z_2)$. Now using Theorem 5.5 in [3], we have

$$\det D_R \mid \det(I-A_1z_1-A_2z_2)$$

(i) $\rightarrow$ (iii) denote by $\mathscr{M}$ and $\mathscr{N}$ the sets of points where (11) and (12) have not full rank. Then by Theorem 2 we obtain

$$\mathscr{S} = \mathscr{M} \cup \mathscr{N} \supseteq \mathscr{V}(W).$$

To prove the converse inclusion, assume, by contradiction, $(z_{10},z_{11}) \in \mathscr{M}$ and $(z_{10},z_{20}) \notin \mathscr{V}(W)$ and consider a straight line

$$z_1 \triangleq z_{10} + \alpha t\ , \quad z_2 = z_{20} + \beta t \tag{24}$$

with α and β such that on that line, $\det D_R$ is not identically zero and $[N_R' \ \ D_R']$ has full rank. Then the restriction to (24) of $N_R D_R^{-1}$, given by

$$N_R(z_{10}+\alpha t,\ z_{20}+\beta t)D_R^{-1}(z_{10}+\alpha t,\ z_{20}+\beta t) \triangleq N_R(t)D_R^{-1}(t)$$

is a 1D right coprime matrix fraction. Also, in the expression of the transfer matrix restricted to (24), given by

$$C(I-A_1(z_{10}+\alpha t) - A_2(z_{20}+\beta t))^{-1}(B_1(z_{10}+\alpha t) + B_2(z_{20}+\beta t)) =$$
$$= C\, V^{-1}(t)U(t) \ ,$$

the matrices V(t) and U(t) are not right coprime since by assumption $[V(0) \quad U(0)]$ does not have full rank.

Consequently det $D_R(t)$ is proper divisor of det V(t), that contradicts the assumption

$$\det D_R(z_1,z_2) = \det(I-A_1z_1-A_2z_2)$$

(iii) → (ii) since $\mathscr{S} = \mathscr{V}(W)$, $\mathscr{S}$ is a finite cardinality set. This implies that $\mathscr{M}$ and $\mathscr{N}$ are also finite cardinality sets and hence $(C,(I-A_1z_1-A_2z_2))$ and $((I-A_1z_1-A_2z_2), (B_1z_1+B_2z_2))$ are factor coprime.

At this point a natural question arises as whether there exist realizations having the property that $\mathscr{S}$ coincides with $\mathscr{V}(W)$. As we shall see, such realizations do exist and the proof of Theorem 5 will also provide an explicit construction for them. In fact it will be shown how to obtain realizations where at least one of matrices (11) and (12) has full rank for every (z_1,z_2) (so Σ turns out to be controllable and/or causally reconstructible) and $\mathscr{V}(W)$ coincides with the subset where the other matrix does not have full rank.

Theorem 5. Let $N_RD_R^{-1}$ be a strictly proper, right coprime MFD. Then there exists a 2D system $\Sigma = (A_1,A_2,B_1,B_2,C)$ that realizes $N_RD_R^{-1}$, i.e. $W_\Sigma = N_RD_R^{-1}$, such that

(i) $\quad [I-A_1z_1-A_2z_2 \quad B_1z_1+B_2z_2]$ has full rank on $\mathbb{C}\times\mathbb{C}$

(ii) $\quad \begin{bmatrix} I-A_1z_1-A_2z_2 \\ C \end{bmatrix}$ has full rank on $\mathbb{C}\times\mathbb{C} \setminus \mathscr{V}(W)$.

proof. there is no restriction in assuming $D_R(0,0) = I_m$. Denote by k_i, $i = 1,2,\ldots,m$ the column degree of the i-th column of

$$\begin{bmatrix} N_R \\ D_R \end{bmatrix},$$

that is the degree of the polynomial of maximal degree in the i-th column.

We can write

$$D_R = I_m - D_{HT}\Psi$$
$$N_R = N_{HT}\Psi$$

where

$$\Psi' = \left[\begin{array}{cccccc|ccc|cccccc} z_2^{k_1} & z_1 z_2^{k_1-1} & \cdots & z_1^{k_1} \cdots z_2 & z_1 & & 0 & \cdots & & 0 & \cdots & & & & \\ \hline & & & & & & \ddots & & & & & & & & \\ 0 & \cdots & & & & & & \cdots & 0 & z_2^{k_m} & z_1 z_2^{k_m-1} & \cdots & z_1^{k_m} \cdots z_2 & z_1 & \end{array}\right]$$

$$D_{HT} = \begin{bmatrix} D_{11} & \cdots & D_{1m} \\ \\ D_{m1} & \cdots & D_{mm} \end{bmatrix}, \quad N_{HT} = \begin{bmatrix} N_{11} & \cdots & N_{1m} \\ \\ N_{p1} & \cdots & N_{pm} \end{bmatrix}$$

and D_{ij} and N_{ij} are row-vectors whose elements are the coefficients of the (i,j) indexed polynomial in $-D_R + I_m$ and N_R.

Introduce now the following matrices

$$A_{10}^{(h)} = \left[\begin{array}{cccc} M_h & & & \\ & M_{h-1} & & \\ & & \ddots & \\ & & & M_2 \\ \hline 0 & \cdots & & 0 \\ 0 & \cdots & & 0 \end{array}\right], \quad B_1^{(h)} = \left[\begin{array}{c} 0 \\ \vdots \\ 0 \\ \hline 0 \\ 1 \end{array}\right]$$

$$A_{20}^{(h)} = \left[\begin{array}{cccc} N_h & & & \\ & N_{h-1} & & \\ & & \ddots & \\ & & & N_2 \\ \hline 0 & \cdots & & 0 \\ 0 & \cdots & & 0 \end{array}\right], \quad B_2^{(h)} = \left[\begin{array}{c} 0 \\ \vdots \\ 0 \\ \hline 1 \\ 0 \end{array}\right]$$

with

$$M_j = \begin{bmatrix} 0 & 0 & \cdots & 0 \\ & & I_j & \end{bmatrix}$$

$$N_j = \begin{bmatrix} 1 & 0 & \cdots & 0 \\ & & 0_j & \end{bmatrix}$$

and define

$$A_{10} = \text{diag}\,[A_{10}^{(k_1)} \quad A_{10}^{(k_2)} \quad .... \quad A_{10}^{(k_m)}]$$

$$A_{20} = \text{diag}\,[A_{20}^{(k_1)} \quad A_{20}^{(k_2)} \quad .... \quad A_{20}^{(k_m)}]$$

$$B_1 = \text{diag}\,[B_1^{(k_1)} \quad B_2^{(k_2)} \quad .... \quad B_2^{(k_m)}]$$

$$B_2 = \text{diag}\,[B_2^{(k_1)} \quad B_2^{(k_2)} \quad .... \quad B_2^{(k_m)}]$$

It is a matter of simple computation to show that

$$(I-A_{10}z_1-A_{20}z_2)^{-1}(B_1z_1+B_2z_2) = \Psi$$

Assuming now

$$\mathscr{A}_0 = A_{10}z_1 + A_{20}z_2, \qquad \mathscr{B} = B_1z_1 + B_2z_2$$

$$\mathscr{A} = \mathscr{A}_0 + \mathscr{B}\,D_{HT}$$

we have

$$(I-\mathscr{A})^{-1}\mathscr{B} = (I-\mathscr{A}_0-\mathscr{B}\,D_{HT})^{-1}\mathscr{B} =$$

$$= ((I-\mathscr{B}\,D_{HT}(I-\mathscr{A}_0)^{-1})(I-\mathscr{A}_0))^{-1}\mathscr{B} =$$

$$= (I-\mathscr{A}_0)^{-1}(I-\mathscr{B}\,D_{HT}(I-\mathscr{A}_0)^{-1})^{-1}\mathscr{B} =$$

$$= (I-\mathscr{A}_0)^{-1}\mathscr{B}\,(I-D_{HT}(I-\mathscr{A}_0)^{-1}\mathscr{B})^{-1} =$$

$$= \Psi(z_1,z_2)(I-D_{HT}\Psi)^{-1} = \Psi\,D_R^{-1}$$

Since

$$N_RD_R^{-1} = N_{HT}\Psi D_R^{-1} = N_{HT}(I-\mathscr{A})^{-1}\mathscr{B} =$$

$$= N_{HT}\,(I-(A_{10}+B_1D_{HT})z_1 - (A_{20}+B_2D_{HT})z_2)^{-1}(B_1z_1+B_2z_2)$$

the matrices $A_1 = A_{10}+B_1D_{HT}$, $A_2 = A_{20}+B_2D_{HT}$, $B_1,B_2,C = N_{HT}$ furnish a realization of $N_RD_R^{-1}$.

The realization obtained in this way is controllable. In fact

$$\text{rank}[I-\mathscr{A} \,|\, \mathscr{B}] = \text{rank}[I-\mathscr{A}_0-\mathscr{B}D_{HT} \,|\, \mathscr{B}] =$$

$$= \text{rank}[I-\mathscr{A}_0 \,|\, \mathscr{B}] =$$

$$= \text{rank}[\text{diag}\{I-A_{10}^{(k_i)}z_1-A_{20}^{(k_i)}z_2,\ i=1...m\} \,|\, \text{diag}\{B_1^{(k_i)}z_1+B_2^{(k_i)}z_2,\ i=1...m\}]$$

is full for every (z_1,z_2) since

$$[I-A_{10}^{(k_i)}z_1-A_{20}^{(k_i)}z_2 \mid B_1^{(k_i)}z_1+B_2^{(k_i)}z_2], \quad i=1,2,\ldots,m$$

has full rank for every (z_1,z_2).

We have still to prove that $\mathscr{V}(W)$ is the set of points where

$$\begin{bmatrix} I-A_1z_1-A_2z_2 \\ C \end{bmatrix}$$

does not have full rank. Because of Theorem 4 it is enough to prove that $\det(I-A_1z_1-A_2z_2)=\det D_R$. This follows from the identities

$$\det D_R = \det(I-D_{HT}\Psi) = \det(I-\Psi D_{HT})$$

and

$$\det(I-\mathscr{A}) = \det(I-\mathscr{A}_o-\mathscr{B}D_{HT}) =$$

$$= \det(I-\mathscr{A}_o)\det[I-(I-\mathscr{A}_o)^{-1}\mathscr{B}D_{HT}] =$$

$$= \det(I-\Psi D_{HT})$$

CONTROLLER DESIGN

Consider the 2D system of fig. 1.

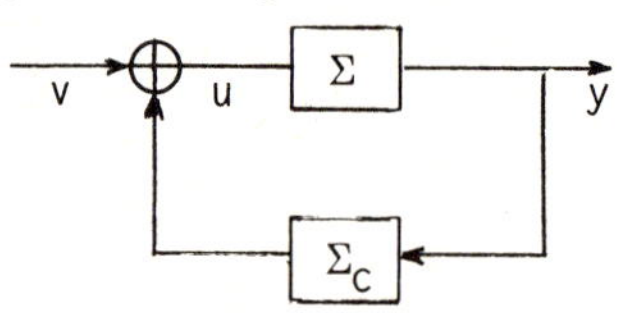

fig. 1

Let $N_RD_R^{-1}$ be a right coprime MFD of the strictly proper transfer matrix of $\Sigma=(A_1,A_2,B_1,B_2,C)$. Hence, by Theorem 4,

$$\det(I-A_1z_1-A_2z_2) = h(z_1,z_2)\det D_R \tag{25}$$

Also, let $R_L^{-1}S_L$ be a left MFD of the proper transfer matrix of $\Sigma_c=(F_1,F_2,G_1,G_2,H,J)$ with

$$\det(I-F_1z_1-F_2z_2) = \det R_L$$

Let x and x_c denote the local state vectors of Σ and Σ_c and assume $\bar{x}' = [x' \; x_c']$. It is easy to check that the dynamics of the local state of the feedback system is given by

$$\bar{x}(h+1,k+1) = \bar{A}_1\bar{x}(h,k+1)+\bar{A}_2\bar{x}(h+1,k)+\bar{B}_1 v(h,k+1)+\bar{B}_2 v(h+1,k)$$

where

$$\bar{A}_1 = \begin{bmatrix} A_1+B_1JC & B_1H \\ G_1C & F_1 \end{bmatrix}, \quad \bar{A}_2 = \begin{bmatrix} A_2+B_2JC & B_2H \\ G_2C & F_2 \end{bmatrix}$$

$$\bar{B}_1 = \begin{bmatrix} G_1 \\ 0 \end{bmatrix}, \quad \bar{B}_2 = \begin{bmatrix} G_2 \\ 0 \end{bmatrix}$$

Then

$$\det(I-\bar{A}_1z_1-\bar{A}_2z_2) = h(z_1,z_2)\det(R_LD_R+S_LN_R) \tag{26}$$

We shall now investigate to what extent it is possible to assign arbitrarily the characteristic polynomial (26) of the feedback system and/or the associated variety, by selecting an appropriate Σ_C.

Obviously, the factor $h(z_1,z_2)$ in (26), that descends from the existence of common factors in at least one of the matrix fractions $C(I-A_1z_1-A_2z_2)^{-1}$ and $(I-A_1z_1-A_2z_2)^{-1}(B_1z_1+B_2z_2)$, cannot be influenced by any dynamic output feedback.

It remains to analyze how $\det(R_LD_R+S_LN_R)$ changes as we vary R_L and S_L with the constraint that they characterize a proper transfer matrix. Once we have selected R_L and S_L, it is possible to construct a 2D system that realizes $R_L^{-1}S_L$ (see for instance the construction in Theorem 5).

Using the Binet-Cauchy formula, $\det(R_LD_R+S_LN_R)$ can be expressed as the sum of the products of all possible minors of maximal order, q_i, $i=1,2,\dots,\nu$, of $[S_L\ R_L]$ into the corresponding minors of the same order, m_i, $i=1,2,\dots,\nu$, of $[N_R'\ D_R']'$, that is

$$\det(R_LD_R + S_LN_R) = \sum_{1}^{\nu} q_im_i \tag{27}$$

Clearly all polynomials that are obtained from (27) as R_L and S_L vary, vanish on $\mathcal{V}(W)$, since they belong to the ideal $(m_1,m_2,\dots,m_\nu)$ generated by the minors of maximal degree of $[N_R'\ D_R']$.

<u>Theorem 6.</u> Let $p\in(m_1,m_2,\dots,m_\nu)$. Then there exist R_L and S_L such that

$$R_LD_R + S_LN_R = p\ I_m \tag{28}$$

proof. Denote by M_i, $i=1,2,\dots,\nu$ the submatrix of maximal order of $[N_R'\ D_R']'$ correspondig to m_i. Then there exist constant matrices L_i and K_i such that

$$M_i = L_iD_R + K_iN_R$$

So we have

$$m_iI_m = (\mathrm{Adj}\ M_i)M_i = (\mathrm{Adj}\ M_i)L_iD_R + (\mathrm{Adj}\ M_i)K_iN_R, \quad i=1,2,\dots,\nu \tag{29}$$

Now, assuming $p = \Sigma_{1}^{\nu} q_i m_i$, from (29) we obtain

$$p\, I_m = \Sigma_{1}^{\nu} q_i m_i I_m = (\Sigma_{1}^{\nu} q_i (\text{Adj } M_i) L_i) D_R + (\Sigma_{1}^{\nu} q_i (\text{Adj } M_i) K_i) N_R$$

that coincides with (28) once we have taken

$$R_L = \Sigma_{1}^{\nu} q_i (\text{Adj } M_i) L_i, \qquad S_L = \Sigma_{1}^{\nu} q_i (\text{Adj } M_i) K_i$$

As a direct consequence of the Theorem above and of Hilbert Nullstellensatz, we have that given any polynomial p vanishing on $\mathscr{V}(W)$, there exist R_L, S_L and a positive integer r such that

$$\det(R_L D_R + S_L N_R) = p^r \tag{30}$$

It is easy to check that there are cases where r cannot be unitary. As an example, consider the following right coprime matrix fraction

$$N_R D_R^{-1} = [(z_1 - z_2)^2 \;\vdots\; 0] \begin{bmatrix} (1-z_1)^2 & 0 \\ 0 & 1 \end{bmatrix}^{-1} = W$$

We have $m_1 = (1-z_1)^2$, $m_2 = (z_1 - z_2)^2$, so that $(W) = (1,1)$. The polynomial $(1-z_2)^2$ vanishes on (W) but does not belong to (m_1, m_2), so that $(1-z_2)^2 \neq \det(R_L D_R + S_L N_R)$ for every R_L and S_L.

The requirement that (R_L, S_L) characterizes a proper transfer matrix is equivalent to impose that p in (30) satisfies $p(0,0) = 0$. In fact, since $N_R(0,0) = 0$ and $D_R(0,0) = I_m$, we have

$$\det[R_L(0,0) \;\vdots\; S_L(0,0)] \begin{bmatrix} D_R(0,0) \\ N_R(0,0) \end{bmatrix} = \det R_L(0,0)$$

so that $p(0,0) \neq 0$ is equivalent to $\det R_L(0,0) \neq 0$ that is to the possibility of realizing $R_L^{-1} S_L$ by means of a 2D proper system.

In some applications we are interested in designing a compensator Σ_c such that the resulting feedback system is finite memory, i.e. $\det(I - \bar{A}_1 z_1 - \bar{A}_2 z_2) \equiv 1$. Such a compensator is called dead-beat compensator.

The existence of a dead-beat compensator depends on the following facts:

(i) the transfer matrix of Σ should make possible to solve the equation

$$\det(R_L D_R + S_L N_R) = 1 \tag{31}$$

for any right factor coprime MFD $N_R D_R^{-1} = W$.

(ii) The matrices A_1, A_2, B_1, B_2 and C of Σ should be selected so that the polynomial h in (25) is unitary.

Now, equation (31) is solvable if and only if $[N_R' \; D_R']$ has full rank for every (z_1,z_2), that is if N_R and D_R are zero coprime. In this case by Theorem 6 the equation

$$R_L D_R + S_L N_R = I_m$$

has solution and hence (31) too.

The condition h=1 implies that $(C,\ I-A_1z_1-A_2z_2)$ are right factor coprime and $(I-A_1z_1-A_2z_2,\ B_1z_1+B_2z_2)$ are left factor coprime.

If we recall the controllability and causal reconstructibility criteria /1, 2/, the existence of a dead-beat controller can be related both to internal structural properties of Σ and to external properties of $N_R D_R^{-1}$. To this purpose we have the following Theorem.

Theorem 7. Let $\Sigma = (A_1,A_2,B_1,B_2,C)$ be a realization of the right coprime matrix fraction $N_R D_R^{-1}$. Then the following properties are equivalent:

(i) there exists a dead-beat compensator

(ii) Σ is controllable and causally reconstructible

(iii) $(C,\ I-A_1z_1-A_2z_2)$ are right factor coprime, $(I-A_1z_1-A_2z_2,\ B_1z_1+B_2z_2)$ are left factor coprime, N_R and D_R are right zero coprime.

REFERENCES

1. M. Bisiacco, State and output feedback stabilizability of 2D systems, to appear in IEEE Trans. on Circuits and Systems (1985).

2. M. Bisiacco, On the structure of 2D observers, to appear in IEEE Trans. on Autom. Control (1985).

3. S. Kung et al., New results in 2D systems theory. Part I and part II. Proc. of IEEE, 65, 6, (1977), 861-72 and 945-61.

4. E. Fornasini, G. Marchesini, Doubly indexed dynamical systems: state space models and structural properties, Math. Systems Theory, 12, 1, (1978),59-72.

Frequency Domain and State Space Methods for Linear Systems
C.I. Byrnes and A. Lindquist (Editors)

CANONICAL FORMS

Shmuel Friedland*

Institute of Mathematics
Hebrew University
Jerusalem, ISRAEL

Department of Mathematics
Cornell University
Ithaca, NY 14853

1. INTRODUCTION

Let $\mathfrak{G}$ be an algebraic group acting on an irreducible variety $\mathfrak{V}$. Then there exists a natural classification of the orbits of $\mathfrak{G}$. See for example Rosenlicht [1956], Friedland [1983] and Friedland [1985]. That is there exists a stratification of $\mathfrak{V}$ to a union of quasi-irreducible sets

$$(1.1) \qquad \mathfrak{V} = \bigcup_{1}^{p} \mathfrak{V}_i$$

such that the following conditions are satisfied:

(i) each $\mathfrak{V}_i$ is an invariant under the action of $\mathfrak{G}$;

(ii) for each $\mathfrak{V}_i$ there exists a set of rational functions $\psi_1^{(i)},\ldots,\psi_{m_i}^{(i)}$ which are well defined constants on each orbit in $\mathfrak{V}_i$. Moreover the values of these rational functions separate the orbits in $\mathfrak{V}_i$. In Friedland [1983] we carried out this classification for the simultaneous similarity of matrices. That is for the orbits

$$(1.2) \qquad \mathrm{orb}(A_0,\ldots,A_k) = \{(B_0,\ldots,B_k),\ B_j = T^{-1} A_j T \in M_n,\ j = 0,\ldots,k,\ T \in GL_n\} .$$

In Friedland [1985] we carried out the above classification for linear dynamical systems. That is we considered the orbits of the triple of matrices (F,G,H) under the action of the general linear group:

$$(1.3) \qquad \mathrm{orb}(F,G,H) = \{(T^{-1}FT,\ T^{-1}G, HT),\ T \in GL\} .$$

However, in many instances one really prefers a classification of orbits by means of canonical forms. Of course, the model example of a such classification is the Jordan canonical form. The purpose of this paper to show that such classification exists once we accept the idea that canonical forms exist on each components of certain stratification (1.1) of $\mathfrak{V}$. In what follows we restrict

*Partially supported by the NSF grant MCS 83-00842.

ourselves to the field of complex numbers $\mathbb{C}$. Technical modifications of our arguments can be carried over to any algebraically closed field F. Our main result is

THEOREM 1. Let $\mathfrak{V}$ be an irreducible variety in $\mathbb{C}^m$. Then there exists a decomposition (1.1) of $\mathfrak{V}$ to disjoint union of quasi-varieties $\mathfrak{V}_i$ which are also manifolds such that in addition to the conditions (i) and (ii) the following condition holds.

(iii) for each $\mathfrak{V}_i$ there exist a subset $\mathfrak{S}_i \subset \{1,\ldots,m\}$ and fixed numbers $a_j^{(i)}$, $j \in \mathfrak{S}_i$ such that each orbit in $\mathfrak{V}_i$ has precisely μ_i distinct points $\xi^{(r)}$ of the form

$$\xi^{(r)} = (\xi_1^{(r)},\ldots,\xi_m^{(r)}), \quad \xi_j^{(r)} = a_j^{(r)} ,$$

$$j \in \mathfrak{S}_i, \; r = 1,\ldots,\mu_i . \tag{1.4}$$

Moreover the points $\xi^{(r)}$, $r = 1,\ldots,\mu_i$, depend continuously on the orbits.

Thus the points $\xi^{(r)}$, $r = 1,\ldots,\mu_i$, are the canonical forms of the orbit. Since Zariski closure of any orbit is an irreducible variety, the number μ_i is the degree of this variety (= the degree of the orbit) which lies in $\mathfrak{V}_i$. In particular, we can not expect μ_i to be equal to 1.

Finally we analize the Jordan canonical form in the frame of Theorem 1. As expected, for many components $\mathfrak{V}_i$ the corresponding degree $\mu_i > 1$. The simplicity of the Jordan forms can be explained by the very simple relations between the different points $\xi^{(r)}$.

More precisely, there exists a Galois group $\mathfrak{G}_i$ corresponding to the component $\mathfrak{V}_i$ which shifts one canonical form $\xi^{(r)}$ to the other one.

2. MAIN RESULT

As usual denote by $\mathbb{C}[x_1,\ldots,x_m]$ and $\mathbb{C}(x_1,\ldots,x_m)$ the ring of polynomials and the field of rational functions in m variables with the coefficients in $\mathbb{C}$. A set $\mathfrak{V} \subset \mathbb{C}^m$ is called a variety if it is a zero set of a finite number of polynomials. That is

$$\mathfrak{V} = \{v, \; \phi_i(v) = 0, \; \phi_i \in \mathbb{C}[x_1,\ldots,x_m], \; i = 1,\ldots,p\} . \tag{2.1}$$

The codimension of $\mathfrak{V}$ is the number of algebraically independent polynomials among $\phi_1,\ldots,\phi_p$. Equivalently, there exists a subvariety $\mathfrak{U}$ of $\mathfrak{V}$ such that $\mathfrak{V}$-$\mathfrak{U}$ is a manifold of dimension $d = \dim \mathfrak{V}$. A subset $\mathfrak{W}$ of $\mathfrak{V}$ is called quasi-variety if it is a complement of some subvariety $\mathfrak{U}$ in $\mathfrak{V}$. That is $\mathfrak{W}$ is an open set in $\mathfrak{V}$ in Zariski topology - which identifies closed sets in $\mathbb{C}^m$ with varieties. For any set $\mathfrak{X} \subset \mathbb{C}^m$ let $\mathrm{cl}(\mathfrak{X})$ denote Zariski closure $\mathfrak{X}$. That is $\mathrm{cl}(\mathfrak{X})$ is the smallest variety $\mathfrak{V}$ which contains $\mathfrak{X}$. The dimension and the degree of $\mathfrak{X}$ is defined to be the dimension and the degree of $\mathrm{cl}(\mathfrak{X})$. A variety $\mathfrak{V}$ is called irreducible if it is not a union of two strict subvarieties of $\mathfrak{V}$. Otherwise $\mathfrak{V}$ is called reducible. It is well known that any variety is a finite union of irreducible varieties which are unique. Moreover any finite union or infinite intersection of varieties is also a variety. For a set $\mathfrak{V} \subset \mathbb{C}^m$ let $\mathcal{I}(\mathfrak{V})$ be the ideal of polynomials which vanish on $\mathfrak{V}$. Then by $\mathbb{C}[\mathfrak{V}]$ we denote the quotient ring $\mathbb{C}[x_1,\ldots,x_m]/\mathcal{I}(\mathfrak{V})$. That is $\mathbb{C}[\mathfrak{V}]$ can be viewed as the ring of polynomials restricted to $\mathfrak{V}$. Assume that $\mathfrak{V}$ is an irreducible variety. Then $\mathbb{C}[\mathfrak{V}]$ is an integral domain. In that case let $\mathbb{C}(\mathfrak{V})$ be the field of quotients of $\mathbb{C}[\mathfrak{V}]$. In other words $\mathbb{C}(\mathfrak{V})$ is the field of rational functions restricted to $\mathfrak{V}$. Let ξ_i be the restriction of the polynomial x_i to $\mathfrak{V}$. Assume for simplicity of notation that

$$\xi_i \neq 0,\ i = 1,\ldots,n, \quad \xi_i = 0,\ i = n+1,\ldots,m .$$

Then $\mathbb{C}[\mathfrak{V}]$ is generated by $\xi_1,\ldots,\xi_n$. That is any $\phi \in \mathbb{C}[\mathfrak{V}]$ is a polynomial in $\xi_1,\ldots,\xi_n$. Thus $\mathbb{C}(\mathfrak{V})$ is generated by $\xi_1,\ldots,\xi_n$, i.e. any $\theta \in \mathbb{C}(\mathfrak{V})$ is a rational function in $\xi_1,\ldots,\xi_n$. As usual let M_n be the set of $n \times n$ matrices with the entries in $\mathbb{C}$. Let us identify M_n with $\mathbb{C}^{n^2}$. Then GL_n - the general linear group of dimension n is a quasi-variety since it is a complement of the variety of singular matrices (the zero set of the determinant). A group $\mathfrak{G}$ is called algebraic if it is isomorphic to a subgroup $\mathfrak{S}$ of GL_n where $\mathfrak{S}$ is a quasi-variety and a manifold. In particular $\mathrm{cl}(\mathfrak{S})$ is irreducible.

Assume that $\mathfrak{G}$ is a group which acts on an irreducible variety $\mathfrak{V}$. That is there exists a map

$$\pi : (\mathfrak{G},\mathfrak{V}) \to \mathfrak{V}$$

which preserve the group structure. Thus, if

$$\pi(\gamma,v) = \gamma(v)$$

then

$$\iota(v) = v, \quad \lambda(\gamma(v)) = (\lambda\gamma)(v)$$

where ι is the identity of $\mathfrak{G}$. Denote by $\mathbb{C}[\mathfrak{V}]^{\mathfrak{G}}$ and $\mathbb{C}(\mathfrak{V})^{\mathfrak{G}}$ the subring and the subfield of $\mathbb{C}[\mathfrak{V}]$ and $\mathbb{C}(\mathfrak{V})$ of all functions which are constant on the orbits. That is

$$\mathbb{C}[\mathfrak{V}]^{\mathfrak{G}} = \{\phi, \phi \in \mathbb{C}[\mathfrak{V}], \quad \phi(\pi(v)) = \phi(v), \ v \in \mathfrak{V}, \ \pi \in \mathfrak{G}\}$$

$$\mathbb{C}[\mathfrak{V}]^{\mathfrak{G}} = \{\psi, \psi \in \mathbb{C}(\mathfrak{V}), \ \psi(\pi(v)) = \psi(v), \ v \in \mathfrak{V}, \ \pi \in \mathfrak{G}\}.$$

The ring of invariants $\mathbb{C}[\mathfrak{V}]^{\mathfrak{G}}$ does not have to be finitely generated as was shown by Nagata. However, the field of invariants $\mathbb{C}(\mathfrak{V})^{\mathfrak{G}}$ is finitely generated, say by $\psi_1,\ldots,\psi_q$. See for example Fogarty [1969]. We say that ψ_i, $i = 1,\ldots,q$ are regular on the $\mathrm{orb}(v)$, $v \in \mathfrak{V}$, if there exists a point $w \in \mathrm{orb}(v)$ such that each rational function ψ_i is well defined on w. That is either the denominator or the numerator of ψ_i do not vanish. It can be shown that:

THEOREM 2. *Let* $\mathfrak{G}$ *be an algebraic group acting on irreducible variety* $\mathfrak{V} \subset \mathbb{C}^m$. *Then* $\mathbb{C}(\mathfrak{V})^{\mathfrak{G}}$ *is finitely generated and there exist generators* $\psi_1,\ldots,\psi_q$ *such that the following holds. There exists an open quasi-variety* $\mathfrak{X}$ *in* $\mathfrak{V}$ $(= \mathrm{cl}(\mathfrak{X}))$ *which is invariant under the action of* $\mathfrak{G}$ $(\mathfrak{G}(\mathfrak{X}) = \mathfrak{X})$ *such that* $\psi_1,\ldots,\psi_q$ *are regular for any* $x \in \mathfrak{X}$ *and the values* $\psi_1(\mathrm{orb}(x)),\ldots,\psi_q(\mathrm{orb}(x))$ *determine the orbit of* x.

In fact, in the proof given in Friedland [1984] the quasi-variety $\mathfrak{X}$ is the set of all orbits of the maximal dimension d^* and the maximal degree δ^* (among all orbits of dimension d^*). Let $\mathfrak{W}$ be the complement of $\mathfrak{X}$ in $\mathfrak{V}$. Since $\mathfrak{X}$ is a quasi-variety $\mathfrak{W}$ is a subvariety of $\mathfrak{V}$. So we can apply Theorem 2 to each irreducible component of $\mathfrak{W}$. Continuing this procedure we deduce the existence of the decomposition (1.1) which satisfies the properties (i)-(ii).

PROOF OF THEOREM 1. We assume the notations and the results of Theorem 2. We call a point $x \in \mathfrak{X}$ a regular point if the ranks of the Jacobians of the set of functions $(\phi_1,\ldots,\phi_p)$ and $(\psi_1,\ldots,\psi_p, \psi_1,\ldots,\psi_q)$ are maximal at x. That is

$$\operatorname{rank} J(\phi_1,\ldots\phi_p)(x) = m-d, \ d = \dim \mathfrak{V},$$

(2.2)

$$\operatorname{rank} J(\phi_1,\ldots,\phi_p,\psi_1,\ldots,\psi_q)(x) = m-d^* .$$

Let $a \in \mathfrak{X}$ be a regular point. Whence the matrix $J(\phi_1,\ldots,\phi_p,\psi_1,\ldots,\psi_q)(a)$ has a nonsingular submatrix of order $m-d^*$. That is, there exists a set $\mathfrak{S} \subset \{1,\ldots,m\}$ of cardinality d^* with the following properties. Let

$$x(a,\mathfrak{S}) = (x_1,\ldots,x_m), \ x_i = a_i, \ i \in \mathfrak{S} \tag{2.3}$$

Thus we can identify $x(a,\mathfrak{S})$ with $y \in \mathbb{C}^{m-d^*}$. By renaming ϕ's and ψ's if necessary we may assume

$$\operatorname{rank} J(\phi_1,\ldots,\phi_{m-d})(x(a,\mathfrak{S})) = m-d,$$

(2.4)

$$\operatorname{rank} J(\phi_1,\ldots,\phi_{m-d},\psi_1,\ldots,\psi_{d-d^*})(x(a,\mathfrak{S})) = m-d^*.$$

Consider the system of polynomial equations

$$\phi_i(x(a,\mathfrak{S})) = 0, \ i = 1,\ldots,d$$

(2.5)

$$\psi_j(x(a,\mathfrak{S})) = \kappa_j, \ j = 1,\ldots,d-d^* \ .$$

Use the implicit function theorem to deduce that for $x(a,\mathfrak{S})$ and $\kappa = (\kappa_1,\ldots,\kappa_{d-d^*})$ in the neighborhood of a and $(\psi_1(a),\ldots,\psi_{d-d^*}(a))$ respectively the above system has a unique solution $\xi(a,\kappa)$. Theorem 2 implies that locally for distinct values of κ the points $\xi(a,\kappa)$ lie on distinct orbits. Note that locally $\xi(a,\kappa)$ are on the quasi-variety $\mathfrak{X}$ since the rank of the first matrix in (2.4) is maximal. The same conclusions hold if we replace a by any point b in the neighborhood of a. Using the fact that ψ's are rational functions we can rewrite (2.5) as a system of polynomial equations which depends on the parameters a_i, $i \in \mathfrak{S}$ and κ. Since for an open set of b's and κ's we have a local unique solution $\xi(b,\kappa)$ it then follows that for most of the points in $\mathfrak{X}$, i.e., except a subvariety $\mathfrak{U}$ of $\mathfrak{V}$, the system (2.5) has μ_1 distinct solutions in $\mathfrak{V}-\mathfrak{U}$. See for example Friedland [1983, Section 3] for the precise version of the needed arguments. Let $\mathfrak{V}_1 = \mathfrak{V}-\mathfrak{U}$. Clearly $\mathfrak{V}_1$ is invariant under the action of $\mathfrak{G}$. Theorem 2 implies that all these distinct solutions are lying on the same orbit. To simplify the notation we may assume that the point $a \in \mathbb{C}^m$ is one of such b's. Hence we proved the theorem for $\mathfrak{V}_1$. Let $\mathfrak{W}$ be the complement of $\mathfrak{V}_1$ in $\mathfrak{V}$. So $\mathfrak{W}$ is an invariant subvariety of $\mathfrak{V}$ which satisfies

(2.6) $$\dim \mathfrak{W} < \dim \mathfrak{V}.$$

Next apply the above arguments to the irreducible components of $\mathfrak{W}$. The inequality (2.6) shows that our procedure ends in a finite number of steps. So we get the decomposition (1.1) which satisfy the assumptions (i)-(iii).

3. JORDAN CANONICAL FORM

Let $\mathfrak{V} = M_n$ and assume that the action of GL_n is given by (1.2) with $k=0$. That is orb(A) is the set of all similar matrices to A. In that case each orbit is uniquely characterized by the Jordan canonical form of A. Let $D(A,r)$ be the ball of radius r centered at A in M_n. A manifold $\mathfrak{W} \subset M_n$ is called stable if for any $A \in \mathfrak{W}$ all matrices $B \in \mathfrak{W} \cap D(A,r)$ have the same type of the Jordan canonical form as A. That is if A has s distinct eigenvalues $\lambda_1(A),\ldots,\lambda_s(A)$ with the multiplicities $\nu_1,\ldots,\nu_s$ then each

$B \in \mathfrak{W} \subset M_n$ has s distinct eigenvlaues $\lambda_1(B),\ldots,\lambda_s(B)$ with the multiplicities $\nu_1,\ldots,\nu_s$. Furthermore, the number of Jordan blocks and their sizes corresponding to the eigenvalues of A and B in the Jordan canonical forms of A and B respectively are identical. It is easy to see that M_n splits to disjoint unions

$$M_n = \bigcup_{1}^{P_n} \mathfrak{v}_i \tag{3.1}$$

where each $\mathfrak{v}_i$ is maximally stable manifold. Thus $\mathfrak{v}_1$ is the manifold of all matrices with pairwise distinct eigenvalues and $\mathfrak{v}_{P_n}$ is the manifold of all scalar matrices λI. It is possible to compute P_n explicitly but we shall not discuss this problem.

As before we identify M_n with $\mathbb{C}^{n^2}$. Clearly on $\mathfrak{v}_1$ the values of the polynomials

$$\psi_r(X) = \mathrm{tr}(X^r), \quad r = 1,\ldots,n, \quad X = (x_{ij}) \tag{3.2}$$

determine a unique orbit. As $\mathfrak{v}_1$ is dense in M_n it is not difficult to show that $\psi_1,\ldots,\psi_n$ are the generators of the ring of the invariant polynomials $\mathbb{C}[M_n]$ and the field of rational functions $\mathbb{C}(M_n)$. Let $A = (a_{ij})$ be a diagonal matrix with pairwise distinct eigenvalues. Denote by $\mathfrak{S}$ the set of all pairs (i,j), $i \neq j$. Then the Jordan canonical forms on $\mathfrak{v}$ can be obtained by considering the system

$$\begin{aligned} &\mathrm{tr}(X^r) = \kappa_r, \quad r = 1,\ldots,n, \\ &x_{ij} = 0 \ (= a_{ij}), \ (i,j) \in \mathfrak{S}, \ X = (x_{ij}). \end{aligned} \tag{3.3}$$

Clearly, the number of distinct solutions of the above system is precisely $n!$ for $X \in \mathfrak{v}_1$. That is $\mu_1 = n!$ which is the degree of each orbit in $\mathfrak{v}$.

Let $\mathfrak{G}_1$ be the symmetric group of $n \times n$ permutation matrices. Then $\mathfrak{G}_1$ acts on M_n as a subgroup of GL_n. That is $A \to P^{-1}AP$, $P \in \mathfrak{G}_1$. Thus, if $\xi^{(r)}$ is a diagonal matrix which is similar to the given $A \in \mathfrak{v}_1$ then all other diagonal matrices $\xi^{(k)}$ which are similar to A are of the form $P^{-1}\xi^{(r)}P$, $P \in \mathfrak{G}_1$. Hence, all the Jordan canonical forms on $\mathfrak{v}_1$ are identified by the action of the Galois group $\mathfrak{G}_1$.

We now show that $\mathfrak{G}_k$ is a certain subgroup of $\mathfrak{G}_1$. Let $\mathfrak{v}_k$ be maximally stable manifold. Then each $A \in \mathfrak{v}_k$ has s pairwise distinct eigenvalues $\lambda_1(A),\ldots,\lambda_s(A)$. To each $\lambda_q(A)$ correspond a fixed Jordan canonical form. That is the number of the Jordan blocks and their sizes are fixed and do not depend on $A \in \mathfrak{V}_q$. Let $\mathfrak{S}$ the set of pairs (i,j), $i \neq j$. We chose the numbers a_{ij}, $(i,j) \in \mathfrak{S}$ to be equal to 0 or 1 such that for some choice of the diagonal elements the resulting A is in the Jordan canonical form and $A \in \mathfrak{V}_q$. It is straightforward to show that a Jordan canonical form on $\mathfrak{V}_q$ is given by the system

$$(3.4) \qquad \begin{aligned} &\mathrm{tr}(X^r) = \kappa_r, \ r = 1,\dots,s, \\ &x_{ij} = a_{ij}, \ (i,j) \in \mathfrak{S}. \end{aligned}$$

We can permute in the above canonical form any two eigenvalues λ_u and λ_v which have the identical Jordan blocks. This action is realized by the action of the corresponding subgroup $\mathfrak{G}_k \subset \mathfrak{G}_1$.

We close this paper by pointing out that the choice of the parameters a_{ij}, $(i,j) \in \mathfrak{S}$ is crucial even in this case. For simplicity we restrict ourselves to $\mathfrak{V}_1$. Suppose we do not choose $A \in \mathfrak{V}_1$ to be the diagonal matrix. As in Theorem 1 we consider the system

$$(3.5) \qquad \mathrm{tr}(X^r) = \kappa_r, \ r = 1,\dots,n, \quad x_{ij} = a_{ij}, \ (i,j) \in \mathfrak{S}$$

Then according to Friedland [1977] there exists an open quasi-variety $\mathfrak{X}$ in M_n such there are precisely $n!$ distinct solutions $X^{(j)}$ to (3.5). Now the relation between these different solutions $X^{(j)}$ is quite complicated. In general the quasi-variety $\mathfrak{X}$ would be strictly contained in $\mathfrak{V}_1$,

The above arguments show that there may exist best choices of the points a and sets $\mathfrak{S}$ appearing in Theorem 1.

REFERENCES

[1] Fogarty, J. [1969], Invariant Theory, Benjamin, New York.

[2] Friedland, S. [1977], Inverse eigenvalue problems, Linear Algebra and Applications 17, 15-51.

[3] Friedland, S. [1983], Simultaneous similarity of matrices, Advances in Mathematics, 50, 189-265.

[4] Friedland [1985], Classification of linear systems, Proc. A.M.S. Conf. on Linear Algebra and Systems Theory, Contemporary Mathematics 1985 (to appear).

[5] Rosenlicht, M. [1956], Some basic theorems on algebraic groups, Amer. J. Math. 78, 401-443.

[6] Shafarevich, R. [1974], Basic Algebraic Geometry, Springer-Verlag, New York.

Frequency Domain and State Space Methods for Linear Systems
C.I. Byrnes and A. Lindquist (Editors)

ON VARIETIES OF INVARIANT SUBSPACES

William H. Gustafson
and
Clyde Martin

Department of Mathematics
Texas Tech University
Lubbock, Texas 79409
U.S.A.

We study the k-dimensional invariant subspaces of a nilpotent linear transformation over an arbitrary field as an algebraic subvariety of the appropriate Grassmannian variety. The equations of this subvariety are discovered, and we show that the invariant subspaces affording a fixed Jordan form make up a smooth variety.

INTRODUCTION

Let V be a t-dimensional vector space over a field K. The <u>projective space</u> P(V) is the set of equivalence classes [v] of nonzero vectors in V, where v and αv are equivalent whenever $0 \neq \alpha\varepsilon K$. When $V = K^t$, the row t-vectors, an element of P(V) is thus given by <u>homogeneous coordinates</u> $[a_1,\ldots,a_t]$, where not all the a_i are zero, and $[a_1,\ldots,a_t] = [\alpha a_1,\ldots,\alpha a_t]$ when $0 \neq \alpha\varepsilon K$. For $1 \leq i \leq t$, the set U_i of points with $a_i \neq 0$ can be identified with K^{t-1} by $\phi_i : [a_1,\ldots,a_t] \to a_i^{-1}(a_1,\ldots,a_{i-1},a_{i+1},\ldots,a_t)$. The U_i's cover P(V), and the restriction of $\phi_j \circ\phi_i^{-1}$ to $\phi_i(U_i \cap U_j)$ is a rational map, whence $P(K^t)$ forms an abstract algebraic variety.

Let M be an n dimensional K-vector space, with a chosen basis $e_1,\ldots,e_n$. For $1 \leq k \leq n$, the k-fold exterior product $\wedge^k M$ (c.f[3]) has a basis $\{e_{i_1}\wedge\ldots\wedge e_{i_k} | 1 \leq i_1 < \ldots < i_k \leq n\}$, and so can be identified with K^t, where $t = \binom{n}{k}$. If N is a k-dimensinal subspace of M, with basis $f_1,\ldots,f_k$, the element $f_1\wedge\ldots\wedge f_k$ of $\wedge^k M$ is almost determined by N; if a different basis is used, this element is multiplied by the determinant of the basis change matrix. Hence,

N determines $[f_1\wedge\ldots\wedge f_k]\varepsilon P(\bigwedge^k M)$ uniquely. The corresponding point $[(p_{i_1\ldots i_k})]\varepsilon P(K^t)$ can be obtained by taking for $p_{i_1\ldots i_k}$ the minor determinant formed from rows $i_1,\ldots,i_k$ of the matrix (a_{ij}), where $f_j = \sum_{j=1}^{n} a_{ij}e_j$. The $p_{i\ldots i_k}$ are the Plücker coordinates of N. The mapping $N \to [f_1\wedge\ldots\wedge f_k]$ is known to be a one-to-one mapping from the set of all k-dimensional subspaces of M onto a quadratic closed algebraic subvariety $G^k(M)$ in $P(K^t)$; it is the <u>Grassmannian variety</u> (see [2] for further details).

Henceforth, let M be equipped with a fixed nilpotent endomorphism T. Following Shayman [4], who studied the real and complex cases, we will examine the set $S_T(k)$ of points in $G^k(M)$ corresponding to T-invariant subspaces. We show first how to find the defining equations of $S_T(k)$; then we show that the elements of $S_T(k)$ having a fixed Jordan structure form a smooth variety.

EQUATIONS FOR $S_T(k)$

The basis $e_1,\ldots,e_n$ of M can be chosen so that for each i, $T(e_i)$ is either zero or e_j, with $j > i$. Let $f_j = \sum a_{ij}e_i$, $1 \le j \le k$ be linearly independent vectors in M. The k x k minors of $A = (a_{ij})$ give a set of Plücker coodinates $(p_{i_1\ldots i_k})$ of the subspace N spanned by $f_1,\ldots,f_k$. N is T-invariant if and only if it is invariant under the invertible transformation I + T (I = identity map of M). Plücker coordinates $(p'_{i_1\ldots i_k})$ for (I + T)N are given by the k x k minors of the matrix obtained from A by adding row i to row j, whenever $T(e_i) = e_j$. By elementary properties of determinants, it follows that $p'_{i_1\ldots i_k}$ is obtained from $p_{i_1\ldots i_k}$ by adding a (possibly zero) sum $q(i_1,\ldots,i_k)$ of $p_{j_1\ldots j_k}$'s whose subscripts preceed $i_1\ldots i_k$ in the lexicographic order. So, if we form the $p_{i_1\ldots i_k}$ into a column vector Π in lexicographic order, and similarly form Π' from the $p'_{i_1 \ldots i_k}$, we have $\Gamma\Pi = \Pi'$, where Γ is a lower triangular t x t matrix with ones on the main diagonal. Now, N is T-invariant if and only if $[p_{i_1 \ldots i_k}] = [p'_{i_1 \ldots i_k}]$, i.e. $\Pi' = \alpha\Pi$, for some $0 \neq \alpha \varepsilon K$. But then α is an eigenvalue of Γ, whence $\alpha = 1$, and $\Pi = \Pi'$. Thus, $S_T(k)$ is given by the vanishing of the linear forms $q(i_1,\ldots,i_k)$ above, together with the quadratic equations defining $G^k(M)$.

Let us work out the example where M is four-dimensional, T is given by $e_1 \to e_2 \to 0$, $e_3 \to e_4 \to 0$, and k = 2. Let f_1, f_2 span a two-dimensional subspace N of M. Expressing f_1, f_2 in terms of $e_1,\ldots,e_4$ gives a 4 x 2 matrix

$$A = \begin{pmatrix} r_1 \\ r_2 \\ r_3 \\ r_4 \end{pmatrix},$$

where r_i is a row 2-vector. The Plücker coordinates $[p_{12},\ldots,p_{34}]$ of N are given by $p_{ij} = \begin{vmatrix} r_i \\ r_j \end{vmatrix}$. Those of (I + T)N are given by the 2 x 2 minors of

$$\begin{pmatrix} r_1 \\ r_1+r_2 \\ r_3 \\ r_3+r_4 \end{pmatrix}$$

Thus,
$$p'_{12} = \begin{vmatrix} r_1 \\ r_1+r_2 \end{vmatrix} = \begin{vmatrix} r_1 \\ r_2 \end{vmatrix} = p_{12}$$
$$p'_{13} = p_{13}$$
$$p'_{14} = \begin{vmatrix} r_1 \\ r_3+r_4 \end{vmatrix} = \begin{vmatrix} r_1 \\ r_3 \end{vmatrix} + \begin{vmatrix} r_1 \\ r_4 \end{vmatrix} = p_{13} + p_{14}$$
$$p'_{23} = p_{13} + p_{23}$$
$$p'_{24} = p_{13} + p_{14} + p_{23} + p_{24}$$
$$p'_{34} = p_{34}.$$

The condition for T-invariance of N is the vanishing of q(1,4) =

$q(2,3) = p_{13}$ and $q(2,4) = p_{13} + p_{14} + p_{23}$. Now, $G^2(M)$ is defined by the single equation

$$p_{12}\,p_{34} - p_{13}\,p_{24} + p_{14}\,p_{23} = 0,$$

so $S_T(2) = \{[x,0,y,-y,z,w] \in P(K^6) \mid y^2 = xw\}$. Since there are four parameters, one equation, and the coordinates are homogeneous, this variety has dimension two. Easy calculations show that $P = [0,0,0,0,1,0]$ is the only singular point, and that if $K = \mathbb{R}$, $S_T(k)$ is a pinched torus. The invariant subspace corresponding to P is that spanned by e_2 and e_4, i.e. the kernel of T. (The subspace N corresponding to a point of $S_T(k)$ is easily found. For, if f_1, f_2 is a basis for N, as above, then $a_1 e_1 + \ldots + a_4 e_4 \in N$ if and only if the 4 x 3 matrix $\begin{pmatrix} r_1 & a_1 \\ r_2 & a_2 \\ r_3 & a_3 \\ r_4 & a_4 \end{pmatrix}$ has rank two, i.e. its 3 x 3 minors vanish. These minors are determined by expansion along the last column, giving homogeneous equations for the a_i's with p_{ij}'s as coefficients.

Hence, the subspace corresponding to $[x,0,y,-y,z,w]$ is described by the nullspace of

$$\begin{pmatrix} -y & 0 & x & 0 \\ z & -y & 0 & x \\ w & 0 & -y & 0 \\ 0 & w & -z & -y \end{pmatrix}).$$

When P is removed, the remaining points of $S_T(k)$ correspond to the invariant subspaces on which T has Jordan form $\begin{pmatrix} 0 & 0 \\ 1 & 0 \end{pmatrix}$. Looking, for example, at the point $Q = [1,0,0,0,0,0,]$, one sees that the points $[1,0,y,-y,z,y^2]$ form a Zariski neighborhood of Q isomorphic to the yz-plane. In this way, one sees that $S_T(k)\backslash\{P\}$ is a smooth (open) subvariety; we generalize this in the next section.

NONSINGULARITY UNDER FIXED JORDAN FORM

In this section, we present an analogue of Theorem 7 of Shayman [4], valid for arbitrary fields. Let $T^S = 0 \neq T^{S-1}$. M can be viewed as a module over $R = K[T]$, which is isomorphic to $K[X]/(X^S)$.

R is a local ring with J = (T) as maximal ideal; any other ideal of R equals J^u, for some u = 0,...,S (where J^0=R). The T-invariant subspaces of M are the R-submodules, and the restrictions of T to two such submodules N and N' have the same Jordan form if and only if N and N' are R-isomorphic.

For a fixed R-dimensional R-submodule N of M, let S(N,M) be the subset of $S_T(k)$ corresponding to submodules of M isomorphic to N. We will exhibit for each PεS(N,M) a neighborhood in S(N,M) isomorphic to an affine space. Thus, we prove

Theorem: S(N,M) is nonsingular

We just remark that S(N,M) can be obtained as the quotient space resulting from acting on the open set Mono(N,M) of monorphisms in $Hom_R(N,M)$ by the open set Aut(N) of bijections in $Hom_R(N,M)$. Aut(N) is a linear algebraic group. Since this group action is free, the dimension of S(N,M) is that of $Hom_R(N,M)/Hom_R(N,N)$. On the other hand, the exact sequence

$$0 \to N \to M \overset{\theta}{\to} M/N \to 0$$

gives rise to the exact sequence

$$0 \to Hom_R(N,N) \to Hom_R(N,M) \overset{\theta^*}{\to} Hom_R(N,M/N),$$

in which θ_*, given by $\theta_*(f) = \theta f$ is not in general surjective. We will parametrize a neighborhood of N in S(N,M) by Im $\theta_* \cong Hom_R(N,M)/Hom_R(N,N)$.

The theory of the Jordan form shows that as R-modules, M and N each decompose as direct sums of cyclic R-modules, which have the form R/J^u. These cyclic modules are indecomposable, since each can be identified with its own endomorphism ring, and all R/J^u are local rings. Via the ring homomorphism K[X]→R given by f(X)→f(T), M can be viewed as K[X]-module. Let τ:F→M be a K[X]-surjection, where F is K[X]-free of least possible rank. Some (necessarily free) submodule F' of F is carried onto N by τ. By the Invariant Factor

Theorem [1,p.97], there are a free basis $f_1,\ldots,f_\ell$ of F and elements $d_1,\ldots,d_m$ of K[X], with $m \leq \ell$, so that $d_1F_1,\ldots,d_mF_m$ is a basis of F'. Letting $\varepsilon_i = \tau(f_i)$, one gets a decomposition of M as a direct sum $I_1 \oplus \ldots \oplus I_\ell$, with $I_i = R\varepsilon_i$. Further, $N = I_1' \oplus \ldots \oplus I_m'$, where $I_i' \subseteq I_i$. From the list of ideals of R given earlier, it follows that there are integers $\mu_1,\ldots,\mu_m$ with $I_i' = RT^{\mu_i}\varepsilon_i$. Thus, N and M have "aligned" decompositions. Let $J^{\beta_t} = \mathrm{Ann}_R(\varepsilon_t)$, the R-annihilator of ε_t. Then $\mathrm{Ann}(T^{\mu_t}\varepsilon_t) = J^{\beta_t-\mu_t}$, and we assume that $I_1',\ldots,I'_m$ are ordered so that $i \leq j$ implies $\beta_i - \mu_i \geq \beta_j - \mu_j$. Now let $f \varepsilon \mathrm{Hom}_R(N,M)$; we examine $\theta_*(f) = \theta f$. Let $f(T^{\mu_t}\varepsilon_t) = \sum g_{kj}T^k\varepsilon_j$, with $g_{kj} \varepsilon K$. Since $f(T^{\mu_t}\varepsilon_t)$ must be annihilated by $T^{\beta_t-\mu_t}$, g_{kj} can be nonzero only when $\beta_t - \mu_t + k \geq \beta_j$. Now, $T^k\varepsilon_j \varepsilon N$ if and only if $j \leq m$ and $k \geq \mu_j$. So, letting $\bar{\varepsilon}_j = \theta(\varepsilon_j)$, we have

$$\theta f(T^{\mu_t}\varepsilon_t) = \sum g_{kj}T^k\bar{\varepsilon}_j,$$

and g_{kj} can be nonzero only when $g_{kj} \varepsilon S_t$, where

$$S_t = \{(k,j) \mid \beta_t - \mu_t + k \geq \beta_j, \text{ and either } j > m \text{ or } \mu_j > k\}.$$

Define $\tilde{f}$: N → M by

$$\tilde{f}(T^{\mu_t}\varepsilon_t) = \sum_{(k,j)\varepsilon S_t} g_{kj}\, T^k\varepsilon_j.$$

The corrspondence $\theta f \leftrightarrow \tilde{f}$ is clearly bijective. Observe that when $(k,j)\varepsilon S_t$, we must have $j < t$. For if $j \leq m$, then $\mu_j > k$, whence $\beta_t - \mu_t + \mu_j > \beta_j$, and so $j > t$ by the ordering of $I_1',\ldots,I_m'$. From this it follows easily that f':N→M, given by $f'(n) = n + \tilde{f}(n)$ is an R-monomorphism. Let $N_{\theta f}$ be its image, $[N_{\theta f}]$ the corresponding point on S(N,M). We claim the map $\theta f \to [N_{\theta f}]$ gives an isomorphism of $\mathrm{Im}\theta^*$ to a neighborhood of $[N] = [N_{\theta\circ 0}]$.

We have to examine this map in terms of the Plücker coordinates. A basis for $N_{\theta f}$ is obtained as follows: first get a basis of N by taking the vectors $T^{\mu_t}\varepsilon_t$, together with their nonzero iterated images $T^{\mu_t+\sigma}\varepsilon_t$ under T. Then add to each $\sum g_{kj}T^{\sigma+k}\varepsilon_t$. Since these added terms come after $T^{\mu_t+\sigma}\varepsilon_t$ in the obvious ordering of a basis for M (left-to-right among the I_i, top-down within any I_i), the wedge product

$$\bigwedge_{t,\sigma}(T^{\mu_t+\sigma}\varepsilon_t + \sum g_{kj}T^{\sigma+k}\varepsilon_t)$$

has $\bigwedge_{t,\sigma}(T^{\mu_t+\sigma}\varepsilon_t)$ as leading term (relative to lexicographic order), and each g_{kj} appears as a coefficient in the natural basis. All other coordinates are polynomials in the g_{kj}, and we are done.

As an example, let M be 5-dimensional, with T given by $e_2 \to e_1 \to 0$, $e_4 \to e_3 \to 0$, $e_5 \to 0$. Let N be the submodule spanned by e_1, e_2 and e_3. A typical R-homomorphism $f: N \to M$ has matrix

$$\begin{pmatrix} a & b & f \\ 0 & a & 0 \\ c & d & g \\ 0 & c & 0 \\ 0 & e & h \end{pmatrix},$$

and θf is given by $\begin{pmatrix} 0 & c & 0 \\ 0 & e & h \end{pmatrix}$, using the basis $e_4 + N$, $e_5 + N$ in M/N. The map f' then is given by

$$\begin{pmatrix} 1 & 0 & 0 \\ 0 & 1 & 0 \\ c & 0 & 1 \\ 0 & c & 0 \\ 0 & e & h \end{pmatrix},$$

and the columns of this matrix span $N_{\theta f}$. In S(N,M), we have [N] = [1,...,0] and $[N_{\theta f}]$ = [1,0,h, - c,-e,ch,0, - ch,0,c^2h].

REFERENCES

[1] Curtis, C.W. and Reiner, I., Representation theory of finite groups and associative algebras (Interscience, New York, 1962).

[2] Hodge, W.V.D. and Pedoe, D., Methods of algebraic geometry, volume II (Cambridge University Press, Cambridge 1968).

[3] MacLane, S. and Birkhoff, G., Algebra, second edition (Macmillan, New York, 1979).

[4] Shayman, M., On the variety of invariant subspaces of a finite-dimensional linear operator, Trans. Amer. Math. Soc. 274 (1982) 721-747.

Frequency Domain and State Space Methods for Linear Systems
C.I. Byrnes and A. Lindquist (Editors)

ROSENBROCK SYSTEM THEORY FOR MATRICES OVER BEZOUT DOMAINS

Malcolm C. Smith

DFVLR
Institute for Flight Systems Dynamics
Oberpfaffenhofen, D-8031 Wessling
Federal Republic of Germany

Basic results of Rosenbrock system theory will be shown to carry over to the case of matrices with elements in a Bezout domain. For the ring of holomorphic functions the extended definition of strict system equivalence will be shown to concide with the usual one for polynominal systems. A discussion of the case where the ring is no longer a Bezout domain will be given.

INTRODUCTION

The purpose of this paper is to show how the basic ideas of Rosenbrock system theory, together with some later additions, carry over to the case of matrices with elements in a Bezout domain **R**. The level of generality of the fractional representation approach to feedback system analysis and design ([1], [4]) will be seen to carry over to include system representation of the more general form $G = VT^{-1}U + W$. Furthermore, it will be shown that notions such as strict system equivalence, minimality and reducibility can be correspondingly extended. The theory allows two definitions of the concept of internal stability to be compared. If **R** is an elementary divisor domain (this is in fact the case for all known Bezout domains) then the results of the Rosenbrock theory pertaining to the Smith and Smith-McMilan form will also be seen to hold. For the case when the elements of **R** are holomorphic functions the generalisations of the theory of decoupling zeros will be considered. Also a consistency result will be proved showing that the extended definition of strict system equivalence coincides with the usual one for polynominal systems. Finally, the possibility of developing Rosenbrock's system theory under weaker conditions on **R** will be considered.

PRELIMINARIES

Let **R** be an integral domain. For $r_1 \ldots . r_n \in \mathbf{R}$ we say $c \in \mathbf{R}$ is a <u>greatest common divisor</u> (gcd) of $r_1 \ldots r_n$ if $c | r_i$ and $d | c$ for any other common divisor d of $r_1 \ldots . r_n$. $r_1 \ldots . r_n \in \mathbf{R}$ are said to be <u>coprime</u> if 1 is a greatest common divisor of the r_i's. **R** is called a <u>greatest common divisor domain</u> (GCDD) if any pair (and hence any finite set) of elements has a greatest common divisor.

R is called a <u>Bezout domain</u> (BD) if every finitely generated ideal of **R** is principal. This is a generalisation of a principal ideal domain (PID) because of the possibility of infinitely generated ideals. A necessary and sufficient condition for **R** to be a BD is that (i) **R** is a GCDD and (ii) for any coprime $a,b \in \mathbf{R}$ there exists $p,q \in \mathbf{R}$ such that $pa + qb = 1$. Further details about GCDD's and BD's can be found in [5].

A square matrix over **R** is called <u>unimodular</u> if its determinant is a unit in **R**. Two matrices X and Y over **R** (with the same number of rows r and the sum of the columns at least r) have an r x r matrix C as <u>common left divisior</u> if $X = CX_1$ and $Y = CY_1$. C is a <u>greatest common left divisor</u> (gcld) if it is a common left divisor and $C = DC_1$ for any other common left divisor D of X and Y. If X and Y have a non-singular gcld then the gcld's of X and Y are unique up to multiplication on the right by a unimodular matrix. If the identity matrix I is a gcld of X and Y then we say X and Y are <u>left coprime</u>. The concepts of (greatest) common right divisors and right coprimeness are similarly defined.

<u>Theorem 1</u> ([3])

Given any r x c matrix A (where $r \leq c$) over a Bezout domain **R** there exists a unimodular matrix M such that

$$AM = [C\ 0]$$

where C is r x r (in fact it is possible to make C triangular). □

Using theorem 1 we can show that greatest common left (right) divisors always exist for matrices over Bezout rings. Indeed, given X and Y as above and writing $A = [X\ Y]$, theorem 1 allows us to find P and Q such that $C = XP + YQ$.

We can also deduce the following corollary. For $A = [X\ Y]$ this gives seven equivalent conditions for X and Y to be left coprime. Corresponding conditions exist for right coprimeness. **F** denotes the quotient field of **R**.

<u>Corollary 1</u>

Let A be any r x c matrix (where $r \leq c$) over a Bezout domain **R**. The following conditions are equivalent:

(C1) (a) There exists a unimodular matrix M such that $AM = [I\ 0]$,

(b) A can be included as the first r rows of a unimodular matrix,

(C2) (a) There exists a c x r matrix B such that $AB = I_r$,

(b) For all $x \in \mathbf{R}^r$ there exists $y \in \mathbf{R}^c$ such that $x = Ay$,

(C3) $A^T z \in \mathbf{R}^c$ for $z \in \mathbf{F}^r$ implies $z \in \mathbf{R}^r$,

(C4) The r x r minors of A are coprime,

(C5) If $A = CA_1$ where C is square then C is unimodular.

<u>Proof</u> Using only the fact that **R** is an integral domain the following implications can be established: (C1a)$\Leftrightarrow$(C1b)$\Rightarrow$(C2a)$\Leftrightarrow$(C2b)$\Rightarrow$(C3)$\Rightarrow$(C4)$\Rightarrow$(C5). Theorem 1 shows (C5)$\Rightarrow$(C1). □

R is called an <u>elementary divisor domain</u> (EDD) if (i) **R** is a BD and (ii) for any non-zero, coprime a,b,c, ε **R** there exists s,t ε **R** such that sa and sb + tc are coprime. It can be shown [3] that **R** is an elementary divisor domain if and only if for any matrix A whose elements come from **R** there exist unimodular matrices M_1 and M_2 such that $M_1\ A\ M_2$ = pdiag($d_1, d_2, \ldots d_r$) where $d_i | d_{i+1}$. For the next four sections we assume that **R** is a BD and for some results that **R** is an EDD. It should be noted however that at present no rings are known which are BD's and not EDD's (see [16] and the references therein).

Two examples of EDD's of interest in control theory are given below ((a) is in fact a Euclidean domain).

(a) $R_\Omega(s)$ - the (real) rational functions with no poles in a subset Ω of the extended complex plane $\mathbb{C}$. (If Ω is a proper subset of $\mathbb{C}_\infty$ then Ω must exclude either ∞ or at least one point of the real axis).

(b) $H(\Omega)$ (resp. $H_S(\Omega)$) - the set of functions holomorphic in a region of the extended complex plane (resp. the subset of $H(\Omega)$ with $f(s)^* = f(s^*)$).

BASIC SYSTEM THEORY

This section is indebted to the work of Coppel ([6], [7]) who showed how the realisation theory of Rosenbrock and the equivalence theory of Rosenbrock and Fuhrmann could be extended to the case of matrices whose elements come from a PID. Using corollary 1 the extension to a BD is immediate. We now summarise the main definitions and results.

The matrix G ε $\mathbf{F}^{p\times m}$ is said to have a <u>realisation</u> {T,U,V,W}, where T, U, V and W are matrices over **R** with T being r x r, if $G = VT^{-1}U + W$. The realisation {T,U,V,W} is associated with (i) the <u>system matrix</u>

$$\begin{bmatrix} T & -U \\ V & W \end{bmatrix}$$

and (ii) an input space $\mathbf{U} = \mathbf{R}^m$, an output space $\mathbf{Y} = \mathbf{F}^p/\mathbf{R}^p$ and the state space $\mathbf{X} = T^{-1}\mathbf{R}^r/\mathbf{R}^r$ which is a finitely generated torsion module. For a given G, r may vary according to the realisation. Notice that the so-called matrix fraction decompositions (MFD's) of G are special realisations in which W = 0 and either U = I or V = I. The following facts can be established:

$$T^{-1}U \text{ maps } \mathbf{U} \text{ onto } \mathbf{X} \Leftrightarrow T \text{ and } U \text{ are left coprime},$$

$$V \text{ maps } \mathbf{X} \text{ one-to-one into } \mathbf{Y} \Leftrightarrow T \text{ and } V \text{ are right coprime}.$$

The realisation $\{T,U,V,W\}$ is said to be irreducible if T, U are left coprime and T, V are right coprime. The following result can be deduced from theorem 1.

Theorem 2

Any $G \in \mathbf{F}^{p \times m}$ has an irreducible realisation. □

The definition of equivalence due to Fuhrmann [9] can be expressed as follows [7]: Two realisations $\{T_i,U_i,V_i,W_i\}_{i=1,2}$ (where T_i is $r_i \times r_i$) are strict system equivalent (sse, ~) if $V_1T_1^{-1}U_1 + W_1 = V_2T_2^{-1}U_2 + W_2$ and there is an isomorphism $\phi : \mathbf{X}_1 \to \mathbf{X}_2$ such that the following diagram commutes

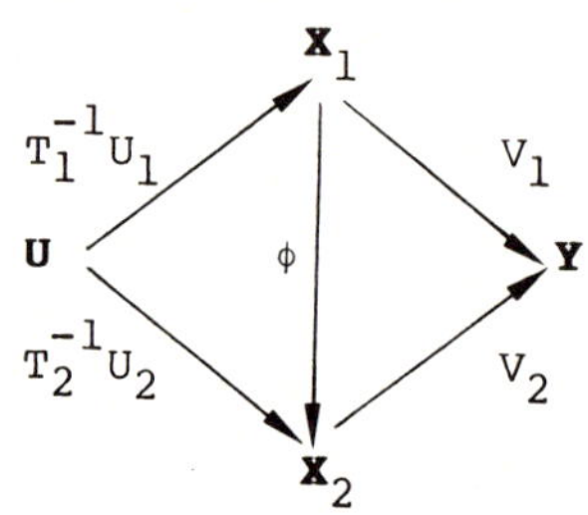

The following result can be established:

Theorem 3 [9]

Two realisations $\{T_i,U_i,V_i,W_i\}_{i=1,2}$ are strict system equivalent if and only if there exist N_i,Y_i of appropriate dimension such that

$$\begin{bmatrix} N_2 & 0 \\ Y_2 & I_p \end{bmatrix} \begin{bmatrix} T_1 & -U_1 \\ V_1 & W_1 \end{bmatrix} = \begin{bmatrix} T_2 & -U_2 \\ V_2 & W_2 \end{bmatrix} \begin{bmatrix} N_1 & Y_1 \\ 0 & I_m \end{bmatrix} \qquad (1)$$

where T_2,N_2 are left coprime and T_1,N_1 are right coprime. □

The original notion of strict system equivalence given by Rosenbrock [8] is characterised in the next theorem.

Theorem 4

Two realisations $\{T_i,U_i,V_i,W_i\}_{i=1,2}$ are strict system equivalent if and only if there exists an $r \geq \max(r_1,r_2)$, unimodular matrices M_1 and M_2 and matrices X_1 and X_2 such that

$$\begin{bmatrix} M_2 & 0 \\ X_2 & I_p \end{bmatrix} \begin{bmatrix} I_{r-r_1} & 0 & 0 \\ 0 & T_1 & -U_1 \\ 0 & V_1 & W_1 \end{bmatrix} = \begin{bmatrix} I_{r-r_2} & 0 & 0 \\ 0 & T_2 & -U_2 \\ 0 & V_2 & W_2 \end{bmatrix} \begin{bmatrix} M_1 & X_1 \\ O & I_m \end{bmatrix} \quad (2)$$

Proof See [10], [11]. □

For the important class of realisations - matrix fraction decompositions - the following result holds.

Theorem 5 ([11] p. 564)

If the two realisations $\{D_i, N_i, I_p, 0\}_{i=1,2}$ are strict system equivalent then there exists a unimodular matrix U such that

$$N_1 = UN_2, \qquad D_1 = UD_2$$

A corresponding result holds for right MFD's. □

The following is a basic result on irreducible realisations which is originally due to Rosenbrock. Rosenbrock's proof does not generalise to our case and instead we describe an approach based on the work of Coppel [6].

Theorem 6

Two irreducible realisations $\{T_i, U_i, V_i, W_i\}_{i=1,2}$ of G are strict system equivalent.

Proof As in [6] theorem 9 we can find A_1, B_1 right coprime so that $T_1B_1 - U_1A_1 = 0$ with A_1 non-singular. Thus

$$\begin{bmatrix} U_1 & 0 \\ -W_1 & I \end{bmatrix} \begin{bmatrix} A_1 & -I \\ C_1 & 0 \end{bmatrix} = \begin{bmatrix} T_1 & -U_1 \\ V_1 & W_1 \end{bmatrix} \begin{bmatrix} B_1 & 0 \\ 0 & I \end{bmatrix}$$

where $C_1 = W_1A_1 + V_1B_1$. Hence $\{T_1, U_1, V_1, W_1\} \sim \{A_1, I, C_1, 0\}$ from theorem 3. Since T_1, V_1 are right coprime so are A_1, C_1 (use theorem 4 or proceed as in [6]). Similarly we can find A_2, C_2 left coprime such that $\{T_2, U_2, V_2, W_2\} \sim \{A_2, C_2, I, 0\}$. Since $C_1A_1^{-1} = A_2^{-1}C_2$ we have

$$\begin{bmatrix} C_2 & 0 \\ 0 & I \end{bmatrix} \begin{bmatrix} A_1 & -I \\ C_1 & 0 \end{bmatrix} = \begin{bmatrix} A_2 & -C_2 \\ I & 0 \end{bmatrix} \begin{bmatrix} C_1 & 0 \\ 0 & I \end{bmatrix}$$

Thus $\{A_1,I,C_1,0\} \sim \{A_2,C_2,I,0\}$. Hence $\{T_1,U_1,V_1,W_1\} \sim \{T_2,U_2,V_2,W_2\}$. □

Using theorem 1 and theorem 6 we can deduce the following result.

Corollary 2

$\{T,U,V,W\}$ is an irreducible realisation of G if and only if for any other realisation $\{T_1,U_1,V_1,W_1\}$ of G $\det T | \det T_1$. □

In conclusion we remark that if **R** is an EDD then G admits reduction G Smith-McMillan form. The theorem of Rosenbrock ([8] p. 111) relating the Smith-McMillan form of G to the Smith form of T and the system matrix also generalises.

ON STRICT SYSTEM EQUIVALENCE FOR HOLOMORPHIC SYSTEM MATRICES

Theorem 5 shows that the natural definition of equivalence for matrix fraction decompositions (multiplication by a unimodular matrix) is precisely equivalent to the notion of strict system equivalence. A result of Rosenbrock ([8] p. 56) in the same spirit establishes that state-space systems which are strict system equivalent are system similar. For polynominal (rational) system matrices a corresponding question arises: are the equivalence classes induced by ~ changed when the ring $\mathbf{R} = R_\Omega(s)$ is extended to $\mathbf{R} = H_S(\Omega)$? The following theorem will now be proved which shows that the answer is no.

Theorem 7

Let Ω be a region of the extended complex plane and let $\{T_i,U_i,V_i,W_i\}_{i=1,2}$ be realisations of G having elements in $R_\Omega(s)$. If the realisations are equivalent over $H_S(\Omega)$ they are equivalent over $R_\Omega(s)$ i.e. there exist choices of N_i and Y_i in (1) with entries in $R_\Omega(s)$.

Proof Suppose N_i and Y_i are holomorphic matrices satisfying theorem 3. Then for any $r_2 \times r_1$ holomorphic matrix A the following equations are satisfied

$$(N_2 + T_2A)T_1 = T_2(N_1 + AT_1), \tag{3a}$$
$$-(N_2 + T_2A)U_1 = T_2(Y_1 - AU_1) - U_2, \tag{3b}$$
$$(Y_2 + V_2A)T_1 + V_1 = V_2(N_1 + AT_1), \tag{3c}$$
$$-(Y_2 + V_2A)U_1 + W_1 = V_2(Y_1 - AU_1) + W_2, \tag{3d}$$

that is, (1) holds with

$$N_1 \to N_1' = N_1 + AT_1,$$

$$N_2 \to N_2' = N_2 + T_2A,$$

$$Y_1 \to Y_1' = Y_1 - AU_1,$$

$$Y_2 \to Y_2' = Y_2 + V_2A.$$

Notice, that T_1, N_1' are right coprime and T_2, N_2' are left coprime. If we can show that there always exist an A over $H_S(\Omega)$ such that N_i' and Y_i' have elements in $R_\Omega(s)$ then the proof is complete. We consider the matrix $N_1T_1^{-1}$. Evidently there are a finite number of (distinct) locations $p_1,\ldots,p_\ell$ in Ω whre $N_1T_1^{-1}$ is not holomorphic. Let L_i denote the proper part of the Laurent expansion of $N_1T_1^{-1}$ at p_i and write

$$A = \sum_{i=1}^{\ell} L_i - N_1T_1^{-1}.$$

By contruction, A has elements in $H_S(\Omega)$. Thus $N_1' = N_1 + AT_1$, has no poles in Ω. However

$$N_1' = \sum_{i=1}^{\ell} L_iT_1$$

and thus has elements in $R_\Omega(s)$. If we multiply (3a) on the right by T_1^{-1} we conclude that N_2' is rational. Similarly from (3b) and (3c) we deduce that Y_i' are rational. □

Example

The following two system matrices ([8], p. 107) are strictly system equivalent (over the polynomials)

$$\begin{bmatrix} (s+2)^2 & -1 \\ s+1 & 0 \end{bmatrix} \qquad \begin{bmatrix} (s+2)^2 & s+1 \\ -1 & 0 \end{bmatrix}$$

That is we can find polynomials m(s) and n(s) such that

$$\begin{bmatrix} m(s) & 0 \\ n(s) & 1 \end{bmatrix} \begin{bmatrix} (s+2)^2 & -1 \\ s+1 & 0 \end{bmatrix} = \begin{bmatrix} (s+2)^2 & s+1 \\ -1 & 0 \end{bmatrix} \begin{bmatrix} m(s) & n(s) \\ 0 & 1 \end{bmatrix} \quad (4)$$

with m(s) having no zeros at -2 (e.g. m(s) = -(s+1), n(s) = 0). If we try to make m(s) unimodular (i.e. a constant) the equation

$$0 = m(s) + (s+1) + n(s)(s+2)^2$$

implies n = m = 0 and so (4) cannot be satisfied. This is a case where to apply the Rosenbrock criterion for equivalence (theorem 4) the system matrices must first be extended.

On the other hand (4) can be satisfied with m(s) unimodular if we allow holomorphic choices of m(s) and n(s), e.g.

$$m(s) = -e^{-(2+s)}$$
$$n(s) = \frac{-s-1-e^{-(2+s)}}{(s+2)^2}.$$

It is not always true that allowing the transforming matrices to be in $H_S(\Omega)$ removes the need for extension of the system matrices (although this is the case over $H(\Omega)$). This can be seen if the (1,1) element in the above example is replaced by s(s+2). (4) then implies that m(0) = -1 and m(-2) = 1 which is impossible over $H_S(\Omega)$.

DECOUPLING ZEROS

Let **R** = $H_S(\Omega)$. Let {T,U,V,W} be a realisation of G and let $[T\ -U]M = [B\ 0]$ where M is unimodular. B is a gcld of T and U and is unique up to postmultiplication with a unimodular matrix. We define the <u>input decoupling zeros</u> to be the complete set of roots of detB = 0 - $\{\beta_1,\beta_2,\ldots\}$ - where the entries are repeated according to the multiplicity of the zeros. Note that there may be infinitely many. The <u>output decoupling zeros</u> - $\{\gamma_1,\gamma_2,\ldots\}$ - are similarly defined. Now let $T = BT_1$ and $U = BU_1$. Then the realisation $\{T_1,U_1,V,W\}$ of G has no input decouping zeros. Let $\{T_1,U_1,V,W\}$ have output decoupling zeros $\{\theta_1,\theta_2,\ldots\}$ - note that these are independent of the choice of B. Since

$$\begin{bmatrix} T \\ V \end{bmatrix} = \begin{bmatrix} B & 0 \\ 0 & I \end{bmatrix} \begin{bmatrix} T_1 \\ V \end{bmatrix}$$

every rxr minor of $[T_1^t\ \ V^t]$ divides the corresponding rxr minor of $[T^t\ \ V^t]$ and so $\{\theta_i\} \subseteq \{\gamma_i\}$. We define the <u>input-output decoupling zeros</u> - $\{\delta_1,\delta_2,\ldots\}$ - of {T,U,V,W} by

$$\{\delta_i\} = \{\gamma_i\} - \{\theta_i\}.$$

By definition $\{\delta_i\} \subseteq \{\gamma_i\}$. Clearly every rxr minor of $[T^t\ \ V^t]$

divides the corresponding rxr minor of $[T_1^t \; V^t]$ times detB and so $\{\theta_i, \beta_i\} \supseteq \{\gamma_i\}$. Hence $\{\delta_i\} \subseteq \{\beta_i\}$.

Now let C be a gcrd of T_1 and V:

$$T_1 = T_2 C,$$
$$V = V_2 C.$$

Then the realisation $\{T_2, U_1, V_2, W\}$ of G has no output decoupling zeros. It is easily seen that T_2, U_1 are left coprime so the realisation also possesses no input decoupling zeros and is thus irreducible. If $\{\eta_i\}$ denotes the set of zeros of detT(s) and $\{\alpha_i\}$ the set of zeros of $\det T_2(s)$ then since $T = BT_2C$ we have

$$\{\eta_i\} = \{\alpha_i, \beta_i, \theta_i\} = \{\alpha_i, \beta_i \gamma_i\} - \{\delta_i\}. \tag{5}$$

It is of course possible to extract the input and output decoupling zeros from {T,U,V,W} to reduce to a minimal realisation in a variety of ways. For example, we could first remove all the output decoupling zeros and then all the remaining input decoupling zeros. More generally still we could extract alternately input and output decoupling zeros, removing them all in a finite number of steps. We can now see that the input-output decoupling zeros so defined will always be given by $\{\delta_i\}$. This follows from theorem 6 (which shows that $\{\alpha_i\}$ is well-defined - independent of the particular irreducible realisation) and (5).

INTERCONNECTED SYSTEMS

If two systems G and H are connected in series (or parallel), a system matrix can be written down for the combined system in a natural way. Rosenbrock's proofs ([8] p. 126, 131) that the resulting system matrices are equivalent for different (but equivalent) representations of the individual systems carry over easily to the case of system matrices over a Bezout domain.

A composite feedback configuration of particular interest in control is shown below, where H represents a compensator and G

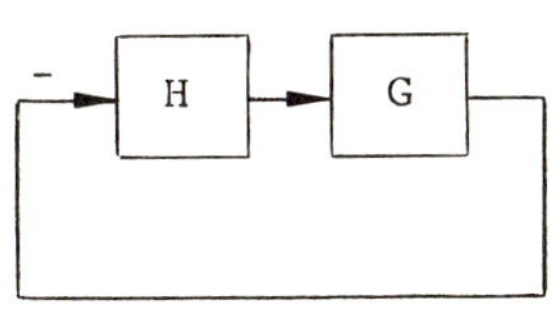

the plant. The following theorem generalises a result of Anderson and Gevers ([12], Theorem 5). Although valid for any **R**, the result has its most meaningful interpretation when **R** is a characterisation of the "desired" or "stable" transfer-functions e.g. proper with poles in a certain region. The result can be viewed as connecting algebraically the notions of internal stability in [1] and [8].

Theorem 8

The following are equivalent:

(1) $GH(I + GH)^{-1}$ has elements in **R** and any open loop system matrix (with G and H having irreducible realisations) is irreducible,

(2) The matrix

$$S = \begin{bmatrix} (I + GH)^{-1} & -G(I + HG)^{-1} \\ H(I + GH)^{-1} & (I + HG)^{-1} \end{bmatrix}$$

has elements in **R**. □

The proof of this theorem is facilitated by the following lemma (cf. [13] corollary to Theorem 2).

Lemma 1
Let $A \in \mathbf{R}^{c\times r}$, $B \in \mathbf{R}^{r\times r}$ and $C \in \mathbf{R}^{r\times d}$ with $\det B \neq 0$. The following conditions are equivalent:

(a) $AB^{-1}C \in \mathbf{R}^{c\times d}$, A and B right coprime, B and C left coprime,

(b) $B^{-1} \in \mathbf{R}^{r\times r}$.

Proof Assume (a) holds. Clearly $AB^{-1}[B\ C] \in \mathbf{R}^{c\times(d+r)}$. Using the fact that B and C are left coprime it follows that $AB^{-1} = T \in \mathbf{R}^{c\times r}$. Hence

$$\begin{bmatrix} B \\ A \end{bmatrix} = \begin{bmatrix} I \\ T \end{bmatrix} B.$$

Since A and B are right coprime B is unimodular hence (b). If (b) holds then (a) follows easily. □

Proof of theorem 8. Any two open-loop system matrices (i.e. realisations of the series connection GH) in which G and H have irreducible realisations are sse. Thus (1) is self-consistent and we may take irreducible representations $G=ND^{-1}$ and $H=Q^{-1}P$. An open-loop system matrix is as follows

$$\begin{bmatrix} 0 & Q & -P \\ D & -I & 0 \\ N & 0 & 0 \end{bmatrix}$$

which is irreducible if and only if N,QD are right coprime P,QD are left coprime. Now $GH(I+GH)^{-1} = G(I+HG)^{-1}H = N(QD+NP)^{-1}P$. Thus (1) is equivalent to conditions (a) of lemma 1 with A = N, B = QD+NP and C = P. Thus (1) is equivalent to $(QD+NP)^{-1} \in \mathbf{R}^{m\times m}$. Now let $\hat{S}$ be the matrix obtained from S by replacing the (1,1) block with $I-(I+GH)^{-1}$ and by negating the (1,2) block. Clearly (2) holds if and only if $\hat{S}$ has all elements in **R**. Now

$$\hat{S} = \begin{bmatrix} N \\ D \end{bmatrix} (QD+PN)^{-1}[P \quad Q],$$

hence (2) is equivalent to condition (a) of lemma 1 with $A = [N^t \; D^t]^t$, $B = (QD+NP)^{-1}$ and $C = [P \; Q]$. Thus (2) holds if and only if $(QD+NP)^{-1} \in \mathbf{R}^{m\times m}$, which establishes the equivalence of (1) and (2).

CONCLUDING REMARKS

In the preceeding sections we have seen how many of the basic ideas and results of Rosenbrock's system theory generalise to matrices over Bezout domains. For example, the theory carries over to the case of holomorphic system matrices.

As is remarked in [4] it is by no means the case that all important rings in system theory are Bezout domains. Two rings which are not Bezout are H^∞ and $k[z_1,z_2,\ldots,z_n]$ where k is a field and n>2. In each case corollary 2 no longer holds.

In general, for **R** an integral domain, the following implications do hold:

$$(C1a)\Leftrightarrow(C1b)\Rightarrow(C2a)\Leftrightarrow(C2b)\Rightarrow(C3)\Rightarrow(C4)\Rightarrow(C5).$$

If **R** is a GCCD then additionally (C4)⇒(C3).

Since $k[z_1,z_2,\ldots,z_n]$ is a UFD, and hence a GCDD, (C4)⇒(C3). The Quillen-Suslin theorem (formerly known as the Serre conjecture) establishes that (C2)⇒(C1) - for an illuminating tutorial account of this theory see [14]. (C3)⇏(C2) for n > 2. For n=2, (C5)⇒(C4), however the implication fails for n > 3 [13].

In general it is clear that we have several possible definitions of coprimeness at our disposal. In the subsequent discussion we will say: X and Y are (C_i) left coprime if A = [X Y] satisfies condition (C_i) of corollary 1 ((C_j) right coprimeness is similarly defined) and {T,U,V,W} is (i,j) irreducible if T,U are (C_i) left coprime and T,V are (C_j) right coprime.

The maps $T^{-1}U$ and V defined in the third section are in general onto and one-to-one (respectively) if and only if {T,U,V,W} is (2,3) irreducible. For any ring containing a pair of coprime elements which generate a non-principal ideal (this is the case for H^∞ and $k[z_1,\ldots,z_n]$) (C2) and (C3) coprimeness are not equivalent, and a surprising lack of symmetry appears in the above characterisation of irreducibility. This situation is apparent in previous work, for example [15]. Here $\mathbf{R} = R[s,\lambda]$ and the class of "state-space" realisations $\{sI-A(\lambda),B(\lambda),C(\lambda),0\}$, where $A(\lambda),B(\lambda)$ and $C(\lambda)$ are polynomial matrices, is considered. The realisation is said to be controllable if the module $\mathbf{B} + A\mathbf{B} + \ldots + A^{n-1}\mathbf{B}$ is equal to $R[\lambda]^m$ (where $\mathbf{B} = \mathrm{Im}B(\lambda)$), observable if $\bigcap_{i=0}^{n-1}$ kernel CA^i is the zero module and canonical if it is both controllable and observable. Clearly the realisation is controllable if and only if there exists $Y_i(\lambda)$ such that $BY_o + ABY_1 + \ldots + A^{n-1}BY_{n-1} = I$, which is true if and only if $sI - A(\lambda),B(\lambda)$ are (C2) left coprime (see [6] theorem 14). Now the realisation is observable if and only if

$$\begin{bmatrix} C \\ CA \\ \cdot \\ \cdot \\ \cdot \\ CA^{n-1} \end{bmatrix} v = 0$$

for $v \in R[\lambda]^n$ implies $v=0$. This condition holds if and only if the observability matrix has full rank over $R(\lambda)$, which is equivalent to the existence of rational matrices $X_i(\lambda)$ such that $X_oC + X_1CA + \ldots + X_{n-1}CA^{n-1} = I$. Thus the realisation is observable if and only if there are matrices $P(s,\lambda)$, $Q(s,\lambda)$ which are polynomial in s and λ, such that

$$P(s,\lambda)(sI-A(\lambda)) + Q(s,\lambda)C(\lambda) = t(\lambda)I \qquad (6)$$

where $t(\lambda)$ is a polynomial in λ. Evidently (6) holds if and only if the gcd of the pxp minors of $[sI - A(\lambda)^t \; C(\lambda)^t]$ divides $t(\lambda)$. (The "if" part here is not immediate. A procedure for constructing a P and Q is given in [13], theorem 2). Note however that $\det(sI - A(\lambda))$ is monic in s and thus the pxp minors of $[sI - A(\lambda)^t \; C(\lambda)^t]$ must be coprime. We conclude that $\{sI-A(\lambda),B(\lambda),C(\lambda),0\}$ is observable if and only if $sI-A(\lambda),C(\lambda)$ are (C3) right coprime. We have thus proved the following:

<u>Proposition 1</u>

Let $\mathbf{R} = R[s,\lambda]$. The realisation $\{sI-A(\lambda),B(\lambda),C(\lambda),0\}$ is canonical if and only if it is (2,3) irreducible. □

This same type of asymmetry seems to appear in the characterisation of strict system equivalence. Two realisations are sse in the sense of Fuhrmann if their state spaces are isomorphic and the diagram of fig.1 commutes. In matrix terms, there must exist N_i and Y_i such that the identity (1) holds with T_2, N_2 (C2) left coprime and T_1, N_1 (C3) right coprime. In this case we will say that the two realisations are (2,3) sse. Although there is an apparent lack of symmetry, it is in fact possible to show the following:

<u>Proposition 2</u>

For any **R**, if two realisations are (2,3) sse they are (1,1) sse.

<u>Proof</u> By definition there are matrices P_1, Q_1 over **R** and P_2, Q_2 over **F** such that

$$\begin{bmatrix} T_2 & -N_2 \\ P_2 & Q_2 \end{bmatrix} \begin{bmatrix} P_1 & T_1 \\ -Q_1 & N_1 \end{bmatrix} = \begin{bmatrix} I & 0 \\ ? & I \end{bmatrix}$$

and thus

$$\begin{bmatrix} T_2 & -N_2 \\ P_3 & Q_3 \end{bmatrix} \begin{bmatrix} P_1 & T_1 \\ -Q_1 & N_1 \end{bmatrix} = \begin{bmatrix} I & 0 \\ 0 & I \end{bmatrix} \tag{7}$$

for some matrices P_3, Q_3 over **F**. The matrices on the l.h.s of (7) are square, and thus

$$\begin{bmatrix} P_1 & -T_1 \\ -Q_1 & N_1 \end{bmatrix} \begin{bmatrix} T_2 & -N_2 \\ P_3 & Q_3 \end{bmatrix} = \begin{bmatrix} I & 0 \\ 0 & I \end{bmatrix}.$$

It follows that

$$\begin{bmatrix} T_1 \\ N_1 \end{bmatrix} \begin{bmatrix} P_3 & Q_3 \end{bmatrix}$$

has elements in **R**. Thus P_3, Q_3 have elements in **R** and the matrices in (7) are both unimodular. □

ACKNOWLEDGEMENTS

This work was supported by the Royal Society (European Science Exchange Programme) and by the Alexander von Humboldt-Stiftung.

REFERENCES

[1] Desoer, C.A., Lin, R.W., Murray, J. and Saeks, R., 1980, Feedback system design: the fractional representation approach to analysis and synthesis, IEEE Trans. AC, 25 (1980) 399-412.

[2] MacDuffee, C.C., The theory of matrices (Springer-Verlag, Berlin, 1933).

[3] Kaplansky, I., Elementary divisors and modules, Trans. AMS 66, (1949) 464-491.

[4] Vidyasagar, M., Schneider, H. and Francis, B.A., Algebraic and topological aspects of feedback stabilization, IEEE Trans. AC, 27 (1982) 880-894.

[5] Kaplansky, I., Commutative rings (University of Chicago Press, Chicago, 1974).

[6] Coppel, W.A., Matrices of rational functions, Bull. Austral. Math. Soc., 11 (1974) 89-113.

[7] Coppel, W.A., Some remarks on strict system equivalence, International Symposium on Mathematical Theory of Networks and Systems, Western Periodical Company, Santa Monica, 4 (1981) 33-34.

[8] Rosenbrock, H.H., State space and multivariable theory (Nelson, London, 1970).

[9] Fuhrmann, P.A., On strict system equivalence and similarity, Int. J. Control, 25 (1977) 5-10.

[10] Levy, B., Kung, S.Y., Morf, M. and Kailath, T., A unification of system equivalence definitions, Proc. IEEE Conference on Decision and Control (New Orleans) (1977).

[11] Kailath, T., Linear Systems (Prentice-Hall, Englewood Cliffs, 1980).

[12] Anderson, B.D.O. and Gevers, M.R., On multivariable pole-zero cancellations and the stability of feedback systems, IEEE Trans. CAS, 28 (1981) 830-833.

[13] Youla, D.C. and Gnavi, G., Notes on n-dimensional system theory, IEEE Trans. CAS, 26 (1979) 105-111.

[14] Youla, D.C. and Pickel, P.F., The Quillen-Suslin theorem and the structure of n-dimensional elementary polynomial matrices, IEEE Trans. CAS, 31 (1984) 513-518.

[15] Morse, A.S., Ring Models for Delay-Differential Systems, Automatica, 12 (1976) 529-531.

[16] Larsen, M.D., Lewis, W.J. and Shores, T.S., Elementary divisor rings and finitely presented modules, Trans. Amer. Math. Soc. 187 (1974) 231-248.

Frequency Domain and State Space Methods for Linear Systems
C.I. Byrnes and A. Lindquist (Editors)
Elsevier Science Publishers B.V. (North-Holland), 1986

REALIZATION THEORY AND MATRIX FRACTION REPRESENTATION FOR LINEAR SYSTEMS OVER COMMUTATIVE RINGS

Li Tiejun

Department of Computer and System Science
Nankai University
Tianjin
CHINA

In this paper the realization theory and matrix fraction representations for linear systems over commutative rings are discussed over generally commutative rings. A new concept--the cyclicity of linear input and output maps(LIOM) is introduced. By means of this concept, a necessary and sufficient condition of realizability of a LIOM is obtained. Another main result in this paper is that we obtain the matrix fraction representations of transfer function matrices for LIOM.

1. INTRODUCTION

The realization theory and matrix fraction representations for linear systems over commutative rings have been studied by some authors [1-3]. However, for matrix fraction representations, [1] studied the problem only over the principal ideal domains and [2] restricted the problem on the finite generated projective state modules(referred to as projection realization). In this paper, these problems are discussed over the generally commutative rings.

Throughout this paper R will denote a commutative ring with identity. R^n denotes the finite free R-module having the natural basis $e_1, e_2, \cdots, e_n$, where $e_i = (0,\cdots,0,1,0\cdots,0)^T$ with the 1 in the i-th position and T denotes the transpose. By a discrete-time linear system over R we mean the following equations :

$$\Sigma: \begin{array}{l} x(t+1) = Fx(t) + Gu(t) \\ y(t) = Hx(t) \end{array} \tag{1.1}$$

where $t \in Z$(the set of integers), the state $x \in X$, a finite generated R-module, the inputs $u \in R^m$ and the output $y \in R^p$. We denote the system (1.1) by $\Sigma=(X,F,G,H)$, for short.

The input space and the output space of (1.1) are defined as follows, respectively

$$\Omega = \{u: Z \to R^m \mid u(t)=0, \ \forall t>0 \text{ and } t<T\leq 0\} \tag{1.2}$$

where the integer T possibly dependent on u, and

$$\Gamma = \{\gamma: Z \to R^p \mid \gamma(t)=0, \ \forall t \leq 0\} \tag{1.3}$$

Let $R^m[z]$ denote the set of polynomials in z with the coefficients in R^m. It is clear that $R^m[z]$ is a finite free $R[z]$-module having the basis $e_1, e_2, \cdots, e_m$. Let $z^{-1}R^p[[z^{-1}]]$ denote the set of formal power seriers in z^{-1} with the coefficients in R^p and vanishing constant term, i.e.,

$$z^{-1}R^p[[z^{-1}]] = \left\{ \sum_{t=1}^{\infty} u_t z^{-t} \,\middle|\, u_t \in R^p \right\} \tag{1.4}$$

We define scalar multiplication o of the elements of $R[z]$ and $z^{-1}R^p[[z^{-1}]]$ as follows: for $a(z) = \sum_{i=0}^{k} a_i z^i \in R[z]$ and $\gamma = \sum_{t=1}^{\infty} u_t z^{-t} \in z^{-1}R^p[[z^{-1}]]$,

$$a(z)\circ\gamma=\sum_{t=1}^{\infty}\sum_{i=0}^{k}a_i u_{t+i}z^{-t}. \tag{1.5}$$

Then $z^{-1}R^p[[z^{-1}]]$ is a $R[z]$-module. We denote also the element $\sum_{t=1}^{\infty}u_t z^{-t}$ by $(u_1, u_2, \cdots)$. It is clear thet $\Omega\cong R^m[z]$ and $\Gamma\cong z^{-1}R^p[[z^{-1}]]$, where $\cong$ denotes the R-module isomorphism.

It is not difficult to check that the input-output map f_Σ of the system (1.1) is a $R[z]$-module homomorphism from Ω (or $R^m[z]$) into Γ (or $z^{-1}R^p[[z^{-1}]]$).

Definition 1.1. A map $f: R^m[z]\to z^{-1}R^p[[z^{-1}]]$ is called a linear input-output map (LIOM) if f is a $R[z]$-module homomorphism, i.e., for any $a(z)\in R[z]$ and $u\in R^m[z]$ we have

$$f(a(z)u)=a(z)\circ f(u). \tag{1.6}$$

It is clear that f is determined completely by $f(e_k)$, $k=1,2,\cdots,m$. From (1.5) it is not difficult to show that

$$f(z^i e_k)=z^i\circ f(e_k)=(f(e_k)_{i+1}, f(e_k)_{i+2}, \cdots) \tag{1.7}$$

The matrix

$$Z_f=(f(e_1), f(e_2), \cdots, f(e_m)) \tag{1.8}$$

is called the transfer function matrix of f.

This paper is organized as follows. In section 2 we first introduce an important concept for LIOM--cyclicity. By means of this concept, a necessary and sufficient condition of the realizability of a LIOM is obtained. Some results about the $R[z]$-submodule $\Delta=QR^m[z]$ of $R^m[z]$ are contained in the section 3 and the main result is to show that for a realizable LIOM there exists a reachable realization (X,g,h), where $X=R^m[z]/\Delta$, $\Delta=QR^m[z]$ with Q is a regular polynomial matrix over $R[z]$. These results will be applied in section 4. In section 4, we obtain the matrix fraction representation of transfer function matrices for LIOM.

2. THE CYCLICITY AND REALIZABILITY OF LIOM

Definition 2.1. A LIOM $f: R^m[z]\to z^{-1}R^p[[z^{-1}]]$ is called realizable if there exist a $R[z]$-module X which is a finite generated R-module and two $R[z]$-module homomorphism g and h such that the following diagram:

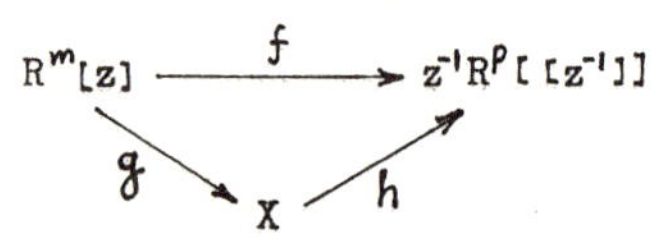

(2.1)

is commutative.

(X,g,h) is called a realization of f. (X,g,h) is called reachable if g is surjective, observable if h is injective and canonical if it is both reachable and observable. We can construct a discrete-time linear system $\Sigma=(X,F,G,H)$ over R by (X,g,h) such that the input-output map f_Σ of Σ equals f.

Definition 2.2. An infinite sequence $\gamma=(\gamma_1, \gamma_2, \cdots)$ in R^p is called cyclic if there exists a polynomial with the leading coefficient being identity in R $a(z)=z^s+a_{s-1}z^{s-1}+\cdots+a_1z+a_0\in R[z]$ such that

$$\gamma_{s+i}+a_{s-1}\gamma_{s+i-1}+\cdots+a_1\gamma_{i+1}+a_0\gamma_i=0, \quad i=1,2,\cdots.$$

We call the $a(z)$ cyclic polynomial of γ.

Now we introduce an important concept for LIOM--cyclicity as follows.
Definition 2.3. A LIOM $f: R^m[z]\to z^{-1}R^p[[z^{-1}]]$ is called cyclic if each $f(e_k)$ is cyclic, $k=1,2,\cdots,m$.

Lemma 2.1. A polynomial $a(z)\in R[z]$ with the leading coefficient being identity in R is the cyclic polynomial of $f(e_k)$ if and only if $a(z)e_k\in\ker(f)$.
Proof: Let $a(z)=z^s+a_{s-1}z^{s-1}+\cdots+a_1z+a_0$. From (1.7) we have

$$f(a(z)e_k)=a(z)\circ f(e_k)=z^s\circ f(e_k)+a_{s-1}z^{s-1}\circ f(e_k)+\cdots+a_1 z\circ f(e_k)+a_0\circ f(e_k)$$
$$=(f(e_k)_{s+1}, f(e_k)_{s+2},\cdots)+a_{s-1}(f(e_k)_s, f(e_k)_{s+1},\cdots)+$$
$$+a_1(f(e_k)_2, f(e_k)_3,\cdots)+a_0(f(e_k)_1, f(e_k)_2,\cdots).$$

By this equality the lemma can be proved easily.

From this lemma we can prove a main result in this paper as follows.

Theorem 2.1. A LIOM f is realizable if and only if f is cyclic.
Proof: Assume that f is cyclic. Let $a_k(z)=z^{s_k}+a^{(k)}_{s_k-1}z^{s_k-1}+\cdots+a^{(k)}_1 z+a^{(k)}_0$ be the cyclic polynomial of $f(e_k)$, $k=1,2,\cdots,m$. We construct a $R[z]$-submodule Δ of $R^m[z]$ as follows:

$$\Delta=\left\{\sum_{k=1}^{m} b_k(z)a_k(z)e_k \;\middle|\; b_k(z)\in R[z],\ k=1,2,\cdots,m\right\} \tag{2.2}$$

From lemma 2.1., $\Delta\subset\ker(f)$. We can show that the quotient $R[z]$-module $X=R^m[z]/\Delta$ is a finite generated R-module [4]. Now we define two $R[z]$-module homomorphism g and h as follows:

$$g:\ R^m[z]\longrightarrow X\ :\ u\longmapsto u+\Delta=\bar{u} \tag{2.3}$$

$$h:\ X\longrightarrow z^{-1}R^p[[z^{-1}]]\ :\ \bar{u}=u+\Delta\longmapsto f(u) \tag{2.4}$$

It is not difficult to prove that (X,g,h) is a realization of f. Conversely, if f is realizable. Let (X,g,h) is a realization of f. We define R-module homomorphisms F,G and H as follows:

$$F:\ X\longrightarrow X\ :\ x\longmapsto zx \tag{2.5}$$

$$G:\ R^m\longrightarrow X\ :\ e_k\longmapsto ge_k,\ k=1,2,\cdots,m \tag{2.6}$$

$$H:\ X\longrightarrow R^p:\ x\longmapsto (hx)_1 \tag{2.7}$$

where $(hx)_1$ denotes the first component of hx. Then we can prove that the characteristic polynomial $\det(zI-F)\in R[z]$ of the matrix F is the cyclic polynomial of each $f(e_k)$, $k=1,2,\cdots,m$. Therefore f is cyclic.

Corollary. If a LIOM f is realizable, then there exist reachable and canonical realizations of f. The realization (X,g,h) is reachable then $X=R^m[z]/\ker(g)$ and canonical if and only if $X=R^m[z]/\ker(f)$.

3. THE STATE MODULE OF A REALIZABLE LIOM

Let Q be an mxm polynomial matrix over $R[z]$, then it is clear that $QR^m[z]$ is a $R[z]$-submodule of $R^m[z]$. If R is a field, Fuhrmann had shown that each $R[z]$-submodule of $R^m[z]$ must be the form as $QR^m[z]$. This important fact is the fundamental tool of the polynomial model approach to linear systems over a field which was developed principally by Fuhrmann [5,6]. However, this fact does not hold for the commutative rings. Even so, we remain to be able to establish the relation in between the realizability and matrix fraction representation of transfer function matrix of LIOM. The purpose of this section is to study the construction of the state module of a realizable LIOM which we will need in the section4 for obtaining the matrix fraction representations of the transfer function matrix.

Definition 3.1. An mxm polynomial matrix Q is called regular if its determinant detQ is a polynomial with the leading coefficient being identity in R.

Lemma 3.1. Let Q_1 and Q_2 be mxm polynomial matrices and $\Delta_i=Q_iR^m[z]$, $i=1,2$. Then we have:
(1) $\Delta_1\subset\Delta_2$ if and only if there exists an mxm polynomial matrix S such that $Q_1=Q_2S$.
(2) $\Delta_1=\Delta_2$ if and only if there exists an mxm unimodular matrix U (i.e., detU is an unit in R) such that $Q_1=Q_2U$.
Proof: (1). If $Q_1=Q_2S$. then $Q_1R^m[z]=Q_2SR^m[z]$. Since $SR^m[z]\subset R^m[z]$, we have $\Delta_1\subset\Delta_2$. Conversely, if $\Delta_1\subset\Delta_2$, i.e., $Q_1R^m[z]\subset Q_2R^m[z]$, then $Q_1e_i=Q_2s_i$, $s_i\in R^m[z]$,

$i=1,2,\cdots,m$. We have therefore $Q_1=Q_1I=Q_2S$, where $S=(s_1,s_2,\cdots,s_m)$.
(2). If $Q_1=Q_2U$, by (1) we have $\Delta_1\subset\Delta_2$. Since U is unimodular, U has an inverse U^{-1} over $R[z]$. Thus we have $Q_2=Q_1U^{-1}$, by (1) therefore $\Delta_2\subset\Delta_1$, hance $\Delta_1=\Delta_2$. Conversely, if $\Delta_1=\Delta_2$, by (1) there exist two polynomial matrices S and T over R such that $Q_1=Q_2S$ and $Q_2=Q_1T$. Therefore we have $Q_1=Q_1TS$ and $Q_2=Q_2ST$. Hance we obtain $\det Q_1=\det Q_1\det(TS)$, thus $\det(TS)=1$, therefore we can assert that T and S are unimodular.

Lemma 3.2. Let Q be mxm polynomial matrix, then we have $(\det Q)R^m[z]\subset QR^m[z]$.
Proof: We notice that in the commutative rings equality $(\det Q)I=Q\operatorname{adj}Q$ remains to be true. From the fact and lemma 3.1 we obtain $(\det Q)R^m[z]\subset QR^m[z]$.

Theorem 3.1. If Q is an mxm regular polynomial matrix and $\Delta=QR^m[z]$, then the quotient $R[z]$-module $X=R^m[z]/\Delta$ is a finite generated R-module.[4].

Theorem 3.2. Assume that a LIOM f is realizable, then there exists a regular polynomial matrix Q such that $\Delta=QR^m[z]\subset\ker(f)$.
Proof: Since f realizable, f is cyclic. Let $a_k(z)$ be the cyclic polynomial of $f(e_k)$, $k=1,2,\cdots,m$. Let Δ be as (2.2), then $\Delta\subset\ker(f)$ and $\Delta=QR^m[z]$, where Q is as follows:

$$Q=\begin{pmatrix} a_1(z) & & & \\ & a_2(z) & & \\ & & \ddots & \\ & & & a_m(z)\end{pmatrix}.$$

Then Q is a regular polynomial matrix.

Summarize to above two theorems we obtain
Theorem 3.3. Assume that a LIOM f is realizable, then there exists a realization (X,g,h) of f, where $X=R^m[z]/\Delta$, $\Delta=QR^m[z]$ and Q is a regular polynomial matrix.

4. MATRIX FRACTION REPRESENTATIONS OF TRANSFER FUNCTION MATRICES OF LIOM

Let $R^p((z^{-1}))$ be the set of Laurent series in z^{-1} with the coefficients in R^p, i.e., $R^p((z^{-1}))=\{\sum_{t=t_0}^{\infty}u_tz^{-t}\,|\,u_t\in R^p\}$. $R^p((z^{-1}))$ is a free $R((z^{-1}))$-module having the basis $e_1,e_2,\dots,e_p$. It is not difficult to show that $R^p((z^{-1}))/R^p[z]\cong z^{-1}R^p[[z^{-1}]]$, where $\cong$ denotes the $R[z]$-module isomorphism. Therefore there exists a cononical $R[z]$-module homomorphism from $R^p((z^{-1}))$ onto $z^{-1}R^p[[z^{-1}]]$, we denote it by π. Assume that f : $R^m[z]\longrightarrow z^{-1}R^p[[z^{-1}]]$ is a LIOM. We now introduce a map as follows:

$$\bar{f}:R^m((z^{-1}))\longrightarrow R^p((z^{-1}))\ :\ \bar{f}(u)=\sum_{t\in Z}f(s(z^tu))_1z^{-(t+1)} \qquad (4.1)$$

where s is a map from $R^m((z^{-1}))$ into $R^m[z]$ defined by

$$s(\sum_{t=t_0}^{\infty}u_tz^{-t})=\sum_{t=t_0}^{0}u_tz^{-t}. \qquad (4.2)$$

The map $\bar{f}$ is called the extended map of f. It is easy to check that $\bar{f}$ is a $R((z^{-1}))$-module homomorphism.

Lemma 4.1. Let f and $\bar{f}$ is as above, then we have the following commutative diagram

$$\begin{array}{ccc} R^m((z^{-1})) & \xrightarrow{\bar{f}} & R^p((z^{-1})) \\ i\uparrow & & \downarrow\pi \\ R^m[z] & \xrightarrow[f]{} & z^{-1}R^p[[z^{-1}]] \end{array} \qquad (4.3)$$

i.e.,

$$f=\pi\cdot\bar{f}\cdot i \qquad (4.4)$$

where i is the identity map.
Proof: For any $u\in R^m[z]$, it is clear that $i(u)=u$. Since $z^tu\in R^m[z]$, $s(z^tu)=z^tu$. By the definition of scalar multiplication o, we have

$$(z^t\circ f(u))_1=\begin{cases} f(u)_{t+1}, & t\geqslant 0 \\ 0, & t<0 \end{cases},$$

where $f(u)_{t+1}$ denotes the coefficient of $z^{-(t+1)}$ in $f(u)$. By these facts we have

$$\begin{aligned}\pi\cdot\bar f\cdot i(u) = \pi\cdot\bar f(u) &= \pi\Big(\sum_{t\in Z} f(s(z^t u))_1\, z^{-(t+1)}\Big)\\ &= \pi\Big(\sum_{t\in Z}(z^t u)_1\, z^{-(t+1)}\Big)\\ &= \pi\Big(\sum_{t\in Z}(z^t \circ f(u))_1\, z^{-(t+1)}\Big)\\ &= \pi\Big(\sum_{t=0}^{\infty} f(u)_{t+1} z^{-(t+1)}\Big)\\ &= \pi\Big(\sum_{t=1}^{\infty} f(u)_t z^{-t}\Big) = \pi f(u) = f(u).\end{aligned}$$

Lemma 4.2. $f(e_k) = \bar f(e_k)$, $k=1,2,\cdots,m$.
Proof: From lemma 4.1, it is not difficult to show this equality.

Let Q be an mxm polynomial matrix, we can regard Q as a map from $R^m[z]$ into itself. Moreover we can also regard Q as a map from $R^m((z^{-1}))$ into itself and we denote it by $\bar Q$. Then we have the following commutative diagram:

$$\begin{array}{ccc} R^m((z^{-1})) & \xrightarrow{\ \bar Q\ } & R^m((z^{-1})) \\ i\uparrow & & \uparrow i \\ R^m[z] & \xrightarrow[\ Q\ ]{} & R^m[z] \end{array} \tag{4.5}$$

We now use again Q to define a map P as follows:

$$P:\ R^m((z^{-1})) \longrightarrow R^p((z^{-1})):\quad Pu = \bar f\,\bar Q u \tag{4.6}$$

i.e., the following diagram is commutative:

$$\overbrace{R^m((z^{-1})) \xrightarrow{\ Q\ } R^m((z^{-1})) \xrightarrow{\ \bar f\ } R^p((z^{-1}))}^{P} \tag{4.7}$$

Moreover we have
Lemma 4.3. The following diagram is commutative:

$$\begin{array}{ccccc} & & \overset{P}{\frown} & & \\ R^m((z^{-1})) & \xrightarrow{\ \bar Q\ } & R^m((z^{-1})) & \xrightarrow{\ \bar f\ } & R^p((z^{-1})) \\ i\uparrow & & \uparrow i & & \downarrow \pi \\ R^m[z] & \xrightarrow[\ Q\ ]{} & R^m[z] & \xrightarrow[\ f\ ]{} & z^{-1}R^p[[z^{-1}]] \\ & & \underset{\pi P}{\smile} & & \end{array} \tag{4.8}$$

Because $R^m((z^{-1}))$ and $R^p((z^{-1}))$ are all finite free $R((z^{-1}))$-modules, thus the map P can be represented by an pxm matrix with the entries in $R((z^{-1}))$, i.e.,

$$P = \sum_{t=t_0}^{\infty} P_t z^{-t} \tag{4.9}$$

where P_t are all pxm matrices over R.

Assume that a LIOM $f: R^m[z] \longrightarrow z^{-1}R^p[[z^{-1}]]$ is realizable, then f has a reachable realization (X,g,h), where $X=R^m[z]/\Delta$, $\Delta = QR^m[z]$ and Q is a regular polynomial matrix. For the matrix Q, we then can define $\bar Q$ and P according to (4.5) and (4.6).

We first have the following important result.

Theorem 4.1. The map P maps submodule $R^m[z]$ of $R^m((z^{-1}))$ into submodule $R^p[z]$ of $R^p((z^{-1}))$.
Proof: For any $u \in R^m[z]$, from theorem 3.2, we have $Qu \in \ker(f)$, i.e., $f(Qu)=0$. Then we have

$$\pi Pu=\pi\bar{f}(\bar{Q}u)=\pi\bar{f}(\bar{Q}iu)=\pi\bar{f}i(Qu)=f(Qu)=0.$$

This implies that $Pu \in R^p[z]$, $\forall u \in R^m[z]$.

Therefore P is a polynomial matrix, so (4.9) becomes

$$P=\sum_{t=t_0}^{0} P_t z^{-t} \tag{4.10}$$

Now we can prove the other main result of this paper.

Theorem 4.2. A LIOM f is realizable if and only if its transfer function matrix Z_f has the following fraction representation:

$$Z_f=PQ^{-1}. \tag{4.11}$$

Proof: Assume that we have $Z_f=PQ^{-1}$. Because $f(e_i)=\bar{f}(e_i)$, we have $\bar{f}=P\bar{Q}^{-1}$, hance $\bar{f}\bar{Q}=P$. Because $\pi P=fQ$, thus $f=\pi PQ^{-1}$. Let $Q^{-1}=(r_{ij}/s_{ij})$, we take $a_i=[s_{1i},s_{2i},\cdots,s_{mi}]$ then we have $f(a_i(z)e_i)=\pi PQ^{-1}a_i(z)e_i=0$. This implies that f is cyclic, therefore f is realizable. Conversely, if f is realizable, then it has a reachable realization (X,g,h), where $X=R^m[z]/\Delta$, $\Delta=QR^m[z]$ and Q is regular. Then from theorem 4.1. and $f(e_i)=\bar{f}(e_i)$ we can obtain $P=Z_fQ$, therefore $Z_f=PQ^{-1}$.

Definition 4.1. Let P and Q be pxm and mxm polynomial matrices, respectively. P and Q is called right coprime if every mxm polynomial matrix S satisfing $P=P_1S$ and $Q=Q_1S$ is necessarily unimodular; P and Q is called strong right coprime if there exist mxp and mxm matrices A and B such that $AP+BQ=I_m$.

It is well known that if R is a field then the right coprime and strong right coprime is equivalent. However, the fact does not hold for commutative rings.

Theorem 4.3. Assume that a LIOM f is realizable and (X,g,h) is a reachable realization of f with $X=R^m[z]/\Delta$, where $\Delta=QR^m[z]$ and Q is regular. The transfer function matrix $Z_f=PQ^{-1}$. Then we have
(1). If (X,g,h) is observable then **P and** Q is right coprime.
(2). If P and Q is strong right coprime, then (X,g,h) is observable.
Proof: (1). From $Z_f=PQ^{-1}$ we have $Z_f=P_1Q_1^{-1}$, thus $f=\pi P_1Q_1^{-1}$, i.e., $fQ_1=\pi P_1$. For any $u \in R^m[z]$, then we have $fQ_1u=\pi P_1u=0$, thus $\Delta_1=Q_1R^m[z] \subset \ker(f)$. Because (X,g,h) is observable, thus $\Delta=\ker(f)$, then we have $\Delta=\Delta_1$.
(2). If P and Q is strong right coprime, thus there are mxp and mxm matrices A and B such that $AP+BQ=I_m$, from $f=\pi PQ^{-1}$, for any $u \in \ker(f)$, $0=f(u)=\pi PQ^{-1}u$, thus $PQ^{-1}u \in R^p[z]$. Because A is a polynomial matrix, thus we have $APQ^{-1}u \in R^m[z]$, i.e., $(I_m+BQ)Q^{-1}u \in R^m[z]$. Therefore $Q^{-1}u \in R^m[z]$, so we have $u \in QR^m[z]=\Delta$. This implies that $\ker(f) \subset \Delta$, thus we obtain $\Delta=\ker(f)$.

Therefore, Z_f of a realizable LIOM f admits a right coprime polynomial matrix fraction representation.

REFERENCES

[1] Conte,G. and Perdon, A.M., Systems over a principal ideal domain, A polynomial model approach, SIAM J. Control & Optimization, 20(1982),112-124.
[2] Khargoneker, P.P., On matrix fraction representation for linear systems over commutative rings, Ibed., 20(1985), 172-197.
[3] Rouchaleau,y. and Wyman,B.F., Linear dynamical systems over integral domains, J. Computer and System Sciences, 9(1974), 129-142.
[4] Li Tiejun, Realization theory for linear systems over commutative rings (in

Chinese), Control Theory and Applications, 2(1985),112-116.
[5] Fuhrmann,P.A., Algebraic system theory: an analysis point of view, J.Franklin Inst., 301(1976),521-540.
[6] ——, Simulation of linear systems and fractorization of matrix polynomials , Internat. J. Control, 28(1978) 689-705.

Frequency Domain and State Space Methods for Linear Systems
C.I. Byrnes and A. Lindquist (Editors)

A COMPLETE DESCRIPTION OF SELF-ADJOINT DETERMINANTAL REPRESENTATIONS OF REAL IRREDUCIBLE CUBICS

Victor Vinnikov

Department of Mathematics and Computer Sciences
Ben-Gurion University of the Negev
Beer-Sheva, Israel

We obtain a complete classification of self-adjoint determinantal representations of real irreducible cubics. The case of real rational cubics have been treated previously [13]; we treat mainly the case of real elliptic cubics. The complete set of canonical representations is topologically equivalent either to a punctured circle or to a disjoint union of a circle and a punctured circle. We investigate elementary transformations [7,8,9] which allow us to obtain from a given self-adjoint determinantal representation all the non-equivalent representations within the same component. Finally, we introduce and study a certain notion of positivity (or definiteness).

1. INTRODUCTION

Let $F(x_0,x_1,x_2)$ be a (complex) algebraic curve of degree n. A determinantal representation $U = U(x_0,x_1,x_2)$ of F is a matrix of order n whose entries are linear in x_0,x_1,x_2:

$$U(x_0,x_1,x_2) = x_1A + x_2B + x_0C \tag{1}$$

(A,B,C- complex matrices of order n)
satisfying $\det U(x) \equiv cF(x)$ $(c \in C,\ c \neq 0)$

Determinantal representations have been previously investigated in [3,4,12]. It is shown there that a general complex curve possesses a determinantal representation.

Suppose now F is a real curve (i.e. all its coefficients are real). A self-adjoint determinantal representation $U(x_0,x_1,x_2) = x_1A + x_2B + x_0C$ of F is a determinantal representation satisfying $U^*(x) = U(x)$ for $x_0,x_1,x_2 \in R$ (i.e. A,B,C are complex hermitian matrices) and so that $\det U(x) = cF(x)$ with $c > 0$.

Two self-adjoint determinantal representations U_1, U_2 of a real curve F of degree n are called equivalent if there exists a complex matrix $P \in GL(n,C)$ so that:

$$U_2 = PU_1P^* \tag{2}$$

Study and classification of self-adjoint determinantal representations of a real curve is an interesting problem by itself, it is equivalent to the algebraic problem of the simultaneous classification of triples of hermitian matrices, and as shown by Livšic [9], it has a deep connection with the colligation theory for pairs of commuting operators in a Hilbert space. In this paper we investigate and classify self-adjoint determinantal representations of real irreducible cubics.

Before proceeding further we introduce some additional general concepts. Let $U(x)$ be an arbitrary (not necessarily self-adjoint) determinantal representation of a complex curve F of degree n; we define the corresponding vector bundle $\varepsilon(x)$ along the curve by

$$\varepsilon(x) = \ker U(x) \tag{3}$$

It is easy to see [3,9,13] that if $\tilde{x}$ is a regular point of F, then the vector bundle at it is one-dimensional: $\dim \varepsilon(\tilde{x}) = 1$.

Let now $U_1(x)$, $U_2(x)$ be self-adjoint determinantal representations of a real curve F of degree n, the corresponding vector bundles $\varepsilon_1(x)$, $\varepsilon_2(x)$ are called equivalent if there exists a complex matrix $T \in GL(n,C)$ so that

$$\varepsilon_2(x) = \varepsilon_1(x)T \tag{4}$$

Clearly, the equivalence of self-adjoint determinantal representations implies the equivalence of corresponding vector bundles.

We proceed now to recall some facts about real irreducible cubics (see [2,5,11,14]).

2. REAL IRREDUCIBLE CUBICS

A complex irreducible curve $f(x,y)$ is called rational if there exist complex rational functions $\varphi(t)$, $\psi(t)$ satisfying:

a) For almost all $t_0 \in C$ (i.e., all but a finite number) the point $(\varphi(t_0),\psi(t_0))$ belongs to the curve: $f(\varphi(t_0),\psi(t_0)) = 0$;

b) For almost all points (x_0,y_0) belonging to the curve there exists a unique $t_0 \in C$, so that $x_0 = \varphi(t_0)$, $y_0 = \psi(t_0)$.

For rational curves the genus $g = 0$, and the Riemann surface of the curve is homeomorphic to the closed complex plane, i.e. to the sphere.

We recall that an elliptic function is a doubly periodic meromorphic function (see [1]). A complex irreducible curve $f(x,y)$ is called elliptic if there exist elliptic functions $\varphi(u)$, $\psi(u)$ satisfying:

a) For almost all u_0 in the period parallelogram, the point $(\varphi(u_0),\psi(u_0))$ belongs to the curve: $f(\varphi(u_0),\psi(u_0)) = 0$;

b) For almost all points (x_0,y_0) belonging to the curve there exists a unique u_0 in the period parallelogram, so that $x_0 = \varphi(u_0)$, $y_0 = \psi(u_0)$.

For elliptic curves the genus $g = 1$, and the Riemann surface of the curve is homeomorphic to the torus.

Now, a real irreducible cubic either is non-singular (smooth), in which case it is elliptic, or it has a unique singular point of multiplicity 2, in which case it is rational. We have presented a complete classification of self-adjoint determinantal representations of real rational cubics in the previous meeting of this conference (see [13]), so in this paper we are going to concentrate on real elliptic (i.e. smooth) cubics.

We remind that complex curves can be brought into different forms by a projective change of coordinates in the complex (projective) plane. Since we are dealing with real curves, we shall consider only real changes of coordinates - which have real coefficients and leave the real part of the complex plane invariant; they are given by:

$$[x_0',x_1',x_2'] = [x_0,x_1,x_2]A \tag{5}$$

where $A \in GL(3,R)$.

Suppose that the change of coordinates (5) transforms a real curve $F(x_0,x_1,x_2)$ into the real curve $F'(x_0',x_1',x_2')$. Then it clearly induces a one-to-one real linear correspondence between self-adjoint determinantal representations of F and those of F'. It is also clear that the equivalence of self-adjoint determinantal representations is preserved under such a correspondence. Hence, instead of studying self-adjoint determinantal representations of the original curve F, we may study those of F'.

It can be shown that any real smooth cubic can be brought, by a real projective change of coordinates, to the following normal form:

$$y^2 = g(x) \tag{6}$$

where $g(x)$ is a (real) polynomial of degree 3 without multiple roots. $g(x)$ must have at least one real root, and by performing an additional real shift in coordinates we may assume that this real root is 0. Thus the final normal form for real smooth cubics (under the real projective change of coordinates) is:

$$y^2 = x(x + \theta_1)(x + \theta_2) \tag{7}$$

where θ_1, θ_2 are either distinct real numbers, different from 0, or complex conjugates. In projective coordinates:

$$x_2^2x_0 - x_1(x_1 + \theta_1x_0)(x_1 + \theta_2x_0) = 0 \tag{8}$$

We shall denote this cubic by $F_{\theta_1\theta_2}(x_0,x_1,x_2)$. Using the Weirstrass p-function with appropriate (real) invariants, it is not difficult to write down explicitly the parametrization of $F_{\theta_1\theta_2}$ by elliptic functions.

By the remarks of the previous paragraph we may limit ourselves now to the investigation of self-adjoint determinantal representations of $F_{\theta_1\theta_2}$.

3. CANONICAL FORMS FOR DETERMINANTAL REPRESENTATIONS

To start the classification of self-adjoint determinantal representations we have to obtain a complete set of canonical forms for them, so that each self-adjoint determinantal representation of $F_{\theta_1\theta_2}$ is equivalent to exactly one canonical representation.

Thus given any self-adjoint determinantal representation of $F_{\theta_1\theta_2}$

$$\det(x_1A + x_2B + x_0C) = cF_{\theta_1\theta_2}(x_0,x_1,x_2) \qquad (c > 0) \tag{9}$$

we want to reduce it to the canonical form by the group action:

$$x_1A + x_2B + x_0C \to P(x_1A + x_2B + x_0C)P^* , \qquad P \in GL(3,C) \tag{10}$$

Alternatively we may consider the group action:

$$(A,B,C) \to (PAP^*, PBP^*, PCP^*) , \qquad P \in GL(3,C) \tag{11}$$

on triples (A,B,C) of hermitian matrices of order 3 satisfying (9).

We proceed now in two steps. At the first step, using the theory of indefinite inner-product spaces [6,10], we reduce simultaneously the above matrices A,B to the canonical form. More precisely, we define an indefinite metric in C^3 by:

$$[h,g] = (hA,g) \tag{12}$$

where $(\cdot, \cdot)$ is the usual scalar product. It turns out that the simultaneous reduction of the matrices A,B to the canonical form by the transformation

$A \to PAP^*$, $B \to PBP^*$ is equivalent to the reduction to the canonical form of the matrix of a certain self-adjoint linear transformation in the vector space C^3 endowed with the indefinite metric (12).

At the second step we find a matrix P preserving the canonical form for A,B, and reduce the matrix C to the canonical forms by the transformation $C \to PCP^*$. Because of the lack of space we omit all further details of the reduction procedure (which is very similar to the corresponding procedure in the case of real rational cubics [13]) and just state the final result:

Theorem 1: The complete set of canonical forms for self-adjoint determinantal representations of $F_{\theta_1\theta_2}$ is given explicitly by:

$$x_1\begin{bmatrix} 0 & 0 & 1 \\ 0 & 1 & 0 \\ 1 & 0 & 0 \end{bmatrix} + x_2\begin{bmatrix} 0 & 1 & 0 \\ 1 & 0 & 0 \\ 0 & 0 & 0 \end{bmatrix} + x_0\begin{bmatrix} t & d & c \\ -d & b & 0 \\ c & 0 & -1 \end{bmatrix} = \begin{bmatrix} tx_0 & dx_0+x_2 & cx_0+x_1 \\ -dx_0+x_2 & bx_0+x_1 & 0 \\ cx_0+x_1 & 0 & -x_0 \end{bmatrix} \tag{13}$$

where b is a real parameter and

$$c = \frac{\theta_1+\theta_2-b}{2}, \quad t = \theta_1\theta_2 + \frac{b-\theta_1-\theta_2}{2} \cdot \frac{\theta_1+\theta_2+3b}{2}, \tag{14}$$

$$d = \pm i\sqrt{b(b-\theta_1)(b-\theta_2)}, \quad d \text{ - purely imaginary}. \quad \square$$

Of course, since d is purely imaginary, b must take its values in certain intervals on the real axis; the precise structure of these intervals depends on whether θ_1, θ_2 are real or complex conjugate.

4. TOPOLOGICAL STRUCTURE OF THE SET OF CANONICAL REPRESENTATIONS

Our objective now is to clarify the topological structure of the set of canonical representations (13), i.e. the structure of the set of self-adjoint determinantal representations of $F_{\theta_1\theta_2}$ modulo the equivalence relation.

The canonical representation (13) is determined by two numbers b,d ($b \in R$, $d \in iR$) satisfying

$$(-d)^2 = -b(-b+\theta_1)(-b+\theta_2) \tag{15}$$

Consider the point $(-b,-d) \in C^2$. It is an affine point whose first coordinate is real and the second one is purely imaginary, and it is situated on the curve

$$y^2 = x(x+\theta_1)(x+\theta_2) \tag{16}$$

i.e. on $F_{\theta_1\theta_2}$. We get thus a natural homeomorphism between the set of canonical representations (13) and the set of all affine points on $F_{\theta_1\theta_2}$ whose first coordinate is real and the second one is purely imaginary.

The set of these points on $F_{\theta_1\theta_2}$ looks differently according to whether the roots θ_1, θ_2 are real or complex conjugate. The precise picture is given by Figures 1-2, where, of course, the x-axis is real, while the y-axis is purely imaginary.

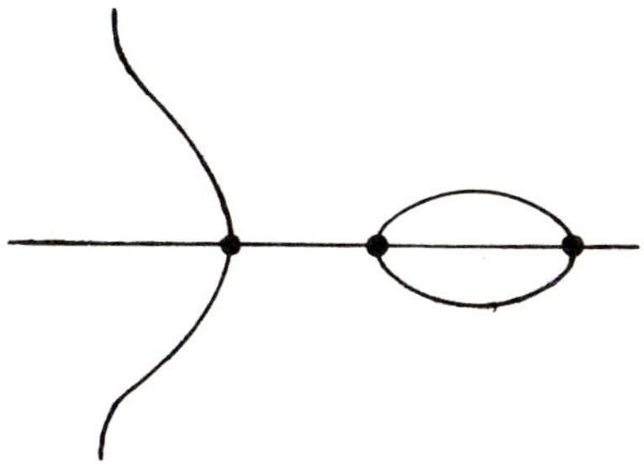

$\theta_1, \theta_2 \in R$

Figure 1

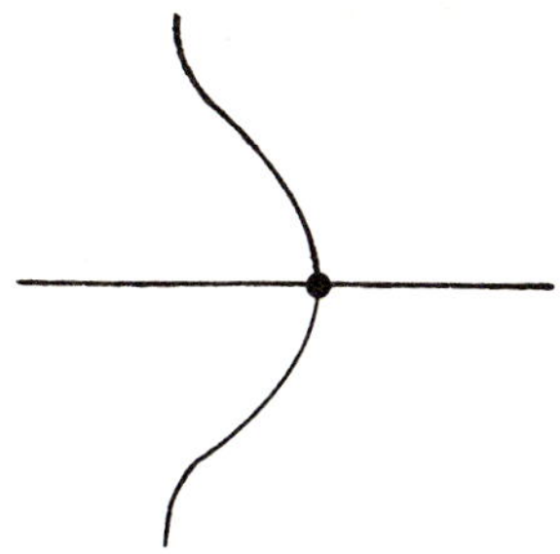

$\theta_1 = \bar{\theta}_2 \notin R$

Figure 2

We summarize:

Theorem 2: There is a natural homeomorphism between the set of canonical forms (13) for self-adjoint determinantal representations of $F_{\theta_1\theta_2}$ and the set of affine points on $F_{\theta_1\theta_2}$ whose first coordinate is real and the second one is purely imaginary. Consequently, the set of canonical representations (13) is topologically equivalent to:

a) $\theta_1, \theta_2 \in R$: a disjoint union of a punctured circle and a circle;

b) $\theta_1 = \bar{\theta}_2 \notin R$: a punctured circle. □

We see that the set of canonical representations of $F_{\theta_1\theta_2}$ contains in any case a non-compact component; if $\theta_1, \theta_2 \in R$ it contains an additional component which is compact. The meaning of these two components will become clear later.

5. CLASS OF DIVISORS OF THE VECTOR BUNDLE AND COMPLETE SYSTEM OF INVARIANTS

To complete the classification of self-adjoint determinantal representations of $F_{\theta_1\theta_2}$ we have to find a complete system of invariants for them, so that given two representations we will be able to tell whether they are equivalent or not. To this end we shall use the notion of the divisor (see [2,5,11,14]).

First, let $R(x_0,x_1,x_2)$ be a rational function in x_0, x_1, x_2 (whose numerator and denominator are homogeneous of the same degree). We remind that we can assign multiplicities to its zeroes and its poles along the curve $F_{\theta_1\theta_2}$ (this can be done in several equivalent ways: we can use general methods of algebraic geometry, or we can consider the rational function along the curve as a meromorphic function on the corresponding Riemann surface, or we can use the fact that $F_{\theta_1\theta_2}$ is an elliptic curve and substitute into $R(x)$ the parameterization of $F_{\theta_1\theta_2}$ by elliptic functions).

Let now $U(x)$ be a self-adjoint determinantal representation of $F_{\theta_1\theta_2}$, and let $\varepsilon(x)$ be the corresponding vector bundle. Let $u(x)$ be a rational section of $\varepsilon(x)$ (i.e. a representative $u(x) \in \varepsilon(x)$ whose entries are rational functions in x_0,x_1,x_2 ; a rational section can always be constructed using the algebraic complements of a column in $U(x)$). We consider zeroes and poles of $u(x)$ along $F_{\theta_1\theta_2}$ and define their multiplicities in the natural way: the multiplicity of T as a zero (resp. as a pole) of $u(x)$ is defined as the minimal (resp. as the maximal) between the multiplicities of T as a zero (resp. as a pole) of the entries of $u(x)$.

Denote by $Z_1,\dots,Z_k$ the zeroes and by $P_1,\dots,P_\ell$ the poles of $u(x)$ along $F_{\theta_1\theta_2}$ and let $n_1,\dots,n_k$ and $m_1,\dots,m_\ell$ be the corresponding multiplicities. The divisor of $u(x)$ is by definition the following formal sum:

$$\operatorname{div}(u(x)) = \sum_i n_i Z_i - \sum_j m_j P_j \tag{17}$$

The divisor of a rational function $R(x)$ is defined similarly.

In general, a <u>divisor</u> D on $F_{\theta_1\theta_2}$ is any (finite) formal sum of points T_i on $F_{\theta_1\theta_2}$ with integer coefficients r_i: $D = \sum_i r_i T_i$. Clearly, the divisors on $F_{\theta_1\theta_2}$ form an (additively written) abelian group. The sum of the coefficients of the divisor D is called the <u>degree</u> of D: $\deg D = \sum_i r_i$.

A divisor D on $F_{\theta_1\theta_2}$ is said to be equivalent to 0: $D \equiv 0$, if there exists a rational function $R(x)$ so that $D = \operatorname{div}(R(x))$. A necessary condition for this is $\deg D = 0$; it is not a sufficient conditon, since the cubic $F_{\theta_1\theta_2}$ is elliptic, and not rational. Two divisors D_1, D_2 on $F_{\theta_1\theta_2}$ are called (linearly) equivalent: $D_1 \equiv D_2$, if $D_1 - D_2 \equiv 0$.

We return now to our vector bundle $\varepsilon(x)$ and consider two rational sections $u(x)$, $v(x)$ of it. Since $F_{\theta_1\theta_2}$ is smooth, $\varepsilon(x)$ is one-dimensional, thus there exists a rational function $R(x)$ so that:

$$u(x) = R(x)v(x) \tag{18}$$

Hence,

$$\operatorname{div}(u(x)) = \operatorname{div}(R(x)) + \operatorname{div}(v(x)) \equiv \operatorname{div}(v(x)) \tag{19}$$

We conclude that the vector bundle $\varepsilon(x)$ determines uniquely a class of equivalent divisors on $F_{\theta_1\theta_2}$, that we shall denote by $C\ell \operatorname{div}(\varepsilon(x))$. Clearly, when we multiply a rational section of $\varepsilon(x)$ by a constant non-singular matrix (of order 3), its zeroes and poles along $F_{\theta_1\theta_2}$ are preserved, including multiplicities. It follows that the class of divisors is invariant under the equivalence of vector bundles.

Consequently, we have found an invariant for self-adjoint determinantal representations of $F_{\theta_1\theta_2}$. We note that our considerations were completely general; for self-adjoint determinantal representations of any real irreducible curve, we can define the class of divisors of the corresponding vector bundle, and it is invariant under the equivalence of self-adjoint determinantal representations.

Returning to the self-adjoint determinantal representations of $F_{\theta_1\theta_2}$, we can check directly, that for different canonical representations (13), the corresponding vector bundles possess different classes of divisors. Thus we obtain

<u>Theorem 3</u>: U being a self-adjoint determinantal representation of $F_{\theta_1\theta_2}$, the class of divisors of the corresponding vector bundle forms the complete system of invariants for U (under the equivalence of self-adjoint determinantal

representations). □

We note that using the parametrization of $F_{\theta_1\theta_2}$ by elliptic functions and Abel's theorem for elliptic curves (see [1,2]), we can write down explicit conditions for the equivalence of divisors on $F_{\theta_1\theta_2}$. Hence Theorem 3 gives us an explicit way to check whether two self-adjoint determinantal representations of $F_{\theta_1\theta_2}$ are equivalent, and to construct, for the given representation, its canonical form (13).

6. ELEMENTARY TRANSFORMATIONS

In this section we consider an elementary transformation of self-adjoint determinantal representations of a real curve, first suggested by Livšic [9] and Kravitsky [7,8]. We show then that given a self-adjoint determinantal representation of $F_{\theta_1\theta_2}$, we can obtain from it by elementary transformations all the non-equivalent representations within the same component (recall the structure given by Theorem 2).

Let $f(x_1,x_2)$ be a real irreducible curve of degree n and take a self-adjoint determinantal representation of it:

$$\det(x_1\sigma_2-x_2\sigma_1+\gamma) = cf(x_1,x_2) \qquad (c > 0) \tag{20}$$

(we shall use in this section affine coordinates, and denote them by (x_1,x_2) instead of (x,y)). In addition to the corresponding vector bundle $\varepsilon(x) = \ker(x_1\sigma_2-x_2\sigma_1+\gamma)$, we shall consider now also the vector bundle $\varphi(x) = \operatorname{coker}(x_1\sigma_2-x_2\sigma_1+\gamma)$. Both $\varepsilon(x)$ and $\varphi(x)$ are one-dimensional, except perhaps at the singular points of f.

Let $\lambda = (\lambda_1,\lambda_2)$, $\bar{\lambda} = (\bar{\lambda}_1,\bar{\lambda}_2)$ be two conjugate points on f, and suppose they are non-singular and different (i.e. $\lambda,\bar{\lambda} \notin R^2$). Take $\hat{v} \in \varepsilon(\lambda)$, $\hat{u} \in \varphi(\bar{\lambda})$:

$$\begin{cases} \hat{v}(\lambda_1\sigma_2-\lambda_2\sigma_1+\gamma) = 0 \\ (\bar{\lambda}_1\sigma_2-\bar{\lambda}_2\sigma_1+\gamma)\hat{u} = 0 \end{cases} \tag{21}$$

Since the representation $x_1\sigma_2-x_2\sigma_1+\gamma$ is self-adjoint, of course

$$\hat{u} = q\hat{v}^* \quad (q \in C,\ q \neq 0) \tag{22}$$

From (21) it follows that:

$$(\lambda_1-\bar{\lambda}_1)\hat{v}\sigma_2\hat{u} = (\lambda_2-\bar{\lambda}_2)\hat{v}\sigma_1\hat{u} \tag{23}$$

Unless $\hat{v}\sigma_1\hat{u} = \hat{v}\sigma_2\hat{u} = 0$, it follows from (23) that we can normalize $\hat{v}$, $\hat{u}$ so that:

$$\begin{cases} \lambda_1-\bar{\lambda}_1 = \hat{v}\sigma_1\hat{u} \\ \lambda_2-\bar{\lambda}_2 = \hat{v}\sigma_2\hat{u} \end{cases} \tag{24}$$

The pair of conjugate points λ, $\bar{\lambda}$ on f is called <u>admissible</u> with respect to the representation $x_1\sigma_2-x_2\sigma_1+\gamma$ if $\hat{v}\sigma_1\hat{u}$, $\hat{v}\sigma_2\hat{u}$ are not both 0. From (21), (22) we conclude that the pair λ, $\bar{\lambda}$ is unadmissible when:

$$\hat{v}\sigma_1\hat{v}^* = \hat{v}\sigma_2\hat{v}^* = \hat{v}\gamma\hat{v}^* = 0 \tag{25}$$

i.e. when $\hat{v} \in \varepsilon(\lambda)$ is a common self-orthogonal vector of σ_1, σ_2, γ ($h \in C^n$ is a self-orthogonal vector of a $n\times n$ hermitian matrix A if

$(hA,h) = hAh^* = 0)$.

Suppose now that the pair λ, $\bar{\lambda}$ is admissible with respect to $x_1\sigma_2 - x_2\sigma_1 + \gamma$ and that $\hat{v}$, $\hat{u}$ are normalized so that (24) holds. Define:

$$\tilde{\gamma} = \gamma + \sigma_1(\hat{u}\hat{v})\sigma_2 - \sigma_2(\hat{u}\hat{v})\sigma_1 \tag{26}$$

From the normalization conditions (24) it follows that in (22) q is purely imaginary, hence $\tilde{\gamma}$ is hermitian. It is a straightforward although tedious matter to verify the following relations:

$$(x_1\sigma_2 - x_2\sigma_1 + \tilde{\gamma})\left(I - \frac{(\hat{u}\hat{v})}{\lambda_k - x_k}\sigma_k\right) = \left(I - \sigma_k \frac{(\hat{u}\hat{v})}{\lambda_k - x_k}\right)(x_1\sigma_2 - x_2\sigma_1 + \gamma), \tag{27}$$

$$k = 1,2 \qquad (x_k \neq \lambda_k)$$

Recall now that for any two square matrices K_1, K_2

$$\det(I + K_1K_2) = \det(I + K_2K_1) \tag{28}$$

Thus by taking determinants it follows from (27) that:

$$\det(x_1\sigma_2 - x_2\sigma_1 + \tilde{\gamma}) = \det(x_1\sigma_2 - x_2\sigma_1 + \gamma) = cf(x_1, x_2) \tag{29}$$

Hence $x_1\sigma_2 - x_2\sigma_1 + \tilde{\gamma}$ is a new self-adjoint determinantal representation of f; we say that it was obtained from the original representation $x_1\sigma_2 - x_2\sigma_1 + \gamma$ by the elementary transformation (26) based upon the admissible pair of conjugate points λ, $\bar{\lambda}$.

To explore the connection between the original representation and the new one, we consider the corresponding vector bundles $\varepsilon(x) = \ker(x_1\sigma_2 - x_2\sigma_1 + \gamma)$, $\tilde{\varepsilon}(x) = \ker(x_1\sigma_2 - x_2\sigma_1 + \tilde{\gamma})$. Let us define:

$$T_k(x) = I - \sigma_k \frac{(\hat{u}\hat{v})}{\lambda_k - x_k}, \qquad k = 1,2 \tag{30}$$

$T_k(x)$ is defined for $x_k \neq \lambda_k$, and it is easily seen that for $x_k \neq \bar{\lambda}_k$ it is invertible and

$$T_k^{-1}(x) = I + \sigma_k \frac{(\hat{u}\hat{v})}{\bar{\lambda}_k - x_k} \tag{31}$$

From (27) it follows now that for $x_k \neq \lambda_k, \bar{\lambda}_k$

$$\varepsilon(x)T_k^{-1}(x) = \tilde{\varepsilon}(x) \tag{32}$$

It should be noted that (if $x_1 \neq \bar{\lambda}_1$, $x_2 \neq \bar{\lambda}_2$) $T_1^{-1}(x)$ and $T_2^{-1}(x)$ coincide when restricted to the vector bundle $\varepsilon(x)$.

Suppose now $v(x)$ is a rational section of $\varepsilon(x)$, then

$$\tilde{v}(x) = v(x)T_k^{-1}(x) \tag{33}$$

is a rational section of $\tilde{\varepsilon}(x)$. From (30), (31) it can be seen that:

$$\operatorname{div}(\tilde{v}(x)) = \operatorname{div}(v(x)) + \lambda - \bar{\lambda} \tag{34}$$

(see [9] for the details). Thus we have

Theorem 4 [9]: $C\ell\ \mathrm{div}(\tilde{\varepsilon}(x)) = C\ell\ \mathrm{div}(\varepsilon(x)) + C\ell\{\lambda-\bar{\lambda}\}$. □ (35)

If the curve f has genus $g > 0$ (i.e. if it is not rational), then for any two distinct points A, B on it, $A-B \not\equiv 0$. Hence $C\ell\ \mathrm{div}(\tilde{\varepsilon}(x)) \neq C\ell\ \mathrm{div}(\varepsilon(x))$, and by the results of the previous section the new representation $x_1\sigma_2-x_2\sigma_1+\tilde{\gamma}$ is not equivalent to the original $x_1\sigma_2-x_2\sigma_1+\gamma$. The natural question is: given a self-adjoint determinantal representation of f, how many other (non-equivalent) representations can we obtain from it by elementary transformations based on admissible pairs of conjugate points?

For the case of our cubic $F_{\theta_1\theta_2}$ we can answer this question completely. Let U_1, U_2 be two self-adjoint determinantal representations of $F_{\theta_1\theta_2}$, we shall say that they can be bridged by an elementary transformation, if there exists an elementary transformation based on an admissible pair of conjugate points, that transforms U_1 into a representation equivalent to U_2 (this definition is actually symmetric in U_1, U_2). We recall that the set of canonical representations (13) of $F_{\theta_1\theta_2}$ consists, in general, of two connected components.

Theorem 5: Two canonical representations (13) of $F_{\theta_1\theta_2}$ can be bridged by an elementary transformation if and only if they both belong to the same component (in the set of canonical representations). □

The proof of this theorem is based on Theorems 3-4; it is performed through explicit calculations with elliptic functions, using the parametrization of $F_{\theta_1\theta_2}$.

7. DEFINITE AND INDEFINITE REPRESENTATIONS

We want to consider now a certain notion of positivity, or definiteness, which appears quite naturally in the context of self-adjoint determinantal representations. This notion will also enable us to understand the partition of the set of canonical representations (13) into two components.

A self-adjoint determinantal representation $U = x_1A+x_2B+x_0C$ of a real curve F of degree n is called definite if there exist $c_0,c_1,c_2 \in R$ so that $c_1A+c_2B+c_0C > 0$, indefinite otherwise. Clearly, equivalent representations are definite or indefinite simultaneously. We note that the question whether three given hermitian matrices admit a positive definite real linear combination is interesting by itself.

It is an important fact that elementary transformations preserve definiteness. Indeed, using the notations of the previous section, it follows from (27) that for $\xi_1,\xi_2 \in R$ the matrices $\xi_1\sigma_2-\xi_2\sigma_1+\gamma$ and $\xi_1\sigma_2-\xi_2\sigma_1+\tilde{\gamma}$ are congruent; hence the representations $x_1\sigma_2-x_2\sigma_1+\gamma$, $x_1\sigma_2-x_2\sigma_1+\tilde{\gamma}$ are simultaneously definite or indefinite.

We consider now the set of canonical representations (13) of $F_{\theta_1\theta_2}$ and we recall that it contains, in general, the non-compact component and the compact one.

Theorem 6: In the set of canonical representations (13) of $F_{\theta_1\theta_2}$, the non-compact component consists precisely of indefinite representations, and the compact component consists precisely of definite representations. □

Hence $F_{\theta_1\theta_2}$ always possesses indefinite self-adjoint determinantal representations; if the roots $\theta_1,\theta_2 \in R$, then it possesses also definite representations.

Outline of Proof of Theorem 6: First let $U = x_1A+x_2B+x_0C$ be a canonical representation in the non-compact component. Using the parametrization of $F_{\theta_1\theta_2}$ by elliptic functions, it can be seen that there exist pairs of conjugate points on $F_{\theta_1\theta_2}$ which are unadmissible with respect to U. By (25), the coefficient matrices A, B, C possess a common self-orthogonal vector, hence they admit no positive definite real linear combination and U is an indefinite representation.

Consider now the compact component. Any two canonical representations there can be bridged by an elementary transformation, which preserves definiteness. Hence we pick up a certain canonical representation in the compact component and check directly that it is definite; it follows that all the representations in the compact component are definite, and the theorem is proved.

From this proof we obtain the following interesting result:

Corollary: A self-adjoint determinantal representation $U = x_1A+x_2B+x_0C$ of $F_{\theta_1\theta_2}$ either is definite, or the coefficient matrices A, B, C possess a common self-orthogonal vector in the corresponding vector bundle. □

Finally we make some remarks on self-adjoint determinantal representations of real rational cubics. For them also, the set of canonical representations always contains a non-compact component, while in some cases it contains an additional component which is compact. It turns out that Theorem 6 and the above Corollary hold true also for self-adjoint determinantal representations of real rational cubics. All these statements can be verified directly using the results presented in [13].

I wish to express my most deep thanks to Prof. Livšic for his help and guidance.

REFERENCES

[1] Ahiezer, N.I., Elements of the Theory of Elliptic Functions, Izdat. "Nauka", Moscow, 1970 (Russian).

[2] Clemens, C.H., A Scrapbook of Complex Curve Theory, Plenum Press, New-York, 1980.

[3] Cook, A.J., Thomas, A.D., Line bundles and homogeneous matrices, Quart. J. Math. Oxford (2), 30 (1979), 423-429.

[4] Dixon, A.C., Note on the reduction of a ternary quartic to a symmetrical determinant, Proc. Comb. Phil. Soc. 2 (1900-1902), 350-351.

[5] Fulton, W., Algebraic Curves, Benjamin, Reading, Massachusets, 1974.

[6] Gohberg, I., Lancaster, P. and Rodman, L., Matrices and Indefinite Scalar Products, Birkhäuser-Verlag, 1983.

[7] Kravitsky, N., Regular colligations for several commuting operators in Banach space, Integral Equations and Operator Theory, 6 (1983), 224-249.

[8] Kravitsky, N., On determinantal representation of algebraic curves, Mathematical Theory of Networks and Systems, Proceedings of the MTNS-83 International Symposium, Beer-Sheva, Israel, June 1983 (Springer-Verlag, Berlin, Heidelberg, 1984).

[9] Livšic, M.S., Cayley-Hamilton theorem, vector bundles and divisors of commuting operators, Integral Equations and Operator Theory, 6 (1983), 250-273.

[10] Maltzev, A.I., Principles of Linear Algebra, Gosudarst. Izdat. Tehn.-Teor. Lit., Moscow-Leningrad, 1948 (Russian).

[11] Mumford, D., Algebraic Geometry I, Springer-Verlag, Berlin, Heidelberg, 1976.

[12] Room, T.G., The Geometry of Determinantal Loci, Cambridge University Press, Cambridge, 1938.

[13] Vinnikov, V., Determinantal representations of real cubics and canonical forms of corresponding triples of matrices, Mathematical Theory of Networks and Systems, Proceedings of the MTNS-83 International Symposium, Beer-Sheva, Israel, June 1983 (Springer-Verlag, Berlin, Heidelberg, 1984).

[14] Walker, R.J., Algebraic Curves, Princeton, New Jersey, 1950.

Frequency Domain and State Space Methods for Linear Systems
C.I. Byrnes and A. Lindquist (Editors)

LOCAL AND GLOBAL LINEAR SYSTEM THEORY

B.F. Wyman
Dept. of Math. O.S.U.
Columbus, Ohio

G. Conte
Ist. Mat.Univ. Genova
Genova, Italy

A.M. Perdon
Ist. Mat. Univ. Genova
Genova, Italy

The notion of linear dynamical system over a field k is investigated from two different points of view, called respectively "local" and "global". The first deals with the properties of the system described by the zeros and the poles at a single specified frequency and is modeled on the realization theory at infinity of non proper transfer functions. The second is a generalization of the classical case, in which finite zeros and poles are considered, and deals with systems having modes in an arbitrarily specified fixed set (stability region). The local and global theories so developed form a coherent algebraic framework for the study of linear systems.

1. INTRODUCTION

The aim of this paper is to develop, in a general framework, a "local theory" and a "global theory" for linear dynamical systems over a field k together with local-global results. The models we have in mind are, for the global case, the classical realization theory of proper transfer functions (or finite poles theory) and the finite zeros theory developed in [18] and, for the local case, the realization at infinity and the theory of poles and zeros at infinity considered in [4] and [5]. Assuming for the moment that k is the field of complex numbers $\mathbb{C}$, by a generalization of the above model in the local case we mean a theory of poles and zeros at a fixed arbitrary point of the Riemann sphere $\mathbb{C}\cup\{\infty\}$ together with a realization theory concerning the modes at a single point. On the other hand, in the global case we consider systems and realization problems in which the modes are restricted to lie in a fixed region of the Riemann sphere which excludes at least one point. For example, if we are interested in stability questions, the region will be the one corresponding to the finite points with negative real part. In the global case the zeros and the poles under interest will be those lying in the fixed region. Then, remarking that the points of the Riemann sphere correspond to the valuations of $\mathbb{C}(z)$, a further generalization is obtained by substituting to $\mathbb{C}\cup\{\infty\}$ the set of all valuation of k(z) for an arbitrary field k.

Essentially, the main motivation in developing this theory is to construct a set of algebraic tools (namely the local and the global pole and zero modules) which are useful in dealing with various factorization and design problems. Examples of these techniques can be found in [6] and in [9].

The paper is organized as follows. In Sections 2 and 3 we describe the rings of transfer functions and the modules over them which will be used. The description is given in terms of valuations of k(z) and the basic algebraic results concerning the structure of finitely generated torsion modules over the above rings. In Section 4 we develop a local system theory defining what we mean by a local system at v, where v is a valuation of k(z), and introducing the notion of local realization and local pole and zero module at v. In Section 5 we introduce a global system theory based on the notion of S-system, where S is any set of valuations of k(z) omitting at least one valuation, and we show that localizing one reobtains the local theory. Moreover we state the Local-Global Principle which relates the minimality of the S-realization to the minimality of the local realizations obtained by localization. The pole and zero modules are introduced and briefly described. In the Appendix we state some technical results which are used through the paper.

2. RINGS OF TRANSFER FUNCTIONS

Let k be a field called the "field of scalars". We consider the field k(z) of rational functions in the variable z over k, called transfer functions. We denote the ring of polynomials in k(z) by k[z] and we wish to consider subrings O of k(z) which generalize k[z] in a reasonable sense.

The rings which we have in mind are illustrated by the ring of stable transfer functions in $\mathbb{C}(z)$:

$$O_{stable} = \{ a(z)/b(z) : \mathrm{Re}(\lambda) < 0 \text{ if } b(\lambda) = 0 \}$$

and the ring of proper transfer functions

$$O_{\infty} = \{ a(z)/b(z) : \deg(b(z)) \geq \deg(a(z)) \}$$

where in each case a(z) and b(z) denote polynomials.

We will restrict our attention to subrings O of K(z) such that $O \supset k$ as a subfield and every rational function can be written as a quotient of members of O. These rings can be described in a uniform way in terms of valuations of k(z). In the complex case, valuations of $\mathbb{C}(z)$ correspond to points on the Riemann sphere, namely complex numbers and the point at infinity. The valuation attached to a point counts zeros or poles of a rational function at that point. For general k the set of all valuations on k(z) is a little more complicated, but it can still be listed explicitly. The valuation "at infinity' is defined just as in the complex case, namely $\mathrm{ord}_{\infty}(a(z)/b(z)) = \deg(b(z)) - \deg(a(z))$ for any rational function f(z) = a(z)/b(z), a(z) and b(z) in k[z]. The "finite" valuations are in one-to-one correspondence with the irreducible polynomials of k[z]. If q(z) is such a polynomial, the corresponding valuation is defined as follows. If f(z) lies in k(z), factor f(z) as $f(z) = q(z)^n(a(z)/b(z))$ where n is positive, negative or zero and a(z) and b(z) are polynomials prime to q(z). Then ord f(z) = n . If deg(q(z)) = d, we say that the corresponding valuation has degree d. The degree of the valuation at infinity is taken equal to one.

If v is any valuation of k(z), the corresponding valuation ring is given by

$$O_V = \{ f(z) \text{ in } k(z) : ord_V f(z) \geq 0 \} ;$$

that is, O_V is the set of all rational functions which have no pole at v. Each O_V is a ring containing k, and the field of fractions of O_V is the whole field k(z). The units, or invertible elements of O_V, are those u(z) in k(z) such that $ord_V u(z) = 0$. The set of all non-units of O_V

$$m_V = \{ f(z) \in O_V : ord_V f(z) > 0 \}$$

is the unique maximal ideal of O_V.

The valuation rings are atoms from which more complicated rings of transfer functions will be constructed.

Let S be an arbitrary set of valuations of k(z) which omits at least one valuation. Define

$$O_S = \{ f(z) \in k(z) : ordv(f(z)) \geq 0 \text{ for all } v \text{ in } S \}$$

We say loosely that O_S is the ring of rational functions regular at every point of S. If S consists of all valuations except v_∞, then $O_S = k[z]$, since the polynomials are exactly the functions with poles only at infinity. The ring k[z] is used to study systems with only finite poles, the classical case. If k is the real or complex field, and S contains all valuations corresponding to Hurwitz polynomials (roots with negative real part), then O_S consists of functions which have only unstable poles. and will be used to study systems with only stable modes.Similarly, the ring of stable transfer functions corresponds to the set of anti-Hurwitz polynomials (see also, for example, [12],[14],[16]. Note carefully that denominators of functions in O_S must have no roots at points corresponding to valuations in the set S. The overall intuition is that modes which should appear in the linear system under investigation should be included in the set S, and they will thereby be omitted from the allowable denominators in the ring O_S. See Section 5 for further discussion.

Note that since each valuation ring O_V contains the scalars k, the ring O_S contains k for any set of valuations. If S contains all valuations of k(z), then O_S consists of constants alone, since a non-constant rational function has at least one pole. If S omits at least one valuation, as we are assuming, then k(z) is the field of fractions of O_S.

<u>For technical convenience, we assume that S omits at least one valuation of degree one.</u> This assumption is standard in the literature and guarantees that O_S is a principal ideal domain.

3. MODULES OVER RINGS OF TRANSFER FUNCTIONS

In this section we review some properties of modules over the rings of transfer functions discussed in the previous section. In the following we will be concerned with three types of modules: finitely generated free modules for inputs, finitely generated torsion modules for states, and (infinitely generated) torsion divisible modules for outputs.

According to the standard structure theory of finitely generated modules over a principal ideal domain O [2] each f.g. O-module M can be written as a direct sum $M = F \oplus T$, where $F \cong O^r$ is free of finite rank and T is a f.g. torsion module. A torsion module admits a decomposition into "primary" or " modal" components. Since such a decomposition plays a crucial role in what follows, we recall it here in some detail.

Suppose O is a principal ideal domain with field of fractions K. Let T be a torsion O-module, so that for each t in T there is a nonzero a in O such that at = 0 in T. Since T is not necessarily f.g., there may be not a single a in O that kills all of T. Next, let $\mathbf{m}$ be a nonzero prime (hence maximal) ideal of O, and let q be any generator of $\mathbf{m}$. The $\mathbf{m}$-primary component of T is given by

$$T_{\mathbf{m}} = \{ t \in T : q^r t = 0 \text{ for some } r \geq 0 \}$$

It follows easily that $T_{\mathbf{m}}$ is an O-submodule of T independent of the choice of q. If $\mathbf{m}$ and $\mathbf{m}'$ are distinct maximal ideals of O, a coprimeness argument ([2],p.132) shows that $T_{\mathbf{m}} \cap T_{\mathbf{m}'} = (0)$. Furthermore, $T = \Sigma\, T_{\mathbf{m}}$. If T is f.g., then all but a finite number of the components $T_{\mathbf{m}}$ will be zero. We summarize this result in the following

3.1 Primary Decomposition Theorem . Suppose that O is a principal ideal domain and T is a torsion O-module. For each maximal ideal $\mathbf{m}$ of O, let $T_{\mathbf{m}}$ be the $\mathbf{m}$-primary component of T. Then T is isomorphic to a direct sum

$$T \cong \oplus\, T_{\mathbf{m}}.$$

It will be helpful to discuss the submodules $T_{\mathbf{m}}$, which may be called <u>modal components</u>, in some more detail. Given O and $\mathbf{m}$ we can define the localization of O at $\mathbf{m}$ as a certain ring of fractions in the field K. Thus ([1], chapter 3)

$$O_{\mathbf{m}} = \{ a/b \in K : a,b \in O ; b \notin \mathbf{m} \}.$$

One can also define the localization of a module T as $T_{\mathbf{m}} = O_{\mathbf{m}} \otimes_O T$, or more concretely as

$$T_{\mathbf{m}} = \{ t/b : t \in T, b \notin \mathbf{m} \},$$

subject to the equivalence relation $t/b = t'/b'$ if $b''(b't - bt') = 0$ for some b" outside **m**. See [1], chap.3. The notation is reasonable, since the **m**-primary component $T_{\mathbf{m}}$ is the same as the localization at **m** for torsion modules . We apply these ideas to rings of the form O_S, where S is a set of valuations. The maximal ideals of O_S are precisely

$$\mathbf{m}^S{}_v = \{ f \in O_S : \mathrm{ord}_v(f) > 0 \},$$

and the localization of O_S with respect to $\mathbf{m}^S{}_v$ is just the valuation ring O_v. We can choose a generator q for $\mathbf{m}^S{}_v$ in O_S which also serves as a generator for $\mathbf{m}_v$ in O_v. In this case the Primary Decomposition Theorem goes as follows :

<u>3.2 COROLLARY</u> If T is a torsion O_S- module, then T is a direct sum

$$T = \oplus_{v\in S} T_v$$

where, for each $v \in S$, T_v is an O_v-module given by

$$T_v = \{ t \in T : q^r{}_v t = 0 \text{ for some } r \geq 0 \}.$$

Furthermore T_v is the localization of T at the maximal ideal of O_S corresponding to the valuation v.

It remains to study the structure of torsion O_v-modules. If T is f.g., then by the structure theory it is sufficient to consider the case of a single cyclic module, i.e. $T = O_v/\mathbf{m}^n{}_v$. Then, if d is the degree of the valuation v, T is a K-vector space of dimension dn. Choose a generator q of **m** and define a linear transformation $j: T \to T$ by $J(t) = qt$. The map J is a nilpotent K-linear transformation with $J^n = 0$, since q^n lies in $\mathbf{m}^n$. If $d = 1$, then $\dim(T) = n$ and J is similar to a single Jordan block; in this case, knowledge of J determines the O_S-module structure of T uniquely. However, if $d > 1$, the situation is more complicated. Each module $O_v/\mathbf{m}^n{}_v$ is a vector space over $k(v) = O_v/\mathbf{m}_v$. The O_v-module structure of T is then specified by its k(v)-structure and the Jordan form of J considered as a k(v)-linear transformation.

The only infinitely generated modules we consider in this paper are modules of the form $k(z)/O_v$ and direct sums of these modules. The module

$k(z)/O_v$ is torsion, since if $f(z)$ lies in $k(z)$ with $\mathrm{ord}_v(f) = -n$, then $q^n{}_v f(z)$ lies in O_v. It is not f.g., since if it were we could find a single n which would work for all f in $k(z)$.

Accordingly to the Primary Decomposition Theorem, for any S, the O_S-module $k(z)/O_S$ can be split into modal components. In this case we have from 3.2

$$k(z)/O_S = \oplus_{v\in S}\, k(z)/O_v \,.$$

These modules will be discussed in detail and related to systems outputs in the next section.

4. LOCAL SYSTEM THEORY

In this section we study linear systems "locally" using module theory over valuation rings O_v.

Suppose v is a valuation of $k(z)$ of degree $d \geq 1$ and O_v is the corresponding valuation ring. Let $q(z)$ denote a fixed generator of the maximal ideal $\mathbf{m}_v$, such that $q(z)$ is an irreducible polynomial of degree d, or $q(z) = 1/z$ with $d = 1$, if $v = \infty$.

4.1 DEFINITION A local system at v, $\Sigma_v = (X, U, Y, B, C)$, consists of finite dimensional k-vector spaces X (the state space), U and Y (respectively the input and output spaces), together with an O_v-module structure on X.

The O_v-module structure gives explicit endomorphisms of X as follows:

(1) the generator $q(z)$ defines a nilpotent endomorphism $J: X \to X$ by $Jx = q(z)x$;

(2) If $v \neq \infty$ then z lies in O_v and a standard dynamical map $A: X \to X$ can be defined by $Ax = zx$ for all x in X.

These two maps are obviously related by $J = q(A)$, and since J is nilpotent, this implies that the eigenvalues of A must be roots of $q(z)$. Dynamical interpretations in terms of difference equations and Z-transform theory are given as follows:

Finite valuations:

$$x(t+1) = Ax(t) + B\,u(t) \qquad\qquad y(z) = C(zI - A)^{-1}Bu(z)$$

$$y(t) = Cx(t) \qquad\qquad G_v(z) = C(zI - A)^{-1}B$$

The valuation at infinity

$$Jx(t+1) = x(t) + Bu(t) \qquad\qquad y(z) = C(zJ - I)^{-1}Bu(z)$$

$y(t) = Cx(t) \qquad\qquad G_\infty(z) = C(zJ - I)^{-1}B$

Here the two transfer functions $G_v(z)$ and $G_\infty(z)$ can be considered as $k(z)$-linear maps : $U(z) \to Y(z)$.

The natural set of inputs for a local system at v, corresponding to the classical $k[z]$- module $U[z]$, is the free O_v-module $\Omega_v = O_v \otimes_k U$. If $U \equiv k^m$, then we identify $\Omega_v U = O_v{}^m$. The usual injection $i(u) = 1 \otimes u$, $i : U \to \Omega_v U$ corresponds to the inclusion $k^m \to O_v{}^m$. The familiar <u>universal mapping property</u> holds:

to every O_v-module X and every k-linear map B: $U \to X$, there exists a unique O_v-module map $\tilde{B}$: $\Omega_v U \to X$ such that $\tilde{B}|_U = B$. In other words, the diagram

(4.2)

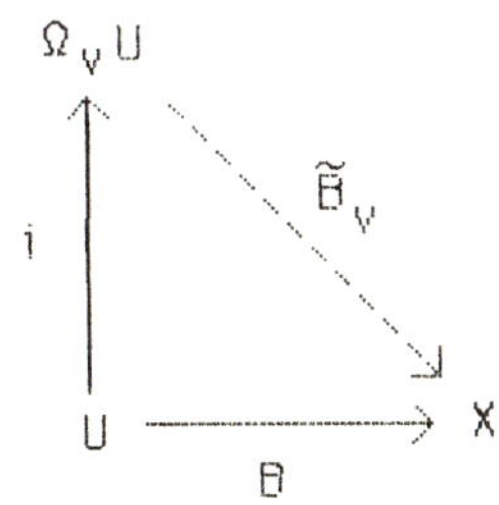

commutes. In fact $\tilde{B}(p(z) \otimes u) = p(z)Bu = p(A)Bu$.

The precise form of the output module depends on whether v is a finite or infinite valuation. Suppose first that $v \neq \infty$, and define $\Gamma_v Y = Y(z)/\Omega_v Y$. If $Y \equiv k^p$, then $\Gamma_v Y$ can be identified with $(k(z)/O_v)^p$, alternatively $\Gamma_v Y \equiv k(z)/O_v \otimes Y$. To dualize the input mapping diagram above, we need to define a map $\rho_v : \Gamma_v Y \to Y$, and it suffices to construct a map $\rho_v : k(z)/O_v \to k$. The map ρ_v is induced from the map T: $k(z) \to k$ given by

$$T(f(z)) = \mathrm{Tr}_{\,k(v)/k} \mathrm{Res}_v f(z)dz.$$

See the Appendix for a discussion of the local residue map and the construction of ρ_v. The <u>output mapping property</u> then states:

for every O_v-module X and every k-linear map C: $X \to Y$, there exists a unique O_v-module map $\tilde{C}_v$: $X \to \Gamma_v Y$ such that $\rho_v \circ \tilde{C}_v = C$. In other words, the diagram

(4.3) 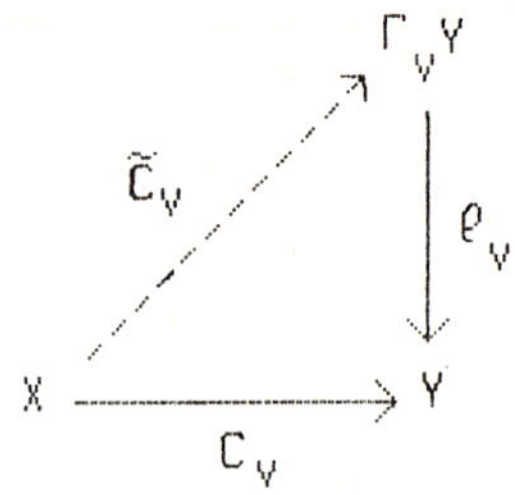

commutes.
In fact, when v is a finite valuation, $\tilde{C}_v$ is given by the familiar expression $\tilde{C}_v x = (zI - A)^{-1}x$, but the verification of $\rho_v \tilde{C} = C$ is a little involved if the degree d > 1. Again see the Appendix.
Following [4] for the point at infinity, we define $\Gamma_\infty Y = Y(z)/z^{-1}\Omega_v Y$. The extra factor of z^{-1} is used to capture the constant (or time t = 0) term of an output sequence. A typical member of $\Gamma_\infty Y$ has the form
$y(z) = y_0 + y_1 z + \cdots + y_r z^r$ for some $r \geq 0$, and $\rho_\infty(y) = -y_0$. (Alternatively, $\rho_\infty(y) = \mathrm{Res}_\infty(z^{-1}y(z))dz$). The same mapping property holds, as above, with $\tilde{C}x = C(zJ - I)^{-1}x$.

To summarize, a local system Σ_v = =(X, U, Y, B, C) gives rise to a commutative local input/output diagram of O_v-modules (valid for finite v and $v=\infty$)

(4.4) 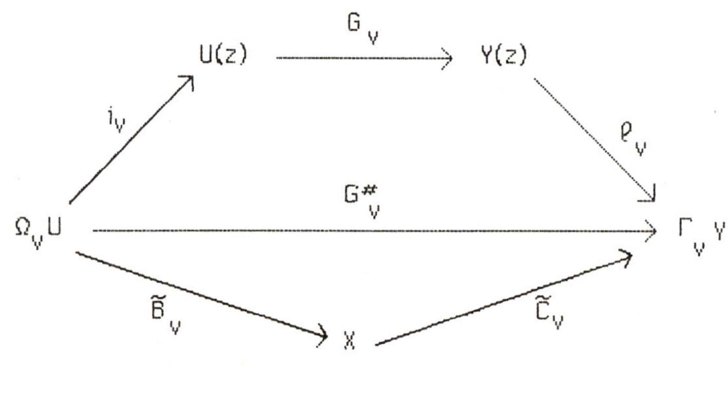

Here G_v is the mapping defined by Z-transform considerations above, $\tilde{B}_v$, $\tilde{C}_v$ are the maps stemming from the mapping properties, and $G^\#_v = \tilde{C}_v \tilde{B}_v$. The maps i, p are obvious inclusion and projection maps. The commutativity of the top trapezoid is a straightforward consequence of the explicit formulas for $\tilde{B}_v$ and $\tilde{C}_v$.

As usual, we say that Σ_v is (locally) reachable if $\tilde{B}_v$ is epic, (locally)

<u>observable</u> if $\tilde{C}_v$ is monic, and (<u>locally</u>) <u>minimal</u> if it is both reachable and observable.

The present machinery allows the development of a local realization theory. Suppose given any transfer function, or k(z)-linear transformation, $G(z)$: $U(z) \rightarrow Y(z)$. Define $G^\#_v = p_v G i_v$, so $G^\#_v : \Omega_v U \rightarrow \Gamma_v Y$ is an O_v-module homomorphism. Define next $X = X_v(G)$ by $X = \Omega_v U/\ker G^\#_v$. Then X is the state space of a minimal local system defined by the triangle

(4.5)

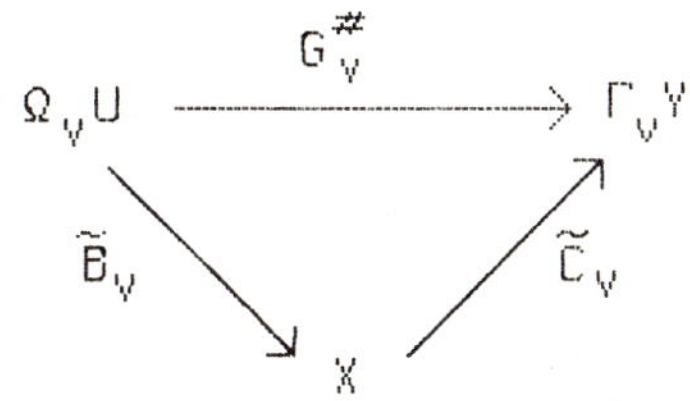

where $\tilde{B}_v$ is the natural projection and $\tilde{C}_v$ is such that $\tilde{C}_v \tilde{B}_v = G^\#_v$.
Explicitly, the dynamic endomorphismas A and J come from the action of O_v on X, B is defined as $\tilde{B}_v|_U$ and $C = \rho_v \circ \tilde{C}_v$. When v is finite,

$$X_v(G) = \Omega_v(U)/(G^{-1}(\Omega_v Y) \cap \Omega_v U) \equiv (G(\Omega_v U) + \Omega_v Y)/\Omega_v Y,$$

so that, essentially, $X_v(G)$ consists of the input sequences which have no poles at v but produce responses having poles at v. We can therefore call $X_v(G)$ the <u>local pole module at v</u> of G(z). Its dimension over k is equal to the number of independent modes at v of the system represented by G(z).
When $v = \infty$, $X_\infty(G)$ contains a non dynamical component related to the possible constant term of G(z). Therefore $X_\infty(G)$ cannot, in general, be interpreted as the module of poles at infinity, but, this one, denoted by $P_\infty(G)$, has to be defined, accordingly to the above discussion, as

$$P_\infty(G) = \Omega_\infty U/(G^{-1}(\Omega_\infty Y) \cap \Omega_\infty U) \equiv (G(\Omega_\infty U + \Omega_\infty Y)/\Omega_\infty Y.$$

If $X_\infty(G) = \oplus_{i=1,\dots,m} O_\infty / z^{-n_i} O_\infty$ is a decomposition of $X_\infty(G)$ into a direct sum of cyclic submodules, then it is not difficult to show that the analogous decomposition for $P_\infty(G)$ is $P_\infty(G) = \oplus_{i=1,\dots,m} O_\infty / z^{-n_i+1} O_\infty$ (see [5] for this and further results).

To compare the original transfer function $G(z)$:$U(z) \rightarrow Y(z)$ with $G^\#_v$, note that any $f(z) \in k(z)$ can be written uniquely as $f(z) = a(z)/q^s(z) + b(z)$ for some $s \geq 0$, $a(z) \in k[z]$ of degree less than ds and $\mathrm{ord}_v b(z) \geq 0$. We define a map $\sigma_v : k(z)/O_v \rightarrow k(z)$ by $\sigma_v([f(z)]) = a(z)/q^s(z)$, which extends to a map $\sigma_v : \Gamma_v Y \rightarrow Y(z)$. Now, a k-basis $u_1, \dots, u_m$ of U serves as an O_v-basis of $\Omega_v U$ and as a k(z)-basis of U(z), then, given $G^\#_v : \Omega_v U \rightarrow \Gamma_v Y$, we can define $G_v : U(z) \rightarrow$

$Y(z)$ by $G_v(u_i) = \sigma_v G^*{}_v(u_i)$. It follows easily that $G(z) = G^*{}_v(z) + H_v(z)$, where $H_v(z)$ is a transfer function whose local realization at v is zero. That is, $H_v(z)$ has no poles at v.

Besides the notion of local pole module we have described, we can introduce the notion of local zero module at v of G(z), denoted by $Z_v(G)$. Following the general lines of [18], we consider, in defining $Z_v(G)$, the input sequences which produces responses having no poles at v and we ignore, at the same time, those belonging to Ker G + $\Omega_v U$. In this way we take into account, only the input sequences whose poles at v fail to appear in the response. In formula, we have

$$Z_v(G) = (G^{-1}(\Omega_v Y) + \Omega_v U)/(\mathrm{Ker}\, G + \Omega_v U).$$

For $v=\infty$, $Z_\infty(G)$ is described in details in [5]. Various application of $Z_\infty(G)$ to factorization problems, such as the causal factorization, the exact model matching and the pseudo-inversion problem, can be found in [6],]7], [8], [9].These results extend to $Z_v(G)$ with almost no change.

5. GLOBAL SYSTEM THEORY

In this Section, we consider linear systems whose modes are restricted to lie in a fixed (but arbitrarily specified) set. This theory includes the classical case, in which any finite mode may appear, as well as the theory of "stable systems" in which a distinguished set of modes, called stable, is identified. If the point at infinity is included in the specified set, then the infinite frequency behaviour of systems may also be studied.

Let S be a set of valuations of k(z) which omits at least one valuation of degree one, and consider the ring O_S of Section 2.

5.1 DEFINITION An S-system $\Sigma_v = (X, U, Y; B,C)$ consists of finite dimensional k-vector spaces X, U and Y, k-linear maps $B : U \to X$ and $C : X \to Y$, and an O_S-module structure on X.

Since X is finite dimensional, it must be a f.g. torsion O_S-module. According to the Primary Decomposition Theorem we can write X as direct sum of modal pieces

$$X = \oplus_{v \in S} X_v$$

where only a finite number of the X_v are nonzero. Each X_v is an O_v-module, giving rise to a dynamic endomorphism $A_v : X_v \to X_v$ (induced by z for v finite) and $J_\infty : X_\infty \to X_\infty$ (induced by 1/z).

Furthermore $B: U \to X$ can be composed with the appropriate natural projections $X \to X_v$ to give $B_v: U \to X_v$, and $C: X \to Y$ can be composed with the natural injections $X_v \to X$ to give $C_v : X_v \to Y$. To simplify notation, write the nonzero components as $X_i = X_{v_i}$ for $i = 1,2, \cdots, r$ (say) and index A, B and C accordingly. If v_∞ appears in S, we write X_∞, J_∞, B_∞ and C_∞. The local transfer functions of the last section glue together to give a global transfer function $G_S(z)$ given by

$$G_S(z) = \begin{bmatrix} C_1 & C_2 & \cdots & C_r & C_\infty \end{bmatrix} \begin{bmatrix} zI - A_1 & & & & \\ & zI - A_2 & & & \\ & & \cdots\cdots & & \\ & & & zI - A_r & \\ & & & & zJ_\infty - I \end{bmatrix}^{-1} \begin{bmatrix} B_1 \\ B_2 \\ \cdots \\ B_r \\ B_\infty \end{bmatrix}$$

where the component at infinity may be absent.

For some purposes we lump the finite modes together and write

$$X = X_{finite} \oplus X_\infty$$

This leads to a transfer function of the form

$$G_S(z) = \begin{bmatrix} C_f & C_\infty \end{bmatrix} \begin{bmatrix} zI - A_f & \\ & zJ_\infty - I \end{bmatrix}^{-1} \begin{bmatrix} B_f \\ B_\infty \end{bmatrix}$$

such that, if they appear, J_∞ is nilpotent and A_f is a matrix whose characteristic polynomial $\chi(A_f)$ has the property that $\mathrm{ord}_v(\chi(A_f)) > 0$ only for v in S. Loosely speaking, A_f has eigenvalues only in the region defined by S. It is often convenient to write

$$C = [\, C_\infty \quad C_f \,] \qquad B = \begin{bmatrix} B_\infty \\ B_f \end{bmatrix} \qquad A = \begin{bmatrix} I & 0 \\ 0 & A_f \end{bmatrix} \qquad E = \begin{bmatrix} J & 0 \\ 0 & I \end{bmatrix}$$

so that $G_S(z)$ appears in "descriptor" or "singular" form

$$G_S(z) = C\,(zE - A)^{-1} B$$

See [4], [15] for this formulation.

The development of an abstract realization theory requires input and output O_S-modules with the familiar mapping properties. These modules can be defined as follows.

For inputs, let $\Omega_S U = O_S \otimes_k U$. Suppose given a k-linear map $B : U \to X$. Then B extends to a unique O_S-module homomorphism $\tilde{B}_S : \Omega_S U \to X$ defined by $\tilde{B}_S(f \otimes u) = f \cdot Bu$ for all $u \in U$ and $f \in O_S$.

For outputs, define $\Gamma_S Y = Y(s)/\Omega'_S Y$ where $\Omega'_S Y = \Omega_S Y$ if v_∞ does not belong to S, and $\Omega'_S Y = \Omega_S Y \cap z^{-1}\Omega_\infty Y$ otherwise. This slight complication is necessary because of the special form of $\Gamma_\infty Y$.

By the Primary Decomposition Theorem $\Gamma_S Y = \oplus_{v \in S} \Gamma_v$ so we can define $\rho_S : \Gamma_S \to Y$ by $\rho_S(\gamma) = \Sigma_{v \in S}\, \rho_v(\gamma)$. For every k-linear map $C : X \to Y$ there is a unique O_S-module map $\tilde{C}_S : X \to \Gamma_S Y$ such that $\rho_S \cdot \tilde{C}_S = C$, namely $C_S = \oplus_v \tilde{C}_v$. In terms of these modules, the S-system $\Sigma_S = (X, U, Y; B, C)$ gives a commutative realization triangle

(5.2)

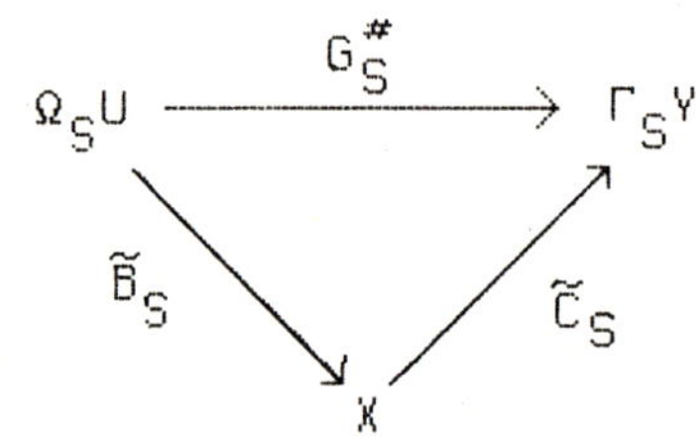

where by definition $G^{\#}_S = \tilde{C}_S \cdot \tilde{B}_S$. We say that Σ_S is <u>S-reachable</u> if $\tilde{B}_S$ is epic, <u>S-observable</u> if $\tilde{C}_S$ is monic, and <u>S-minimal</u> if it is both reachable and observable.

We would like to establish a "local-global" principle for the realization triangle. Localization at the maximal ideal $\mathbf{m}_v$ described earlier in this paper, is a functor which transforms O_S-modules and O_S-homomorphisms to O_v-modules and O_v-homomorphisms. See [1], chap.3 for more details. Free modules localize to free modules, so that $\Omega_S U$ localizes to $\Omega_v U$. Torsion modules localize to their $\mathbf{m}_v$-primary pieces, so that X_S localizes to X_v and $\Gamma_S Y$ localizes to $\Gamma_v Y$. The localization of the realization triangle is

(5.3)

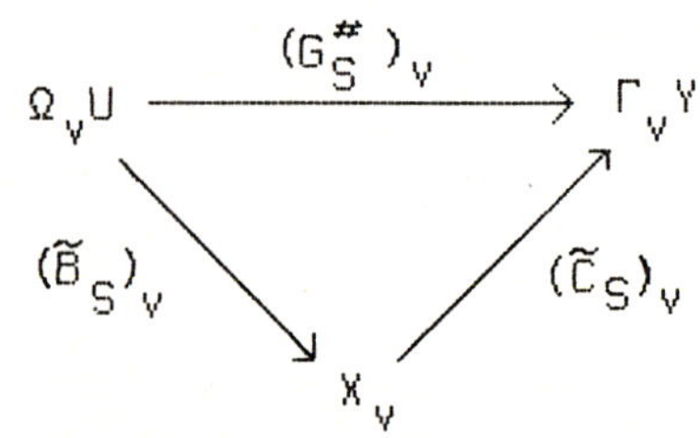

Since the module localizations have been identified, our problem is to identify the mappings. In fact, this localized triangle <u>coincides</u> with the

local realization triangle of Section 4. We omit the details which are straightforward but a little tedious. The ingredients for the proof are the uniqueness properties of $\tilde{B}_v$ and $\tilde{C}_v$ and the fact that (for torsion modules at least) localization of a morphism is just the restriction of the map to the corresponding $\mathbf{m}_v$-primary pieces. Furthermore, a global map of O_S-modules is epic or monic iff all its localizations are epic or monic, respectively. ([1], p.40).

This allows us to summarize the facts as follows:

5.4 Local-Global Principle for Linear Systems Consider an S-systems $\Sigma_S = (X_S, U, Y; B, C)$. For each v in S, the corresponding local system at v is given by $\Sigma_v = (X_v, U, Y; B_v, C_v)$ where X_v is the $\mathbf{m}_v$-primary component of X_S and B_v and C_v are obtained by projecting or restricting B and C to this component. The S-realization triangle for Σ_S localizes to the local realization triangle for Σ_v. The system Σ_S is S-reachable (S-observable) if and only if Σ_v is locally reachable (locally observable) for every v in S.

This completes the basic work on the restricted input/output theory of a given S-system. To reverse the process, we would like to do realization theory.

Given a k(z)-linear map $G(z) : U(z) \to Y(z)$ we would like to find a minimal S-system which produces the S-part of G(z) in an appropriate sense. Consider the standard injection $i_S : \Omega_S U \to U(z)$ and the standard projection $p_S : Y(z) \to \Gamma_S Y$, and define $G^*_S = p_S \cdot G(z) \cdot i_S$, so $G^*_S : \Omega_S U \to \Gamma_S Y$ is an O_S-module homomorphism. Define next the O_S-module X_S by $X_S = \Omega_S U/\mathrm{Ker} G^*_S$, or (equivalently via a natural isomorphism), $X_S = \mathrm{Im}\, G^*_S \subseteq \Gamma_S Y$. Then X_S is the state space of a minimal S-system defined by a triangle as in (5.2) where, G^*_S being given, $\tilde{B}_S : \Omega_S U \to X_S$ is the natural projection, and $\tilde{C}_S : X_S \to \Gamma_S Y$ is such that $\tilde{C}_S \cdot \tilde{B}_S = G^*_S$. Just as in the local case, we can costruct B by restricting $\tilde{B}_S$ to U, while C comes as $C = \rho_S \cdot \tilde{C}_S$.

The minimal S-system $\Sigma_S = (X_S, U, Y; B, C)$ so constructed is unique in a strong sense [17]. Furthermore, the Local-Global Principle assures that each local realization $\Sigma_v = (X_v, U, Y; B_v, C_v)$ is a localization of Σ_S and is itself minimal. Thus, Σ_S can be constructed from the direct sum of finitely many local minimal realizations.

The S-pole module $P_S(G)$ and the S-zero module $Z_S(G)$ are defined, analogously to the local case, by

$$P_S(G) = \Omega_S U/(G^{-1}(\Omega_S Y) \cap \Omega_S U) \equiv (G(\Omega_S U) + \Omega_S Y)/\Omega_S Y$$

and by

$$Z_S(G) = (G^{-1}(\Omega_S Y) + \Omega_S U)/(\mathrm{Ker}\ G + \Omega_S U).$$

If $v=\infty$ does not appear in S, $P_S(G)$ coincides with $X_S(G)$. In particular, if S consists of all finite valuations, $P_S(G)$ is the state space of the classical realization theory (see also [18]). In this case, moreover, $Z_S(G)$ is the zero module introduced first in [18]. Applications of these notions to factorization problems are in [18], [20], [6].
Clearly, the localization at any v of $Z_S(G)$ coincides with $Z_v(G)$, while the localization at any finite v of $P_S(G)$ coincides with $X_v(G)$ and the localization at $v=\infty$ of $P_S(G)$ coincides with $P_\infty(G)$.

APPENDIX - Residues and Outputs

Assume that k is a perfect field (i.e. k admits no inseparable extensions). This hypothesis is satisfied if k has characteristic zero or if k is finite. Consider a valuation ring $O_v \subset k(z)$, let q(z) be a generator of the maximal ideal of O_v, and let $k(v) = O_v/\mathbf{m}_v$ be the residue class field. Then every rational function f(z) in k(z) can be expanded in a formal Laurent series in q over k(v):

$$f(z) = c_{-n}q^{-n} + \cdots + c_{-1}q^{-1} + c_0 + c_1 q + \cdots + c_r q^r + \cdots\cdots$$

with $c_i \in k(v)$ (see [3]). Now, every differential form ω of k(z) can be written in the form $\omega = f(z)dz$ and the <u>residue</u> is defined by

$$\mathrm{Res}_v f(z)dz = c_{-1}.$$

The crucial facts about the residue are, first, it is independent of the choice of q(z), and, second, it satisfies the <u>residue formula:</u>

$$\Sigma_v \mathrm{Tr}_{k(v)/k} \mathrm{Res}_v f(z)dz = 0$$

for all f(z) in k(z), where the sum runs over <u>all</u> valuations v of k(z).

Assume next that v is a finite valuation corresponding to an irreducible polynomial q(z) of degree $d \geq 1$. Define $T_v : k(z) \to k$ by

$$T_v(f(z)) = \mathrm{Tr}_{k(v)/k} \mathrm{Res}_v f(z)dz = \mathrm{Tr}_{k(v)/k} \mathrm{Res}_v (f(z)dq/q'(z)),$$

using the fact that $dq = q'(z)dz$.

Since k is perfect, q(z) is a separable polynomial and $\mathrm{ord}_v q'(z) = 0$. It follows that if f(z) lies in O_v, so does $f(z)/q'(z)$, and $(f(z)/q'(z))dz$ has residue 0. Thus, T_v kills O_v, and therefore T_v induces a k-linear mapping $\rho_v : k(z)/O_v \to k$. If $Y = k^p$, $\Gamma_v Y = (k(z)/O_v)^p$, so we can extend ρ_v coordinatewise to $\rho_v : \Gamma_v Y \to Y$.

<u>THEOREM</u> Let v be a valuation of k(z) different from ∞. If X is any finitely generated torsion O_v-module, and $C : X \to Y$ is any k-linear transformation, then there exists a unique O_v-module homomorphism $\tilde{C} : X \to \Gamma_v Y$ such that $\rho_v \cdot \tilde{C} = C$. In fact, if $A : X \to X$ is defined by $Ax = zx$, then $\tilde{C} = C(zI - A)^{-1}x$ (mod $O_v{}^p$).

Proof. Existence and uniqueness follow from abstract nonsense. See [11]. The important point here is the explicit formula. We must show that

$$\rho_v C(zI - A)^{-1}x = Cx, \text{ for all } x \text{ in } X.$$

It suffices to consider the case Y = X and C = identity, in which case the result follows from a computational lemma.

<u>LEMMA</u> Suppose X is a finite dimensional O_v-module as above, and z in O_v induces the endomorphism $A : X \to X$, $Ax = zx$. Then

$$Tr_{k(v)/k} \, Res_v(zI - A)^{-1}dz = \mathbf{I}.$$

Proof of Lemma. Recall that q(A) is nilpotent since q(z) generates $\mathbf{m}_v$. Therefore the characteristic polynomial of A is $q(z)^r$ for some r. If q(z) has degree d, then X has dimension n = dr over k. Using the classical adjoint, we obtain that

$$q(z)^r(zI - A)^{-1} = adj(zI - A) = B_0(z) + B_1(z)q(z) + .. + B_{r-1}(z)q(z)$$

where each $B_i(z)$ is a polynomial matrix of degree less than r. In fact,

$$B_{r-1}(z) = z^d I + z^{d-1}B^1{}_{r-1} + \cdots$$

since the leading term of adj(zI - A) is $z^n I$. Thus

$$(zI - A)^{-1}dz = ((zI - A)^{-1}/q'(z))\, dq =$$

$$= (B_{r-1}(z)/q'(z))\, dq/q + (B_{r-2}(z)/q'(z))dq/q^{\,2} + \cdots$$

If β is a root of q(z) in k(v), we have

$$Res_v(zI - A)^{-1}dz = B_{r-1}(\beta)/q'(\beta) =$$

$$= \beta^d/q'(\beta) \cdot \mathbf{I} + \beta^{d-1}/q'(\beta)\, B^1{}_{r-1} + \cdots$$

for various constant matrices $B^i{}_{r-1}$. By a standard identity [13]

$$Tr_{k(v)/v} \; \beta^i / q'(\beta) = \begin{cases} 1 \text{ if } & i = d \\ 0 \text{ if } & 0 \le i \le d-1 \end{cases}$$

and we obtain immediately $\quad Tr_{k(v)/k} \, Res_v \, (zI - A)^{-1} \, dz = \mathbf{I}$.

Consider next the classical case O = k[z], corresponding to S = { all finite valuations }. Then $\Gamma_S k = k(z)/k[z]$, and we have to competing definitions of the critical mapping $\rho_S : \Gamma_S k \to k$. Classically an element γ of $\Gamma_S k$ has a unique strictly proper representative $y(z) = y_1 z^{-1} + y_2 z^{-2} + \ldots$, and $\rho_S(\gamma) = y_1$. The local-global approach gives $\rho_S(\gamma) = \Sigma \, \rho_v(\gamma)$, summing over all finite valuations. In fact, these two definitions coincide, since

$$\Sigma \rho_v(\gamma) = \Sigma T_v(y(z)) \ (v \text{ finite}) = \Sigma \ Tr_{k(v)/k} Res_v \ y(z)\, dz = - Res_\infty \ y(z)\, dz$$

by the residue formula.

Now, using $d(z^{-1}) = -z^{-2}dz$, we compute $Res_\infty \ y(z)\, dz = -y_1$, as required.

REFERENCES

[1] Atiyah, M.F. and Mac Donald, I.G. , Introduction to Commutative Algebra (Addison-Wesley, Reading ,1969)

[2] Hartley, B. and Hawkles, T.O., Rings, Modules and Linear Algebra, (Chapman and Hall, London, 1970)

[3] Chevalley, C., Introduction to Theory of Algebraic Function of One Variable, (A.M.S. , 1951)

[4] Conte, G. and Perdon, A.M., Generalized state space realization of non proper rational transfer functions, Systems & Control Letters, 1 (1982) 112-124

[5] Conte, G. and Perdon, A.M., Infinite zero module and infinite pole module, Proc. VII Int. Conf. on Analysis and Opt. of Systems, Nice, Lecture Notes in Control and Information Science, Springer-Verlag 62 (1984) 302-315

[6] Conte , G. and Perdon, A.M., Zero module and factorization problems, to appear in A.M.S. Series in Contemporary Mathematics (1985)

[7] Conte, G. and Perdon, A.M., On the causal factorization problem, to appear in IEEE Trans. Aut. Control (1985)

[8] Conte, G. and Perdon, A.M., On the minimum delay problem, Systems & Control Letters, 5 (1985) 213-215

[9] Conte, G. and Perdon, A.M., Zeros of cascade composition, to appear in Proceedings of MTNS 85, Stockholm (1985)

[10] Hartshorne, R., Algebraic Geometry, (Graduate Texts in Mathematics, Springer-Verlag , 1977)

[11] Hartshorne, R., Local Cohomology, (Lecture Notes in Mathematics, Springer-Verlag , 41, 1966)

[12] Khargonekar, P. and Ozgüler, A.B., Systems theoretic and algebraic aspects of the rings of stable and proper rational functions, Linear Algebra and Appl., 66 (1985)

[13] Lang, S., Algebra, (Addison-Wesley, Reading, 1965)

[14] Pernebo, L., An algebraic theory of controllers for linear multivariable systems- Part I: Structure Matrices and feedforward design", IEEE Trans. on Aut. Control, AC-26 (1981)

[15] Verghese, G., Infinite-frequency behaviour in generalized dynamical systems, Ph. D. Thesis, Dept. Electrical Eng., Stanford Univ. (1978)

[16] Vardulakis, A. and Karcanias, N., Structure, Smith-MacMillan form and coprime MFD of a rational matrix inside a region $P = \Omega \cup \{\infty\}$, Int. J. on Control, 38 (1983) 927-958

[17] Wyman, B., Time-varying linear discrete time systems: realization

theory", Studies in Foundations and Combinatorics, G. Rota ed., Academic Press (1978)

[18] Wyman, B. and Sain, M., The zero module and essential inverse systems IEEE Trans. Circuit and Systems, CAS-28 (1981) 112-126

[19] Wyman, B. and Sain, M., Internal zeros and the system matrix, Proc. 20th Allerton Conf. on Communication, Control and Computing, Urbana (1982) 153-158

[20] Wyman, B. and Sain, M., The zero module of a minimal realization, Linear Algebra and Appl., 50 (1983) 621-637

3. APPLICATIONS

Frequency Domain and State Space Methods for Linear Systems
C.I. Byrnes and A. Lindquist (Editors)

MODAL CONTROL AND VIBRATIONS

Dr. A. N. Andry, Jr.
Director of Engineering
Teledyne Solid State Products
Hawthorne, CA, USA

Dr. E. Y. Shapiro
General Manager, Special Products
HR Textron
Valencia, CA, USA

Dr. K. M. Sobel
Senior Scientist
Lockheed California Company
Burbank, CA, USA

ABSTRACT

The use of output feedback is considered as a means of shaping the response of a vibrating system. Of primary importance is the nature of the underlying eigenstructure (eigenvalues/eigenvectors) of the system. The technique has proven to be a valuable one in the preliminary design stage of synthesizing a feedback controller in order to alter the vibrational characteristics of a system without resorting to a costly redesign.

1. Motivating Example

An important use of active control might be to dispense with the need for a costly redesign of a physical system. A typical example might be the use of a feedback control system to attain a desired set of natural frequencies and/or to reshape mode shapes of a structure rather than adding and/or redistributing mass and stiffness.

Along these lines, we consider a simple motivating example. Assume we have an undamped structure whose model has been discretized and is described by the familiar equation

$$M\underline{\ddot{x}}(t) + K\underline{x}(t) = \underline{0} \tag{1}$$

where $\underline{x}$ is an n-vector of "displacements" and M, K are the mass and stiffness matrices, respectively. For the sake of argument, also assume that M, K are symmetric and positive definite.

Given this background, (1) has "natural frequencies" $\{\omega_i^2\}$, i = 1, 2, ..., n and mode shapes $\{\underline{w}_i\}$, i = 1, 2, ..., n, [1]. Suppose these frequencies and/or mode shapes are unacceptable and that acceptable sets are given by $\{\Omega_i^2\}$ and $\{\underline{W}_i\}$, i = 1, 2, ..., n. The question is: how does one achieve these acceptable sets?

A standard approach to this problem might be to determine modifications to the M and K matrices necessary to achieve the acceptable sets and then attempt to realize the model. These changes in M and K, quite possibly very expensive to actually realize, are reflected in new mass and stiffness matrices, M_0 and K_0, respectively, and a new set of desired equations of motion,

$$M_0\ddot{\underline{x}}(t) + K_0\underline{x}(t) = \underline{0} \qquad \text{or} \tag{2'}$$

$$\ddot{\underline{x}}(t) + M_0^{-1}K_0\underline{x}(t) = \underline{0} \tag{2}$$

A different, and possibly cheaper, approach might be to modify the natural frequencies and mode shapes by addition of a feedback control system. This control system would be driven by outputs taken from measurement of a dynamic behavior of the original system. Assuming that we have m infinite bandwidth actuators and r infinite bandwidth sensors, then the equations describing the augmented system are given by

$$M\ddot{\underline{x}}(t) + K\underline{x}(t) = B\underline{u}(t) \tag{3}$$

$$\underline{y}(t) = C\underline{x}(t) \tag{4}$$

where $\underline{u}$ is the m-vector of force inputs, B is an (nxm) matrix reflecting the manner in which forces are introduced, $\underline{y}$ is the r-vector of outputs, and C is an (rxn) matrix reflecting the manner in which measurements are taken. Assuming we wish to implement constant output feedback, we seek an (mxr) matrix, F, such that the control

$$\underline{u}(t) = -F\underline{y}(t) \tag{5}$$

when applied yields

$$M\underline{\ddot{x}}(t) + K\underline{x}(t) = -BFC\underline{x}(t) \quad \text{or} \tag{6}$$

$$M\underline{\ddot{x}}(t) + (K+BFC)\underline{x}(t) = \underline{0} \quad \text{or} \tag{7'}$$

$$\underline{\ddot{x}}(t) + M^{-1}(K+BFC)\underline{x}(t) = \underline{0} \tag{7}$$

In order for (7) and (2) to be equivalent, we must have

$$M^{-1}(K+BFC) = M_0^{-1}K_0 \tag{8}$$

From physical considerations (to be discussed later), we assume B and C possess full rank; invoking this assumption and carrying out matrix arithmetic, we calculate F to be

$$F = (B^TB)^{-1}B^TM[M_0^{-1}K_0-M^{-1}K]C^T(CC^T)^{-1} \tag{9}$$

Thus, for this example, if F is implemented, the closed loop system will have the same characteristics as the modified system. We note that the gains that appear in the feedback matrix depend on

a. The "differences" between the original mass and stiffness matrices and the desired mass and stiffness matrices;

b. The B, C matrices which reflect how and where the forces are applied (B) and how and which measurements are taken (C).

We also note that the gains in F may be altered by the choices of inputs (types and position) and measurements (types and location). Hence, as stated previously, it may be possible to use a control system to ensure desired response qualities rather than a structural alteration which may incur larger redesign costs as well as performance penalties.

We now proceed in more detail.

2. Assumptions and Problem Statement - Modal Control

We consider the following problem. Assume that we have a system described by the set of differential equations given as

$$M\ddot{\underline{q}}(t) + D\dot{\underline{q}}(t) + K\underline{q}(t) = \underline{F}(t) \tag{10}$$

where

i) M, D, and K are real square matrices of dimension n;

ii) M is of full rank;

iii) $\underline{q}(t)$, $\underline{F}(t)$ are the generalized forces, respectively

Typically, if (10) results from an elastic system, there may be additional symmetry and/or matrix structure assumptions concerning M, D, and K. For this presentation, these additional assumptions are not required.

To imbed this system in a control setting, we assume that the vector of generalized forces, $\underline{F}(t)$, may be written as

$$\underline{F}(t) = B_0\underline{u}(t). \tag{11}$$

The matrix B_0, termed the input matrix, is an (nxm) real valued matrix. The term $\underline{u}(t)$ is an m-vector of inputs, called the control. For this paper, we assume that the inputs result from actuators possessing an infinite bandwidth; the more realistic case of finite bandwidth actuators will be dealt with in a later paper.

In addition, we assume that a variety of measurements are being taken. These measurements are a linear combination of the generalized displacements $\underline{q}$ and/or velocities $\dot{\underline{q}}$ and are expressed as

$$\underline{y}(t) = C_0\underline{q}(t) + C_1\dot{\underline{q}}(t) \tag{12}$$

where dim $\underline{y}$ = r and C_0 and C_1 are real valued matrices of appropriate dimension.

Two final assumptions are:

$$\begin{aligned} &\text{rank } [B_0] = m \\ &\text{rank } [C_0 \; C_1] = r \end{aligned} \tag{13}$$

The justification of these last two statements is straightforward. In practical problems, no two control inputs will serve precisely the same function and, except for redundancy requirements, no two sensors will give precisely the same measurement. Essentially, these rank conditions attest to the independence of both inputs and outputs of the system.

We may now rewrite (10) along with (12) as our describing equations:

$$\begin{aligned} &\underline{M}\ddot{\underline{q}}(t) + \underline{D}\dot{\underline{q}}(t) + \underline{K}\underline{q}(t) = B_0\underline{u}(t) \\ &\underline{y}(t) = C_0\underline{q}(t) + C_1\dot{\underline{q}}(t) \end{aligned} \tag{14'}$$

Recalling that M is of full rank, we rewrite (14') as

$$\begin{aligned} &\ddot{\underline{q}}(t) + M^{-1}D\dot{\underline{q}}(t) + M^{-1}K\underline{q}(t) = M^{-1}B_0\underline{u}(t) \\ &\underline{y}(t) = C_0\underline{q}(t) + C_1\dot{\underline{q}}(t) \end{aligned} \tag{14}$$

We put these equations into state space form by letting $B = M^{-1}B_0$, $\underline{x}_1 = \dot{\underline{q}}$ and $\underline{x}_2 = \underline{q}$ to obtain

$$\begin{aligned} \begin{bmatrix} \dot{\underline{x}}_1(t) \\ \dot{\underline{x}}_2(t) \end{bmatrix} &= \begin{bmatrix} -M^{-1}D & -M^{-1}K \\ I_n & 0 \end{bmatrix} \begin{bmatrix} \underline{x}_1(t) \\ \underline{x}_2(t) \end{bmatrix} + \begin{bmatrix} B \\ 0 \end{bmatrix} \underline{u}(t) \\ \underline{y}(t) &= \begin{bmatrix} C_1 & C_0 \end{bmatrix} \begin{bmatrix} \underline{x}_1(t) \\ \underline{x}_2(t) \end{bmatrix} \end{aligned} \tag{15}$$

Trivially, rank $\begin{bmatrix} B \\ 0 \end{bmatrix} = m$.

Corresponding to this 2n dimensional system are sets of eigenvalues $\{\lambda_i\}$, $i = 1, 2, \ldots, 2n$ and eigenvectors $\{\underline{v}_i\}$, $i = 1, 2, \ldots, 2n$. In relation to the motivating example, each λ_i has a magnitude related to the square root of the appropriate ω_i^2 while each eigenvector $\underline{v}_i$ is related to the appropriate $\underline{w}_i$ in the following way:

$$\underline{v}_i = \begin{bmatrix} \lambda_i \underline{w}_i \\ \\ \underline{w}_i \end{bmatrix}$$

Given this background and the previous assumptions, we state the modal control problem:

Given a self conjugate set of scalars, $\{\lambda_i^d\}$, $i = 1, 2, \ldots, r$ and a corresponding self conjugate set of vectors, $\{\underline{v}_i^d\}$, $i = 1, 2, \ldots, r$, find a real (mxr) matrix F such that r of the eigenvalues of the closed loop system are precisely those of the set $\{\lambda_i^d\}$ with corresponding eigenvectors the set $\{\underline{v}_i^d\}$

whenever

$$\underline{u}(t) = F\underline{y}(t)$$

From a physical standpoint, we are trying to use output feedback to control frequencies and, if necessary, damping ratio and mode shapes.

3. Eigenvector ("Mode Shape") Assignability

Before focusing on the feedback gain matrix necessary to achieve the desired sets, it is necessary to study the nature of the eigenvectors of the closed loop systems. As we shall see, these eigenvectors will play a crucial role in the feedback gain calculation.

To carry out this analysis, recall equations (15) and the associated relevant assumptions. Given these statements as our background, we proceed with the technical analysis of

(1) characterizing eigenvectors (hence, mode shapes) which can be assigned as closed loop eigenvectors; and

(2) determining the best possible approximation of a desired closed loop eigenvector that cannot be achieved. The cases of full and partial specifications are treated.

3a. Characterization of $\underline{v}_i$

We begin by considering the closed loop system written in the following way:

$$\dot{\underline{x}}(t) = \begin{bmatrix} -M^{-1}D & -M^{-1}K \\ I_n & 0 \end{bmatrix} \underline{x}(t) + \begin{bmatrix} B \\ 0 \end{bmatrix} F \begin{bmatrix} C_1 & C_0 \end{bmatrix} \underline{x}(t) \tag{16}$$

Assume we are given $\{\lambda_i\}$, i = 1, 2, ..., r as the desired eigenvalues and that $\underline{v}_i$ is the eigenvector corresponding to λ_i. Then

$$\lambda_i\underline{v}_i = \begin{bmatrix} -M^{-1}D & -M^{-1}K \\ I_n & 0 \end{bmatrix} \underline{v}_i + \begin{bmatrix} B \\ 0 \end{bmatrix} F\, C\, \underline{v}_i \tag{17}$$

where

$$C = [C_1 \quad C_0]$$

Thus,

$$\underline{v}_i = \begin{bmatrix} \lambda_i I_n + M^{-1}D & M^{-1}K \\ -I_n & \lambda_i I_n \end{bmatrix}^{-1} \begin{bmatrix} B \\ 0 \end{bmatrix} F\, C\, \underline{v}_i \tag{18}$$

where it is tacitly assumed that λ_i is not an eigenvalue of the matrix

$$\begin{bmatrix} -M^{-1}D & M^{-1}K \\ I_n & 0 \end{bmatrix} \quad \text{i.e.,}$$

we are desiring a set of eigenvalues <u>different</u> from the original set.

Analyzing (18), we define the m-vector $\underline{m}_i$ as

$$\underline{m}_i = F\, C\, \underline{v}_i \tag{19}$$

and (18) becomes

$$\underline{v}_i = \begin{bmatrix} \lambda_i I_n + M^{-1}D & M^{-1}K \\ -I_n & \lambda_i I_n \end{bmatrix}^{-1} \begin{bmatrix} B \\ 0 \end{bmatrix} \underline{m}_i \qquad (20)$$

Using standard techniques, (20) reduces to

$$\underline{v}_i = \begin{bmatrix} \lambda_i(\lambda_i^2 I_n + \lambda_i M^{-1}D + M^{-1}K)^{-1}B \\ (\lambda_i^2 I_n + \lambda_i M^{-1}D + M^{-1}K)^{-1}B \end{bmatrix} \underline{m}_i \qquad (21)$$

We simplify notation by letting

$$G_i = (\lambda_i^2 I_n + \lambda_i M^{-1}D + M^{-1}K)^{-1}B \qquad (22)$$

Thus,

$$\underline{v}_i = \begin{bmatrix} \lambda_i G_i \\ G_i \end{bmatrix} \underline{m}_i \qquad (23)$$

The form of (23) is not surprising given the form of the definition of the state vector. Basically, equation (23) states that the eigenvector $\underline{v}_i$ which is associated with the eigenvalue λ_i must lie in the subspace spanned by the columns of

$$\begin{bmatrix} \lambda_i G_i \\ G_i \end{bmatrix} \qquad (24)$$

We note that the dimension of this subspace is precisely m - the number of independent control inputs to the system. The implication is obvious and consistent with engineering intuition - the number of control variables determines the dimension of the subspace in which the achievable eigenvectors must reside; i.e., the more inputs one has, the more flexibility one has in shaping the eigenvector. The orientation of this subspace depends upon i) the parameters in M, D, and K, and ii) the desired eigenvalue, λ_i.

3b. A Best Possible $\underline{v}_i^d$

It is a rare case when the desired eigenvector lies within the span of the matrix

$$L_i = \begin{bmatrix} \lambda_i G_i \\ G_i \end{bmatrix} \tag{25}$$

Typically, the desires of the designer do not translate to eigenvectors which lie exactly in the prescribed subspace and, hence, cannot be achieved exactly. In this usual situation, we find the "best possible" achievable eigenvector; i.e., the vector in the prescribed subspace that is, in the least squares sense, closest to the desired eigenvector.

Pictorially,

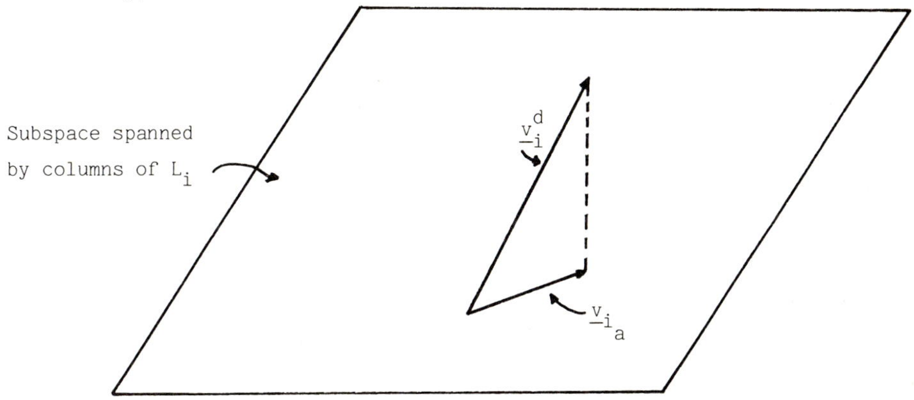

Figure 1
Best Achievable Eigenvector, $\underline{v}_{i_a}$

Mathematically, it is easily computed that

$$\underline{v}_{i_a} = L_i (L_i^T L_i)^{-1} L_i^T \underline{v}_i^d \tag{26}$$

where L_i is defined as before.

Due to possible ill conditioning, care must be taken when inverting $L_i^T L_i$. A Cholesky decomposition or a singular value decomposition is advised at this stage [5].

We re-emphasize the importance of the subspace spanned by L_i and its implications.

(1) If a desired eigenvector $\underline{v}_i^d$ is nearly orthogonal to the subspace spanned by the columns of L_i, then there is little hope of achieving this mode. At this point, the only recourse to the designer would be changing the corresponding λ_i. A different choice of λ_i may result in an orientation of the column space of L_i that would "capture" a larger component of the desired eigenvector. Choosing a different λ_i may not be unreasonable given that in many cases, performance specifications are given in terms of maximum and minimum values for frequencies, damping ratios, rise times, overshoots, settling times, etc. Thus, some flexibility is afforded here naturally and it allows natural performance tradeoffs to be considered by the designer.

(2) The only way to increase the "size" of the subspace in order to more closely approximate the desired eigenvector, is to increase the number of independent control inputs. As stated previously, this is entirely consistent with engineering intuition.

3c. Partial Specification of $\underline{v}_i^d$

In the previous sections, it was assumed that the desired eigenvector was totally specified by the designer. If one is dealing with a large model or a model of a large structure, total specification of the vector may be unwarranted (if not impossible). An example might be that of a large flexible structure where a localized modification may be all that is necessary to achieve the desired results.

To handle such a situation, we again consider our eigenvector and assume that it has the following structure:

$$\underline{v}_i^d = \begin{bmatrix} v_{i1} \\ x \\ x \\ v_{ij} \\ x \\ x \\ v_{i,2n} \end{bmatrix}$$

where

(1) v_{ij} are designer specified components;

(2) x is an unspecified component;

(3) the desired structure is consistent with the form of eigenvector given by (24).

Proceeding, we define, as in [4], a reordering operator as follows:

$$\begin{bmatrix} \underline{v}_i^d \end{bmatrix}^{R_i} = \begin{bmatrix} \underline{l}_i \\ \\ \underline{d}_i \end{bmatrix}$$

where $\underline{l}_i$, $\underline{d}_i$ are the specified and unspecified subvectors of $\underline{v}_i^d$, respectively.

We also reorder the rows of L_i to conform with the reordered elements of $\underline{v}_i^d$, i.e.,

$$\begin{bmatrix} \lambda_i G_i \\ \\ G_i \end{bmatrix}^{R_i} = \begin{bmatrix} L_i \\ \\ D_i \end{bmatrix} \tag{27}$$

If we proceed in precisely the same manner as in Section 3b. with $\underline{l}_i$ replacing $\underline{v}_i^d$ and $\tilde{L}_i$ replacing L_i (in a certain manner), we obtain

$$\underline{v}_{i_a} = L_i(\tilde{L}_i^T\tilde{L}_i)^{-1}\tilde{L}_i^T\underline{l}_i \tag{28}$$

A tacit assumption for this formula is that the dimension of the desired vector is less than the row dimension and rank of $\tilde{L}_i$. If this is not the case, the formula is trivially modified.

In conclusion, we remain mindful of the natures of a desired eigenvector $\underline{v}_i^d$, of an eigenvector $\underline{v}_i$, and of an achievable eigenvector, $\underline{v}_{i_a}$. In all that follows (e.g., the calculation of the feedback gain matrix), achievable eigenvectors are used for computation. We now turn to the feedback gain calculation.

4. Feedback Gain Calculation

From the problem statement, we seek a control law of the form

$$\underline{u}(t) = F\underline{y}(t).$$

Expanding this expression using equation (15), we have

$$\begin{aligned}\underline{u}(t) &= F\,[C_1 \quad C_0]\begin{bmatrix}\underline{x}_1(t)\\ \underline{x}_2(t)\end{bmatrix}\\ &= [FC_1 \quad FC_0]\begin{bmatrix}\underline{x}_1(t)\\ \underline{x}_2(t)\end{bmatrix}\end{aligned} \tag{29}$$

Using (29) in conjunction with (15) again, we have

$$\begin{bmatrix}\dot{\underline{x}}_1(t)\\ \dot{\underline{x}}_2(t)\end{bmatrix} = \begin{bmatrix}-M^{-1}D & -M^{-1}K\\ I_n & 0\end{bmatrix}\begin{bmatrix}\underline{x}_1(t)\\ \underline{x}_2(t)\end{bmatrix} + \begin{bmatrix}BFC_1 & BFC_0\\ 0 & 0\end{bmatrix}\begin{bmatrix}\underline{x}_1(t)\\ \underline{x}_2(t)\end{bmatrix} \tag{30}$$

Before going any further, a few points are in order concerning (30). Let

$$G = \begin{bmatrix}BFC_1 & BFC_0\\ 0 & 0\end{bmatrix}$$

If tr G = 0, then, in a straightforward manner,

$$\mathrm{tr}\begin{bmatrix}-M^{-1}D & -M^{-1}K\\ I_n & 0\end{bmatrix} = \mathrm{tr}\left[\begin{bmatrix}-M^{-1}D & -M^{-1}K\\ I_n & 0\end{bmatrix} + G\right] \quad \text{i.e.,}$$

if tr G = 0, then the sum of the open loop eigenvalues remain constant under feedback. For the trace of G to be zero, we must have tr $(BFC_1) = 0$. Practically speaking, either we do not measure velocities ($C_1 = 0$) or velocities are not fed back (the structure of F mandates $FC_1 = 0$ and hence, $BFC_1 = 0$). A simplistic statement might be that the "total damping" within the system remains "constant" under feedback if tr G = 0.

Returning to (30) and considering the related eigenvalue problem, if λ_i is an eigenvalue and $\underline{v}_i$ is the corresponding eigenvector, then

$$\lambda_i \underline{v}_i = \begin{bmatrix} -M^{-1}D & -M^{-1}K \\ I_n & 0 \end{bmatrix} \underline{v}_i + \begin{bmatrix} BFC_1 & BFC_0 \\ 0 & 0 \end{bmatrix} \underline{v}_i \tag{31}$$

Since $\underline{v}_i$ is a vector of dimension 2n, we partition it conformally with respect to the partitions of the matrices. Thus,

$$\underline{v}_i = \begin{bmatrix} \underline{z}_i \\ \underline{w}_i \end{bmatrix}$$

where dim $\underline{z}_i$ = dim $\underline{w}_i$ = n. The $\underline{z}_i$ term is related to the generalized velocities while the $\underline{w}_i$ term corresponds to the generalized displacements. Using this partition,

$$\lambda_i \begin{bmatrix} \underline{z}_i \\ \underline{w}_i \end{bmatrix} = \begin{bmatrix} -M^{-1}D & -M^{-1}K \\ I_n & 0 \end{bmatrix} \begin{bmatrix} \underline{z}_i \\ \underline{w}_i \end{bmatrix} + \begin{bmatrix} BFC_1 & BFC_0 \\ 0 & 0 \end{bmatrix} \begin{bmatrix} \underline{z}_i \\ \underline{w}_i \end{bmatrix} \tag{32}$$

Expanding this equation yields

$$\begin{aligned} \lambda_i \underline{z}_i &= -M^{-1}D\underline{z}_i - M^{-1}K\underline{w}_i + BFC_1\underline{z}_i + BFC_0\underline{w}_i \\ \lambda_i \underline{w}_i &= \underline{z}_i . \end{aligned} \tag{33}$$

Combining the equations of (33), we have

$$\lambda_i^2 \underline{w}_i = -M^{-1}D\lambda_i\underline{w}_i - M^{-1}K\underline{w}_i + BFC_1\lambda_i\underline{w}_i + BFC_0\underline{w}_i \tag{34}$$

Equation (34) must hold for each eigenvalue/eigenvector pair. More importantly, it will hold for the r pairs that we are trying to modify in accordance with the problem statement. In the compact form of matrix notation,

$$W\Lambda^2 = -M^{-1}DW\Lambda - M^{-1}KW + BFC_1W\Lambda + BFC_0W \tag{35}$$

where

$$W = [\underline{w}_1 \underline{w}_2 \underline{w}_3 \ . \ . \ . \ \underline{w}_r],$$

$$\Lambda = \text{diag}\ [\lambda_1, \lambda_2, \lambda_3, \ . \ . \ ., \lambda_r].$$

Rewriting (35) after matrix arithmetic yields

$$W\Lambda^2 + M^{-1}DW\Lambda + M^{-1}KW = BF[C_1W\Lambda + C_0W] \tag{36}$$

Since B is of full rank, we may rewrite (36) as

$$(B^TB)^{-1}B^T[W\Lambda^2 + M^{-1}DW\Lambda + M^{-1}KW] = F[C_1W\Lambda + C_0W] \tag{37}$$

Ultimately, we wish to extract F from equation (37). From a mathematical standpoint, this inverse will exist as long as the only point of intersection of the null space of the output matrix

$$[C_1 \quad C_0]$$

and the column space of the matrix

$$\begin{bmatrix} W\Lambda \\ W \end{bmatrix}$$

is the origin.

From a practical standpoint, the inverse can be taken only when there is a link between the observations we are making and the eigenvector we are trying to control. More explicitly, recall that the matrix W is populated by the achievable eigenvectors that lie in the column space spanned by L_i. Furthermore, the form of the L_i depends crucially on the input matrix B. Thus, the strong link is between what is being measured and how the control enters the system. This line of reasoning suggests, especially for large space structures, the necessity for the proximity of sensors and actuators; i.e., collocation.

Proceeding as if the required inverse exists, we solve for F as

$$F = (B^TB)^{-1}B^T[W\Lambda^2 + M^{-1}DW\Lambda + M^{-1}KW]\,[C_1W\Lambda + C_0W]^{-1} \tag{38}$$

Obviously, once we have chosen our desired eigenvalues $\{\Lambda\}$, and our eigenvectors $\{W\}$, F follows easily.

Summary

The concept of modal control of vibrating systems using output feedback is seen to be one that is very powerful. Depending on a given physical system, the design can decide among a number of options which contain inherent performance tradeoffs.

In honesty, as users of the above technique, an in-depth knowledge of the system to be controlled is required. The reader must note that once a designer enters the realm of output feedback, stability is not mathematically guaranteed. Hence, care must be exercised in its use.

Finally, the technique has proven successful in studying flutter problems [6].

References

1. Meirovitch, L., Analytical Methods in Vibrations, Macmillan Company, London, 1967.

2. Andry, A. N., Shapiro, E. Y., and Chung, J. C., "Eigenstructure Assignment for Linear Systems," IEEE Transactions on Aerospace and Electronic Systems, Vol. AES-19, No. 5, September 1983, pp. 711-729.

3. Srinathkumar, S., Eigenvalue/Eigenvector Assignment Using Output Feedback," IEEE Trans. Automat. Contr., AC-23, 1, 1978, pp. 79-81.

4. Andry, A. N., Shapiro, E. Y., and Chung, J. C., "Modal Control for a Class of Linear, Discrete, Vibrating Mechanical Systems," Proceedings of the 1st International Modal Analysis Conference, Orlando, Florida, November 1982, pp. 430-435.

5. Wilkinson, J. H., The Algebraic Eigenvalue Problem, Oxford University Press, New York, 1965.

6. Garrard, W. L., and Liebst, B. S., "Active Flutter Suppression Using Eigenspace and Linear Quadratic Design Techniques," AIAA Paper 83-2222-CP, Proceedings 1983 AIAA Guidance and Control Conference, August 1983.

Frequency Domain and State Space Methods for Linear Systems
C.I. Byrnes and A. Lindquist (Editors)

MODAL CONTROL OF ROBOT ARMS WITH JOINT ELASTICITY

AWOBAMISE, B.O., KATENDE, J. and ALOS, A.J.
Electrical Engineering Department
University of Lagos
Lagos, Nigeria.

A multiple joint robot arm is a complex, highly non-linear multivariable system with interaction forces and cross couplings among the joint variables. Movements of robots are task dependent and in most applications the robots must be capable of tracking continuous trajectories well. In this paper, in order to achieve the best possible tracking, an approach which employs a combination of two strategies is proposed. The robot global work space is subdivided into subspaces to permit linearization of system equations and modal control is then deployed to yield a simple controller with high accuracy and non - interacting joint control. The proposed scheme is applied to the control of robot arms with joint elasticity and simulation studies carried out for the manipulation of various payloads along a prescribed path are presented.

INTRODUCTION

The robot control problem basically involves the maintenance of the end-effector at a fixed location (e.g. position and orientation) and/or the need to make it follow accurately a prescribed path in spite of disturbances to the robot system. The design of such a high performance controller is usually fraught with difficulties. Firstly industrial robots are generally described by a set of complex non-linear second order differential equations [1,2] resulting in a multivariable system with interaction forces and cross - couplings among the joint variables. Secondly, for large displacements, the system parameters can vary considerably primarily due to the varying moments of inertia as the link inclinations change during the motion. Moreover, the payload also influences the inertia characteristics of the robot arm. These difficulties have been addressed by the deployment of various controllers, e.g. localized joint servomechanisms (which neglect system dynamics) [3], controllers based on the computed torque method (in which the joint actuating torques are computed on-line by the controller computer using the robot's dynamic model) [4,5] and various adaptive control schemes [6,7,8]. The major disadvantage of most of these schemes either relates to the computational effort which is usually too severe or to the problem of signal interaction which generally results in insufficient tracking accuracy.

An approach consisting of two strategies is employed in this paper to tackle the above problems. On the one hand, the robot's global work space is subdivided into subspaces in each of which the nonlinear system dynamics are modelled by a corresponding set of linear differential equations with parameter variations within the range of the subspace. A modal controller, which reduces the structure of the complex multivariable system into a number of decoupled first order systems, is then evaluated off-line for a specific subspace. The approach is demonstrated by designing a control system for the robot of Fig.1, whose dynamic model incorporates the effects of elasticity at the joints. Results of simulation studies are then presented showing the performance of the designed scheme with respect to payload variations.

LINEARIZATION AND WORK SPACE CONSIDERATIONS

Consider a nonlinear system described by,

$$\dot{x} = f(x,\alpha,u) \tag{1}$$

with $x \in R^n$ output/state vector

$u \in R^m$ input vector

$\alpha \in \Omega(x)$ a compact set of system parameters which are also functions of x.

In the case of an industrial robot, the system parameters, α, are not only functions of x, but also depend on the payload. Linearization of system (1) is usually carried out about an operating point, which takes into account some working conditions. If the working conditions change, the linearized system equations are not valid for the actual operating point due to the deviations from the linearization point. Thus for a robot arm, operating point variations constitute further parametric changes in the linearized system equations. This problem can be overcome if the control system can cope with changes of the nonlinear system due to the variations in the work space, and the payload. Furthermore, to avoid large parameter variations due to large displacements, the global work space is subdivided into subspaces [9].

Let the global work space, W_G, of system (1) be defined by

$$W_G = \{x : x_{\ell b} \leqslant x \leqslant x_{ub}\}, \quad (\ell b = \text{lower bound and ub} = \text{upper bound}) \tag{2}$$

and subdivide it into subspaces W^i with

$$W_G = \bigcup_{i=1}^{N} W^i \tag{3}$$

with

$$W^i = \{x : x^i_{\ell b} \leqslant x \leqslant x^i_{ub}\} \tag{4}$$

Then, system (1) can be represented in each subspace W^i by a set of linearized system equations with nxm transfer function matrix

$$H^i(s,a^{ij}) \tag{5}$$

for every set point $x^j \in W^i$; $j = \overline{1,k}$,

where

- s the Laplace operator
- a^{ij} scalar vector of parameters of the transfer function matrix.
- k the number of fixed parameter set points x^j.

The choice of scalar k depends on the admissible error ε between the real nonlinear system and the linearized one measured by

$$||x(t)_{nonlinear} - x(t)_{linear}|| < \varepsilon \text{ for all t and u(t)} \tag{6}$$

and

$$x(o)_{nonlinear} = x(o)_{linear}$$

The resulting set of linear transfer function matrices represents system (1) with an accuracy determined by ε.

Using this linearized system a robust modal controller [10] is designed. For each subspace W^i the designed controller C^i can be evaluated with a prescribed closed-loop behaviour for all parameter variations a^{ij}; $j = \overline{1,k}$, so that C^i will be switched whenever the system operates in W^i. The whole approach is summarized as follows:

(i) Linearize the nonlinear system **at** N*k set points obtain $H^r(s,a^r)$, $r = \overline{1,N*k}$

(ii) Determine for each subspace W^i, $i = \overline{1,N}$, the corresponding transfer matrix $H^i(s,a^{ij})$, $i = 1,N$, $j = \overline{1,k}$.

(iii) Evaluate C^i for each W^i using modal analysis and synthesis.

MODAL ANALYSIS (A REVIEW).

Modal control techniques typically aim at reducing a complex interconnected system to a parallel combination of first-order subsystems. These techniques utilize modal analysis in order to manipulate the system's eigenvalues or poles for control objectives such as

(i) stabilization of the unstable modes existing in the system's dynamic equations, by new pole assignments and,

(ii) speeding up of the transient response of the system by introducing state variable feedback control loops which cause the eigenvalues of the uncontrolled system to be arbitrarily shifted to more negative values.

A necessary condition for the above objectives to be achievable is that the system must be state or mode controllable [10]. To illustrate modal analysis consider a linear time - invariant dynamical system described by the equations

$$\dot{x} = Ax + Bu$$
$$y = C^T x \tag{7}$$

with

$x \in R^n$ state vector

$y \in R^r$ output vector

$u \in R^m$ input vector

$A \in R^{nxn}$ system matrix

$B \in R^{nxm}$ input matrix

$C^T \in$ rxn output matrix

The transfer function relating u and x is given by

$$H(s) = C^T(sI - A)^{-1}B = \frac{C^T adj(sI-A)B}{det(sI-A)} \tag{8}$$

Thus the poles of H(s) are determined by the eigenvalues of the system matrix A. It is required to suitably shift these poles to new locations using the state variable feedback control law

$$u = Kx + Gv \tag{9}$$

where v is the new external input, G is a nonsingular mxm constant feed forward matrix and K is an mxn constant feedback gain matrix.

The resulting feedback system is depicted in Fig.2 and is described by

$$\dot{x} = (A + BK)x + BGv$$
$$y = C^T x \tag{10}$$

with a closed - loop transfer function matrix of the form

$$H_c(s) = C^T(sI-A - BK)^{-1}B = \frac{C^T adj(sI-A-BK)B}{det(sI-A-BK)} \tag{11}$$

where the characteristic polynomial

$$det\ [sI - A - BK] = p_c(s) \tag{12}$$

has arbitrary roots.

The problem of achieving arbitrary pole locations of $H_c(s)$ is that of selecting constant matrix K so that (A+BK) has arbitrary eigenvalues. The system pair is said to be completely mode controllable if and only if the controllability matrix

$$\phi_c = [B \vdots AB \vdots A^2B \vdots ---- \vdots A^{n-1}B] \tag{13}$$

has rank n. A matrix, P, which puts pair (A,B) into the standard controllable form can be obtained using the analysis in [9]. Thus applying the transformation

$$\hat{x} = Px \tag{14}$$

eqn.(10) becomes

$$\begin{aligned}\hat{\dot{x}} &= (PAP^{-1} + PBKP^{-1})\hat{x} + PBGv \\ &= (\hat{A} + \hat{B}\hat{K})\hat{x} + \hat{B}Gv\end{aligned} \tag{15}$$

where

$\hat{K} = KP^{-1}$

$\hat{A} = (\hat{a}_{ij})$ is a block matrix of the form $\begin{bmatrix} \hat{A}_{11} & \cdots & \hat{A}_{1m} \\ \vdots & & \\ \hat{A}_{m1} & & \hat{A}_{mm} \end{bmatrix}$

with the (m+1) x (m+1) diagonal blocks $\hat{A}_{ii}$ in the companion form

$$\hat{A}_{ii} = \begin{bmatrix} o & 1 & \cdots & o \\ \vdots & o & \cdots & o \\ & \vdots & & \\ o & o & \cdots & 1 \\ -\hat{a}_{ii,o} & -\hat{a}_{ii,1} & & -\hat{a}_{ii,m} \end{bmatrix} \tag{16}$$

and the (m+1) off-diagonal blocks in the form

$$\hat{A}_{ij} = \begin{bmatrix} o & --- & o \\ \vdots & & \vdots \\ o & & o \\ -\hat{a}_{ij,o} & & -\hat{a}_{ij,m} \end{bmatrix} \quad \text{for } i \neq j \tag{17}$$

Also $\hat{B} = (\hat{b}_{ij})$ is a nxm matrix given by

$$\hat{B} = \begin{bmatrix} \tilde{B}_1 \\ \tilde{B}_2 \\ \vdots \\ \tilde{B}_m \end{bmatrix} \tag{18}$$

where each (m+1) x m block $\tilde{B}_i$ is given by

$$\tilde{B}_i = \begin{bmatrix} - - - - - \overset{o}{-} - - - - - - - - - - \\ o - - - 1 \; \hat{b}_{i,i+1} - - - \hat{b}_{i,m} \end{bmatrix} \tag{19}$$

$$\uparrow \text{ i-th column}$$

Matrix G is therefore chosen so as to make all $\hat{b}_{ij} \neq 1$ in (19) to be zero, i.e.

$$\bar{B} = \hat{B}G = \begin{bmatrix} \bar{B}_1 \\ \vdots \\ \bar{B}_m \end{bmatrix}, \qquad \bar{B}_i = \begin{bmatrix} - - - - \overset{o}{-} - - - - - \\ o --- 1 \; o - - - o \end{bmatrix} \tag{20}$$

$$\uparrow \text{ i-th column}$$

Thus G is obtained by arranging its inverse to be given by

$$G^{-1} = \begin{bmatrix} g_1 \\ g_2 \\ \vdots \\ g_m \end{bmatrix} \tag{21}$$

where each g_i is given by the last row of $\hat{B}_i$ in (19).

Substituting for $\hat{K} = G\bar{K}$, eqn.(15) becomes

$$\dot{\hat{x}} = (\hat{A} + \bar{B}\,\bar{K})\,\hat{x} + \bar{\mathbf{B}}\mathbf{v} \tag{22}$$

Matrix $\bar{K}$ is chosen in such a way that $(\hat{A} + \bar{B}\,\bar{K})$ is in block diagonal form with the elements $\hat{a}_{ii,j}$, $j = \overline{o,m}$ in the diagonal blocks of $\hat{A}$ replaced by the coefficients of the desired closed loop characteristic polynomial. From the structure of $\hat{\mathbf{A}}$, matrix $\bar{K}$ is of the form

$$\bar{K} = \begin{bmatrix} \bar{K}_1 & - - - & \bar{K}_m \end{bmatrix} \tag{23}$$

where

$$\bar{K}_i = \begin{bmatrix} \vdots & \vdots & & \vdots \\ \hat{a}_{(i-2)i,o} & \hat{a}_{(i-2)i,1} & \cdots & \hat{a}_{(i-2)i,m} \\ \hat{a}_{(i-1)i,o} & \hat{a}_{(i-1)i,1} & \cdots & \hat{a}_{(i-1)i,m} \\ (\hat{a}_{ii,o}-p_{ii,o}) & (\hat{a}_{ii,1}-p_{ii,1}) & \cdots & (\hat{a}_{ii,m}-p_{ii,m}) \\ \hat{a}_{(i+1)i,o} & \hat{a}_{(i+1)i,1} & \cdots & \hat{a}_{(i+1)i,m} \\ \hat{a}_{(i+2)i,o} & \hat{a}_{(1+2),1} & \cdots & \hat{a}_{(i+2),m} \\ \vdots & \vdots & & \vdots \end{bmatrix} \tag{24}$$

and $p_{ii,j}$, $j = o,1,2,\ldots. m$ are the coefficients of the characteristic polynomials in terms of the desired eigenvalues. Substituting for $\bar{B}$ and $\bar{K}$ in (22) shows that the system is decoupled and has the desired eigenvalues.

The actual feedback gain matrix is given by

$$K = G\bar{K}P \tag{25}$$

ROBOT CONTROL SCHEME DESIGN

In general the nonlinear dynamic model for an n degrees-of-freedom robot arm with elastic joints may be derived using the Lagrange formulation [2,11],

$$\Gamma = \frac{d}{dt}\left[\frac{\partial L}{\partial \dot{q}}\right] - \frac{\partial L}{\partial q} \tag{26}$$

where Γ is nx1 vector of the joint actuating torques, $q,\dot{q}$ are 2nx1 vectors representing angular positions and rates and L is the Lagrangian given by

$$L = T(q,\dot{q}) - P(q) \tag{27}$$

where T is the total kinetic energy, and P is the total potential energy. The kinetic energy is given by

$$T(q,\dot{q}) = \tfrac{1}{2}\, \dot{q}^T M(q)\dot{q} \tag{28}$$

where $M(q)$ is a symmetric positive definite 2nx2n inertia matrix , for joint j, the elastic potential energy is given by

$$E_j(q) = \tau_j(q_{i+1}-q_i/N_j) \qquad i = 2_j-1 \tag{29}$$

where τ_j is the torsional constant for joint j, N_j is the gear ratio. For $i = 2j-1$, q_i and q_{i+1} are respectively, the angular displacements of the motor shaft and link j, with respect to link j-1. If the potential energy of the system due to gravity loading on link j is g_j, then the total potential energy due to elasticity and gravity effects is given by

$$P(q) = \sum_{j=1}^{2n} (E_j(q) + g_j(q) \tag{30}$$

substituting (27-30) into (26) and rearranging, gives

$$\ddot{q} = M^{-1}(q)\ [-c(q,\dot{q}) - \gamma(q) + \Gamma] \tag{31}$$

where $\gamma(q)$ is 2nx1 vector representing gravity and elasticity forces, and the 2nx1 vector, $c(q,\dot{q})$, represents Coriolis and Centrifugal forces. The i-th component of vector c is given by [2]

$$c_i = \tfrac{1}{2}\left[\frac{\partial b_i}{\partial q} + \frac{\partial b_i^T}{\partial q} - \frac{\partial M(q)}{\partial q_i}\right] \tag{32}$$

where b_i denotes the i-th column of matrix M(q).

In order to synthesize the modal controller, (31) is expressed in state-space form as

$$\dot{x} = f(x) + B(x)\ u \tag{33}$$

where

$$x = \left(\frac{q}{\dot{q}}\right),\ f(x) = \left[\begin{array}{c}\dot{q}\\ \hline -M^{-1}(c+\gamma)\end{array}\right],\ B = \left[\begin{array}{c}0\\ \hline M^{-1}\end{array}\right] \text{ and, } u = \Gamma \tag{34}$$

For the robot of Fig.1, with n = 3, the rotor variables, denoted by q_1, q_3 and q_5 are internal to the joint, and so the link positions denoted by q_2, q_4 and q_6 may be used to define the global work space,

$$W_G: \begin{array}{rcll} 0 \leq & q_2 & \leq \pi & \text{rad} \\ 0 \leq & q_4 & \leq \pi/2 & \text{rad} \\ 0 \leq & q_6 & \leq -\pi/2 & \text{rad} \\ -2 \leq & \dot{q}_i & \leq 2 & \text{rad/s} \qquad i = 2,4,6 \end{array} \tag{35}$$

To illustrate the approach, the work space was linearized in the subspace defined by [11] W^1 at thirty different points, where

$$W^1: \begin{array}{rcll} 0 \leq & q_2 & \leq 2.2 & \text{rad} \\ 0.4 \leq & q_4 & \leq 1.2 & \text{rad} \\ -0.8 \leq & q_6 & \leq -1.2 & \text{rad} \\ 0 \leq & \dot{q}_2 & \leq 1.9 & \text{rad/s} \\ -0.9 \leq & \dot{q}_4 & \leq 1.7 & \text{rad/s} \\ -1.6 \leq & \dot{q}_6 & \leq 1.4 & \text{rad/s} \end{array} \tag{36}$$

For linearization of f(x) in (33), let x_d be some nominal trajectory in subspace W^1, and ε be small excursions of x from x_d so that it is possible to write

$$x = x_d + \varepsilon \tag{37}$$

Substituting for x into the state equation (33) and using a Taylor expansion yields

$$\dot{x}_d + \dot{\varepsilon}_d = \{ f(x_d) + A(x_d)\,\varepsilon + h\} + B(x_d)\,u \tag{38}$$

where term h is an expression whose value is supposed to be small with respect to ε and A is a Jacobian matrix of f, with respect to x, whose (i,j)-th element is given by

$$A_{ij} = \left.\frac{\partial f_i(x)}{\partial x_j}\right|_{x=x_d} \qquad \begin{array}{l} i = 1,12 \\ j = 1,12 \end{array} \tag{39}$$

Thus neglecting h, ε approximately satisfies the linear equation

$$\dot{\varepsilon} = A\varepsilon + Bu \tag{40a}$$

$$\varepsilon = x - x_d \tag{40b}$$

where $A = A(x_d)$ and $B = B(x_d)$, are 12x12 and 12x3 matrices respectively.

Assuming the desired operating state vector is

$$x_d^T = \left[q_1^d\,,\ q_2^d,\ q_3^d,\ q_4^d,\ q_5^d,\ q_6^d,\ 0,\ 0,\ 0,\ 0,\ 0,\ 0, \right] \tag{41}$$

Then the deviation of the actual state of the system from the operating point is given by (40b), with components

$$\left.\begin{array}{l} \varepsilon_i = q_i - q_i^d \\ \varepsilon_{i+6} = \dot{q}_i \end{array}\right\} \quad i = 1,6 \tag{42}$$

Thus the actual modal control scheme is that shown in Fig.3, in which feedforward block $u_g(x_d)$ was introduced to compensate for the effects of gravity on links 2 and 3.

The structures of the robot of Fig.1 carry a nominal payload of 5kg. A dc electric motor situated at each joint actuates the links through a gear box with reduction ratio 160. A major difficulty in the choice of matrix K relates to the choice of a suitable set of eigenvalues to render the system insensitive to parameter variations in the work space. The choice used here is that arrived at via simulation in [11] .

SIMULATION RESULTS AND CONCLUSIONS

The robot was simulated transferring various payloads along a trajectory represented in Fig.1 by the straight 1.5 m line between points A and B within the subspace defined by (36). The velocity profile of the end-effector was planned in such a way that 3s are used to move from A to B. During the first and third seconds the end-effector has constant acceleration and deceleration, respectively, of $0.75 m/s^2$.

Simulation results are shown in Figs 4, and 5 which show the control torques developed by the motors, and the tracking error components - σ (the lateral displacement from the path) and Δ (the delay along the path) for payloads of 5 kg and

10kg, respectively. It can be seen that the tracking error is dominated by the delay, Δ, along the path, so that the actual displacement σ from the curve is relatively very small. Fig.6 also shows the performance of the system when the torsional constants are reduced by 32% of the nominal values. In this case the lateral displacement from the path is pronounced.

In all cases, σ is less than 0.6mm. It is also observed that when the payload is 10kg (twice the nominal value), the tracking error remains constant at the end of the trajectory. Thus to eliminate steady state error, there is need to incorporate an observer as suggested in [10].

Results illustrate that the modal controller with suitable assignment of eigenvalues can be combined with the concept of dividing the workspace into subspaces to yield a robust control system.

REFERENCES

1. Hollerbach,J.M., A recursive Lagrangian formulation of manipulator dynamics and a comparable study of dynamics formulation complexity, IEEE Trans. Systems Man and Cybernetics, vol. SMC-10, no.11 (Nov.1980) pp.730-736.

2. Cesareo,G.,Nicolo,F. and Nicosia,S., DYMIR: A code for generating dynamic models of robots, Rome University Industrial Robotics Lab, Report no.02-83 (July 1983).

3. Whitney,D.E., Resolved motion rate control of manipulators and human prostheses, IEEE Trans Man-Machine Systems, vol. MMS-10,(June 1969) pp. 47-53.

4. Luh,J.Y.S., Conventional controller design for industrial robots - A tutorial, IEEE Trans.Systems Man and Cybernetics, vol SMC-13,no.3,(May/June 1983) pp. 298 - 316.

5. Lentini,D.,Nicolo,F.,Nicosia,S. and Vecchio,L.,Decoupling and Nonlinearities compensation in robot control systems, Report no.R.80-08 (April 1980),University of Rome, Instituto di Automatica.

6. Dubowsky,S.,and Desforges,D.T., The application of model reference adaptive control to robotic manipulators, Trans. ASME,J. Dynamic systems, measurements and control, vol.101 (Sept. 1979).

7. Nicolo,F. and Katende,J., A robust MRAC for industrial robots,Proc.IASTED 2nd Int. Symp. on Robotics and Automation,June 1983, Switzerland, pp. 64-67.

8. Katende,J.,Awobamise,B.O. and Alos,A.J., An adaptive scheme for robot trajectory control, IASTED 4th Int. Symp. Modelling Identification and control - MIC '85, Switzerland, Feb. 1985.

9. Ersu,E. and Neddermeyer,W., An approach via robust linear multivariable system design methods to nonlinear manipulator control, IASTED 2nd Int.Symp. on Robotics and Automation, (June 1983), Switzerland, pp. 160 - 163.

10. Isidori,A., Systemi di controllo, Siderea Rome, (1979).

11. Katende,J., Computer algorithms for non-linear control of industrial robots, Ph.D. Thesis, University of Lagos, 1985.

ACKNOWLEDGEMENT

The authors would like to thank the Department of Computer Science and Systems Engineering, University of Rome for the facilities provided and are also grateful to Professor F. Nicolo for his valued help in the development work leading up to this paper.

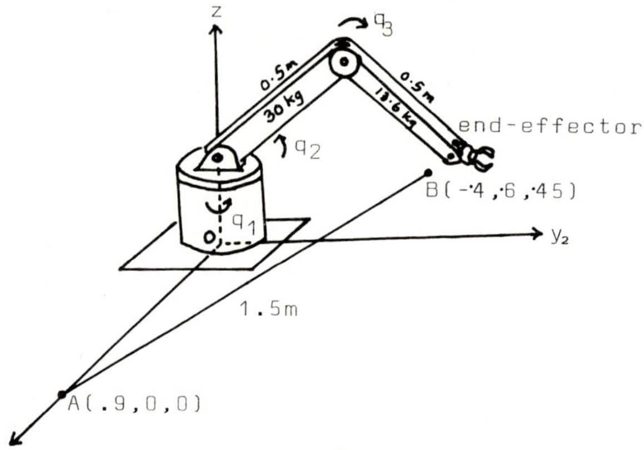

Fig.1 Three dof robot arm and linear trajectory AB

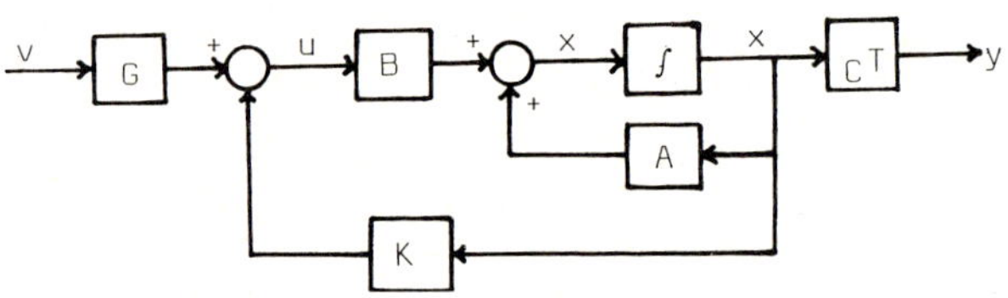

Fig 2. General modal control block diagra

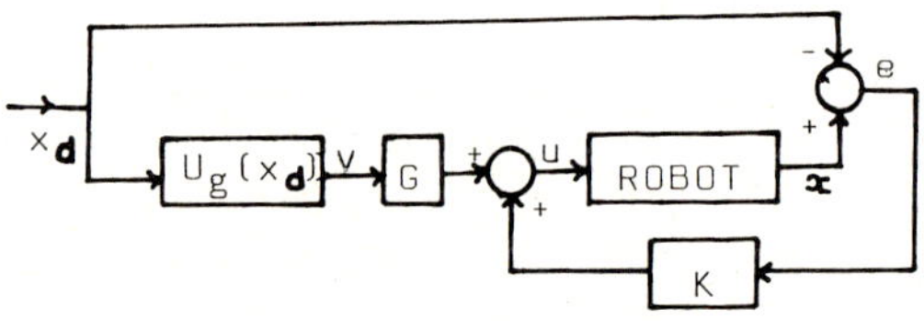

Fig.3 Actual modal control scheme

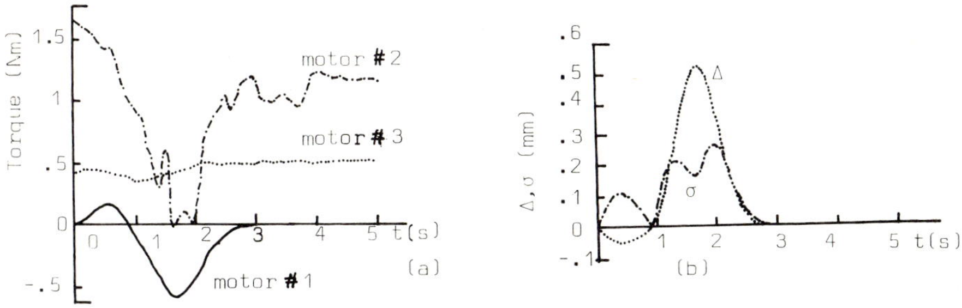

Fig.4 Trajectory tracking: nominal payload and torsional coefficients
(a) motor developed torques
(b) components of the tracking error

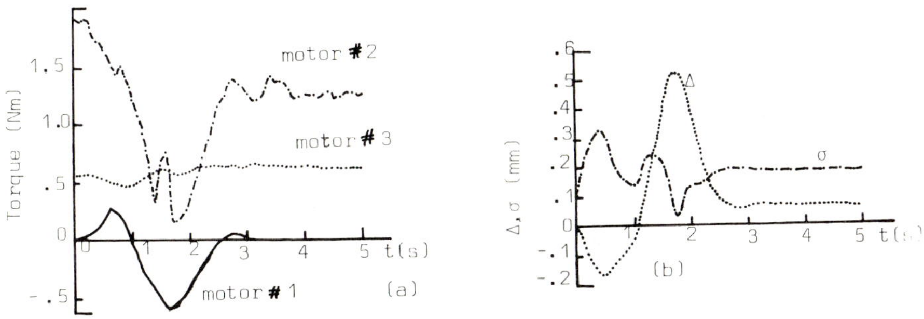

Fig.5 Trajectory tracking: payload = 10kg; nominal torsional coefficient
(a) motor developed torques
(b) components of the tracking error

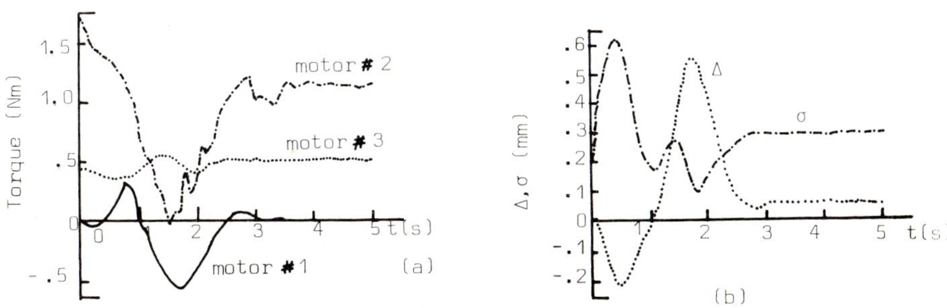

Fig.6 Trajectory tracking: nominal payload-torsional coefficients reduced to 70% of nominal values
(a) motor developed torques
(b) components of the tracking error

Frequency Domain and State Space Methods for Linear Systems
C.I. Byrnes and A. Lindquist (Editors)

SOME RESULTS ABOUT A LEONTIEFF-TYPE MODEL

J.W. Nieuwenhuis

Department of Econometrics
University of Groningen
The Netherlands

Abstract. Let A be an invertible linear operator from $\mathbb{R}^n$ to $\mathbb{R}^n$ and $\mathbb{R}^n_+$ the nonnegative orthant in $\mathbb{R}^n$. In this paper we will study properties of the (largest) closed convex cone $K \subseteq \mathbb{R}^n_+$ such that $AK \subseteq K$. This cone naturally arises when studying a Leontieff-type model such as $x_{t+1} = Ax_t$, $x_t \in \mathbb{R}^n_+$, $t \geqq 0$. We will give sufficient conditions such that K is polyhedral. We will indicate that there are relations with some questions in linear systems theory.

MOTIVATION AND INTRODUCTION

Our motivation comes from a remarkable result (to be published) by J.C. Willems. In order to state this result we need some notation. Z_+ will denote the nonnegative integers and $(\mathbb{R}^q)^{Z_+}$ all the functions from Z_+ to $\mathbb{R}^q$, i.e., elements of $(\mathbb{R}^q)^{Z_+}$ are sequences and each element of such a sequence is just an element of $\mathbb{R}^q$. By σ we denote the shift-operator, i.e., let $w \in (\mathbb{R}^q)^{Z_+}$ then σw is defined by $(\sigma w)(t) := w(t+1)$, hence σ is a linear operator from $(\mathbb{R}^q)^{Z_+}$ to $(\mathbb{R}^q)^{Z_+}$. Further we define $\sigma^n := \sigma(\sigma^{n-1})$. By $R(\sigma)$ we denote an $(r \times q)$ matrix with polynomials in σ as entries. Let us endow $(\mathbb{R}^q)^{Z_+}$ with the topology of pointwise convergence, i.e. $w^i \to \bar{w}$ iff $w^i(t) \to \bar{w}(t)$ in the norm-topology, $\forall t \in Z_+$.

Then the following result holds true.

THEOREM (by J.C. Willems, submitted for publication Ref. 1) Let $B \subseteq (\mathbb{R}^q)^{Z_+}$ be a nonempty set. Then the following two statements are equivalent.

1) $\sigma B \subseteq B$, B is closed, B is a linear subspace.
2) There is an integer $r \in Z_+$ and an $(r \times q)$ polynomial matrix $R(\sigma)$ such that $B = \{w | R(\sigma)w = 0\}$

As we are used to work in surroundings where <u>in</u>equalities play a dominant role the following question naturally arises:

QUESTION. Under what conditions (if any) is it possible to give a similar characterization of closed shift invariant convex cones B in $(\mathbb{R}^q)^{Z_+}$, i.e., such that $\sigma B \subseteq B$.

In order to get some feeling for this problem we studied in some detail the Leontieff-system, well-known in mathematical economics,

$$\begin{aligned} x_{t+1} &= Ax_t \\ x_t &\geqq 0 \end{aligned} \qquad t \in Z_+,$$

where A is supposed to be an invertible linear mapping from $\mathbb{R}^n$ to $\mathbb{R}^n$.

Let us define the set $K \subseteq \mathbb{R}^n_+$, the nonnegative orthant in $\mathbb{R}^n$, by $K := \{x \in \mathbb{R}^n \mid A^t x \geqq 0,\ \forall t \in Z_+\}$. This set K evidently is a closed convex cone, i.e. $\alpha K + \beta K \subseteq K$, $\forall \alpha, \beta \in \mathbb{R}_+$. Now suppose we could prove that K actually is polyhedral, i.e., there are vectors $\{k^1, k^2, \ldots, k^m\} \subseteq K$ such that

$$k \in K \Leftrightarrow k = \Sigma \lambda_i k^i \text{ for some } \lambda_i \in \mathbb{R}_+ ,$$

then we could, by a well-known result in convex analysis, represent K as follows:

$$K = \{x \in \mathbb{R}^n \mid \tilde{B}x \leqq 0\},$$

for some linear operator $\tilde{B}$ from $\mathbb{R}^n$ to $\mathbb{R}^p$ for some $p \in Z_+$.

In this case $\{x_{t+1} = Ax_t,\ x_t \geqq 0,\ t \in Z_+\}$ is equivalent to $\{x_{t+1} = Ax_t,\ \tilde{B}x_t \leqq 0,\ t \in Z_+\}$, or $x_{t+1} - Ax_t \leqq 0$, $-x_{t+1} + Ax_t \leqq 0$, $\tilde{B}x_t \leqq 0$, or

$$\begin{pmatrix} I, & -A \\ -I, & A \\ 0, & \tilde{B} \end{pmatrix} \begin{pmatrix} x_{t+1} \\ x_t \end{pmatrix} \leqq 0,$$

so we could represent this Leontieff-system by $\{w \mid R(\sigma)w \leqq 0\}$, where w and $R(\sigma)$ are defined accordingly.

It will appear in the sequel that the situation is not that simple. In the first place K does not need to be polyhedral, in the second place we could only find sufficient conditions such that K is polyhedral.

At the end of this paper we will mention some more problems we encountered in our attempt to give a characterization of translation-invariant closed convex cones in $(\mathbb{R}^q)^{Z_+}$.

We will then also indicate some more relations with the existing literature in linear systems theory and especially with realization theory.

The further line of this paper is as follows. First we give some results from the theory of nonnegative matrices (a bad name as it covers more matrices). Then we will give our own results. Some boring proofs will be postponed to the appendix. After that we give the promised relations with realization theory and nearly at the end we indicate some relations with mathematical economics.

Cones invariant under linear operators.

First of all some notions we will use in the sequel. Whenever we use the capital Q we will denote a proper cone in $\mathbb{R}^n$, $n > 1$.
DEFINITION. Q is pointed whenever $Q \cap (-Q) = \{0\}$. Q is solid whenever Q has a nonempty interior, int Q. Q is called a proper cone whenever it is a closed pointed solid cone.

There is an extensive theory about linear operators $L: \mathbb{R}^n \to \mathbb{R}^n$ such that they leave invariant a given proper cone Q, i.e., such that $LQ \subseteq Q$. In case $LQ \subseteq Q$ we denote this by $L \in \pi(Q)$. In the sequel we also need the following notation. Let $L: \mathbb{R}^n \to \mathbb{R}^n$ be a linear operator, then $\lambda_1, \lambda_2, \ldots, \lambda_k$ will denote its eigenvalues. The multiplicity of λ_i is denoted by m_i and in order to avoid confusion we explicitly state that $\sum_{i=1}^{k} m_i = n$.

By $\rho(L)$ we denote the spectral radius of L, i.e.,

$$\rho(L) := \max_{i=1,\ldots k} |\lambda_i|, \text{ where}$$

$|\lambda_i|$ is the modulus of the eigenvalue λ_i.

The following results are well-known, see for instance Ref. 2.
THEOREM 1. If $L \in \pi(Q)$ then
(a) $\rho(L)$ is an eigenvalue
(b) if λ is an eigenvalue of L such that $|\lambda| = \rho(L)$ then the degree of $\lambda \leqq$ the degree of $\rho(L)$.
(c) Q contains an eigenvector of L corresponding to $\rho(L)$.
Here the degree of an eigenvalue λ is the multiplicity of λ in the minimal polynomial of Q, and this is equivalent to saying that it is the largest multiplicity m_j such that $\lambda = \lambda_j$, $j \in \{1,2,\ldots k\}$.

The other way round we have the following result, see also Ref. 2.
THEOREM 2. If $\rho(L)$ is an eigenvalue of L and if deg $\rho(L) \geqq$ degree λ for every eigenvalue λ such that $|\lambda| = \rho(L)$ then L leaves a proper cone invariant.
The proof of this result goes by construction and a variant of it will be used in the sequel of this paper.
A useful subclass of $\pi(Q)$ is the collection of so-called irreducible matrices.
DEFINITION. $L \in \pi(Q)$ is called irreducible if no eigenvalue of L lies on the boundary of Q.
This definition is equivalent to the following.
THEOREM 3. $L \in \pi(Q)$ is irreducible iff L has (up to scalar multiples) exactly one eigenvector in Q, and iff further this vector is in int Q, the interior of Q.
In case $L \in \pi(Q)$ is irreducible a good deal more can be said about the eigenvalue-structure of L.
THEOREM 4. If $L \in \pi$ is irreducible then $\rho(L)$ is a simple eigenvalue and any other eigenvalue with the same modulus is also simple and there is an eigenvalue corresponding to $\rho(L)$ in int Q and no other eigenvector lies in Q.

The matrix $L \in \pi(Q)$ is called positive if $L(Q \setminus \{0\}) \subseteq \text{int } Q$. In that case we know the following.
THEOREM 5. If $L \in \pi(Q)$ is positive then $\rho(L)$ is a simple eigenvalue greater than the magnitude of any other eigenvalue and an eigenvector corresponding to $\rho(L)$ lies in int Q.
A matrix $L \in \pi(Q)$ is called primitive if there exists a natural number

such that L^m is positive. One can prove that a primitive matrix necessarily is irreducible. Within the class of irreducible matrices one can characterize the "primitives" completely by means of their eigenvalue structure as follows.

THEOREM 6. An irreducible matrix $L \in \pi(Q)$ is primitive if and only if $\rho(L)$ is greater in magnitude than any other eigenvalue.

All the results above are about matrices that leave invariant a given cone, where we are faced with the opposite; given a matrix A we should prove that a given A-invariant cone has desirable properties.

Some answers will be given in the next part.

About the largest cone in $\mathbb{R}^n_+$ invariant under a linear operator A.

The structure of this part is as follows.

As before A will denote an invertible linear mapping from $\mathbb{R}^n$ to $\mathbb{R}^n$, and $K := \{x \mid A^t x \geqq 0,\ t \in Z_+\}$.

First of all we will explore the situation in case n = 2. Secondly we will give some results in $\mathbb{R}^n$. At the end of this part we will formulate some open problems.

Case n = 2.

The following result gives a characterization of the cone K.

THEOREM 7. $[(K\setminus\{0\}) \subseteq \text{int}\ \mathbb{R}^2_+] \Rightarrow \text{int}\ K = \phi$.

Before proving this result, notice that this statement is equivalent to the following.

One can only have that:

i) $K = \{0\}$.

ii) K is exactly one ray.

iii) int $K \neq \phi$ and bd $K \cap (\text{bd}\ R^2_+\setminus\{0\}) \neq \phi$.

PROOF OF THEOREM 7. Unfortunately we have to consider several cases.

a) Let A be such that $\lambda_1 = 1$ and $\lambda_2 = 2$ with corresponding eigenvectors $(-1,2)^T$ and $(-1,3)^T$ then, applying Theorem 1, it follows that $K = \{0\}$.

b) Let A be such that $\lambda_1 = 2$ with eigenvector $(1,1)^T$ and $\lambda_2 = 3$ with eigenvector $(1,-1)^T$, then, again applying Theorem 1, it easily follows that $K = \{x \in \mathbb{R}^2_+ \mid x = \mu(1,1)^T,\ \mu \leqq 0\}$.

c) It remains to prove statement iii). The proof goes by contradiction, hence, suppose that $K\setminus\{0\} \subseteq \text{int}\ \mathbb{R}^2_+$. In this case there are vectors $a_1, a_2 \in \in \text{int}\ \mathbb{R}^2_+$ such that

$$K = \{x \mid x = \lambda a_1 + \lambda a_2,\ \lambda \geqq 0,\ \mu \geqq 0\}.$$

We now know that $Aa_1 \in \text{bd}K$, the boundary of K, and $Aa_2 \in \text{bd}K$.

Otherwise it would follow that there is a point in the neighbourhood of say a_1, not in K, but still in $\mathbb{R}^2_+$ such that the image of this point under A is in K, and this contradicts the definition of K.

As A is invertible only the following situations are possible:

α) $Aa_1 = \eta_1 a_1$, $Aa_2 = \eta_2 a_2$, $\eta_1 > 0$, $\eta_2 > 0$.

β) $Aa_1 = \zeta_1 a_2$, $Aa_2 = \zeta_2 a_1$, $\zeta_1 > 0$, $\zeta_2 > 0$.

First we consider case α.

Assume that $\eta_2 \cdot \eta_1^{-1} \leqq 1$. Take $\bar{\varepsilon} > 0$ such that $a_1 - \varepsilon a_2 \in \text{int}\ \mathbb{R}^2_+$, $\forall \varepsilon \in [0, \bar{\varepsilon}]$.

By construction $a_1-\varepsilon a_2 \notin K$. But $A(a_1-\varepsilon a_2) = \eta_1 a_1-\varepsilon\eta_2 a_2 = \eta_1(a_1-\varepsilon\eta_2\eta_1^{-1}a_2)$ and $\varepsilon\eta_2\eta_1^{-1} \in [0,\bar{\varepsilon}]$ in case $\varepsilon \in [0,\bar{\varepsilon}]$, hence $A^t(a_1-\varepsilon a_2) \in \text{int } \mathbb{R}^2_+$, $\forall t \in Z_+$ and this contradicts the definition of K, and therefore α) cannot exist. Finally we consider case β.

We assume that $\zeta_2.\zeta_1^{-1} \leqq 1$ and we take $\bar{\varepsilon} > 0$ such that $a_1 - \varepsilon a_2 \in \text{int } \mathbb{R}^2_+$

$a_2 - \varepsilon a_1 \in \text{int } \mathbb{R}^2_+$, $\forall\varepsilon \in [0,\bar{\varepsilon}]$.

Now $A(a_1-\varepsilon a_2) = \zeta_1 a_2 - \varepsilon\zeta_2 a_1 = \zeta_1(a_2-\varepsilon\zeta_2\zeta_1^{-1}a_1)$ and

$$A^2(a_1-\varepsilon a_2) = \zeta_1\zeta_2 a_1 - \varepsilon\zeta_2\zeta_1 a_2 = \zeta_1\zeta_2(a_1-\varepsilon a_2),$$

hence $A^t(a_1-\varepsilon a_2) \in \text{int } \mathbb{R}^2_+$, $\forall t \in Z_+$ and again we have a contradiction, and this implies this theorem to hold.

In $\mathbb{R}^2$ evidently every convex closed cone is polyhedral, but in $\mathbb{R}^n$, $n \geqq 3$ this is no longer true, even more, K does not need to be polyhedral.

An example showing that K does not need to be polyhedral.

We define $A: \mathbb{R}^3 \Rightarrow \mathbb{R}^3$ as follows.
First of all we take the plane $\{(x,y,z)/z = 0\}$ and define A on this plane as follows. Let φ be an irrational multiple of the number π, then $A\binom{x}{y} := \begin{pmatrix}\cos\varphi & -\sin\varphi\\ \sin\varphi & \cos\varphi\end{pmatrix}\binom{x}{y}$. Hence A rotates points.

Further we define $A(1,1,1) := (1,1,1)$. Now we defined A completely. It is now not hard to see that K contains the cone $\tilde{K}$ generated by $\{(x,y,z)/z = 1$ and $(1-x)^2+(1-y)^2 = 1\}$. Actually $K = \tilde{K}$ by construction, for take a point outside $\tilde{K}$, then eventually this point will be rotated outside $\mathbb{R}^3_+$. This follows from some results in number theory.

Notice that we still have that int $K \neq \phi$ and that K has points in common with the nontrivial boundary of $\mathbb{R}^3_+$.

Now we come to some sufficient conditions which guarantee that K is polyhedral.

Sufficient conditions such that K is polyhedral.

The following result is a building-block in our main result but deserves explicit mentionning.

THEOREM 8. Suppose that $\rho(L)$ is an eigenvalue, say, $\rho(L) = \lambda_1$ and suppose further that $\lambda_1 > |\lambda_i|$, $i \geqq 2$ and that an eigenvector belonging to λ_1 is in int $\mathbb{R}^n_+$, then int $K \neq \phi$.

PROOF. We choose a Jordan canonical basis of A in $\mathbb{C}^n$ as follows:

$$Ax_{ij} = \lambda_i x_{ij} + x_{i,j-1},\ x_{i0} = 0,\ 1 \leqq i \leqq k,\ 1 \leqq j \leqq m_i.$$

By assumption we have that $x_{11} \in \text{int } \mathbb{R}^n_+$, and hence we may take $x_{1j} \in \mathbb{R}^n_+$, $j \in \{1,\ldots,m_1\}$. This can be seen as follows. Take an $\varepsilon > 0$ and define $\tilde{x}_{11} := \varepsilon x_{11}$ and $\tilde{x}_{1j} := \tilde{x}_{1,j-1}.\varepsilon^{-1} + \varepsilon x_{1j}$. Then we conjecture that

$A\tilde{x}_{1j} = \lambda_1\tilde{x}_{1j} + \tilde{x}_{1,j-1}$ and further that for ε sufficiently small $\tilde{x}_{1j} \in \text{int } \mathbb{R}^n_+$, $\forall j$.

Of course we define $\tilde{x}_{i0} := 0$. By easy calculations it follows that $A\tilde{x}_{1j} = \lambda_1\tilde{x}_{1j} + \tilde{x}_{1,j-1}$ for $j = 1,2$. Now suppose the result to be true for $j = 1,2,\ldots,r-1$. Then we start calculating $A\tilde{x}_{1r}$.

$A\tilde{x}_{1r} = A(\tilde{x}_{1,r-1}\cdot\varepsilon^{-1}+\varepsilon x_{1r}) = \varepsilon^{-1}(\lambda_1\tilde{x}_{1,r-1}+\tilde{x}_{1,r-2}) + \varepsilon A x_{1r}$

$= \varepsilon^{-1}(\lambda_1\tilde{x}_{1,r-1}+\tilde{x}_{1,r-2}) + \varepsilon\lambda_1 x_{1r} + \varepsilon x_{1,r-1}$

$= \lambda_1(\varepsilon^{-1}\tilde{x}_{1,r-1}+\varepsilon x_{1r}) + \varepsilon^{-1}\tilde{x}_{1,r-2} + \varepsilon x_{1,r-1}$

$= \lambda_1\tilde{x}_{1r} + \tilde{x}_{1,r-1}$.

From the definition of $\tilde{x}_{ij}$ it follows that

$$\tilde{x}_{1,r} = \varepsilon^{2-r}x_{11} + \varepsilon^{3-r}x_{12} + \ldots + x_{1,r-1} + \varepsilon x_{1,r}.(*),\ r \geqq 2.$$

As m_1 is finite and x_{11} is supposed to be a vector in int $\mathbb{R}^n_+$ it follows that for ε sufficiently small $x_{1,j} \in$ int int $\mathbb{R}^n_+$, $\forall j$.

The n vectors x_{ij} form a basis for $\mathbb{C}^n$.

As this fact is important in our proof we should check whether span $\{x_{1,j},\ j = 1,2,\ldots,m_1\}$ = span $\{\tilde{x}_{1,j},\ j = 1,2,\ldots,m_1\}$. From statement (*) it immediately follows that this is indeed the case. In the sequel we now assume that all x_{1j}, $j = 1,2\ldots m_1$ are in the interior of $\mathbb{R}^n_+$.

Unfortunately we need a further rescaling of the vectors $\{x_{ij}\}$. To that end we assume that $|\lambda_2|$ is the second largest in modulus, and we define $\delta := \lambda_1 - |\lambda_2| > 0$.

Further we define $\hat{x}_{i1} := \delta x_{i1}$, $\hat{x}_{i2} = \delta x_{i1} + \delta^2 x_{i2}$, $\hat{x}_{i3} := \delta x_{i1} + \delta^2 x_{i2} + \delta^3 x_{i3}$ etc. Then it is easy to prove that $A\hat{x}_{ij} = \lambda_i\hat{x}_{ij} + \delta\hat{x}_{i,j-1}$, $\hat{x}_{i0} = 0$, $1 \leqq i \leqq k$, $1 \leqq j \leqq m_i$. Notice that all the vectors $\hat{x}_{1,j}$ remain in the interior of $\mathbb{R}^n_+$. Let c be an arbitrary number greater than zero.

Then we define the set $K(c) \subseteq \mathbb{R}^n$ by:

$$K(c) := \{x \in \mathbb{R}^n \mid x = \sum_{i=1}^{k}\sum_{j=1}^{m_i}\alpha_{ij}\hat{x}_{ij},\ |\alpha_{ij}| \leqq c\alpha_{ij},\ j \leqq m_1$$

$$|\alpha_{ij}| \leqq c\alpha_{1m_1},\ j \geqq m_1$$

$$\alpha_{ij} \in \mathbb{R} \text{ when } \hat{x}_{ij} \in \mathbb{R}^n \text{ and}$$

$$\alpha_{ij} = \bar{\alpha}_{pq} \text{ when } \hat{x}_{ij} = \hat{x}_{pq}\}.$$

The overbars denote complex conjugation. As by construction all the vectors $\hat{x}_{1j} \in \text{int } \mathbb{R}^n_+$ it easily follows that $K(c) \subseteq \mathbb{R}^n_+$, for c sufficiently small. We further conjecture that K(c) is an A-invariant solid cone. It was for this reason that we needed the final rescaling. As for small numbers c the set K(c) is in K we are done with the proof when we succeed in proving our conjecture. Suppose first that all the vectors x_{ij} are $\mathbb{R}^n$. Then K(c) is the image under an invertible linear mapping from $\mathbb{R}^n$ to $\mathbb{R}^n$ of the set

$$P := \{(\alpha_{11},\dots,\alpha_{1m_1},\dots,\alpha_{km_k}) \in \mathbb{R}^n \mid |\alpha_{ij}| \leqq c\alpha_{1j},\ j \leqq m_1$$
$$|\alpha_{ij}| \leqq c\alpha_{1m_1},\ j \geqq m_1\}$$

In this case P is polyhedral convex cone in $\mathbb{R}^n$ and $(\underbrace{1,1,\dots 1}_{m_1},0,0\dots 0)$ is an interior point of this set.

Hence K(c) is a solid polyhedral cone. In case not very vector x_{ij} is in $\mathbb{R}^n$ the analysis is a bit more complicated.

Suppose x_{ij} has an imaginary part, say $x_{ij} = y_{ij} + iz_{ij}$ with $z_{ij} \in \mathbb{R}^n \neq 0$, then all such y_{ij} and z_{ij} together with the real vectors x_{ij} are n linearly independent vectors over $\mathbb{R}$ and form a basis for $\mathbb{R}^n$.

It is not hard to show that K(c) is the image under an invertible linear operator from $\mathbb{R}^n$ to $\mathbb{R}^n$ of a nonpolyhedral solid cone, and again we find that K(c) is a solid cone.

So there remains to be proved that K(c) is invariant under A. To that end we take $x = \Sigma\Sigma\alpha_{ij}\hat{x}_{ij}$. Call the coefficient of $\hat{x}_{ij}$ in Ax $\hat{\alpha}_{ij}$ then it follows that:

$$\hat{\alpha}_{ij} = \lambda_i\alpha_{ij} + \delta\alpha_{i,j+1} \qquad 1 \leqq j \leqq m_i - 1$$

$$\hat{\alpha}_{ij} = \lambda_i\alpha_{ij} \text{ when } j = m_i$$

We now have to check whether the inequalities in the definition of K(c) do hold. This exercise is left to the reader who will notice in doing this that the rescaling of x_{ij} in $\hat{x}_{ij}$ and the introduction of the factor δ is essential in proving this result. By this we completed the proof.

The construction of the cone K(c) using the vectors $\hat{x}_{ij}$ is essentially due to Vandergraft, see Ref. 2.
It is possible to construct other cones in a similar way such that they are polyhedral when the conditions in the theorem (or even weaker conditions) are fulfilled.
THEOREM 9. Suppose that in addition to the assumptions in Theorem 8 there is only one eigenvector in $\mathbb{R}^n_+$ and that $m_1 = 1$, then A is primitive with respect to K.
PROOF. From Theorem 8 it follows that K is solid, so there remains to prove the existence of an integer m such that $A^m(K \setminus \{0\} \subseteq \text{int } K$. From the proof of Theorem 8 it follows that the eigenvector is in the interior of K(c), and hence in the interior of K. But this implies according to Theorem 3 that A is irreducible with respect to K. From Theorem 6 it now follows that A is K-positive and this had to be proven.

Without the addition that $m_1 = 1$ this result no longer holds. This can be seen as follows. The eigenvector x_{11} is in K. In order that A is irreducible and afortiori K-positive it is necessary that $x_{11} \in \text{int } K$. Suppose now that $m_1 \geqq 2$, then there is a vector x_{12} linearly independent from x_{11} such that $Ax_{12} = \lambda_1 x_{12} + x_{11}$. It is not hard to prove that

$$A^k(x_{11}+\varepsilon x_{12}) = \lambda_1^k(x_{11}+\frac{k\varepsilon}{\lambda_1}x_{11}+\varepsilon x_{12}) \notin \mathbb{R}^n_+ \text{ for } \varepsilon < 0 \text{ and } k \text{ large enough.}$$

But this implies that $x_{11} \in \text{bd } K$, it even implies that there is no solid cone Q such that A is irreducible with respect to Q.

The next result makes clear why K-positivity of A is important in our story.

THEOREM 10. Suppose that A is K-primitive, then K is polyhedral.

We think it to be a surprising result; one could consider K-primitiveness of A as a kind of contraction condition, so contraction in this sense implies polyhedrality.
Before proving this result we state some auxiliary results, to be proved in the appendix.

LEMMA 1. Let $Q \subseteq \mathbb{R}^n$ be a closed convex cone with nonempty interior. Let $\text{bd } Q \subseteq \bigcup_{i=1}^p H_i$ where H_i is a hyperplane in $\mathbb{R}^n$ and let $H_i \cap \text{int } Q = \phi$, $i = 1,2,\ldots p$, then Q is polyhedral.

LEMMA 2. Let $Q \subseteq \mathbb{R}^n$ be a closed convex cone with $\text{int } Q \neq \phi$. Let $\text{bd } Q \subseteq \bigcup_{i=1}^p H_i$, H_i is a hyperplane in $\mathbb{R}^n$. Let $H_1 \cap \text{int } Q \neq \phi$, then $\text{bd } Q \subseteq \bigcup_{i=2}^p H_i$.

Now we give the <u>proof of Theorem 10</u>.

Let K be such that $(\text{bd } K) \cap \text{int } \mathbb{R}^n_+ = \phi$, then we claim that $K = \mathbb{R}^n_+$ and hence polyhedral. For suppose to the contrary that $K \neq \mathbb{R}^n_+$, then there is a point $\bar{x} \in \mathbb{R}^n_+$ such that $\bar{x} \notin K$. Take an arbitrary $\bar{y} \in \text{int } K$, and consider $x(\lambda) := \lambda\bar{x} + (1-\lambda)\bar{y}$. It is easy to see that $x(\lambda) \in \text{int } \mathbb{R}^n_+$, $\forall\lambda \in (0,1)$ and further there is a $\bar{\lambda} \in (0,1)$ such that $x(\bar{\lambda}) \in \text{bd } K$ and we arrived at a contradiction.
Hence we may assume that there is a $\bar{k} \in (\text{bd } K) \cap \text{int } \mathbb{R}^n_+$. As A is K-positive there is an integer m such that $A^m(K\setminus\{0\} \subseteq \text{int } K$. We claim that there is an integer $i \leqq m-1$ such that $A^i\bar{k} \in \text{bd } \mathbb{R}^n_+$. For suppose to the contrary that $A^i\bar{k} \in \text{int } \mathbb{R}^n_+$, $i = 1,2,\ldots m-1$. By assumption we know that $A^m\bar{k} \in \text{int } K$. We now take open neighbourhoods U_i of the origin in $\mathbb{R}^n$ such that

$$A^i(\bar{k}+U_i) \subseteq \text{int } \mathbb{R}^n_+, \quad i = 0,2,\ldots m-1$$

$$A^i(\bar{k}+U_i) \subseteq \text{int } K \quad , \quad i = m$$

We define $U := \bigcap_i U_i$, then U is an open neighbourhood of the origin in $\mathbb{R}^n$ such that $A^i(\bar{k}+U) \subseteq \mathbb{R}^n_+$, $\forall i \in Z_+$. As $(\bar{k}+U) \cap \mathbb{R}^n_+ \cap K^c \neq \phi$, where K^c denotes the complement of K a contradiction arises and our claim is correct.

We define the hyperplanes H^i_j by

$$H^i_j := \{x \in \mathbb{R}^n \mid A^i x \in \{\tilde{x} \mid \tilde{x}_j = 0\}\}, \quad i = 0,1,2\ldots m-1, \quad j = 1,2,\ldots n.$$

As A is invertible H_j^i is indeed a hyperplane for every i and j. Notice that we did prove that

$$\text{bd } K \subseteq \bigcup_{i,j} H_j^i$$ and applying Lemma 1 and

Lemma 2 leads to the desired result.

We would like to remark that the proof reveals even more about the structure of K; it has points in common with $(\text{bd } \mathbb{R}_+^n)\setminus\{0\}$, a situation as in the case n = 2.

Relations with Realizationtheory.

Let a sequence $\{W_k\}_{k=1}^{\infty} \subseteq \mathbb{R}_+$ be given such that $W_k = HF^{k-1}G$ where H is a $(1 \times n)$, F a $(n \times n)$ and G a $(n \times 1)$ real matrix. Then one could ask when it is possible to realize the nonnegative sequence in a nonnegative way i.e. such that $W_k = \tilde{H}\tilde{F}^{k-1}\tilde{G}$ for <u>nonnegative</u> matrices $\tilde{H}$, $\tilde{F}$ and $\tilde{G}$. This question is addressed in Ref. 3. Related problems are discussed in Ref. 4 and Ref. 5.

The next result from Ref. 3 indicates the relation with the previous parts of this paper.

THEOREM 11. Let $\{W_k\}_1^{\infty}$ be a nonnegative sequence.

Let (F_c,G_c,H_c) be the realization as descibed below. If there exists a polyhedral cone $Q \subseteq \mathbb{R}_+^n$ with $H_c \in Q$ and $QF_c \subseteq Q$, then $\{W_k\}_1^{\infty}$ can be realized in a nonnegative way. Here (F_c,G_c,H_c) is a <u>minimal</u> realization with

$$F_c := \begin{pmatrix} 0 & 0 & \cdot & \cdot & \cdot & -a_n \\ 1 & 0 & \cdot & \cdot & \cdot & \\ \cdot & & & & & \cdot \\ \cdot & & & & & \cdot \\ \cdot & & & & & \cdot \\ 0 & 0 & \cdot & \cdot & 1 & -a_1 \end{pmatrix} \qquad \begin{aligned} H_c &= (W_1,W_2,\ldots,W_n \\ G_c &= (1,0,\ldots,0)^T. \end{aligned}$$

We refer to Ref. 3 for a proof. In that paper is a sufficient condition in terms of eigenvalues such that the conditions in Theorem 11 do hold. All the eigenvalues have to be real in that condition. An alternative condition will be presented now. Denote, as before, the eigenvalues of F_c by $\lambda_1,\lambda_2,\ldots,\lambda_k$. We now have to work with <u>left-eigenvectors</u>. It is easy to see that these vectors are

$$e^i := (1,\lambda_i,\lambda_i^2,\ldots,\lambda_i^{n-1}),\ i = 1,2\ldots,k$$

Now we present the alternative condition.

THEOREM 12. Let $\lambda_1 = \rho(F_c)$ and $m_1 = 1$. Let there be no other <u>positive</u> eigenvalue and let λ_1 be such that $\lambda_1 > |\lambda_i|$, i = 2,...k. Then there is a polyhedral cone $Q \subseteq \mathbb{R}_+^n$ with $H_c \in Q$ and $QF_c \subseteq Q$, and hence $\{W_k\}_1^{\infty}$ can be realized in a nonnegative way.

PROOF. From Theorem 8 it follows that int $K \neq \phi$, and the assumptions together with Theorem 9 lead to the conclusion that F_c is primitive with respect to K, and hence K is polyhedral as follows from Theorem 10.
As K is the largest F_c invariant cone in $\mathbb{R}^n_+$ it has to be that $H_c \in K$ and we are done with the proof.

So we are a little bit closer to a complete understanding of the problem of nonnegative realizability.
Finally we will tell something about the origins of the problem discussed in the field of mathematical economics.

Leontieff-models in mathematical economics.

One of the simplest models in input-output analysis is the following

$$\text{LE}\begin{cases} x_{t+1} = [I+\tilde{B}^{-1}(I-\tilde{A})]x_t, & t \in Z_+, \\ x_t \geqq 0 \end{cases}$$

where $\tilde{A}$ is a so-called $(n \times n)$ matrix of input-coefficients, $\tilde{A} \geqq 0$, and $\tilde{B}$ is an $(n \times n)$ matrix of capital-coefficients.
For more details and economic background we refer to Ref. 6. Within this economic context it is reasonable to assume that $(I-\tilde{A})^{-1} > 0$, i.e., all the entries of this matrix are positive, and further that $\tilde{B}$ is non-singular and $\tilde{B} \geqq 0$.
Let us denote by ρ_1 the Frobenius root of $(I-\tilde{A})^{-1}\tilde{B}$. The other eigenvalues of $(I-\tilde{A})^{-1}\tilde{B}$ are indicated by ρ_i.

Further it is assumed that $[I+\tilde{B}^{-1}(I-\tilde{A})]$ is nonsingular and that the eigenvalues $\eta_i = 1 + \rho_i^{-1}$ of this matrix are such that $\eta_1 \neq |\eta_i|$, $i = 2,3,\ldots n$. For simplicity we assume that all the eigenvalues are different, hence there are n eigenvectors $k_1,k_2,\ldots k_n$. What can be said about these eigenvectors?
In the first place it are precisely the eigenvectors of the matrix $(I-\tilde{A})^{-1}\tilde{B}$.
By assumption we have that $(I-\tilde{A})^{-1}\tilde{B} > 0$ and hence ρ_1 is a simple eigenvalue, greater than the modulus of any other eigenvalue and $k_1 \in \text{int } \mathbb{R}^n_+$ and no other eigenvector is in $\mathbb{R}^n_+$. This follows from Theorem 1.3 and Theorem 1.4 in Ref. 2.

In the sequel we will relate our results concerning polyhedrality of K to the concept of relative stability in input-output analysis.

An important solution to (LE) is $x_t = c\eta_1^t k_1$ where $c > 0$.
DEFINITION. Model (LE) is called relatively stable if

$$\lim_{t\to\infty} \frac{\|x_t - c\eta_1^t k_1\|}{\|x_t\|} = 0$$

where (x_t) is an arbitrary solution to LE with initial condition $\geqq 0$ and not equal to zero. And it now appears, see Ref. 6, that the following result is true.

THEOREM 13. Model (LE) (with all the assumptions made) is relatively stable if and only if $\eta_1 = 1 + \rho_1^{-1}$ is greatest in modulus among all η_i's.

However, it is not hard to see that this eigenvalue condition also implies that $A := [I+\widetilde{B}^{-1}(I-\widetilde{A})]$ has a solid polyhedral cone K. This follows from Theorem 9 and Theorem 10 as a little exercise shows. Under the condition that all eigenvalues η_i are different (and this is the case in Tsukui's book Ref. 6) the fact that $AK \subseteq K$ for a solid cone K implies that $\rho(A) = 1 + \rho_1^{-1}$. So under the assumptions imposed on the model (LE) relative stability is equivalent to the existence of a solid polyhedral cone $K \subseteq \mathbb{R}^n_+$ such that $AK \subseteq K$ and such that there is no point outside K in $\mathbb{R}^n_+$ that forever (under A) remains in $\mathbb{R}^n_+$!

And maybe an economic interpretation can be given to the polyhedrality of K, but we do not know.

Final remarks.

Our first hope that we could characterize closed convex cones $\mathcal{B} \subseteq (\mathbb{R}^q)^{Z_+}$ in a manner similar to J.C. Willems' result is not fulfilled.

The situation seems to be even more hopeless, for consider the following example of a closed cone $\mathcal{B} \subseteq (\mathbb{R}^q)^{Z_+}$.

Example.

Take a collection $\{Q_i\}_{i=1}^{\infty}$ of closed convex cones in $\mathbb{R}^2_+$ with $Q_{i+1} \subseteq Q_i$, $\forall i \geq 1$ and such that $\bigcap_i Q_i \neq Q_j$, $\forall j$. A moment thought reveals that it is not hard to choose this collection further in such a way it seems impossible to describe $\widetilde{\mathcal{B}} := \{w := (w_t) \mid w_t \in Q_t\}$ with finitely many object different from $\widetilde{\mathcal{B}}$, although $\widetilde{\mathcal{B}}$ is a shift-invariant closed convex cone in $(\mathbb{R}^2)^{Z_+}$.

In the case of our simple model $x_{t+1} = Ax_t$, $x_t \geq 0$ we can describe it also by $x_{t+1} = Ax_t$, $x_t \in K$, and a similar expression seems impossible in the example above.

Concerning the cone K there remain a number of unsolved questions, to name a few:

1) give necessary <u>and</u> sufficient conditions such that K is polyhedral.
2) in case K is <u>solid</u>, is it true that K has nontrivial points in common with the boundary of $\mathbb{R}^n_+$?

Appendix.

PROOF OF LEMMA 1. By assumption we have that $H_i = \{x \in \mathbb{R}^n \mid x_i^* x = 0\}$ for some $x_i^* \in \mathbb{R}^n \setminus \{0\}$. As H_i has nothing in common with the interior of Q it follows that we may take without loss of generality

$$Q \subseteq \bigcap_{i=1}^{p} \{x \mid x_i^* x \geq 0\} =: \tilde{Q}$$

We claim that $Q = \tilde{Q}$ and hence that Q is polyhedral.

To prove that claim we take an element $\bar{x} \notin Q$ and an element $\bar{k} \in \text{int } Q$.

Now there is a $\bar{\lambda} \in (0,1)$ such that $\bar{\lambda}\bar{x} + (1-\bar{\lambda})\bar{k} \in \text{bd } Q$ and hence, for some $j \in \{1,2,\ldots,p\}$,

$$x_j^*(\bar{\lambda}\bar{x}+(1-\bar{\lambda})\bar{k}) = 0 \qquad (*)$$

As $\bar{k} \in \text{int } Q$ it follows that $x_j^*\bar{k} > 0$, for suppose to the contrary that $x_j^*\bar{k} = 0$, then $x_j^*(\bar{k}+u) \geq 0$, $\forall u \in U$, where U is a neighbourhood of the origin such that $\bar{k}+U \subseteq Q$, hence $x_j^* = 0$, a contradiction. But this implies, together with (*) that $x_j^*\bar{x} < 0$, and hence $\bar{x} \notin \tilde{Q}$ and we are done with this proof.

The proof of Lemma 2 is unfortunately tedious (bus easy).

PROOF OF LEMMA 2. Again we will take $H_i = \{x \in \mathbb{R}^n \mid x_i^* x = 0\}$, where $x_i^* \in \mathbb{R}^n \setminus \{0\}$. We will take an element $\bar{k} \in (\text{bd } Q) \cap H_1$ and prove that $\bar{k} \in \bigcup_{i=2}^{p} H_i$. Essentially that we do in the sequel is to construct a sequence of points $\{k_b(\lambda) \mid \lambda \downarrow 0\}$ such that $x_1^*(k_b(\lambda)) > 0$ and $k_b(\lambda) \in \text{bd } K$ and further $\lim_{\lambda \downarrow 0} k_b(\lambda) = \bar{k}$. As $\bigcup_{i=2}^{p} H_i$ is the union of finitely many closed sets it will follow that $\bar{k} \in \bigcup_{i=2}^{p} H_i$.

We take a point $\hat{k} \in H_1 \cap \text{int } Q$ and we define $\tilde{k} := 2^{-1}\bar{k} + 2^{-1}\hat{k} \in H_1 \cap \text{int } Q$.

We also take a point $k_+ \in \text{int}$ such that $x_1^* k_+ > 0$.

Further we define $k(\lambda) := (1-\lambda)\tilde{k} + \lambda k_+ \in \text{int } Q$, $\forall \lambda \in [0,1]$, and

$k(\lambda,\mu) := (1-\mu)\hat{k} + \mu k(\lambda)$, $\mu k(\lambda)$, $\mu \geq 1$, $\lambda \in [0,1]$.

We claim that for λ sufficiently small the following holds:

$\mu(\lambda) := \sup \{\mu \mid (\lambda,\mu) \in Q\}$ is finite and

$k(\lambda,\mu(\lambda)) =: k_b(\lambda) \in \text{bd } Q$.

We will prove this. Notice that by construction $k(\lambda,1) \in \text{int } Q$. Take a fixed $\lambda \in [0,1]$ and suppose that there is a sequence $\{\mu_i \to +\infty\}$ such that $k(\lambda,\mu_i) \in Q$, then it would follow from the closedness of Q that $k(\lambda) - \hat{k} \in Q$. Suppose this to be true for every $\lambda > 0$, then $\tilde{k}-\hat{k} = 2^{-1}\bar{k} + 2^{-1}\hat{k} - \hat{k} = 2^{-1}\bar{k} - 2^{-1}\hat{k} \in Q$ and this would imply that $\bar{k} \in \text{int } Q$ but by assumption $\bar{k} \in \text{bd } Q$. Up till now we proved that $\mu(\lambda)$ is finite for λ sufficiently small.

As Q is closed it easily follows that $k_b(\lambda) \in Q$.

By definition of $\mu(\lambda)$ we even have that $k_b(\lambda) \in \text{bd } Q$ (hence the subscript b).

Now we claim that $x_1^* k_b(\lambda) > 0$, $\forall \lambda \in (0,1)$ and further that $\lim_{\lambda \downarrow 0} k_b(\lambda) = \bar{k}$.

First we prove that $x_1^* b_k(\lambda) > 0$, $\lambda > 0$.

By construction $k_b(\lambda) = (1-\mu(\lambda))\hat{k} + \mu(\lambda)k(\lambda)$ with $x_1^*k(\lambda) > 0$, $\forall\lambda \in [0,1]$, and $x_1^*\hat{k} = 0$, hence indeed $x_1^*k_b(\lambda) > 0$, for all $\lambda > 0$ sufficiently small. Suppose not to the contrary that there is a sequence $\{\lambda_i\}$ with $\lambda_i \downarrow 0$ and a sequence $\{\mu(\lambda_i)\}$ with $\mu(\lambda_i) \to +\infty$, then $\bar{k} \in \text{int } Q$, a contradiction. This can be seen as follows.

$$(1-\mu(\lambda_i))\hat{k} + \mu(\lambda_i)((1-\lambda_i)\tilde{k}+\lambda_i k_+) \in \text{bd } Q, \text{ hence}$$

$$-\hat{k} + \tilde{k} \in \text{bd } Q, \text{ or } \bar{k}-\hat{k} \in \text{bd } Q, \text{ but}$$

$\hat{k} \in \text{int } Q$ and hence $\bar{k} \in \text{int } Q$, a contradiction.

Therefore we may assume that $\lambda_i \downarrow 0$ implies that $\mu(\lambda_i) \to \bar{\mu}$, and this leads to the conclusion that

$$(1-\bar{\mu})\hat{k} + \bar{\mu}\tilde{k} \in \text{bd } Q, \text{ implying that}$$

$$(1-\bar{\mu})\hat{k} + \bar{\mu}(2^{-1}\bar{k}+2^{-1}\hat{k}) \in \text{bd } Q, \text{ or}$$

$$(1-2^{-1}\bar{\mu})\hat{k} + 2^{-1}\bar{\mu}\bar{k} \in \text{bd } Q.$$

We now claim that $\bar{\mu} = 2$. Suppose to the contrary that $\bar{\mu} \neq 2$, then we distinguish two possibilities:

(a) $\bar{\mu} > 2$, but then $(2^{-1}\bar{\mu}-1)\hat{k} \in \text{int } Q$ and hence $\bar{k} \in \text{int } Q$, a contradiction.

(b) $1 \leqq \bar{\mu} < 2$ leads again to a contradiction, hence $\bar{\mu} = 2$ implying that $\lim_{\lambda \downarrow 0} k(\lambda,\mu(\lambda)) = \bar{k}$ and we are done with the proof.

REFERENCES:

[1] Willems, J.C., From Time Series to Linear System, Part I, 1985 submitted for publication.

[2] Berman, A., Plemmons, J., Nonnegative matrices in the mathematical sciences, Academic Press (1979).

[3] Nieuwenhuis, J.W., About nonnegative realizations, Systems and Control Letters vol 1, (1982) pp 283-287.

[4] Maeda, H., Kodama, S., Positive realization of difference equations, IEEE, trans- on circuits and systems, (1981) pp 39-47.

[5] Ohta, Y., Maeda, H., Kodama, S., Reachability, observability and realizability of continuous-time positive systems, SIAM J. control and optimization, vol 22, (1982) pp 171-180.

[6] Tsukui, J., Murakami, Y., Turnpike optimality in Input-Output Systems, North-Holland (1979).

Frequency Domain and State Space Methods for Linear Systems
C.I. Byrnes and A. Lindquist (Editors)

FLAT TURN CONTROL LAW DESIGN USING EIGENSTRUCTURE ASSIGNMENT

Dr. K. M. Sobel
Senior Scientist
Lockheed California Company
Burbank, California
U.S.A.

Dr. E. Y. Shapiro
General Manager
Special Products Division
HR Textron
Valencia, California
U.S.A.

Dr. A. N. Andry
Director of Engineering
Teledyne Solid
State Products
Hawthorne, California
U.S.A.

A flat turn capability will be important for advanced aircraft of the future. This specialized control mode will provide for a reduced pilot workload during a bombing task. A modern control design is suggested which utilizes eigenstructure assignment. This method allows for direct gain computation based upon decoupling the lateral acceleration mode from both the dutch roll and roll mode.

INTRODUCTION

Advanced aircraft such as control configured vehicles (CCV) provide the capability for implementing specialized modes for bombing, strafing, and air-to-air combat. The control objectives for these advanced modes include the ability to command a chosen variable without causing significant motion in other variables.

For the longitudinal dynamics of an aircraft such decoupled flight modes (DFM) include direct lift control (DLC), pitch pointing, and vertical translation. For the lateral dynamics the corresponding modes are direct sideforce control, yaw pointing, and lateral translation.

In this paper, we consider the direct sideforce mode. This mode is characterized by a lateral acceleration command accompanied by no change in bank angle or sideslip angle. The result of such a maneuver is to cause a change in both heading (direction of the fuselage) and lateral flight path (direction of the velocity vector of the c.g.). This is the so-called flat turn maneuver which will enhance the bombing capability of the aircraft.

The design methodology to be utilized is eigenstructure assignment as described in [1]. Application of eigenstructure assignment to DFM design was first proposed in [2] where the longitudinal dynamics were considered. In this paper, we design a direct sideforce or flat turn control law by choosing the desired eigenvectors such that the lateral acceleration mode is decoupled from both the dutch roll and the roll mode. We obtain an acceptable design in which several elements of the output feedback gain matrix are constrained to be zero. By not feeding back every output to every control surface, we increase the reliability of the control system.

THEORETICAL BACKGROUND

Consider the linear time invariant system described by

$$\underline{\dot{x}} = A\underline{x} + B\underline{u} \tag{1}$$

$$\underline{y} = C\underline{x} \tag{2}$$

with the assumptions

$$\text{rank} \quad [B] \quad = \quad m \tag{3}$$

and

$$\text{rank} \quad [C] \quad = \quad r \tag{4}$$

Under the above assumptions the problem can be stated as follows:

Given a set of desired eigenvalues $\{\lambda_i^d\}$, i = 1, 2, ..., r and a corresponding set of desired eigenvectors $\{\underline{v}_i^d\}$, i = 1, 2, ..., r, find a real m x r matrix F such that the eigenvalues of A + BFC contain $\{\lambda_i^d\}$ as a subset, and the corresponding eigenvectors of A + BFC are "close" to the respective members of the set $\{\underline{v}_i^d\}$.

The solution to this problem, which is given in [1], is summarized below. Consider the closed loop system.

$$\dot{\underline{x}} = (A + BFC)\, \underline{x} \tag{5}$$

For an eigenvalue/eigenvector pair, λ_i and $\underline{v}_i$

$$(A + BFC)\, \underline{v}_i = \lambda_i\, \underline{v}_i \tag{6}$$

or

$$\underline{v}_i = (\lambda_i\, I-A)^{-1}\, BFC\, \underline{v}_i \tag{7}$$

Define the m-vector $\underline{m}_i$ as

$$\underline{m}_i = FC\, \underline{v}_i \tag{8}$$

Then (8) becomes

$$\underline{v}_i = (\lambda_i\, I-A)^{-1} B\underline{m}_i \tag{9}$$

The importance of (9) is that the eigenvector $\underline{v}_i$ must be in the subspace spanned by the columns of $(\lambda_i\, I-A)^{-1}B$. This subspace is of dimension m which is the number of independent control variables. Therefore, the number of control variables determines the dimension of the subspace in which the achievable eigenvectors must reside. The orientation of the subspace is determined by the open loop parameters described by A, B and the desired closed loop eigenvalue λ_i. We conclude that if we choose an eigenvector $\underline{v}_i$ which lies precisely in the subspace spanned by the columns of $(\lambda_i\, I-A)^{-1}B$, it will be achieved exactly.

In general, however, a desired eigenvector, $\underline{v}_i^d$ will not reside in the prescribed subspace and hence, cannot be achieved. Instead, a "best possible" choice for an achievable eigenvector is made. This "best possible" eigenvector is the projection of $\underline{v}_i^d$ onto the subspace spanned by the columns of $(\lambda_i \ I-A)^{-1}B$.

To further complicate the situation, complete specification of $\underline{v}_i^d$ is neither required nor known in most practical situations. When the designer is interested in only certain elements of the eigenvector, we assume that the desired eigenvector has a structure given by

$$\underline{v}_i^d = \left[v_{i1}, X, X, X, X, v_{ij}, X, X, X, v_{in}\right]^T$$

where

v_{ij} = designer specified components

X = unspecified components.

We define a reordering operator $\left\{ \ \right\}^{R_i}$ as follows [3]:

$$\left\{\underline{v}_i^d\right\}^{R_i} = \begin{bmatrix} \underline{\ell}_i^d \\ \underline{d}_i \end{bmatrix} \tag{10}$$

where

$\underline{\ell}_i^d$ is the vector of specified components of $\underline{v}_i^d$

$\underline{d}_i$ is the vector of unspecified components of $\underline{v}_i^d$

We begin the computation of an achievable eigenvector, $\underline{v}_i^d$ by defining

$$L_i = (\lambda_i \ I-A)^{-1}B \tag{11}$$

An achievable eigenvector must reside in the required subspace and hence,

$$\underline{v}_i^a = L_i \ \underline{z}_i \tag{12}$$

We reorder the rows of L_i to conform with the reordered components of $\underline{v}_i^d$. Thus, as shown in [3], we have

$$\left\{L_i\right\}^{R_i} = \begin{bmatrix} \tilde{L}_i \\ D_i \end{bmatrix} \tag{13}$$

To find the value of z_i corresponding to the projection of $\underline{\ell}_i^d$ onto the "achievability subspace," we choose $\underline{z}_i$ which minimizes

$$J = || \underline{\ell}_i^d - \underline{\ell}_i^a ||^2 = || \underline{\ell}_i^d - L_i \underline{z}_i ||^2 \tag{14}$$

Upon setting the first derivative of J with respect to $\underline{z}_i$ equal to zero we obtain,

$$\underline{z}_i = \left(\tilde{L}_i^T \tilde{L}_i \right)^{-1} \tilde{L}_i^T \underline{\ell}_i^d \tag{15}$$

and

$$\underline{v}_i^a = L_i \left(\tilde{L}_i^T \tilde{L}_i \right)^{-1} \tilde{L}_i^T \underline{\ell}_i^d \tag{16}$$

As shown in [1], the feedback gain matrix is computed by using a transformation to obtain a system equivalent to (1) - (2). This transformed system is described by $(\tilde{A}, \tilde{B}, \tilde{C})$ where

$$\tilde{A} = T^{-1} A T \tag{17}$$

$$\tilde{B} = T^{-1} B = \begin{bmatrix} I_m \\ \hline 0 \end{bmatrix} \tag{18}$$

$$\tilde{C} = C T \tag{19}$$

Under this transformation,

$$\underline{x} = T \underline{\tilde{x}} \tag{20}$$

$$\lambda_i = \tilde{\lambda}_i$$

$$\underline{v}_i^a = T \underline{\tilde{v}}_i^a \tag{22}$$

We partition $\underline{\tilde{v}}_i^a$ and $\tilde{A}$ conformally with $\tilde{B}$ to obtain

$$\underline{\tilde{v}}_i^a = \begin{bmatrix} \underline{\tilde{s}}_i \\ \hline \underline{\tilde{w}}_i \end{bmatrix} \tag{23}$$

$$\tilde{A} = \begin{bmatrix} \tilde{A}_1 \\ \hline \tilde{A}_2 \end{bmatrix}$$

Then, as shown in [1], the feedback gain matrix is computed by using

$$F = \left(\tilde{S} - \tilde{A}_1 \tilde{V}\right) \left(\tilde{C} \tilde{V}\right)^{-1} \tag{24}$$

where

$$\tilde{S} = \left[\lambda_1 \underline{\tilde{s}}_1, \lambda_2 \underline{\tilde{s}}_2, \ldots, \lambda_r \underline{\tilde{s}}_r\right] \tag{25}$$

and

$$\tilde{V} = \left[\underline{\tilde{v}}_1^a, \underline{\tilde{v}}_2^a, \ldots, \underline{\tilde{v}}_r^a\right] \tag{26}$$

In order to constrain certain elements of F to be zero, we write (24) as follows [1]:

$$F \tilde{C} \tilde{V} = \tilde{S} - \tilde{A}_1 \tilde{V} \tag{27}$$

Define

$$\Omega = \tilde{C} \tilde{V} \tag{28}$$

$$\Psi = \tilde{S} - \tilde{A}_1 \tilde{V} \tag{29}$$

Then, (27) becomes

$$I_m F \Omega = \psi \tag{30}$$

We rewrite (30) in terms of the Kronecker product and a row stacking operator [1],

$$\left[I_m \times \Omega^T\right] \text{STACK}(F) = \text{STACK}(\Psi) \tag{31}$$

Letting $\underline{f}_i$ denote the i-th row of F and $\underline{\psi}_i$ denote the i-th row of Ψ, we obtain

$$\begin{bmatrix} \Omega^T & 0_m & \cdots & 0_m \\ 0_m & \Omega^T & \cdots & 0_m \\ \cdot & & & \cdot \\ \cdot & & & \cdot \\ \cdot & & & \cdot \\ & & \cdots & \Omega^T \end{bmatrix} \begin{bmatrix} \underline{f}_1^T \\ \underline{f}_2^T \\ \cdot \\ \cdot \\ \cdot \\ \underline{f}_m^T \end{bmatrix} = \begin{bmatrix} \underline{\psi}_1^T \\ \underline{\psi}_2^T \\ \cdot \\ \cdot \\ \cdot \\ \underline{\psi}_m^T \end{bmatrix} \tag{32}$$

From (32), we observe that each row of the feedback gain matrix, $\underline{f}_i$, can be computed independently of the other rows.

Now consider the equation for the i-th row given by

$$\Omega^T \underline{f}_i^T = \underline{\psi}_i^T \tag{33}$$

If we want to constrain f_{ij} to be zero, then we delete f_{ij} from $\underline{f}_i^T$ and delete the j-th column of Ω^T. We solve the reduced problem given by

$$\hat{\Omega}^T \hat{\underline{f}}_i^T = \underline{\psi}_i^T \tag{34}$$

where $\hat{\Omega}^T$ is the matrix Ω^T with its j-th column deleted and $\hat{\underline{f}}_i^T$ is the vector $\underline{f}_i^T$ with its j-th element deleted.

EXAMPLE

The model of the Flight Propulsion Control Coupling (FPCC) lateral dynamics is described by:

$$\underline{x} = [\beta, \phi, p, r, \psi + \beta, \delta_r, \delta_a, \delta_c, n_{ypi}]^T$$

$$\underline{u} = [\delta_{rc}, \delta_{ac}, \delta_{cc}]^T$$

$$\underline{y} = [r, \beta, p, \phi, n_{ypi}]^T$$

where

β = sideslip angle

ϕ = bank angle

p = roll rate

r = yaw rate

$\psi+\beta$ = lateral directional flight path

δ_r = rudder deflection

δ_a = aileron deflection

δ_c = canard deflection

n_{ypi} = integrated lateral acceleration at pilot station

The matrix A is given by

$$
\begin{bmatrix}
-.106\text{E }01 & 0.32\text{E-}01 & 0.190\text{E-}02 & -.994\text{E }00 & 0.0 & 0.114\text{E }00 & 0.0 & 0.749\text{E-}01 & 0.0 \\
0.0 & 0.0 & 0.100\text{E }01 & 0.0 & 0.0 & 0.0 & 0.0 & 0.0 & 0.0 \\
-.209\text{E }02 & 0.0 & -.348\text{E }01 & 0.352\text{E }01 & 0.0 & 0.515\text{E }01 & 0.273\text{E }02 & -.362\text{E }01 & 0.0 \\
0.239\text{E }02 & 0.0 & 0.419\text{E }00 & -.151\text{E }01 & 0.0 & -.122\text{E }02 & 0.512\text{E }00 & 0.494\text{E }01 & 0.0 \\
-.106\text{E }01 & 0.302\text{E-}01 & 0.190\text{E-}02 & 0.580\text{E-}02 & 0.0 & 0.114\text{E }00 & 0.0 & 0.749\text{E-}01 & 0.0 \\
0.0 & 0.0 & 0.0 & 0.0 & 0.0 & -.200\text{E }02 & 0.0 & 0.0 & 0.0 \\
0.0 & 0.0 & 0.0 & 0.0 & 0.0 & 0.0 & -.200\text{E }02 & 0.0 & 0.0 \\
0.0 & 0.0 & 0.0 & 0.0 & 0.0 & 0.0 & 0.0 & -.200\text{E }02 & 0.0 \\
-.218\text{E }02 & -.148\text{E-}02 & 0.254\text{E }00 & -.522\text{E }00 & 0.0 & -.211\text{E }01 & 0.239\text{E }00 & 0.464\text{E }01 & 0.0
\end{bmatrix}
$$

and the matrix B is given by

$$\begin{bmatrix} 0.0 & 0.0 & 0.0 \\ 0.0 & 0.0 & 0.0 \\ 0.0 & 0.0 & 0.0 \\ 0.0 & 0.0 & 0.0 \\ 0.0 & 0.0 & 0.0 \\ 0.200E\ 02 & 0.0 & 0.0 \\ 0.0 & 0.200E\ 02 & 0.0 \\ 0.0 & 0.0 & 0.200E\ 02 \\ 0.0 & 0.0 & 0.0 \end{bmatrix}$$

The desired eigenvectors were chosen such that the lateral acceleration mode would be decoupled from both the dutch roll mode and the roll mode. This choice yields the following:

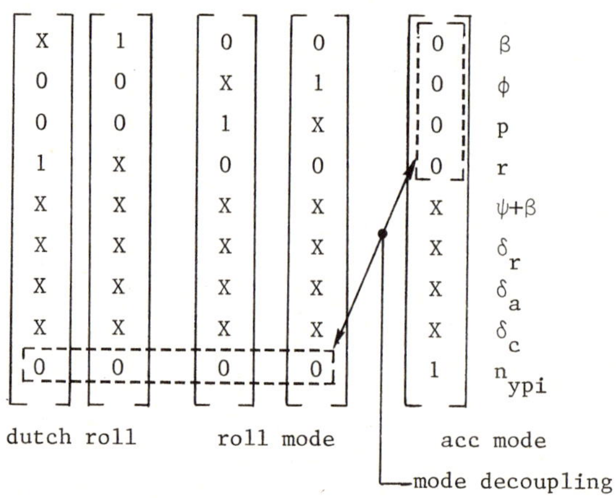

Several designs are summarized in Table 1. Design No. 1 is characterized by an output feedback gain matrix without any gain constraints. The time histories (not shown) exhibit a lateral acceleration which achieves its 1g commanded value while the steady state sideslip angle and bank angle are described by $|\beta_{ss}| = 0.44$ deg and $|\phi_{ss}| = 0.014$ deg, respectively.

Design No. 2 is characterized by the constraints that bank angle and roll rate shall not be fed back to either the rudder or canard. In this design, the number of gains is reduced by more than 25 percent as compared to design No. 1. The time histories (not shown) exhibit almost the same behavior as in design No. 1.

TABLE 1
CONTROL LAW SUMMARY

	Achievable Eigenvalues	Feedback Gain Matrix						Response	Magnitudes
Design #1	$\lambda_{1,2} = -3.000 \pm j2.002$	r	β	p	φ	n_{ypi}		$\lvert\delta_r\rvert \leq 8.03$	$\lvert\beta\rvert \leq .44$
	$\lambda_{3,4} = -1.998 \pm j2.007$	-.0473	-4.05	-.00624	.0229	.275	δ_r	$\lvert\delta_a\rvert \leq .975$	$\lvert\phi\rvert \leq .014$
	$\lambda_5 = -3.002$	.162	-.811	.0939	.405	.0424	δ_a	$\lvert\delta_c\rvert \leq 17.77$	
	$\lambda_{6,7} = -18.36 \pm j0.3677$	.101	-6.53	.0305	-.0542	.696	δ_c		
	$\lambda_8 = -16.33$								
Design #2	$\lambda_{1,2} = -3.068 \pm j1.831$	r	β	p	φ	n_{ypi}		$\lvert\delta_r\rvert \leq 8.03$	$\lvert\beta\rvert \leq .45$
	$\lambda_{3,4} = -2.001 \pm j1.997$	-.0472	-4.05	0	0	.275	δ_r	$\lvert\delta_a\rvert \leq .975$	$\lvert\phi\rvert \leq .015$
	$\lambda_5 = -2.997$	.162	-.811	.0939	.405	.0424	δ_a	$\lvert\delta_c\rvert \leq 17.77$	
	$\lambda_{6,7} = -17.62 \pm j0.4105$	.101	-6.52	0	0	.695	δ_c		
	$\lambda_8 = -17.66$								
Design #2I	$\lambda_{1,2} = -3.069 \pm j1.801$	r	$\beta - 0.1\int\beta dt$	p	φ	n_{ypi}		$\lvert\delta_r\rvert \leq 7.45$	$\lvert\beta\rvert \leq .42$
	$\lambda_{3,4} = -1.997 \pm j2.009$	-.0927	-3.86	0	0	.273	δ_r	$\lvert\delta_a\rvert \leq .975$	$\lvert\phi\rvert \leq .18$
	$\lambda_5 = -2.999$	.153	-.773	.0939	.405	.0420	δ_a	$\lvert\delta_c\rvert \leq 17.77$	
	$\lambda_{6,7} = -17.61 \pm j0.6761$	.0274	-6.22	0	0	.692	δ_c		
	$\lambda_8 = -17.46$								
Design #3	$\lambda_{1,2} = -3.961 \pm j1.804$	r	β	p	φ	n_{ypi}		$\lvert\delta_r\rvert \leq 8.60$	$\lvert\beta\rvert \leq .58$
	$\lambda_{3,4} = -1.151 \pm j2.038$	-.0472	-4.05	0	0	.275	δ_r	$\lvert\delta_a\rvert \leq 1.09$	$\lvert\phi\rvert \leq 2.58$
	$\lambda_5 = -3.096$	0	0	.0958	.416	0	δ_a	$\lvert\delta_c\rvert \leq 18.91$	
	$\lambda_{6,7} = -17.86 \pm j0.3909$	.101	-6.52	0	0	.695	δ_c		
	$\lambda_8 = -17.01$								

Design No. 2I is characterized by feeding back proportional plus integral sideslip angle described by

$$y_2 = \beta - 0.1 \int \beta dt$$

The time histories for this design are shown in Figure 1. Sideslip angle attains a maximum of 0.42 deg, but it approaches zero as the time into the maneuver increases. The bank angle attains a maximum value of 0.18 deg which is acceptable although it is larger than for designs No. 1 and No. 2. Further, we observe that both heading and lateral flight path have achieved a change in excess of 17 deg during the 10 sec maneuver. We conclude that design No. 2I is acceptable because it achieves heading and flight path changes with insignificant variations in sideslip angle and bank angle.

Design No. 3 is characterized by the configuration and constraints of design No. 2 with the additional constraint that only bank angle and roll rate may be fed to the ailerons. The time histories (not shown) exhibit a bank angle variation of 2.58 deg which is excessive.

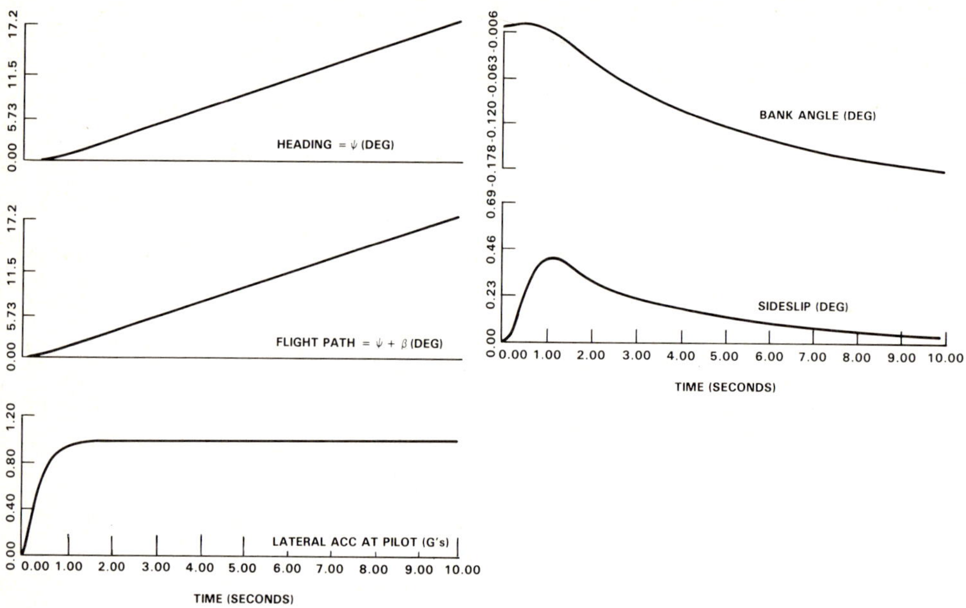

Figure 1
Time Responses (Design No. 2I)

REFERENCES

1. Andry, A. N., Shapiro, E. Y., Chung, J. C., "On Eigenstructure Assignment for Linear Systems," IEEE Transactions on Aerospace and Electronic Systems, Vol. AES-19, No. 5, September 1983

2. Sobel, K. M. and Shapiro, E. Y., "A Design Methodology for Pitch Pointing Flight Control Systems," AIAA Journal of Guidance, Control, and Dynamics, Vol. 8 , No. 2, March - April 1985

3. Harvey, C. H. and Stein, G., "Quadratic Weights for Asymptotic Regulator Properties," IEEE Transactions on Automatic Control, Vol. AC-23, pp. 378-387, June 1978

Frequency Domain and State Space Methods for Linear Systems
C.I. Byrnes and A. Lindquist (Editors)

STRUCTURE SELECTION METHODS APPLIED TO POWER SYSTEM STABILIZER DESIGN

L.H.Y.Wong
Engineering Faculty
University of Regina
Saskatchewan, Canada

J.D.Katzberg
Engineering Faculty
University of Regina
Saskatchewan, Canada

P.Daval
Sask Power Research
Regina
Saskatchewan, Canada

A structure selection technique has been used to select [1] and design [2] a power system stabilizer for Poplar River Unit 2 of the Saskatchewan Power Corporation. This method predicted that feedback of frequency or terminal voltage with power would give near optimal performance. It also predicted that these configurations would give considerably improved performance over the traditional approaches of using frequency [3] or power [4] alone. These predictions were verified by extensive simulations using a large nonlinear model of the complete Saskatchewan Power System. Further it was demonstrated by the same simulation that the selected stabilizer designs provided good damping both for small disturbances and large faults.

HISTORY AND MOTIVATION

Linear state feedback controllers coupled with state estimators obtained by solving the linear quadratic Gaussian problem have many good features such as guaranteed stability, good step responses and insensitivity to noise and plant variations. Over the years many authors have suggested using these methods to solve various power systems problems, for example [5] and [6] apply optimal control methods to stabilizer design. The complicated large nature of power systems and their fast response times encourage the use of such methods.

However power system utility management is very conservative and suspicious of control systems they don't understand. They prefer the tried and true. Reliability and system security are more important than cost or performance. The selection of control structure, the choice of which measurements to feedback to which control inputs, is based on experience and confirmed by simulation. Most utilities use single loop frequency response design methods to get a starting estimate for their controller parameters. The designs are then tuned up on very detailed and large system simulations. Given the very high system security requirements it is important that simulation runs be done for all possible fault conditions to insure the performance of the control system will be adequate.

J.D.Katzberg has been interested in the design and selection of simple controller structures since his doctoral work in the early 70's [7]. He has produced methods for computing optimal gains for reduced structures which use the property that the quadratic cost can be expressed as a quadratic form in the controller gains which are free to be optimized. Linear discrete time [1,2,7], continuous time [8], or function space [9] models may be used for the system. He has produced methods for further expanding the quadratic cost function into a tiangular form in terms of controller gains for linear systems described by discrete time [1], continuous time [8], or function space [9] models. This tiangular form may be used to estimate the effect of each gain and to order the gains by their effect. Use of these methods indicated, that frequently controllers can be reduced to just a few gains and still give near optimal performance. J.D. Katzberg was therefore encouraged to try these approaches on some real industrial problems.

A power system is large and geographically distributed. The requirement of a centralized controller that the measurements be transmitted to a central location where the control calculations can be performed and that the control inputs be transmitted from the central location to the various actuators can cause severe telecommunications difficulties. In addition a multitude of feedback links increase the cost and reduce the reliability. Rather than using a complete linear quadratic controller it would be preferable to use local feedback, that is feedback of measurements taken near the actuators. If such a controller could not provide adequate performance one might wish to try transmitting a few essential variables. Simple control systems are easier to sell to management. J.D.Katzberg spoke to P.Daval about these matters to determine if some collaboration might be possible. P.Daval, a research engineer with The Saskatchewan Power Corporation, expressed some interest.

In the spring of 1983 L.H.Y. Wong indicated to J. D. Katzberg that he would like to do an M.Sc project on applied digital control. J.D. Katzberg then contacted P. Daval who indicated Saskatchewan Power were now very interested in stability improvement by excitation control. Recently, when a 40MW load had tripped, two 300MW thermal generating units had oscillated against each other. The Saskatchewan Power System has a capacity of about 2500MW.

During the remainder of 1983 and early 1984 L.H.Y. Wong produced linear state space models suitable for linear quadratic regulator design, designed and simplified the stabilizer and computed approximate step responses using the simplified design models. He produced a package of programs for his IBM PC using Pascal to do these calculations. In June and July 1984 L. Wilson of The Saskatchewan Power Corporation tested the stabilizer's performance on a detailed nonlinear simulation of the Saskatchewan Power System using the Power Technologies, Inc. power system simulator. Figures 1 to 8 have been selected from the results produced.

MODELING DESIGN AND SIMULATION

Poplar River Unit 2 was modeled as a single machine against an infinite bus. A fourth order linear generator model suggested by Anderson and Fouad [10] was transformed so the measured variables, shaft speed, electric power, terminal voltage and field voltage, were states. Model parameters were computed from data provided by P. Daval.

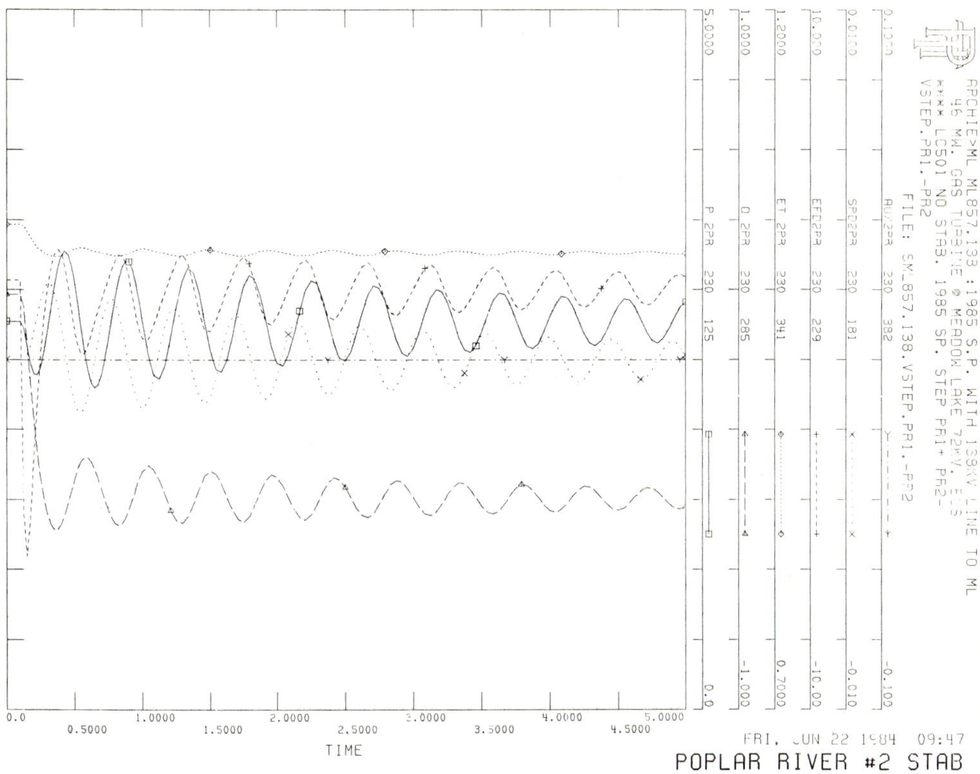

Figure 1: The Response of Poplar River Unit Two without a Stabilizer to a Voltage Disturbance.

It is intended to implement the final design in a microprocessor based product. This will allow the stabilizer to conveniently communicate with the control room and central control computer. It will also permit the controller to be easily changed or expanded in the future.

A discrete time approach seemed most natural. The state space model was discretized, the optimal linear quadratic gains were computed, and the step responses checked on a linear simulation. When quadratic costs that produced good step responses were found, the discrete structure selection method published in [1] was used to order the gains in terms of their effect. It was found power was the most important variable but using it alone has a 40% increase in cost over using either speed or terminal voltage with power with reoptimized gains by the method in [2]. The most interesting controllers were then tested on Saskatchewan Power's complete system simulation. Proportional feedback of power and terminal voltage was found to be most satisfactory.

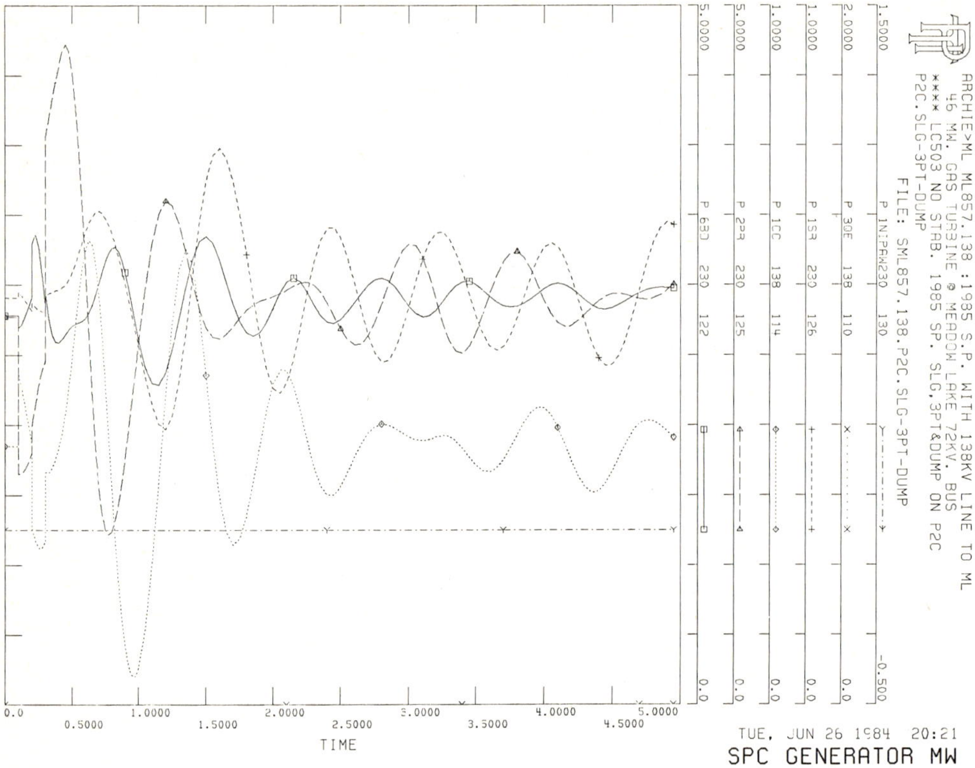

Figure 2: The Response of Poplar River Unit Two without a Stabilizer to a Line Fault.

Figure 1 and 3 are the responses, without and with the power and terminal voltage stabilizer respectively, for the Poplar Unit 2 when a small disturbance occurs. The time scale is in seconds and the other scales are in appropriate per unit quantities. The 2PR extension on the variable names designates Poplar River Unit Two.

The abbreviated variable names represent:

AUX	-	Stabilizer Output (per unit)
SPD	-	Speed or Angular Velocity (per unit variation about the operating point)
EFD	-	Field Voltage (per unit)
ET	-	Terminal Voltage (per unit)
Q	-	Reactive Power (100 MVAR per unit)
P	-	Electrical Power (100 MW per unit)

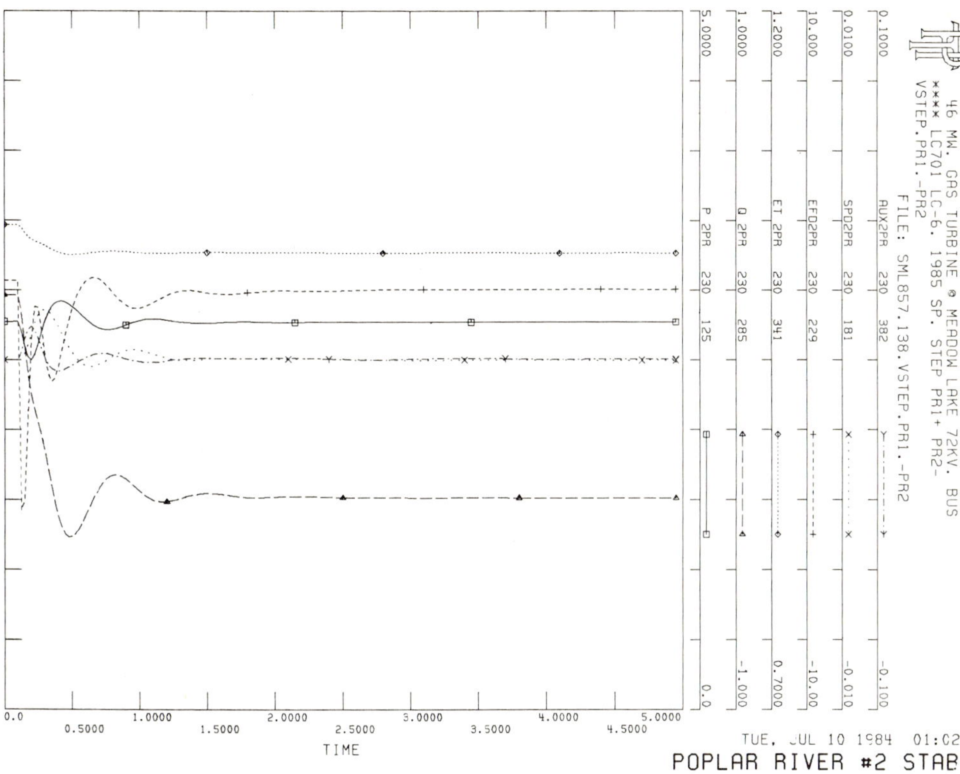

Figure 3: The Response of Poplar River Unit Two with a Proportional Power and Terminal Voltage Stabilizer to a Voltage Disturbance.

Figure 2 and 4 are the power responses, without and with the power and terminal voltage stabilizer respectively, for the largest generators at the main stations in the system when a major line from the Poplar River Station to Regina is lost. Again the time is in seconds and the power is in units of 100MW. The stabilizer is on the unit labled 2PR. It can be clearly seen that both for small and large disturbances the stabilizer improves system performance significantly.

The mnemonics for the generators mean:

1NIPAW	-	Nipawin Unit 1
3QE	-	Queen Elizabeth Unit 3
1SR	-	Squaw Rapids Unit 1
1CC	-	Coteau Creek Unit 1
2PR	-	Poplar River Unit 2
6BD	-	Boundary Dam Unit 6

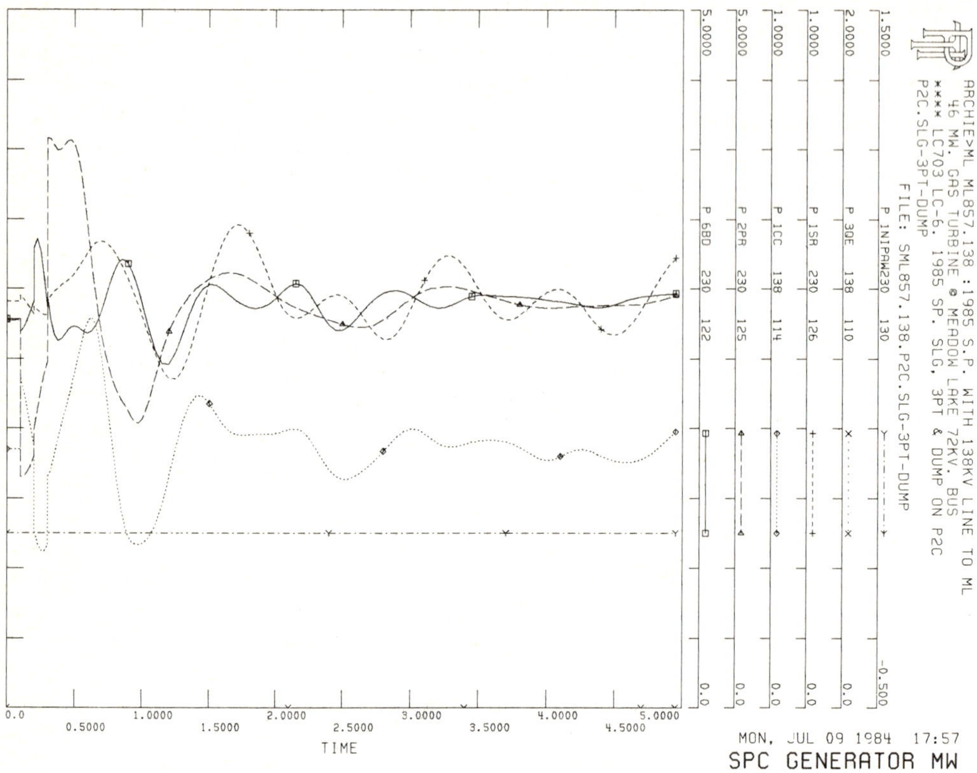

Figure 4: The Response of Poplar River Unit Two with a Proportional Power and Terminal Voltage Stabilizer to a Line Fault.

THE LINEAR MODELS

Saskatchewan Power Poplar River Unit Two is a 300MW Hitachi Generator powered by a three stage turbine fed from a Combustion Engineering lignite fired boiler. The data available limited us to the four state model of a synchronous machine connected to a transmission line. The model suggested by Anderson and Fouad was transformed so the measurable variables: speed, terminal voltage, electrical power and field voltage were states. The control input was a modulating voltage applied to the excitation control system.

Four linear models were formed by linearizing the nonlinear model about nominal, heavy load, underexcited, and long line operating points. The discrete time models were then formed from the continuous time models. A sampling time of 8.33ms = 1/120Hz was selected to minimize the error caused by zero sequence effects. These computations were done by a program package written in Pascal on an IBM Personal Computer.

LINEAR DESIGN

The package of programs written for the IBM Personal Computer also included programs for linear quadratic regulator design, Kalman-Bucy filters, specific optimal design including proportional integral derivative controllers[2] and programs to assess the contribution to cost reduction of the various controller gains[1]. This package included programs for linear simulation as well.

The first stage was to design a linear quadratic regulator then check its step responses on a linear simulation. Choosing the quadratic cost matrices to be diagonal with the on diagonal elements the inverse square of the maximum acceptable deviations gave a very good starting point. However to get satisfactory damping of power it was necessary to increase the damping on that variable by a factor of ten. Designs were done for all four operating points.

The importance of the gains in these controllers was then assessed by expanding the cost function for these discrete time systems in a triangular array as described in [1]. With a nominal load the feedback gains in inverse order of importance are speed, field voltage, terminal voltage and power. With a heavy load the feedback gains in inverse order of importance are terminal voltage, field voltage, speed and power. The system model therefore effects the controller structure. In either case two gains could be removed with little effect but when the third gain was removed there was about a 40% increase in expected cost.

When a specific optimal controller which used terminal voltage and power was computed for the heavy load case the increase in quadratic cost was only 5% over the speed and power controller. Speed measurement can pick up and amplify torsional oscillations. The speed transducers are normally placed within a bearing. This makes the speed instrumentation expensive and difficult to maintain. These factors make the use of terminal voltage and power feedback preferable to speed and power.

As a microprocessor was to be used for control, complex calculations could be performed with no increase in hardware complexity. The main cost is in the measurement of the variables fed back. Added dynamics have a small effect on cost. Various Kalman-Bucy filters with optimal regulators were then designed, simulated and simplified. Similarly various specific optimal proportional integral derivative regulators were then designed, simulated and simplified. Controllers were considered good if they had a near optimal expected cost and the linear simulation produced good step responses for all four operating conditions. The use of terminal voltage and power as inputs to a Kalman filter with optimal quadratic gain regulator and to a P.I.D. controller were shown to give slightly improved linear system performance. Eleven controllers were selected for evaluation on Sask Power's detailed nonlinear simulation.

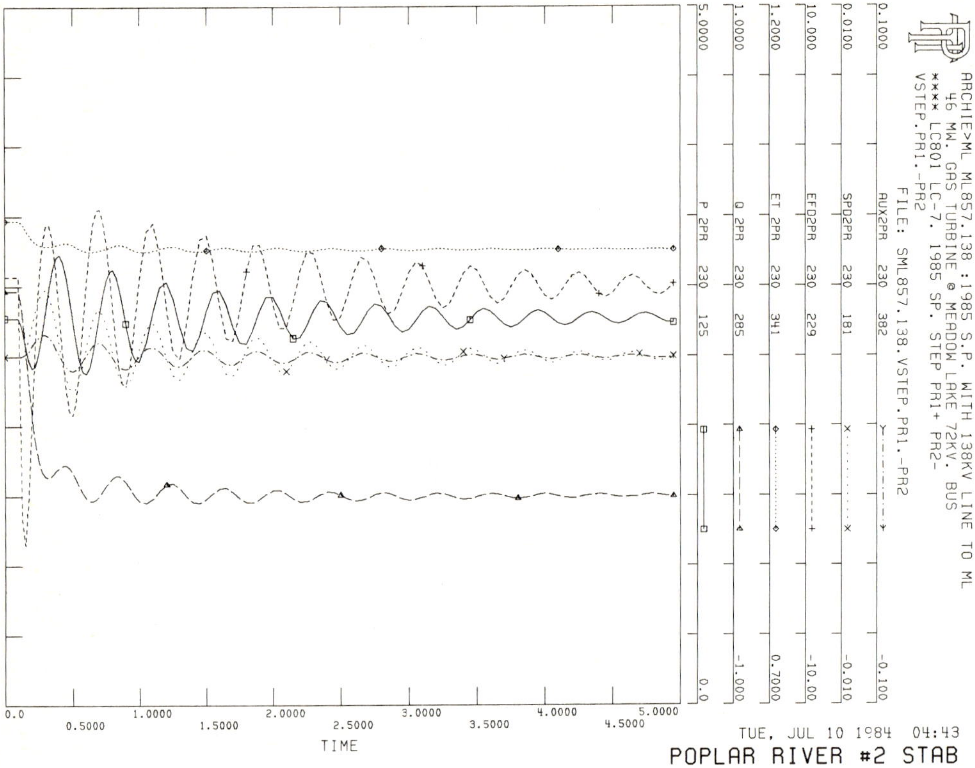

Figure 5: The Response of Poplar River Unit Two with the SPC Speed Stabilizer to a Voltage Disturbance.

NONLINEAR SIMULATION

These controllers were compared against each other, an SPC speed stabilizer independently designed and no control. On this simulation the controllers were band limited to 0.2 to 2.5 Hertz by use of filters to avoid effecting load following at low frequencies or exciting torsional shaft modes at higher frequencies. The detailed nonlinear simulation plots 140 important variables from across the network. This model is used to simulate faults anywhere in the network.

Two disturbances located near the controller were selected for this evaluation:

1. A 2% drop in reference voltage for Poplar River 2 and a 2% increase in reference voltage for Poplar River 1. This causes these units to oscillate against each other as in the original problem.

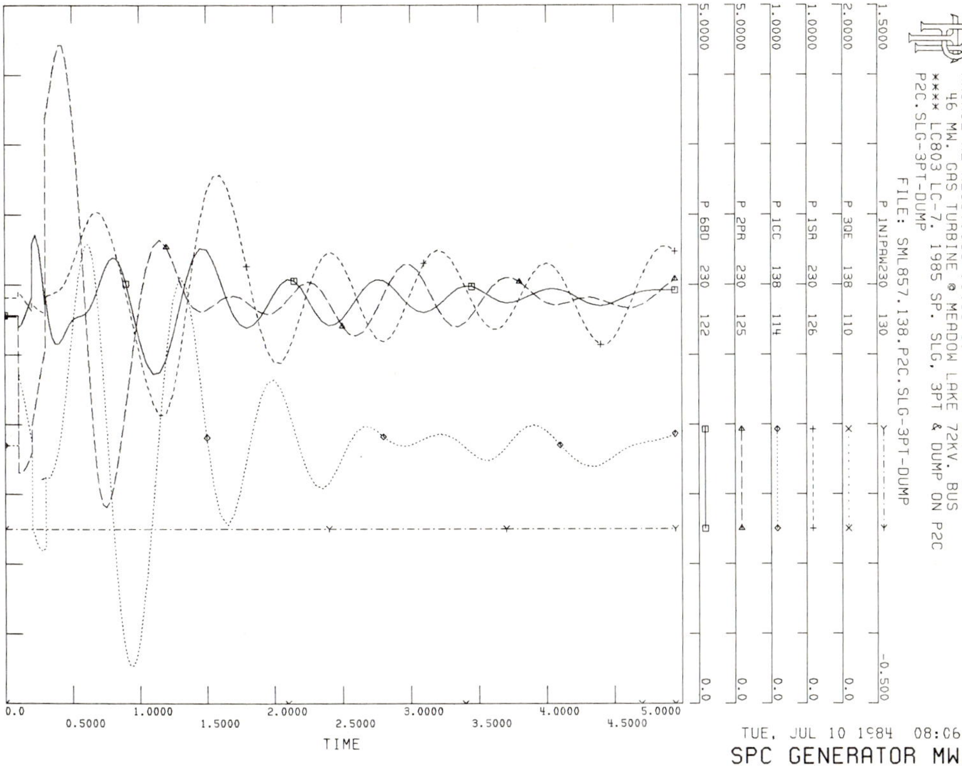

Figure 6: The Response of Poplar River Unit Two with the SPC Speed Stabilizer to a Line Fault.

2. A single line to ground fault between Poplar River and Regina. This is followed by a three pole trip (the line opens). Then as the remaining line cannot handle both generators Poplar River Unit 1 is automatically shut down. 300MW is lost to the system and as the figures indicate this disturbance is felt throughout the system.

To reduce the data to be considered to a length suitable for this paper we will only consider the variables associated with Poplar River Unit 2 for disturbance 1, which is essentially local. As disturbance 2 is a fault which effects stations throughout the system, we have selected the power flows of the largest generators in the major stations in the system for that fault. Three controllers of special interest and no control are considered. Details of the other controllers considered and the other variables plotted may be found in [11].

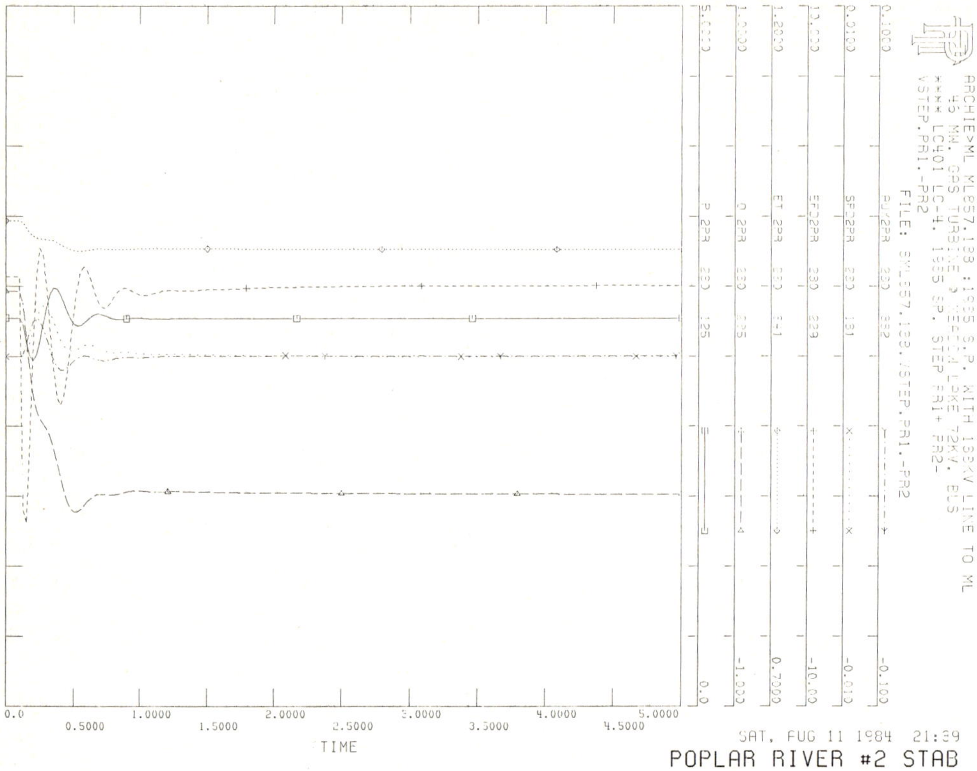

Figure 7: The Response of Poplar River Unit Two with the Linear Quadratic Gaussian Optimal Power Controller to a Voltage Disturbance.

Figures 5 and 6 are the responses of the SPC speed controller to disturbances 1 and 2 respectively. The SPC speed controller is an Ontario Hydro design retuned for the SPC system. It can be seen by comparing figures 1 with 5 and 2 with 6 that the damping is improved over the use of no control at all. However a comparison of figure 3 with 5 and 4 with 6 indicates that the performance of the SPC speed stabilizer is inferior to band limited proportional feedback of power and terminal voltage. The SPC speed stabilizer had better performance than any speed stabilizer we designed using optimal or specific optimal methods. This verifies the cost improvement analysis which indicated speed feedback on its own would not produce near optimal performance.

Figures 7 and 8 are the responses for the linear quadratic Gaussian optimal power controller to disturbances 1 and 2 respectively. A comparison of figures 3 and 7 indicate the performance of the optimal power controller and the proportional feedback of power and terminal voltage are equally good for this small disturbance.

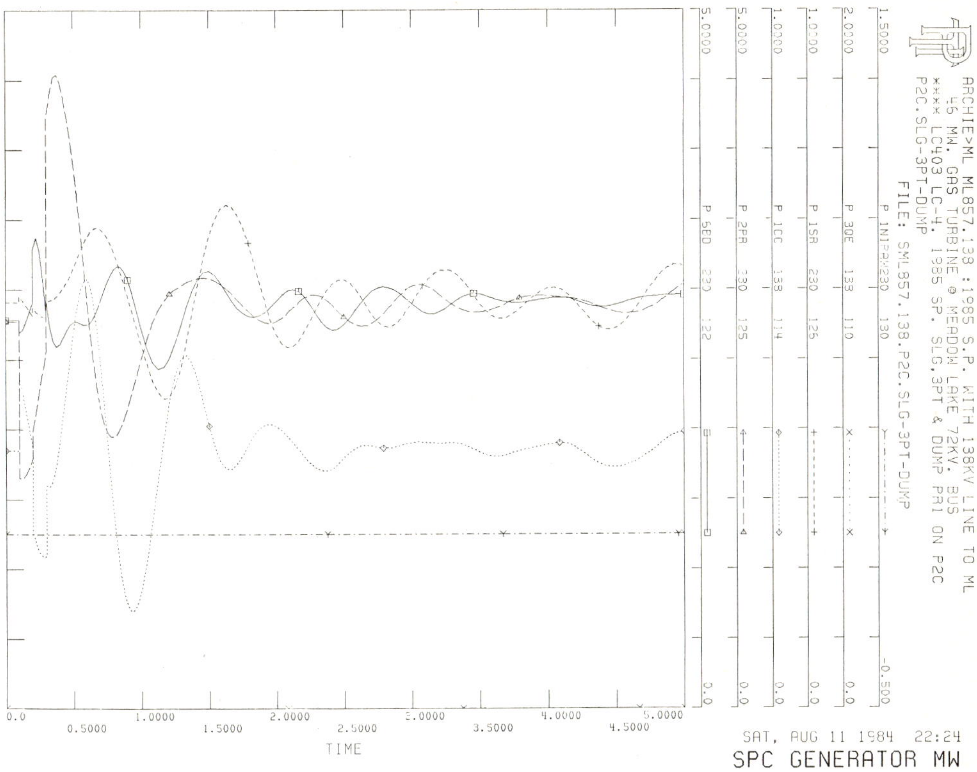

Figure 8: The Response of Poplar River Unit Two with the Linear Quadratic Gaussian Optimal Power Controller to a Line Fault.

However it can be seen from Figure 8 that the peak instantaneous power drawn from Poplar River Unit 2 is 450MW with the optimal power controller. Figure 4 indicates the peak power drawn from that generator when the specific optimal power and terminal voltage controller is used is only 400MW and the disturbance is at least as well damped. This leads us to conclude that a band limited proportional feedback of power and terminal voltage optimally tuned is the best damping control system. The superior performance of the proportional controller over the dynamic power controller can be explained intuitively by noting that the dynamic controller design uses phase relationships between variables that will not be true for a large disturbance which produces a nonlinear response.

CONCLUSIONS

In this design we have used a controller structure selection technique to predict that proportional feedback of power and terminal voltage can produce near optimal performance. Detailed nonlinear simulation indicated that this was indeed a superior controller for both small disturbances and large faults. This stabilizer should be seriously considered when selecting damping controllers for generating units.

The potentially good controller structures can be determined analytically. This should be done early in the design.

REFERENCES

[1] Katzberg, J.D. and Johnson, H.H., Good Feedback Structure Selection, I.E.E.E. Transaction on Automatic Control AC-26 No. 4 (1981) 946-948.

[2] Katzberg, J.D., Structured Feedback Control of Discrete Linear Stochastic Systems with Quadratic Costs, I.E.E.E. Transactions on Automatic Control AC-22 No. 2 (April 1977) 232-236.

[3] Kundur, P. et al, Power System Stabilizers for Thermal Units: Analytical Techniques and On-Site Validation, I.E.E.E. Power Engineering Society Winter Meeting, New York (February 1980) Paper F80-227-9.

[4] deMello, F.P. et al, Practical Approaches to Supplementary Stabilizing From Accelerating Power, I.E.E.E. Transactions on Power Apparatus and Systems PAS-97 (Sept./Oct. 1978).

[5] Moussa, H.A.M. and Yu, Y.N., Optimal Power System Stabilization Through Excitation and/or Governor Control, I.E.E.E. Power Engineering Society Winter Meeting, New York (1971) Transactions Paper 71 TP 581-PWR.

[6] Chen, J.Y.H. and Kusic, G.L, Dynamic Stability and Excitation Control of Directly-Coupled Multimachine Systems, I.E.E.E. Transactions on Power Apparatus and Systems PAS-103 (February 1984).

[7] Katzberg, J.D., Sub-Optimal Control of Discrete Stochastic Processes, Ph.D. Thesis, Department of Computing and Control, Imperial College of Science and Technology, University of London (November 1973).

[8] Katzberg, J.D. and Johnson, H.H., Simplified Controllers for Continuous Linear Stochastic Systems with Quadratic Costs, American Control Conference, Arlington, Virginia (1982) 990-995.

[9] Katzberg, J.D. and Johnson, H.H., A Function Space Approach to the Design of Simplified Controllers for Linear Systems with Quadratic Costs, Large Engineering System 4, Calgary, Alberta (1982) 445-448.

[10] Anderson, P.M. and Fouad, A.A., Power System Control and Stability, Vol. 1 (The Iowa University Press, Ames, Iowa, 1977).

[11] Wong, L.H.Y., Computer-Aided Design of a Microprocessor Based Power System Stabilizer, M.Sc. Thesis, Faculty of Engineering, University of Regina (October 1984).

4. ESTIMATION AND STOCHASTIC CONTROL

Frequency Domain and State Space Methods for Linear Systems
C.I. Byrnes and A. Lindquist (Editors)

A STOCHASTIC REALIZATION APPROACH TO THE DISCRETE-TIME MAYNE-FRASER SMOOTHING FORMULA

Faris A. Badawi
Department of Mathematical Sciences
University of Petroleum & Minerals
Dhahran, SAUDI ARABIA

Anders Lindquist
Division of Optimization and
System Theory
Royal Institute of Technology
Stockholm, SWEDEN

This paper is aimed at providing a theory of smoothing for linear discrete-time stochastic systems in the context of stochastic realization theory. The basic idea is to embed the given system in a class of similar systems all having the same output and the same Kalman filter. A somewhat extended version of this class contains two models which completely determine the smoothing estimate. The discrete-time setting introduces difficulties related to invertibility which are not present in continuous-time.

INTRODUCTION

Consider a linear stochastic system

$$(S)\begin{cases} x(t+1) = F(t)x(t) + B(t)w(t); & x(0) = \xi & \text{(1a)} \\ y(t) = H(t)x(t) + D(t)w(t); & y(0) = 0 & \text{(1b)} \end{cases}$$

defined on $[0,T] := 0,1,2,\ldots,T$, where x is an n-dimensional <u>state</u> process, y is an m-dimensional <u>output</u> process, w is a p-dimensional zero mean white noise sequence satisfying

$$E\{w(t)\} = 0 \quad \text{and} \quad E\{w(t)w(s)'\} = I\delta_{st}, \tag{2}$$

ξ is zero mean random vector with finite covariance matrix $N := E\{\xi\xi'\}$ and uncorrelated with w. The matrices F, B, H and D are time-varying matrices with F invertible for all $t \in [0,T]$.

The <u>fixed-interval smoothing problem</u> is to find the linear least-squares estimate $\hat{x}(t)$ of $x(t)$ given $\{y(s);\ s \in [0,T]\}$ i.e., the wide-sense conditional expectation [1]

$$\hat{x}(t) = \hat{E}\{x(t)|y(s);\ s \in [0,T]\}. \tag{3}$$

This problem has generated a rather extensive literature [2-21] and is of considerable importance in applications.

In this paper, we shall study this problem employing concepts and techniques from the <u>stochastic realization theory</u> developed in [22-39]. In fact, all the well-known smoothing formulas found in [2-21] have natural interpretations in stochastic realization setting. Here we shall only consider a <u>Mayne-Fraser two-filter</u> formula [5,6], on which topic a large number of papers have been written [7-9, 12-18]. The basic idea of our approach is to embed the model (1) in a class of models $\mathcal{S}$ all having the same output process y and the same Kalman-Bucy filter. The class $\mathcal{S}$ will be shown to contain an element S_*, which together with another element S^* (not belonging to $\mathcal{S}$ in general) contains all the information on y needed to estimate x. One of the major obstacles in developing this theory is the fact that the state covariance function $P(t):= E\{x(t)x(t)'\}$, which satisfies

$$P(t+1) = F(t)P(t)F(t)' + B(t)B(t)'; \quad P(0) = N, \tag{4}$$

is in general not positive definite. It is worth noting that in the smoothing literature, conditions to insure invertibility of P are either ignored [20] or mistakenly assumed to hold on all of $[0,T]$ as a consequence of complete contrability. This is clearly incorrect (see e.g. [40]). Hence, at times we shall apply the generalized Moore-Penrose pseudo-inverse $P^{\#}$ of P whenever $P \not> 0$, leading to certain nontrivial complications.

This paper is a shortened version of Chapter 2 in [41](and whenever a proof has been omitted, it can be found in [41])and is the discrete-time version of our continuous-time papers [42,43].

NOTATIONS

Let H be the Hilbert space of all centered stochastic variables with finite second order moments. If u is a p-dimensional stochastic vector process with components in H, define $H(u)$ to be the subspace spanned by $\{u_1,u_2,\dots,u_p\}$. Then, for each stochastic vector process $\{z(t);\ t \in [t_0,t_1]\}$ and $t \in [t_0,t_1]$, define $H_t(z)$ to be $H(z(t))$; $H(z)$, $H_t^-(z)$ and $H_t^+(z)$ will denote the closed linear hulls in H of all subspaces $H_s(z)$ such that $s \in [t_0,t_1]$, $[t_0,t]$ and and $[t,t_1]$ respectively. Given $\eta \in H$ and a subspace $H_1 \subset H$, $\hat{E}\{\eta|H_1\}$ will be the orthogonal projection of η onto H_1, i.e., the wide sense conditional mean [1]. We shall write $\hat{E}\{\eta|u\}$ in place of $\hat{E}\{\eta|H(u)\}$. The process η is called a wide sense Markov process if $\hat{E}\{\eta(t)|H_s^-(\eta)\} = \hat{E}\{\eta(t)|\eta(s)\}$ for $s \leq t$ or equivalently, $\hat{E}\{\eta(s)|H_t^+(\eta)\} = \hat{E}\{\eta(s)|\eta(t)\}$, i.e., the Markov property is independent of the time direction.

FORWARD AND BACKWARD REALIZATION

Let the output process y be defined as before. Any system of type (1) (with $\xi_i \in H$, for $i = 1,2,\dots,n$, w satisfying (2) and $\xi_i \in H(w)$ for all i) having y as its output is called a realization of y. Clearly, the components of x, y and w belong to H.

The purpose of this section is to introduce the two models S_* and $\bar{S}_*$, the knowledge of which determines the frame space (to be defined later), which in turn contains the smoothing estimate.

It is well known that the linear least-squares estimate

$$x_*(t) = \hat{E}\{x(t) \mid H_{t-1}^-(y)\} \tag{5}$$

of the state process x of S is generated on $[0,T]$ by the <u>Kalman-Bucy filter</u>

$$x_*(t+1) = F(t)x_*(t) + B_*(t)R_*(t)^{-\frac{1}{2}}[y(t) - H(t)x_*(t)]; \quad x_*(0) = 0, \tag{6a}$$

where the <u>Kalman gain</u> function B_* is given by

$$B_* = [FQ_*H' + BD']R_*^{-\frac{1}{2}}, \tag{6b}$$

$$R_* = HQ_*H' + DD' \tag{6c}$$

and the <u>error covariance matrix</u>

$$Q_*(t) = E\{[x(t) - x_*(t)][x(t) - x_*(t)]'\} \tag{6d}$$

is the solution of

$$Q_*(t+1) = F(t)Q_*(t)F(t)' - B_*(t)B_*(t)' + B(t)B(t)'; \quad Q_*(0) = N \tag{6e}$$

which is a matrix Riccati difference equation when (6b) is inserted.

As we shall see shortly, there are other realizations of y which have (6a) as their Kalman-Bucy filter. Hence define $\mathcal{S}$ to be the class of all realizations S of y such that $R(t) := D(t)D(t)' > 0 \;\forall\; t \in [0,T]$ and such that the corresponding Kalman-Bucy filter is given by (6a), that is, it has the same matrix functions F, H and $K_* := B_*R_*^{-\frac{1}{2}}$ as those of (6a) and consequently the same estimates $\{x_*(t);\, t \in [0,T+1]\}$.

The sequence $\{w_*(t);\, t \in [0,T]\}$ defined by

$$w_*(t) = R_*(t)^{-\frac{1}{2}}[y(t) - H(t)x_*(t)] \tag{7}$$

is called the <u>innovation process</u>. It is a normalized white noise satisfying (2) and characterized by the property $H_t^-(w_*) = H_t^-(y)$ for all $t \in [0,T]$. Combining (6a) and (7), we obtain the model

$$(S_*)\begin{cases} x_*(t+1) = F(t)x_*(t) + B_*(t)w_*(t); & x_*(0) = 0 \\ y(t) = H(t)x_*(t) + R_*(t)\, w_*(t), & \end{cases} \tag{8a}$$

which clearly belongs to $\mathcal{S}$. It can be immediately seen that the covariance matrix $P_*(t) := E\{x_*(t)x_*(t)'\}$ of $x_*(t)$ satisfies

$$P_*(t+1) = F(t)P_*(t)F(t)' + B_*(t)B_*(t)'; \quad P_*(0) = 0, \tag{8b}$$

and that

$$Q_* = P - P_* \,. \tag{8c}$$

Clearly, the matrices F, H and R are the same for all $S \in \mathcal{S}$, while B, D and P will differ over the class $\mathcal{S}$. Furthermore, the matrix function

$$G := FPH' + BD' \tag{9a}$$

is invariant over $\mathcal{S}$ since it can be easily written

$$G = FP_*H' + B_*R_*^{-\frac{1}{2}} \,. \tag{9b}$$

Our next task is to derive a backward realization $\bar{S}$ for each $S \in \mathcal{S}$. We shall begin by restricting our attention to the subclass $\mathcal{S}_+$ (which can be shown to be nonempty) consisting of all $S \in \mathcal{S}$ for which $P > 0$.

Let $S \in \mathcal{S}_+$. Then $\bar{x}(t-1) := P(t)^{-1}x(t)$ is defined for every $t \in [0,T]$ and satisfies the backward realization of $\{y(t);\, t \in [0,T]\}$

$$\begin{cases} \bar{x}(t-1) = F(t)'\bar{x}(t) + \bar{B}(t)\bar{w}(t); & \bar{x}(T) = \bar{\xi} \\ y(t) = G(t)'\bar{x}(t) + \bar{D}(t)\bar{w}(t) & \end{cases} \tag{10}$$

where $\bar{\xi} = P(T+1)^{-1}x(T+1)$,

$$\bar{B}(t) = -P(t)^{-1}F(t)^{-1}B(t)[I - B(t)'P(t+1)^{-1}B(t)] \,, \tag{11}$$

$$\bar{D}(t) = [D(t) - H(t)F(t)^{-1}B(t)][I - B(t)'P(t+1)^{-1}B(t)] \tag{12}$$

and $\bar{w}$ is a p-dimensional normalized white noise sequence satisfying (2) and the condition $H_t^-(\bar{w}) \perp H_t^+(\bar{x})$ for $t \in [0,T]$ and is given by

$$\bar{w}(t) = [I - B(t)'P(t+1)^{-1}B(t)]^{\frac{1}{2}}[w(t) - B(t)'F'(t)^{-1}P(t)^{-1}x(t)] \tag{13}$$

The covariance function $\bar{P}(t) := E\,\bar{x}(t)\bar{x}(t)'$ is given by

$$\bar{P}(t-1) = P(t)^{-1}$$

and satisfies the Liapunov equation

$$\bar{P}(t-1) = F'(t)\bar{P}(t)F(t) + \bar{B}(t)\bar{B}(t)'; \quad \bar{P}(T) = P(T+1)^{-1}. \tag{14}$$

In analogy with the forward setting, the least-square estimate

$$\bar{x}_*(t) = E\{\bar{x}(t) \mid H^+_{t+1}(y)\} \tag{15}$$

is generated by the backward Kalman-Bucy filter:

$$\bar{x}_*(t-1) = F(t)'\bar{x}_*(t) + \bar{B}_*(t)\bar{R}_*(t)^{-\frac{1}{2}}[y(t) - G(t)'\bar{x}_*(t)]; \; \bar{x}_*(T) = 0 \tag{16a}$$

where

$$\bar{B}_* = [F'\bar{Q}_*G + \bar{B}\bar{D}']\bar{R}_*^{-\frac{1}{2}}, \tag{16b}$$

$$\bar{R}_* = G'\bar{Q}_*G + \bar{D}\bar{D}', \tag{16c}$$

the error covariance matrix function

$$\bar{Q}_*(t) = E\{[\bar{x}(t) - \bar{x}_*(t)][\bar{x}(t) - \bar{x}_*(t)]'\} \tag{16d}$$

being the solution of the dual matrix Riccati equation

$$\bar{Q}_*(t-1) = F(t)'\bar{Q}_*(t)F(t) - \bar{B}_*(t)\bar{B}_*(t)' + \bar{B}(t)\bar{B}(t)'; \; \bar{Q}_*(T) = P(T+1)^{-1} \tag{16e}$$

with $\bar{B}_*$ given by (16b).

The backward innovation process $\bar{w}_*$ defined by

$$\bar{w}_*(t) = \bar{R}_*(t)^{-\frac{1}{2}}[y(t) - G'(t)\bar{x}_*(t)] \tag{17}$$

is a normalized white noise satisfying (2) and the condition $H^+_t(\bar{w}_*) = H^+_t(y)$. The covariance matrix $\bar{P}_*(t) := E\{\bar{x}_*(t)\bar{x}_*(t)'\}$ satisfies the backward Liapunov equation

$$\bar{P}_*(t-1) = F'(t)\bar{P}_*(t)F(t) + \bar{B}_*(t)\bar{B}_*(t)'; \quad \bar{P}_*(T) = 0 \tag{18}$$

Now, in the same way as above, define $\bar{\mathcal{S}}$ to be the class of all backward realizations $\bar{S}$ having (16) as their backward Kalman-Bucy filter, and let $\bar{\mathcal{S}}_+$ be the subclass consisting of those $\bar{S} \in \bar{\mathcal{S}}$ for which $\bar{N} := E\{\bar{\xi}\bar{\xi}'\} > 0$. In the same way as in the forward setting, it is seen that the realization

$$(\bar{S}_*) \begin{cases} \bar{x}_*(t-1) = F'(t)\bar{x}_*(t) + \bar{B}_*(t)\bar{w}_*(t); & \bar{x}_*(T) = 0 \\ y(t) = G'(t)\bar{x}_*(t) + \bar{R}_*(t)^{\frac{1}{2}}\bar{w}_*(t) \end{cases} \tag{19}$$

belongs to $\bar{\mathcal{S}}$, and that the class $\bar{\mathcal{S}}$ is uniquely defined in terms of the invariants F, H, G and Λ_0, and therefore, the backward counterpart $\bar{S}$ of any $S \in \mathcal{S}_+$ belongs to $\bar{\mathcal{S}}$. In particular, since $P(T+1)$ is positive definite and since $\bar{P}(T) = P(T+1)^{-1}$, $\bar{S} \in \bar{\mathcal{S}}_+$.

It is clear that there is a complete symmetry between forward and backward realizations. In particular, the subclasses $\mathcal{S}_+$ and $\bar{\mathcal{S}}_+$ are in one-one correspondence. Therefore, one easily constructs a forward realization from a backward one upon defining $x(t) = \bar{P}(t-1)^{-1}\bar{x}(t-1)$.

In the sequel, we shall be interested in obtaining the backward (forward) counterpart of S_* $(\bar{S}_*)$. As is clear by now, for this we need to invert the matrices

$P_*(t)$ in (8b) and $\bar{P}_*(t-1)$ in (18). However, since $P_*(0) = 0$ and $\bar{P}_*(T) = 0$, P_*^{-1} and $\bar{P}_*^{-1}$ will not exist on the whole interval $[0,T]$; they may not even exist on part of it for that matter. Therefore, we introduce the following

Definition 1. Let t_* be the smallest t such that $P_*(t) > 0$; if there is no such t on $[0,T]$, set $t_* := T+1$. Similarly, let $\bar{t}_*$ be the largest t such that $\bar{P}_*(t-1) > 0$; if there is no such t, set $\bar{t}_* = -1$.

As is clear from the definition above, t_* and $\bar{t}_*$ might lie outside the interval $[0,T]$. However, if we impose some more conditions on the class $\mathcal{S}$, we can guarantee that t_* and $\bar{t}_*$ belong to $[0,T]$.

Definition 2. The class $\mathcal{S}$ is said to be minimal if there is no realization of type (1), the state process of which has dimension smaller than n.

Lemma 3. $\mathcal{S}$ is minimal if and only if $\bar{\mathcal{S}}$ is minimal. And, if $\mathcal{S}$ is minimal, $t_* \leq T$ and $\bar{t}_* \geq 0$.

Of course, there is no guarantee that there is a $t \in [0,T]$ for which both $P_*(t)$ and $\bar{P}_*(t-1)$ are positive definite.

Definition 4. The class $\mathcal{S}$ is said to be regular if $t_* \leq \bar{t}_*+1$, i.e., for each $t \in [0,T]$, either $P_*(t) > 0$ or $\bar{P}_*(t-1) > 0$ (or both), and is said to be irregular otherwise.

In fact, the regularity property of the class $\mathcal{S}$ depends to a certain extent on the length of the interval $[0,T]$; for if T is sufficiently large compared with n, $\mathcal{S}$ will be regular, since then the controllability gramian will eventually become positive definite; the same holds in the backward direction. If $T < n$, we will encounter irregularity; then we do not have minimality either.

Now, we are ready to justify introducing the processes x_* and $\bar{x}_*$. It follows from (5) and (15) that

$$H_t(x_*) = \hat{E}\{H_t(x) | H_{t-1}^-(y)\} \subset H_{t-1}^-(y)$$

and

$$H_{t-1}(\bar{x}_*) = \hat{E}\{H_t(x) \mid H_t^+(y)\} \subset H_t^+(y),$$

for all $t \in [0,T]$. Define the orthogonal complements.

$$N_t^- := H_{t-1}^-(y) \ominus H_t(x_*) \quad \text{and} \quad N_t^+ := H_t^+(y) \ominus H_{t-1}(\bar{x}_*).$$

Then, $H(y)$ can be decomposed as

$$H(y) = N_t^- \oplus H_t^0 \oplus N_t^+ , \tag{20}$$

where H_t^0 is the frame space [35,36,38]

$$H_t^0 = H_t(x_*) \vee H_{t-1}(\bar{x}_*) \tag{21}$$

for all $t \in [0,T]$.

Lemma 5. Let x be the state process of a realization $S \in \mathcal{S}$. Then

$$H_t(x) \subset H_t^0 \oplus [H(y)]^\perp \tag{22}$$

for all $t \in [0,T]$.

Proof. See the proof of Lemma 7 in [42]. □

Notice that therefore the smoothing estimate (3) will always be contained in the frame space, hence its importance.

It will be more convenient in the sequel to express the frame space H_t^0 in a somewhat more symmetric form. To this end, we shall construct the forward counterpart of $\bar{S}_*$. However, since $\bar{P}_*$ is not positive definite for every $t \in [0,T]$, the previous argument of constructing a backward realization from a forward one cannot be reversed on the whole interval. Therefore, we apply the generalized Moore-Penrose pseudo-inverse $P^\#$ of any matrix P.

Now, define the process

$$x^*(t) = \bar{P}_*(t-1)^\# \bar{x}_*(t-1), \tag{23}$$

for each $t \in [0,T]$. Then, premultiplying both sides of (23) by $\bar{P}_*(t-1)$ and using the fact that $PP^\# x = x$, we obtain

$$\bar{x}_*(t-1) = \bar{P}_*(t-1)x^*(t). \tag{24}$$

Hence $H_t(x^*) = H_{t-1}(\bar{x}_*)$ and consequently, we have the following symmetric expression

$$H_t^0 = H_t(x_*) \vee H_t(x^*) \tag{25}$$

for the frame space.

THE MODEL S^*

The aim of this section is to construct and study the properties of the model S^* whose state process is x^* defined by (23). As we mentioned earlier, since $\bar{S}_* \notin \bar{S}_+$, the previous forward-backward construction cannot be applied here. However, the basic idea of the derivation is the same, but the results will be somewhat different.

The following proposition, the proof of which can be found in [14], gives the required representation for S^*.

Proposition 6. Let $\bar{S}_*$ and x^* be given by (19) and (23) respectively and define Θ^* as

$$\Theta^*(t) = I + \bar{B}_*(t)'F(t)^{-1}\bar{P}_*(t)^\# F(t)'^{-1}\bar{B}_*(t) \tag{26}$$

Then x^* is the state process of the following model, which is the forward counterpart of $\bar{S}_*$.

$$(S^*) \begin{cases} x^*(t+1) = \Pi(t+1)F(t)x^*(t) + \Pi(t+1)B^*(t)w^*(t); \; x^*(0) = \bar{P}_*(-1)^\# \bar{x}_*(-1) \\ y(t) = \tilde{H}(t)x^*(t) + R^*(t)^{\frac{1}{2}} w^*(t)^{\frac{1}{2}}, \end{cases} \tag{27}$$

where

$$B^*(t) = -\bar{P}_*(t)^\# F(t)'^{-1}\bar{B}_*(t)[I - \bar{B}_*(t)'\bar{P}_*(t-1)^\# \bar{B}_*(t)]\Theta^*(t)^{\frac{1}{2}} \tag{28a}$$

$$R^*(t)^{\frac{1}{2}} = [\bar{R}_*(t)^{\frac{1}{2}} - \tilde{H}(t)\bar{P}_*(t-1)^\# \bar{B}_*(t)]\Theta^*(t)^{\frac{1}{2}}, \tag{28b}$$

$$\tilde{H}(t) = H(t)F(t)^{-1}\Pi(t+1)F(t), \tag{28c}$$

w^* is a white noise satisfying (2) and the forward property $H_t^+(w^*) \perp H_t^-(x^*)$ given by

$$w^*(t) = \Theta^*(t)^{-\frac{1}{2}}[\bar{w}_*(t) - \bar{B}_*(t)'F(t)^{-1}\bar{P}_*(t)^\# \bar{x}_*(t), \tag{29}$$

and the $n \times n$-matrix function Π is defined by

$$\Pi(t) := P^{*}(t)P^{*}(t)^{\#}, \tag{30}$$

where P^{*} is the covariance function of x^{*}, i.e., $P^{*}(t) = E\{x^{*}(t)x^{*}(t)'\}$, which satisfies the Liapunov type equation

$$P^{*}(t+1) = \Pi(t+1)F(t)P^{*}(t)F(t)'\Pi(t+1) + B^{*}(t)B^{*}(t)'; \; P^{*}(0) = \bar{P}_{*}(-1)^{\#}. \tag{31}$$

The matrix function Π defined by (30) will play an important role in what follows. It follows from the definition of pseudo-inverse that $\Pi^2 = \Pi$ and that $\Pi = \Pi'$, hence Π is an orthogonal projection. Also, we have

$$\Pi(t)x^{*}(t) = x^{*}(t) \tag{32}$$

for all $t \in [0,T+1]$. Finally, if S is minimal, $\bar{S}$ is also minimal and there exists a $\bar{t}_{*} \in [0,T]$ such that $P^{*}(t) > 0$ for all $t \in [0,\bar{t}_{*}]$ (Lemma 3). Then $\Pi \equiv I$ on $[0,\bar{t}_{*}]$; in this case, x^{*} will satisfy a recursion of type (1a), and the following relation

$$P_{*}(t) \leq P(t) \leq P^{*}(t) \tag{33}$$

holds for each such t. (In fact, $P_{*}(t) \leq P(t)$ holds for all $t \in [0,T]$.)

Finally, the following three lemmas the proof of which can be found in [41], will be needed in the sequel.

<u>Lemma 7</u>. Let x be the state process and P the state covariance function of any $S \in S_{+}$ and let x_{*} and x^{*} be defined by (5) and (23) respectively. Then

$$E\{x(t)x_{*}(t)'\} = P_{*}(t) \tag{34}$$

and

$$E\{x(t)x^{*}(t)'\} = P(t)\Pi(t). \tag{35}$$

for all $t \in [0,T+1]$.

<u>Lemma 8</u>. Assume that S is regular. Let x_{*} and x^{*} be defined by (6) and (27) respectively. Then, for all $t \in [0,T+1]$,

$$\hat{E}\{x^{*}(t) \mid x_{*}(t)\} = \Pi(t)x_{*}(t) \tag{36}$$

and

$$E\{x_{*}(t)x^{*}(t)'\} = P(t)\Pi(t). \tag{37}$$

<u>Lemma 9</u>. Let $S \in S_{+}$ and let Q^{*} be the covariance function of $\Pi(x^{*} - x)$. Then

$$Q^{*} = \Pi(P^{*} - P)\Pi. \tag{38}$$

If in addition S is regular, then

$$E\{[x - x_{*}][x^{*} - x]'\Pi\} = 0. \tag{39}$$

A MAYNE-FRASER TYPE SMOOTHING FORMULA

Given the state process x and the output process y of a model (1), the smoothing problem consists in determining the <u>smoothing estimate</u>

$$\hat{x}(t) = \hat{E}\{x(t) \mid H(y)\} \tag{40}$$

of x and the smoothing error covariance

$$\Sigma(t) = E\{[x(t) - \hat{x}(t)][x(t) - \hat{x}(t)]'\} \tag{41}$$

for all $t \in [0,T]$. In this section, we shall derive a smoothing formula for the case that S is regular. At the end of the section, we shall present a conjecture which, if true, would allow us to remove this regularity assumption. Our approach will utilize an orthogonal decomposition of the frame space H_t^0 to be given below.

In view of Lemma 5, $\hat{x}(t) \in H_t^0$. In order to obtain a formula for $\hat{x}$, we need to decompose the frame space H_t^0 perpendicularly. To this end, we introduce the process z defined by

$$z(t) = x^*(t) - \Pi(t)x_*(t) \tag{42}$$

for all $t \in [0,T+1]$.

Lemma 10. Assume that the class S is regular. Let z be defined by (42), and let $Q(t) := E\{z(t)z(t)'\}$. Then

$$H_t^0 = H_t(x_*) \oplus H_t(z) \tag{43}$$

for all $t \in [0,T+1]$. Moreover,

$$Q = \Pi(P^* - P_*)\Pi \tag{44}$$

and

$$Q = \Pi(Q_* + Q^*)\Pi \tag{45}$$

where Q_* and Q^* are given by (8c) and (38) respectively and P is the state covariance of any $S \in S$.

Proof. Since S is regular, Lemma 8 implies that the components of z are orthogonal to those of x_*, and therefore (43) follows from (25). Next, since $x^* = \Pi x^*$, $z = \Pi(x^* - x_*)$. Then (24) is a direct consequence of (37). Finally, write $Q = \Pi(P^* - P + P - P_*)\Pi$ to obtain (45).

It is easy to prove the relations in the following (see [41]).

Lemma 11. Let z be given by (42) and Q be its covariance function. Then

$$Q = \Pi Q = Q\Pi = \Pi Q \Pi , \tag{46a}$$

$$Q^{\#} = \Pi Q^{\#} = Q^{\#}\Pi = \Pi Q^{\#}\Pi \tag{46b}$$

and

$$QQ^{\#} = Q^{\#}Q = \Pi. \tag{46c}$$

Now, we are ready to present the main result of this paper.

Theorem 12. Assume that the class S is regular. Let x be the state process of any S of class S_+. Then the smoothing estimate (40) is given by

$$\hat{x}(t) = [I - Q_*(t)Q(t)^{\#}]x_*(t) + Q_*(t)Q^{\#}(t)x^*(t) \tag{47}$$

and the error covariance (41) by

$$\Sigma(t) = Q_*(t) - Q_*(t)Q(t)^{\#}Q_*(t) \tag{48}$$

for all $t \in [0,T+1]$.

Proof. Since $\hat{x}(t) = \hat{E}\{x(t) \mid H_t^0\}$ (Lemma 5), (43) yields

$$\hat{x}(t) = \hat{E}\{x(t) \mid x_*(t)\} + \hat{E}\{x(t) \mid z(t)\},$$

which upon using (5) can be written as

$$\hat{x}(t) = x_*(t) + E\{x(t)z(t)'\}Q(t)^{\#}z(t).$$

In view of (34) and (35), $E\{x(t)z(t)'\} = Q_*(t)\Pi(t)$. Since $z = \Pi z$, (47) follows from the above relation noting (46). To prove (48), observe that

$$x - \hat{x} = (I - Q_*Q^{\#})(x - x_*) + Q_*Q^{\#}(x - x^*).$$

Replacing $Q^{\#}$ by $\Pi Q^{\#}\Pi$, and noting that the two terms above are, in view of (39), orthogonal, we obtain

$$\Sigma = Q_* - Q_*\Pi Q^{\#}\Pi Q_* - Q_*\Pi Q^{\#}\Pi Q_* + Q_*\Pi Q^{\#}\Pi Q_*\Pi Q^{\#}\Pi Q_* + Q_*\Pi Q^{\#}\Pi Q^*\Pi Q^{\#}\Pi Q_* ,$$

which, in view of (45) and (46), yields (48).

To obtain the discrete-time version of the Mayne-Fraser formula [5,6], the following lemma is needed.

Lemma 13. Let $S \in \mathcal{S}_+$. Then, for each $t \in [0,T+1]$, $Q_*(t)$ and $\Sigma(t)$ are positive definite and satisfy

$$\Sigma(t)^{-1} = Q_*(t)^{-1} + Q^*(t)^{\#} \tag{49}$$

where Q^* is given by (38).

Proof. Let P be the covariance function of S. Since $S \in \mathcal{S}_+$, $Q_*(0) = P(0) - P_*(0) = N > 0$, consequently $Q_*(t) > 0$ for all $t \in [0,T+1]$. An analogus backward argument implies that $\Sigma(t) \geq 0$ for all $t \in [0,T+1]$ since $\Sigma(t) \geq \Sigma(T+1) = Q_*(T+1) > 0$. Next, it follows from (48) that

$$I - Q_*Q^{\#} = \Sigma Q_*^{-1} \tag{50}$$

which is nonsingular for all $t \in [0,T+1]$. Then applying the matrix inversion lemma to (48), it is seen that $\Sigma^{-1} = Q_*^{-1} + Z$, where

$$Z = Q^{\#}(I - Q_*Q^{\#})^{-1} = (I - Q^{\#}Q_*)^{-1}Q^{\#} \tag{51}$$

Therefore, it just remains to show that $Z = Q^{*\#}$. Since π is a projection, (45) yields

$$Q^* = Q - \Pi Q_*\Pi . \tag{52}$$

In view of Lemma 11, (52) and the first of relations (51) yield $Q^*Z = \Pi$, which is symmetric. Likewise, using the second of relations (51), we obtain $ZQ^* = \Pi$. Then Lemma 11 implies that $Q^*ZQ^* = Q^*$ and $ZQ^*Z = Z$ also. Consequently $Z = Q^{*\#}$.

Now, we are in a position to state the Mayne-Fraser two-filter formula.

Theorem 14. Let $\mathcal{S}$ be regular and let $S \in \mathcal{S}_+$. Then the smoothing estimate $\hat{x}$ of the state x of S is given by

$$\hat{x}(t) = \Sigma(t)[Q_*(t)^{-1}x_*(t) + Q^*(t)^{\#}x^*(t)] \tag{53}$$

where x_*, x^* and Σ are given by (6), (27) and (49) respectively.

Proof. It follows from (49) and (50) that $Q_*Q^\# = I - \Sigma Q_*^{-1} = \Sigma(\Sigma^{-1} - Q_*^{-1}) = \Sigma Q^{*\#}$. This together with (50) yields (53) when inserted in (48).

CONCLUSION

Although the regularity assumption imposed on the class S is not a very stringent one (especially if T is large enough in comparison with n), it would be interesting to see if it can be removed. Actually, this assumption was introduced only to obtain relation (36). We believe that (36) holds without the regularity assumption.

Moreover, the other smoothing formulas, e.g. the Brayson-Frazier formula [2] and the Rauch,et al [3] formula were obtained in [41] by using a procedure based on an orthogonal decomposition of $H(y)$ (rather than the frame space), which does not require that S be regular.

ACKNOWLEDGEMENT

The first author would like to thank the University of Petroleum and Minerals for supporting this research.

REFERENCES

[1] J.L. Doob, Stochastic Processes. (Wiley, 1953).

[2] A.E. Bryson and M. Frazier, Smoothing for linear and nonlinear dynamic systems, TDR 63-119, Aero. Sys. Div., (1963), Wright-Patterson Air Force Base, Ohio, 353-364.

[3] H.E. Rauch, F. Tung, and C.T. Striebel, Maximum likelihood estimates of linear dynamic systems, AIAA J., Vol.3, No.8, August 1965, 1445-1450.

[4] J.S. Meditch, On optimal fixed point smoothing, Int. J. Control, Vol.6, No.2(1976), 189-199.

[5] D.Q. Mayne, A solution of the smoothing problem for linear dynamic systems, Automatica, Vol.4(1966), 73-92.

[6] D.C. Fraser, A new technique for optimal smoothing of data, Sc.D. Dissertation, MIT, Cambridge, MA (1967).

[7] R.K. Mehra, On optimal and suboptimal linear smoothing, Proc. Nat. Electr. Conf., XXIV(1968), 119-124.

[8] D.C. Fraser and J.E. Potter, The optimum linear smoother as a combination of two optimum linear filters, IEEE Trans. Automatic Control, AC-14(1969), 387-390.

[9] T. Kailath and P.A. Frost, An innovations approach to least-squares estimation -- Part II: linear smoothing in additive noise, IEEE TRans. Automatic Control, AC-13(1968), 655-660.

[10] L.E. Zachrisson, On optimal smoothing of continuous-time Kalman processes, Information Sciences, I(1969), 143-172.

[11] A. Lindquist, A theorem on duality between estimation and control for linear stochastic systems with time delay, J. Mathematical Analysis and Applications, Vol.37, No.2(1972), 516-536.

[12] D.G. Lainiotis, Partitioned linear estimation algorithms; discrete case, IEEE Trans. Automatic Control, AC-20(1975), 255-257.

[13] D.G. Lainiotis and K.S. Govindaraj, A unifying approach to linear estimation via the partitioned algorithms, II: discrete models, Proc. of the 1975 IEEE Decision and Control Conf.

[14] T. Kailath and L. Ljung, Backwards Markovian models for second order stochastic process, IEEE Trans. Information Theory, IT-22(1976), 488-491.

[15] B. Friedlander, T. Kailath, and L. Ljung, Scattering theory and linear least squares estimation - II: discrete-time problems, J. Franklin Inst., Vol.301(1976), 71-82.

[16] L. Ljung and T. Kailath, A unified approach to smoothing formulas, Automatica, Vol.12(1976), 147-157.

[17] G.S. Sidhu and U.B. Desai, New smoothing algorithms based on reversed-time lumped models, IEEE Trans. Automatic Control, AC-21(1976), 538-541.

[18] ________________________, Reversed-time lumped models, smoothing algorithms and Fredholm resolvents, Proc. of the JHU Conf. on Information Science and Systems, Johns Hopkins Univ., Baltimore, MD, March-April, 1976.

[19] J.S. Meditch, A survey of data smoothing for linear and non-linear dynamic systems, Automatica, Vol.9(1973), 151-162.

[20] J.E. Wall, Jr., Control and estimation for large-scale systems having spatial symmetry, Ph.D. Thesis, MIT (1978).

[21] P. Faurre, M. Clerget and F. Germain, Operateurs Rationnels Posififs (Dunod, 1979).

[22] A. Lindquist, M. Pavon, and G. Picci, Recent trends in stochastic realization theory, Prediction Theory and Harmonic Analysis: The Pesi Masani Volume, V. Mandrekar and H. Salehi (eds.), North Holland, 1983.

[23] A. Lindquist and G. Picci, On the stochastic realization problem, SIAM J. Control and Optimization 17(1979), 365-389.

[24] ________________________, A state-space theory for stationary stochastic processes, Proc. 21st Midwest Symposium on Circuits and Systems, Ames, Iowa, August 1978.

[25] ________________________, A Hardy space approach to the stochastic realization problem, Proc. 1978 Conf. Decision and Control, San Diego, 933-939.

[26] ________________________, Realization theory for multivariate stationary Gaussian processes I: State space construction, Proc. 4th Intern. Symp. Math. Theory of Networks and Systems, July 1979, Delft, Holland, 140-148.

[27] ________________________, Realization theory for multivariate stationary Gaussian processes II: State space theory revised and dynamical representations of finite dimensional state spaces, Proc. 2nd Intern. Conf. on Information Sciences and Systems, Patras, Greece, July 1979, Riedel Publ. Co., 108-129.

[28] ________________________, State space models for Gaussian stochastic processes, Stochastic System: The Mathematics of Filtering and Identification and Applications, M. Hazewinkel and J.C. Willems, Eds., Reidel Publ.Co, 1981.

[29] A. Lindquist and G. Picci, On a condition for minimality of Markovian splitting subspaces, Systems and Control Letters 1(1982), 264-269.

[30] ________________________, Infinite dimensional stochastic realizations of continuous-time stationary vector processes, Topics in Operators and Systems, H. Dym and I. Gohberg, eds., Birkhauser Verlag, 1984.

[31] ________________________, Forward and backward semimartingale representations for stationary increment processes, Stochastics (to be published).

[32] G. Picci, Stochastic realization of Gaussian processes, Proc. IEEE 64(1976), 112-122.

[33] M. Pavon, Stochastic realization and invariant directions of the matrix Riccati equation, SIAM J. Control and Optimization 18(1980), 155-180.

[34] G. Ruckebusch, Representations Markoviennes de processus Gaussians Stationaires, Ph.D. Thesis, Univ. of Paris VI, 1975.

[35] ____________, Representations markoviennes de processus gaussiens stationaires, C.R. Acad. Sc. Paris, Series A, 282(1976), 649-651.

[36] ____________, Factorisations minimales de densites spectrales et representations markoviennes, Proc. 1re Colloque AFCET-SMF, Palaiseau, France, 1978.

[37] ____________, A state space approach to the stochastic realization problem, Proc. 1978 Intern. Symp. Circuits and Systems, New York.

[38] ____________, Theorie geometrique de la representation markovienne, These de doctorat d'etat, Univ. Paris IV, 1980.

[39] ____________, On the structure of minimal Markovian representations, Nonlinear Stochastic Problems, R. Bucy and J.M.F. Moura (eds.), Reidel, 1983.

[40] R.E. Kalman, Contributions to the theory of optimal control, Bol. Soc. Mat. Mexicana, 5(1960), 102-119.

[41] F.A. Badawi, Structures and algorithms in stochastic realization theory and the smoothing algorithm, Ph.D. Thesis, University of Kentucky, (January, 1980).

[42] F.A. Badawi, A. Lindquist, and M. Pavon, A stochastic realization approach to the smoothing problem, IEEE Trans. Automatic Control, AC-24(1979), 878-888.

[43] ____________________________________, On the Mayne-Fraser Formula and stochastic realization theory for nonstationary stochastic systems, Proc. 1979 Conf. Decision and Control, Fort Leuderdale, Florida, December 1979.

[44] P. Faurre, M. Clerget and F. Germain, Opêrateurs Rationnels Positifs, Dunod 1979.

NOTE
A stochastic realization approach to smoothing similar to that of [41](and that of [42] and [43]) is also used in [44]. However, the main result of [44] is incorrect, the error having to do with the invertibility questions to which a large portion of our paper is devoted.

Frequency Domain and State Space Methods for Linear Systems
C.I. Byrnes and A. Lindquist (Editors)

HARMONIZABLE PROCESSES and NON-STATIONARY PREDICTION THEORY

P. E. Caines* and C. Houdré**

* Canadian Institute for Advanced Research and
Department of Electrical Engineering
McGill University, Montréal.

** Department of Mathematics, McGill University, Montréal

Abstract

The prediction problem for stochastic processes may be stated as follows: given observations on an m component stochastic process $\{\ldots x_{n-1}, x_n\}$ up to time n, what is the best linear least squares estimate of the value of a second p component process y at time $n+\tau$?

If the process $\begin{bmatrix} x \\ y \end{bmatrix}$ is wide sense stationary we may obtain the well known solution to this problem (in both the time and frequency domains) via the Fourier representation of the covariance matrices of $\begin{bmatrix} x \\ y \end{bmatrix}$ and the use of harmonic analysis on the circle.

For discrete time non-stationary processes, a generalization of the time domain analysis of the stationary prediction problem has been given by Rissanen and Barbosa [1969, 1973]. In their theory, linear estimation problems are solved for non-stationary processes which belong to a class M consisting of processes whose covariance operators are realized by infinite matrices in the Banach algebra $B(\ell_p^2) = \{A : \ell_p^2 \to \ell_p^2; \|A\|_B < +\infty\}$.

In this paper, we relate the framework of Rissanen and Barbosa to the theory of harmonizable processes (see [Cramér, 1951], [Loève, 1960]). Then, using the techniques of harmonic analysis on the torus, we derive results concerning the action of time varying linear systems on harmonizable processes and, moreover, obtain an explicit formula (in terms of two complex variables) for the linear non-stationary filter for such processes.

0. Introduction

The objective of this work is the analysis of a class of non-stationary processes and systems via the methods of harmonic analysis. We shall consider the class of harmonizable processes introduced by Loevè; these constitute a natural generalization of wide sense stationary (WSS) processes in the sense that their covariance functions have a "frequency domain" representation as a Fourier transform of a measure with bounded variation on the torus. We examine the action of a certain class S of linear time varying (LTV) systems on these processes and prove that if u is harmonizable and if ξ is an LTV system in S, then $y = \xi u$ is harmonizable. Next we study the prediction problem and give a frequency domain formula for the optimum filter for harmonizable processes; this filter corresponds to the one originally derived by Rissanen and Barbosa [1969, 1973] for a class of non-stationary time domain filtering problems. In the wide sense stationary case it reduces to the Wiener filter.

I. Foundations

All processes will have zero mean and finite second moments.

Definition 1. Let $(\Omega, \mathcal{B}, P)$ be a probability space. Let $x = \{x_n\}_{n \in \mathbf{Z}}$ be a $\mathbf{C}^p$ valued discrete time parameter process. Then $\{x_n\}_{n \in \mathbf{Z}}$ is said to be harmonizable if it has the representation

$$x_n = \int_{-\pi}^{\pi} e^{in\theta} d\varsigma(\theta) \qquad n \in \mathbf{Z}$$

where ς is a $\mathbf{C}^p$ valued random process on $[-\pi, \pi]$.

In general the spectral representation process ς does not have orthogonal increments on $[-\pi, \pi]$, but by the Cramér representation theorem (Cramér and Leadbetter, 1967), x is wide sense stationary if and only if ς has orthogonal increments.

The type of integral in Definition 1 has to be interpreted as a quadratic mean Lebesgue-Stieltjes integral; a sufficient condition for the existence of

$$\int_a^b f(\theta) d\varsigma(\theta) \qquad f : [a,b] \to \mathbf{C}^p \quad f \text{ continuous}$$

is
$$\int_a^b \int_a^b f(\theta) dF(\theta, \psi) f(\psi)^* < +\infty$$

where $*$ denotes complex conjugate transpose, with $dF(\theta, \psi) = \mathrm{E} d\varsigma(\theta) d\varsigma(\psi)^*$, a complex measure with bounded variation on $[a,b] \times [a,b]$.

A well known fact about WSS processes is that the shift invariance of their second moments allows the representation of their covariance function as the Fourier transform of a positive measure i.e.

$$\mathrm{E}(x_{n+k} x_k^*) = R(n) = \int_{-\pi}^{\pi} e^{-in\theta} dF_x(\theta)$$

with $dF_x(\theta) = \mathrm{E} d\varsigma_x(\theta) d\varsigma_x(\theta)^*$. Conversely any positive measure dF gives rise to a sequence $\{R(n)\}_{n \in \mathbf{Z}}$ which is positive definite in the sense that $\sum_{n,m \in \mathbf{Z}} a_n R(n-m) a_m^* \geq 0$. This is the Herglotz-Bochner theorem.

For harmonizable processes we have the following:

Theorem 1: (Loève) A stochastic process x is harmonizable if and only if its covariance function R satisfies $R(n,m) \triangleq \mathrm{E}(x_n x_m^*) = \int_{-\pi}^{\pi} \int_{-\pi}^{\pi} e^{-in\theta} e^{im\psi} dF(\theta, \psi), n, m \in \mathbf{Z}$, where $F(\theta, \psi) = \mathrm{E} d\varsigma(\theta) d\varsigma(\psi)^*$ is a complex measure with bounded variation on the torus $[-\pi, \pi] \times [-\pi, \pi]$.

So a harmonizable process is a generalization of a WSS process in which the "frequency domain" representation is taken with respect to an arbitrary process in the former case and to an orthogonal increment process in the latter, and whose spectral distribution is respectively a complex measure on the torus and a positive measure distributed on the diagonal.

II. System Theory

Let $\ell_p^2 = \ell^2(\mathbf{C}^p) = \{x = \{x_n\}_{n \in \mathbf{Z}} \quad x_n = \begin{pmatrix} x_n^1 \\ \vdots \\ x_n^p \end{pmatrix}$ with $\sum_{i=-\infty}^{+\infty} tr(x_i x_i^*) < +\infty\}$

Then ℓ_p^2 is an Hilbert space with the usual inner product

$$(x, y)_{\ell_p^2} = \sum_{i=-\infty}^{+\infty} tr(x_i y_i^*) = \sum_{i=-\infty}^{+\infty} (x_i, y_i)_{\mathbf{C}^p}.$$

Following Rissanen and Barbosa [1969, 1973] (see also the exposition in Chapter 2 of [Caines, 1985]), let $B(\ell_p^2, \ell_m^2)$ be the set of linear operators from ℓ_p^2 to ℓ_m^2 with finite operator norm i.e.

$$B(\ell_p^2, \ell_m^2) = \{A : \ell_p^2 \to \ell_m^2 : \|A\|_B = \sup_{\|x\|_{\ell_p^2} \leq 1} \|Ax\|_{\ell_p^2} < +\infty\}$$

Let us set

$$B(\ell^2_p) \,\Delta\, \{A : \ell^2_p \to \ell^2_p : \; \|A\|_B < +\infty\}$$

then $B(\ell^2_p)$ is a Banach algebra and we have

$$\forall A, B \in B(\ell^2_p): \qquad \|AB\|_B \leq \|A\|_B \|B\|_B.$$

To begin we shall restrict attention to operators on the B spaces which can be represented by infinite sequences of $m \times p$ matrices, that is to say, operators A which can be represented as

$$A = \begin{bmatrix} \ddots & & & \\ & A_{i-1,i-1} & A_{i+1,i} & \\ & A_{i,i-1} & A_{i,i} & \\ & & & \ddots \end{bmatrix}$$

In this case $x \in \ell^2_p$ is mapped into $Ax \in \ell^2_m$, where $(Ax)_i \,\Delta\, \sum_{j=-\infty}^{+\infty} A_{ij} x_j$.

Moreover, if $A \in B(\ell^2_p, \ell^2_m)$ and $B \in B(\ell^2_n, \ell^2_p)$, then $AB \in B(\ell^2_n, \ell^2_m)$ with $(AB)_{ij} = \sum_{k=-\infty}^{+\infty} A_{ik} B_{kj}$.

An important feature of operators in $B(\ell^2_p, \ell^2_m)$ which can be represented by an infinite array is that their rows and their columns are square summable; to be precise

$$\sum_{i=-\infty}^{+\infty} tr A_{ik} A^*_{ki} \leq \|A\|_B \quad < \infty$$

$$\sum_{k=-\infty}^{+\infty} tr A_{ki} A^*_{ki} \leq \|A\|_B \quad < \infty.$$

This feature allows the action of such operators on orthogonal bounded variance process i.e. if w is such that

$$\mathrm{E} w_n w^*_m = C_{n,m} \delta_{n,m}; \quad C_{n,m} > 0, \quad \forall n, m \in \mathbf{Z}$$

and

$$\mathrm{E} w^*_n w_n < C < +\infty \qquad \forall n \in \mathbf{Z}$$

where δ is the Kronecker symbol.

Then we define

$$(Aw)_n \,\Delta \lim_{N,N' \to +\infty} \sum_{k=-N'}^{N} A_{nk} w_k. \qquad \text{(q.m. limit)}$$

this limit exists since

$$\mathrm{E} \|\sum_{k=M}^{N} A_{nk} w_k\|^2 \leq C \sum_{j=M}^{N} tr A_{jk} A^*_{jk} \to 0 \quad \text{as} \quad N, M \to \infty$$

and we write

$$(Aw)_n \,\Delta \sum_{k=-\infty}^{+\infty} A_{nk} w_k.$$

In general we are interested in the action of operators $A \varepsilon B(\ell_p^2)$ on harmonizable processes, but to facilitate the analysis let's define the class $S \quad B(\ell_p^2)$ as follows:

Definition 2. An operator $A \varepsilon B(\ell_p^2)$ is said to belong to the class S if the following two conditions are satisfied

(i) $$\int_{-\pi}^{\pi} \sum_n \sum_p A_{n p} e^{-in\theta} e^{ip\psi} d\theta = \mu_p(\psi), \quad p \varepsilon \mathbf{Z}$$

(ii) $$\int_{-\pi}^{\pi} \sum_n \sum_p A_{n,p} e^{-in\theta} e^{ip\psi} d\psi = Z_n(\theta), \quad n \varepsilon \mathbf{Z}$$

where μ and Z are respectively a measure and a function on the $[-\pi, \pi]$.

The next result describes the action of operators on harmonizable processes in terms of their spectral representation processes and spectral distributions.

Theorem 2. Let u be an orthogonal bounded variance harmonizable process, let ξ a system represented via an infinite array $A \varepsilon S$ then $y = Au$ is harmonizable with

$$F_y(\theta, \psi) = \int_{-\pi}^{\theta} \int_{-\pi}^{\psi} \int_{-\pi}^{\pi} \int_{-\pi}^{\pi} m(\theta'', \psi') dF_u(\theta'', \theta') m(\theta', \psi'')^* d\psi' d\psi''$$

Proof

We have

$$y_n \Delta (Au)_n = \sum_{p=-\infty}^{+\infty} A_{n,n-p} u_{n-p}$$

and since u is harmonizable

$$= \sum_{-\infty}^{+\infty} A_{n,n-p} \int_{-\pi}^{\pi} e^{-i(n-p)\theta} d\varsigma_u(\theta)$$

$$= \int_{-\pi}^{\pi} \sum_{p=-\infty}^{+\infty} A_{n,n-p} e^{-i(n-p)\theta} d\varsigma_u(\theta)$$

because $A \varepsilon S$.

On the other hand for $A \varepsilon S$

$$\int_{-\pi}^{\pi} \sum_{n=-\infty}^{+\infty} \sum_{p=-\infty}^{+\infty} A_{n,n-p} e^{ip\theta} e^{in\psi} e^{ik\psi} d\psi = \begin{cases} Z_n(e^{i\theta}) & \text{for } k = n \\ 0 & \text{otherwise} \end{cases}$$

and so

$$y_n = \int_{-\pi}^{\pi} \int_{-\pi}^{\pi} m(\theta, \psi) e^{-in\psi} d\psi e^{-in\theta} d\varsigma_u(\theta)$$
$$\Delta \int_{-\pi}^{\pi} e^{-in\psi} d\varsigma_y(\psi)$$

where

$$\varsigma_y(\psi) = \int_{-\pi}^{\psi} \int_{-\pi}^{\pi} m(\theta, \psi') d\varsigma_u(\theta) d\psi'$$

hence y_n is harmonizable.

Furthermore the spectral distribution of y and u are related in the following way:

$$F_u(\theta,\psi) = \mathrm{E}\left(\int_{-\pi}^{\theta}\int_{-\pi}^{\pi} m(\theta'',\psi')d\varsigma_u(\theta'')d\psi'\right)\left(\int_{-\pi}^{\psi}\int_{-\pi}^{\pi} m(\theta',\psi'')d\varsigma_u(\theta')d\psi''\right)^*$$

$$= \int_{-\pi}^{\theta}\int_{-\pi}^{\psi}\int_{-\pi}^{\pi}\int_{-\pi}^{\pi} m(\theta'',\psi')\mathrm{E}d\varsigma_u(\theta'')d\varsigma_u(\theta')^* m(\theta',\psi'')^* d\psi'$$

since

$$\mathrm{E}\left(\int_a^b f(\theta)d\varsigma_u(\theta)\right)\left(\int_a^b g(\psi)d\varsigma_u(\psi)\right)^* = \int_a^b\int_a^b f(\theta)\mathrm{E}d\varsigma_u(\theta)d\varsigma_u(\psi)^* g(\psi)^*$$

and because

$$\mathrm{E}d\varsigma_u(\theta'')d\varsigma_u(\theta')^* = dF_u(\theta'',\theta')$$

this can be rewritten as

$$F_y(\theta,\psi) = \int_{-\pi}^{\theta}\int_{-\pi}^{\psi}\int_{-\pi}^{\pi}\int_{-\pi}^{\pi} m(\theta'',\psi')dF_u(\theta'',\theta')m(\theta',\psi'')^* d\psi' d\psi''$$

or as

$$dF_y(\theta,\psi) = \int_{-\pi}^{\pi}\int_{-\pi}^{\pi} m(\psi',\theta)dF_u(\psi'.\theta')m(\theta',\psi)^* d\theta d\psi.$$

□

As a particular case of this result we obtain the convolution representation of linear time invariant (LTI) systems, as in

$$y_n = \sum_{j=-\infty}^{+\infty} Z_j u_{n-j}, \qquad n \varepsilon \mathbf{Z}.$$

Further, if u is harmonizable, then y is harmonizable with

$$dF_y(\theta,\psi) = \sum_{j=-\infty}^{+\infty} Z_j e^{ij\theta} dF_u(\theta,\psi) \sum_{k=-\infty}^{+\infty} Z_k^* e^{-ik\psi}.$$

This is a generalization of the classical formula for the output y of a LTI system with transfer function $Z(e^{i\theta}) = \sum_{-\infty}^{\infty} Z_j e^{ij\theta}$ whose input u is a WSS process, namely

$$dF_y(\theta) = Z(e^{i\theta})dF_u(\theta)Z^*(e^{i\theta})$$

III. Prediction Theory

Let

$$B^+(\ell_p^2) = \{A \varepsilon B(\ell_p^2) \quad \text{s.t.} \quad A_{ij} = 0; i > j \quad i,j \varepsilon \mathbf{Z}\}$$
$$B^-(\ell_p^2) = \{A \varepsilon B(\ell_p^2) \quad \text{s.t.} \quad A_{ij} = 0; i < j \quad i,j \varepsilon \mathbf{Z}\}$$
$$M_p = \{u : u = Aw; A \varepsilon B(\ell_p^2) \text{ and } w \text{ an orthogonal bounded variance process}\}$$
$$M_p^+ = \{u : u = Aw; A \varepsilon B^+(\ell_p^2) \text{ and } w \text{ an orthogonal bounded variance process}\}$$

and let's introduce the following definition.

Definition 3: Let x a p component process with bounded variance. Let $R^x_{i,j} \triangleq Ex_ix_j^*$ $i,j \in \mathbf{Z}$. If $R^x \in B(\ell^2_p)$ i.e. $\|R^x\|_B < +\infty$ then R^x is called the covariance operator of x.

For a covariance operator, R^x is such that $(R^x)^* \in B(\ell^2_p)$, where $((R^x)^*u,v)_{\ell^2_p} \triangleq (u, R^x v)_{\ell^2_p}$. Further define

$$[U^m x]_n = x_{n+m} \qquad \forall x \in M_+, \quad \forall n,m \in \mathbf{Z}$$
$$(U^m A)_{i,j} = A_{i+m,j} \qquad \forall i,j,m \in \mathbf{Z}, \quad \forall A \in B(\ell^2_p)$$
$$(AH^m)_{i,j} = A_{i,j+m}. \qquad \forall i,j,m \in \mathbf{Z}, \quad \forall A \in B(\ell^2_p)$$

and

$$([A]_+)_{i,j} = \begin{cases} A_{i,j} & j \le i \\ 0 & j > i \end{cases}$$

Consider the prediction of $x \in M_+$, given $y \in M_+$ via the non-stationary filter H^m as follows

$$\widehat{x}_{n+m|n} = \sum_{i=-\infty}^{n} H^m_{n,i} y_i.$$

Now $\widehat{x}_{n+m|n}$ is characterized by

$$x_{n+m} - \widehat{x}_{n+m|n} \perp H^y_n = \overline{\mathrm{Sp}}\{\dots, y_n\}$$

which implies

$$(*) \sum_{i=-\infty}^{n} H^m_{n,i} R^y_{i,k} - R^{xy}_{n+m,k} = 0 \quad \forall k \le n \quad \forall n \in \mathbf{Z}$$

where R^y is the covariance operator of $y \in M^+$, R^{xy} is the cross covariance of x and y.

So as in Rissanen and Barbosa [1969, 1973], (see Caines [1985]) the prediction problem is reduced to solving the Wiener-Hopf equation $(*)$. To solve this equation, we have the following:

Theorem 3 (Rissanen-Barbosa) $R \in B(\ell^2_p), R = R^*R > 0$ then $\exists A \in B^+(\ell^2_p)$ with $A^{-1} \in B^+(\ell^2_p)$ s.t. $R = AA*$.

But since $R^y \in B(\ell^2_p), R^y = GG^*$ with G and $G^* \in B^+(\ell^2_p)$, hence $(*)$ can be rewritten as

$$[H^m GG^* - U^m R^{xy}]_{nk} = 0 \quad \forall k \le n \quad \forall n \in \mathbf{Z}$$

i.e.

$$H^m GG^* - U^m R^{xy} \in B^-(\ell^2_p)$$

Hence

$$H^m G - U^m R^{xy} G^{*-1} \in B^-(\ell^2_p)$$

but since

$$G \in B^+(\ell^2_p) \qquad H^m G \in B^+(\ell^2_p)$$

and so

$$H^m G - [U^m R^{xy} G^{*-1}]_+ = 0$$

which gives the non-stationary Wiener-Hopf Equation

$$H^m = [U^m R^{xy} G^{*-1}]_+ G^{-1} \in B^+.$$

This is the Rissanen-Barbosa filter in the time domain.

When all the operators involved may be concatenated to yield harmonizable processes, the frequency domain expression for this filter is

$$H(\theta,\psi) = \int_{-\pi}^{\pi}\left[\int_{-\pi}^{\pi}\Phi_{xy}(\theta',\theta)G_y^{*}(\theta'),\theta''d\theta'\right]_+ G_y(\theta''\psi)d\theta''$$

where

$$A(\theta,\psi) = \sum_{n=-\infty}^{\infty}\sum_{p=0}^{\infty} A_{n,n-p}e^{-i(n-p)\theta}e^{-in\psi}$$

for any $A \varepsilon S$ and $G_y(\cdot,\cdot)$ is the spectral factor of $F_y(\theta,\theta')$ given by

$$dF_y(\theta,\psi) = \int_{-\pi}^{\pi}\int_{-\pi}^{\pi} G_y(\theta'',\theta)d\theta''d\theta' G_y^{*}(\theta',\psi)d\theta d\psi.$$

This formula will be derived in a later paper.

References

Caines, P.E. "Linear Stochastic Systems" Lecture Notes, Dept. Elec. Eng. McGill Univ. Montreal, (to be published by John Wiley) 1985.

Cramér, H. "A Contribution to the Theory of Stochastic Processes", Proceedings of the 2nd Berkeley Symposium on Mathematical Statistics and Probability, Berkeley and Los Angeles, University of California Press, 1951, pp. 329-399.

Cramér, H. and Leadbetter, M.R. "Stationary and Related Stochastic Processes". John Wiley, N.Y.C. 1967.

Loève, M. *Probability Theory*, Van Nostrand Publishers, 1960.

Rissanen, J. "A Fast Algorithm for Optimum Linear Predictors", *IEEE Trans. Automatic Control*, Vol. AC-18, p.555, Oct. 1973.

Rissanen, J. and Barbosa, L. "Properties of Infinite Covariance Matrices and Stability of Optimum Predictors", Information Sciences, Vol. 1, pp. 221-234, 1969.

Frequency Domain and State Space Methods for Linear Systems
C.I. Byrnes and A. Lindquist (Editors)

SINGULAR OPTIMAL FILTERING

Aysin Kitapci, Edmond Jonckheere, and Leonard Silverman

Department of Electrical Engineering-Systems
University of Southern California
Los Angeles, California 90089-0781
U.S.A.

In this paper, a system theory approach to the linear-quadratic-gaussian singular filtering problem is developed. The singular filter processes the output of a (A,B,C,D)-system driven by a white noise and constructs the "best" estimate of its state. The problem is said to be singular whenever D does not have full row rank. The approach developed here is system theoretic, in the sense that it identifies the crucial role played by the weakly unobservable and the strongly reachable subspaces of (A,B,C,D). Further, the state space of (A,B,C,D) is decomposed according to these subspaces. Finally, from this state space decomposition, it is obvious that the optimal filter is the cascade of a strong observer, that differentiates the output, and a reduced-order, nonsingular Kalman-Bucy filter.

I. INTRODUCTION

We consider a standard linear state space system

$$\Sigma: \quad \begin{aligned} \dot{x} &= Ax + Bn \\ y &= Cx + Dn \end{aligned} \tag{1}$$

driven by a stationary, zero-mean, Gaussian white noise $n(\cdot)$ with unit intensity:

$$En(t)n^T(\tau) = I\,\delta(t-\tau)$$

All along the paper, we will implicitly assume that $\dot{x} = Ax + Bn$ should be rather interpreted as the Ito stochastic differential equation $dx = Ax\,dt + Bd\beta$, where β is the Brownian or Wiener process with covariance $E\beta(t)\beta^T(\tau) = I \min(t,\tau)$. We keep the usual white noise formulation for notational convenience.

The purpose of this paper is to construct the filter that takes $y(\cdot)$ and synthesizes the best estimate $\hat{x}(\cdot)$ of the state $x(\cdot)$ of the system. As long as D has full row rank, this is the standard Kalman-Bucy filtering problem. However, when D fails to be of full row rank, this becomes the singular filtering problem, a problem whose solution has been "targeted" over the past few years.

Much of the results presented here rely on geometric insight. The weakly unobservable subspace $\mathcal{V}$ and the strongly reachable subspace $\mathcal{W}$ of (A,B,C,D) are shown to play a crucial role. Along $\mathcal{V}^{\perp}$, the filter error can be made to go to zero arbitrarily fast by means of a strong observer that differentiates the

---

This research was supported by JSEP Grant F49620-85-C-0071.

measurement data. In contrast, $\mathcal{W}$ is that part of the state which is inherently corrupted by a nonsingular estimation error when a causal filter of the Kalman-Bucy type is used to reconstruct the state. More concretely, the geometry of $\mathcal{V}$ and $\mathcal{W}$ reveals a realization of (A,B,C,D) as the cascade of a strongly observable and a properly invertible subsystems. Therefore, the state of the strongly observable subsystem is reconstructed with arbitrary accuracy by means of a strong observer that differentiates the output, while the state of the properly invertible subsystem is estimated by means of a nonsingular Kalman-Bucy filter that takes the measurement $y(\cdot)$ as well as the output of the strong observer, thereby resulting in a cascade realization of the singular filter.

The present paper follows in the footstep of the geometric approach to the linear-quadratic singular regulator problem developed by Hautus and Silverman (1983) and Willems, Kitapci and Silverman (1985). In an independent effort, Schumacher (1985) developed another approach to singular filtering. In contrast to the present approach that involves the original system data (A,B,C,D), Schumacher's approach rather involves a system model equivalent data $(A,\bar{B},C,\bar{D})$ defined from the factorization of the linear matrix inequality (an idea apparently formulated for the first time in the context of linear-quadratic singular problems in Jonckheere (1981)). Systems (A,B,C,D) and $(A,\bar{B},C,\bar{D})$ are said to be system model equivalent if they generate the same Wiener-Hopf equations. From this observation, it follows that (A,B,C,D) and $(A,\bar{B},C,\bar{D})$ are equivalent as far as the Wiener filter and associated stationary phenomena are concerned, but the equivalence breaks down when it comes to the Kalman-Bucy filter and associated transient phenomena. Because our singular filter is derived from the original data (A,B,C,D), it allows a treatment of the transient phenomena occuring at $t = 0$. These transient phenomena occur in the strong observer that cancels the $\mathcal{V}^{\perp}$-component of the initial error arbitrarily fast. These transient phenomena are unaccounted for in Schumacher's approach.

To avoid misinterpretations, we make the following disclaimer: This is a system theoretic approach; no deep probabilistic interpretation is attempted here.

The present paper is a short version of a more complete paper to appear elsewhere; for the complete proofs, the reader is referred to Kitapci (1985).

II. GEOMETRIC PREREQUISITIES

In this section, we develop the geometric insight that will lead to the state space decomposition that will itself reveal the structure of the singular filter. Two subspaces play a central role:

II.1. The Largest Null-Output (A,B)-Invariant, or Controlled Invariant, Subspace $\mathcal{V}$; In Other Words, The Weakly Unobservable Subspace

This subspace $\mathcal{V}$ of (A,B,C,D) has received various equivalent motivations and definitions, most of them inspired from the control problem $\dot{x} = Ax + Bu$, $y = Cx + Du$, where $y(\cdot)$ is some output-to-be-controlled and $u(\cdot)$ some control input. Three equivalent definitions are provided here:

i) $\mathcal{V} = \{x_0 : y(t) = 0, \forall t \geq 0$, for some regular input $u(\cdot)\}$

ii) $\exists$ state feedback F $\ni$

$$(A + BF)\mathcal{V} \subseteq \mathcal{V}$$

$$(C + DF)\mathcal{V} = 0$$

$$\text{iii)}\quad \mathcal{V}_{i+1} = \begin{pmatrix} A \\ C \end{pmatrix}^{-1} \left\{ \begin{pmatrix} \mathcal{V}_i \\ 0 \end{pmatrix} + \operatorname{Im} \begin{pmatrix} B \\ D \end{pmatrix} \right\}$$

$$\mathcal{V}_0 = \mathcal{X}$$

Definition i) is the intuitive notion of weak unobservability of x_0: The initial state $x_0 \in \mathcal{V}$ does not generate output in the future for some appropriately selected control input. From Definition i), it is easily seen that $\mathcal{V}$ is (A,B)-invariant or controlled invariant. System is said to be strongly observable when $\mathcal{V} = 0$. As a corollary, strong observability means that the state can be reconstructed without knowledge of the input. From Definition ii), it follows that (A,B,C,D) is strongly observable iff (A+BF, C+DF) is observable for all state feedbacks F.

The complete filtering interpretation of $\mathcal{V}$, i.e., the construction of a strong observer to reconstruct part of the state without knowledge of the input, is postponed to Section II.4.

II.2. The Smallest Input-Containing (C,A)-Invariant, or Conditioned-Invariant, Subspace $\mathcal{W}$; In Other Words, The Strongly Reachable Subspace

Again, there are various definitions and interpretations of this subspace:

i) $\mathcal{W} = \{x_0 : x(o) = x_0$ for some impulsive control $u(t),\ t \leq 0$, such that $y(t) = 0,\ t \leq 0\}$

ii) $\exists$ output injection K $\ni$

$$(A + KC)\mathcal{W} \subseteq \mathcal{W}$$

$$\operatorname{Im}(B + KD) \subseteq \mathcal{W}$$

$$\text{iii)}\quad \mathcal{W}_{i+1} = (A \quad B)\left\{ \left\{ \begin{pmatrix} \mathcal{W}_i \\ \mathcal{U} \end{pmatrix} \right\} \cap \operatorname{Ker}(C \quad D) \right\}$$

$$\mathcal{W}_0 = 0$$

Definition i) is the notion of strong reachability for the system $\dot{x} = Ax + Bu$, $y = Cx + Du$; the state x_0 in $\mathcal{W}$ can be reached by some impulsive control $u(\cdot)$ while at the same time keeping the output $y(\cdot)$ at zero. It is rather easily seen from this definition that $\mathcal{W}$ is (C,A)- or conditioned-invariant.

II.3. Left Invertibility

Another important concept in singular filtering is left invertibility. The system $\Sigma = (A,B,C,D)$ is said to be left invertible if $D + C(sI-A)^{-1}B$ is invertible as a rational matrix, or, more intuitively, if the input $u(\cdot)$ can be recovered from the output $y(\cdot)$. The following lemma will prove useful.

Lemma 1: $\Sigma = (A,B,C,D)$ is left invertible iff $\mathcal{V} \cap \mathcal{W} = 0$ and $\operatorname{Ker}\begin{bmatrix} B \\ D \end{bmatrix} = 0$.

Proof: See Hautus and Silverman (1983).

II.4. Filtering Interpretation of Weakly Unobservable Subspace $\mathcal{V}$ and Construction of Strong Observer

The filtering interpretation of $\mathcal{V}$ has been a leading motivation behind the early development of the geometric theory (Basile and Marro (1969) and Rappaport and Silverman (1971).) Here, this filtering interpretation is provided very explicitly by resorting to the structure algorithm (Silverman (1969)).

Let F be a state feedback such that

$$(A + BF)\mathcal{V} \subseteq \mathcal{V}$$

$$(C + DF)\mathcal{V} = 0$$

Such a feedback F is said to be maximally unobservable, because it makes the unobservable subspace of (A + BF, C + DF) as large as possible. With this feedback F, we define the following noise transformation:

$$n = Fx + \eta$$

where η is, in general, colored. This noise transformation acts as a state feedback F on the system Σ:

$$\Sigma^F : \begin{aligned} \dot{x} &= (A + BF)x + B\eta \\ y &= (C + DF)x + D\eta \end{aligned}$$

Kalman's observability decomposition yields

$$\Sigma^* : \begin{aligned} \begin{pmatrix} \dot{x}_1 \\ \dot{x}_2 \end{pmatrix} &= \begin{pmatrix} A^*_{11} & A^*_{12} \\ 0 & A^*_{22} \end{pmatrix} \begin{pmatrix} x_1 \\ x_2 \end{pmatrix} + \begin{pmatrix} B_1 \\ B_2 \end{pmatrix} \eta \\ y &= (0 \quad C^*_2) \begin{pmatrix} x_1 \\ x_2 \end{pmatrix} + D\eta \end{aligned}$$

(A star (*) is a short way to denote a quantity transformed by the state feedback F.) The unobservable subspace of Σ^*, that has been made as large as possible by appropriate choice of F, is nothing other than $\mathcal{V}$:

$$\mathcal{V} = \left\{ \begin{pmatrix} x_1 \\ 0 \end{pmatrix} \right\}.$$

As far as the process $y(\cdot)$ is concerned, it can be equally generated by the subsystem

$$\Sigma^*_{so} : \begin{aligned} \dot{x}_2 &= A^*_{22} x_2 + B_2 \eta \\ y &= C^*_2 x_2 + D\eta \end{aligned}$$

This system is strongly observable because $\mathcal{V}(\Sigma^*_{so}) = 0$. To make it left invertible, we decompose the noise space as follows (see Lemma 1):

$$\eta = \eta_1 \oplus \eta_2 \quad , \quad \eta_2 = \mathrm{Ker}\begin{bmatrix} B_2 \\ D \end{bmatrix}$$

This yields the system

$$\Sigma^*_{sol} : \quad \begin{aligned} \dot{x}_2 &= A^*_{22}x_2 + B_{21}\eta_1 \\ y &= C^*_2\, x_2 + D_1\eta_1 \end{aligned} \qquad (2)$$

which is strongly observable and left invertible. Left invertibility means that the input noise $\eta_1(\cdot)$ can be recovered from the output $y(\cdot)$. Therefore, a perfect estimate of x_2 is to be expected. Concretely, this is done as follows: Applying the left structure algorithm to the system Σ^*_{sol} yields a sequence of transformed outputs terminating in

$$\begin{pmatrix} \bar{y}_\alpha \\ \tilde{y}_\alpha \end{pmatrix} = \begin{pmatrix} \bar{C}^*_{2\alpha} \\ \tilde{C}^*_{2\alpha} \end{pmatrix} x_2 + \begin{pmatrix} \bar{D}_{1\alpha} \\ 0 \end{pmatrix} \eta_1 \qquad (3)$$

$(\bar{y}^T_\alpha \tilde{y}^T_\alpha)^T$ is obtained from y through nonsingular transformations and α differentiations. Because Σ^*_{sol} is left invertible, the structure algorithm terminates with a square nonsingular matrix $\bar{D}_{1\alpha}$. Therefore, Equation (3) can be solved for η_1 and this gives

$$\eta_1 = -\bar{D}^{-1}_{1\alpha}\bar{C}^*_{2\alpha}x_2 + \bar{D}^{-1}_{1\alpha}\,\bar{y}_\alpha$$

Substituting for η_1 in (2) has the effect of transforming Σ^*_{sol} by the state feedback $-\bar{D}^{-1}_{1\alpha}\bar{C}^*_{2\alpha}$:

$$\Sigma^{**}_{sol} : \quad \begin{aligned} \dot{x}_2 &= (A^*_{22} - B_{21}\bar{D}^{-1}_{1\alpha}\bar{C}^*_{2\alpha})x_2 + B_{21}\bar{D}^{-1}_{1\alpha}\,\bar{y}_\alpha \\ y &= (C^*_2 - D_1\bar{D}^{-1}_{1\alpha}\bar{C}^*_{2\alpha})x_2 + D_1\bar{D}^{-1}_{1\alpha}\,\bar{y}_\alpha \end{aligned}$$

Since the system Σ^*_{sol} is strongly observable, Σ^{**}_{sol} is observable. Therefore, one can construct the observer for Σ^{**}_{sol}:

$$\begin{aligned} \dot{\hat{x}}_2 = {} & (A^*_{22} - B_{21}\bar{D}^{-1}_{1\alpha}\bar{C}^*_{2\alpha})\hat{x}_2 + B_{21}\bar{D}^{-1}_{1\alpha}\,\bar{y}_\alpha \\ & + K_2(y - (C^*_2 - D_1\bar{D}^{-1}_{1\alpha}\bar{C}^*_{2\alpha})\hat{x}_2 - D_1\bar{D}^{-1}_{1\alpha}\,\bar{y}_\alpha) \end{aligned}$$

The dynamics of this observer can be made arbitrarily fast by appropriate choice of the output injection K_2. This observer for Σ^{**}_{sol} is a strong observer for Σ^*_{sol}:

$$\dot{\hat{x}}_2 = A^{***}_{22}\,\hat{x}_2 + K_2 y + L_2\bar{y}_\alpha$$

where

$$A_{22}^{***} \;:=\; A_{22}^{*} - B_{21}\bar{D}_{1\alpha}^{-1}\bar{C}_{2\alpha}^{*} - K_2(C_2^{*} - D_1\bar{D}_{1\alpha}^{-1}\bar{C}_{2\alpha}^{*})$$

$$L_2 \;:=\; -K_2D_1\bar{D}_{1\alpha}^{-1} + B_{21}\bar{D}_{1\alpha}^{-1}$$

and the eigenvalues of A_{22}^{***} are freely assignable. If we define the error

$$e := x - \hat{x}$$

we get

$$\dot{e}_2 = A_{22}^{***} e_2$$

and the spectrum of A_{22}^{***} is arbitrarily fast. Therefore, along $\mathscr{V}^{\perp}$, the filter error goes to zero arbitrarily fast. To be more precise, if $e_2(o)$ is considered deterministic, it can be shown that

$$\forall \varepsilon > 0 \;, \quad \exists K_2 \ni \| e_2 \|_{L^1[0,\infty)} < \varepsilon \,.$$

On the other hand, if $e_2(o)$ is a random vector, the covariance kernel $\Gamma(\cdot,\cdot)$ of $e_2(\cdot)$ is "small" in the following sense:

$$\forall_{i,j};\ \forall \varepsilon > 0 \;;\ \exists K_2 \ni \int_0^\infty\!\!\int_0^\infty |\Gamma_{ij}(t,\tau)|\,dt\,d\tau < \varepsilon \;.$$

II.5 Filtering Interpretation of Strongly Reachable Subspace $\mathscr{W}$ and Rationale Behind the Cascade Decomposition of the Singular Filter

The role of the strongly reachable subspace $\mathscr{W}$ in singular filtering can be understood as follows: Consider the causal estimator:

$$\dot{\hat{x}} = A\hat{x} + K(y - C\hat{x}) \tag{4}$$

which yields the error

$$\dot{e} = (A - KC)e + (B - KD)n \,,$$

regardless of whether or not the problem is singular. Choose an output injection K such that

$$(A - KC)\mathscr{W} \subseteq \mathscr{W}$$

$$\mathrm{Im}\,(B - KD) \subseteq \mathscr{W}$$

and consider Kalman's controllability decomposition of the error system:

$$\begin{pmatrix} \dot{e}_1 \\ \dot{e}_2 \end{pmatrix} = \begin{pmatrix} A_{11}^{*} & A_{12}^{*} \\ 0 & A_{22}^{*} \end{pmatrix}\begin{pmatrix} e_1 \\ e_2 \end{pmatrix} + \begin{pmatrix} B_1 \\ 0 \end{pmatrix} n$$

(A_{11}^{*}, B_1) is controllable, and the "star" is a short notation to denote a coefficient matrix that has been transformed by the output injection K. Clearly

$$\mathscr{W} = \left\{ \begin{pmatrix} e_1 \\ 0 \end{pmatrix} \right\}$$

Therefore, the following interpretation of $\mathcal{W}$ emerges

$$e(o) \in \mathcal{W} \quad \Rightarrow \quad e(t) \in \mathcal{W} \quad ;$$

$$E\, e_1 e_1^T > 0$$

The above means that with a causal filter of the form (4), an initial error e(o) in $\mathcal{W}$ cannot be compensated for, and, in fact, the error variance along $\mathcal{W}$ is and will remain positive definite.

In view of the above limitation, it appears necessary to estimate the largest possible subspace of $\mathcal{W}$ by means of a (noncausal) strong observer. What can be estimated this way is $\mathcal{V}^{\perp}$. Therefore, the ultimate solution of the singular filtering problem depends on how $\mathcal{V}$ and $\mathcal{W}$ are intertwined. This is the topic of the next section.

The above reasoning is the leading intuition behind our cascade decomposition of the singular filter: A strong observer estimates with arbitrary accuracy $\mathcal{V}^{\perp}$, and $\mathcal{V}$ is estimated by means of a nonsingular Kalman-Bucy filter.

Our motivation for looking at the subspaces $\mathcal{V}$ and $\mathcal{W}$ in order to clarify the singular filtering problem strongly contrasts with Schumacher's. Schumacher defines a system model equivalent problem $\overline{\Sigma} = (A,\overline{B},C,\overline{D})$ such that $\mathcal{V}(\overline{\Sigma}) \cap \mathcal{W}(\overline{\Sigma}) = 0$, from which it follows that the singular filter consists of the parallel connection of a filter that reconstructs x (mod $\mathcal{V}$) and another that reconstructs x (mod $\mathcal{W}$).

III. STATE SPACE DECOMPOSITION AND STRUCTURE OF THE SINGULAR FILTER

Before decomposing the state space of $\Sigma = (A,B,C,D)$ according to the weakly unobservable and the strongly reachable subspaces, we operate some simplifying transformations on the data (A,B,C,D). These transformations by no means will affect the problem.

Consider the original filtering problem (1), and let us decompose the input noise space and the output measurement space as follows:

$$\mathcal{N} = \mathcal{N}_1 \oplus \mathcal{N}_2 \qquad \mathcal{N}_2 := \text{Ker } D$$

$$\mathcal{Y} = \mathcal{Y}_1 \oplus \mathcal{Y}_2 \qquad \mathcal{Y}_1 := \text{Im } D$$

Therefore, the problem becomes

$$\Sigma : \quad \begin{aligned} \dot{x} &= Ax + B_1 n_1 + B_2 n_2 \\ y_1 &= C_1 x + n_1 \\ y_2 &= C_2 x \end{aligned} \tag{5}$$

The noise n_1 can be eliminated by making use of $n_1 = y_1 - C_1 x$ and this gives

$$\Sigma_2^* : \quad \begin{aligned} \dot{x} &= (A - B_1 C_1)x + B_1 y_1 + B_2 n_2 \\ y_2 &= C_2 x \end{aligned}$$

In the above y_1 is a known input to the system that does not affect the nature of the filtering problem. The fact that there is no fundamental difference between the problem $\Sigma = (A,B,C,D)$ and the problem $\Sigma_2^* = (A - B_1C_1, B_2, C_2)$ is asserted by the following:

Lemma 2:

$$\mathcal{V}(\Sigma) = \mathcal{V}(\Sigma_2^*)$$
$$\mathcal{W}(\Sigma) = \mathcal{W}(\Sigma_2^*)$$
$$\Sigma \text{ detectable} \Rightarrow \Sigma_2^* \text{ detectable} .$$

Proof: See Kitapci (1985).

The absence of direct transmission in $\Sigma_2^* = (A - B_1C_1, B_2, C_2)$ is believed to simplify the problem with respect to $\Sigma = (A,B,C,D)$. In particular, in the state space of Σ_2^*, $\mathcal{V}$ is the largest controlled invariant subspace contained in the kernel of C_2, while $\mathcal{W}$ is the smallest conditioned invariant subspace containing the image of B_2.

Exploiting this last interpretation of $\mathcal{W}$, there exists an output injection K_2 such that

$$(A - B_1C_1 - K_2C_2)\mathcal{W} \subseteq \mathcal{W}$$
$$\text{Im } B_2 \subseteq \mathcal{W}$$

With this output injection, we transform Σ_2^* as follows:

$$\Sigma_2^{**} : \quad \begin{aligned} \dot{x} &= (A-B_1C_1-K_2C_2)x + B_1y_1 + K_2y_2 + B_2n_2 \\ y_2 &= C_2x \end{aligned}$$

Again, there is no difference between $\Sigma_2^* = (A-B_1C_1, B_2, C_2)$ and $\Sigma_2^* = (A-B_1C_1-K_2C_2, B_2, C_2)$ as asserted by the following:

Lemma 3:

$$\mathcal{V}(\Sigma_2^*) = \mathcal{V}(\Sigma_2^{**})$$
$$\mathcal{W}(\Sigma_2^*) = \mathcal{W}(\Sigma_2^{**})$$
$$\Sigma_2^* \text{ detectable} \Rightarrow \Sigma_2^{**} \text{ detectable} .$$

Proof: See Kitapci (1985).

The filtering problems Σ and Σ_2^{**} are thus fully equivalent and from now on we will work with Σ_2^{**} to elucidate the structure of the singular filter.

We choose a system of coordinates for the state space X of Σ_2^{**} as follows:

$$X = X_1 \oplus X_2 \oplus X_3 \oplus X_4 \oplus X_5 \tag{6}$$
$$\mathcal{W} = X_1 \oplus X_2 \oplus X_3$$
$$\mathcal{V} = X_3 \oplus X_4$$

$$\text{Im } B_2 = X_2 \oplus X_3$$

This is illustrated in the diagram of Figure 1. In this coordinate system, the coefficient matrices of Σ_2^{**} take a highly structured form that will reveal the structure of the singular filter.

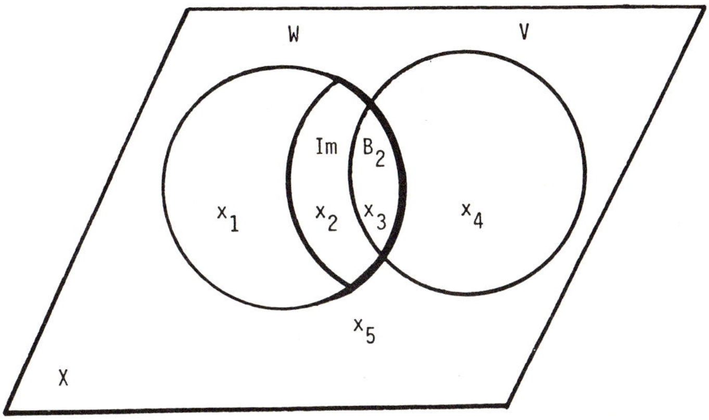

Figure 1

Theorem 1 (State Space Decomposition): In the coordinate system (6), the state space equations of $\Sigma_2^{**} = (A-B_1C_1-K_2C_2, B_2, C_2)$ have the form

$$\dot{x} = \begin{bmatrix} A_{11} & A_{12} & 0 & 0 & A_{15} \\ A_{21} & A_{22} & A_{23} & A_{24} & A_{25} \\ A_{31} & A_{32} & A_{33} & A_{34} & A_{35} \\ 0 & 0 & 0 & A_{44} & A_{45} \\ 0 & 0 & 0 & 0 & A_{55} \end{bmatrix} x + \begin{bmatrix} 0 \\ B_{22} \\ B_{32} \\ 0 \\ 0 \end{bmatrix} n_2 + [B_1 \quad K_2] \begin{bmatrix} y_1 \\ y_2 \end{bmatrix} \tag{7}$$

$$y_2 = [C_{21} \quad C_{22} \quad 0 \quad 0 \quad C_{25}]\, x$$

Further, the matrix $\begin{bmatrix} B_{22} \\ B_{32} \end{bmatrix}$ has full row rank.

Proof: See Kitapci (1985).

From the structure of Σ_2^{**} as revealed by the partitioned state space equations (7), it follows that Σ_2^{**} is the interconnection of the following two systems:

$$\Sigma_{2so}^{**}: \quad \begin{bmatrix} \dot{x}_1 \\ \dot{x}_2 \\ \dot{x}_5 \end{bmatrix} = \begin{bmatrix} A_{11} & A_{12} & A_{15} \\ A_{21} & A_{22} & A_{25} \\ 0 & 0 & A_{55} \end{bmatrix} \begin{bmatrix} x_1 \\ x_2 \\ x_5 \end{bmatrix} + \begin{bmatrix} 0 \\ B_{22} \\ 0 \end{bmatrix} n_2$$

$$+ \begin{bmatrix} B_{11} & K_{12} \\ B_{21} & K_{22} \\ B_{51} & K_{52} \end{bmatrix} \begin{bmatrix} y_1 \\ y_2 \end{bmatrix} + \begin{bmatrix} 0 \\ A_{23}x_3 + A_{24}x_4 \\ 0 \end{bmatrix}$$

$$y_2 = [C_{21} \quad C_{22} \quad C_{25}]\, x$$

$$\begin{bmatrix} \dot{x}_3 \\ \dot{x}_4 \end{bmatrix} = \begin{bmatrix} A_{33} & A_{34} \\ 0 & A_{44} \end{bmatrix} \begin{bmatrix} x_3 \\ x_4 \end{bmatrix} + \begin{bmatrix} B_{32} \\ 0 \end{bmatrix} n_2 + \begin{bmatrix} B_{31} & K_{32} \\ B_{41} & K_{42} \end{bmatrix} \begin{bmatrix} y_1 \\ y_2 \end{bmatrix}$$

$$\Sigma_{2pi}^{**}: \quad + \begin{bmatrix} A_{31}x_1 + A_{32}x_2 + A_{35}x_5 \\ A_{45}x_5 \end{bmatrix}$$

$$\dot{x}_2 = \begin{bmatrix} A_{23} & A_{24} \end{bmatrix} \begin{bmatrix} x_3 \\ x_4 \end{bmatrix} + B_{22}n_2 + \begin{bmatrix} B_{21} & K_{22} \end{bmatrix} \begin{bmatrix} y_1 \\ y_2 \end{bmatrix}$$

$$+ A_{21}x_1 + A_{22}x_2 + A_{25}x_5$$

Each subsystem has a very specific structure:

<u>Theorem 2:</u> Σ_{2so}^{**}, considered as the system from $n_2(\cdot)$ to $y_2(\cdot)$, is strongly observable.

<u>Proof:</u> Easy.

<u>Theorem 3:</u> Σ_{2pi}^{**} considered as the system from $n_2(\cdot)$ to $\dot{x}_2(\cdot)$ is properly invertible. Σ_{2pi}^{**} is further detectable from $\dot{x}_2(\cdot)$ if Σ_2^{**} is detectable from $y_2(\cdot)$.

<u>Proof:</u> See Kitapci (1985).

From Theorems 2 and 3, the intuition leading to the structure of the singular filter should be clear: Since Σ_{2so}^{**} is <u>strongly observable</u>, the state components x_1, x_2, and x_5 can be almost perfectly reconstructed despite the noise $n_2(\cdot)$. Once x_1, x_2, and x_5 are identified, Σ_{2pi}^{**} is a <u>nonsingular Kalman-Bucy filtering</u>

problem, because B_{22} has full row rank and Σ_{2pi}^{**} is detectable from $\dot{x}_2(\cdot)$. This suggests a cascade realization of the singular filter: a strong observer followed by a Kalman-Bucy filter.

There are, however, several technicalities that must be worked out before this intuition can be made rigorous. A first trouble is the driving force $A_{23}x_3 + A_{24}x_4$ on Σ_{2so}^{**}, that acts as a feedback from Σ_{2pi}^{**} to Σ_{2so}^{**} and hence destroys the cascade structure. However, $A_{23}x_3 + A_{24}x_4$ is easily eliminated by means of the noise transformation

$$n_2 = -B_{22}^{-R}(A_{23}x_3 + A_{24}x_4) + \eta_2 \tag{8}$$

B_{22}^{-R} is the right inverse of B_{22}, which is guaranteed to exist in view of Theorem 1. After this transformation, Σ_{2so}^{**} becomes

$$\Sigma_{2so}^{***}: \quad \begin{bmatrix} \dot{x}_1 \\ \dot{x}_2 \\ \dot{x}_5 \end{bmatrix} = \begin{bmatrix} A_{11} & A_{12} & A_{15} \\ A_{21} & A_{22} & A_{25} \\ 0 & 0 & A_{55} \end{bmatrix} \begin{bmatrix} x_1 \\ x_2 \\ x_5 \end{bmatrix} + \begin{bmatrix} 0 \\ B_{22} \\ 0 \end{bmatrix} \eta_2 + \begin{bmatrix} B_{11} & K_{12} \\ B_{21} & K_{22} \\ B_{51} & K_{52} \end{bmatrix} \begin{bmatrix} y_1 \\ y_2 \end{bmatrix}$$

$$y_2 = [C_{21} \quad C_{22} \quad C_{25}] \begin{bmatrix} x_1 \\ x_2 \\ x_5 \end{bmatrix}$$

The noise transformation (8) by no means affects strong observability of Σ_{2so}^{**}. Therefore, we have the following:

<u>Corollary 1:</u> Σ_{2so}^{***}, considered as a system with input $\eta_2(\cdot)$ and output $y_2(\cdot)$, is strongly observable.

Now, it is clear that Σ is the <u>true cascade</u> of the <u>strongly observable</u> subsystem Σ_{2so}^{***} and the properly invertible subsystem Σ_{2pi}^{**}.

The next difficulty is that, although Σ_{2pi}^{**} is a <u>nonsingular</u> Kalman-Bucy filtering problem with state <u>detectable</u> from the observations, it is not guaranteed that the optimal filter is asymptotically stable, i.e., does not have poles on the $j\omega$-axis. Molinari (1977) correctly pointed out that optimal closed-loop poles on the imaginary axis can occur, even in a <u>nonsingular</u> regulator problem. However, the following is true.

<u>Theorem 4</u> (<u>Optimal Filtering with Stability Constraint</u>): Let Σ be detectable. Then the optimal Kalman-Bucy filter of Σ_{2pi}^{**} is asymptotically stable if and only if the matrix A_{44} has no imaginary axis eigenvalues.

<u>Proof:</u> See Kitapci (1985).

The last technical difficulty is that the "known" inputs to Σ_{2pi}^{**}, i.e., x_1, x_2, and x_5, are reconstructed by the strong observer and are hence contaminated by an arbitrarily fastly decaying noise; see Section II.4. However, if the condition of Theorem 4 for asymptotic stability of the Kalman-Bucy filter is verified, then the response of the Kalman-Bucy filter to this noise is also arbitrarily small; see Section II.4. However, the input $\dot{x}_2$ to the Kalman-Bucy filter is contaminated with the derivative of a rapidly decaying noise, and the effect of this last noise is more difficult to assess. Let $K(t,\tau)$ be the (bounded) kernel of the Kalman-Buch filter of Σ^{**}_{2pi} and let $\Phi(t,\tau)$ be the filter's state transition matrix. The optimal state estimates are given by

$$\begin{bmatrix} \hat{x}_3(t) \\ \hat{x}_4(t) \end{bmatrix} = \Phi(t,o) \begin{bmatrix} \hat{x}_3(o) \\ \hat{x}_4(o) \end{bmatrix} + \int_0^t K(t,\tau)\, \dot{x}_2(\tau)\, d\tau \tag{9}$$

However, $\dot{x}_2$ is corrupted by an arbitrarily fastly decaying noise $e_{\dot{x}_2} = \dot{e}_{x_2}$, and to mitigate the damage created by this noise, we will rather, in practice, use the following "modified" filter:

$$\begin{bmatrix} \tilde{x}_3(t) \\ \tilde{x}_4(t) \end{bmatrix} = \Phi(t,o) \begin{bmatrix} \hat{x}_3(o) \\ \hat{x}_4(o) \end{bmatrix} + \int_\varepsilon^t K(t,\tau)(\dot{x}_2(\tau) + e_{\dot{x}_2}(\tau))\, d\tau \tag{10}$$

It remains to see whether the error

$$\begin{bmatrix} \hat{x}_3(t) - \tilde{x}_3(t) \\ \hat{x}_4(t) - \tilde{x}_4(t) \end{bmatrix} = \int_0^\varepsilon K(t,\tau)\, \dot{x}_2(\tau)\, d\tau + \int_\varepsilon^t K(t,\tau) e_{\dot{x}_2}(\tau)\, d\tau$$

is "small." This is made precise in the following lemma:

Lemma 4: Let the conditions for stability of the Kalman-Bucy filter of Σ^{**}_{2pi} be verified. Let $\Gamma(t,\tau)$ be the covariance kernel of $\int_0^\varepsilon K(t,\tau)\dot{x}_2(\tau)\, d\tau$. Then $\int_0^\infty \int_0^\infty |\Gamma_{ij}(t,\tau)|\, dt\, d\tau \to 0$ as $\varepsilon \to 0$.

Proof: See the full paper.

Lemma 5: Let the conditions for asymptotic stability of the Kalman-Bucy filter of Σ^{**}_{2pi} be verified. Let $\Gamma(t,\tau)$ be the covariance kernel of $\int_\varepsilon^t K(t,\tau)e_{\dot{x}_2}(\tau)\, d\tau$, where e_{x_2} is the error of the strong observer with gain K_2 used to reconstruct x_2. Then, $\forall\, \varepsilon' > 0$, there exist an $\varepsilon > 0$, small enough, and a strong observer gain $K_2(\varepsilon,\varepsilon')$, stiff enough, such that $\int_0^\infty \int_0^\infty |\Gamma_{ij}(t,\tau)|\, dt\, d\tau < \varepsilon'$.

Proof: See the full paper.

Intuitively, the modified filter does not take the observation at $\tau = 0$, so as to avoid the transients due to the differentiation of the very quickly decaying error of the strong observer. Further, the output of the modified filter can be made arbitrarily close to that of the optimal filter, so that we will rather use the modified filter in the final implementation.

From the above, the full structure of the singular filter can be completely elucidated, a result which we formulate as the following:

<u>Theorem 5</u> (<u>Main Result</u>): Consider the filtering problem (5) with (A,C) detectable and assume that A_{44} has no $j\omega$-axis poles.

The system Σ can be represented as the cascade of the strongly observable subsystem Σ_{2so}^{***} and the properly invertible system Σ_{2pi}^{**}. Therefore, the singular filter consists of the cascade of a strong observer for Σ_{2so}^{***}, the construction of which is detailed in Section II.4 and is not repeated here, and the "modified" Kalman-Bucy filter (10) for the properly invertible, detectable subsystem Σ_{2pi}^{**}. This is schematically shown in Figure 2. Further, the modified and optimal Kalman-Bucy filters can be made arbitrarily close by adjusting the small parameter ε and the strong observer gain according to Lemma 5.

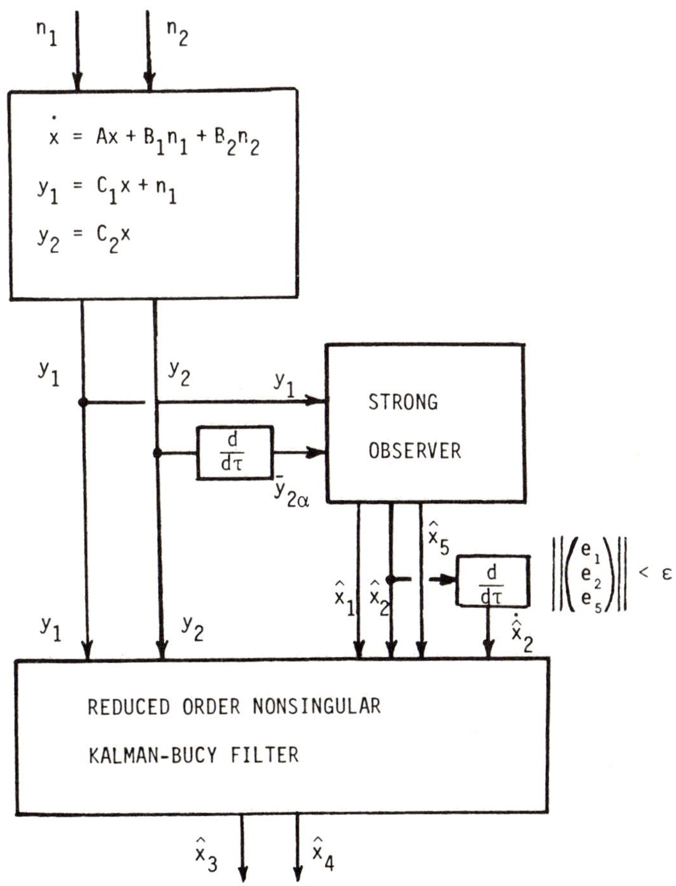

Figure 2

IV. CONCLUSION

It is believed that the present paper has provided a fairly exhaustive system theory analysis of the linear-quadratic-gaussian singular filtering problem. The deeper probabilistic interpretations have not been addressed here and are postposed to a further paper.

REFERENCES

[1] Basile, G. and Marro, G., Controlled and conditioned invariant subspaces in linear systems theory, J. Opt. Th. Appl. 3 (1969), 306-315.

[2] Basile, G. and Marro, G., On the observability of linear, time-invariant systems with unknown inputs, J. Opt. Th. Appl. 3 (1969), 410-415.

[3] Hautus, M.L.J., Strong detectability and observers, Linear Algebra and Applications 50 (1983).

[4] Hautus, M.L.J. and Silverman, L.M., System structure and singular control, Linear Algebra and Applications 50 (1983), 369-402.

[5] Jonckheere, E., On the existence of a negative semidefinite, antistabilizing solution to the discrete-time algebraic Riccati equation, IEEE Trans. Automatic Control, AC-26 (1981), 707-712.

[6] Kalman, R. E. and Bucy , R. S., New results in linear filtering and prediction theory, Trans. ASME, Ser. D, J. Basic Engineering 83 (1961), 95-107.

[7] Kitapci, A., System structure and singular problems, Ph.D. dissertation, Department of Electrical Engineering-Systems, University of Southern California, Los Angeles, USA (1985).

[8] Molinari, B. P., The time invariant linear quadratic optimal control problem, Automatica 13 (1977), 347-357.

[9] Rappaport, D. and Silverman, L. M., Structure and stability of discrete time optimal systems, IEEE Trans. Automat. Contr. AC-16 (1971), 227-233.

[10] Schumacker, J. M., A geometric approach to the singular filtering problem, Report OS-R8502, Department of Operations Research and System Theory, Centre for Mathematics and Computer Science, Amsterdam, The Netherlands (March 1985).

[11] Silverman, L. M., Inversion of multivariable linear systems, IEEE Trans. Automatic Control, AC-14 (1969), 270-276.

[12] Silverman, L. M., Discrete Riccati equations: alternative algorithms, asymptotic properties, and system theory interpretations, in: C. T. Leondes (ed.), Control and Dynamic Systems 12 (Academic, New York, 1976).

[13] Wiener, N., Extrapolation, interpolation and smoothing of stationary time series (The MIT Press, Cambridge, Massachusetts, 1949).

[14] Willems, J. C., Kitapci, A. and Silverman, L. M., Singular optimal control, SIAM Journal of Control and Optimization (1985).

[15] Wonham, W. M., Linear multivariable control: a geometric approach (Springer, New York, 1979).

Frequency Domain and State Space Methods for Linear Systems
C.I. Byrnes and A. Lindquist (Editors)

OBSERVATION CONTROL IN AUTOMATIC SYSTEMS

N.A. Kuznetsov

Institute for Control Sciences
Moscow, USSR

The aim of this paper is to present a systematic description of recent results in an application of the theory of estimation and control under incomplete observation to the solution of various problems of automatic control for systems with controlled observations. Control of observations implies that the measuring devices are appropriately chosen and time distribution of the observation intervals and their duration as well as the priorities in processing the input information are established. The problems of chosing the optimal control by switching the observation channels have also been investigated and a method for their solution has been developed. Besides, joint optimization problems of control for a dynamic system and the process of observation are under a detailed consideration.

INTRODUCTION

Progress of digital computing technology has led to extensive use of computers in control. Hence the possibility of utilizing more complicated mathematical models of controlled systems and observations, which both may be more relevant to actual processes, appears in the synthesis of an optimal control law. On the one hand the quality of the control system functioning is improved, but on the other hand, the cost of data processing greatly increases. Thus the optimality of the estimates obtained at the output of the corresponding filters should be considered with due regard for the cost of rudder drives as well.

The use of digital devices in control loops of a continuous-time system requires time quantization. Even at the stage of designing a control system the time intervals should be determined both between measurements of the system output coordinates and between the moments of generating control signals.

Depending on the choice of the sampling technique it is possible to consider the following problems of synthesis of optimal control.

THE FIRST PROBLEM

The first problem is the classical one. Let us assume that the state of the system to be controlled is described by the vector linear stochastic equation:

$$d\Theta_t = c(t)u(t)dt + a(t)\Theta_t dt + b(t)dW_t$$

where for each $t \in [0,T]$ Θ_t denotes the K -dimentional state vector; $u(t)$ is the control vector; W_t is a standard Wiener process; $c(t)$, $a(t)$ and $b(t)$ are matrices with dimensions $K \times \tau$, $K \times K$ and $K \times K$, respectively whose elements are deterministic bounded time functions. The random vector $\xi(t_j)$ presents the 1-dimentional process of observation:

$$\xi(t_j) = A(t_j)\Theta_{t_j} + B(t_j)\varepsilon_{t_j} ,$$

where ε_{t_j} denotes the Gaussian sequence of vectors with ℓ independent components such as $E\{\varepsilon_m(t)\} = 0$, $E\{\varepsilon_m^2(t)\} = 1$, $(m = 1, 2, \ldots \ell)$ and $E\{\cdot\}$ denotes an expectation operator. Matrices $A(t)$ and $B(t)$ are of $\ell \times K$ and $\ell \times \ell$ dimentions. Under the term "observation" we consider the system's outputs which are measured and immediately entered into the computer at a definite time t_j . If the measuring device works continuously, "observation" means only discrete enter of the appropriate current value in the computer. The initial value Θ_0 of vector Θ_t is a Gaussian vector with given moments m_0 and γ_0 which are independent of the Wiener process W_t and the sequence of ε_t .

It is essential that in this problem the distribution of t_j on [0, T] is supposed to be chosen in advance. Vector $u(t)$ is considered to be a step-function, changing at times t_j . It is chosen on the basis of entering discrete observations in order to minimize the following mean square performance index (loss functional):

$$\min_u J_1 = \min_u E\{\Theta_T^* H \Theta_T + \int_0^T (\Theta_t q(t)\Theta_t + u^*(t)r(t)u(t))dt\}$$

where $q(t)$, $r(t)$ and H are deterministic constrained symmetrical nonnegative matrices.

This problem is known to be solved in the following way. First of all, the proper discrete process is constructed according to the continuous time process Θ_t and the quadratic functional is replaced by its digital analog. The resulting discrete optimal control problem can be solved in a usual way. The structure of the designed system is represented by Figure 1.

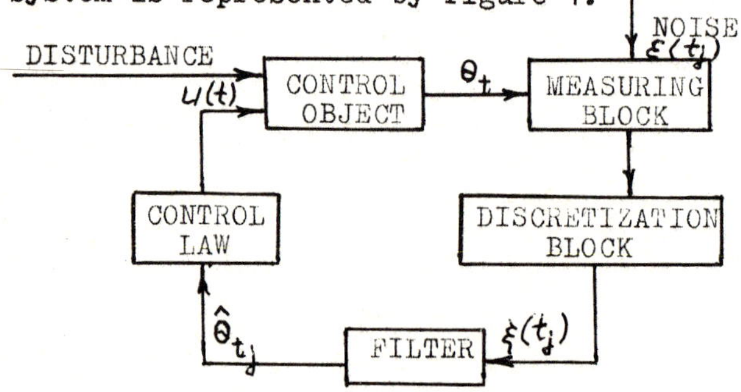

Figure 1

THE SECOND PROBLEM

This problem differs from the first one in that the observation process may be controlled. So at any time the state vector of the system can be observed either through all observation channels or through only some of them (if any). Such new statement of the problem is possible only if the expenditure on carrying out an individual observation is comparable with the payoff caused by the associated increase of accuracy of estimates in terms of the profit obtained in the overall value of the performance criteria.

A mathematical model describing the observation process differs from the above classical one:

$$\xi(t_{j+1}) = A(\alpha, t_j)\Theta_{t_j} + B(t_j)\varepsilon_{t_{j+1}}, \qquad \xi(t_0) = 0$$

where

$$A(\alpha, t_j) = \begin{pmatrix} \alpha_1(t_j)\, A_1(t_j) \\ \alpha_2(t_j)\, A_2(t_j) \\ \cdots \\ \alpha_\ell(t_j)\, A_\ell(t_j) \end{pmatrix}$$

Matrix A depends here on the so-called observation control vector $\alpha(t_j) = [\alpha_1(t_j), \alpha_2(t_j), \ldots, \alpha_\ell(t_j)]^*$, which corresponds to ℓ scalar observation trunks. The components $\alpha_i(t_j)$, $i = \overline{1,\ell}$, $j = \overline{0,N-1}$ of the observation control vector are constrained:

$$\alpha_i(t_j) = \{0, \text{ or } 1\}$$

The condition $\alpha_i(t_j) = 1$ means that the i-th output coordinate is observed through the i-th scalar observation trunk at time t_j.

The quality of the system is characterized by the value of its performance functional:

$$\min_{u,\alpha} \tilde{J}_2 = \min_{u,\alpha} [\tilde{J}_1 + \sum_{j=0}^{N-1} F(t_j)\alpha^*(t_j)]$$

which is practically the same as in the first problem but an additional term is included describing the cost of the observations and their processing.

It is required to choose the control vector $u(t)$ and the observation control vector $\alpha(t)$ in order to minimize the losses. As in the first case, the solution of this problem starts with constructing a discrete process in compliance with the process Θ_t and the appropriate sampling of the loss functional. But here an additional problem arises, that of finding the optimal observation control vector $\alpha(t_j)$ at each $t_j \in [0,T]$.

It turns out that this additional problem may be separated from the main problem of finding $u(t)$ and, according to [1] the optimal observation control may be computed in advance. After solving the additional problem the main problem can be solved in a usual way.

The structure of the designed system is represented by Figure 2.

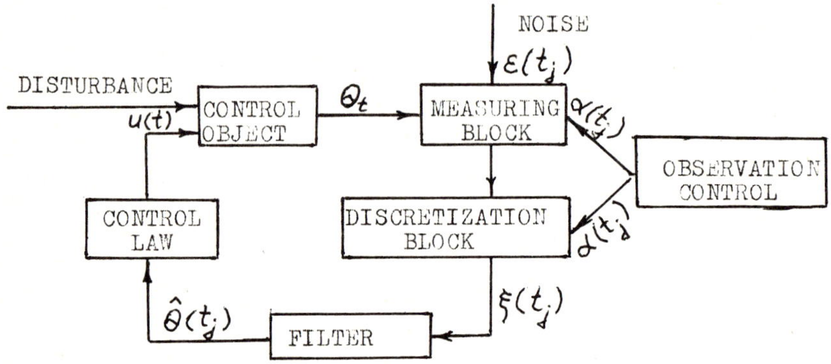

Figure 2

THE THIRD PROBLEM

The third problem differs from the second one because it contains another kind of control, the control of times t_j at which control signals $u(t)$ can be changed. In this case at any time t_j either all the components of the system control vector or only some of them can be changed, the rest of the components remaining the same. Moreover, it is possible that at any t_j all components of $u(t)$ remain unchanged. The need for the optimal choice of times t_j is dictated by the following reasons. In order to improve the performance of the system the control actions should be computed as frequently as possible, but this increases expenditure on computing these controls and more importantly, their implementation (for example, too frequent switch of the actuators to change the angular position of the vehicle rudders). To avoid such technically unrealizable solution it is necessary to include in the loss functional the cost of computing, generating and implementing the control actions.

So the complete form of mathematical model of the system is:

$$d\theta_t = [c(t)\beta(t)u(t) + C(t)(I - \beta(t)\,\eta(t)]dt + a(t)\theta_t dt + b(t)dW_t \,,$$

where all the components of vectors $u(t)$, $\eta(t)$ and matrix $\beta(t)$ are piecewise-constant time functions so that $u(t) = u(t_j)$, $\eta(t) = \eta(t_j)$, $\beta(t) = \beta(t_j)$ for any $t \in [t_j, t_{j+1})$ and

$$\eta(t_{j+1}) = \beta(t_j)u(t_j) + (I - \beta(t_j))\,\eta(t_j), \quad \eta(t_0) = 0.$$

The diagonal matrix $\beta(t)$ denotes the control of times of changing the system state control actions. The possibility to control these time instants is realized in the mathematical model by means of expanding the state vector so that it includes vector $\eta(t)$. Thus, the total state vector in the third problem consists of the couple $\{\Theta_t, \eta(t)\}$.

The elements $\beta_{ii}(t)$ of the control matrix $\beta(t)$ are restricted:

$$\beta_{ii}(t) = \{0, \text{ or } 1\}, \qquad t \in [t_j, t_{j+1})$$

Under this assumption, each coordinate of vector $\eta(t)$ satisfies the requirement:

$$\eta_i(t_{j+1}) = \begin{cases} u_i(t_j) & \text{if } \beta_{ii}(t_j) = 1 \\ \eta_i(t_j) & \text{if } \beta_{ii}(t_j) = 0 \end{cases}$$

It follows from the initial condition $\eta(t_0)$ and the above requirement that $\eta_i(t_K) = 0$, $K = \overline{0, j+1}$ as long as $\beta_{ii}(t_\ell) = 0$, $\ell = \overline{0, j}$. If $\eta_i(t_{j+1}) = u_i(t_K)$ then t_K is the last time instant which precedes t_{j+1} when $\beta_{ii}(t_K) = 1$. Thus the new vector of system state control is $[\beta(t)u(t) + (I - \beta(t))\eta(t)]$ (I denotes the identity matrix) instead of $u(t)$ in the above two problems.

With $\beta_{ii}(t) = 1$, $t \in [t_j, t_{j+1})$ the value of the i-th component of the system control action equals $u_i(t_j)$, while with $\beta_{ii}(t) = 0$, $t \in [t_j, t_{j+1})$ it is equal to the value of the same component of the control action affecting the system during the preceding half-interval $[t_{j-1}, t_j)$. The problem is to choose the control of the system $u(t^j)$, the control of times of changing the system control actions $\beta(t)$ and the observation control $\alpha(t)$, using the set of observations $\xi(t_j)$ in order to minimize the loss functional:

$$\min_{u,\alpha,\beta} \tilde{J}_3 = \min_{u,\alpha,\beta} [\tilde{J}_2 + \sum_{j=0}^{N-1} sp(G(t_j)\beta(t_j))]$$

As it was proved in [2] the given problem of joint optimization can be reduced to the solution of the binary optimization problem of finding optimal $\alpha(t)$ and $\beta(t)$ and then solving the problem for $u(t)$. The solution of the binary problem defines the optinal times at which the optimal system state control signals are to be changed.

The optimal state control of the system can then be evaluation in a usual way. The structure of the designed system is represented by Figure 3.

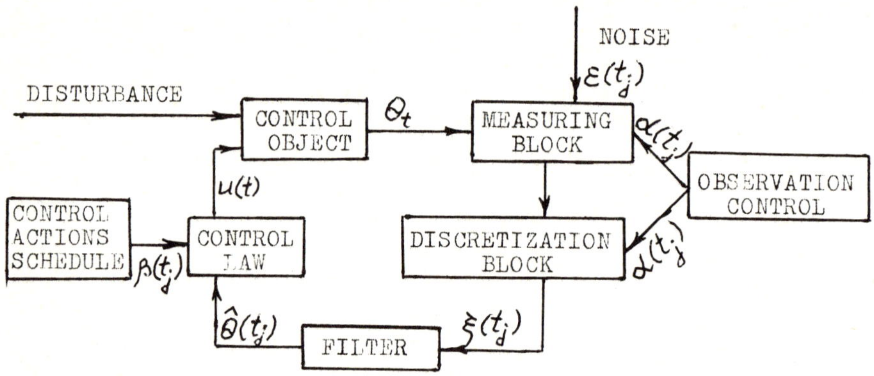

Figure 3

These results are contained in the forthcoming book with the same title written by the author and his coworkers F.N. Grigor'ev and A.P. Serebrovskiy. The book is intended for scientific and technical audience engaged in designing the control in stochastic dynamic systems. It may also be used by university teachers and postgraduates specializing in related professional fields.

REFERENCES:

1 Butov, A.A., Kuznetsov, N.A., et all, Generalized observations control in problems of stochastical optimization, Preprints 8-th Triennial World Congress IFAC, (Kyoto, Japan, August 24-28, 1981), Vol.VI, p.185-190.

2 Kuznetsov, N., Grigor'ev, F., Serebrovskiy, A., On the synthesis of the optimal discrete control system for continuous-time object, Preprints 9-th World Congress IFAC, (Budapest, Hungary, July 2-6, 1984), Vol.IX, p.136-141.

Frequency Domain and State Space Methods for Linear Systems
C.I. Byrnes and A. Lindquist (Editors)

A LJAPUNOV EQUATION AND SUBOPTIMAL STRATEGIES FOR STOCHASTIC JUMP PROCESSES

M. MARITON , P. BERTRAND
Laboratoire des Signaux et Systèmes
C.N.R.S. - E.S.E.
Plateau du Moulon
91190 GIF-SUR-YVETTE - France

INTRODUCTION

Control for a class of stochastic linear systems, subject to sudden jumps in parameter values, is considered. When the performance index is quadratic the optimal solution is known, and can be computed in a closed-loop form. However it requires a feedback that switches whenever a jump occurs. Since, in practice, such a jump might be difficult to detect, the use of a non-switching feedback must be considered, but some measure of the inevitable loss of optimality should be given. This communication shows first how for a suboptimal strategy one may compute the criterion value in terms of the solutions of a set of coupled Ljapunov equations, and from then how one may compare different non-switching strategies.

1°) A class of linear systems with jump parameters

Control of systems with random parameters described as Markovian jump processes was first considered by Krasovskii and al. [1] in 1961. Here the presentation will be restricted to the class of linear systems, the description of which is now given.

The system state is defined by $(x^T(t) \ , \ r(t))^T$ where $x(t) \in R^n$ is the usual plant state, and $r(t)$, the plant mode, takes discrete values in a finite set $S = \{ 1, 2, \ldots N \}$.

The dynamical evolutions are respectively given by

$$\text{plant state : } x(t) = A(r(t))x(t) + B(r(t))\, u(t) \tag{1}$$

with $u(t)$ a control input of dimension m and $A(r(t))$, $B(r(t))$ matrices of suitable dimensions

$$\text{plant mode : } \text{Prob}\,\{r(t+\Delta) = j \mid r(t) = i\} = \begin{cases} \pi_{ij}\,\Delta + o(\Delta) & i \neq j \\ 1 + \pi_{ii}\Delta + o(\Delta) & i = j \end{cases} \tag{2}$$

the jump transition density matrix being noted as $\Pi = (\pi_{ij})_{\substack{i = 1,N \\ j = 1,N}}$

when $rt() = i$, the pair $[A(r(t))\ ,\ B(r(t))]$ will take the form $[A_i\ ,\ B_i]$. An initial plant mode distribution P achieves the system description :

$$\text{Prob}\,\{[A(t{=}o)\ ,\ B(t{=}o)] = [\,A_i\ ,\ B_i]\} = p_i \qquad i = 1,N$$

After the initial work of Krasovskii and al. [1], the study of such dynamic systems was pursued by Wonham and Sworder [12], [5], and was focused on the optimal control problem for a quadratic performance index. In a series of papers, Sworder gave more complete results to include uncertainty [3] and state dependance in the densities π_{ij}[7], or plant state discontinuities [6].

These efforts were fruitfull enough to result in a real scale implementation on a solar thermal process [9].

More recently these models were proposed by Willsky [10] to represent large scale power systems. Here the above model (1), (2), is specialized to describe the various modes of operation in an energy production network. The plant mode $r(t)$ follows the connection/disconnection of units, tie-lines ruptures... The name "hybrid model" was proposed to recall the two-fold system state ($x \in R^n$, $r \in S$).

In the context of redundant multiplex control, Siljak [4] also used a hybrid model to represent the plant dynamics as controllers fail or recover.

2°) Quadratic optimal control

The optimal control of the system (1), (2), was studied independantly by Sworder [5] from a maximum principle point of view, and by Wonham [11] from a dynamic programming point of view.

This section recalls their main results.

The optimality is defined with respect to the expected value of a quadratic index

$$J(u, t_o, x(t_o), r(t_o)) = E\{\frac{1}{2}\int_{t_o}^{T}(x(z)^TQx(z) + u(z)^TRu(z))dz|t_o,x(t_o),r(t_o)\} \quad (3)$$

where the weighting matrices $Q > 0$ and $R > 0$ could be mode dependant ($Q = Q(r(t))$, $R = R(r(t))$).

The optimal control can be expressed in a closed-loop form by

$$u^*(x(t), r(t)) = -R^{-1} B_i^T K_i(t)\, x(t) \quad \text{for} \quad r(t) = i \quad (4)$$

where the K_i, $i = 1,N$ matrices are the positive solutions of a set of coupled Riccati equations

$$\begin{cases} -\dot{K}_i = A_i^TK_i + K_iA_i - K_iB_iR^{-1}B_i^TK_i + \sum_{j=1}^{N} \pi_{ij}K_j + Q & i = 1,N \\ \text{with } K_i(T) = 0 \end{cases} \quad (5)$$

It is important to note that a full state dependant feedback is obtained here

$$u^* = U(x,r) \quad (6)$$

While, for a purely deterministic system $[A(t), B(t)] = [A_i, B_i]\ \forall t$ the corresponding Riccati equation would be

$$-\dot{K}_i = A_i^TK_i + K_iA_i - K_iB_iR^{-1}B_i^TK_i + Q \quad (7)$$

The only difference between (7) and (5) comes from the term $\sum_{j=1}^{N} \pi_{ij} K_j$ that couples the equations of the various modes.

Several methods are available to solve (5). As noted by Sworder the problem is far easier if the transition matrix Π is lower or upper triangular since the system can then be solved in sequence starting with the independant equation ($i = 1$ or $i = N$). When not in that case, Wonham proposed a quasilinearization technique. The principle of homotopy can also be employed and an algorithm derived to compute the solution [2].

3°) The suboptimal criterion equation

As noted above, the mode must be measured or reconstructed in order to implement the optimal control (4). This might not be possible or the mode observer might be to costly to implement. One consequently has to consider the performance of the system under a non-switching feeback strategy

$$u = U(x) \tag{8}$$

The control (8) could be viewed as a partial feedback (x) of the full process state (x, r).

Among such strategies, linear ones are of particular interest, since for

$$u = \Gamma x \tag{9}$$

the suboptimal criterion can be analytically computed. The expression of this suboptimal criterion is the main result of this communication and is given by the following theorem.

Theorem

The performance (3) of the system (1), (2) operated under the control (9) is given by

$$J = \frac{1}{2} x(t_o)^T E\{L(t_o, r(t_o))\} x(t_o)$$

where the L matrices are the solutions of the set of coupled Ljapunov equation

$$\begin{cases} -\dot{L}_i = \tilde{A}_i^T L_i + L_i \tilde{A}_i + \Gamma^T R \Gamma + \sum_{j=1}^{N} \pi_{ij} L_j + Q \\ L_i(T) = 0 \end{cases} \qquad i = 1,N \tag{10}$$

where $\tilde{A}_i = A_i + B_i \Gamma$ and $L_i(t) = L(t, r(t))$ when $r(t) = i$

The proof which is straightforward and requires the evaluation of the conditionnal expectation (3), is presented along the lines of [5].

Proof :

For a final time T, the cost-function is written as

$$J = \frac{1}{2} x(t_o)^T E\{L(t_o, r(t_o))\} x(t_o) \tag{11}$$

where

$$L(t) = \int_{t_o}^{t} x^T (Q + \Gamma^T R \Gamma) x \, dz \tag{12}$$

The integral in (12) can be differentiated in two ways. First one has

$$E\{\dot{L}(t) | r(t) = i\} = \lim_{\Delta \to 0} \frac{E\{L(t+\Delta) | r(t)=i\} - E\{L(t) | r(t)=i\}}{\Delta} \tag{13}$$

For the system described by (1), (2) one can further write

$$E\{L(t+\Delta) | r(t)=i\} = L_i(t+\Delta) + \Delta \sum_{j=1}^{N} \pi_{ij} L_j(t) + o(\Delta) \tag{14}$$

so that (13) becomes

$$E\,\{\dot{L}(t)|r(t) = i\} = \dot{L}_i(t) + \sum_{j=1}^{N} \pi_{ij}L_j \qquad (15)$$

Second, (12) is differentiated with respect to its upper-bound

$$\dot{L} = -\tilde{A}L - L\tilde{A}^T - Q - \Gamma^T R \Gamma \qquad (16)$$

Equation (16) averaged for $r(t) = i$ yields

$$E\,\{\dot{L}(t)|r(t) = i\} = -\tilde{A}_i{}^T L_i - L_i\tilde{A}_i - \Gamma^T R\Gamma - Q \qquad (17)$$

Equalling (17) and (15) then gives (10) and completes the proof.

4°) Application : comparison of non-switching strategies

Once the designer has decided to give up optimality and to implement a non-switching feedback, an adequate choice of the gain matrix of (8) is required. The above theorem provides a mean to compare different choices.

To be more precise consider the following illustrative example of a scalar plant with two modes

mode 1 $\dot{x} = -x + u$

mode 2 $\dot{x} = u$

Two cases are studied for the transition matrix

case 1 $\Pi = \begin{pmatrix} 0 & 0 \\ 1 & -1 \end{pmatrix}$ (mode 1 is a trapping mode)

case 2 $\Pi = \begin{pmatrix} -1 & 1 \\ 1 & -1 \end{pmatrix}$ (symetric unstopped transitions)

The initial mode is distributed under Prob $\{r(t_o) = 1\} = p$

With the previous notations and $x(t_o) = 2$, the suboptimal criterion is expressed by

$$J = p l_1 + (1-p)\, l_2 \qquad (18)$$

where l_1 and l_2 depends on the non-switching gain γ as indicated by (10)

The following reasonnable strategies were considered :

Switching optimal strategy S* $u = k_1^* x$ if $r = 1$

$= k_2^*$ if $r = 2$

with $k^*_{i=1,2}$ solution of (5)

Non-switching strategies

NS1* $u = k_1^* x \quad \forall r$

NS2* $u = k_2^* x \quad \forall r$

NS1 $u = k_1 x \quad \forall r$

with k_1 solution of (7) for $i = 1$

NS2 $u = k_2 x \quad \forall r$

with k_2 solution of (7) for $i = 2$

For the two cases each strategy, performances are compared by the degree of suboptimality μ defined as

$$\mu = \frac{J - J^*}{J^*} \tag{19}$$

were J is the suboptimal criterion, and J^* the optimal on :

$$J^* = \frac{1}{2} x (t_0)^T E \{K (t_0, r(t_0)\} x (t_0) \tag{20}$$

which reduces in this example to

$$J^* = pk_1^* + (1-p)k_2^*$$

Ineach case table 1 gives the regulator gains and table 2 the corresponding suboptimality degrees

GAINS	NS1*	NS2*	NS1	NS2
Case 1	0.414	0.790	0.414	1.000
Case 2	0.520	0.830	0.414	1.000

Table 1 Regulator Gains

p / Strategy	1.	0.75	0.5	0.25	0.	1.	0.75	0.5	0.25	0.
NS1*	0.	3.8	6.4	8.3	9.8	2.	12.	20.	26.	31.
NS2*	9.6	6.6	4.6	3.1	1.9	4.	42.	71.	94.	112.
NS1	0.	3.8	6.4	8.3	9.8	7.	13.	18.	22.	24.
NS2	20.8	14.8	10.7	7.8	5.5	10.	80.	133.	176.	210.

Table 2 Suboptimality Degrees

Table 2 shows that for every p the best among the non-switching strategies that were considered performs very close to the optimal switching strategy. The next logical step is then to obtain the optimal non-switching strategy. This result will appear as [13].

CONCLUSION

The distribution of this communication is to show how to compute the suboptimal quadratic performance resulting from a non-switching linear feedback, applied to a linear dynamic system subject to sudden changes in parameter values.

This leads to a matrix Ljapunov equation of stochastic control that is akin to the Riccati one as in the deterministic case. An application is given to show its usefulness when comparing different non-switching strategies. As cautionned by Sworder [8], consideration of the average value of the stochastic index is not sufficient and work is underway to characterize higher order statistics.

REFERENCES

[1] KRASOVSKII, N.N., and LIDSKII, E.A., Analytical design of controllers in systems with random attributes, I-III, Aut. Remote Control n° 22, 1961, pp 1021-1025, pp 1141-1146, pp 1289-1294.

[2] MARITON, M., and BERTRAND, P., A homotopy algorithm for solving coupled Riccati equations, to appear Optimal Control Application & Methods, 1985.

[3] PIERCE, B.D., and SWORDER, D.D., Bayes and Minimax controllers for a linear system with stochastic jump parameters, IEEE Trans. AC-16, n° 4, Aug. 1971, pp 300-307.

[4] SILJAK, D.D., Reliable control using multiple control systems, Int. J. Control, vol. 31, n° 2, 1090, pp 303-329.

[5] SWORDER, D.D., Feedback control of a class of linear systems with jump parameters, IEEE Trans. AC-14, n° 1, Feb. 1969, pp 9-14.

[6] SWORDER, D.D., Control of jump parameter systems with discontinuous state trajectories, IEEE Trans. AC-17, n° 5, Oct. 1972, pp 740-741.

[7] SWORDER, D.D., and ROBINSON, V.G., Feedback regulators for jump parameter systems with state and control dependant transition rates, IEEE Trans. AC-18, n° 4, Aug. 73, pp 355-360.

[8] SWORDER, D.D., and CHOI, L.L., Stationnary cost densities of optimally controlled stochastic systems, IEEE Trans. AC-21, n° 4, Aug. 1976, pp 492-499.

[9] SWORDER, D.D., and ROGERS, R.O., An LQ - solution to a control problem associated with a solar thermal central receiver, IEEE Trans. AC-28, n° 10, Oct. 1983, pp 971-978.

[10] WILLSKY, A.S., and LEVY, B.C., Stochastic stability research for complex power systems, Report ET-76-C-01-2295 DOE Contract, LIDS, MIT, 1979.

[11] WONHAM, W.M., On a matrix Riccati equation of stochastic control, SIAM J. Control, vol. 6, n° 4, 1968, pp 681-697.

[12] WONHAM, W.M., Random differential equations in control theory, in Probabilistic methods in applied mathematics, A.T. Bharucha-Reid Ed, Academic Press, New York, 1970.

[13] MARITON, M., and BERTRAND, P., Non-switching control strategies for continuous time Jump linear Quadratic Systems, to appear CDC 1985.

Frequency Domain and State Space Methods for Linear Systems
C.I. Byrnes and A. Lindquist (Editors)

CYCLOSTATIONARY IN THE DIGITAL REGULATION OF CONTINUOUS TIME SYSTEMS

Darrell Williamson and Kus Kadiman
Department of Systems Engineering,
Australian National University,
Canberra, A.C.T. 2600
Australia

The consequences for control and state prediction in the presence of cyclostationary (or periodic) statistics caused by digital control are examined. A minimax quadratic performance criterion which takes into account the periodic statistical effects in both states and disturbance noise is defined and optimized. The resulting state feedback law minimizes the maximum intersample error. In particular a new design (the minimax variance regulator) is given which provides significant improvement over the classical minimum variance regulator. For state prediction it is shown that synchronous measurement and control is not optimal.

1. INTRODUCTION

For the control of continuous systems it is generally more practical and advantageous to use digital control especially if the desired control law is to be adaptive. From an analytical point of view, the simplest solution is to generate the control law as a pulse amplitude modulated signal whose period is constant. Specifically, consider the minimal time invariant plant

$$\dot{x}(t) = Ax(t) + Bu(t) + \omega(t) \tag{1.1a}$$

$$y(t) = Cx(t) + \eta(t)$$

where y(t) is the measured variable with $\omega(t)$, $\eta(t)$ zero mean independent wide sense stationary (WSS) 'white noise' disturbance and measurement processes with respective covariances

$$r_\omega(0) = \Omega \quad ; \quad r_\eta(0) = \Lambda \tag{1.1b}$$

The control signal u(t) is given by

$$u(t) = \sum_{k=-\infty}^{\infty} u(k)\, p(t-kT) \; ; \quad p(t) = \begin{cases} 1 & ; t\varepsilon[0,T) \\ 0 & \text{otherwise} \end{cases} \tag{1.2}$$

The sequence of amplitude levels $\{u_k\}$ is generated by a digital computer and the signal u(t) is constructed by a digital/analog converter and holding circuit.

At the controlling instants kT, the system (1.1) has an equivalent linear time invariant description in which the corresponding discrete disturbance and measurement processes are WSS. The usual approach is to base the design of the discrete control sequence on this system description. Often (as in the case of linear regulation) the resulting control sequence is (asymtotically) WSS. However the signal u(t) in (1.2) will then exhibit periodic statistical properties, and as a consequence so will the system state x(t) and output y(t). This special class of nonstationary processes is termed cyclostationary [1] or periodically stationary [2,3]. Cyclostationary processes have long been an interest and concern in communications theory but the consequences for digital control and estimation have received little attention.

In section 2, we define a cyclostationary process and consider some system properties. In particular we show that in system modelling, the disturbance noise $\omega(t)$ should normally be modelled as a cyclostationary process if digital control is to be implemented. In section 3, we consider the representation of cyclostationary processes and develop explicit results which characterize two classes of first and second order processes. In section 4, we define the minimax regulator for discrete control of continuous systems. As a particular example we consider minimax output variance regulation (MMVR) which is the appropriate generalization of the classical minimum variance regulation (MVR) problem [7]. We demonstrate that the ill conditioning often associated with MVR is due in part to the cyclostationary nature of the output. In particular we provide a useful suboptimal solution to the MMVR problem which offers significant improvement over MVR. The consequences of cyclostationary disturbance noise in prediction are explored in section 5. We show that optimal prediction is not generally achieved by synchronous control and output sampling.

2. CYCLOSTATIONARY PROCESSES

Definition 2.1 A random process $\{x(t)\}$ is wide sense cyclostationary (WSCS) with period T if the mean function $\overline{x(t)}$ is periodic, and the autocorrelation function $r_x(s,t)$ is jointly periodic of period T; that is

$$\overline{x(t)} = \overline{x(t+T)} \quad \forall t \qquad (2.1)$$
$$r_x(s,t) = r_x(s+T,t+T) \quad \forall s,t$$

More generally, a cyclostationary process is one in which the probability density function (pdf) f_x and the joint pdf f_{xy} satisfy the conditions

$$f_{x(t)}(z) = f_{x(t+T)}(z) \quad \forall t,z \qquad (2.2)$$
$$f_{x(t)x(s)}(z_1,z_2) = f_{x(t+T)x(s+T)}(z_1,z_2) \quad \forall s,t,z_1,z_2$$

Since the performance measures to be considered only involve second order statistics, WSCS is sufficient for our purposes. Simple manipulations reveal the following two results.

Lemma 2.1
Assume the discrete process $\{u(k)\}$ is WSS with correlation sequence $\{r_u(m)\}$.Then the signal $u(t)$ in (1.2) is WSCS of period T with correlation function

$$r_u(t+\Delta,t) = \sum_{m=-\infty}^{\infty} r_u(m)\, q(t,\Delta+mT) \ ; \ 0 < \Delta < T \qquad (2.3a)$$

where

$$q(t,\Delta) = \begin{cases} 1 & ; \ kT \leqq t \leqq (k+1)T-\Delta \\ 0 & ; \ \text{otherwise} \end{cases} \qquad (2.3b)$$

Lemma 2.2
Consider the continuous system (1.1) with control law (1.2) where $\{\omega(t)\}$ and $\{n(t)\}$ are zero mean WSS. Then if $\{u_k\}$ is zero mean WSS, the continuous process $\{x(t)\}$ is zero mean WSCS of period T. More generally, if $\{\omega(t)\}$ is WSCS of period P with mT = nP (n,m integer with n/m irreducible), then $\{x(t)\}$ is WSCS of period nm.

Process (or disturbance) noise $\omega(t)$ is often used to represent the uncertainty which exists in developing appropriate system models. Model reduction techniques [13,14] seek low order models in order to simplify the digital control algorithm and in this case, the resulting disturbance noise should be modelled as a WSCS process. In order to see this, suppose A,B,C in (1.1) are partitioned according to

$$A = \begin{bmatrix} A_r & A_{rt} \\ A_{tr} & A_t \end{bmatrix} \qquad B = \begin{bmatrix} B_r \\ B_t \end{bmatrix} \qquad C' = \begin{bmatrix} C'_r \\ C'_t \end{bmatrix}$$

where $x' = (x'_r, x'_t)$. If the states x_r defined the reduced order model then

$$\dot{x}_r(t) = A_r x_r(t) + B_r u(t) + \omega_r(t)$$
$$y_r(t) = C_r x_r(t) + \eta_r(t)$$

where

$$\omega_r(t) = A_{rt} x_t(t) + \omega(t) \quad ; \quad \eta_r(t) = C_t x_t(t) + \eta(t)$$

However by lemma 2.2, $x(t)$ hence $x_r(t)$ and as well as $\omega_r(t)$ and $\eta_r(t)$ are WSCS of period T. The consequences of cyclostationary process noise in optimal prediction are explored in section 5.

Henceforth for simplicity and without any loss of generality we assume the sampling period T=1. The system model from (1.1) and (1.2) is then described for $0<\Delta\leq 1$ by

$$x(k+\Delta) = F_\Delta x(k) + G_\Delta u(k) + \omega_\Delta(k) \tag{2.4a}$$
$$y(k+\Delta) = Cx(k+\Delta) + \eta_\Delta(k)$$

where

$$F_\Delta = e^{A\Delta} \quad ; \quad G_\Delta = \int_0^\Delta e^{A(\Delta-\sigma)} B d\sigma \tag{2.4b}$$

$$\omega_\Delta(k) = \int_0^\Delta e^{A(\Delta-\sigma)} \omega(\sigma+k) d\sigma$$

Using (1.1b),the covariance matrices of the zero mean independent WSS discrete processes $\{\omega_\Delta(k)\}$ and $\{\eta_\Delta(k)\}$ are given by W_Δ and Λ where

$$W_\Delta = \int_0^\Delta e^{A(\Delta-\sigma)} \Omega e^{A'(\Delta-\sigma)} d\sigma \tag{2.5}$$

For each Δ, the process $\{x(k+\Delta\}$ is WSS. The cyclostationary characteristics of $\{x(t)\}$ may therefore be calculated directly from (2.4) and (2.5). That is, define

$$P_\Delta = \xi\{x(k+\Delta)x'(k+\Delta)\} \quad ; \quad U = \xi\{u(k)u'(k)\} \tag{2.6a}$$

then

$$P_\Delta = F_\Delta P_0 F'_\Delta + G_\Delta U G'_\Delta + W_\Delta \tag{2.6b}$$

where P_0 $(=P_1)$ is given by $\Delta = 1$ in (2.6). The problem of discrete control has traditionally been treated by considering only P_0 while ignoring the periodic properties of P_Δ. These characteristics are the statistical equivalent of the deterministic phenomena of intersample ripple [6].

3. REPRESENTATION OF CYCLOSTATIONARY PROCESSES

Definition 3.1 A cyclostationary process $\omega(t)$ with period 1 is said to have a translational series representation [4] of order L if there exists a set of L deterministic basis functions $\{\phi_p : 1\leq p\leq L\}$ and a set of L jointly WSS random sequences $\{\{\alpha_{np}\} : 1\leq p\leq L\}$ such that

$$\xi\{|\omega(t)-\sum_{n=-\infty}^{\infty}\sum_{p=1}^{L}\alpha_{np}\phi_p(t-n)|^2\} = 0 \tag{3.1}$$

In the context of this paper we consider a state space translational representation of $\omega(t)$ order L as defined by

$$\begin{aligned}\dot{z}(t) &= Dz(t) + \sum_{k=-\infty}^{\infty} e(k)p(t-k) \quad ; \quad z(t) \varepsilon R^L \\ \omega(t) &= Hz(t)\end{aligned} \tag{3.2}$$

where D is strictly stable and $\{e(k)\}$ is a zero mean vector white noise process with covariance matrix E. The covariance matrix $\Omega_\Delta = \Omega_{\Delta+1}$ of $\{\omega(t)\}$ for $\Delta\varepsilon[0,1]$ is therefore given by

$$\Omega_\Delta = H\Phi_\Delta H' \quad ; \quad \Phi_\Delta = \Psi_\Delta\Phi_0\Psi_\Delta' + \Gamma \quad ; \quad \Psi_\Delta = e^{D\Delta} \tag{3.3}$$

where

$$\Gamma = \int_0^\Delta e^{D(\Delta-\sigma)}Ee^{D'(\Delta-\sigma)}d\sigma$$

Note that Φ_0 $(=\Phi_1)$ is given by $\Delta = 1$ in (3.3).

Example 3.1

Consider the first order representation

$$\dot{\omega}(t) = -\beta\omega(t) + \beta\sum_k e_k p(t-k) \quad ; \quad \xi(e_k^2) = \frac{1-e^{-\beta}}{(1-e^{-\beta})^2} \tag{3.4}$$

Then $\Omega_0 = \Omega_1 = 1$, $\Omega_\Delta = r_\omega(\Delta,\Delta)$ and for $0\leq t,\ s<1$

$$r_\omega(t,s) = e^{-\beta(t+s)} + \frac{(1-e^{-\beta t})(1-e^{-\beta s})(1-e^{-2\beta})}{(1-e^{-\beta})^2} \tag{3.5}$$

It follows that $\Omega_\Delta < 1 \ \forall\ 0<\Delta<1$ and that a minimum occurs at Δ_1 where

$$e^{-\beta\Delta_1} = \frac{1+e^{-\beta}}{2} \quad ; \quad \Omega_{\Delta_1} = \frac{1+e^{-\beta}}{2} \tag{3.6}$$

Note that $\Delta_1 \to 0$ as $\beta \to \infty$. Plots of Ω_Δ are illustrated in Fig 3.1.

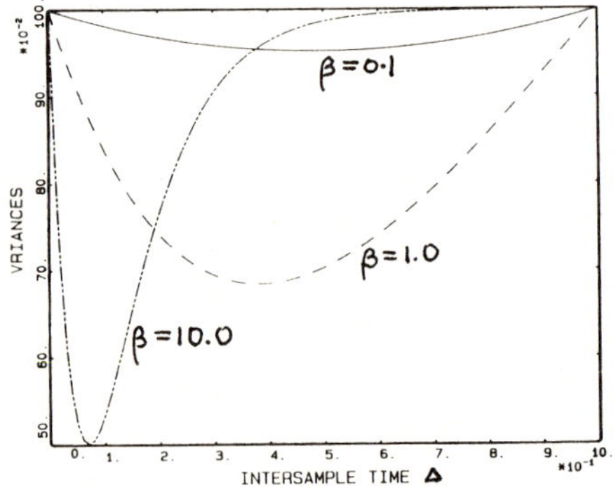

Fig 3.1 The Covariance (Ω_Δ) for 1st Order Systems

Example 3.2

Consider the second order representation

$$\ddot{\omega}(t) + 2\zeta w_n\dot{\omega}(t) + w_n^2\omega(t) = \sum_k e_k p(t-k) \quad ; w_n,\zeta>0 \tag{3.7}$$

where for each pair (ζ,w_n) the covariance $E=E(\zeta,w_n)$ of $\{e_k\}$ is normalised such that Ω_0 $(=\Omega_1)$ = 1. Plots of Ω_Δ are illustrated in Fig 3.2. For high damping ξ, Ω_Δ in Fig 3.2a are similar to the curves in Fig. 3.1 with $\omega_n=2\beta$. It is postulated for an Lth order overdamped system (3.2) that the characteristics of Ω_Δ are described by $\omega_n=L\beta$. This approximation is currently under investigation. For low damping ξ and high natural frequency ω_n, the characteristics of Ω_Δ can be quite complex as evidenced in Fig. 3.2b whereas for low ω_n in Fig. 3.2c the characteristics are more sinusoidal. Further discussion for low ω_n is given below.

The cyclostationary process $\omega(t)$ in (3.2) can also be represented in harmonic series [4,12]. That is,

$$\omega(t) = \sum_{k=-\infty}^{\infty} a_k(t)\exp(j2\pi kt) \qquad \forall\, t\varepsilon(-\infty,\infty)$$

$$a_k(t) = \int_{-\infty}^{\infty} \frac{\sin\pi(t-\tau)}{\pi(t-\tau)}\,\omega(\tau)\exp(-j2\pi k\tau)d\tau \tag{3.8}$$

The autocorrelation function $r_\omega(t,s)$ is given by

$$r_\omega(t,s) = \sum_{n=-\infty}^{\infty}\sum_{m=-\infty}^{\infty} r_{nm}(t-s)\exp(j2\pi(mt-ns)) \tag{3.9}$$

$$r_{nm}(t-s) \mathrel{\hat{=}} \xi\{a_n(t)a_m^*(s)\}$$

Since the components of $\omega(t)$ are individually WSS but not jointly WSS (due to the exponential factors) it follows [4] that the zero mean WSCS process $\{\omega(t)\}$ is WSS if and only if the representors $\{a_k(t)\}$ are uncorrelated. The covariance function $\Omega_\Delta = r_\omega(\Delta,\Delta)$ can be expressed as

$$\Omega_\Delta = \alpha_0 + \sum_{k=1}^{\infty} \alpha_k\cos(2\pi k\Delta+\phi_k) \tag{3.10}$$

A first order approximation is given by $\hat{\Omega}_\Delta = \alpha_0 + \alpha_1\cos(2\pi\Delta+\phi_1)$ where α_0, α_1 and ϕ_1 are calculated by evaluating Ω_Δ in (3.3) for three values of Δ (say Δ = 0,0.3 and 0.7). For the second order representation (3.2), the functions α_0, α_1 and ϕ_1 of damping ζ and natural frequency w_n are illustrated in Fig 3.3. Numerical results have revealed (see Fig 3.2) that $\hat{\Omega}_\Delta$ is a good approximation of Ω_Δ for $w_n < \pi T^{-1}$ and $\zeta < 1$. (sampling period T).

4. MINIMAX QUADRATIC REGULATION

Definition 4.1 Given the continuous plant (1.1), the discrete control minimax quadratic regulator is given by

$$\min_{\{u_k\}} \left\{ \max_{0\leq\Delta\leq 1} J_\Delta \right\} \tag{4.1a}$$

where for $Q \geq 0$ and $R \geq 0$

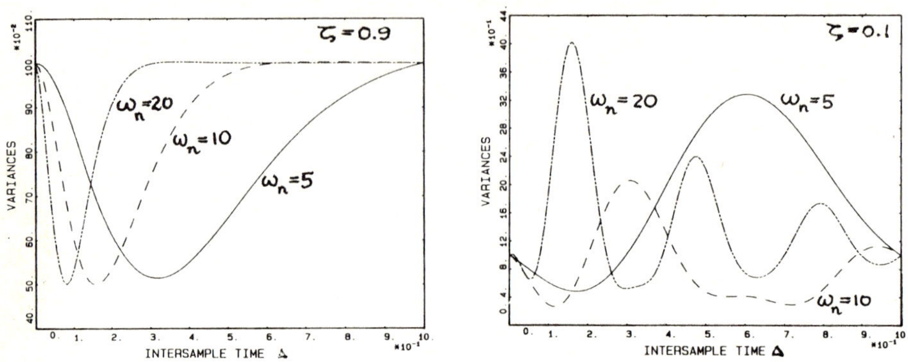

(a) (b)

Fig 3.2 Covariance (Ω_Δ) for 2nd Order System

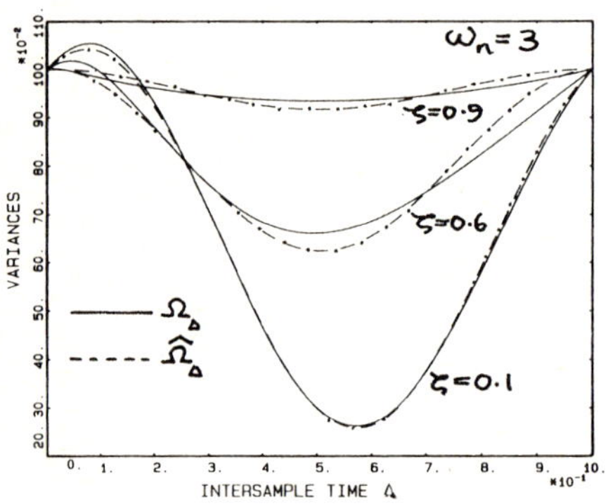

Fig 3.2c

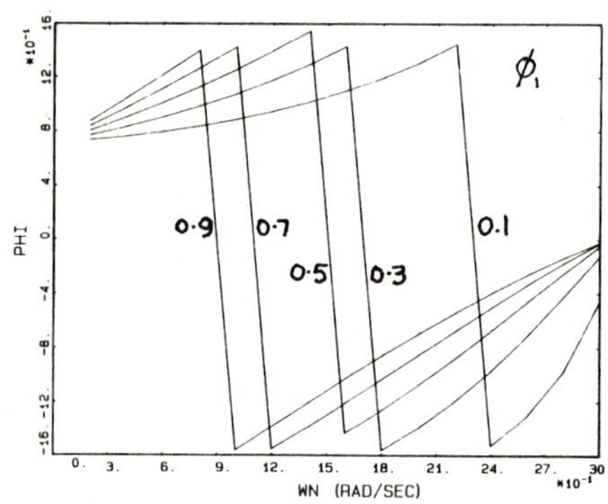

Fig 3.3a

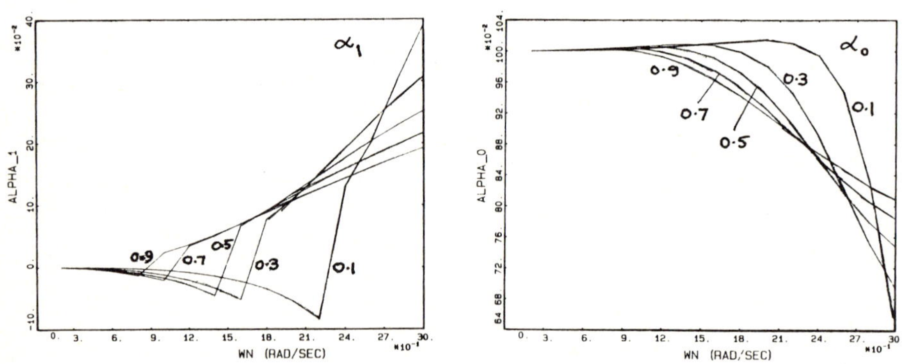

(b) (c)

Fig 3.3 Coefficients of $\hat{\Omega}_\Delta$

$$J_\Delta = \lim_{\substack{n\to\infty \\ m\to\infty}} \frac{1}{m+n} \xi\{ \sum_{k=-n}^{m} x(k+\Delta)'Qx(k+\Delta) + u(k)'Ru(k)\} \tag{4.1b}$$

From (2.4) and (2.5) after completing squares we have the following result.

Lemma 4.1
The discrete equation (2.4) for $\Delta=1$ can be rewritten as

$$x_{k+1} = \tilde{F}_\Delta x_k + G_1\tilde{u}_k + \omega_k \; ; \; \tilde{u}_k = u_k + R_\Delta^{-1}S_\Delta x_k \tag{4.2a}$$

where

$$\tilde{F}_\Delta = F_1 - G_1R_\Delta^{-1}S_\Delta \; ; \; R_\Delta = R + G_\Delta'QG_\Delta \; ; \; S_\Delta = G_\Delta'QF_\Delta \tag{4.2b}$$

Furthermore

$$J_\Delta = \xi\{x_k'Q_\Delta x_k + \tilde{u}_k'R_\Delta\tilde{u}_k\} + tr[QW_\Delta] \tag{4.3a}$$

where tr denotes trace, and

$$Q_\Delta = \tilde{F}_\Delta'Q\tilde{F}_\Delta - S_\Delta'R_\Delta^{-1}S_\Delta \tag{4.3b}$$

For each Δ, J_Δ can be minimized by a linear feedback regulator

$$\tilde{u}(k) = K_\Delta x(k) \; ; \; K_\Delta = (R_\Delta + G_1'P_\Delta G_1)^{-1}G_1'P_\Delta\tilde{F}_\Delta \tag{4.4}$$

where P_Δ (≥ 0) satisfies the discrete algebraic Riccati equation (ARE)

$$P_\Delta = \tilde{F}_\Delta'P_\Delta\tilde{F}_\Delta + Q_\Delta - \tilde{F}_\Delta'P_\Delta G_1(R_\Delta + G_1'P_\Delta G_1)^{-1}G_1'P_\Delta\tilde{F}_\Delta \tag{4.5}$$

Definition 4.2 A solution $P_\Delta \geq 0$ of the ARE (4.5) is said to be a strong solution if $\tilde{F}_\Delta + G_1K_\Delta$ has all eigenvalues inside or on the unit circle. If no roots are on the unit circle then the solution is termed stabilizing.

Theorem 4.1 [15]
Consider the ARE (4.5), and define

$$Q_\Delta = D_\Delta D_\Delta' \tag{4.6}$$

Provided $\{\tilde{F}_\Delta, G_1\}$ is stabilizable, then
(i) the strong solution exists and is unique.
(ii) if $\{\tilde{F}_\Delta, D_\Delta\}$ is detectable, the strong solution is the only nonnegative definite solution.
(iii) if $\{\tilde{F}_\Delta, D_\Delta\}$ has no unobservable mode on the unit circle, the strong solution coincides with the stabilizing solution.
(iv) if $\{\tilde{F}_\Delta, D_\Delta\}$ has an unobservable mode on the unit circle then although the strong solution exists, there is no stabilizing solution.
(v) if $\{\tilde{F}_\Delta, D_\Delta\}$ has an unobservable mode inside or on the unit circle, then the strong solution is not positive definite.
(vi) if $\{\tilde{F}_\Delta, D_\Delta\}$ has an unobservable mode outside the unit circle, then besides the strong solution, there is at least one other nonnegative definite solution.

Note that if $\{F_1, D_1\}$ is observable then $\{\tilde{F}_\Delta, D_\Delta\}$ is unobservable for at most a finite number of values of $0<\Delta<1$. In the first order case $D_1\neq 0$ implies $D_\Delta\neq 0$ for $0\leq\Delta<1$, and so the observability of $\{\tilde{F}_\Delta, D_\Delta\}$ is guaranteed in this case. The general case is under investigation.

Conjecture 4.1 Suppose $\{F_1, D_1\}$ is observable, then $\{\tilde{F}_\Delta, D_\Delta\}$ is observable for all $0<\Delta<1$.

For a fixed Δ, the minimum cost $\hat{J}_\Delta$ in lemma 4.1 is given by

$$\hat{J}_\Delta = tr[P_\Delta(Q_\Delta + K_\Delta' R_\Delta K_\Delta) + QW_\Delta] \tag{4.7}$$

Lemma 4.2

A necessary condition for (K_Δ, Δ^*) to be a minimax solution in (4.1) is that

$$\text{(i)}\quad \frac{\partial \hat{J}_\Delta}{\partial \Delta} = 0 \quad ; \quad \frac{\partial^2 \hat{J}_\Delta}{\partial \Delta^2} < 0 \quad \text{for } 0 < \Delta^* < 1$$

$$\text{(ii)}\quad \left.\frac{\partial \hat{J}_\Delta}{\partial \Delta}\right|_{\Delta^*=1} \geq 0 \quad \text{or} \quad \left.\frac{\partial \hat{J}_\Delta}{\partial \Delta}\right|_{\Delta^*=0} \leq 0$$

4.1 Minimax Variance regulation

For remainder of this section we consider a single input single output (SISO) plant (1.1). The minimax variance regulator (MMVR) is defined by

$$\min_{\{u_k\}} \left(\max_{0 \leq \Delta \leq 1} Y_\Delta \right) \quad ; \quad Y_\Delta = \xi\{y^2_{k+\Delta}\} \tag{4.8}$$

The classical [7] minimum variance regulator (MVR) is defined by

$$\min_{\{u_k\}} (Y_1) \tag{4.9}$$

For MVR, the resulting control signal $\{u_k\}$ is very often too large. One possible solution [8] is to minimize Y_1 subject to the constraint $\xi\{u^2_k\} \leq \varepsilon$ on the input variance. However an appropriate $\varepsilon > 0$ must first be chosen to guarantee the existence of a stabilizing solution. Placing magnitude constraints (although physically justifiable) on the input leads to a difficult if not intractible nonlinear dynamic programming solution [9]. Another approach to this problem is to implicitly restrict the input variance by using a quadratic loss function with weighting on the input [10]. However, the choice of the weighting on $\{u(k)\}$ is then somewhat artificial. A more recent approach [11] introduces secondary quadratic objective function besides the primary variance cost in order to achieve satisfactory control. On the other hand, the criterion (4.8) is well justified. The largest value of Y_Δ indeed limits the system performance. The effects of sampling in the MVR has also been considered by Souza and Goodwin [16].

Lemma 4.3

Consider a SISO plant (1.1), (1.2). Then with

$$Q = C'C \quad ; \quad R = 0 \tag{4.10a}$$

in (4.2),(4.3) we have

$$Q_\Delta = 0 \quad ; \quad R_\Delta = (CG_\Delta)^2 \quad ; \quad S_\Delta = (CG_\Delta)(CF_\Delta) \tag{4.10b}$$

Furthermore, $R_\Delta > 0$ for $0 < \Delta < 1$.

Note that since (1.1) has no pure time delay, $R_1 = (CG_1)^2 \neq 0$. The property of R_Δ in (4.10b) then follows from (2.4b).

If $\tilde{F}_\Delta$ is asymptotically stable, then the optimal control law which minimizes Y_Δ is given by $\tilde{u}(k) = 0$; that is

$$u(k) = - R_\Delta^{-1} S_\Delta x(k)$$

otherwise one must solve the ARE (4.5). Unfortunately in view of Theorem 4.1, since $Q_\Delta = 0$, the ARE (4.5) is equally difficult to solve for all $0 < \Delta < 1$. In particular, numerical difficulties will be experienced in attempting to solve (4.5) iteratively using a recursive Riccati equation [15]. This problem can be overcome by using <u>modified</u> index of the following form

$$\min_{\{u(k)\}} \left(\max_{0\leq\Delta\leq 1} \tilde{Y}_\Delta \right) \tag{4.11}$$

$$\tilde{Y}_\Delta = \alpha Y_1 + \beta Y_\Delta \quad ; \alpha \geq 0 \ , \ \beta \geq 0 \ , \ 0 < \Delta < 1$$

where Y_Δ is defined by (4.8). A little algebraic manipulation after completing squares yields the next result.

<u>Lemma 4.4</u>

Define

$$I_\Delta \triangleq \alpha[F_1x_k + G_1u_k]'C'C[F_1x_k + G_1u_k] + \beta[F_\Delta x_k + G_\Delta u_k]'C'C[F_\Delta x_k + G_\Delta u_k] \tag{4.12}$$

Then

$$I_\Delta = \alpha\beta r_\Delta^{-1}(q_\Delta x_k)^2 + r_\Delta \tilde{u}^2_k$$

where

$$\tilde{u}_k = u_k + r_\Delta^{-1}s_\Delta x_k \ ; \ r_\Delta = \alpha(CG_1)^2 + \beta(CG_\Delta)^2 \tag{4.13}$$

$$q_\Delta = (CG_1)(CF_\Delta) - (CG_\Delta)(CF_1) \ ; \ s_\Delta = \alpha(CG_1)(CF_1) + \beta(CG_\Delta)(CF_\Delta)$$

From lemma 4.1 and Theorem 4.2 we can now state the optimal design for the modified index.

<u>Theorem 4.2</u>

Consider a minimal SISO time invariant system (1.1) and control law (1.2) for a sampling period T = 1. Then for any $0 < \Delta < 1$, the performance index $\tilde{Y}_\Delta$ defined in (4.11) for $\alpha > 0$, $\beta > 0$ is minimized by

$$u_k = (K_\Delta - r_\Delta^{-1}s_\Delta)x_k$$

where K_Δ is given by (4.4) and (4.5), r_Δ and s_Δ by (4.13) and $Q_\Delta = q_\Delta q_\Delta'$. The closed loop system is asymtotically stable provided $\{F_\Delta, q_\Delta\}$ is observable.

The optimal selection Δ^* of Δ which achieves the minimax solution for the modified performance measure (4.1) is still under investigation. A useful practical estimate $\hat{\Delta}$ for Δ^* is obtained from the MVR solution. That is, let $r_y(\Delta)$ be the covariance of $\{y_{k+\Delta}\}$ corresponding to the MVR, and define

$$r_y(\hat{\Delta}) = \max_{0\leq\Delta\leq 1} r_y(\Delta) \tag{4.14}$$

The following example demonstrates this design procedure.

<u>Example 4.1</u>

Consider a second order system (1.1) where

$$A = \begin{bmatrix} 0 & 1 \\ 0 & -0.1 \end{bmatrix} \quad B = \begin{bmatrix} 0 \\ 1 \end{bmatrix} \quad C' = \begin{bmatrix} 1 \\ 0 \end{bmatrix}$$

and

$$\Omega = \begin{bmatrix} 1 & 0 \\ 0 & 0 \end{bmatrix} \quad \Lambda = 1$$

The control laws are defined by (1.2) where $u(k) = K_\Delta x(k)$. For MVR (ie. $\Delta = 1$),

$K_1 = [2.0697\ 1.6971]$. The maximum of the corresponding output covariance $r_y(\Delta)$ occurs at $\hat{\Delta} = 0.53$. The corresponding gains for the approximate MMVR solution are given by

$$K_{0.53} = [1.5799\ 1.7091] \quad \text{for } \alpha = 1 \quad , \beta = 1$$
$$K_{0.53} = [1.4445\ 1.6304] \quad \text{for } \alpha = 0.01 \quad , \beta = 1$$

The output variances $r_y(\Delta)$ corresponding to the MVR design and the MMVR design for the modified index (4.11) are illustrated in Fig 4.1a. In Fig 4.1b, the corresponding variances U of the piecewise constant control laws (1.2) are given by

$$U = r_v(0)$$

The signal $v(t) = K_\Delta x(t)$ is a 'fictitious' control signal since $u(t) = v(t)$ only at the sampling instants k. However the characteristics of the covariance $r_v(\Delta)$ of $\{v_{k+\Delta}\}$ is of interest in Fig 4.1b. In particular the minimum values of $r_v(\Delta)$ coincide for all control laws.

5. OPTIMAL STATE PREDICTION

Suppose now that only the output $y(t_k)$ is available for measurement at the periodic measuring instants

$$t_k = k + \delta \tag{5.1}$$

for some $0 \leqq \delta \leqq 1$. The discrete control minimax quadratic regulator now involves the design of an optimal state predictor. The main issue of this section is to optimize the choice of δ given that the disturbance noise $\omega(t)$ and/or measurement noise $\eta(t)$ are WSCS. We shall see that the separation principle holds so that the control value $\tilde{u}(k+1)$ in (4.4) can be replaced by

$$\tilde{u}(k+1) = K_\Delta \hat{x}(k+1 | k+\delta) \tag{5.2}$$

where the predicted value of the state is based on measurements $y(t_k)$. Since the measurement and control instants are not necessarily synchronous, the steady state prediction equations must be solved across the time instants k at which the control changes as detailed below.

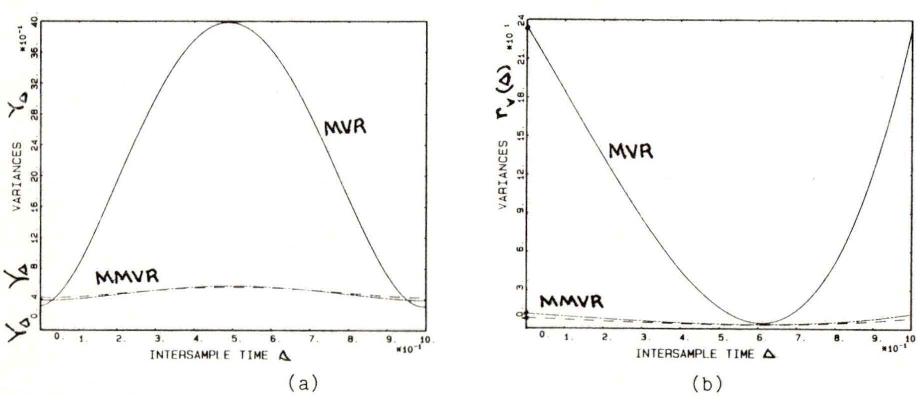

(a) (b)

Fig 4.1 The Covariance of Output (Y_Δ) and The 'Fictitious' Control Signal ($r_v(\Delta)$)

Lemma 5.1

Consider the predictor equations

$$\hat{x}(k+1|k+\delta) = F_{1-\delta}\hat{x}(k+\delta|k-1+\delta) + G_{1-\delta}u(k) + K_\delta[y(k+\delta) - C\hat{x}(k+\delta|k-1+\delta)] \quad (5.3a)$$

$$\hat{x}(k+1+\delta|k+\delta) = F_\delta\hat{x}(k+1|k+\delta) + G_\delta u(k+1) \quad (5.3b)$$

where F_Δ, G_Δ are defined by (2.4b) and the gain K_δ is to be determined. Then the reconstruction error $\varepsilon = x - \hat{x}$ is governed by

$$\varepsilon(k+1) = (F_{1-\delta} - K_\delta C)\varepsilon(k+\delta) - K_\delta\eta_\delta(k) + \tilde{\omega}_{1-\delta}(k) \quad (5.4a)$$

and

$$\varepsilon(k+1+\delta) = (F_1 - F_\delta K_\delta C)\varepsilon(k+\delta) + v_\delta(k+1) - F_\delta K_\delta\eta_\delta(k) \quad (5.4b)$$

where

$$v_\delta(k+1) = \omega_\delta(k+1) + F_\delta\tilde{\omega}_{1-\delta}(k) \quad (5.5a)$$

The process $\omega_\Delta(k)$ is defined by (2.4b) while

$$\tilde{\omega}_\Delta(k) = \int_0^\Delta e^{A(\Delta-\tau)}\omega(\tau+k+1-\Delta)d\tau \quad (5.5b)$$

We now suppose the continuous disturbance $\omega(t)$ and measurement noise $\eta(t)$ are independent zero mean WSCS of period 1 with respective covariance functions

$$\Omega_t = r_\omega(t,t) \quad ; \quad \Lambda_t = r_\eta(t,t) \quad (5.6)$$

As explained in section 2, this circumstance would arise after model reduction. The variances of the resulting WSS discrete processes $\{\omega_\delta(k)\}$ and $\{\eta_\delta(k)\}$ are given by

$$W_\delta = \int_0^\delta\int_0^\delta e^{A(\delta-\sigma)}r_\omega(\sigma,\tau)e^{A'(\delta-\sigma)}d\sigma d\tau \quad ; \quad \Lambda_\delta \quad (5.7)$$

while the covariance matrix V_δ of the zero mean process $\{v_\delta(k)\}$ defined in (5.5a) can be calculated according to

$$V_\delta = W_\delta + F_\delta\tilde{W}_{1-\delta}F'_\delta + F_\delta X'_\delta + X_\delta F'_\delta \quad (5.8)$$

where

$$\tilde{W}_\delta \triangleq \xi\{\tilde{\omega}_\delta(k)\tilde{\omega}'_\delta(k)\} = \int_0^\delta\int_0^\delta e^{A(\delta-\sigma)}r_\omega(\sigma-\delta,\tau-\delta)e^{A'(\delta-\tau)}d\tau d\sigma$$

and

$$X_\delta \triangleq \xi\{\omega_\delta(k+1)\tilde{\omega}'_{1-\delta}(k)\} = \int_0^\delta\int_0^{1-\delta} e^{A(\delta-\sigma)}r_\omega(\sigma,\tau+\delta)e^{A'(1-\delta-\tau)}d\tau d\sigma$$

Note that $V_1 = V_0 = W_1$. Both $\delta = 0$ and $\delta = 1$ correspond to making measurements synchronous with the control changes. For $\delta = 0$ we predict one period ahead whereas for $\delta = 1$ we 'instantaneously' estimate. Furthermore, if $\omega(t)$ is in fact WSS and uncorrelated, then $r_\omega(t,s+\delta) = 0$ unless $t = s+\delta$, so from (5.8)

$$X_\delta = 0 \quad ; \quad V_\delta = W_1 \quad \forall \; 0 < \delta < 1$$

Optimal state prediction when both $\omega(t)$ and $\eta(t)$ are independent WSS and uncorrelated is therefore achieved with $\delta = 1$ (In practice some finite delay between the measurement instant and control change is necessary in order to allow time t_c for the calculations to be performed in which case $\delta = 1 - t_c$).

For WSCS measurement and/or disturbance noise, synchronous measurement and control is not necessarily optimal.

Theorem 5.1

Consider the continuous-time plant (1.1) where measurements $y(k+\delta)$; $0<\delta<1$ are made periodically. Assume $\{\omega(t)\}$ and $\{\eta(t)\}$ are independent zero mean WSCS processes of period 1 with covariances $\Omega_\delta \geq 0$ and $\Lambda_\delta > 0$ respectively. Then the optimal steady state predictor which minimize the prediction error $x(k+1)-\hat{x}(k+1|k+\delta)$ is described by (5.4) where the gain K_δ satisfies

$$K_\delta = F_{1-\delta}P_\delta C'(\Lambda_\delta + CP_\delta C')^{-1} \tag{5.9a}$$

$$P_\delta = F_1 P_\delta F_1' + V_\delta - F_1 P_\delta C'(\Lambda_\delta + CP_\delta C')^{-1}CP_\delta F_1' \tag{5.9b}$$

$$Q_\delta = \bar{F}_{1-\delta}P_\delta \bar{F}_{1-\delta}' + K_\delta \Lambda_\delta K_\delta' + \tilde{W}_{1-\delta} \quad ; \quad \bar{F}_{1-\delta} = F_{1-\delta} - K_\delta C \tag{5.10}$$

The covariance P_δ is the steady state covariance of the state estimation error $e(k+\delta)$ in (5.4b) at the measuring instants (5.2). The covariance Q_δ of the prediction error at the control instants is achieved by propagating P_δ according to (5.10). Optimal prediction is achieved by

$$\min_{0\leq\delta\leq 1} \operatorname{tr}[Q_\delta] \tag{5.11}$$

Example 5.1

Consider the first order plant model

$$\dot{x}(t) = -\alpha x(t) + u(t) + \omega(t)$$

$$y(t) = x(t) + \eta(t)$$

where $\{\eta(t)\}$ is zero mean and WSS with unity covariance and $\{\omega(t)\}$ is WSCS of period 1 as defined by the first order representation (3.4) of example 2.1. Plots of V_δ, $\tilde{W}_{1-\delta}$, P_δ and Q_δ for $\alpha=2$ and $\beta=1$ are illustrated in Fig 5.1. The optimal measuring instant $\delta^*=0.83$.

It is clear from (4.1) and (5.11) that the optimal selection of Δ for control design, and the optimal selection of δ for predictor design are independent. It is also clear that the optimal design in section 4 can be extended to WSCS disturbance and measurement - simply replace W_Δ and N_Δ in (2.5) by W_δ in (5.7) and Λ_δ.

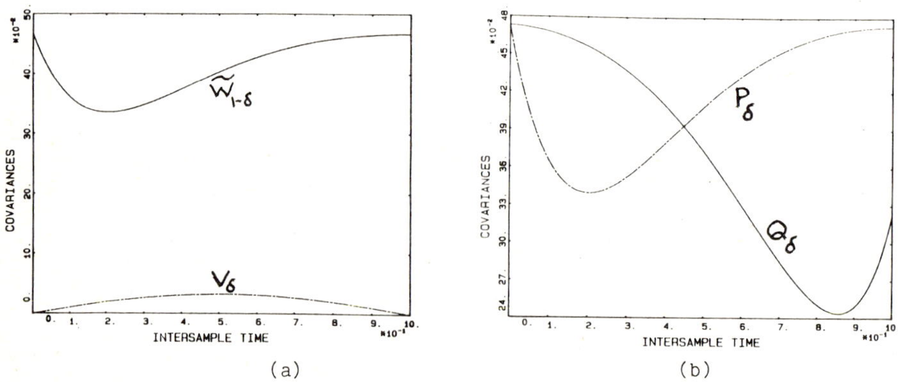

Fig 5.1 Noise Variances V_δ, $\tilde{W}_{1-\delta}$ and Error Variances Q_δ, P_δ

REFERENCES

1. W.R. Bennet, "Statistics of Regenerative Digital Transmission", Bell Sup.Tech. Journal, vol 37, pp 1501-1542, Nov. 1958.
2. A. Papoulis, "Probability, Random Variables, and Stochastic Processes" McGraw Hill, 1965.
3. L. Franks, "Signal Theory", Prentice Hall, 1969.
4. W.A. Gardner and L.E. Franks, "Characterization of Cyclostationary Random Signal Processes," IEEE Trans. on Information Theory, vol IT-21, no 1, pp 4-14, Jan 1975.
5. K.L. Jordan, "Discrete Representations of Random Signals," PhD dissertation, MIT, Cambridge, 1961.
6. J.T. Tou, "Digital and sampled - data control systems," McGraw Hill, 1959.
7. K.J. Aström, "Introduction to Stochastic Control Theory," Academic Press, 1970.
8. G.E. Box and G.M. Jenkins "Time Series Analysis, Forecasting and Control," Holden Day, 1970.
9. H.T. Toivenen "Suboptimal Control of Linear Discrete Stochastic Systems with Linear Input Constraints", IEEE Trans. Auto. Control, vol AC-28, Feb. 1983, pp 246-248.
10. K.J. Aström, V. Borisson, L. Ljung and B. Wittenmark, "Theory and Applications of Selftuning Regulators", Automatica, vol 13,1977 pp457-476.
11. H. T. Toivenen "A Multiobjective Linear Quadratic Gaussian Problem", IEEE Trans. Auto. Control vol AC-29, no 3, March 1984, pp 279-280.
12. H. Ogura "Spectral Representation of A Periodic Nonstationary Random Process", IEEE Transaction on Information Theory, vol IT-17, March 1971, pp 143-149
13. R.E Skelton, "Cost Decomposition of Linear Systems with Application to Model Reduction", Int. J. Control, vol. 32, No 6, pp 1031-1055, 1980.
14. B.C Moore, "Principal Component Analysis in Linear Systems: Controllability, Observability and Model Reduction", IEEE Trans. A to. Control, vol. AC-26, no 1, pp 17-32, Feb. 1981.
15. S.W Chan, G.C Goodwin and K.S Sin, "Convergence Properties of The Riccati Difference Equation in Optimal Filtering of Nonstabilizable Systems", IEEE Transaction on A tomatic Control, vol AC-29, Feb 1984, pp 110-118
16. C.E. de So za and G.C. Goodwin,"Robustness Effects of Sampling in Model Reference and Minimum Variance Control", Proc. 22nd IEEE CDC , San Antonio, Texas, Dec 1983,pp997-1001

5. GENERALIZED SYSTEMS

Frequency Domain and State Space Methods for Linear Systems
C.I. Byrnes and A. Lindquist (Editors)

CONSISTENT INITIAL CONDITIONS FOR LINEAR TIME VARYING SINGULAR SYSTEMS

Stephen L. Campbell[1]

Department of Mathematics
North Carolina State University
Raleigh, North Carolina 27695-8205

A fairly direct method is given for calculating the consistent initial conditions for a wide class of solvable implicit linear time varying systems $A(t)x'(t) + B(t)x(t) = f(t)$ with $A(t)$ singular.

INTRODUCTION

Singular linear systems of differential equations

$$A(t)x'(t) + B(t)x(t) = f(t) \tag{1.1}$$

with $A(t)$ singular on the entire interval of interest arise in a variety of mechanical, electrical, economic, and control systems [1], [2], [6]. Generally three methods have been proposed for solving (1.1) numerically. The first was implicit backward differences [11], [12]. The second is to repeatedly differentiate and algebraically manipulate the constraints and derive an explicit nonsingular differential equation [14] [15]. Recently a more general approach has been developed in [5], [7]. However, all approaches have suffered from the lack of any good general method of determining the consistent initial conditions for (1.1), that is, those initial conditions x_o such that there is a solution to the initial value problem

$$A(t)x'(t) + B(t)x(t) = f(t),\ x(t_o) = x_o \tag{1.2}$$

This paper will provide a characterization of the consistent initial conditions for a large class of singular systems (1.2). The key ideas come from [5], [7], [8]. The characterization is then used to develop a numerical algorithm for the computation of the consistent initial conditions.

THEORETICAL RESULTS

Suppose that (1.1) is solvable on a finite interval I and A, B are n×n. That is, for every sufficiently smooth f (n-times differentiable will suffice) there exists a smooth solution and solutions for a given f are defined on all of I and are uniquely determined by their value at any point. Also if f, A, B are m-times differentiable, $m \geq n$, then the solution x is at least m+1-n times differentiable. Define the matrices A_i, B_i and vectors x_i, f_i by

$$A(t) = \sum_{i=0} A_i t^i,\ B(t) = \sum_{i=0} B_i t^i$$
$$F(t) = \sum_{i=0} F_i t^i,\ x(t) = \sum_{i=0} x_i t^i \tag{2.1}$$

[1]Research supported in part by the National Science Foundation under Grant No. DMS-8318026 and the Air Force Office of Scientific Research under Grant No. 84-0240.

where the expansions in (2.1) are taken to be finite Taylor approximations if A, B, x or f is not real analytic. Substituting the expansions (2.1) into (1.1) gives for any integer $j \geq 1$ (in what follows j is assumed less than the number of terms in the expansions in (2.1));

$$\begin{bmatrix} A_o & 0 & \cdot & \cdot \\ A_1 + B_o & 2A_o & \cdot & \cdot \\ \cdot & \cdot & \cdot & 0 \\ A_{j-1} + B_{j-2} & 2A_{j-2} + B_{j-3} & \cdot & jA_o \end{bmatrix} \begin{bmatrix} x_1 \\ \cdot \\ \cdot \\ x_j \end{bmatrix} = \begin{bmatrix} f_o \\ \cdot \\ \cdot \\ f_{j-1} \end{bmatrix} - \begin{bmatrix} B_o \\ \cdot \\ \cdot \\ B_{j-1} \end{bmatrix} x_o \tag{2.2}$$

or

$$\mathcal{A}_j \underline{x}_j = \underline{f}_j - \mathcal{B}_j x_o \tag{2.3}$$

The matrix $\mathcal{A}_j$ is singular since A_o is. The system (2.2) is the basis of the numerical methods developed in [5], [7] except that there the $\mathcal{A}_j$, $\mathcal{B}_j$, $\underline{f}_j$ depend on t. For any $j \geq 1$ and given $\underline{f}_j$ define

$$\mathcal{M}_j[f] = \{x_o | \mathcal{A}_j \underline{x} = \underline{f}_j - \mathcal{B}_j x_o \text{ is consistent}\}. \tag{2.4}$$

Alternatively

$$\mathcal{M}_j[f] = \{x_o | \mathcal{B}_j x_o \in \mathcal{R}(\mathcal{A}_j) + \underline{f}_j\} \tag{2.5}$$

where $\mathcal{R}$ denotes the range (column space) of a matrix. For a solvable system (1.1), let $C[f]$ = {consistent initial conditions for (1.1)}. Clearly for any $j > 1$ the linear manifolds $C[f]$ and $\mathcal{M}_j[f]$ satisfy

$$C[f] \subset \mathcal{M}_j[f]. \tag{2.6}$$

Since $C[f]$ and $\mathcal{M}_j[f]$ are linear manifolds, (translates of linear subspaces) to establish equality in (2.6) it suffices to show they have the same dimension. If (1.1) is solvable, then $\mathcal{M}_j[f]$ is nonempty for any sufficiently smooth f. Thus solvability implies that

$$\mathcal{R}(\mathcal{A}_j) + \mathcal{R}(\mathcal{B}_j) = \mathbb{C}^{jn} \tag{2.7}$$

for all $j \geq 1$. Suppose that

$$\text{rank}(\mathcal{A}_j) = r_j$$

$$\text{nullity}(\mathcal{A}_j) = n_j$$

so that $r_j + n_j = jn$. Let $\{\psi_1, \ldots, \psi_s\}$, $s = n_j$ be a basis of $\mathcal{R}(\mathcal{A}_j)^\perp$. Multiplying (2.3) by ψ_i^* (* denotes conjugate transpose) gives

$$\psi_i^* \mathcal{B}_j x_o = \psi_i^* f_j, \quad i = 1, \ldots, s. \tag{2.8}$$

But the vectors $\{\psi_i^* \mathcal{B}_j, \ldots, \psi_s^* \mathcal{B}_j\}$ are linearly independent. For if $\Sigma \alpha_i \psi_i^* \mathcal{B}_j = 0$, then $\Sigma \bar{\alpha}_i \psi_i$ is a nonzero vector perpendicular to both $\mathcal{R}(\mathcal{A}_j)$ and $\mathcal{R}(\mathcal{B}_j)$ which contradicts (2.7). Since the vectors $\{\psi_i^* \mathcal{B}_j, \ldots, \psi_s^* \mathcal{B}_j\}$ are linearly independent, the equations (2.8) show that

$$\dim \mathcal{M}_j[f] \leq n - n_j \tag{2.9}$$

We have arrived then at the point where

PROPOSITION 2.1. To establish that $M_j[f] = C[f]$ it suffices to show that

$$\dim C[f] = n - n_j. \tag{2.10}$$

Note that (2.10) includes the possibility that $n = n_j$. Proposition 2.1 allows us to utilize several recent results in obtaining characterizations of initial conditions. In all of these results the required differentiability need only hold on a neighborhood of the origin.

THEOREM 2.1. Suppose that (1.1) is solvable and A, B are constant. Then the pencil $\lambda A + B$ is regular [9]. Let k be the index of the pencil. Then if f is m-times differentiable, $m \geq 2k$, then for $j \leq m+1-k$

$$M_j[f] = C[f] \tag{2.11}$$

if and only if $j \geq k+1$.

PROOF. Suppose that $j \geq k+1$. Then from the proof of Theorem 4.3 in [5] we have that the nullity n_j of A_j is the same as the size of the nilpotent block in the Jordan form of $(\lambda A+B)^{-1}A$ which from [1] is $n - \dim C[f]$. Thus (2.11) follows from Proposition 2.1. On the other hand, if $j \leq k$, then the same argument shows that $n_j < n_k$. □

Of more general importance recall that in [4] the standard canonical form (SCFr) of (1.1) was defined to be

$$\begin{bmatrix} G_{11} & 0 & \cdot & \cdot & 0 \\ 0 & 0 & \cdot & \cdot & \cdot \\ 0 & G_{32} & \cdot & \cdot & \cdot \\ \cdot & \cdot & \cdot & \cdot & \cdot \\ 0 & G_{r2} & \cdot & G_{r,r-1} & 0 \end{bmatrix} \begin{bmatrix} z_1' \\ \cdot \\ \cdot \\ \cdot \\ z_r' \end{bmatrix} = \begin{bmatrix} H_{11} & 0 & \cdot & \cdot & 0 \\ H_{21} & I & \cdot & \cdot & \cdot \\ \cdot & 0 & \cdot & \cdot & \cdot \\ \cdot & \cdot & \cdot & \cdot & 0 \\ H_{r1} & 0 & \cdot & 0 & I \end{bmatrix} \begin{bmatrix} z_1 \\ \cdot \\ \cdot \\ \cdot \\ z_r \end{bmatrix} + \begin{bmatrix} g_1 \\ \cdot \\ \cdot \\ \cdot \\ g_r \end{bmatrix} \tag{2.12}$$

where G_{11} is invertible.

THEOREM 2.2. Suppose that (1.1) is solvable on $[0,t_1]$ and there exists an interval $[0,\varepsilon_0)$ with $\varepsilon_0 > 0$ and m-times continuously differential matrix functions P(t), Q(t) such that on $[0,\varepsilon_0)$ the transformations $x = Q(t)z$ and left multiplication by P(t) transforms (1.1) on $[0,\varepsilon_0)$ to SCFr where the G_{ij}, H_{ij} are m-times continuously differentiable. If f is m-times continuously differentiable, $m \geq 2r-2$, then

$$M_j[f] = C[f] \quad \text{if} \quad r \leq j \leq m-r+2$$

PROOF. From Propositions 4.2 and 4.3 of [5] we have that the rank and nullity of A_j are unchanged by the transformations P,Q. Thus we may work directly with the form (2.12). But then the proof of Theorem 2.2 may be easily adapted. The m-differentiability insures that the solutions are at least (m-r+1)-times differentiable near $t_0 = 0$. □

It is important to note that Theorem 2.2 does not require (1.1) to have the form SCFr on all of $[0,t_1]$ but only on a small neighborhood of zero. Combining Theorem

2.2 with the results of [10] leads to the following simple criterion.

THOEREM 2.3. If (1.1) is solvable on $[0,t_1]$ and A(t), B(t) are real analytic on any neighborhood of zero, then for any 2n-times differentiable f.

$$M_{n+1}[f] = C[f].$$

PROOF. If (1.1) is solvable on $[0,t_1]$ and A(t), B(t) are real analytic on an interval $[0,\varepsilon_o] \subset [0,t_1]$ then, it is shown in [10, Theorem 2] that (1.1) can be transformed to SCFr with $r \leq n$. □

However, probably the most general result is the following:

THEOREM 2.4. Suppose that (1.1) is solvable on I and that on some subinterval $\tilde{I} = [0,\varepsilon] \subset I$, that A, B are 2n-times continuously differentiable on $\tilde{I}$. Suppose also that not only is (1.1) solvable but that solutions of the associated homogeneous equation are 2n+1-times continuously differentiable on $\tilde{I}$. Then (2.11) holds if j = n+1.

PROOF. Under the assumptions of Theorem 2.4, it is shown in [8] that (2.10) holds for j = n+1. Again the differentiability of A, B insures that the solutions are at least n+1-times differentiable near zero. □

Theorem 2.4 includes a large variety of nonanalytic solvable systems and systems which cannot be put into SCF. It also includes all of the solvable systems with real analytic coefficients. Finally, we give the following result derived from Theorem 4.4 of [5].

THEOREM 2.5. For the system

$$A(t)x'(t) + x(t) = f(t) \tag{2.14}$$

suppose (2.14) is solvable on $[0,t_1]$ with A, f, 4-times continuously differentiable. Suppose that rank A^3 = rank A^2 and is constant on an interval $[0,\varepsilon] \subset [0,t]$. Define P,Q,N on $[0,\varepsilon]$ by $P = A^DA$, $Q = I - P$, $N = AQ$ where A^D denotes the Drazin inverse of A [8]. Suppose that $I - N'$, $I \pm 2N'$ are invertible on $[0,\varepsilon]$ and rank $(Q(I - N')Q)$ = rank $([(I - 2N')Q)$ = rank Q. Then $M_3[f] = C[f]$.

Some necessary conditions for solvability are given in [8] as well as necessary and sufficient conditions for the real analytic case. These conditions are given directly in-terms of A_j, B_j and do not require coordinate changes.

In closing this section we give a simple example to illustrate why some smoothness of A, B is required.

EXAMPLE 2.1. Let ϕ be a differentiable function such that $\phi' \neq -1$ for all t. Then the solution of

$$\begin{bmatrix} \phi & \phi^2 \\ -1 & -\phi \end{bmatrix} \begin{bmatrix} x' \\ y' \end{bmatrix} + \begin{bmatrix} x \\ y \end{bmatrix} = \begin{bmatrix} f \\ g \end{bmatrix}$$

is formally

$$x = -\phi y + f + \phi g$$

$$y = (1 + \phi')^{-1}(g + f' + (\phi g)')$$

Notice that in general y will have one less degree of smoothness than ϕ and f.

NUMERICAL PROCEDURES

Suppose that (1.1) is solvable and we are in one of the situations where $M_j[f] = C[f]$ described earlier. From Section 2 we have

PROPOSITION 3.1. To find $M_j[f]$ if $M_j[f] = C[f]$ when (1.1) is solvable and (2.10) holds, it suffices to compute a matrix W such that $A_j^{*}W = 0$, rank W = nullity of A_j^* and then solve

$$W^* B_j x_o = W^* \underline{f}_j . \tag{3.1}$$

The solutions will form an n-nullity (A_j) dimensional manifold.

There are numerous ways to compute W. One of the numerically most reliable (and most expensive) would be to compute a singular value decomposition (SVD) of A_j

$$A_j = U\Sigma V = [U_1 \quad U_2]\begin{bmatrix} \Sigma_1 & 0 \\ 0 & 0 \end{bmatrix}\begin{bmatrix} V_1 \\ V_2 \end{bmatrix}$$

where U,V are unitary, Σ_1 is a diagonal matrix with entries $\sigma_i \geq 0$ [1]. Then one can take $W = U_2$. Since W has orthonormal columns, the matrix $U_2^* B_j$ should be as well conditioned as the problem allows.

If the methods of [5], [7] are being used to numerically integrate (1.1), then on the first time step it is desirable to determine the numerical rank of A_j and solve (2.3). The system (3.1) may be derived as a by-product of these calculations. A reasonable way to do this is with an (SVD) [13], [16]. There is no need to save V_1, V_2, U_1. If only one f is involved, then there is no need to save U_2^* either. Rather it may be accumulated onto B_j, $\underline{f}_j$. The resulting system (3.1) is n×n whereas the original A_j was jn × nj. In addition this rank determination for A_j also gives us the rank of W^*B_j. Since the rank of W^*B_j is known, it is reasonable to solve the underdetermined system (3.1) by a QR algorithm with column pivoting [13]. The forming of (3.1) and its solution will take less computational effort then the original SVD.

A somewhat less reliable method, but one that often works well is to just do a QR decomposition on A_j as

$$A_j = QR = [Q_1 \; Q_2]\begin{bmatrix} R_{11} & R_{12} \\ 0 & 0 \end{bmatrix}$$

where Q is unitary, R_{11} is invertible and upper triangular. Then we may take $W = Q_2$. In the numerical solution of (1.1) by the methods of [5], [7] a modified QR is used which in selecting pivot columns never interchanges the first n columns with the last (j-1)n columns. It probably should be pointed out that pivoting the last (j-1)n columns is not just desirable but necessary for the systems considered here. We have run fairly simple 2-dimensional solvable examples where failure to do this pivoting leads to very ill-conditioned systems.

For well-conditioned problems one may simply find a basis $\{\psi_1,\ldots,\psi_s\}$ of solutions of $A_j^* \psi = 0$ by (for example) Gaussian elimination (with modified column pivoting) to the row echelon form and then set $W = [\psi_1,\ldots,\psi_s]$.

CONCLUSION

We have presented an explicit, easy to compute characterization of the consistent initial conditions of $Ax' + Bx = f$ which is valid for all known types of solvable systems. The method does not require any coordinate changes or deflation procedures and is applied directly to an algebraic system made up of the Taylor coefficients of the known A, B, f at the initial time.

REFERENCES

[1] Campbell, S. L., Singular Systems of Differential Equations (Pitman, 1980).

[2] Campbell, S. L., Singular Systems of Differential Equations II (Pitman, 1982).

[3] Campbell, S. L., Index two linear time-varying singular systems of differential equations, SIAM J. Alg. Disc. Meth., 4 (1983) 237-243.

[4] Campbell, S. L., One canonical form for higher-index linear time varying singular systems, Circuits Systems & Signal Processing, 2 (1983) 311-326.

[5] Campbell, S. L., The numerical solution of higher index linear time-varying singular systems of differential equations, SIAM J. Sci. Stat. Comp., 6 (1985) 334-348.

[6] Campbell, S. L., Non-linear time varying generalized state-space systems: an overview, Proc. 23rd IEEE-SIAM Conference on Decision & Control (1984). 268-273.

[7] Campbell, S. L., Rank deficient least squares and the Numerical Solution of Linear Singular Implicit Systems of Differential Equations, Linear Algebra & Its Role in Systems Theory, Contemporary Mathematics (to appear).

[8] Campbell, S. L., A general form for solvable linear time varying singular systems of differential equations, preprint, 1985.

[9] Campbell, S. L., and Meyer, C. D., Jr., Generalized Inverses of Linear Transformations (Pitman, 1979).

[10] Campbell, S. L. and Petzold, L. R., Canonical forms and solvable singular systems of differential equations, SIAM J. Alg. Disc. Meth., 4 (1983) 517-521.

[11] Gear, C. W., Simultaneous numerical solution of differential-algebraic equations, IEEE Trans. Circuit Theory CT-18 (1971) 89-95.

[12] Gear, C. W. and Petzold, L. R., ODE Methods for the solution of differential/algebraic systems, SANDIA-Report SAND 82-8051 (1982).

[13] Golub, G. H. and Van Loan, C. F., Matrix Computations (Johns Hopkins University Press, Baltimore, 1983).

[14] Rheinboldt, W. C., Differential-algebraic systems as differential equations on manifolds, Math. Comp. 43 (1984) 473-482.

[15] Silverman, L. M., Inversion of multivariable linear systems, IEEE Trans. Aut. Control AC-14 (1969) 270-276.

[16] Stewart, G. W., Rank degeneracy, SIAM J. Sci. Stat. Comp., 5 (1984) 403-413.

Frequency Domain and State Space Methods for Linear Systems
C.I. Byrnes and A. Lindquist (Editors)

STRONG SYSTEM EQUIVALENCE

W. A. Coppel

Department of Mathematics
Institute of Advanced Studies
Australian National University
Canberra, A.C.T.
AUSTRALIA

The notion of (strict) system equivalence of polynomial realizations was introduced by Rosenbrock and has been further studied by other authors. Verghese has posed the problem of defining a stronger concept, which will take into account the behaviour of the system at infinity, and has given such a definition for the special case of generalized state-space systems. A (time-domain) solution of Verghese's problem in the general case will be given.

The notion of (strict) *system equivalence* of polynomial realizations was introduced by Rosenbrock [7],[8] and has been studied also by Fuhrmann [4], Pernebo [6], Hinrichsen and Prätzel-Wolters [5], and Coppel [3]. Verghese [9] has posed the problem of defining a stronger concept, which will take into account the behaviour of the system at infinity, and has given such a definition for the special case of generalized state-space systems. A solution of Verghese's problem in the general case will be given here. We simply present the definitions and refer for full details to Anderson, Coppel and Cullen [1], and Coppel and Cullen [2]. We also present a formulation of Verghese's concepts of strong controllability and strong observability and, for purposes of comparison, of the corresponding "ordinary" concepts.

We are given a system of linear differential equations of the form

$$\begin{aligned} T(D)x &= U(D)u \\ y &= V(D)x + W(D)u \end{aligned} \tag{1}$$

where $D = d/dt$ and $T(s)$, $U(s)$, $V(s)$, $W(s)$ are polynomial matrices of size $n{\times}n, n{\times}m, p{\times}n, p{\times}m$ with $T(s)$ non-singular. We are interested in its *distributional* solutions. Furthermore we choose to restrict attention to distributions of the form

$$z(t) = \xi(t)H(t) + c_0\delta(t) + c_1\delta^{(1)}(t) + \ldots + c_h\delta^{(h)}(t) ,$$

where $\xi(t)$ is an exponential polynomial, $H(t)$ is Heaviside's unit step function, c_k is a constant and $\delta^{(k)}$ is the k-th derivative of the delta function. We can write

$$z(t) = \xi(t)H(t) + p(D)\delta(t) ,$$

where $p(s) = c_0 + c_1 s + \ldots + c_h s^h$ is a polynomial. We will call $z_\infty(t) = p(D)\delta(t)$ the *impulsive part* of $z(t)$. Given two such distributions z and $\tilde{z}$, we will say that z is *quasi-equal* to $\tilde{z}$, and write $z \equiv \tilde{z}$, if z and $\tilde{z}$ have the same impulsive part.

The distributional derivative of $z(t)$ has the same form, in fact

$$Dz(t) = (D\xi(t))H(t) + [\xi(0) + Dp(D)]\delta(t) .$$

Convolution products also have the same form; in particular, for any complex number λ ,

$$e^{\lambda t}H(t) * p(D)\delta(t) = re^{\lambda t}H(t) + q(D)\delta(t) ,$$

where $q(s) = [p(s) - p(\lambda)]/(s - \lambda)$ is the quotient and $r = p(\lambda)$ the remainder in the division of $p(s)$ by $s - \lambda$. The Laplace transform of $z(t)$ is

$$\hat{z}(s) = \hat{\xi}(s) + p(s) ,$$

where $\hat{\xi}(s)$ is the classical Laplace transform of $\xi(t)$. Moreover, for an arbitrary polynomial $\ell(s)$ the Laplace transform of $\ell(D)z(t)$ is $\ell(s)\hat{z}(s)$.

It is easily shown that for any impulsive input $u(t) = \ell(D)\delta(t)$, where $\ell(s) \in \mathbb{C}^m[s]$, the polynomial system (1) has a unique distributional solution

$$x(t) = \xi(t)H(t) + p(D)\delta(t) ,$$

where $\xi(t)$ is an exponential polynomial and $p(s) \in \mathbb{C}^n[s]$. Moreover $\xi(t)$ is a solution of the homogeneous equation

$$T(D)\xi = 0 . \qquad (2)$$

We will say that the solution $\xi(t)$ of (2) is the *regular part* of the solution of (1) with the impulsive input $u(t)$.

The polynomial system (1) is *controllable* if every solution of (2) is the regular part of the solution of (1) for some impulsive input $u(t)$. It is *observable* if $\xi = 0$ is the only solution of (2) for which $V(D)\xi = 0$.

The polynomial system (1) is *system equivalent* to the polynomial system

$$\begin{aligned} \tilde{T}(D)\tilde{x} &= \tilde{U}(D)u \\ y &= \tilde{V}(D)\tilde{x} + \tilde{W}(D)u \end{aligned} \qquad (3)$$

if the transfer matrices are the same, i.e.

$$W(s) + V(s)\, T^{-1}(s)\, U(s) = \tilde{W}(s) + \tilde{V}(s)\, \tilde{T}^{-1}(s)\, \tilde{U}(s) ,$$

and there exists a one-to-one linear map $\varphi\colon \xi(t) \to \tilde{\xi}(t)$ of all solutions of the homogeneous equation (2) onto all solutions of the homogeneous equation

$$\tilde{T}(D)\tilde{\xi} = 0 \qquad (4)$$

such that

(i) $\varphi\,[D\xi(t)] = D[\varphi\,\xi(t)]$,

(ii) $V(D)\,\xi(t) = \tilde{V}(D)\,[\varphi\,\xi(t)]$,

(iii) if $\xi(t)$ is the regular part of the solution of (1) with some impulsive input $u(t)$, then $\varphi\,\xi(t)$ is the regular part of the solution of (3) with the same input.

We consider next the corresponding results for behaviour at infinity. A distribution (x,u,y) will be said to be a *quasi-solution* of the polynomial system (1) if

$$u \equiv 0,\ T(D)x \equiv U(D)u,\ y \equiv V(D)x + W(D)u .$$

The impulsive part $(x_\infty,\ 0,\ y_\infty)$ of a quasi-solution will be called an *impulsive solution.* We draw attention to the fact that in general y_∞ is not uniquely determined by x_∞ .

If (x,u,y) is a quasi-solution of (1), then so also is the convolution product $e^{\lambda t}H(t) * (x,u,y)$, for any complex number λ. It follows that if $(x_\infty, 0, y_\infty)$ is an impulsive solution of (1), then so also is the impulsive part $\Delta_\lambda(x_\infty, 0, y_\infty)$ of $e^{\lambda t}H(t) * (x_\infty, 0, y_\infty)$.

The polynomial system (1) is *controllable at infinity* if every impulsive solution of (1) is the impulsive part of a solution (and not merely of a quasi-solution). It is *observable at infinity* if it has no nontrivial impulsive solution $(x_\infty, 0, y_\infty)$ with $y_\infty = 0$.

The polynomial system (1) is *system equivalent at infinity* to the polynomial system (3) if the transfer matrices are the same and there exists a one-to-one linear map $\psi: (x_\infty, 0, y_\infty) \to (\tilde{x}_\infty, 0, \tilde{y}_\infty)$ of all impulsive solutions of (1) onto all impulsive solutions of (3) such that

(i) $\psi[\Delta_\lambda(x_\infty, 0, y_\infty)] = \Delta_\lambda[\psi(x_\infty, 0, y_\infty)]$ for any $\lambda \in \mathbb{C}$,

(ii) $\tilde{y}_\infty = y_\infty$,

(iii) if $(x_\infty, 0, y_\infty)$ is the impulsive part of the solution of (1) with input $u(t)$, then $\psi(x_\infty, 0, y_\infty)$ is the impulsive part of the solution of (3) with the same input.

The polynomial system (1) is *strongly controllable* if it is both controllable and controllable at infinity. It is *strongly observable* if it is both observable and observable at infinity.

The polynomial systems (1) and (3) are *strongly system equivalent* if they are both system equivalent and system equivalent at infinity.

In [1] system equivalence at infinity is defined in the frequency domain, in ways analogous to the definitions by Rosenbrock and Fuhrmann of ordinary system equivalence. It is shown also that anv polynomial system is strongly system equivalent to a generalized state-space system, and that two generalized state-space systems are strongly system equivalent if and only if they are constant system equivalent. In [2] system equivalence at infinity is given a module-theoretic characterization, together with the dynamical interpretation outlined above.

REFERENCES

[1] B.D.O. Anderson, W.A. Coppel and D.J. Cullen, "Strong system equivalence (I)", *J. Austral. Math. Soc. Ser. B.* (to appear).

[2] W.A. Coppel and D.J. Cullen, "Strong system equivalence (II)", *J. Austral. Math. Soc. Ser. B.* (to appear).

[3] W.A. Coppel, "Some remarks on strict system equivalence", *Proc. 5th Internat. Symp. Math. Theory of Networks and Systems,* Santa Monica (1981), 33-34.

[4] P.A. Fuhrmann, "On strict system equivalence and similarity", *Intenat. J. Control* 25 (1977), 5-10.

[5] D. Hinrichsen and D. Prätzel-Wolters, "Solution modules and system equivalence", *Internat. J. Control* 32 (1980), 777-802.

[6] L. Pernebo, "Notes on strict system equivalence", *Internat. J. Control* 25 (1977), 21-38.

[7] H.H. Rosenbrock, *State-space and multivariable theory* (Nelson, London, 1970).

[8] H.H. Rosenbrock, "The transformation of strict system equivalence", *Internat. J. Control* 25 (1977), 11-19.

[9] G.C. Verghese, "Infinite-frequency behavior in generalized dynamical systems", Ph.D. dissertation, Stanford University, 1978.

Frequency Domain and State Space Methods for Linear Systems
C.I. Byrnes and A. Lindquist (Editors)

EQUIVALENCE OF GENERALISED STATE-SPACE SYSTEMS: NEW RESULTS

G. E. Taylor

Electronic Engineering Department,
Hull University,
Hull,
U. K.

A. C. Pugh

Mathematics Department,
The University of Technology,
Loughborough, Leics.
U. K.

A system matrix transformation for systems in generalised state-space form is developed and an interpretation in terms of mappings between sets having important dynamical meaning is given.

INTRODUCTION

In linear systems theory, systems whose defining equations are in generalised state-space form have received much recent interest (Rosenbrock, 1984; Verghese et at, 1981; Bosgra and Van der Weiden, 1981). The main reason for this interest is that such systems represent the simplest form of system equations which can exhibit simultaneously significant finite and infinite frequency behaviour. It is necessary therefore when transforming such systems, that the permitted transformations should not only preserve the usually considered finite frequency structure but also that associated with the point at infinity. A number of transformations which accomplish this have been proposed but with all these transformations the set of invariants include rather more than are necessary to characterise the system's dynamic behaviour, so that the resulting equivalence classes are artificially fragmented. The notion of "strong equivalence" proposed by Verghese et al (1981) is the nearest to overcoming this limitation but unfortunately it suffers the defect that no closed form expression for it in terms of the system matrices is apparent. It is the purpose of this paper to indicate how such difficulties may be overcome and to develop the necessary closed form expression of the equivalence that exists between two generalised state space systems exhibiting identical finite and infinite frequency behaviour.

TRANSFORMATIONS OF REGULAR MATRIX PENCILS

Before considering the case of generalised state space systems it is instructive to consider the transformations of the underlying matrix pencils. Within the generalised state-space system matrix there are a number of interesting matrix pencils the most important of which is the pole pencil.

$$D(s) = sE-A \quad . \tag{1}$$

In constructing any system matrix transformation it is essential that the pole pencil is transformed in a wholly proper manner. Since such a pencil is regular (Gantmacher, 1959) we will concern ourselves primarily with the equivalence of regular matrix pencils and in this section only those results of direct interest to the present discussion will be given. For a more complete analysis of transformations of the various matrix pencils relevant to generalised state-space systems matrices reference is made to Pugh, Hayton and Fretwell (1985).

Let $P_0(r)$ denote the class of $(r\times r)$ regular matrix pencils where the integer $r > 0$ is variable. We then say

Definition 1: Two pencils $D_1(s) = sE_1-A_1, D_2(s) = sE_2-A_2$ in $P_o(r)$ are said to be COMPLETELY EQUIVALENT (c.e.) in case . constant matrices N_1, N_2 such that

$$D_1(s)N_2 = N_1D_2(s) \ , \tag{2}$$

where

$$(sE_1-A_1 \quad N_1) \ , \quad \begin{bmatrix} -N_2 \\ sE_2-A_2 \end{bmatrix} \quad \text{have no finite nor infinite zeros.} \tag{3}$$

It is noted that the usual requirements of strict equivalence (s.e.) (Gantmacher, 1959) that N_1, N_2 be square and nonsingular are generalised by (3). This imposes conditions on the way the transformation of complete equivalence acts with respect to points in the finite part of the complex plane and the point at infinity. Although this is also true of the requirements of strict equivalence the conditions (3) do this in a much less restrictive way since (2) now allows matrices of different dimensions to be related as the following result shows

Lemma 1: Two regular pencils obtained from each other by a trivial deflation/inflation are (c.e.)

Proof: $D_1(s) = sE-A$ and $D_2(s) = \begin{bmatrix} I & 0 \\ 0 & sE-A \end{bmatrix}$ are related by

$$\text{Deflation:} \quad \begin{bmatrix} I & 0 \\ 0 & sE-A \end{bmatrix} \begin{bmatrix} 0 \\ I \end{bmatrix} = \begin{bmatrix} 0 \\ I \end{bmatrix} \begin{bmatrix} sE-A \end{bmatrix} ,$$

$$\text{Inflation:} \quad \begin{bmatrix} sE-A \end{bmatrix} \begin{bmatrix} 0 & I \end{bmatrix} = \begin{bmatrix} 0 & I \end{bmatrix} \begin{bmatrix} I & 0 \\ 0 & sE-A \end{bmatrix} , \tag{4}$$

and the conditions (3) are clearly satisfied.
Another technical result which is useful in the sequel is

Lemma 2: Two pencils $D_1(s), D_2(s) \in P_o(r)$ are (c.e.) iff any pencils obtained from them by (s.e.) are also (c.e.).

Proof: Suppose $D_1'(s), D_2'(s)$ are obtained from $D_1(s), D_2(s)$ respectively by (s.e.) then

$$D_1(s) = L_1D_1'(s)L_2^{-1} \ , \qquad D_2 = M_1D_2'(s)M_2^{-1} \ , \tag{5}$$

with L_1, L_2, M_1, M_2 square and nonsingular. If $D_1(s), D_2(s)$ are (c.e.) then

$$D_1'(s)L_2^{-1}N_2M_2 = L_1^{-1}N_1M_1D_2'(s) \ . \tag{6}$$

This transformation is in the form of (c.e.) and the appropriate relative primeness conditions are readily verified in view of (3).

Theorem 1: The transformation of (c.e.) preserves the finite and infinite zeros of a regular* pencil sE-A.

Proof: (*This result also holds for singular pencils (Pugh et al. 1985)). The result for the finite zeros is due to Pugh and Shelton (1978) but a simpler proof is due to Smith (1981). For the infinite zeros

$$D_1(s)N_2 = N_1D_2(s) \ , \tag{7}$$

where $D_1(s) = sE_1-A_1, D_2(s) = sE_2-A_2$ and the conditions (3) hold. Let

$$\left.\begin{aligned} \left[\tfrac{1}{\omega} E_1-A_1 \quad N_1\right] &= T_1(\omega)^{-1}\left[\underline{D}_1(\omega) \quad \underline{N}_1(\omega)\right] \\ \begin{bmatrix} -N_2 \\ \tfrac{1}{\omega} E_2-A_2 \end{bmatrix} &= \begin{bmatrix} -\underline{N}_2(\omega) \\ \underline{D}_2(\omega) \end{bmatrix} T_2^{-1}(\omega) \end{aligned}\right\} , \qquad (8)$$

be relatively prime factorisations then (3) imply that

$$\left[\underline{D}_1(\omega) \quad \underline{N}_1(\omega)\right] ; \qquad \begin{bmatrix} -\underline{N}_2(\omega) \\ \underline{D}_2(\omega) \end{bmatrix} , \qquad (9)$$

have full rank for all finite ω.

Now putting $s = 1/\omega$ into (7) and substituting from (8) yields

$$\underline{D}_1(\omega)\underline{N}_2(\omega) = \underline{N}_1(\omega)\underline{D}_2(\omega) \ . \qquad (10)$$

Equations (10) and (9) may be used in the same way as (2) and (3) to show that $\underline{D}_1(\omega)$, $\underline{D}_2(\omega)$ have the same zero structure at $\omega = 0$.

It now remains to prove that $\underline{D}_1(\omega)$ and $\underline{D}_2(\omega)$ are numerators of $D_1(1/\omega)$ and $D_2(1/\omega)$ and as such (when $\omega = 0$) represent the infinite zero structure of $D_1(s)$ and $D_2(s)$ respectively. Now if $\nu(.)$ denotes the least order and $\delta(.)$ denotes the McMillan degree of the indicated matrix, it follows since $D_1(s)$ is polynomial and N_1 is constant, that

$$\begin{aligned} \nu(\tfrac{1}{\omega}E_1-A_1 \quad N_1) &= \delta(sE_1-A_1 \quad N_1) \\ &= \delta(sE_1-A_1) \qquad (11) \\ &= \nu(\tfrac{1}{\omega} E_1-A_1) \ . \end{aligned}$$

Hence $\frac{1}{\omega} E_1 - A_1 = T_1^{-1}(\omega)\underline{D}_1(\omega)$, is a relatively left prime factorisation. i.e. $\underline{D}_1(\omega)$ is a numerator of $D_1(\frac{1}{\omega})$. In a similar way it follows that $\underline{D}_2(\omega)$ is a numerator of $D_2(\frac{1}{\omega})$ and the result follows.

Now it is known that to each finite elementary divisor of a certain degree there corresponds a finite zero of exactly that degree while to every infinite elementary divisor of degree k there corresponds an infinite zero of degree k − 1, and vice versa. (Verghese et al, 1981). In these terms theorem 1 yields

<u>Corollary</u>: (c.e.) preserves all finite elementary divisors and those infinite elementary divisors of degree two or more. (c.e.) does not preserve the infinite elementary divisors of degree 1.

It is not obvious that (c.e.) is a transformation of equivalence but the above corollary may be used to formally establish this.

<u>Theorem 2</u>: (c.e.) is an equivalence relation on the set $P_o(r)$.

<u>Proof</u>: The reflexivity is obvious since if $D(s) \in P_o(r)$ then

$$I.D(s) = D(s).I$$

with the properties (3) holding.

For symmetry let $K(D_1), K(D_2)$ denote the Kronecker standard forms (Gantmacher,

1959) of D_1, D_2 respectively. Since any pencil is (s.e.) to this form and since D_1, D_2 are (c.e.) it follows by application of lemma 2 that $K(D_1), K(D_2)$ are (c.e.). Thus by the Corollary to theorem 1 $K(D_1), K(D_2)$ have the same finite elementary divisors and the same infinite elementary divisors of degree two or more. Hence $K(D_1), K(D_2)$ differ only in the number of infinite elementary divisors of degree 1. Thus

$$K(D_i) = \begin{bmatrix} sI-A' & 0 & 0 \\ 0 & I_{p_i} & 0 \\ 0 & 0 & I-sJ \end{bmatrix}, \quad i = 1,2 . \tag{12}$$

Now $$K(D_2)M = MK(D_1) , \tag{13}$$

with an obvious notation

$$M = \begin{bmatrix} I & 0 & 0 \\ 0 & 0_{p_2,p_1} & 0 \\ 0 & 0 & I \end{bmatrix} . \tag{14}$$

It is readily verified that the relation (13) satisfies the conditions (3) and hence $K(D_2)$ and $K(D_1)$ are (c.e.). Application of Lemma 2 then shows that D_2 and D_1 are (c.e.) which establishes symmetry.

For transitivity suppose that $D_1, D_2, D_3 \in P_0(r)$ and that D_1, D_2 and D_2, D_3 are (c.e.). By Lemma 2 $K(D_1), K(D_2)$ and $K(D_2), K(D_3)$ are (c.e.) and hence

$$K(D_1)N = NK(D_2) \; ; \qquad K(D_2)L = LK(D_3) , \tag{15}$$

with N and L having the same form as M of (14) with appropriate modifications. From (15)

$$K(D_1)NL = NLK(D_3) , \tag{16}$$

with NL having a similar form to (14). It is then easily verified that $K(D_1), K(D_3)$ are (c.e.). Hence by Lemma 2 D_1, D_3 are (c.e.) as required.

Another important result concerning (c.e.) of regular pencils in Kronecker form is

<u>Theorem 3</u>: Let $K(D_1), K(D_2)$ be (c.e.) regular pencils in Kronecker standard form. If $K(D_1), K(D_2)$ have the same dimension then the transforming matrices N_1, N_2 in (2) are square and nonsingular.

<u>Proof</u>: Since $K(D_1), K(D_2)$ are (c.e.)

$$\begin{bmatrix} sI_{r_1}-A_1 & 0 \\ 0 & I-sJ_1 \end{bmatrix}\begin{bmatrix} N_{21} & N_{22} \\ N_{23} & N_{24} \end{bmatrix} = \begin{bmatrix} N_{11} & N_{12} \\ N_{13} & N_{14} \end{bmatrix}\begin{bmatrix} sI_{r_2}-A_2 & 0 \\ 0 & I-sJ_2 \end{bmatrix} , \tag{17}$$

with the appropriate form of conditions (3) holding. As the pencils have equal dimension there has been no trivial inflation or deflation and hence the corollary to Theorem 1 may be applied to show that

$$J_1 = J_2 \; ; \qquad r_1 = r_2 = n, \quad \text{say} . \tag{18}$$

Considering (block) element 1,2 in (18)

$$(sI-A_1)N_{22} = N_{12}(I-sJ_2) , \tag{19}$$

that is

$$N_{22} = N_{12}J_2 \quad , \tag{20}$$

$$-A_1N_{22} = N_{12} \quad . \tag{21}$$

Eliminating N_{12} between (20) and (21)

$$N_{22} = - A_1N_{22}J_2 \quad . \tag{22}$$

Let p be the dimension of the first Jordan block in J_2 and let n_i $i=1,2,\ldots,p$ denote the first p columns of N_{22}, then the first p columns of (22) become

$$(n_1\ n_2 \ldots n_p) = -(0\ An_1\ An_2 \ldots An_{p-1}) \quad . \tag{23}$$

$$\left.\begin{array}{lll} \text{Thus} & n_1 = 0 & \\ \text{and} & n_i = An_{i-1} = 0 & i=2,3,\ldots,p \end{array}\right\} \quad . \tag{24}$$

The argument above may be repeated for each Jordan block in J_2 in turn and so

$$N_{22} = 0 \quad , \tag{25}$$

and hence from (21)

$$N_{12} = 0 \quad . \tag{26}$$

An analogous argument may be used to show

$$N_{23} = N_{13} = 0 \quad . \tag{27}$$

Now consider (block) elements 1,1 in (17)

$$(sI-A_1)N_{21} = N_{11}(sI-A_2) \quad , \tag{28}$$

with conditions (3) holding so that N_{11}, $sI-A_1$ are relatively left prime and hence

$$\text{rank}(N_{11},\ A_1N_{11},\ldots,A_1^{n-1}N_{11}) = n \quad . \tag{29}$$

Equating constants in (28)

$$A_1N_{21} = N_{11}A_2 \quad , \tag{30}$$

and equating powers of s

$$N_{21} = N_{11} \quad . \tag{31}$$

From (29), (30), (31)

$$\text{rank}(N_{11},N_{11}A_2,\ldots,N_{11}A_2^{n-1}) = n \quad . \tag{32}$$

Clearly the matrix $(N_{11},N_{11}A_2,\ldots,N_{11}A_2^{n-1})$ may be reduced by invertible operations to $(N_{11},0,\ldots,0)$. Hence the condition (29) requires that N_{11} has rank n. Thus since N_{11} is nxn it is nonsingular.

Finally consider (block element 2,2 in (18)

$$(I-sJ)N_{24} = N_{14}(I-sJ) \quad , \tag{33}$$

with conditions (3) holding so that

$$\begin{bmatrix} I-sJ \\ N_{24} \end{bmatrix}$$

has no infinite zeros (the special form I-sJ has no finite zeros). A numerator for $\begin{bmatrix} I-sJ \\ N_{24} \end{bmatrix}$ is $\begin{bmatrix} \text{diag.}(T_{p_i}) \\ N_{24} \end{bmatrix}$ where T_{p_i} are matrices of the form:

$$T_{p_i} = \begin{bmatrix} s & 1 & 0 & \cdots & \cdots & 0 \\ 0 & s & 1 & & & \\ \cdot & & \ddots & \ddots & & \\ \cdot & & & \ddots & \ddots & \\ 0 & & & & s & 1 \\ 0 & 0 & 0 & \cdots & 0 & 1 \end{bmatrix} \Big\updownarrow p_i \, , \qquad \text{(34)}$$

$$\longleftarrow p_i \longrightarrow$$

with p_i, the dimensions of the corresponding Jordan block in J. Let $N' = (n_1', n_2' \ldots n_p')$ be a matrix comprising the first p_1 columns of N_{24}. Since

$\begin{bmatrix} 1-sJ \\ N_{24} \end{bmatrix}$ has no infinite zeros $\begin{bmatrix} \text{diag.}(T_{p_i}) \\ N_{24} \end{bmatrix}$ has full rank at s = 0 and hence it is

hence it is clear from the form of T_{p_1} that $n_1' \neq 0_1$. Equating constants in (33)

$$N_{24} = N_{14} \, . \qquad (35)$$

and equating powers of s

$$JN_{24} = N_{14}J \, . \qquad (36)$$

Assume N_{24} has less than full rank then there exists a nonzero vector ξ such that

$$\begin{bmatrix} N_{24} \\ JN_{24} \\ J^2N_{24} \\ \vdots \\ J^{p_1-1}N_{24} \end{bmatrix} \xi = 0 \, , \qquad (37)$$

substituting into (37) from (35) and (36) and considering only the first p_1 columns of $N_{24} = (N', N'')$ together with the corresponding elements of $\xi(\xi_1, \xi_2, \ldots, \xi_{p_1})$

$$\begin{bmatrix} N' \\ N'J \\ \vdots \\ N'J^{p_1-1} \end{bmatrix} \begin{bmatrix} \xi_1 \\ \xi_2 \\ \vdots \\ \xi_{p_1} \end{bmatrix} = 0 \, , \qquad (38)$$

that is

$$\begin{bmatrix} n_1' & n_2' \cdots\cdots n_{p_1}' \\ & n_1' \cdots\cdots n_{p_1-1}' \\ 0 & \ddots \\ & & n_1' \end{bmatrix} \begin{bmatrix} \xi_1 \\ \xi_2 \\ \vdots \\ \xi_{p_1} \end{bmatrix} = 0 \ . \tag{39}$$

Considering the last (block) row of (39)

$$n_1' \xi_{p_1} = 0 \ , \tag{40}$$

and since $n_1' \neq 0$, we have $\xi_{p_1} = 0$.

Substituting for ξ_{p_1} in the penultimate (block) row of (39)

$$\left. \begin{aligned} n_1' \xi_{p_1-1} &= 0 \\ \text{i.e.} \qquad \xi_{p_1-1} &= 0 \end{aligned} \right\} \tag{41}$$

Proceeding in this way and considering each of the Jordan blocks in J in turn it may be shown that $\xi = 0$ which contradicts the initial assumption that N_{24} is singular. It follows that N_{24} is nonsingular and collecting all the results together (18) reduces to

$$\begin{bmatrix} sI-A_1 & 0 \\ 0 & I-sJ \end{bmatrix} \begin{bmatrix} N_{21} & 0 \\ 0 & N_{24} \end{bmatrix} = \begin{bmatrix} N_{21} & 0 \\ 0 & N_{24} \end{bmatrix} \begin{bmatrix} sI-A_2 & 0 \\ 0 & I-sJ \end{bmatrix} , \tag{42}$$

with N_{21}, N_{24} nonsingular, thus proving the theorem.

The above theorem has important corollaries which describe the way in which the transformation of (c.e.) is a generalisation of other well-known matrix pencil transformations. Application of Lemma 2 to theorem 3 gives

<u>Corollary 1</u>: If $D_1(s), D_2(s) \in P_o(r)$ are of the same dimension than $D_1(s)$ and $D_2(s)$ are (c.e.) if and only if they are (s.e.).

For the special case of pencils in the conventional form sI-A we note the following result obtained directly from (42).

<u>Corollary 2</u>: $sI-A_1$, $sI-A_2 \in P_o(r)$ are (c.e.) if and only if they are similar.

A MATRIX TRANSFORMATION FOR GENERALISED STATE-SPACE SYSTEMS

In the case of matrix pencils it has been seen that the transformation of (c.e.) possesses the kind of properties of interest. It not only preserves the finite and infinite frequency structure of any pencil to which it is applied but also it is capable of transforming between pencils at different dimensions. It is natural therefore to consider the systems theory extension of (c.e.).

Let $P(m, \ell)$ denote the class of $(r+m)\times(r+\ell)$ generalised state-space system matrices of the form

$$P(s) = \begin{bmatrix} sE-A & B \\ -C & D \end{bmatrix} , \tag{43}$$

where the integer $r > 0$ is variable. As is usual in this context it is assumed that

$$|sE-A| \neq 0 . \tag{44}$$

The equivalence induced by (c.e.) on the class $P(m,\ell)$ is then as follows.

<u>Definition 2</u>: $P_1(s), P_2(s) \in P(m,\ell) \in P(m,\ell)$ are said to be COMPLETELY SYSTEM EQUIVALENT (c.s.e.) if constant matrices N_1, N_2, X, Y such that

$$\begin{bmatrix} sE_1-A_1 & B_1 \\ -C_1 & D_1 \end{bmatrix}\begin{bmatrix} N_2 & Y \\ 0 & I \end{bmatrix} = \begin{bmatrix} N_1 & 0 \\ X & I \end{bmatrix}\begin{bmatrix} sE_2-A_2 & B_2 \\ -C_2 & D_2 \end{bmatrix} , \tag{45}$$

where the matrices

$$\begin{bmatrix} sE_1-A_1 & N_1 \end{bmatrix} ; \quad \begin{bmatrix} -N_2 \\ sE_2-A_2 \end{bmatrix} \tag{46}$$

have no finite or infinite zeros.

It is noted that earlier studies (Verghese et al, 1981) of generalised state-space systems have approached the problem of defining equivalence in an algorithmic way, the elementary operations permitted being described essentially as in the following definition.

<u>Definition 3</u>: The operations embodied in the transformation

$$\begin{bmatrix} sE_1-A & B_1 \\ -C_1 & D_1 \end{bmatrix}\begin{bmatrix} N_2 & Y \\ 0 & I \end{bmatrix} = \begin{bmatrix} N_1 & 0 \\ X & I \end{bmatrix}\begin{bmatrix} sE_2-A_2 & B_2 \\ -C_2 & D_2 \end{bmatrix} , \tag{47}$$

where N_1, N_2 are square and nonsingular are called OPERATIONS OF STRONG EQUIVALENCE.

This notion of equivalence of generalised state-space systems is then summarised as follows.

<u>Definition 4</u>: (Verghese et al, 1981): $P_1(s), P_2(s) \in P(m,\ell)$ are said to be STRONGLY EQUIVALENT (str. eq.) if after some sequence of operations of strong equivalence and after each system matrix has been deflated as far as possible, the resulting system matrices are related by operations of strong equivalence.

Unfortunately a closed form expression for (str. eq.) involving the system matrices $P_1(s), P_2(s) \in P(m,\ell)$ was not given in Verghese et al (1981). It will now be shown that (c.s.e.) remedies this omission. A first result is

<u>Lemma 3</u>: $P_1(s), P_2(s) \in P(m,\ell)$ are (c.s.e.) iff the generalised state-space system matrices obtained from them by operations of strong eqivalence are obtained (c.s.e.).

<u>Proof</u>: The proof of Lemma 2 is simply adapted.

As a first step to establishing the precise connection between the two notions of equivalence it will now be formally proved that (c.s.e.) is an equivalence relation.

Theorem 4: (c.s.e.) is an equivalence relation on the set $P(m,\ell)$.

Proof: For reflexivity the transforming matrices N_1, N_2 are taken as unit matrices and X, Y as zero and conditions (46) are easily verified.
For transitivity suppose $P_1(s), P_2(s), P_3(s) \in P(m,\ell)$ are related as

$$\begin{bmatrix} sE_1-A_1 & B_1 \\ -C_1 & D_1 \end{bmatrix}\begin{bmatrix} N_1 & Y_1 \\ 0 & I \end{bmatrix} = \begin{bmatrix} N_2 & 0 \\ X_2 & I \end{bmatrix}\begin{bmatrix} sE_2-A_2 & B_2 \\ -C_2 & D_2 \end{bmatrix}, \tag{48}$$

$$\begin{bmatrix} sE_2-A_2 & B_2 \\ -C_2 & D_2 \end{bmatrix}\begin{bmatrix} N_3 & Y_3 \\ 0 & I \end{bmatrix} = \begin{bmatrix} N_4 & 0 \\ X_4 & I \end{bmatrix}\begin{bmatrix} sE_3-A_3 & B_3 \\ -C_3 & D_3 \end{bmatrix}, \tag{49}$$

combining (48), (49) gives

$$\begin{bmatrix} sE_1-A_1 & B_1 \\ -C_1 & D_1 \end{bmatrix}\begin{bmatrix} N_1N_3 & N_1Y_3+Y_1 \\ 0 & I \end{bmatrix} = \begin{bmatrix} N_2N_4 & 0 \\ X_2N_4+X_4 & I \end{bmatrix}\begin{bmatrix} sE_3-A_3 & B_3 \\ -C_3 & D_3 \end{bmatrix}. \tag{50}$$

This has the form of (c.s.e.). The properties (46) follow from Theorem 2.

For symmetry assume that $P_1(s), P_2(s) \in P(m,\ell)$ are (c.s.e.). Then by Lemma 3, $P_i(s)$ may be replaced by $P_i'(s)$ in which the pole pencil is $K(sE_i-A_i)$ (i=1,2). It is thus only necessary to show that $P_2'(s), P_1'(s)$ are (c.s.e.).

Let $\ddot{P}_i(s)$ be the totally deflated form of $P_i'(s)$ (i=1,2). The deflation of $P_2'(s)$ to $\ddot{P}_2(s)$ and the inflation of $\ddot{P}_1(s)$ to $P_1'(s)$ are both transformations of (c.s.e.) by Lemma 1. Hence by the transitivity property of (c.s.e.). It follows that $\ddot{P}_1(s), \ddot{P}_2(s)$ are (c.s.e.). That is

$$\left[\begin{array}{cc|c} sI-A & 0 & \bar{B}_1 \\ 0 & I-sJ & \hat{B}_1 \\ \hline -\bar{C}_1 & -\hat{C}_1 & \hat{D}_1 \end{array}\right]\begin{bmatrix} N_2 & Y \\ 0 & I \end{bmatrix} = \begin{bmatrix} N_1 & 0 \\ X & I \end{bmatrix}\left[\begin{array}{cc|c} sI-A & 0 & \bar{B}_2 \\ 0 & I-sJ & \hat{B}_2 \\ \hline -\bar{C}_2 & -\hat{C}_2 & \hat{D}_2 \end{array}\right]. \tag{51}$$

Application of Theorem 3 shows that N_2, N_1 and hence

$$\begin{bmatrix} N_2 & Y \\ 0 & I \end{bmatrix}, \quad \begin{bmatrix} N_1 & 0 \\ X & I \end{bmatrix}$$ are square and invertible so that

$$\begin{bmatrix} N_1^{-1} & 0 \\ -XN_1^{-1} & I \end{bmatrix}\left[\begin{array}{cc|c} sI-A & 0 & \bar{B}_1 \\ 0 & I-sJ & \hat{B}_1 \\ \hline -\bar{C}_1 & -\hat{C}_1 & \hat{D}_1 \end{array}\right] = \left[\begin{array}{cc|c} sI-A & 0 & \bar{B}_2 \\ 0 & I-sJ & \hat{B}_1 \\ \hline -\bar{C}_2 & -\hat{C}_2 & \hat{D}_2 \end{array}\right]\begin{bmatrix} N_2^{-1} & -YN_2^{-1} \\ 0 & I \end{bmatrix}. \tag{52}$$

That is $\ddot{P}_2(s)$ is c.s.e. to $\ddot{P}_1(s)$. Further application of transitivity then gives the required result.

The connection between (str.eq.) and (c.s.e.) may now be formally established.

Theorem 5: $P_1(s), P_2(s) \in P(m,\ell)$ are (str. eq.) iff they are (c.s.e.).

Proof: From Lemma 1 and Lemma 3 respectively it is seen that both trivial inflation/deflation and operations of strong equivalence may be checked by transformations of (c.s.e.). It follows from the transitivity property of (c.s.e.) that if $P_1(s), P_2(s)$ are (str. eq) they are also (c.s.e.). Conversly suppose that $P_1(s), P_2(s)$ are (c.s.e) then $P_i(s)$ may be replaced by $P_i'(s)$ i=1,2 as in the proof of symmetry above. $P_1'(s)$ and $P_2'(s)$ are related by trivial inflation/deflation together with the operations (51) which are clearly operations of strong equivalence. It follws that $P_1'(s), P_2'(s)$ and hence $P_1(s), P_2(s)$ are (str. eq.) completing the proof.

The above theorem shows that (str. eq.) and (c.s.e.) are equivalent notions and as such (c.s.e.) provides a closed form matrix statement of the requirements of (str. eq.). The next results show that (c.s.e.) preserves the dynamic structure of the system, and is a consequence of the note on thoerem 1.

Theorem 6: (c.s.e.) leaves invariant the following sets
- (i) the finite and infinite transmission zeros
- (ii) the finite and infinite transfer function poles
- (iii) the finite and infinite system zeros
- (iv) the finite and infinite system poles
- (v) the finite and infinite decoupling zeros.

Theorem 6 shows that (c.s.e.) leaves the pole zero structure of the system unchanged. Other parameters of importance to the dynamical behaviour of the system are the input (resp. output) dynamical indices defined as the row (resp. column) minimal indices of the singular pencil $(sE-A \quad B) \left\{ \text{resp.} \begin{bmatrix} sE-A \\ -C \end{bmatrix} \right\}$ (Rosenbrock and Hayton, 1974). The following result shows that these are also invariant under (c.s.e.).

Theorem 7: (c.s.e.) leaves invariant the system input and output dynamical indices

Proof: Let $P_1(s), P_2(s) \in P(m, \ell)$ be (c.s.e.) and let $\delta_{i1} \leq \delta_{i2} \leq \ldots \leq \delta_{i\ell}$ be the input dynamical indices of $P_i(s)$ i=1,2.

Consider

$$\begin{bmatrix} sE_i-A_i & B_i \end{bmatrix} \begin{bmatrix} \xi(s) \\ u(s) \end{bmatrix} = 0 \qquad i=1,2\ , \tag{53}$$

and let $\begin{bmatrix} \xi_{21}(s) \\ u_{22}(s) \end{bmatrix}$ be a solution of (53) for i=2 of largest degree then $\begin{bmatrix} \xi_{21}(s) \\ u_{21}(s) \end{bmatrix}$ has degree δ_{21} (Gantmacher, 1959). Since $P_1(s), P_2(s)$ are (c.s.e.) it is readily seen that

$$\begin{bmatrix} \xi_{11}(s) \\ u_{11}(s) \end{bmatrix} = \begin{bmatrix} N_2 & Y \\ 0 & I \end{bmatrix} \begin{bmatrix} \xi_{21}(s) \\ u_{22}(s) \end{bmatrix}, \tag{54}$$

is a solution of (53) for i=1. The tranforming matrix in (54) is constant so $\begin{bmatrix} \xi_{11}(s) \\ u_{11}(s) \end{bmatrix}$ has degree no greater than $\begin{bmatrix} \xi_{21}(s) \\ u_{21}(s) \end{bmatrix}$ that is

$$\delta_{11} \leq \delta_{12}\,. \tag{55}$$

By the symmetry property of Theorem 4, however, $P_2(s), P_1(s)$ are (c.s.e.) and the argument above may be repeated to show

$$\delta_{12} \leq \delta_{11} . \tag{56}$$

It follows from (55) and (56) that

$$\delta_{11} = \delta_{12} . \tag{57}$$

Now among those solutions of (53), i=2, which are independent of $\begin{bmatrix} \xi_{21}(s) \\ u_{21}(s) \end{bmatrix}$ let $\begin{bmatrix} \xi_{22}(s) \\ u_{22}(s) \end{bmatrix}$ be one having lowest degree. Then the degree of $\begin{bmatrix} \xi_{22}(s) \\ u_{22}(s) \end{bmatrix}$ is δ_{22}.

$$\begin{bmatrix} \xi_{12}(s) \\ u_{12}(s) \end{bmatrix} = \begin{bmatrix} N_2 & Y \\ 0 & I \end{bmatrix} \begin{bmatrix} \xi_{22}(s) \\ u_{22}(s) \end{bmatrix} , \tag{58}$$

is certainly a solution of (53) for i=1, it will have degree δ_{12} if it is independent of $\begin{bmatrix} \xi_{11}(s) \\ u_{11}(s) \end{bmatrix}$. Suppose that $\begin{bmatrix} \xi_{1i}(s) \\ u_{1i}(s) \end{bmatrix}$ i=1,2 are linearly dependent, that is suppose there exist constant nonzero α_1, α_2 such that

$$\alpha_1 \begin{bmatrix} N_2 & Y \\ 0 & I \end{bmatrix} \begin{bmatrix} \xi_{21}(s) \\ u_{21}(s) \end{bmatrix} + \alpha_2 \begin{bmatrix} N_2 & Y \\ 0 & I \end{bmatrix} \begin{bmatrix} \xi_{22}(s) \\ u_{22}(s) \end{bmatrix} = 0 , \tag{59}$$

from consideration of the second (block) row in (59)

$$\alpha_1 u_{21}(s) + \alpha_2 u_{22}(s) = 0 . \tag{60}$$

Recalling that $\begin{bmatrix} \xi_{2i}(s) \\ u_{2i}(s) \end{bmatrix}$ i=1,2 are linearly independent solutions of (53), i=2 it follows from (60) that there exists a non-zero vector $\xi'(s) = \alpha_1 \xi_1(s) + \alpha_2 \xi_2(s)$ such that

$$(sE_2 - A_2)\xi'(s) = 0 , \tag{61}$$

(61) contradicts the regularity of the pole pencil $sE_2 - A_2$. Hence $\begin{bmatrix} \xi_{1i}(s) \\ u_{1i}(s) \end{bmatrix}$ i=1,2 are linearly independent, $\begin{bmatrix} \xi_{12}(s) \\ u_{12}(s) \end{bmatrix}$ has degree δ_{12} and so

$$\delta_{12} \leq \delta_{22} . \tag{62}$$

The symmetry of c.s.e. is again used to show that equality holds in (62). The argument may be repeated to show that all the input dynamical indices of the system are invariant under (c.s.e.). Invariance of the output dynamical indices may be shown in the same way by considering solutions of

$$\left[\xi^t(s), y^t(s)\right] \begin{bmatrix} sE_i - A_i \\ -C_i \end{bmatrix} = 0 \quad i=1,2 . \tag{63}$$

A MAPPING INTERPRETATION OF (C.S.E.)

The transformation (c.s.e.) described in Section 3 above was derived as a closed form matrix transformation which retains input/output behaviour, (finite and infinite) pole and zero structure and input and output dynamical indices – that is it preserves all the properties fundamental to the dynamic behaviour of the system. An alternative and intuitively more attractive way of defining equivalence of two systems is now presented. Space precludes the inclusion of complete proof of the results in this section – details can be found in Hayton et al (1984). The system to be considered is described by a set of first order, linear differential equations Laplace transformed with nonzero initial conditions to give

$$\begin{bmatrix} sE-A & B \\ -C & D \end{bmatrix} \begin{bmatrix} x(s) \\ u(s) \end{bmatrix} = \begin{bmatrix} Ex(0-) \\ -y(s) \end{bmatrix} . \tag{64}$$

Equivalence is then defined in terms of maps between

(i) the solution/input vector pairs $[x_1^t(s), u^t(s)]^t$, $[x_2^t(s), u^t(s)]^t$ and

(ii) the restricted initial condition/output vector pairs $[(E_1x_1(0-)^t, y^t(s)]^t$, $[E_2x_2(0-)^t, y^t(s)]^t$.

It is noted from the form of the system equation (64) that the pairs referred to here are fundamental to the system. The system equations themselves are simply mappings to an appropriate set of initial condition/ouput pairs from an appropriate set of solution/input pairs. It is the nature of these two sets that characterise a system, for the solution/input pairs completely describe the controllability properties of the system, while the restricited intial condition/output pairs encapsulate the observability properties. These comments are probably best understood in the context of system equations in the classical state-space form for then the number of initial condition/output pairs which contains a given output describes the extent to which the system is unobservable. A system in state-space form will, of course, be completely observable in case a given output occurs in one and only one initial conditon/output pair. Dual statements may be made concerning the solution/input pairs and complete controllability.

It would appear that to preserve the structure of the pairs in (i) and (ii) above (and hence to preserve the controllability and observability properties of the system) it would be necessary for there to be two bijections, one relating the solution/input vector pairs and the second relating the restricted initial condition/output vector pairs. However this is not necessary and it will be shown that the following definition suffices.

<u>Definition 5</u>: Let $\mathbb{P}_1, \mathbb{P}_2$ be two generalised state-space systems of the form (64). They are said to be fundamentally equivalent (f.e.) if there exist

(i) a constant, injective map

$$\begin{bmatrix} x_2(s) \\ -u(s) \end{bmatrix} = \begin{bmatrix} N & Y \\ 0 & I \end{bmatrix} \begin{bmatrix} x_1(s) \\ -u(s) \end{bmatrix} , \tag{65}$$

and

(ii) a constant, surjective map

$$\begin{bmatrix} E_2x_2(0-) \\ -y(s) \end{bmatrix} = \begin{bmatrix} M & 0 \\ X & I \end{bmatrix} \begin{bmatrix} E_1x_1(0-) \\ -y(s) \end{bmatrix} . \tag{66}$$

Following Pernebo (1977) Verghese et al (1981) have proposed a similar definition to 5 which refers only to the mapping (65) with the requirement that it be a bijection. However to specify two maps as in Definition 5 is a recognition of the importance from a system viewpoint of the pairs referred to in (65) and (66). Further, the splitting of the bijection requirement in the manner indicated permits the duality of the controllability and observability concepts contained within the pairs to be fully exploited. The first result formally establishes the bijectivity of the above maps.

Theorem 8: Let $\mathbb{P}_1, \mathbb{P}_2$ be two systems in generalised state-space form. Then the maps of (65), (66) are both bijective.

Proof: Let t_1(=rank E_1), t_2(=rank E_2) be the dimensions of the vector spaces $\{E_1x_1(0-)\}$, $\{E_2x_2(0-)\}$. From (66) there exists a surjective map between the two spaces showing that

$$t_1 \geqslant t_2 \ . \tag{67}$$

A second, injective map between the spaces may be constructed (75) together with the system equation (64) and from this

$$t_1 \leqslant t_2 \ . \tag{68}$$

Thus the two spaces are of equal dimension and proof of the theorem follows.

If the pole pencils of the systems are in Kronecker form then (f.e.) acts in a decoupled fashion as is shown by

Theorem 9: Let $\mathbb{P}_1, \mathbb{P}_2$ be (f.e.) generalised state-space systems with pole pencils in Kronecker form then the subsystems

$$\begin{bmatrix} sI-\bar{A} & \bar{B} \\ -\bar{C} & 0 \end{bmatrix}\begin{bmatrix} x(s) \\ -u(s) \end{bmatrix} = \begin{bmatrix} x(0-) \\ -y(s) \end{bmatrix}; \quad \begin{bmatrix} I-sJ & \hat{B} \\ -\hat{C} & \hat{D} \end{bmatrix}\begin{bmatrix} \eta(s) \\ -u(s) \end{bmatrix} = \begin{bmatrix} -J\eta(0-) \\ -y(s) \end{bmatrix} \tag{69}$$

(referred to as the state-space and impulsive subsystems respectively) are (f.e.).

Proof: The proof is similar to that of Theorem 3 with each (block) element of the maps being considered individually.

The relationship between the state-space subsystems may be further specialised to provide the expected connection with the conventional state-space matrix transformation.

Corollary: If two generalised state-space systems in Kronecker standard form are (f.e.) then their state-space subsystems are system similar.

The result above actually holds for any pair of systems in the classical state space form which are (f.e.) and is essentially that due to Pernebo (1977).

The main difficulty in deriving a matrix characterisation of (f.e.) in the more general case lies in the fact that the maps whose existence is guaranteed by Definition 5 may not commute in a diagrammatic sense as is demonstrated by

Example: Consider two, identical, undriven systems described by the equations

$$\begin{bmatrix} I & s \\ 0 & I \end{bmatrix}\begin{bmatrix} \eta_1(s) \\ \eta_2(s) \end{bmatrix} = \begin{bmatrix} \eta_1(0-) \\ 0 \end{bmatrix} . \tag{70}$$

Solutions and initial conditons in both systems then take the form $(\eta_1(0-), 0)^t$ and appropriate maps are

$$M = N = \begin{bmatrix} 1 & 0 \\ 0 & 0 \end{bmatrix} , \tag{71}$$

but

$$ME_1 = \begin{bmatrix} 0 & 1 \\ 0 & 0 \end{bmatrix} , \qquad E_2N = \begin{bmatrix} 0 & 0 \\ 0 & 0 \end{bmatrix} . \tag{72}$$

Thus $E_2N \neq ME_1$ that is $(sE_2-A_2)N \neq M(sE_1-A_1)$.

The problem of non-commutability highlighted above may always be resolved by the construction of new maps defining the fundamental equivalence which additionally possess the required commutability property. The following result gives the details.

<u>Lemma 4</u>: If two generalised state-space systems with their pole pencils in Kronecker form are (f.e.) then

(i) there exists a constant map N_1 such that

$$\begin{bmatrix} x_2(s) \\ -u(s) \end{bmatrix} = \begin{bmatrix} N_1 & 0 \\ 0 & I \end{bmatrix} \begin{bmatrix} x_1(s) \\ -u(s) \end{bmatrix} ; \quad \begin{bmatrix} x_2(0-) \\ -y(s) \end{bmatrix} = \begin{bmatrix} N_1 & 0 \\ 0 & I \end{bmatrix} \begin{bmatrix} x_1(0-) \\ -y(s) \end{bmatrix} , \tag{73}$$

are bijections and $(sI-A_2)N_1 = N_1(sI-A_1)$, N_1 square and nonsingular and

(ii) there exists a constant map N_2 such that

$$\begin{bmatrix} N_2(s) \\ -u(s) \end{bmatrix} = \begin{bmatrix} N_2 & Y_2 \\ 0 & I \end{bmatrix} \begin{bmatrix} n_1(s) \\ -u_1(s) \end{bmatrix} ; \quad \begin{bmatrix} -J_2n_2(0-) \\ -y(s) \end{bmatrix} = \begin{bmatrix} N_2 & 0 \\ X_2 & I \end{bmatrix} \begin{bmatrix} -J_1n_1(0-) \\ -y(s) \end{bmatrix} \tag{74}$$

are bijections and $(I-sJ_2)N_2 = N_2(I-sJ_1)$.

<u>Proof</u>: The proof is constructive with suitable N_1, N_2 being generated from maps originally defining the fundamental equivalence.

Lemma 4 is the key to the next result.

<u>Theorem 10</u>: If two generalised state-space systems $\mathbb{P}_1, \mathbb{P}_2$ are (f.e.) then there exist maps M^*, N^* such that:

$$\begin{bmatrix} x_2(s) \\ -u(s) \end{bmatrix} = \begin{bmatrix} N^* & Y \\ 0 & I \end{bmatrix} \begin{bmatrix} x_1(s) \\ -u(s) \end{bmatrix} ; \quad \begin{bmatrix} E_2x_2(0-) \\ -y(s) \end{bmatrix} = \begin{bmatrix} M^* & 0 \\ X & I \end{bmatrix} \begin{bmatrix} E_1x_1(0-) \\ -y(s) \end{bmatrix} , \tag{75}$$

are bijections and $(sE_2-A_2)N^* = M^*(sE_1-A_1)$.

Use of Theorem 10 is the basic tool in proving the main result of this section.

<u>Theorem 11</u>: Two generalised state-space systems $\mathbb{P}_1, \mathbb{P}_2$ are (f.e.) iff they are (c.s.e.).

<u>Proof</u>: Theorem 10 provides a partial matrix characterisation of (f.e.), this is completed by considering the effect of the maps (65), (66) on the input matrix B, the ouput matrix C and the feedfoward term D. The resulting transformation is of the form (2) and is readily shown to satisfy conditon (3). For the converse case it is noted that the matrices in (2) actually form the maps (65) and (66)

and conditions (3) ensure that these maps are respectively injective and surjective. The above theorem unifies the various ideas contained in the several definitions of equivalence of generalised state-space systems discussed in this paper.

REFERENCES

[1] Bosgra, O.H. and Van Der Weiden, A.J.J., "Realisations in generalised state-space form for polynomial system matrices, and the definition of poles, zeros and decoupling zeros at infinity". Int. J. Control, 33, (1981) pp. 393-411.

[2] Gantmacher, F.R., "The Theory of Matrices" (Chelsea Pub. Co., New York, 1959).

[3] Hayton, G.E., Fretwell, P. and Pugh, A.C., "Fundamental Equivalence of Generalised State-space Systems". Research Report 182, Dept. of Maths., Loughborough University of Technology. (1984).

[4] Pernebo, L., "Notes on Strict System Equivalence", Int. J. Control, 25, (1977) pp. 21-38.

[5] Pugh, A.C., Hayton, G.E. and Fretwell, P., "On the Equivalence of Generalised State-space Systems", Research report 186, Dept. of Maths, Loughborough University of Technology (1985).

[6] Pugh, A.C. and Shelton, A.K., "On a New Definition of Strict System Equivalence", Int. J. Control, 27 (1978) pp. 657-672.

[7] Rosenbrock, H.H., "Structural Properties of Linear Dynamical Systems", Int. J. Control, 20 (1974) pp. 191-202.

[8] Rosenbrock, H.H. and Hayton, G.E., "Dynamical Indices of a Transfer Function Matrix", Int. J. Control, 20 (1974) pp. 177-189.

[9] Smith, M.C., "Matrix Functions and Strict System Equivalence", Int. J. Contol, 34 (1981) p. 869.

[10] Verghese, G., Levy, B.C. and Kailath, T., "A Generalised State-space for Singular Systems", I.E.E.E. Trans. Aut. Control, AC-26 (1981) pp. 811-830.

6. INFINITE DIMENSIONAL SYSTEMS

Frequency Domain and State Space Methods for Linear Systems
C.I. Byrnes and A. Lindquist (Editors)

TRANSFER FUNCTIONS OF EXPONENTIAL ORDER OF DISTRIBUTED SYSTEMS

F.M. Callier and J. Winkin

Department of Mathematics
Facultés Universitaires N.D. de la Paix
8, Rempart de la Vierge
B-5000 Namur BELGIUM

A transfer function algebra of fractions, viz. $\hat{\mathbf{F}}(\sigma_0)$, is constructed for modeling possibly unstable distributed linear systems in such a way that a) any transfer function in $\hat{\mathbf{F}}(\sigma_0)$ is holomorphic in $\operatorname{Re} s \geq \sigma_0$, (i.e. is σ_0-stable), iff it is σ_0-exponentially stable, and b) an infinite sum of delays is allowed in the direct input-output transmission of the system. This algebra is, on one hand, an extension of the algebra of proper rational functions such that the exponential order properties of these transfer functions for lumped systems are maintained, and, on the other hand, a restriction of the algebra $\hat{\mathbf{B}}(\sigma_0)$, [1], [2]. $\hat{\mathbf{F}}(\sigma_0)$ can be used for modeling and feedback system design, and fits standard semigroup systems better than $\hat{\mathbf{B}}(\sigma_0)$.

1. INTRODUCTION

Exp., I/O are abbreviations of "exponential(ly)", "input-output". $R\ M^{-1}$ denotes the set $\{f : f = n\,d^{-1}\ ,\ n \in R\ ,\ d \in M\}$. The Laplace transform as well as corresponding sets of Laplace transforms are denoted by a superhat. The symbol $*$ denotes convolution. We report the construction of a transfer function algebra of fractions $\hat{\mathbf{F}}(\sigma_0) := [\hat{\mathbf{E}}(\sigma_0)].[\hat{\mathbf{E}}^{\infty}(\sigma_0)]^{-1}$, (σ_0 is a real number), for distributed systems, [3], where (i) we allow **delay in the direct I/O transmission** of the system, and (ii) σ_0-stability, (i.e. the transfer function $\hat{f}$ is σ_0-stable iff $\hat{f}$ is holomorphic in $\operatorname{Re} s \geq \sigma_0$), is σ_0**-exp.** stability, (viz. exp. stability after multiplication of the impulse response $f(t)$ by $\exp(-\sigma_0 t)$). Because of the latter, $\hat{\mathbf{F}}(\sigma_0)$ fits semigroup systems better than the algebra of fractions $\hat{\mathbf{B}}(\sigma_0) := [\hat{\mathbf{A}}_-(\sigma_0)].[\hat{\mathbf{A}}_-^{\infty}(\sigma_0)]^{-1}$ developed by Callier and Desoer, [1], [2].
Note that $\hat{f} \in \hat{\mathbf{B}}(\sigma_0)$ is σ_0-stable iff $\hat{f} \in \hat{\mathbf{A}}_-(\sigma_0)$, or equivalently, on $\mathbb{R}_+$,

$f(t) = f_a(t) + \Sigma_{i=0}^{\infty} f_i.\delta(t-t_i)$ such that for some $\sigma < \sigma_0$,

(i) $\exp(-\sigma\cdot).f_a(\cdot) \in L_1$,

(ii) $t_0 = 0$ and $t_i > 0$ for $i=1,2,\ldots$, with $\delta(\cdot)$ the delta function, and

(iii) $f_i \in \mathbb{C}$ for $i=0,1,2,\ldots$ such that $\Sigma_{i=0}^{\infty} \exp(-\sigma t_i)\,|f_i| < \infty$;

$\hat{\mathbf{A}}_-^{\infty}(\sigma_0)$ is the multiplicative system[1] of elements of $\hat{\mathbf{A}}_-(\sigma_0)$ that are bounded away from zero at infinity in $\operatorname{Re} s \geq \sigma_0$. It is also reported that $\hat{\mathbf{F}}(\sigma_0)$ is an extension of the transfer function algebra for lumped systems, (viz. the algebra $\mathbb{C}_p(s) := [\mathbf{R}(\sigma_0)].[\mathbf{R}^{\infty}(\sigma_0)]^{-1}$ of proper rational functions, where $\mathbf{R}(\sigma_0)$ is the algebra of σ_0-stable and proper rational functions), maintaining the exponential order properties of the latter; ($\mathbf{R}^{\infty}(\sigma_0)$ is the multiplicative system of elements of $\mathbf{R}(\sigma_0)$ that are nonzero at infinity).

2. σ_0-EXP. STABLE TRANSFER FUNCTIONS, [3]

Observing that $g(\cdot) \in L_1$ does not imply in general that $g(t)$ tends to zero exp. fast as t tends to infinity, we deduce that $\hat{f} \in \hat{\mathbf{A}}_-(\sigma_0)$ does not always imply σ_0- exp. stability : therefore we study first the class $L_\varepsilon(\sigma_0)$ of **σ_0-exp. stable functions.**

2.1. The ideal $\hat{L}_\varepsilon(\sigma_0)$

Definition 2.1. A function[2] $f : \mathbb{R} \to \mathbb{C}$ is said to be **σ_0-exp. stable** (denoted by : $f \in L_\varepsilon(\sigma_0)$) iff $|f(t)| \leq M.\exp(\sigma t)$ a.e. on $\mathbb{R}_+$, for some $\sigma < \sigma_0$ and $M > 0$, or equivalently, iff

$\exp(-\sigma_0 t).f(t)$ is bounded and
tends to zero **exp.** fast as t tends to infinity.□ (1)

We give now the main properties of $\hat{L}_\varepsilon(\sigma_0)$.

Theorem 2.1. [Ideal Property].

a) $\hat{L}_\varepsilon(\sigma_0)$ is **strictly** included in $\hat{\mathbf{A}}_-(\sigma_0)$.
b) $\hat{L}_\varepsilon(\sigma_0)$ is a (proper) **ideal** of $\hat{\mathbf{A}}_-(\sigma_0)$, i.e.

for all $\hat{u} \in \hat{L}_\varepsilon(\sigma_0)$ and $\hat{f} \in \hat{\mathbf{A}}_-(\sigma_0)$, $\hat{y} = \hat{f}\hat{u} \in \hat{L}_\varepsilon(\sigma_0)$.□ (2)

Fact 2.1. [Dynamic Interpretation of [1, Th. 2.2]].
Let $f \in \mathbf{A}_-(\sigma_0)$; $u(t) = \exp(zt).1(t)$ [4], where $\mathrm{Re}\, z \geq \sigma_0$ and $y(t) = (f*u)(t)$. Under these conditions,

$$q(\cdot) := y(\cdot) - \hat{f}(z).\exp(z\cdot).1(\cdot) \in L_\varepsilon(\sigma_0) . \quad (3)$$

Hence, as t tends to infinity, $\exp(-\sigma_0 t).y(t)$ will be attracted **exp.** fast to the weighted stationary waveform $\exp(-\sigma_0 t).\hat{f}(z).\exp(zt).1(t)$, since for some $\sigma < \sigma_0$ and $M > 0$, $\exp(-\sigma_0 t).|y(t)-\hat{f}(z).\exp(zt).1(t)| \leq M.\exp((\sigma-\sigma_0)t)$.□

Comments 2.1. α) For $\sigma_0 = 0$, the blocking property of an unstable transmission zero z of $\hat{f} \in \hat{\mathbf{A}}_-(0)$, (see (3) with $\hat{f}(z) = 0$), is displayed in the same manner as for an exp. stable rational transfer function in $\mathbf{R}(0)$. β) The properties of $\hat{L}_\varepsilon(0)$ are also important to show that a transfer function in $\hat{\mathbf{A}}_-(0)$ has the I/O properties of an exp. stable rational transfer function in $\mathbf{R}(0)$, [e.g. BIBO-stability; L^p-stability for all $p \in [1,\infty]$; $\hat{L}_\varepsilon(\sigma_0)$ interacts well with $\hat{\mathbf{A}}_-(\sigma_0)$: I/O properties of exp. order]. γ) The ideal property, (2), of $\hat{L}_\varepsilon(\sigma_0)$ as well as property (3), play a central role in what follows.□

Now, by adding (as in $\mathbf{A}_-(\sigma_0)$) to each f in $L_\varepsilon(\sigma_0)$ a suitable series of Dirac pulses we define (after applying the Laplace transform) the class $\mathbf{E}(\sigma_0)$ of **σ_0-exp. stable transfer functions.**

2.2. The algebra $\hat{\mathbf{E}}(\sigma_0)$

More precisely, we have the following

Definition 2.2. A distribution $f(\cdot) = f_a(\cdot) + \Sigma_{i=0}^{\infty} f_i.\delta(\cdot-t_i) \in \mathbf{A}_-(\sigma_0)$ is said to be **σ_0-exp. stable** (denoted by : $f \in \mathbf{E}(\sigma_0)$) iff

its functional part f_a is σ_0-exp. stable, (4)

(i.e. $f_a \in L_\varepsilon(\sigma_0)$). We call $\hat{\mathbf{E}}(\sigma_0)$ the subset of σ_0-exp. stable transfer functions of $\hat{\mathbf{A}}_-(\sigma_0)$.

$\hat{\mathbf{E}}^\infty(\sigma_0)$ is the subset of elements of $\hat{\mathbf{E}}(\sigma_0)$ that are bounded away from zero at infinity in $\mathrm{Re}\, s \geq \sigma_0$.□ (4a)

Comment 2.2. A transfer function $\hat{f} \in \hat{E}(\sigma_0)$ has a **strictly proper** part $\hat{f}_a$, (Riemann-Lebesgue), such that $\exp(-\sigma_0 t).f_a(t)$ is bounded by a decreasing exp. : this property is characteristic for rational σ_0-exp. stable transfer functions in $R(\sigma_0)$. Hence the decision to call elements of $\hat{E}(\sigma_0)$ also σ_0-exp. stable.□

The main properties of $\hat{E}(\sigma_0)$ and $\hat{E}^\infty(\sigma_0)$ are the following ...

Theorem 2.2. [Properties of $\hat{E}(\sigma_0)$ and $\hat{E}^\infty(\sigma_0)$].

a) $\hat{E}(\sigma_0)$ is **strictly** included in $\hat{A}_-(\sigma_0)$.

b) $\hat{E}(\sigma_0)$ is a (commutative) subalgebra of $\hat{A}_-(\sigma_0)$. (5)

c) If $\hat{f} \in \hat{L}_\varepsilon(\sigma_0)$ and $(1+\hat{f})$ is invertible in $\hat{A}_-(\sigma_0)$, then

$$\hat{h} = (1+\hat{f})^{-1} \in \hat{E}(\sigma_0) .$$

d) $\hat{E}(\sigma_0)$ is closed under inversion :

$$\hat{f} \in \hat{E}(\sigma_0) \text{ is invertible in } \hat{E}(\sigma_0) \tag{6a}$$

iff

$$\inf\{|\hat{f}(s)| : \operatorname{Re} s \geq \sigma_0\} > 0 , \tag{6b}$$

(i.e. $\hat{f}$ is invertible in $\hat{A}_-(\sigma_0)$).

e) $\hat{E}^\infty(\sigma_0)$ is a multiplicative system of the commutative ring $\hat{E}(\sigma_0)$.□(7)

Comments 2.2. α) The proof of property c) is very easy. Indeed, $h = \delta-(f*h)$, where $f \in L_\varepsilon(\sigma_0)$ and $h \in A_-(\sigma_0)$. Hence $h \in E(\sigma_0)$, because $h_a = -(f*h)$ is in $L_\varepsilon(\sigma_0)$ by the ideal property (2). β) The proof of equivalence (6a)-(6b) shows that under assumption (6b) the singular atomic part $(\Sigma_{i=0}^{\infty} f_i.e^{-st_i} = \hat{f}(s) - \hat{f}_a(s))$ of $\hat{f}$ becomes invertible in $\hat{E}(\sigma_0)$ and uses then for the functional part the ideal property of $\hat{L}_\varepsilon(\sigma_0)$, (Theorem 2.1.b)), and the property of $\hat{E}(\sigma_0)$ dealing with unity feedback, (Theorem 2.2.c)).□

We define and characterize now coprimeness.

Definition 2.3. Let $\hat{f}$ and $\hat{g}$ be in $\hat{E}(\sigma_0)$. The pair $(\hat{f},\hat{g})$ is said to be **$\varepsilon\sigma_0$- coprime** iff

$$\text{there exist } \hat{u} , \hat{v} \in \hat{E}(\sigma_0) \text{ s.t. } \hat{u}\hat{f} + \hat{v}\hat{g} = 1 .\square \tag{8}$$

Condition (8), known as the **Bezout condition,** is crucial for studying e.g. algebras of fractions and feedback compensator design problems, e.g. [1]-[4].

Theorem 2.3. [Characterization of $\varepsilon\sigma_0$-Coprimeness].

Let $\hat{f}$ be in $\hat{E}(\sigma_0)$ and let $\hat{g}$ be in $\hat{E}^\infty(\sigma_0)$.

The pair $(\hat{f},\hat{g})$ is $\varepsilon\sigma_0$-coprime

iff

$$(\hat{f}(s),\hat{g}(s)) \neq (0,0) \text{ for all } s \in \mathbb{C} \text{ s.t. } \operatorname{Re} s \geq \sigma_0 .\square \tag{9}$$

Comments 2.3. α) Since $\hat{E}(\sigma_0) \subset \hat{A}_-(\sigma_0)$ it follows by [2, Facts 1, 2] and [1, Th. 2.1], that condition (9) is equivalent to " $(\hat{f},\hat{g})$ is **σ_0-coprime**" or equivalently :

$$\text{there exist } \hat{u}_1 \text{ and } \hat{v}_1 \text{ in } \hat{A}_-(\sigma_0) \text{ s.t. } \hat{u}_1 \hat{f} + \hat{v}_1 \hat{g} = 1 , \tag{9a}$$

i.e. the Bezout condition is satisfied with respect to $\hat{A}_-(\sigma_0)$. The proof of the theorem shows that this condition also holds with respect to $\hat{E}(\sigma_0)$ by a regularization procedure. β) If (9) is not satisfied, then $(\hat{f},\hat{g})$ can be made to be $\varepsilon\sigma_0$-coprime by removing from $\hat{f}$ and $\hat{g}$ common factors $(s-z)(s-p)^{-1}$ in $R^\infty(\sigma_0)$ associated with the common zeros z of $\hat{f}$ and $\hat{g}$ in $\operatorname{Re} s \geq \sigma_0$.□

This characterization of $\varepsilon\sigma_0$-coprimeness leads to a transfer function algebra of fractions $\hat{\mathbf{F}}(\sigma_0)$, suitable for modeling distributed systems and for feedback.

3. σ_0-EXP. UNSTABLE[3] TRANSFER FUNCTIONS, [3]

Definition 3.1. We define $\hat{\mathbf{F}}(\sigma_0)$ as the (commutative) algebra of fractions of $\hat{\mathbf{E}}(\sigma_0)$ with respect to $\hat{\mathbf{E}}^\infty(\sigma_0)$, i.e.

$$\hat{\mathbf{F}}(\sigma_0) := [\hat{\mathbf{E}}(\sigma_0)].[\hat{\mathbf{E}}^\infty(\sigma_0)]^{-1} .\square \tag{10}$$

Theorem 3.1. [Main Properties of $\hat{\mathbf{F}}(\sigma_0)$].

a) $\hat{\mathbf{F}}(\sigma_0)$ is **strictly** included in $\hat{\mathbf{B}}(\sigma_0)$.

b) $\mathbb{C}_p(s)$, (resp. $\hat{\mathbf{F}}(\sigma_0)$), is a subalgebra of $\hat{\mathbf{F}}(\sigma_0)$, (resp. $\hat{\mathbf{B}}(\sigma_0)$).(11)

c) $\hat{f} \in \hat{\mathbf{F}}(\sigma_0)$ is σ_0-stable
iff (12)
$\hat{f} \in \hat{\mathbf{E}}(\sigma_0)$, (σ_0-**exp.** stable).

d) $\hat{f} \in \hat{\mathbf{F}}(\sigma_0)$
iff (13)
$\hat{f} = \hat{r} + \hat{g}$,
where (i) $\hat{r}$ is completely σ_0-unstable (all poles in $\operatorname{Re} s \geq \sigma_0$) strictly proper rational, and (ii) $\hat{g} \in \hat{\mathbf{E}}(\sigma_0)$, (i.e. $\hat{g}$ is σ_0-exp. stable).$\square$

Comments 3.1. α) Property (13), (resp. (12)), illustrates the fact that $\hat{\mathbf{F}}(\sigma_0)$, (resp. $\mathbf{E}(\sigma_0)$), is an appropriate extension of the algebra of proper rational, (resp. σ_0-exp. stable rational), transfer functions **maintaining exp. order properties.** β) For standard semigroup systems the following facts hold (using (12)-(13) and the results of Jacobson, [5]-[6]) : if (A,B) is σ_0-exp. stabilizable and (C,A) is σ_0-exp. detectable, then (i) the transfer function $\hat{f}(s) = C(sI-A)^{-1}B$ is in $\hat{\mathbf{F}}(\sigma_0)$, ($\underset{+}{\subset}$ $\hat{\mathbf{B}}(\sigma_0)$), and (ii) internal σ_0-exp. stability, (viz. σ_0-exp. stability of the semigroup), is equivalent to external σ_0-exp. stability, (viz. $\hat{f} \in \hat{\mathbf{E}}(\sigma_0) \underset{+}{\subset} \hat{\mathbf{A}}_-(\sigma_0)$). For more details, see [3, especially Section 6].

4. CONCLUSION

Four classes of distributed system transfer functions of exponential order have been defined viz. (i) $\hat{L}_\varepsilon(\sigma_0)$, (σ_0-exp. stable functions, (1)); (ii) $\hat{\mathbf{E}}(\sigma_0)$, (σ_0-exp. stable transfer functions, (4)); (iii) $\hat{\mathbf{E}}^\infty(\sigma_0)$, ("biproper" elements of $\hat{\mathbf{E}}(\sigma_0)$, (4a)), (iv) $\hat{\mathbf{F}}(\sigma_0) := [\hat{\mathbf{E}}(\sigma_0)].[\hat{\mathbf{E}}^\infty(\sigma_0)]^{-1}$, (fractions, (10)). They are

a) restrictions of resp. (i) $\cup \{\hat{L}_{1,\sigma} : \sigma < \sigma_0\}$, (ii) $\hat{\mathbf{A}}_-(\sigma_0)$, (iii) $\hat{\mathbf{A}}_-^\infty(\sigma_0)$, and (iv) $\hat{\mathbf{B}}(\sigma_0) = [\hat{\mathbf{A}}_-(\sigma_0)].[\hat{\mathbf{A}}_-^\infty(\sigma_0)]^{-1}$, and

b) extensions of resp. (i) the class of σ_0-exp. stable strictly proper rational functions, (ii) $\mathbf{R}(\sigma_0)$, (iii) $\mathbf{R}^\infty(\sigma_0)$, and (iv) $\mathbb{C}_p(s) = [\mathbf{R}(\sigma_0)].[\mathbf{R}^\infty(\sigma_0)]^{-1}$ such that the exponential order properties of the latter classes (associated with lumped systems) are maintained.

The algebra $\hat{\mathbf{F}}(\sigma_0)$ can be used for modeling and feedback design, [3]-[6].

REFERENCES

[1] F.M. CALLIER and C.A. DESOER, "An algebra of transfer functions for distributed linear time-invariant systems", IEEE Trans. Circuits Syst., Vol. CAS-25, pp. 651- 663, Sept. 1978; correction : vol. CAS-26, p. 360, May 1979.

[2] F.M. CALLIER and C.A. DESOER, "Simplifications and clarifications on the paper "An algebra of transfer functions for distributed linear time-invariant systems"", IEEE Trans. Circuits Syst., Vol. CAS-27, pp. 320-323, April 1980.

[3] F.M. CALLIER and J. WINKIN, "Distributed system transfer functions of exponential order", Dept. of Mathematics, FNDP Namur, Report 84/18, November 1984; Int. Journal of Control, accepted.

[4] C.N. NETT, C.A. JACOBSON and M.J. BALAS, "Fractional representation theory : robustness results with applications to the finite-dimensional control of a class of linear distributed systems", Proceedings of the 1983 CDC Conference, San Antonio, TX, pp. 268-280, 1983, (esp. Theorem 3.2).

[5] C.A. JACOBSON, "Some aspects of the structure and stability of a class of linear distributed systems", M.Sc. Thesis, ECSE Dept., Rensselaer Polytech. Inst., Troy, NY, May 1984.

[6] C.A. JACOBSON, "Necessary and sufficient conditions for output feedback stabilization of a class of linear distributed systems", ECSE Dept., Rensselaer Polytech. Inst., Troy, NY, June 1984; submitted to SIAM Jour. on control and optimization.

FOOTNOTES

[1] M is a **multiplicative system** of a commutative ring R , if M is a subset of R which does not contain neither the zero nor divisors of zero of R and which is closed under multiplication.

[2] Here all the functions have their support on $\mathbb{R}_+$.

[3] By "unstable", we mean "possibly unstable".

[4] 1(t) is the unit step function.

Frequency Domain and State Space Methods for Linear Systems
C.I. Byrnes and A. Lindquist (Editors)

ON A SYSTEM-THEORETIC APPROACH TO DYNAMICS OF OPEN QUANTUM SYSTEMS

Andrzej Jamiołkowski*

Institute of Physics, N. Copernicus University
Toruń, Poland

This paper discusses the question of the minimal number η of observables $Q_1,\ldots,Q_\eta$ whose expectation values at some instants $t_1,\ldots,t_s$ determine the statistical state of an N-level quantum system and a similar question of the minimal number Θ of instants $t_1,\ldots,t_\Theta$ of measurements which are necessary for reconstructibility of trajectories of the system in question. To give an exact answer to the questions we introduce the following two concepts: index of reconstructibility and index of cyclicity of the generator of time evolution. Some applications of those concepts to Hamiltonian and Gaussian semigroups are discussed.

1. INTRODUCTION

The dynamics of a finite isolated quantum system is usually represented by a one-parameter group of unitary transformations in a complex Hilbert space. In this formalism one assumes that we are mostly concerned with those states of quantum systems (called "pure states") that can be completely specified by assigning state vectors. However, in many physical problems, it is necessary to consider a given quantum system as an open system which interacts with its surroundings. In this case the representation of states of the system in question by state vectors should be replaced by the representation in terms of density operators describing "mixture states". To study the dynamics of such systems (or at least some of its aspects) it appears useful to introduce the notion of a quantum dynamical semigroup of maps (cf., e.g., [1] - [3]). In contrast with the case of a rather idealized situation of strictly isolated systems, time evolution of which is Hamiltonian, the evolution of open quantum systems is governed by properly selected dynamical semigroups and master equations .

In this paper we establish and discuss some effective conditions for reconstructibility of trajectories in a set of all density operators representing states of an open quantum system $\mathcal{L}$. We assume that our knowledge about the system in question is given by the mean values $\langle Q_i(t_j)\rangle = \mathrm{Tr}(Q_i\,\rho(t_j))$ of n self-adjoint operators $Q_1,\ldots,Q_n$ at some instants $t_1<\ldots<t_s$. The operators $Q_1,\ldots,Q_n$ correspond to fixed physical quantities of the system $\mathcal{S}$ and the symbol $\varrho(t_j)$ denotes the state of the system at

---

*) This paper was supported in part by the Alexander von Humboldt Foundation and by the Ministry of Higher Education, Proj.MR.I.7

the moment t_j , where $t_j \in [0,T]$, $T>0$.

Another goal of the present paper is a merging of density-operator and system-theoretic techniques. Both methods have hardly ever been used simultaneously. In applications to non-isolated (open) quantum systems the density-operator techniques have preferentially been used in the cases where the system $\mathcal{S}$ is a quantum system with only a few degrees of freedom, whereas the system-theoretic methods have been especially successful where the system $\mathcal{S}$ is a macroscopic classical system. This has both formal and physical reasons. Specification of the density operator for a given system provides a complete quantum-statistical description. In general, however, even for systems with a small number of degrees of freedom, it is difficult to find $\varrho(0)$ performing measurements only at the moment $t=0$. We intend to show that a combination of both techniques can prove fruitful, each of them contributing its particular advantages.

2. RECONSTRUCTIBILITY OF TRAJECTORIES IN GIBBS SPACES

The time evolution of an open quantum system of finitely many degrees of freedom coupled to infinite systems, usually called reservoirs, can be described by a one-parameter semigroup of transformations [1], [2]. Let $\mathcal{H}$ be the Hilbert space of the system in question ($\dim \mathcal{H} = N < \infty$) and let $\Phi(t) : T(\mathcal{H}) \to T(\mathcal{H})$, $t \in \mathbb{R}^1_+$, be the dynamical semigroup, where $T(\mathcal{H})$ is the real Banach space of all self-adjoint operators on $\mathcal{H}$ under the trace norm $\|\varrho\| := \mathrm{Tr}(\varrho^*\varrho)^{1/2}$. States of the system are described by density operators $\varrho \in \mathcal{P}(\mathcal{H})$ where the set $\mathcal{P}(\mathcal{H})$ (called the Gibbs space) is defined in the following way

$$\mathcal{P}(\mathcal{H}) := \{\varrho \in T(\mathcal{H}) \;;\; \varrho \geq 0,\ \mathrm{Tr}\,\varrho = 1\} . \qquad (2.1)$$

Usually one assumes that 1) $\Phi(t)$ is linear for all $t \in \mathbb{R}^1_+$, 2) $\Phi(t)$ is trace preserving, 3) the dual map $\Phi^*(t)$ is completely positive, 4) $\Phi(t_1)\circ\Phi(t_2) = \Phi(t_1+t_2)$, $t_1, t_2 > 0$ and 5) $\lim \Phi(t) = I$. Since $\Phi(t)$ is a contraction ([2]) it follows from the Hille-Yosida theorem that there exists a linear operator $L : T(\mathcal{H}) \longrightarrow T(\mathcal{H})$ such that $\Phi(t) = \exp(Lt)$ for all $t > 0$ and the evolution of the system in question is governed by the "master equation"

$$\frac{d\varrho(t)}{dt} = L\,\varrho(t) , \qquad (2.2)$$

where $\varrho(t) = \Phi(t)\,\varrho(0)$. The operator L is called the generator of the dynamical semigroup $\{\Phi(t),\ t \in \mathbb{R}^1_+\}$. If $\varrho(0) \in \mathcal{P}(\mathcal{H})$ then the whole trajectory $\{\varrho(t),\ t \in \mathbb{R}^1_+\}$ generated by (2.2) belonges to $\mathcal{P}(\mathcal{H})$, i.e., $\{\varrho(t),\ t \in \mathbb{R}^1_+\} \subset \mathcal{P}(\mathcal{H})$.

The master equation (2.2) defines the assignment

$$\mathbb{R}^1_+ \ni t \longmapsto \varrho(t) \in \mathcal{P}(\mathcal{H}), \qquad (2.3)$$

provided that we know the initial state of the system $\varrho(0) \in \mathcal{P}(\mathcal{H})$. The specification of the state $\varrho(0)$ is given by measurements of mean values of suitably chosen operators. We recall from quantum mechanics that all information on the behaviour of a given system can be expressed in terms of expectation values of suitably chosen observables. From an operational point of view the density matrix $\varrho(0)$ can be defined in the following way. As many operators $Q_1,\ldots,Q_n$ as there are independent parameters in $\varrho(0)$ are chosen and their mean values $\langle Q_1(0)\rangle, \ldots, \langle Q_n(0)\rangle$ are taken as the initial information on the state of the system. The corresponding density matrix can then be determined by solving the set of equations

$$\mathrm{Tr}\,(\varrho(0)\,Q_i) = \langle Q_i(0)\rangle. \tag{2.4}$$

The number of independent parameters which are needed to specify a given density matrix is determined by the dimension of the Hilbert space $\mathcal{H}$. In general, this number is infinite, but it is often finite when only one particular property of the system (the spin, for example) is considered and the dependence on all other variables can be omitted. In the following discussion we will consider the case in which the number of linearly independent vectors in the space $\mathcal{H}$ is equal to N. In this case an N-dimensional density operator is completely specified by N^2-1 real parameters. In conclusion we can say that in order to specify the state $\varrho(0)$ by equalities (2.4) we need the mean values of N^2-1 independent observables $Q_1,\ldots,Q_{N^2-1}$.

Now the following question may arise: what can we say about the assignment (2.3) if the only information on the system in question is given by the mean values

$$\mathrm{Tr}(\varrho(t_j)\,Q_i) = \langle Q_i(t_j)\rangle \tag{2.5}$$

of, say $n < N^2-1$, linearly independent self-adjoint operators $Q_1,\ldots,Q_n$ at some instants $t_1 < \ldots < t_s$, where $t_j \in [0,T]$ for $j=1,\ldots,s$ and $T>0$. In fact, the problem of the reconstruction of trajectories in the Gibbs space $\mathcal{P}(\mathcal{H})$ consists in the establishment of current states of the system in question from known expectation values (2.5).

To be more precise we introduce the following definition:

Definition 1. The N-level quantum system $\mathcal{S}$ is said to be $(Q_1,\ldots,Q_n)$-reconstructible on the interval $[0,T]$ if there exists at least one set of instants $0 \leqslant t_1 < \ldots < t_s \leqslant T$ such that $\varrho(t)$ can be reconstructed by using the assignments of the form

$$[0,T] \ni t_j \longmapsto \langle Q_i(t_j)\rangle = \mathrm{Tr}(Q_i\,\varrho(t_j)), \tag{2.6}$$

where $\varrho(t_j) = \Phi(t_j)\,\varrho(0)$ for $j = 1,\ldots,s$.

The above definition is equivalent to the following one.

Definition 1 . The N-level quantum system $\mathcal{S}$ is said to be $Q_1,\ldots,Q_n$ -reconstructible on the interval $[0,T]$ if for every two trajectories $\{\varrho_1(t),\ t\in[0,T]\}$ and $\{\varrho_2(t),\ t\in[0,T]\}$ with distinct initial states there exists at least one moment $\hat{t}\in[0,T]$ and at least one operator $\hat{Q}_i \in \{Q_1,\ldots,Q_n\}$, such that

$$\mathrm{Tr}(\hat{Q}_i\, \rho_1(\hat{t})) \neq \mathrm{Tr}(\hat{Q}_i\, \rho_2(\hat{t})) \qquad (2.7)$$

Remark 1. It is well known that each measurement of an observable can cause alterations in the state of the system (cf. e.g. [4]). In this paper we consider $\omega = n \cdot s$ identical systems and we assume that at $t = 0$ all these systems are described by the same density operator $\rho(0)$ (we assume that all systems are prepared by identical procedures in a quantum sense). The quantities $\langle Q_i (t_j) \rangle$, where $i = 1,\dots,n$ and $j = 1,\dots,s$ can then be measured independently.

Remark 2. It is important that for the number s of the moments $t_1,\dots,t_s$ we do not formulate any restrictions (of course, apart from that the number s should be finite).

Starting from Def.1 (or equivalently from 1′) one obtains the following criterion for $(Q_1,\dots,Q_n)$-reconstructibility. The N-level open quantum system $\mathcal{S}$ is $(Q_1,\dots,Q_n)$ -reconstructible if and only if

$$\sum_{i=0}^{n} \langle L(m) \mid Q_i \rangle = T^*(\mathcal{H}), \qquad (2.8)$$

where

$$\langle L(m) \mid Q_i \rangle := \mathrm{Span}\{Q_i, L^* Q_i, \dots, (L^*)^{m-1} Q_i\} \qquad (2.9)$$

and $m := \deg(\mu, L)$, $(\mu(\lambda, L)$ denotes the minimal polynomial of the generator L . In (2.8) we assume that $Q_0 := I$.

The subspace $\langle L(m) \mid Q_i \rangle$ is the smallest subspace of $T^*(\mathcal{H})$ which is invariant under L^* and contains Q_i . Of course, this subspace can also be defined as the intersection of all L^* - invariant subspaces of $T^*(\mathcal{H})$ which contain Q_i .

A question of an obvious physical interest is to find the minimal number of observables $Q_1,\dots,Q_n$ for which the N-level quantum system $\mathcal{S}$ with the generator L can be $(Q_1,\dots,Q_n)$-reconstructible. It can be shown that if the time evolution of the system $\mathcal{S}$ is described by the master equation $\dot{\rho}(t) = L\rho(t)$ in the Gibbs space $\mathcal{P}(\mathcal{H})$ then there exists ([5] , [6]) a set of observables $Q_1,\dots,Q_\eta$, where

$$\eta := \max_{\lambda \in \sigma(L)} \{\dim \mathrm{Ker}(\lambda I - L)\}, \qquad (2.10)$$

such that the system $\mathcal{S}$ is $(Q_1,\dots,Q_\eta)$ - reconstructible. Moreover, if we have another set of observables $\tilde{Q}_1,\dots\tilde{Q}_{\tilde{\eta}}$ such that the system $\mathcal{S}$ is $(\tilde{Q}_1,\dots,\tilde{Q}_{\tilde{\eta}})$ - reconstructible, then $\tilde{\eta} \geq \eta$. The number η given by (2.10) we will call the index of cyclicity of the quantum system $\mathcal{S}$.

Let us now consider an open quantum system $\mathcal{S}$ which is reconstructible with respect to observables $Q_1,\dots,Q_n$ (of course, this means that $n \geq \eta$, where η is given by (2.10) . An interesting problem is connected with the number of moments $t_1,\dots,t_s$ of measurements which are necessary for reconstructibility. It may be important to know an answer to the following question: what is the minimal number of moments $t_1, \dots, t_\Theta$ such that the assignments of the form

$$[0,T] \ni t_j \longmapsto \mathrm{Tr}\,(\varrho(t_j)Q_i) \qquad (2.11)$$

where $i=1,\dots,n$, $j=1,\dots,\Theta$ determine uniquely the state $\varrho(0)$. Of course, we should give an answer in terms of the generator L and the observables $Q_1,\dots,Q_n$. It appeares ([7]) that the minimal number of moments $0 \leqslant t_1 < \dots < t_s \leqslant T$ of observations at which we have to measure the expectation values $\langle Q_i(t_j)\rangle$ is given by the equality

$$\Theta := \min \left\{ k \in \mathbb{N} \;;\; \sum_{i=0}^{n}{}' \langle L(k) \mid Q_i \rangle = T^*(\mathcal{H}) \right\}, \qquad (2.12)$$

where

$$\langle L(k) \mid Q_i \rangle = \mathrm{Span}\left\{ Q_i, L^*Q_i, \dots, (L^*)^{k-1}Q_i \right\} \qquad (2.13)$$

and, as before, $m := \deg \mu(\lambda, L)$. The number Θ given by (2.12) we will call the index of reconstructibility of the system $\mathcal{S}$ with respect to the given observables $Q_1,\dots,Q_n$.

Let us observe that for $s < \Theta$ the information about the system in question given by $\langle Q_i(t_j)\rangle$, where $i = 1,\dots,n$ and $j = 1,\dots,s$, is not sufficient to be used for reconstruction of trajectories (in spite of the fact that our system is $(Q_1,\dots,Q_n)$-reconstructible).

3. EXAMPLES

a) Reconstructibility of isolated N-level quantum systems

The time evolution of an isolated quantum system $\mathcal{S}$ is described by the von Neumann equation

$$i\dot{\varrho}(t) = L_0\,\varrho(t) = [H, \varrho(t)], \qquad (3.1)$$

where H stands for the Hamiltonian of the system $\mathcal{S}$. This means that the evolution is given by a one-parameter group of unitary transformations on the Hilbert space $\mathcal{H}$ associated with the system under consideration. The minimal number of observables $Q_1,\dots,Q_\eta$ for which the system $\mathcal{S}$ can be reconstructible (i.e., the index of cyclicity of the generator L_0) is given by the equality ([5])

$$\eta = n_1^2 + n_2^2 + \dots + n_r^2, \qquad (3.2)$$

where

$$n_i := \dim\mathrm{Ker}\,(\lambda_i I - H) \qquad (3.3)$$

for all eigenvalues of H; $\lambda_i \in \sigma(H) = \{\lambda_1,\dots,\lambda_r\}$. We assume that if $i \neq j$ then $\lambda_i \neq \lambda_j$ for all $i,j = 1,\dots,r \leqslant N$.

b) Reconstructibility of Gaussian N-level quantum systems

Consider the problem of the minimal number of observables $Q_1,\dots,Q_\eta$ for reconstructibility of the system $\mathcal{S}$ governed by the so-called Gaussian semigroups

$$\Phi(t)\varrho = (2\pi t)^{-1/2}\int_{-\infty}^{\infty} ds\; e^{-s^2/2t} e^{-iHs}\varrho\, e^{iHs} \tag{3.4}$$

with the generator of the form

$$L\varrho := -\frac{1}{2}[H,[H,\varrho]] = \frac{1}{2}\{[H\varrho,H]+[H,\varrho H]\}, \tag{3.5}$$

where H is a self-adjoint opertor on $\mathcal{H}$; $\dim\mathcal{H} = N$.

To determine the exact value of η (the index of cyclicity of the generator (3.5)) we must find the number of nontrivial invariant factors of the operator L . Let the spectrum of H be

$$\sigma(H) = \{\lambda_1,\dots,\lambda_r\}, \tag{3.6}$$

where $\lambda_1,\dots,\lambda_r$ are all distinct eigenvalues of H and by definition $n_i := \dim\mathrm{Ker}(\lambda_i I - H)$ are their multiplicities. In other words we assume that if $i \neq j$ then $\lambda_i \neq \lambda_j$. Of course, we may also assume that $\lambda_1<\lambda_2<\dots<\lambda_r$. For systems with spectra of the form (3.6) we can formulate the following theorem.

The minimal number of observables $Q_1,\dots,Q_\eta$ for which the system $\mathcal{S}$ can be reconstructible is determined by the formulae

$$\eta = \max\{a, b_1,\dots, b_z\}, \tag{3.7}$$

where

$$z = \begin{cases}\frac{1}{2}(r-1) & \text{when } r \text{ is odd,}\\ \frac{1}{2}(r-2) & \text{when } r \text{ is even,}\end{cases} \tag{3.8}$$

and

$$a := n_1^2 + n_2^2 + \dots + n_r^2, \qquad b_k = 2\sum_{i=1}^{r-k} n_i n_{i+k}, \tag{3.9}$$

for $k = 1,\dots,r-1$. Indeed, let us observe that if $\sigma(H) = \{\lambda_1,\dots,\lambda_r\}$ then the spectrum of the operator L is given by

$$\sigma(L) = \{\nu\in\mathbb{R}^1;\ \nu = (\lambda_i - \lambda_j)^2,\ i,j=1,\dots,r\}. \tag{3.10}$$

The above statement follows from the fact that the operator L can also be represented as

$$L = H^2\otimes I + I\otimes H^2 - 2H\otimes H. \tag{3.11}$$

Since H is a self-adjoint operator therefore the algebraic multiplicity of λ_i (i.e., the multiplicity of λ_i as a root of the characteristic polynomial of H) is equal to the geometric multiplicity of λ_i ; $n_i = \dim \mathrm{Ker}(\lambda_i I - H)$. Of course, we have $n_1+n_2+\ldots+n_r = \dim \mathcal{H}$.

From the equality (3.11) we can see that the multiplicities of the eigenvalues of the operator L are not determined uniquely by multiplicities of $\lambda_i \in \sigma(H)$. But, if we assume that $\lambda_1<\ldots<\lambda_r$ and $\lambda_k = (k-1)\varkappa + \lambda_1$, where $k = 1,\ldots,r$ and $\varkappa = \mathrm{const} > 0$, then the multiplicities of all eigenvalues of L are given by

$$\dim \mathrm{Ker}\left[(\lambda_i - \lambda_j)^2 I - L\right] = b_{|i-j|} \tag{3.12}$$

for $i \neq j$ and $\dim \mathrm{Ker}(L) = n_1^2 + \ldots + n_r^2$ when $i = j$.

The minimal number of observables $Q_1,\ldots,Q_\eta$ for which the system $\mathcal{S}$ can be $(Q_1,\ldots,Q_\eta)$-reconstructible is given by (2.10), so in our case

$$\eta = \max_{\substack{i=1,\ldots,r\\ j=1,\ldots,r}} \left\{\dim \mathrm{Ker}\left[(\lambda_i - \lambda_j)^2 I - L\right]\right\}, \tag{3.13}$$

where $\lambda_i \in \sigma(H)$. Using (3.13) and the inequality $b_k < a$ for $k > z$ we can observe that also without the assumption $\lambda_k=(k-1)\varkappa+\lambda_1$ one obtains

$$\eta = \max\{a, b_1,\ldots,b_z\} \tag{3.14}$$

where z is determined by (3.7). This completes the proof.

Remark 3. It is fairly easy to see that the quantities $\langle Q_i(t_j)\rangle$ where $i = 1,\ldots,n$ and $j = 1,\ldots,s$ give no new information about the state of the system if $j > m = \deg\mu(\lambda,L)$. Hence we can consider only the case $s \leqslant \deg\mu(\lambda,L)$. It is also obvious that if we want to discuss problems connected with indices of reconstructibility we have to assume an explicit form of observables [7].

REFERENCES

[1] G. Lindblad, Commun. Math. Phys. 48 (1976) 119-131

[2] V. Gorini, A. Frigerio, M. Verri, A. Kossakowski, E.C,G. Sudarshan, Rep. Math. Phys. 13 (1978) 149-162

[3] H. Spohn, Rev. Mod. Phys. 53(1980) 569-615

[4] A. Bohm, Quantum Mechanics Springer-Verlag, New York, 1979

[5] A. Jamiołkowski, Int. J. Theor. Phys. 22(1983) 369-376

[6] A. Jamiołkowski, Rep. Math, Phys. 21(1985) 101-109

[7] A. Jamiołkowski, Rep. Math. Phys. (in press).

Frequency Domain and State Space Methods for Linear Systems
C.I. Byrnes and A. Lindquist (Editors)

NEW TECHNIQUES FOR THE CONTROL OF LINEAR INFINITE-DIMENSIONAL SYSTEMS*

E. W. Kamen
Dept. of Electrical Engr.
Univ. of Florida
Gainesville, FL 32611
USA

P. P. Khargonekar
Dept. of Electrical Engr.
Univ. of Minnesota
Minneapolis, MN 55455
USA

A. Tannenbaum
Dept. of Electrical Engr.
McGill University
Montreal H3A 2A7
CANADA

Abstract

Output feedback control of a large class of linear time-invariant infinite-dimensional systems is studied in terms of a left-Bezout fractional representation of the transfer function matrix. The entries of the fractional representation belong to an algebra of stable functions of a complex variable. The design of finite-dimensional stabilizing compensators is approached using a procedure for constructing proper rational approximations to functions analytic in the open right-half plane and continuous on the boundary of the one-point compactification of the closed right-half plane. An explicit construction is then given for computing a Bezout fractional representation for a system consisting of a lumped system in cascade with a pure time delay. For systems with commensurate time delays, a summary is given of recent results on the existence and construction of finite-dimensional stabilizing compensators.

I. INTRODUCTION

Many linear time-invariant infinite-dimensional systems arising in applications can be modeled by a transfer function matrix W(s) whose entries are functions of the complex variable s. Examples of such systems include systems with time delays and systems described by a parabolic partial differential equation.

In the first part of this paper, we consider a large class of multi-input multi-output infinite-dimensional systems for which the transfer function matrix has the left fractional representation

$$W(s) = A^{-1}(s)B(s) \tag{1}$$

In (1), A(s) and B(s) are matrices whose entries are analytic in the open right-half plane H and continuous on the boundary of the compactification $\tilde{H}$ of the closed right-half plane.

In this paper we assume that the transfer function matrix W(s) given by (1) admits a left-Bezout fractional representation; that is, there exist matrices P(s) and Q(s), whose entries are analytic on H and continuous on $\tilde{H}$, such that

*This work was supported in part by the U.S. Army Research Office under Contract No. DAAG29-84-K-0081, and the National Science Foundation under Grant Nos. ECS-8400832 and ECS-8451519.

$$A(s)P(s) + B(s)Q(s) = I \tag{2}$$

The representation of an infinite-dimensional system by a Bezout fractional form (1) has been considered by several authors; in particular, we mention Vidyasagar [1,2,3], Callier and Desoer [4], Vidyasagar, Schneider, and Francis [5], and Nett [6]. In this existing work, fractional forms (1) that satisfy the Bezout identity (2) are said to be coprime. We prefer to use the term "Bezout."

Given an infinite-dimensional system specified by (1), in the next section we consider the existence of a stabilizing compensator. In particular, we would like to know if there is a finite-dimensional stabilizing compensator, since finite-dimensional compensators are easy to implement. Most previous approaches to the design of finite-dimensional compensators are based on the approximation of the given infinite-dimensional system by a lumped system. However, the authors are not aware of any general theorem on the existence of finite-dimensional stabilizing compensators where the proof is based on a finite-dimensional approximation of the given system.

In the design of a finite-dimensional stabilizing compensator, instead of approximating the given system, we can consider approximating the entries of $P(s), Q(s)$ in the Bezout identity (2) by proper rational functions of s. This will result in a finite-dimensional compensator since the transfer function of the stabilizing compensator can be constructed from the rational approximations to $P(s)$ and $Q(s)$. This approach to the design of finite-dimensional stabilizing compensators is considered in Kamen, Khargonekar, and Tannenbaum [7,8]. An explicit method for computing rational approximations is given in [7,8]. For the sake of completeness, the procedure is repeated in this paper.

In the next section, we show that if $W(s)$ has a left-Bezout fractional representation as defined above, there is a finite-dimensional stabilizing compensator. Starting from the Bezout identity (2), we show how to compute a finite-dimensional stabilizing compensator. We then show how to compute a Bezout fractional representation for any system consisting of a lumped system in cascade with a pure time delay. In the last section of the paper, we consider the class of linear systems with commensurate time delays.

II. FRACTIONAL REPRESENTATIONS OVER AN ALGEBRA OF ANALYTIC FUNCTIONS

With C = field of complex numbers, let H (resp. $\bar{H}$) denote the open (resp. closed) right-half plane. Let $\tilde{H} = \bar{H} \cup \{\infty\}$, where $\tilde{H}$ is the one-point compactification of $\bar{H}$. Finally, let $\Gamma_r(\tilde{H})$ denote the Banach algebra of functions from $\tilde{H}$ into C which are analytic on H, continuous on the boundary of $\tilde{H}$, and which have real coefficients (i.e., any power-series expansion about a real point has real coefficients).

A linear time-invariant system with transfer function belonging to $\Gamma_r(\tilde{H})$ is said to be stable. Note that the algebra $\Gamma_r(\tilde{H})$ can be viewed as a generalization of the algebra of proper stable rational functions in s.

Now consider a m-input p-output system with transfer function matrix $W(s)$ given by the left fractional representation

$$W(s) = A^{-1}(s)B(s) \tag{3}$$

where $A(s)$ is a $p \times p$ matrix over $\Gamma_r(\tilde{H})$ with $\det A(s) \neq 0$ (as an element of $\Gamma_r(\tilde{H})$) and $B(s)$ is a $p \times m$ matrix over $\Gamma_r(\tilde{H})$. The fractional representation (3) is said to be Bezout if there exist matrices $P(s), Q(s)$ over $\Gamma_r(\tilde{H})$ such that

$$A(s)P(s) + B(s)Q(s) = I \tag{4}$$

Not every transfer function matrix $W(s)$ given by (3) has a left-Bezout fractional representation. In particular, suppose that $m = p = 1$ and

$$w(s) = \frac{b(s)}{a(s)}$$

where $a(s), b(s)$ are coprime elements of the algebra $\Gamma_r(\tilde{H})$ (i.e., if $a(s)$ and $b(s)$ are divisible by an element $c(s) \in \Gamma(\tilde{H})$, then $c(s)$ is an invertible element of $\Gamma_r(\tilde{H})$). Then if $a(s)$ and $b(s)$ have any common zeros in $\tilde{H}$, $w(s)$ does not have a Bezout fractional representation.

There are algebras (other than $\Gamma_r(\tilde{H})$) for which Bezout fractional representations always exist (see Vidyasagar [3]).

If $W(s)$ does have a left-Bezout fractional representation over $\Gamma_r(\tilde{H})$, we will show that the system can be stabilized by a finite-dimensional compensator. To prove this result, we first need a procedure for approximating the elements of $\Gamma_r(\tilde{H})$ by proper rational functions in s. The basic idea of the approximation scheme given below can be found in [7] (see also [8]).

Let $\bar{\Delta}$ denote the closed unit disc defined by $\bar{\Delta} = \{z \in C: |z| \leq 1\}$. Let $\phi: \tilde{H} \to \bar{\Delta}$ be any conformal equivalence, for example

$$\phi(s) = \frac{s-1}{s+1}$$

Given $f(s) \in \Gamma_r(\tilde{H})$, define

$$g(z) = f(s)\Big|_{s=\phi^{-1}(z)} = f(\phi^{-1}(z)), \quad z \in \bar{\Delta}$$

Since $f(s) \in \Gamma_r(\tilde{H})$, $g(z)$ is analytic on the open unit disc and is continuous on the unit circle. Hence, $g(z)$ has the power-series expansion

$$g(z) = \sum_{i=0}^{\infty} c_i z^i, \quad z \in \bar{\Delta}$$

The coefficients c_i of this expansion can be computed by taking the inverse discrete-time Fourier transform of $g(e^{j\omega})$. For a suitably large positive integer N, the following is a good approximation to the c_i:

$$c_i \cong (1/N) \sum_{r=0}^{N-1} g(e^{j2\pi r/N})e^{-j2\pi ri/N}, \quad i=0,1,2,\ldots,N-1$$

Since

$$g(e^{j2\pi r/N}) = f(\phi^{-1}(e^{j2\pi r/N})),$$

we have that

$$c_i \cong (1/N) \sum_{r=0}^{N-1} f(\phi^{-1}(e^{j2\pi r/N}))e^{-j2\pi ri/N}, \quad i=0,1,\ldots,N-1$$

This expression for the c_i can be evaluated via the FFT.

Now for $n=0,1,2,\ldots,$ define

$$p_n(s) = \sum_{i=0}^{n} \frac{n-i+1}{n+1} c_i(\phi(s))^i$$

Each $p_n(s)$ is an element of $\Gamma_r(\tilde{H})$. Then using standard results on the approximation of functions defined on $\bar{\Delta}$ (see [8]), we have that $p_n(s)$ converges to $f(s)$ uniformly on $\tilde{H}$ as $n \to \infty$.

Using this approximation scheme, we can prove the following result.

Theorem 1: Suppose that $W(s)$ admits a left-Bezout fractional representation over $\Gamma_r(\tilde{H})$. Then there exists a proper finite-dimensional compensator which results in an internally-stable closed-loop system.

Proof: We shall first prove the theorem in the case that $W(s) \to 0$ as $s \to \infty$, where the limit is taken in the right-half plane. Since the fractional representation $W(s) = A^{-1}(s)B(s)$ is Bezout, there exist matrices $P(s),Q(s)$ over $\Gamma_r(\tilde{H})$ such that (4) is satisfied. Using the procedure given above, we can approximate the entries of $P(s),Q(s)$ by proper rational functions in s. This results in proper rational matrices $\hat{P}(s),\hat{Q}(s)$ such that

$$A(s)\hat{P}(s) + B(s)\hat{Q}(s) = \Phi(s)$$

with $\Phi(s)$ uniformly close to I over $\tilde{H}$. In particular, we can choose $\hat{P}(s),\hat{Q}(s)$ so that $\det \Phi(s) \neq 0$ for all $s \in \tilde{H}$. Now since $W(s) \to 0$ as $s \to \infty$, we have that $B(s) \to 0$ as $s \to \infty$, and thus

$$\lim_{s \to \infty} (A(s)\hat{P}(s) + B(s)\hat{Q}(s)) = \lim_{s \to \infty} A(s)\hat{P}(s) = \lim_{s \to \infty} \Phi(s) \cong I$$

Therefore, $A(\infty)\hat{P}(\infty) \cong I$, which implies that $\hat{P}(\infty)$ is invertible. Thus, $\hat{P}^{-1}(s)$ is a proper rational matrix. Taking $\hat{Q}(s)\hat{P}^{-1}(s)$ to be the transfer function matrix of the finite-dimensional compensator, the resulting feedback system can be shown to be internally stable (in the sense of Vidyasagar [3]). Now suppose that $W(s) \not\to 0$ as $s \to \infty$, where again the limit is taken in $\tilde{H}$. Define $\bar{W}(s) = W(s)/(s+1)$. Then $\bar{W}(s) = A^{-1}(s)(B(s)/(s+1))$ is a left-Bezout fractional representation of $\bar{W}(s)$, and by the above arguments, there exist proper rational matrices $\hat{P}(s), \hat{Q}(s)$ such that

$$A(s)\hat{P}(s) + (B(s)/(s+1))\hat{Q}(s) = \Phi(s)$$

with $\Phi(s)$ uniformly close to I over $\tilde{H}$. In addition, since $\bar{W}(s) \to 0$ as $s \to \infty$, we have that $B(s)/(s+1) \to 0$ as $s \to \infty$. It follows that $\hat{P}(\infty)$ is invertible, and thus $\hat{P}^{-1}(s)$ is a proper rational matrix. Taking the transfer function matrix of the compensator to be $\hat{Q}(s)((s+1)\hat{P}(s))^{-1}$, we obtain an internally-stable feedback system. ■

To implement the above approach to compensator design, we need a procedure for constructing Bezout fractional representations over $\Gamma_r(\tilde{H})$ (when these representations exist). In the next section we show how to carry out this construction in the case when the system is a cascade connection of a lumped system and a pure time delay. In the last section of the paper, we consider the class of systems with commensurate time delays.

III. LUMPED SYSTEMS IN CASCADE WITH A PURE DELAY

Let $\mathbb{R}[s]$ denote the ring of polynomials in s with real coefficients. In this section we consider scalar systems $(m = p = 1)$ with transfer function

$$w(s) = \frac{b(s)}{a(s)} e^{-hs} \tag{5}$$

where $a(s), b(s) \in \mathbb{R}[s]$, $a(s)$ is monic with degree n, $\deg b(s) < n$, and $h > 0$. A system with transfer function (5) corresponds to a cascade connection of a lumped system and a time delay of h seconds. As is well known, such systems are very common in applications.

In the following development, we assume that the polynomials $a(s), b(s)$ are coprime, so that there exist polynomials $c(s), d(s)$ such that

$$a(s)c(s) + b(s)d(s) = 1 \tag{6}$$

Note that since $\deg b(s) < \deg a(s)$, it must be true that

$$\deg c(s) < \deg d(s) \tag{7}$$

Our objective is to construct a Bezout fractional representation over $\Gamma_r(\tilde{H})$ for the transfer function $w(s)$ given by (5). A key component of our approach is the definition of a distributed delay: Let $\alpha(t)$ denote the inverse Laplace transform of $1/a(s)$ and define

$$v(t) = \begin{cases} \alpha(t), & 0 \le t \le h \\ 0, & \text{all other } t \end{cases}$$

Clearly, the function $v(t)$ is an analytic function of t for $0 < t < h$. The Laplace transform $V(s)$ of $v(t)$ is analytic on C; that is, $V(s)$ is an entire function of s. By definition of $v(t)$, it follows that we can express $V(s)$ in the form

$$V(s) = \frac{1 - x(s)e^{-hs}}{a(s)} \tag{8}$$

where $x(s)$ is an element of $\mathbb{R}[s]$ with $\deg x(s) < n$. In addition, if $p_1, p_2, \ldots, p_n$ are the zeros of $a(s)$, then

$$1 - x(p_i)e^{-hp_i} = 0 \text{ for all } i \tag{9}$$

The relationship (9) is a result of the fact that $V(s)$ is an entire function of s. From (8), we see that

$$V(s) \to 0 \text{ as } s \to \infty \quad (s \in \tilde{H})$$

Thus $V(s)$ is an element of the algebra $\Gamma_r(\tilde{H})$.

We can view $V(s)$ as the transfer function of a distributed time delay given by the input/output relationship

$$y(t) = \int_{t-h}^{t} v(t-\lambda)u(\lambda)d\lambda$$

Now choose monic polynomials $\phi_1(s), \phi_2(s) \in \mathbb{R}[s]$, both of degree n, and both stable (no zeros in $\bar{H}$). Using long division, we can compute polynomials $q(s), r(s) \in \mathbb{R}[s]$ such that

$$d(s)x(s)\phi_1(s)\phi_2(s) = a(s)q(s) + r(s) \tag{10}$$

where $\deg r(s) < n$. From (10), we have that

$$\deg q(s) = n + \deg d(s) + \deg x(s) \tag{11}$$

We then have the following result.

Theorem 2: The transfer function (5) has a Bezout fractional representation over $\Gamma_r(\tilde{H})$ given by

$$w(s) = \left[\frac{a(s)}{\phi_1(s)}\right]^{-1}\left[\frac{b(s)e^{-hs}}{\phi_1(s)}\right] \tag{12}$$

where

$$\frac{a(s)}{\phi_1(s)}p(s) + \frac{b(s)e^{-hs}}{\phi_1(s)}\gamma(s) = 1 \tag{13}$$

with

$$p(s) = \frac{q(s) + [V(s) + x(s)c(s)e^{-hs}]r(s)}{d(s)x(s)\phi_2(s)} \tag{14}$$

$$\gamma(s) = \frac{r(s)}{\phi_2(s)} \tag{15}$$

Proof: It is obvious that (12) is a fractional representation over $\Gamma_r(\tilde{H})$. We shall verify (13): Multiplying both sides of (6) by $x(s)e^{-hs}$ gives

$$a(s)c(s)x(s)e^{-hs} + b(s)d(s)x(s)e^{-hs} = x(s)e^{-hs}$$

Adding $1 - x(s)e^{-hs}$ to both sides, we get

$$1 - x(s)e^{-hs} + a(s)c(s)x(s)e^{-hs} + b(s)d(s)x(s)e^{-hs} = 1$$

Multiplying both sides by $\phi_1(s)\phi_2(s)$ and using (10), we obtain

$$\frac{[1 - x(s)e^{-hs} + a(s)c(s)x(s)e^{-hs}][a(s)q(s) + r(s)]}{d(s)x(s)} + b(s)[a(s)q(s) + r(s)]e^{-hs} =$$

$$\phi_1(s)\phi_2(s)$$

Using (6) and the relationship $a(s)V(s) = 1 - x(s)e^{-hs}$, we get

$$\frac{a(s)[q(s) + (V(s) + c(s)x(s)e^{-hs})r(s)]}{d(s)x(s)} + b(s)e^{-hs}r(s) = \phi_1(s)\phi_2(s)$$

Dividing both sides by $\phi_1(s)\phi_2(s)$, we obtain (13), with $p(s)$ and $\gamma(s)$ given by (14) and (15). It is clear that $\gamma(s)$ is an element of $\Gamma_r(\tilde{H})$. To prove that $p(s) \in \Gamma_r(\tilde{H})$, first note that by (11),

$$\deg q(s) = \deg(d(s)x(s)\phi_2(s))$$

In addition, combining (7) and (11), we have

$$\deg q(s) > \deg(c(s)x(s)r(s))$$

From the form of $p(s)$ as given by (14), it follows that $p(s)$ converges to a constant as $s \to \infty$ $(s \in H)$. Now to show that $p(s)$ is analytic on H and continuous on the $j\omega$-axis, using (6) and (10) we can first write $p(s)$ in the form

$$p(s) = \frac{[V(s) + c(s)x(s)e^{-hs}]\phi_1(s)\phi_2(s) + b(s)q(s)e^{-hs}}{\phi_2(s)} \tag{16}$$

Since $\phi_2(s)$ is a stable polynomial and $V(s)$ is an entire function of s, from (16) we see that $p(s)$ is analytic on H and continuous on the $j\omega$-axis. Therefore, $p(s) \in \Gamma_r(\tilde{H})$, and the proof is complete. ∎

Suppose that we have computed the fractional representation given in Theorem 2. If we then take the transfer function $w_c(s)$ of the compensator to be

$$w_c(s) = \frac{r(s)d(s)x(s)}{q(s) + [V(s) + c(s)x(s)e^{-hs}]r(s)} \tag{17}$$

the transfer function of the resulting closed-loop system is

$$\frac{[q(s) + (V(s) + c(s)x(s)e^{-hs})r(s)]b(s)e^{-hs}}{\phi_1(s)\phi_2(s)d(s)x(s)}$$

From the arguments given in the proof of Theorem 2, it follows that

$$\frac{q(s) + (V(s) + c(s)x(s)e^{-hs})r(s)}{d(s)x(s)}$$

is an entire function of s. Thus, the closed-loop poles are equal to the zeros of $\phi_1(s), \phi_2(s)$. Since $\phi_1(s)$ and $\phi_2(s)$ were arbitrary, we have a pole-assignability result for infinite-dimensional systems consisting of a cascade connection of of a lumped system and a pure time delay.

To construct a hardware realization of the compensator with transfer function given by (17), it is necessary to realize an h-second time delay. This can be avoided by computing a finite-dimensional stabilizing compensator using the procedure outlined in the previous section. Since $\gamma(s)$ given by (15) is already rational, we only have to compute a proper rational approximation to $p(s)$ given by (14). We should also note that Example 3.6 in [8] is a special case of the design procedure given above.

IV. SYSTEMS WITH COMMENSURATE TIME DELAYS

Given a fixed real number $h > 0$, let $\mathbb{R}[s,e^{-hs}]$ denote the ring of polynomials in s and e^{-hs} with real coefficients. It is well known that $\mathbb{R}[s,e^{-hs}]$ is isomorphic to the ring $\mathbb{R}[x,y]$ of polynomials in the symbols x,y with coefficients in the reals. In this section we consider m-input p-output systems with commensurate time delays given by the $p \times m$ transfer function matrix $W(s) = (w_{ij}(s))$, where

$$w_{ij}(s) = \frac{b_{ij}(s,e^{-hs})}{a_{ij}(s,e^{-hs})}$$

with $a_{ij}(s,e^{-hs}), b_{ij}(s,e^{-hs}) \in \mathbb{R}[s,e^{-hs}]$. The $a_{ij}(s,e^{-hs})$ are monic in s with degree (as a polynomial in s) equal to $\bar{n}$, and $\deg b_{ij}(s,e^{-hs}) < \bar{n}$.

We shall consider the existence and construction of left-Bezout fractional representations over $\Gamma_r(\tilde{H})$ for the above class of systems. The following results are a summary of some of results contained in Kamen, Khargonekar, and Tannenbaum [9].

First, any transfer function matrix W(s) as defined above admits a co-canonical state realization given by the triple $(F(e^{-hs}), G(e^{-hs}), H(e^{-hs}))$ of matrices over the ring $\mathbb{R}[e^{-hs}]$ of polynomials in e^{-hs} (see Rouchaleau and Sontag [10]). By definition of a realization,

$$W(s) = H(e^{-hs})(sI - F(e^{-hs}))G(e^{-hs})$$

where I is the $n \times n$ identity matrix (n = size of F). Co-canonical means that the matrix

$$[G(e^{-hs}) \quad F(e^{-hs})G(e^{-hs}) \quad \cdots \quad F^{n-1}(e^{-hs})G(e^{-hs})]$$

has a right inverse over the field $\mathbb{R}(e^{-hs})$ of rational functions in e^{-hs}, and the matrix

$$\begin{bmatrix} H(e^{-hs}) \\ H(e^{-hs})F(e^{-hs}) \\ \cdot \\ \cdot \\ \cdot \\ H(e^{-hs})F^{n-1}(e^{-hs}) \end{bmatrix}$$

has a left inverse over the polynomial ring $\mathbb{R}[e^{-hs}]$.

The following result can be found in [9].

Theorem 3: Let $(F(e^{-hs}), G(e^{-hs}), H(e^{-hs}))$ be a co-canonical realization of $W(s)$. Then $W(s)$ has a left-Bezout fractional representation over $\Gamma_r(\tilde{H})$ if and only if

$$\text{rank } [sI - F(e^{-hs}) \quad G(e^{-hs})] = n \text{ for all } s \in \bar{H} \tag{18}$$

When the rank condition (18) is satisfied, in [9] the authors give an explicit procedure for computing matrices $X_1(s), X_2(s)$ over $\Gamma_r(\tilde{H})$ such that

$$(sI - F(e^{-hs}))X_1(s) + G(e^{-hs})X_2(s) = I$$

The construction in [9] is based on a particular ring of pure and distributed time delays.

Now let $\phi(s)$ be a monic stable element of $\mathbb{R}[s]$, with the degree of $\phi(s)$ equal to n. Then using Morse's procedure [11], we can compute a $n \times p$ matrix $L(e^{-hs})$ over $\mathbb{R}[e^{-hs}]$ such that

$$\det(sI - F(e^{-hs}) + L(e^{-hs})H(e^{-hs})) = \phi(s)$$

Again from the results in [9], we have the following left-Bezout fractional representation:

$$W(s) = [I - H(sI - F + LH)^{-1}L]^{-1}H(sI - F + LH)^{-1}G$$

with

$$[I - H(sI - F + LH)^{-1}L][I + HX_1] + [H(sI - F + LH)^{-1}G][X_2L] = I$$

Using the approximation scheme discussed in Section II, we can then compute a finite-dimensional stabilizing compensator for the given system.

REFERENCES

[1] Vidyasagar, M., Input-output stability of a broad class of linear time-invariant multivariable feedback systems, SIAM J. Control 10 (1972), pp. 203-209.

[2] Vidyasagar, M., Coprime factorizations and the stability of multivariable distributed feedback systems, SIAM J. Control 13 (1975), pp. 1144-1155.

[3] Vidyasagar, M., Control System Synthesis: A Factorization Approach (MIT Press, Cambridge, 1985).

[4] Callier, F. M. and Desoer, C. A., An algebra of transfer functions for distributed linear time-invariant systems, IEEE Trans. Circuits and Systems CAS-25 (1978), pp. 651-663; Correction: CAS-26 (1979), p. 360.

[5] Vidyasagar, M., Schneider, H. and Francis, B. A., Algebraic and topological aspects of feedback stabilization, IEEE Trans. Automatic Control AC-27 (1982), pp. 880-894.

[6] Nett, C. N., The fractional representation approach to robust linear feedback design: A self-contained exposition, M. S. Thesis, Dept. of Electrical Engr., Rensselaer Polytechnic Institute (April 1984).

[7] Kamen, E. W., Khargonekar, P. P. and Tannenbaum, A., Pointwise stability and feedback control of linear systems with noncommensurate time delays, Acta Appl. Math. 2 (1984), pp 159-184.

[8] Kamen, E. W., Khargonekar, P. P. and Tannenbaum, A., Stabilization of time-delay systems using finite-dimensional compensators, IEEE Trans. Automatic Control AC-30 (January 1985), pp. 75-78.

[9] Kamen, E. W., Khargonekar, P. P. and Tannenbaum, A., Proper stable Bezout factorizations and feedback control of linear time-delay systems, to appear in Int. J. of Control (1985).

[10] Rouchaleau, Y. and Sontag, E. D., On the existence of minimal realizations of linear dynamical systems over Noetherian integral domains, J. Computer and System Sciences 18 (1979), pp. 65-75.

[11] Morse, A. S., Ring models for delay differential systems, Automatica 12 (1976), pp. 529-531.

Frequency Domain and State Space Methods for Linear Systems
C.I. Byrnes and A. Lindquist (Editors)

STABILIZABILITY AND DETECTABILITY FOR INFINITE-DIMENSIONAL LINEAR TIME-VARYING SYSTEMS

Günter Kern

Institut für Mathematik
Technische Universität Graz
Kopernikusgasse 24
A-8010 Graz
Austria

The concepts of stabilizability and detectability are explored for linear time-varying systems in infinite-dimensional Hilbert spaces. We study their duality and an extended version of the lemma of Lyapunov. Finally, we show the equivalence of the given definition with the existence of stabilizing feedback laws.

I. INTRODUCTION

In many control problems the controllability and observability hypothesis have been weakened to stabilizability and detectability and the result are still valid (especially in the time-invariance case). First of all, for time-varying systems the weaker conditions of stabilizability and detectability were not used since the technical issues raised involve nontrivial generalizations of the time-invariant results and on first glance appear formidable. For the time-varying case a lead has been taken in [4] with the introduction of definitions of detectability and stabilizability for discrete-time, time-varying linear systems and some applications to control and filtering problems. In [1] and [2] a generalization of the concepts of detectability and stabilizability to linear, time-varying, discrete-time, finite dimensional systems have been considered, with the intention to obtain their formal duality and invariance under feedback. In fact, the definition says roughly that when a state trajectory is not fast decaying, then that trajectory must be controllable or observable.

In this paper, motivated by [1] and [2], we explore the concepts of detectability and stabilizability to linear time-varying systems in infinite-dimensional Hilbert spaces. We indicate their formal duality and establish some simple consequences. In section III we prove a generalization of the lemma of Lyapunov, which is useful for studying the stability of linear systems in infinite dimensional Hilbert spaces. In section IV we show the equivalence of the definitions of section II with the existence of stabilizing feedback laws of an appropriate form.

II. DEFINITIONS, NOTATIONS AND PRELIMINARY RESULTS

Let X be a separable Hilbert space and consider the linear time-varying system

$$(2.1)\qquad \begin{aligned}\dot{x}(t) &= A(t)x(t) + B(t)u(t) \qquad t \geqq 0\\ z(t) &= C(t)x(t)\end{aligned}$$

where $A \in L^1_{loc}(R_+,L(X))$, $B \in \tau(R_+,L(Y,X))$, $u \in \Omega \subset L^1_{loc}(R_+,Y)$, $C \in \tau(R_+,L(X,Z))$, in which $R_+ = [0,\infty)$, $L(Y,X)$... the Banach space of the bounded linear operators from the Banach space Y to X , $L(X,X) = L(X)$, $L^1_{loc}(R_+,Z)$ ... the space of Z-valued functions defined almost everywhere on R_+

such that these functions are strongly measurable and locally Bochner integrable and finally $\tau(R_+,L(Y,X))$ ... the space of continuous operator valued functions from R_+ to $L(Y,X)$.

The scalar product of a pair of elements $x,y \in X$ will be denoted by $<x,y>$ and its norm by $\|x\| = <x,x>^{1/2}$. The operator norm for an element $A \in L(X,Y)$ is the quantity $\|A\| = \sup\{\|Ax\| \mid x \in X , \|x\| = 1\}$. An operator $H \in L(X)$ is said to be Hermitean if $H = H^*$, where H^* is called the adjoint of H . An operator is said to be positive (nonnegative) if its form $<Hx,x>$ is positive (nonnegative) for any $x \neq 0$. An operator H is said to be uniformly positive if its form $<Hx,x>$ is uniformly positive on the unit sphere $S = \{x \mid \|x\| = 1\}$ in X , i.e. if $\inf\{<Hx,x> \mid \|x\| = 1\} = \alpha > 0$.

Let $U(t,\tau) = U(t)U^{-1}(\tau)$, which will be called the evolution (solving) operator of the differential equation

$$\dot{x}(t) = A(t)x(t) \tag{2.2}$$

since this operator satisfies the system

$$\frac{dZ}{dt} = A(t)Z , \qquad Z(\tau) = I .$$

By the above hypotheses the solution of the Cauchy problem for the homogeneous equation (2.2) can be written in the form

$$x(t) = U(t,\tau)x_\tau , \qquad x(\tau) = x_\tau \tag{2.3}$$

and for the system (2.1) in the form

$$x(t) = U(t,\tau_0)x_0 + \int_{\tau_0}^{t} U(t,s)B(s)u(s)ds . \tag{2.4}$$

Standing assumption: The operator function $A(t)$ is integrally bounded:

$$\sup \int_t^{t+1} \|A(\tau)\| d\tau \leq M , \qquad t \geq 0 . \tag{2.5}$$

<u>Definition 1:</u> Equation (2.2) is said to have the property $B(\nu,N)$, where ν is real and N is positive, if all of its solutions satisfy the estimate

$$\|x(t)\| \leq Ne^{-\nu(t-\tau)}\|x(\tau)\| , \qquad t \geq \tau \geq 0 \tag{2.6}$$

or equivalently

$$\|U(t,\tau)\| \leq Ne^{-\nu(t-\tau)} , \qquad t \geq \tau \geq 0 . \tag{2.7}$$

Clearly, property $B(0,N)$ is equivalent to the uniform stability and $B(\nu,N)$ with $\nu > 0$ is equivalent to the uniform asymptotic or exponential stability of equation (2.2).

We now define the concepts of uniform stabilizability and detectability. Suppose that for any q $(0 < q < 1)$ and any $T > 0$ there exist x_0 and τ_0 such that

$$\|U(t,\tau_0)x_0\| > q\,\|x_0\| \quad , \quad t \in [\tau_0, \tau_0+T] \tag{2.8}$$

then there is no positive ν to estimate the corresponding solution of (2.2) in the form of (2.6) (see [3], Theorem 6.1).
We define the closed subspace

$$X_P(\tau_0) = \overline{\{x(\tau_0) = x_0 \mid x_0 \in X, \|U(t,\tau_0)x_0\| > q\,\|x_0\|\}} \tag{2.9}$$

and decompose the state space X into a direct sum

$$X = X_P \oplus X_P^{\perp}$$

of closed subspaces with the corresponding projections P and $(I-P)$.

Definition 2: System (A,B) is uniformly stabilizable if there exist $\sigma > 0$ and $\delta > 0$ such that

$$< W(\tau_0+\sigma,\tau_0)x_0, x_0 > \;\geq\; \delta\,\|x_0\|^2 \tag{2.1o}$$

for all $x_0 \in X_P$ at any $\tau_0 \geq 0$, where

$$W(\tau_0+\sigma,\tau_0) = \int_{\tau_0}^{\tau_0+\sigma} U(\tau_0,t)B(t)B^*(t)U^*(\tau_0,t)dt \;. \tag{2.11}$$

Definition 3: System (A,C) is uniformly detectable if there exist $\sigma > 0$ and $\delta > 0$ such that

$$< M(\tau_0+\sigma,\tau_0)x_0, x_0 > \;\geq\; \delta\,\|x_0\|^2 \tag{2.12}$$

for all $x_0 \in X_P$ at any $\tau_0 \geq 0$, where

$$M(\tau_0+\sigma,\tau_0) = \int_{\tau_0}^{\tau_0+\sigma} U^*(t,\tau_0+\sigma)C^*(t)C(t)U(t,\tau_0+\sigma)dt \;. \tag{2.13}$$

The above definitions are an infinite-dimensional version of defintions given by B.D.O.Anderson [1],[2]. In fact, the definition says roughly that when a state trajectory is not fast decaying, i.e. (2.8) is satisfied, then that trajectory must be controllable (constructible), i.e. (2.1o) ((2.12)) holds.

The duality between stabilizability and detectability is taken up in the following lemma:

Lemma 1: Make the definitions

$$\tilde{A}(t) = A^*(-t) \;, \quad \tilde{C}(t) = B^*(-t) \;. \tag{2.14}$$

Then

a) $(\tilde{A},\tilde{C})$ is uniformly detectable if and only if (A,B) is uniformly stabilizable.

b) $\dot{x} = A(t)x$ is exponentially stable if and only if $\dot{\tilde{x}} = \tilde{A}(t)\tilde{x}$ is exponentially stable.

Proof is via direct calculation.

We conclude the section by noting two simple consequences of these definitions:

Lemma 2: The system (A,B) is uniformly stabilizable if and only if there exist $\sigma > 0$ and $k > 0$ such that for every $x_0 \in X$ and for any time $\tau_0 \geqq 0$ there exists an input $u \in \Omega$ such that

(i) $x(\tau_0+\sigma;\ \tau_0,x_0,u) \notin X_P(\tau_0+\sigma)$

(ii) $\displaystyle\int_{\tau_0}^{\tau_0+\sigma} \|u(s)\|^2 ds \leq k\ \|x_0\|^2$,

where we consider the projection function

$$P(\tau_0+\sigma) = U(\tau_0+\sigma,\tau_0)\,P\,U(\tau_0,\tau_0+\sigma)$$

and the corresponding subspace $X_P(\tau_0+\sigma) = U(\tau_0+\sigma,\tau_0)PX$.

Proof: Assume that (i) and (ii) hold. Then for $x_0 \in X$ there exists an input $u \in \Omega$ such that

$$x(\tau_0+\sigma) = U(\tau_0+\sigma,\tau_0)\ \{Px_0 + (I-P)x_0 - \int_{\tau_0}^{\tau_0+\sigma} U(\tau_0,t)B(t)u(t)dt\} \tag{2.15}$$

$$P(\tau_0+\sigma)x(\tau_0+\sigma) = U(\tau_0+\sigma,\tau_0)\ \{Px_0 - \int_{\tau_0}^{\tau_0+\sigma} PU(\tau_0,t)B(t)u(t)dt\} = 0\ . \tag{2.16}$$

Now, if (A,B) is not uniformly stabilizable, then there exists $\tau_0 \geqq 0$ such that for every nonincreasing function $\varepsilon : R_+ \rightarrow R_+$ there exists $\hat{x}_0 \in X_P$ with $\|\hat{x}_0\| = 1$ such that

$$< W(\tau_0+\sigma,\tau_0)\hat{x}_0,\hat{x}_0 > < \varepsilon(\tau_0)\ . \tag{2.17}$$

By (2.16) with $Px_0 = \hat{x}_0 \in X_P$ and the Schwarz inequality follows

$$1 = \|\hat{x}_0\|^2 = \int_{\tau_0}^{\tau_0+\sigma} < u(t),\ B^*(t)U^*(\tau_0,t)\hat{x}_0 > dt \leq \tag{2.18}$$

$$\leq \left(\int_{\tau_0}^{\tau_0+\sigma} \|u(t)\|^2 dt\right)^{1/2} < W(\tau_0+\sigma,\tau_0)\hat{x}_0,\hat{x}_0 >^{1/2} \leq \sqrt{k\varepsilon}$$

which is a contradiction since ε can be found arbitrarly small.

If (A,B) is uniformly stabilizable then $W_P(\tau_0+\sigma,\tau_0) = PW(\tau_0+\sigma,\tau_0)P$ are invertible operators on X_P and

$$\|W_P^{-1}(\tau_0+\sigma,\tau_0)\| \leq \frac{1}{m} \qquad \text{for every } \tau_0 \geqq 0\ .$$

Define with $x_0 \in X_P$

$$u(t) = \begin{cases} - B^*(t)U^*(\tau_0,t)PW_P^{-1}(\tau_0+\sigma,\tau_0)x_0 & \text{if } t \in [\tau_0,\tau_0+\sigma] \\ 0 & \text{if } t \notin [\tau_0,\tau_0+\sigma] \end{cases}$$

it is clear that $\int_{\tau_0}^{\tau_0+\sigma} \|u(s)\|^2 ds \leq k \|x_0\|^2$ where $k = \frac{1}{m}$ and

$$P(\tau_0+\sigma)x(\tau_0+\sigma) = U(\tau_0+\sigma,\tau_0) \{Px_0 + P(I-P)x_0 -$$

$$- P \int_{\tau_0}^{\tau_0+\sigma} U(\tau_0,t)B(t)B^*(t)U^*(\tau_0,t)dt\ PW_2^{-1}(\tau_0+\sigma,\tau_0)x_0 = 0$$

which implies (i).

Lemma 3: If the system (A,C) is uniformly detectable, then for the system (2.1) with a zero input

$$\|C(t)x(t)\| \to 0 \text{ as } t \to \infty \Longrightarrow \|x(t)\| \to 0 \text{ as } t \to \infty .$$

The proof can carried out in the same way as in the finite dimensional case, see [1].

III. EXTENDED LEMMA OF LYAPUNOV

In this section we attempt to parallel a result relevant to establishing stability. Lyapunov functions plays in the infinite dimensional case the same important role as in the finite dimensional case, see e.g. [6].

For each system (2.1) we consider the operator differential equation

$$\dot{X}(t) + A^*(t)X(t) + X(t)A(t) = - C^*(t)C(t) . \tag{3.1}$$

Theorem 1: Suppose that (A,C) is uniformly detectable and that there is a bounded nonnegative definite Hermitean solution X satisfying (3.1) for $t \geq \tau_0$, then (2.2) is exponentially stable.

Proof: We shall associate with (2.2) a "Lyapunov-like" function

$$v(t) = < (X(t) + \varepsilon I)x(t),x(t) > \tag{3.2}$$

$$\frac{dv(t)}{dt} = - < (A^*(t)X(t) + X(t)A(t) + C^*(t)C(t))x(t),x(t) > +$$

$$+ < (X(t) + \varepsilon I)A(t)x(t),x(t) > + < (X(t) + \varepsilon I)x(t),A(t)x(t) >$$

$$= - \|C(t)x(t)\|^2 + \varepsilon\frac{d}{dt} \|x(t)\|^2 ;$$

$$v(t+\sigma)-v(t) = \int_t^{t+\sigma} \dot{v}(\tau)d\tau = - < M(t+\sigma,t)x(t),x(t) > + \varepsilon\{\|x(t+\sigma)\|^2 - \|x(t)\|^2\} . \tag{3.3}$$

Two cases arise: Let q and δ be the quantities of definition 3.

Case 1: If $\|U(t+\Theta_{x,t},t)x(t)\| > q\,\|x(t)\|$, $\Theta_{x,t} \in [0,T]$, then under the detectability assumption

$$< M(t,t+\sigma)x(t),\ x(t) > \geqq \delta\|x(t)\|^2$$

so that with the use of (2.5)

$$v(t+\sigma) - v(t) \leqq -\delta\|x(t)\|^2 + \varepsilon\{e^{M(\sigma+1)} - 1\}\,\|x(t)\|^2 \,. \tag{3.4}$$

Hence if ε is sufficiently small, there exists $\eta > 0$ such that

$$v(t+\sigma) - v(t) \leqq -\eta\|x(t)\|^2 \leqq -\frac{\eta}{M_1} < (X(t)+\varepsilon I)x(t),x(t) > = -c'v(t) \tag{3.5}$$

where $X(t) + \varepsilon I \leqq M_1 I$ and $c' = \frac{\eta}{M_1} > 0$.

Case 2: If $\|U(t+\Theta_{x,t},t)x(t)\| \leqq q\,\|x(t)\|$, $\Theta_{x,t} \in [0,T]$

$$v(t+\sigma) - v(t) \leqq \varepsilon(q^2-1)\,\|x(t)\|^2 \leqq \frac{\varepsilon}{M_1}(q^2-1)\,v(t) = -c''v(t) \,, \tag{3.6}$$

where $c'' = \frac{\varepsilon}{M_1}(1-q^2) > 0$.

If $c = \min(c',c'')$, it follows from (3.5) and (3.6) that for both cases

$$v(t+\sigma) - v(t) \leqq -c\,v(t) \,. \tag{3.7}$$

With (3.7) we can carry out in the same way as in [6], page 14 and 15 the remaining part of the proof and obtain that (2.2) is exponentially stable.

IV. CONNECTION WITH THE EXISTENCE OF A STABILIZING STATE FEEDBACK LAW.

In this final section we shall show the equivalence of the definition of section II with the existence of stabilizing feedback laws of appropriate form:

Theorem 2: System (A,B) is uniformly stabilizable if and only if there exist an input $u(t) = -K(t)x(t)$, with $\sup_t \int_t^{t+1} \|K(\tau)\|^2 d\tau < \infty$, such that the closed-loop system

$$\dot{x} = (A(t) - B(t)K(t))x \tag{4.1}$$

is exponentially stable.

Proof: ($\Leftarrow$ by contradiction) Assume there exists an input such that the closed-loop system is exponentially stable, but (A,B) is not uniformly stabilizable. Let $\sigma > 0$, if (A,B) is not uniformly stabilizable then there exists $\tau_0 \geqq 0$ such that for every nonincreasing function $\varepsilon : R_+ \to R_+$ there exists $x_0 \in X_p(\tau_0)$ with $\|x_0\| = 1$ such that

$$(4.2)\qquad \int_{\tau_0}^{\tau_0+\sigma} \|B^*(t)U^*(\tau_0,t)x_0\|^2 dt < \varepsilon(\tau_0)\ .$$

Let $\Psi_k(t,\tau_0)$ the evolution operator associated with the closed-loop-system (4.1). From the obvious equality (see [7])

$$\Psi_k(t,\tau_0) = U(t,\tau_0) - \int_{\tau_0}^{t} U(t,s)B(s)K(s)\Psi_k(s,\tau_0)ds$$

we obtain

$$(4.3)\qquad U(\tau_0,t)\Psi_k(t,\tau_0) = I - \int_{\tau_0}^{t} U(\tau_0,s)B(s)K(s)\Psi_k(s,\tau_0)ds$$

which implies

$$(4.4)\qquad \|\Psi_k(t,\tau_0)\|\ \|U^*(\tau_0,t)x_0\| \geq 1 - p(\sigma)\ \{\int_{\tau_0}^{t} \|B^*(s)U^*(\tau_0,s)x_0\|^2 ds\}^{1/2}$$

where

$$p(\sigma) = \sup_s\ (\int_s^{s+\sigma} \|K(\xi)\|^2 d\xi)^{1/2} \sup_{s\in[\tau_0,\tau_0+\sigma]} \|\Psi_k(s,\tau_0)\| \qquad \sigma > 0\ .$$

Let $\varepsilon = \frac{1}{2p(\sigma)}$, we obtain from (4.2) and (4.4)

$$(4.5)\qquad \|\Psi_k(t,\tau_0)\|\ \|U^*(\tau_0,t)x_0\| \geq 1 - p(\sigma)\ \frac{1}{2p(\sigma)} = \frac{1}{2}\ ,$$

thus we have a contradiction because we assumed that $\Psi_k(t,\tau_0)$ is the evolution operator of an exponentially stable system and the solution of the adjoint equation $U^*(\tau_0,t)x_0$ is right stable (because $x_0 \in X_p(\tau_0)$ and therefore the corresponding solution of (2.2) is left stable, see [3], page 113).

(=>) If system (A,B) is uniformly stabilizable then assumption (i) in [5], page 174 is satisfied and therefore we can show by the same argument as in [5] (Theorem 4, page 179) the existence of stabilizing feedback laws of appropriate form.

REFERENCES

[1] Anderson, B.D.O. and Moore, J.B., Detectability and stabilizability of time-varying discrete-time linear systems, SIAM J. Control and Optimization 19 (1981) 2o-32.

[2] Anderson, B.D.O., Internal and external stability of linear time-varying systems, SIAM J. Control and Optimization 2o (1982) 4o8-413.

[3] Daleckii, J.L. and Krein, M.G., Stability of solutions of differential equations in Banach space, AMS Translations of Math. Monographs, Vol.43, Providence, Rhode Island 1974.

[4] Hager, A.H. and Horowitz, L.L., Convergence and stability properties of the discrete Riccati equation and the associated control and filtering problems, SIAM J. Control and Optimization 14 (1976) 295-312.

[5] Kern, G., Zum Stabilisierungsproblem von linearen nichtautonomen Systemen in Hilbert Räumen, Wissenschaftl. Zeitschrift der Technischen Hochschule Leipzig 3 (1984) 171-182.

[6] Megan, M., On the controllability, stability and Lyapunov differential equation for linear control systems, S.E.F. 39 (1976), Timisoara.

[7] Megan, M. and Reghis, M., Riccati equation, exact controllability and stability, S.E.F. 42 (1977), Timisoara.

Frequency Domain and State Space Methods for Linear Systems
C.I. Byrnes and A. Lindquist (Editors)
Elsevier Science Publishers B.V. (North-Holland), 1986

Some Remarks on Stabilization of Waves and Beams

John Lagnese

Department of Mathematics
Georgetown University
Washington, D.C. 20057

Various boundary feedback stabilization schemes have been proposed in order to stabilize systems modeled by hyperbolic partial differential equations. A brief survey of some of these schemes as they pertain to waves and beams is presented along with a discussion of some of the difficulties that might be encountered in their implementations.

INTRODUCTION

Let Ω be a bounded region in $\mathbb{R}^n$ having a smooth boundary Γ consisting of two parts, Γ_o and Γ_1, with Γ_o nonempty and relatively open in Γ. Γ_o is assumed to either be empty or to have a nonempty interior. A typical control problem of the type to be considered is

(1.1) $\ddot{y} - \Delta_n y = 0$ in $\Omega\times(0,\infty)$,

(1.2) $y = 0$ on $\Gamma_o\times(0,\infty)$,

(1.3) $\dfrac{\partial y}{\partial \nu} = v$ on $\Gamma_1\times(0,\infty)$,

where $\cdot$ denotes time differentiation, Δ_n is the ordinary n-dimensional spatial Laplacian and ν is the unit normal to Γ directed towards the exterior of Ω. The function v is the control through which the system is to be stabilized in some sense. Stability is defined in terms of the asymptotic behavior of the total energy $E(y,t)$ of the system. For solutions of (1.1) this quantity is

$$E(y,t) = \tfrac{1}{2}\int_\Omega (\dot{y}^2 + |\nabla y|^2)dx.$$

We shall be concerned with two types of stability. The first is strong (asymptotic) stability defined as

$$\lim_{t\to\infty} E(y,t) = 0,$$

and the second is uniform (asymptotic) stability:

$$E(y,t) \leq f(t)E(y,0)$$

where

$$\lim_{t\to\infty} f(t) = 0.$$

In this case we refer to $f(t)$ as a decay rate for the system.

In order to obtain an asymptotically stable closed-loop system it is assumed that the velocity is observable on Γ_1:

(1.4) $u = \dot{y}$ on $\Gamma_1\times(0,\infty)$.

A straight-forward calculation shows that

$$\dot{E}(y,t) = \int_{\Gamma_1} \dot{y}v\,d\sigma ,$$

consequently the feedback law

(1.5) $v = -ku$ on $\Gamma_1 \times (0,\infty)$

with $k>0$ assures that the closed-loop system is dissipative. The following result concerning strong stability is well known (see e.g. [1,8,11,13]).

THEOREM 1.1. *Assume* $E(y,0)<\infty$ *and that*

$$k \in C^1(\bar{\Gamma}_1),\ k(x)\geq 0,\ k \not\equiv 0.$$

Then the closed-loop system (1.1) - (1.5) *is strongly stable.*

The above result gives no information about the rate of decay of energy and, indeed, in general there will be no decay rate without restrictions on the geometry of Γ. Conditions on the geometry which assure a decay rate have been given in [2,3,8]. The most general result to date seems to be that in [8] and is as follows.

THEOREM 1.2. *Assume that* $k \in C^1(\bar{\Gamma}_1)$, $k(x)\geq k_o>0$, *and that there is a vector field* $\ell = (\ell_1,\ell_2,\cdots,\ell_n) \in C^2(\bar{\Omega})$ *such that*

$\ell\cdot\nu\leq 0$ on Γ_o,

$\ell\cdot\nu\geq\alpha>0$ on Γ_1,

$(\partial\ell_i/\partial x_j+\partial\ell_j/\partial x_i)$ *is positive definite on* $\bar{\Omega}$.

Then the system (1.1) - (1.5) *admits an exponential decay rate.*

Among the many other stabilization results for waves deserving mention are two due to Lasiecka and Triggiani. In [9] a feedback of the form

$$v = g(x) \int_\Omega w(x)\dot{y}(x,t)dx \quad \text{on } \Gamma\times(0,\infty),$$

acting in the Dirichlet boundary condition

(1.6) $y = v$ on $\Gamma\times(0,\infty)$

is considered, and sufficient conditions on the functions $g\in L^2(\Gamma)$ and $w\in L^2(\Omega)$ which assure strong stabilization of the closed-loop system are given. Theorem 3.1 below is analagous to their result. A second result, similar in spirit to Theorem 1.2, states that there is an exponential decay rate for the system (1.1), (1.6) via the feedback

$$v = -k \frac{\partial}{\partial\nu}(\Delta_n^{-1}\dot{y}) \text{ on } \Gamma\times(0,\infty)\ (k>0)$$

provided the geometry of Γ is suitably restricted (see [10]).

Concerning the stabilization of *beams* by means of boundary feedback, of particular interest is a recent paper by Chen et al [4]. A special case of their main result will now be described. Consider an Euler-Bernoulli beam of length L clamped at the left end and controlled at the right end by means of a point force v_1 and a point bending moment v_2. The system is therefore given by

(1.7) $\ddot{y} + a^4D^4y = 0,\quad 0<x<L,\ t>0,$

(1.8) $y(0,t) = Dy(0,t) = 0,\quad t>0,$

(1.9) $-a^4D^2y(L,t) = v_2,\ a^4D^3y(L,t) = v_1,\quad t>0.$

where $D = \partial/\partial x$. At $x = L$ the observations are assumed to be

(1.10) $u_1 = \dot{y},\ u_2 = D\dot{y},\quad t>0.$

With the energy defined as

$$E(y,t) = \tfrac{1}{2}\int_0^L (|\dot{y}|^2+a^4|D^2y|^2)dx$$

one has

$$\dot{E}(y,t) = -v_2D\dot{y} - v_1\dot{y}\,_{x=L},\ t\ 0,$$

and therefore the closed-loop system defined by

(1.11) $v_1 = k_1 u_1,\ v_2 = k_2 u_2,\ k_1 \geq 0,\ k_2 \geq 0,\ k_1+k_2>0,$

is dissipative. A corollary of the main result of [4] is

THEOREM 1.3. *If $k_1>0$, then the system (1.7) - (1.11) admits an exponential decay rate.*

EFFECT OF TIME DELAY IN THE FEEDBACK LOOP

All of the stabilization schemes of the previous section involve velocity feedback. Any implementation of such schemes would inevitably lead to a time delay in the feedback caused by the necessity of calculating the control action from the observations. In this section a system similar to (1.1) - (1.5) but also containing viscous and elastic damping will be considered. Our purpose is to demonstrate that the introduction of a time delay, no matter how small, into the feedback law (1.4), (1.5) can lead to destabilization of the system. A similar phenomenon has previously been observed by Datko [5] in the case of distributed observations/actuators.

The example to be considered is

(2.1) $\ddot{y} + 2a\dot{y} + a^2 y - D^2 y = 0, \quad 0<x<1,\ t>0,$

(2.2) $y(0,t) = 0,\ Dy(1,t) = v(t),\ t>0.$

When $a>0$, (2.1) contains both viscous and elastic damping, and it is easily shown that the uncontrolled system ($v(t)=0$) is already uniformly stable. Given the observation

(2.3) $u(t) = \dot{y}(1,t),\ t>0,$

and given $\varepsilon>0$ we close the loop by setting

(2.4) $v(t) = -ku(t-\varepsilon),\ t>\varepsilon,$

where k is a positive constant. When $\varepsilon=0$ the system (2.1) - (2.4) is uniformly stable, even if $a = 0$.

The necessary condition that the system (2.1) - (2.4) be uniformly (strongly) stable is that its spectrum σ_ε lie in the half-plane $\mathrm{Re}\omega \leq -\rho<0$ ($\mathrm{Re}\omega<0$). The spectrum can be exhibited by setting $y = e^{\omega t}Y(x)$. The eigenvalues ω and eigenfunctions Y are then determined by the problem

$$D^2Y - (\omega+a)^2 Y = 0,\ 0<x<1,$$

$$Y(0) = 0,\ DY(1) = -k\omega e^{-\varepsilon\omega} Y(1).$$

The following properties of σ_ε have recently been established.

THEOREM 2.1 [6]. Let $K = e^{-2a}$.

(i) *If $0<k<(1-K)/(1+K)$, for each $\varepsilon>0$ there exists $\beta(\varepsilon)>0$ such that*

$$\sigma_\varepsilon \subset \{\omega \mid \mathrm{Re}\omega \leq -\beta\}.$$

(ii) *If $k = (1-K)/(1+K)$, for each $\varepsilon>0$*

$$\sigma_\varepsilon \subset \{\omega \mid \mathrm{Re}\omega<0\}.$$

In addition, there is a sequence S_0 which is dense in $(0,\infty)$ such that for each ε in S_0,

$$\text{distance } (\sigma_\varepsilon,\ \{\omega \mid \mathrm{Re}\omega \geq 0\}) = 0.$$

(iii) *If $k>(1-K)/(1+K)$, there is a dense open subset S_1 in $(0,\infty)$ such that for each ε in S_1,*

$$\sigma_\varepsilon \cap \{\omega \mid \mathrm{Re}\omega>0\} \neq \phi.$$

Thus if $k>(1-K)/(1+K)$, equation (2.1) with boundary conditions (2.2) admits exponentially unstable solutions for arbitrarily small delays in the feedback law (2.3), (2.4), and if $k = (1-K)/(1+K)$ the system can no longer be uniformly stable but may remain strongly stable. In particular, if $a=0$ in (2.1) the closed-loop system will always be unstable for any gain $k>0$ and "almost" arbitrary delays.

In order to compensate for the destabilizing effects of time delays in the case a=0, Skar and Lindner [12] have proposed the following. They first note that solutions of (2.1) are superpositions of left traveling waves and right traveling waves, each having unit velocity. The left traveling waves are reflected at the boundary x=0 with no energy gain or loss and become right traveling waves. The latter are to be attenuated at x=1 via the feedback (2.3), (2.4). Because of the presence of a time delay ε, one might anticipate that waves with frequencies of approximately $1/(2\varepsilon)$ are amplified at x=1 rather than attenuated. In order to avoid this, Skar and Lindner suggest placing an additional sensor at position $x=1-\varepsilon$. Since the waves have velocity one and the feedback calculation introduces a time delay ε, observations at position $1-\varepsilon$ and time $t-\varepsilon$ are fed back at x=1 and time t. More specifically, the form of the proposed feedback law is

$$(2.5)\qquad u_1 = \dot{y}(1,t),\; u_2 = \dot{y}(1-\varepsilon,t)\quad t>0,$$

$$(2.6)\qquad v(t) = -k[u_2(t-\varepsilon)-\tfrac{1}{2}(u_1(t-2\varepsilon)+v(t-2\varepsilon))],$$

and the following result is proved.

THEOREM 2.2 [12]. <u>Suppose</u> $a=0$, $0<k<2$ <u>and</u> $0\leq\varepsilon<\frac{1}{2}$. <u>Then the closed-loop system</u> (2.1), (2.5), (2.6) <u>is uniformly exponentially stable</u>.

CONCENTRATED SENSORS/ACTUATORS

From an engineering point of view, point sensors/actuators such as those considered in the last section, and boundary sensors/actuators in general, are idealizations, since all real sensors will yield measurements of the variables being monitored which are weighted averages over a portion of the spatial region. Similarly, control actions resulting from real actuators affect an open region of the spatial domain and not just a point or the boundary. In this section we consider the effect on stability of replacing point sensors/actuators by ones which influence or are influenced by a neighborhood of the points in question. We refer to such sensors and actuators as "concentrated".

As motivation for the model problem to be treated consider the Euler-Bernoulli beam control problem (1.7) - (1.9) with $v_2=0$ and $v_1 = k^2\dot{y}$, $k\neq 0$. As noted in Theorem 1.3, the resulting closed-loop system is uniformly stable. When that system is written as a variational problem it has the form

$$(3.1)\qquad \int_0^L (\ddot{y}v+a^4D^2yD^2v)\,dx + k^2\int_0^L \xi\dot{y}\,dx \int_0^L \xi v\,dx = 0,$$

$$\forall\, v\in\{v\in H^2(0,L)\,|\,v(0) = Dv(0) = 0\},$$

where $\xi=\delta(L-x)$, the Dirac measure supported at x=L. Use of a concentrated sensor/actuator in place of the boundary sensor/actuator amounts to replacing $\xi=\delta(L-x)$ with an $L^2(0,L)$ approximation. The resulting variational problem (3.1) is the one to be considered here. It is equivalent to the boundary value problem

$$(3.2)\qquad \ddot{y}+a^4D^4y+k^2\xi\int_0^L \xi\dot{y}\,dx = 0,\quad 0<x<L,\; t>0,$$

$$(3.3)\qquad y(0,t) = Dy(0,t) = D^2y(L,t) = D^3y(L,t) = 0,\quad t>0.$$

At issue is how the stability of (3.2), (3.3) depends on the weight function ξ. From a result of Gibson [7] it is known that a decay rate cannot exist for (3.2), (3.3). However, the following has recently been established in joint work with M. Delfour and M. P. Polis.

THEOREM 3.1. <u>Assume</u> $\xi\in L^2(0,L)$. <u>The necessary and sufficient condition that</u> (3.2), (3.3) <u>be strongly stable is that</u>

$$(3.4)\qquad (\xi,\psi)\neq 0$$

<u>for every eigenvector</u> ψ <u>of the problem</u>

$$D^4\psi+\omega^2\psi = 0,\quad 0<x<L,$$

$$\psi(0) = D\psi(0) = D^2\psi(L) = D^3\psi(L) = 0.$$

Suppose now that ξ is an approximation to $\delta(L-x)$. The condition (3.4) places a fairly severe restriction on the allowable choices of ξ. In particular, arbitrarily small changes in the support of ξ may destabilize the system. This may seem counter-intuitive if $\xi(L) \neq 0$ and the support of ξ is small in view of the uniform stability the system enjoys when a point sensor/actuator at $x = L$ is utilized. The following result helps to clarify this apparent anomaly.

THEOREM 3.2. <u>Assume that</u>

$$\xi \geq 0, \ \int_0^L \xi(x)dx = 1, \ \underline{\text{support}}\ \xi = [L\rho, L]$$

<u>where</u> $0 \leq \rho < 1$. <u>Then for</u> t <u>sufficiently large</u>,

$$E(y,t) \leq \frac{C}{t} E(y,0) + \frac{8L^4}{3}(1-\rho^{3/2})E(\dot{y},0). \tag{3.5}$$

Thus it is seen that whether or not (3.2), (3.3) is strongly stable, the energy of the system can be bounded by a sum of two terms: one which tends uniformly to zero as $t \to \infty$ and a residual which depends only on the initial energy $E(\dot{y},0)$ and the support of ξ. As the support converges to the point $x = L$, (3.5) formally gives a decay rate for the energy, in agreement with [4].

REFERENCES

[1] Benchimol, C.D., A note on weak stabilization of contraction semigroups, SIAM J. Control Opt. 16 (1978), 373-379.

[2] Chen, G., Energy decay estimates and exact boundary value controllability for the wave equation in a bounded domain, J. Math. Pures Appl. 58 (1979), 249-274.

[3] _______, A note on the boundary stabilization of the wave equation, SIAM J. Control Opt. 19 (1981), 106-113.

[4] Chen, G., Delfour, M., Krall, A., and Payre, G., Modeling, stabilization and control of serially connected beams, preprint.

[5] Datko, R., Representation of solutions and stability of linear differential-difference equations in a Banach space, J. Differential Equations 29 (1978), 105-166.

[6] Datko, R., Lagnese, J., and Polis, M.P., An example on the effect of time delays in boundary feedback stabilization of wave equations, SIAM J. Control Opt., to appear.

[7] Gibson, J.S., A note on stabilization of infinite dimensional linear oscillators by compact feedback, Ibid. 18 (1980), 311-316.

[8] Lagnese, J., Decay of solutions of wave equations in a bounded region with boundary dissipation, J. Differential Equations 50 (1983), 163-182.

[9] Lasiecka, I., and Triggiani, R., Dirichlet boundary stabilization of the wave equation with damping feedback of finite range, J. Math. Anal. Appl. 97 (1983), 112-130.

[10] _______, Exponential uniform stabilization of the wave equation with $L^2(0,\infty; L^2(\Gamma))$ boundary feedback acting in The Dirichlet boundary conditions, Proc. 24th IEEE CDC, to appear.

[11] Russell, D.L., Controllability and stabilizability theory for linear partial differential equations: Recent progress and open questions, SIAM Rev. 20

(1978), 639-739.

[12] Skar, S.J and Lindner, D.K., On correcting the destabilizing effects of delay in boundary control, Proc. 24th IEEE CDC, to appear.

[13] Slemrod, M., Stabilization of boundary control systems, J. Differential Equations 22 (1976), 402-415.

Frequency Domain and State Space Methods for Linear Systems
C.I. Byrnes and A. Lindquist (Editors)

STRUCTURAL INVARIANTS OF INFINITE DIMENSIONAL SYSTEMS-SYSTEMS WITH DELAYS *

E. B. Lee, W.-S. Lu and S. H. Żak

Department of Electrical Engineering
University of Minnesota
123 Church Street
Minneapolis, Minnesota 55455

The Hermite normal form of a matrix is employed in classification of polynomial matrix pairs (A,B) with respect to similarity transformations. The polynomial (A,B) pairs can be viewed as coming from models for linear systems with time delays. Algorithms for finding canonical forms for such pairs are described.

INTRODUCTION AND PRELIMINARY DEVELOPMENTS

Let R denote the field of real numbers and R[z] the ring of polynomials in the indeterminate z with coefficients in R. Further let R(z) denote the field of rational functions in the indeterminate z with coefficients from R. The polynomial matrix pairs that we will find canonical forms for appear in the systems theory associated with infinite dimensional linear time delay type systems.

Consider a single input, single output (SISO) linear system as modeled by the linear differential-difference equations, (vector-matrix delay operator notation)

$$\dot{x}(t) = A(z)x(t) + b(z)u(t),$$

$$y(t) = c(z)\ x(t).$$

Here $x(t)\varepsilon R^n$, $u(t)\varepsilon R$, and $y(t)\varepsilon R$. The delay operator matrices A(z), b(z), and c(z) are composed of real polynomials in the right shift (delay) operator z which has the property that $zx(t)=x(t-h)$, where $h>0$ is the delay duration. Stated briefly, A=A(z), b=b(z), and c=c(z) are nxn, nx1, and 1xn matrices with entries from the polynomial ring R[z]. Such a system is infinite dimensional since its evolution takes place in a state space such as $M^2 x L^2$ (see [1]).

Consider next the set

$$X = \{(A,b)\varepsilon R[z]^{nxn+nx1} \mid \det[b,Ab,\dots,A^{n-1}b] \not\equiv 0\}$$

Notice that $(A,b)\varepsilon X$ means that (A,b) is an R(z)-controllable pair [2]. On the set X define an equivalence relation via the transformation group

$$G = \{T \mid T\varepsilon R[z]^{nxn},\ \det T\varepsilon R\setminus\{0\}\}.$$

Matrix pairs (A_1,b_1), $(A_2,b_2)\varepsilon X$ are said to be <u>G-equivalent</u>, denoted by ~ (R[z]-similar), on X iff there exists $T\varepsilon G$ such that

$$A_2 = TA_1T^{-1}, \text{ and } b_2 = Tb_1.$$

---

* Research supported in part by the National Science Foundation under Grant Number DMS-8413129.

All the results which are obtained for the set X are applicable to the dual set

$$X_d = \{(A,c) \mid \det[c^T, A^T c^T, \dots, (A^T)^{n-1} c^T] \not\equiv 0\}$$

An equivalence relation on a set X partitions it into its equivalence classes, called cosets [3]. A coset can be characterized by any of its elements. The goal is to find a unique characteristic element, called a canonical form [4], for each coset, i.e., a canonical form corresponding to the coset to which it belongs.

Now a theorem of considerable importance in the derivation of the main results.

Theorem 1 [5], [6]: Given an nxn nonsingular polynomial matrix, by elementary row operations one can convert it to a unique Hermite form D_H, where D_H is upper triangular, with each diagonal element monic and of higher degree than any other element in the same column. If a diagonal element is 1, all other entries in that column will be zero.

Corollary 1: Given an nxn nonsingular polynomial matrix P(z), then unimodular matrix performing elementary row operations on P(z) which brings P(z) into its Hermite form is unique.

Proof: Suppose we have two different unimodular matrices which reduce P(z) to its (unique) Hermite form P_H, i.e., $T_{L1}(z)\, P(z) = P_H(z) = T_{L2}(z)\, P(z)$. P(z) is nonsingular so $T_{L1}(z) = T_{L2}(z)$. □

Consider now an arbitrary element $(A,b) \in X$. Form the controllability matrix

$[A|b] \stackrel{\Delta}{=} [b, Ab, \dots, A^{n-1}b]$ and then find (unique) Hermite form of $[A|b]$:

$$T[A|b] = \begin{bmatrix} \hat{h}_{11} & \hat{h}_{12} & \hat{h}_{13} & \cdots & \hat{h}_{1n} \\ 0 & \hat{h}_{22} & \hat{h}_{23} & \cdots & \hat{h}_{2n} \\ 0 & 0 & \hat{h}_{33} & \cdots & \hat{h}_{3n} \\ \vdots & \vdots & \vdots & & \vdots \\ 0 & 0 & 0 & \cdots & \hat{h}_{nn} \end{bmatrix}$$

$$= [TAT^{-1} \mid Tb] \stackrel{\Delta}{=} [\hat{A}|\hat{b}].$$

It is clear that $\hat{b}$ must be of the form

$$\hat{b} = \begin{bmatrix} \hat{h}_{11} \\ 0 \\ 0 \\ \vdots \\ 0 \end{bmatrix}. \tag{1}$$

Let us now find the structure of $\hat{A}$. The general form of $\hat{A}$ is

$$\begin{bmatrix} \hat{a}_{11} & x & \cdots & x \\ \hat{a}_{21} & x & \cdots & x \\ \hat{a}_{31} & x & \cdots & x \\ \vdots & \vdots & & \vdots \\ \hat{a}_{n1} & x & \cdots & x \end{bmatrix}.$$

Hence, we have to have the following equality satisfied:

$$\begin{bmatrix} \hat{a}_{11}\hat{h}_{11} \\ \hat{a}_{21}\hat{h}_{11} \\ \hat{a}_{31}\hat{h}_{11} \\ \vdots \\ \hat{a}_{n1}\hat{h}_{11} \end{bmatrix} = \begin{bmatrix} \hat{h}_{12} \\ \hat{h}_{22} \\ 0 \\ \vdots \\ 0 \end{bmatrix} .$$

Thus

$$\hat{a}_{11} = \frac{\hat{h}_{12}}{\hat{h}_{11}} , \ \hat{a}_{21} = \frac{\hat{h}_{22}}{\hat{h}_{11}} , \ \hat{a}_{31} = \dots = \hat{a}_{n1} = 0 .$$

Repetition of similar arguments shows that

$$\hat{A} = \begin{bmatrix} \frac{\hat{h}_{12}}{\hat{h}_{11}} & x & \dots & x \\ \frac{\hat{h}_{22}}{\hat{h}_{11}} & x & \dots & x \\ & \frac{\hat{h}_{33}}{\hat{h}_{22}} & \dots & x \\ & & \ddots & \vdots \\ 0 & & \frac{\hat{h}_{n,n}}{\hat{h}_{n-1,n-1}} & x \end{bmatrix} . \tag{2}$$

The transformation matrix $\hat{T}$ is a unimodular one. This means that $\hat{A}$ is a polynomial matrix, i.e., $\hat{h}_{11} \mid \hat{h}_{22}, \ \hat{h}_{22} \mid \hat{h}_{33}, \dots, \hat{h}_{n-1,n-1} \mid \hat{h}_{n,n}$. (3)

The procedure that led us to the pair of matrices (1) and(2) will be referred to later as procedure P1 and the resulting pair $(\hat{A},\hat{b})$ as being in the P1 form.

The set of relations (3) were previously noticed by Watanabe and Ito [7] where these formulas were derived using a different approach.

Theorem 2: If (A,b) is an element of X, then the P1 form of (A,b) is unique.

Proof: By assumption $(A,b)\varepsilon X$, hence $[A|b]$ is nonsingular. By virtue of Theorem 1 Hermite form of such a pair is unique. By Corollary 1 the transformation matrix which brings $[A|b]$ to its Hermite form is also unique, and since this matrix is a similarity transformation which transforms (A,b) into P1 form, this completes the proof of the uniqueness of P1 form. □

Now we will give another procedure that will lead us to the P2 form of a pair $(A,b)\varepsilon X$. Later we will show that both P1 and P2 forms are canonical.

Let $O_1(p,q)$ denote the operation of interchanging rows p and q of A and then columns p and q of A. This corresponds to premultiplication by the elementary matrix $P(p,q)$ obtained by interchanging the pth and qth columns of I_n, and then postmultiplication by the elementary matrix $P^{-1}(p,q)=P(p,q)$.

Next let $O_2(p,q,\alpha(z))$ be the operation of adding polynomial $\alpha(z)$ times row q to row p, and then subtracting $\alpha(z)$ times column p from column q. This corresponds to premultiplication by the elementary matrix

$$P(p,q,\alpha(z)) = I_n + \begin{bmatrix} 0 & \cdots & 0 & \cdots & 0 \\ \vdots & & \vdots & & \vdots \\ 0 & \cdots & \alpha(z) & \cdots & 0 \\ \vdots & & \vdots & & \vdots \\ 0 & \cdots & 0 & \cdots & 0 \end{bmatrix}$$ with $\alpha(z)$ in the pth row and q-th column and then postmultiplication by $P^{-1}(p,q,\alpha(z)) = P(p,q,-\alpha(z))$.

Let $T(z)$ be a product of t elementary matrices of above two types written as

$T = T_t T_{t-1} \cdots T_1$; then form

$\tilde{A} = T_t T_{t-1} \cdots T_1 A T_1^{-1} T_2^{-1} \cdots T_t^{-1}$ and $\tilde{b} = T_t T_{t-1} \cdots T_1 b$.

Consider now the nx(n+1) matrix $[b,A]$ and using the above elementary operations transform it to the form (see reference [8] for details; also [9])

$$\begin{bmatrix} \tilde{b}_1 & \tilde{a}_{11} & \tilde{a}_{12} & \cdots & x & x \\ 0 & \tilde{a}_{21} & \tilde{a}_{22} & \cdots & x & x \\ 0 & 0 & \tilde{a}_{32} & & \vdots & \vdots \\ 0 & 0 & 0 & & x & x \\ \vdots & \vdots & \vdots & & \vdots & \vdots \\ 0 & 0 & 0 & \cdots & \tilde{a}_{n,n-1} & x \end{bmatrix} = [\tilde{b},\tilde{A}] \tag{4}$$

where $\deg \tilde{a}_{i+1,i} > \deg \tilde{a}_{j,i}$ for $j = 1,2,\ldots,i$; $i=1,\ldots,n-1$.

The pair of matrices $(\tilde{A},\tilde{b})$ obtained by applying the above procedure will be referred to as the P2 form of pair (A,b).

Note that

$$[\tilde{A}|\tilde{b}] = \begin{bmatrix} \tilde{b}_1 & \tilde{a}_{11}\tilde{b}_1 & \tilde{a}_{11}^2\tilde{b}_1+\tilde{a}_{12}\tilde{a}_{21}\tilde{b}_1 & \cdots & x \\ 0 & \tilde{a}_{21}\tilde{b}_1 & \tilde{a}_{11}\tilde{a}_{21}\tilde{b}_1+\tilde{a}_{21}\tilde{a}_{22}\tilde{b}_1 & \cdots & x \\ 0 & 0 & \tilde{a}_{21}\tilde{a}_{32}\tilde{b}_1 & \cdots & x \\ 0 & 0 & 0 & & \vdots \\ \vdots & \vdots & \vdots & & \vdots \\ 0 & 0 & 0 & \cdots & \tilde{a}_{21}\tilde{a}_{32}\tilde{a}_{43}\cdots\tilde{a}_{n,n-1}\tilde{b}_1 \end{bmatrix}$$

$$= \begin{bmatrix} \tilde{h}_{11} & \tilde{h}_{12} & \cdots & x \\ 0 & \tilde{h}_{22} & \cdots & x \\ 0 & 0 & & \vdots \\ \vdots & \vdots & & \vdots \\ 0 & 0 & \cdots & \tilde{h}_{nn} \end{bmatrix}. \tag{5}$$

Therefore $\tilde{h}_{ii} | \tilde{h}_{i+1,i+1}$ for $i=1,2,\ldots,n-1$. (6)

Moreover, $\det[\tilde{A}|\tilde{b}] \neq 0$ now implies that $\tilde{b}_1 \neq 0$, $\tilde{a}_{21} \neq 0,\ldots,\tilde{a}_{n,n-1} \neq 0$.

The controllability matrix corresponding to $(\tilde{A},\tilde{b})$ in P2 form is not in general, in its Hermite form as an upcoming example shows.

Theorem 3: If a pair (A,b) is an element of X, then its P2 form is unique.

Proof: Suppose that $(A,b)\sim(\tilde{A},\tilde{b})$, $(A,b)\sim(\tilde{\tilde{A}},\tilde{\tilde{b}})$, and that $(\tilde{A},\tilde{b}),(\tilde{\tilde{A}},\tilde{\tilde{b}})$ are in the P2 form. Then we have $(\tilde{A},\tilde{b})\sim(\tilde{\tilde{A}},\tilde{\tilde{b}})$ i.e., there exist a unimodular matrix, say U, such that

$$U[\tilde{b},\tilde{A}]\left[\begin{array}{c|c}1 & 0\\ \hline 0 & U^{-1}\end{array}\right] = [\tilde{\tilde{b}},\tilde{\tilde{A}}].$$

We have $U\tilde{b} = \tilde{\tilde{b}}$, that is

$$\begin{bmatrix} u_{11} & x\ldots x\\ u_{21} & x\ldots x\\ u_{31} & x\ldots x\\ \vdots & \vdots\quad\vdots\\ u_{n1} & x\ldots x\end{bmatrix}\begin{bmatrix}\tilde{b}_{11}\\ 0\\ 0\\ \vdots\\ 0\end{bmatrix} = \begin{bmatrix}\tilde{\tilde{b}}_{11}\\ 0\\ 0\\ \vdots\\ 0\end{bmatrix}.$$

Hence

$$\begin{bmatrix} u_{11}\\ u_{21}\\ u_{31}\\ \vdots\\ u_{n1}\end{bmatrix}\tilde{b}_{11} = \begin{bmatrix}\tilde{\tilde{b}}_{11}\\ 0\\ 0\\ \vdots\\ 0\end{bmatrix}.$$

This means that $u_{21} = u_{31} = \cdots u_{n1} = 0$ and U has the form

$$\left[\begin{array}{c|c} u_{11} & x\;.\;.\;.\;x\\ \hline 0 & \\ \vdots & X\\ 0 & \end{array}\right].$$

The matrix U is unimodular, therefore u_{11} must be constant and in fact unity to keep $\tilde{b}_{11}$ and $\tilde{\tilde{b}}_{11}$ monic.

Now $U\tilde{A}U^{-1} = \tilde{\tilde{A}}$; that is

$$\begin{bmatrix} 1 & u_{12} & u_{13} & \cdots & u_{1n}\\ 0 & u_{22} & u_{23} & \cdots & u_{2n}\\ 0 & u_{32} & u_{33} & \cdots & u_{3n}\\ \vdots & \vdots & \vdots & & \vdots\\ 0 & u_{n2} & u_{n3} & \cdots & u_{nn}\end{bmatrix}\begin{bmatrix}\tilde{a}_{11} & \tilde{a}_{12} & \cdots & \tilde{a}_{1n}\\ \tilde{a}_{21} & \tilde{a}_{22} & \cdots & \tilde{a}_{2n}\\ 0 & \tilde{a}_{32} & \cdots & \tilde{a}_{3n}\\ \vdots & \vdots & & \vdots\\ 0 & 0 & \cdots & \tilde{a}_{nn}\end{bmatrix}\begin{bmatrix}1 & \bar{u}_{12} & \cdots & \bar{u}_{1n}\\ 0 & \bar{u}_{22} & \cdots & \bar{u}_{2n}\\ \vdots & \vdots & & \vdots\\ 0 & \bar{u}_{n2} & \cdots & \bar{u}_{nn}\end{bmatrix}$$

$$= \begin{bmatrix} x & x & \cdots & x & x & x\\ x & x & \cdots & x & x & x\\ 0 & x & \cdots & x & x & x\\ 0 & 0 & & \vdots & \vdots & \vdots\\ \vdots & \vdots & & & & \\ 0 & 0 & \cdots & 0 & x & x\end{bmatrix}.$$

Therefore

$$\begin{bmatrix} \tilde{a}_{11} + u_{12}\tilde{a}_{21} \\ u_{22}\tilde{a}_{21} \\ u_{32}\tilde{a}_{21} \\ \vdots \\ u_{n2}\tilde{a}_{21} \end{bmatrix} = \begin{bmatrix} \tilde{\tilde{a}}_{11} \\ \tilde{\tilde{a}}_{21} \\ 0 \\ \vdots \\ 0 \end{bmatrix}.$$

This means that we have to have $u_{32} = u_{42} = \ldots = u_{n2} = 0$ because $\tilde{a}_{21} \neq 0$. Both $\tilde{a}_{21}$ and $\tilde{\tilde{a}}_{21}$ are $\neq 0$, U unimodular, hence $u_{22} = 1$, i.e., $\tilde{a}_{21} = \tilde{\tilde{a}}_{21}$. Furthermore, $\deg \tilde{a}_{11} < \deg \tilde{a}_{21} = \deg \tilde{\tilde{a}}_{21} > \deg \tilde{\tilde{a}}_{11}$. Therefore $u_{12} = 0$ and

$$U = \left[\begin{array}{cc:ccc} 1 & 0 & x & \ldots & x \\ 0 & 1 & x & \ldots & x \\ \hdashline 0 & 0 & x & \ldots & x \\ 0 & 0 & \vdots & & \vdots \\ \vdots & \vdots & & & \\ 0 & 0 & x & \ldots & x \end{array}\right].$$

In the same manner find that $U = I_n$ and that the P2 form is unique. □

P1 and P2 forms are different in general as the following example shows.

Example: Consider the pair of matrices

$$\left(A = \begin{bmatrix} z^3 & 1 & 3 \\ z^4 & 0 & 2 \\ 0 & z & 1 \end{bmatrix}, \quad b = \begin{bmatrix} 1 \\ 0 \\ 0 \end{bmatrix}\right). \tag{7}$$

Note that (A,b) is already in the P2 form, i.e., $(A,b) = (\tilde{A},\tilde{b})$. The controllability matrix of pair (7) is

$$[\tilde{A}\,|\tilde{b}] = \begin{bmatrix} 1 & z^3 & z^6+z^4 \\ 0 & z^4 & z^7 \\ 0 & 0 & z^5 \end{bmatrix}.$$

The matrix $[\tilde{A}|\tilde{b}]$ is not in its Hermite form which means that the P1 and P2 forms for pair (7) are different.

Transform $[\tilde{A}|\tilde{b}]$ to its Hermite form via elementary row operations:

$$\begin{bmatrix} 1 & 0 & -z \\ 0 & 1 & -z^2 \\ 0 & 0 & 1 \end{bmatrix} \begin{bmatrix} 1 & z^3 & z^6+z^4 \\ 0 & z^4 & z^7 \\ 0 & 0 & z^5 \end{bmatrix} = \begin{bmatrix} 1 & z^3 & z^4 \\ 0 & z^4 & 0 \\ 0 & 0 & z^5 \end{bmatrix} = T[\tilde{A}|\tilde{b}].$$

Therefore the P1 form is

$$\hat{A} = T\tilde{A}T^{-1} = \begin{bmatrix} z^3 & 1-z^2 & z^2-z+3 \\ z^4 & -z^3 & 2-z^2 \\ 0 & z & z^3+1 \end{bmatrix}, \quad \text{and } \hat{b} = T\tilde{b} = \begin{bmatrix} 1 \\ 0 \\ 0 \end{bmatrix}. \tag{8}$$

CLASSIFICATION OF (A,b) PAIRS

It will be shown using lemma 1 that both P1 and P2 forms are canonical in the algebraic sense. Thus we can characterize the set X by either P1 or P2 forms.

Lemma 1: If $(A_1,b_1),(A_2,b_2)\varepsilon X$ are R[z]-similar then they are related by a unique unimodular transformation matrix.

Proof: Suppose that there are U_1 and U_2 both unimodular such that $A_2=U_1A_1U_1^{-1}$, $b_2=U_1b_1$, and $A_2=U_2A_1U_2^{-1}$, $b_2=U_2b_1$. These relations imply that $U_1[A_1|b_1]=[A_2|b_2]$ and $U_2[A_1|b_1]=[A_2|b_2]$. Subtracting the second equation from the first one get $[U_1-U_2][A_1|b_1] = 0$.
By assumption $\det[A_1|b_1] \neq 0$, hence $U_1=U_2$. □

Consider the set X. A set of canonical forms for the equivalence relation is a subset of X such that there exists to each $(A,b)\varepsilon X$ exactly one (A^*,b^*) in the subset with $(A,b)\sim(A^*,b^*)$. This means that the projection $P:X\to X/\sim$ restricted to the set of canonical forms $C\subset X$ is an isomophism $C\approx X/\sim$. In other words, each coset possesses one and only one canonical form, (all pairs which are R[z]-similar have the same canonical form).

It will now be established that both procedures P1 and P2 give canonical forms.

Theorem 4: The mapping

$$\phi_1 : X \to \hat{C}$$

$$: (A,b) \to (T_1AT_1^{-1},T_1b)$$

is a complete invariant [5, page 493], where T_1 (unimodular matrix) is found by using procedure P1.

Proof: First it will be proven that ϕ_1 is an invariant under R[z]-similarity transformations on X, i.e., $(A_1,b_1)\sim(A_2,b_2)\Rightarrow\phi_1(A_1,b_1)=\phi_1(A_2,b_2)=(\hat{A},\hat{b})$. By Lemma 1 there is exactly one unimodular matrix, say U, such that $A_2=UA_1U^{-1}$, $b_2=Ub_1$, i.e. $U[A_1|b_1]=[A_2|b_2]$. By virtue of Corollary 1, we can find the unique unimodular matrix which brings $U[A_1|b_1]=[A_2|b_2]$ into its Hermite form. Thus this construction gives us similarity transformations which transform (A_1,b_1) and (A_2,b_2) into the same P1 form.

The map ϕ_1 is complete, i.e., $(A_1,b_1)\sim(A_2,b_2) \Leftarrow \phi_1(A_1,b_1) = \phi_2(A_2,b_2)$; for suppose the P1 forms for (A_1,b_1) and (A_2,b_2) are the same, i.e.,

$$T_1A_1T_1^{-1} = \hat{A},\ T_2A_2T_2^{-1} = \hat{A},\ T_1b_1 = \hat{b},\ \text{and}\ ,\ T_2b_2 = \hat{b}.$$

Then $A_2 = (T_2^{-1}T_1)A_1(T_2^{-1}T_1)^{-1}$ and $b_2 = (T_2^{-1}T_1)b_1$, i.e., $(A_1,b_1)\sim(A_2,b_2)$. □

Theorem 5: The mapping

$$\phi_2 : X \to \tilde{C}$$

$$: (A,b) \to (T_2AT_2^{-1},T_2b)$$

is a complete invariant where T_2(unimodular) is found by using procedure P2.

The last two theorems mean that one can fully characterize each coset of the set X/~ under R[z]-similarity transformations in two different ways via P1 or P2 canonical forms. The structures of P1 and P2 forms are quite similar (yet not the same as indicated by the previous example). In both forms operator A is in upper right supertriangular form and operator b has only one nonzero element.

These two forms must be related via an R[z]-similarity transformation because they are both canonical forms. It can be readily shown that the similarity transformation (matrix) relating P1 and P2 forms corresponding to the same coset has an upper triangular form with 1's on the diagonal.

Now attention will be concentrated on the generic case, i.e., the case when $\det[b, Ab, \ldots, A^{n-1}b]$ is a polynomial with distinct zeros. Let us denote the set of pairs (A,b) with this property as X_g. Then we have the following result.

Theorem 6: The mapping $\phi : X_g \to (\bar{A},\bar{b})\{=(\hat{A},\hat{b})=(\tilde{A},\tilde{b})\}$ is a complete invariant, where $X_g = \{(A,b)\,|\det[A|b] = \text{nontrivial polynomial with distinct zeros}\}$,

$$\bar{A}=TAT^{-1} = \begin{bmatrix} 0 & 0 & \ldots & 0 & \bar{h}_{1,n} & x \\ 1 & 0 & \ldots & 0 & \bar{h}_{2,n} & x \\ 0 & 1 & \ldots & 0 & \bar{h}_{3,n} & x \\ 0 & 0 & \ldots & 0 & \bar{h}_{4,n} & x \\ \vdots & \vdots & & \vdots & \vdots & \vdots \\ 0 & 0 & \ldots & 1 & \bar{h}_{n-1,n} & x \\ 0 & 0 & \ldots & 0 & \bar{h}_{n,n} & x \end{bmatrix}, \quad \bar{b}=Tb= \begin{bmatrix} 1 \\ 0 \\ 0 \\ 0 \\ \vdots \\ 0 \\ 0 \end{bmatrix} \tag{9}$$

and the T matrix is a unimodular one which is obtained using either procedure P1 or P2; result is the same.

Proof: First we apply procedure P1 to find a canonical form for the generic case. Next we will show that the effect of procedure P2 is the same as P1.

The Hermite form of $[A|b]$ is

$$\hat{T}[A|b] = \begin{bmatrix} \hat{h}_{11} & & & \\ & \hat{h}_{22} & \cdot & x \\ & & \cdot & \\ 0 & & & \cdot \\ & & & \hat{h}_{nn} \end{bmatrix} = [\hat{A}|\hat{b}] \tag{10}$$

with

$$\hat{h}_{ii} \mid \hat{h}_{i+1,i+1}$$

so $\det[\hat{A}|\hat{b}] = \hat{h}_{11}\,\hat{h}_{22}\cdots\hat{h}_{nn} = \hat{h}_{11}(\hat{h}_{11}h^{*}_{22})(\hat{h}_{11}h^{*}_{22}h^{*}_{33})\cdots$

$$= (\hat{h}_{11})^{n}(h^{*}_{22})^{n-1}(h^{*}_{33})^{n-2}\cdots h^{*}_{nn}\,.$$

Therefore $\hat{h}_{ii} = 1$ for $i = 1,\ldots,n-1$ if $\det[A|b]$ has distinct zeros.

In such a case the Hermite normal form corresponding to $[A|b]$ will be

$$\hat{T}[A|b] = \begin{bmatrix} 1 & 0 & 0 & \ldots & \hat{h}_{1,n} \\ & 1 & 0 & \ldots & \hat{h}_{2,n} \\ & & \cdot & & \vdots \\ 0 & & & \cdot & \hat{h}_{n-1,n} \\ & & & \cdot & \hat{h}_{n,n} \end{bmatrix} = [\hat{A} \mid \hat{b}] \tag{11}$$

(n-1)st column of A

where $\deg\hat{h}_{n,n} > \deg\hat{h}_{i,n}$ for $i = 1,\ldots,n-1$; and in the new basis, the operators (A,b) will have the form (9). Apply procedure P2 to a pair$(A,b)\in X_g$ to get form P2 which is exactly the same as form P1. General structure of (A,b) in P2 form is

$$\tilde{A} = \begin{bmatrix} \tilde{a}_{11} & \tilde{a}_{12} & \cdots & \tilde{a}_{1,n-1} & \tilde{a}_{1n} \\ \tilde{a}_{21} & \tilde{a}_{22} & \cdots & \tilde{a}_{2,n-1} & \tilde{a}_{2n} \\ 0 & \tilde{a}_{32} & \cdots & \tilde{a}_{3,n-1} & \tilde{a}_{3n} \\ 0 & 0 & \cdots & \vdots & \vdots \\ \vdots & \vdots & & & \\ 0 & 0 & \cdots & \tilde{a}_{n,n-1} & \tilde{a}_{n,n} \end{bmatrix}, \quad \tilde{b} = \begin{bmatrix} \tilde{b}_1 \\ 0 \\ 0 \\ \vdots \\ 0 \end{bmatrix}$$

where $\deg \tilde{a}_{i+1,i} > \deg \tilde{a}_{j,i}$, $j=1,2,\ldots,i$ $i=1,\ldots,n-1$. Thus the controllability matrix $[A|b]$ has as its diagonal entries the following elements:

$$\tilde{h}_{ii} = \tilde{b}_1\tilde{a}_{21}\tilde{a}_{32} \cdots \tilde{a}_{i,i-1}, \quad i=1,\ldots,n \quad . \tag{12}$$

By assumption $(A,b)\varepsilon X_g$, therefore $\prod_{i=1}^{n} \tilde{h}_{ii} =$

$$\tilde{b}_1^n(\tilde{a}_{21})^{n-1}(\tilde{a}_{32})^{n-2}\ldots(\tilde{a}_{n-1,n-2})^2\, \tilde{a}_{n,n-1} = \text{polynomial with distinct zeros.} \tag{13}$$

$$\text{Therefore } \tilde{b}_1=1,\ \tilde{a}_{i,i-1}=1,\ i=2,\ldots,n-1 \text{ and } \overline{h}_{n,n} = \tilde{a}_{n,n-1} = \det[A|b] \ . \tag{14}$$

The above set of relations implies that if $(A,b)\varepsilon X_g$, then P2 form of (A,b) is the same as its P1 form.

To prove that ϕ is a complete invariant under R[z]-similarity transformations one can proceed as in the proof of Theorem 4. □

Example: Let us find the Hermite form of the matrix $[A|b]$ for pair

$$\left(A = \begin{bmatrix} z^2 & 0 & z^3 \\ 0 & 0 & 1 \\ 0 & 1 & 0 \end{bmatrix}, \quad b = \begin{bmatrix} -z \\ 0 \\ 1 \end{bmatrix}\right) \tag{15}$$

The unimodular matrix which transforms the controllability matrix $[A|b]$ into its Hermite normal form $[\tilde{A}|\tilde{b}]=[\hat{A}|\hat{b}]$ is

$$T = \begin{bmatrix} 0 & 0 & 1 \\ 0 & 1 & 0 \\ 1 & 0 & z \end{bmatrix},$$

giving

$$TAT^{-1} = \begin{bmatrix} 0 & 1 & 0 \\ 1 & 0 & 0 \\ 0 & z & z^2 \end{bmatrix} = \hat{A} = \tilde{A}, \quad \text{and } Tb = \begin{bmatrix} 1 \\ 0 \\ 0 \end{bmatrix} = \hat{b} = \tilde{b}.$$

Corollary 2: If $\det[A|b] = \text{const} \neq 0$, then

$$(\overline{A},\overline{b}) = \left(\begin{bmatrix} 0 & 0 & \cdots & 0 & x \\ 1 & 0 & \cdots & 0 & x \\ 0 & 1 & \cdots & 0 & x \\ \vdots & \vdots & \ddots & \vdots & \vdots \\ 0 & 0 & \cdots & 1 & x \end{bmatrix}, \begin{bmatrix} 1 \\ 0 \\ 0 \\ \vdots \\ 0 \end{bmatrix} \right) \tag{16}$$

and in such a case the coefficients of the characteristic polynomial constitute the complete system of independent invariants. Thus the set of canonical forms for $X_0 = \{(A,b)\,|\det[A|b] = \text{const} \neq 0\}$ is isomorphic to the module $R^n[z]$.

The complete system of independent invariants when $(A,b)\varepsilon X_0$ constitutes the last column of the augmented controllability matrix of $(\bar{A},\bar{b})$ pair, i.e.,

$$[\bar{A}|\bar{b}]_{aug} = [\bar{b},\bar{A}\bar{b},\ldots,(\bar{A})^{n-1}\bar{b},(\bar{A})^n\bar{b}] = \begin{bmatrix} 1 & 0 & \cdots & 0 & x \\ 0 & 1 & \cdots & 0 & x \\ \vdots & \vdots & & \vdots & \vdots \\ 0 & 0 & & 1 & x \end{bmatrix} \tag{17}$$

Now find a complete system of independent invariants in the more general case.

Lemma 2: Given an nx(n+1) full rank polynomial matrix D(z) in which the first n columns are linearly independent (over R(z)). Then the Hermite (normal) form of such a matrix is unique and there exists one and only one unimodular matrix which brings D(z) into its Hermite form.

Proof: This fact is an easy consequence of Theorem 1 and Corollary 1. □

Now taking into account that the last column of $[A|b]_{aug}$ is $A^n b$ we have some further results (for a comparison in simpler field case see [10], [11], [12] and [13]).

Theorem 7: Given

$$\begin{bmatrix} 1 & 0 & \cdots & h_{1n} & h_{1,n+1} \\ 0 & 1 & \cdots & h_{2n} & h_{2,n+1} \\ \vdots & \vdots & & \vdots & \vdots \\ 0 & 0 & \cdots & h_{nn} & h_{n,n+1} \end{bmatrix} \varepsilon\, R^{nx(n+1)} \tag{18}$$

where $\deg h_{n,n} > \deg h_{i,n}$ for $i=1,\ldots,n$;
and

$$\operatorname*{Span}_{R[z]} \begin{bmatrix} 1 & 0 & \cdots & 0 & h_{1n} \\ 0 & 1 & \cdots & 0 & h_{2n} \\ \vdots & \vdots & & \vdots & \vdots \\ 0 & 0 & \cdots & 1 & h_{nn} \end{bmatrix} = \operatorname*{Span}_{R[z]} \begin{bmatrix} 1 & 0 & \cdots & 0 & h_{1n} & h_{1,n+1} \\ 0 & 1 & \cdots & 0 & h_{2n} & h_{2,n+1} \\ \vdots & \vdots & & \vdots & \vdots & \vdots \\ 0 & 0 & \cdots & 1 & h_{nn} & h_{n,n+1} \end{bmatrix} \tag{19}$$

Then the last two columns of (18), provided (19) holds, constitute the complete set of independent invariants for X_g.

CLASSIFICATION OF (A,B) PAIRS

We now extend the results developed above to the multi-input case. Let $A(z)\varepsilon R^{nxn}[z]$, $B(z)\varepsilon R^{nxm}(z)$, $X_1 = \{(A,B)\varepsilon R^{nxn+nxm}[z] \mid \text{rank } [A|B] = n\}$. For $(A,B)\varepsilon X_1$ with rank $B(z)=m$, pick n independent columns from $[A|B]$ to form

$$Q(z) = [b_1\ Ab_1\ldots A^{p_1-1}b_1\ b_2\ Ab_2\ldots A^{p_2-1}b_2\ldots b_m\ Ab_m\ldots A^{p_m-1}b_m]$$

where for each $1<i<m$, $A^{p_i}b_i$ is a linear combination of $b_1,\ldots A^{p_1-1}b_1,\ldots,b_i,$ $\ldots A^{p_i-1}b_i$ in $R(z)$. There exists a unimodular matrix $T(z)$ such that

$$T(z)Q(z) = \begin{bmatrix} h_{11} & h_{12} & \cdots & h_{1n} \\ & h_{22} & \cdots & h_{2n} \\ 0 & & \ddots & \vdots \\ & & & h_{nn} \end{bmatrix} \equiv H(z) \tag{20}$$

with h_{ii} monic and $\deg h_{ij} < \deg h_{jj}$ for $1 \le i < j \le n$.

Define $\hat{A} = TAT^{-1}$, $\hat{B} = TB = [Tb_1 \ldots Tb_m] \equiv [\hat{b}_1 \ldots \hat{b}_m]$,

$$k_i = \sum_{j=1}^{i} p_j, \text{ and note that}$$

$$H(z) = [\hat{b}_1 \; \hat{A}\hat{b}_1 \ldots \hat{A}^{p_1-1}, \ldots, \hat{b}_m \; \hat{A}\hat{b}_m \ldots \hat{A}^{p_m-1}\hat{b}_m]. \tag{21}$$

Comparing (20) with (21) gives

$$\hat{B} = \begin{bmatrix} h_{11} & x & & x \\ & \cdot & & \cdot \\ 0 & \cdot & & \cdot \\ \cdot & \cdot & & \cdot \\ \cdot & h_{k_1+1,k_1+1} & & \cdot \\ \cdot & & & \cdot \\ \cdot & 0 & & \cdot \\ \cdot & \cdot & & \cdot \\ \cdot & \cdot & & h_{k_{m-1}k_{m-1}} \\ \cdot & \cdot & & \\ \cdot & \cdot & & 0 \\ \cdot & \cdot & & \\ 0 & 0 & & 0 \end{bmatrix} \tag{22}$$

and

$$\hat{A} = \hat{A}HH^{-1} = [\hat{A}\hat{b}_1 \ldots \hat{A}^{p_1}\hat{b}_1, \ldots, \hat{A}\hat{b}_m \ldots \hat{A}^{p_m}\hat{b}_m]H^{-1}$$

$$= H \left[\begin{array}{cccc} \overbrace{\begin{array}{ccccc} 0 & 0 & \cdots & & x \\ 1 & 0 & & & \cdot \\ 0 & 1 & & & \cdot \\ \cdot & \cdot & \ddots & & \cdot \\ 0 & 0 & & 1 & x \end{array}}^{p_1} & \overbrace{\begin{array}{cc} & x \\ & \cdot \\ 0 & \cdot \\ & \cdot \\ & x \end{array}}^{p_2} & & \overbrace{\begin{array}{cc} & x \\ & \cdot \\ & \cdot \\ & \cdot \end{array}}^{p_m} \\ & \begin{array}{ccccc} 0 & 0 & & & x \\ 1 & 0 & & & \cdot \\ 0 & 1 & & & \cdot \\ \cdot & \cdot & \ddots & & \cdot \\ 0 & 0 & & 1 & x \end{array} & & \begin{array}{cc} 0 & \cdot \\ & \cdot \\ & x \end{array} \\ & & \ddots & \\ & 0 & & \begin{array}{ccccc} 0 & 0 & & & x \\ 1 & 0 & & & \cdot \\ 0 & 1 & & & \cdot \\ \cdot & \cdot & \ddots & & \cdot \\ 0 & 0 & & 1 & x \end{array} \end{array}\right] H^{-1}.$$

To compute $\hat{A}$, we partition the upper triangular matrices H and H^{-1} with block diagonal sizes $p_1, p_2, \ldots, p_m$. Thus

$$\hat{A} = \left[\begin{array}{ccc}
\begin{array}{cccc}
x & x & x \cdots & x \\
\frac{h_{22}}{h_{11}} & x & \vdots & \vdots \\
 & x & \vdots & \vdots \\
 & \ddots & \vdots & \vdots \\
0 & & \frac{h_{k_1,k_1}}{h_{k_1-1,k_1-1}} & x
\end{array} & & * \\
 & \begin{array}{cccc}
x & x & x & x \\
\frac{h_{k_1+2,k_1+2}}{h_{k_1+1,k_1+1}} & x & & \\
 & x & & \\
 & & \ddots & \\
0 & & \frac{h_{k_2,k_2}}{h_{k_2-1,k_2-1}} & x
\end{array} & \\
 & \ddots & \\
0 & & \begin{array}{cccc}
x & x & x & x \\
\frac{h_{k_{m-1}+2,k_{m-1}+2}}{h_{k_{m-1}+1,k_{m-1}+1}} & x & \vdots & \vdots \\
 & x & \vdots & \vdots \\
 & & \ddots & \vdots \\
0 & & \frac{h_{n,n}}{h_{n-1,n-1}} & x
\end{array}
\end{array}\right] \tag{23}$$

Bearing in mind that $\hat{A}\epsilon R^{n\times n}[z]$, (23) implies that there exist monic $g_i(z)\epsilon R[z]$ such that $h_{i+1,i+1} = g_i h_{ii}$ for $1\leq i\leq n-1$ with $i\neq k_1,k_2, \ldots,k_m$. We now assume that all zeros of $\det Q(z)$ are distinct which will be referred to later as Hypothesis (H), so that all zeros of $\det H(z)$ are also distinct. Notice that, by (20) and the fact that $h_{i+1,i+1}=g_i h_{ii}$ for $1\leq i\leq n-1$ with $i\neq k_1,k_2,\ldots,k_m$ we have

$$\det H(z) = (h_{11}^{p_1} g_1^{p_1-1} \cdots g_{k_1-2}^{2}\, g_{k_1-1}) \cdots (h_{k_{m-1}+1,k_{m-1}+1}^{p_m}\, g_{k_{m-1}+1}^{p_{m-1}} \cdots g_{n-2}^{2}\, g_{n-1})$$

which with the hypothesis (H) yields

$$h_{11} = g_1 = \cdots = g_{k_1-2} = 1,\ldots,h_{k_{m-1}+1,k_{m-1}+1} = g_{k_{m-1}+1} = \cdots = g_{n-2} = 1.$$

Namely $h_{11} = h_{22} = \cdots = h_{k_1-1,k_1-1} = 1,\ldots,h_{k_{m-1}+1,k_{m-1}+1} = \cdots = h_{n-1,n-1} = 1.$

Hence matrix H(z) in (20) becomes

$$
\left[\begin{array}{ccc:ccc:c:ccc}
1 & & h_{1,k_1} & & & h_{1,k_2} & & & & h_{1,k_n} \\
 & \ddots & \vdots & & 0 & \vdots & & & & \vdots \\
 & 1 & \vdots & & & \vdots & & & & \vdots \\
0 & \cdots & h_{k_1,k_1} & & & x & & & & \vdots \\
\hdashline
 & & & 1 & & x & & & & \vdots \\
 & & & & \ddots & \vdots & & & 0 & \vdots \\
 & & & & 1 & \vdots & & & & \vdots \\
 & & & 0 & \cdots & h_{k_2,k_2} & & & & \vdots \\
\hdashline
 & & & & & & \ddots & & & x \\
 & & 0 & & & & & & & \\
\hdashline
 & & & & & & & 1 & & x \\
 & & & & & & & & \ddots & \vdots \\
 & & & & & & & & 1 & \vdots \\
 & & & & & & & 0 & \cdots & h_{nn}
\end{array}\right] . \tag{24}
$$

The matrix $\hat{B}$ in (22) then becomes a simple matrix with 0's replacing the elements labeled with x's and 1's replacing the elements labeled h_{11}, $h_{k_1+1,k_1+1}, \cdots$ $\ldots, h_{k_{m-1},k_{m-1}}$, which is numbered as equation (25), and a straightforward but quite involved calculation leads to

$$
\hat{A} = \left[\begin{array}{cccc:cccc:c:cccc}
0 & 0 & h_{1,k_1} & x & & & h_{1,k_1} & x & & & & h_{1,n} & x \\
1 & 0 & \vdots & \vdots & & & \vdots & \vdots & & & & \vdots & \vdots \\
0 & 1 & \vdots & \vdots & & 0 & \vdots & \vdots & & & & \vdots & \vdots \\
\vdots & \vdots & \ddots & \vdots & & & \vdots & \vdots & & & & \vdots & \vdots \\
0 & 0 & h_{k_1,k_1} & x & & & \vdots & x & & & & \vdots & \vdots \\
\hdashline
 & & & & 0 & 0 & \vdots & x & & & & \vdots & \vdots \\
 & & & & 1 & 0 & \vdots & \vdots & & & & \vdots & \vdots \\
 & & & & 0 & 1 & \vdots & \vdots & & & & \vdots & \vdots \\
 & & & & \vdots & \vdots & \ddots & \vdots & & & & \vdots & \vdots \\
 & & & & 0 & 0 & h_{k_2,k_2} & x & & & & \vdots & \vdots \\
\hdashline
 & & & & & & & & \ddots & & & \vdots & x \\
\hdashline
 & & & & & & & & & 0 & 0 & \vdots & x \\
 & & & & & & & & & 1 & 0 & \vdots & \vdots \\
 & & 0 & & & & & & & 0 & 1 & \vdots & \vdots \\
 & & & & & & & & & \vdots & \vdots & \ddots & \vdots \\
 & & & & & & & & & 0 & 0 & h_{n,n} & x
\end{array}\right] . \tag{26}
$$

Similar to the single-input case, it is easy to show that under the hypothesis (H), the pair $(\hat{A},\hat{B})$ given in (26),(25, which is (22) modified as indicated above) is canonical in the algebraic sense for the pair $(A,B)\varepsilon X_1$.

The form of $\hat{A}$ as equation (26) is easily verifiable. To see this, we look at the case m=2 for convenience. Note

$$H(z) = \left[\begin{array}{cccc:cccc} 1 & & & x & & & & x \\ & \ddots & & \vdots & & 0 & & \vdots \\ 0 & & 1 & x & & & & x \\ & & & x & & & & \\ \hdashline & & & & 1 & & & x \\ & 0 & & & & \ddots & & \vdots \\ & & & & 0 & & 1 & x \\ & & & & & & & x \end{array}\right] \equiv \left[\begin{array}{c:c} H_1 & H_2 \\ \hdashline 0 & H_3 \end{array}\right].$$

So that

$$H^{-1}(z) = \left[\begin{array}{cccc:cccc} 1 & & & x & & & & x \\ & \ddots & & \vdots & & 0 & & \vdots \\ 0 & & 1 & x & & & & x \\ & & & x & & & & \\ \hdashline & & & & 1 & & & x \\ & 0 & & & & \ddots & & \vdots \\ & & & & 0 & & 1 & x \\ & & & & & & & x \end{array}\right] \equiv \left[\begin{array}{c:c} G_1 & G_2 \\ \hdashline 0 & G_3 \end{array}\right].$$

Thus

$$\hat{A} = H \left[\begin{array}{ccccc:ccccc} 0 & 0 & & & x & & & & & x \\ 1 & 0 & & & \vdots & & & & & \vdots \\ 0 & 1 & & & \vdots & & & 0 & & \vdots \\ \vdots & & \ddots & & \vdots & & & & & \vdots \\ 0 & & & 1 & x & & & & & x \\ \hdashline & & & & & 0 & 0 & & & x \\ & & & & & 1 & 0 & & & \vdots \\ & 0 & & & & 0 & 1 & & & \vdots \\ & & & & & \vdots & & \ddots & & \vdots \\ & & & & & 0 & & & 1 & x \end{array}\right] H^{-1} \equiv H \left[\begin{array}{c:c} A_1 & A_2 \\ \hdashline 0 & A_3 \end{array}\right] H^{-1}$$

$$= \left[\begin{array}{c:c} H_1A_1G_1 & H_1A_1G_2 + (H_1A_2 + H_2A_3)G_3 \\ \hdashline 0 & H_3A_3G_3 \end{array}\right]$$

$$= \left[\begin{array}{cccccc:cccccc}
0&0&&&x&x&&&&&x&x\\
1&0&&&\cdot&\cdot&&&&&\cdot&\cdot\\
0&1&&&\cdot&\cdot&&&&&\cdot&\cdot\\
\vdots&&\ddots&&\cdot&\cdot&&&0&&\cdot&\cdot\\
\vdots&&&1&\cdot&\cdot&&&&&\cdot&\cdot\\
0&\cdots&&0&x&x&&&&&x&x\\ \hdashline
&&&&&&0&0&&&x&x\\
&&&&&&1&0&&&\cdot&\cdot\\
&&&&&&0&1&&&\cdot&\cdot\\
&&0&&&&\vdots&&\ddots&&\cdot&\cdot\\
&&&&&&\vdots&&&1&\cdot&\cdot\\
&&&&&&0&\cdots&&0&x&x
\end{array}\right].$$

Applications of these results are being made in synthesis of stabilizing controllers for the linear delay systems and will be reported on later.

REFERENCES

[1] R. L. Alford and E. B. Lee, Sampled data hereditary systems: linear quadratic theory, to appear IEEE Trans. on Automat. Contr. 1986.

[2] A. W. Olbrot and S. H. Żak, Controllability and observability problems for linear functional-differential systems, Found. of Contr. Eng., 5, 1980, 79-89.

[3] S. MacLane and G. Birkhoff, Algebra, (Macmillan, New York, 1971).

[4] E. B. Lee, S. Neftci, and A. Olbrot, Canonical forms for time delay systems, IEEE Trans. Automat. Contr., AC-27, 1982, 128-132.

[5] T. Kailath, Linear Systems, (Prentice-Hall, Inc., Englewood Cliffs, NJ, 1980).

[6] M. Newman, Integral Matrices, (Academic Press, New York, 1972).

[7] K. Watanabe and M. Ito, A necessary condition for spectral controllability of delay systems on the basis of Finite Laplace Transforms, Int. J. Contr., 39, 1984, 363-374.

[8] E. B. Lee, A. W. Olbrot, and S. H. Żak, Cyclicity of polynomial systems, IEEE Trans. Automat. Contr., AC-27, 1982, 451-453.

[9] W. A. Wolovich, Linear Multivariable Systems, (Springer-Verlag, New York: 1974).

[10] M. J. Denham, Canonical forms for the identification of multivariable linear systems, IEEE Trans. Automat. Contr., AC-19, 1974, 646-656.

[11] V. M. Popov, Invariant description of linear, time-invariant controllable systems, SIAM J. Contr., 10, 1972, 252-264.

[12] J. D. Aplevich, Direct computation of canonical forms for linear systems by elementary matrix operations", IEEE Trans. Automat. Contr., AC-19, 1974, 124-126.

[13] C. I. Byrnes, M. W. Spong, and T-J. Tarn, A several complex variables approach to feedback stabilization of linear neutral delay-differential systems, Math. Sys. Theory, 17, 1984, 97-133.

Frequency Domain and State Space Methods for Linear Systems
C.I. Byrnes and A. Lindquist (Editors)

APPROXIMATION AND REALIZATION OF INFINITE DIMENSIONAL LINEAR SYSTEMS: THE ROLE OF DIRICHLET SERIES

Clyde F. Martin*

Department of Mathematics
Texas Tech University
Lubbock, Texas 79409
U.S.A.

Problems in the approximation and realization of symmetric systems evolving in function spaces of the circle are considered. It is noted that generalized Dirichlet series and their summation properties are naturally related to approximation issues for these systems. It is also noted that there are fundamental obstructions to the uniform approximation of these systems by finite dimensional systems.

INTRODUCTION

Finite dimensional linear systems have held the attention of the control theory community for the last three decades and the theory, while not completely developed, is in a generally quite satisfactory state. Based on this theory the last decade has seen a substantial body of material, outside of the theory of finite dimensional linear systems, being developed to model and control systems which are either nonlinear or nonfinite dimensional or both. Often the techniques being used are direct modificatins of techniques developed for linear systems and the question arises of whether or not linear methods are truly appropriate.

In the case of nonlinear finite dimensional systems there has been a tremendous amount of attention given to the Brockett-Hunt-Meyer-Su results on the feedback linearization of linear analytic systems. While these methods have proven to be very successful in the practical control of aircraft and have attracted, justifiably, an extremely large following they are of limited utility in the larger sense of control theory. It is known from the work of Byrnes [1] and even more elegantly from the work of Gardner and Shadwick [2] that even in the planar case these systems are far from generic. It would suggest that truly nonlinear methods must be developed.

* Supported in part by NASA Grant # NAG 2-203, AFWL Contract No. F29601-85-C-0040, NSA Grant #MDA904-85-H-0009 and a grant from the Texas High Tech Research Fund.

In this paper we want to consider a class of infinite dimensional linear systems and some of the problems both at the mathematical level and at the control theoretic level which arise when finite dimensional techniques are applied to their analysis. The results of this paper are not intended to be very deep nor even particularly fundamental but perhaps will serve to point out some of the pitfalls that can arise when "natural" techniques are applied to a class of simple problems. As in the case of finite dimensional nonlinear systems this would seem to suggest that finite dimensional techniques while useful must be used with discretion rather than with casual familiarity.

The author would like to acknowledge many stimulating discussions with Robert Grossman on the subject of this paper. Many of the ideas of this paper were formulated with him.

THE FORCED HEAT EQUATION

Let S^1 denote the unit circle, μ the normalized Haar measure on the circle and $L^2(S^1)$ the space of square integrable functions on S^1. We consider the following system evolving in $L^2(S^1)$.

$$d/dt(w(x,t)) = d^2/dx^2(w(x,t)) + b(x)u(t)$$
$$y(t) = \int_{S^1} c(x)w(x,t)d\mu,$$

where b(x) and c(x) are in $L^2(S^1)$. We use d/dt instead of the more customary partial derivative symbol for reasons of typographical efficiency. We, of course, assume that the partial derivatives are defined on a dense subset of $L^2(S^1)$ and consequently are unbounded operators. The question of the controllability of this system has been discussed extensively in the context of symmetric systems by Grossman and Martin in [3] and [4]. In this paper we are more concerned with the question of approximation.

We begin by calculating y(t) as a function of u(t) by constructing the fourier coefficients of w(x,t).

$$d/dt\hat{w}(n,t) = -n^2\hat{w}(n,t) + \hat{b}(n)u(t)$$

and hence we have

$$\hat{w}(n,t) = \int_0^t \exp(-n^2(t-s))\hat{b}(n)u(s)ds$$

and it then follows that

$$y(t) = \int_0^t \sum \hat{b}(n)\hat{c}(n)\exp(-n^2(t-s))u(s)ds.$$

We use (^) to denote the fourier transform. The output y is expressed in standard convolutional form

$$y(t) = \int_0^t k(t-s)u(s)ds$$

where k(t) in this case is the classical Dirichlet series

$$H(t) = \sum \hat{b}(n)\hat{c}(n)\exp(-n^2t).$$

The series H(t) converges uniformly and absolutely in the region t>0 and if more traditionally we consider t to be a complex variable we have uniform and absolute convergence in every cone in the right half of the complex plane with vertex at the origin. The assumption that b and c are in $L^2(S^1)$ gives us this convergence. We see that we could relax the assumption on the function c(x) since we really only need that the fourier coefficients of c grow at some exponential order. So in particular if we had polynomial growth of the coefficients we would still have uniform and absolute convergence. From a system theoretic point of view this make a great deal of sense since we are only asking for a observation. If the observer were a bounded linear functional we would, in most cases, be satisfied. We will discuss this in greater detail in a later section.

FINITE DIMENSIONAL APPROXIMATIONS

The problems of approximation of infinite dimensional systems is subtle and care must be taken to ensure that one understands in what sense approximation is being used. Commonly if one starts with an input output relation of the form

$$y(t) = \int_0^t k(t-s)u(s)ds$$

it is assumed that approximation is used in the sense that either the convolutional kernel k(t) is to be approximated or that the Laplace transform of the kernel is approximated by Laplace transforms of functions. In this paper we will confine our attention to the case of the convolution kernel. The approximation of the Laplace transform merely adds complications. We will use the following notation to denote the convolution of k(t) with u(t)

$$y(t) = (k*u)(t).$$

Let $\{k_n(t)\}$ be a sequence converging to k(t) and let

$$y_n(t) = (k_n*u)(t)$$

be the sequence generated by the approximating kernels. From a practical point of view the problem of approximation is to determine how closely the sequence $\{y_n(t)\}$ is to the function y(t). If we were concerned with the problem of a single input function, u(t), the problem of approximation would be quite simple but, of course, the very nature of system theory requires that all inputs be considered. The problem of a single input function is a problem of differential equations not of system theory.

In this paper we are considering the question of finite dimensional approximations. It is worth noting that there is no concept of dimension inherent in the input-output formulation of system theory and dimension only becomes an issue when the system is realized in statespace form. We will say that the sequence $\{k_n(t)\}$ is a <u>finite dimensional</u> approximation if each of the input output pairs

$$y_n(t) = (k_n * u)(t)$$

can be realized with a finite dimensional controllable and observable system of the form

$$(d/dt)x_n(t) = A_n x_n(t) + b_n u(t)$$
$$y_n(t) = c' x_n(t)$$

THE FORCED HEAT EQUATION, CON'T

The representation of the input output behavior of

$$d/dt(w(x,t)) = d^2/dx^2(w(x,t)) + b(x)u(t)$$
$$y(t) = \int_{S^1} c(x)w(x,t)d\mu,$$

by the convolution operator

$$y(t) = \int_0^t \sum \hat{b}(n)\hat{c}(n)\exp(-n^2(t-s))u(s)ds.$$

lends itself readily to the construction of finite dimensional approximations. We will use the truncations of the Dirichlet series

$$H_N(t) = \sum_{n\leq N} \hat{b}(n)\hat{c}(n)\exp(-n^2 t),$$

the so called Dirichlet polynomials. We note that the relations

$$y_N(t) = (H_N * u)(t)$$

can be realized in the form of

$$(d/dt)x_n(t) = A_n x_n(t) + b_n u(t)$$
$$y_n(t) = c' x_n(t)$$

with

$$A_N = \mathrm{Diag}[-N^2,\cdots,-1,0,-1,\cdots,-N^2],$$
$$b'_N = (\hat{b}_{-N},\cdots\cdot,\hat{b}_N),$$
$$c_N = (\hat{c}_{-N},\cdots\cdot,\hat{c}_N).$$

It is clear that A_N is a representation of d^2/dx^2 restricted to the subspace of $L^2(S^1)$ spanned by the complex exponentials $\{\exp(-n^2x): n\leq N\}$ and that b and c are truncations of the fourier series representation of the functions b(x) and c(x) restricted to this subspace.

The general theory of $L^2(S^1)$ combined with the L^2 theory of the Laplace transform would provide a good theory of convergence for the forced heal equation but for the fact that there are inputs of classical and current interest that are not L^2 nor even bounded. One of the classical tests for a control system was the ability of the system to track a ramp input. It is interesting todo the calculation for the simple ramp $u(t) = t$. Sparing the reader the tedium of the actual calculation we find that we have

$$y_N(t) = \quad {}_{n\leq N}(t/n^2 - 1/n^4 + (1/n^4)\exp(-n^2 t)).$$

It is obvious that the sequence converges pointwise to $y(t)$ but because of the term which involves t there is no uniform convergence despite the fact that the Dirichlet polynomials, $H_N(t)$, converge uniformly and absolutely to $H(t)$. As one can imagine the situation becomes worse for functions which are less regular.

DIRICHLET KERNELS

In the previous sections we have examined the input output behavior of a system governed by the forced heat equation and have constructed finite dimensional linear approximations to this system. The finite dimensional approximations were good in the sense that they gave uniform approximations to the convolution kernel and even converged to operators in $L^2(S^1)$. In this section we examine a slightly more general case and will find that the problems become much more subtle.

Let $k(t)$ be represented by a generalized Dirichlet series

$$k(t) = \sum a_n \exp(\mu_n t)$$

where the sequence $\{\mu_n\}$ is a sequence of negative real numbers whose magnitudes are growing without bound. For the time being we will assume that the series converges uniformly and absolutely in any cone in the open right half plane with vertex at the origin. As in the case of the heat kernel we can find a finite dimensional linear approximation to the system defined by $k(t)$. Now however the problem is not as well defined. We must factor the coefficient a_n so that we have

$$a_n = b_n c_n$$

and the sequences $\{b_n\}$ and $\{c_n\}$ are the fourier coefficients of elements of $L^2(S^1)$. Even for some of the classical series this is not well posed. For example, the Riemann zeta function has coefficients 1 and $\mu_n = -\ln(n)$. However we note in passing that even for the heat equation the observation could have been defined in terms of distributions. Thus if the factorization can be accomplished so that the sequence $\{c_n\}$ is the fourier coefficients of a distribution then we will have constructed an adequate approximation.

The question of the convergence of the finite dimensional realizations is much more complicated. In order to discuss this we recall from Grossman and Martin, [3] and [4], the definition of an operator on $L^2(S^1)$ defined in terms of multipliers. Let p be an element of $L^2(S^1)$ with fourier coefficients $\{\hat{p}_n\}$. Given a sequence of complex numbers $\{a_n\}$ we define the operator A by

$$A(p)(x) = \sum a_n \hat{p}_n \exp(-nx).$$

Thus the operator is defined in terms of multiplication of the fourier coefficients. We refer the reader to the monograph of Stein [5] and the related literature for the theory of such operators. We know of course that the operator d^2/dx^2 is of this type with $a_n = -n^2$. Now it is clear that the finite dimensional operators A_n converges to a multiplier operator. The question of the realization of this operator as a more familiar operator of the differential type is naturally enough for the most part quite impossible. Also the question of the domain and range of the operator becomes very subtly involved with the the theory of convergence of Fourier series in $L^2(S^1)$. In fact it is sometimes necessary to resort to the full power of the Toeplitz summation techniques to obtain a reasonable domain and range for such operators.

SUMMATION TECHNIQUES FOR DIRICHLET SERIES

The problem of the interpreting an infinite sum is very old and at one time was a very controversial issue. The problem was "solved" when the definition of convergence was proposed by Cauchy, i.e. a series is convergent if the limit of the partial sums exists. It was, however, recognized that there were series that could be give real interpretations but not by the use of convergence. A familiar example is the theory of **Cesàro** summability for fourier series and the more general but less familiar **Toeplitz** summability. By the beginning of the twentieth century there was a well developed theory of summability of divergent series. For Dirichlet series the problem was a bit more delicate than just the summation of divergent series and rested in the area we now more familiarly refer to as analytic continuation.

Consider the familiar Riemann zeta function, which we will refer to as Z(s). We recall the definition

$$Z(s) = \sum \exp(-\ln(n)s) = \sum (n)^{-s},$$

and note that the series converges for $Re(s) > 1$. It is well known that this function has a pole at s=1 but admits a continuation to the entire punctured plane. On the other hand the function

$$T(s) = (1-2^{1-s})Z(s) = \sum (-1)^{n+1}(n)^{-s}$$

is regular over the entire plane but whose series only converges for $Re(s) > 0$. The series T(s) is in contrast to the theory of convergence for Taylor series which must have a singularity on the radius of convergence. In order to continue such functions past the domain of convergence Riesz [6] introduced the concept of moyennes typiques. We give the basic definitions.

Let $a_n \exp(-\mu_n s)$ be an arbitrary Dirichlet series. We define

$$C_\mu^k(w) = \sum_{\mu_n < w} (w-\mu_n)^k a_n \exp(-\mu_n s),$$

where k is any real number. $C_\mu^k(w)$ is called the **typical means (moyennes typiques) of the first kind of order k.** We say that the series is summable if the limit, as w goes to infinity, of

$$w^{-k} C_\mu^k(w)$$

exists. There is a well developed theory of uniform convergence for the typical means and we refer the reader to the classic treatise of Hardy and Riesz [7] for the details. A typical result is that if one calculates the typical means of order k for the series T(s) we find that we have convergence in the half plane $Re(s) > -k$. The typical means have the effect of not only extending the domain of convergence but in fact accelerate the convergence of the series. It is in this light that summation methods are seeing a small revival in numerical analysis. We will use them to construct finite dimensional approximations to infinite dimensional linear systems.

FINITE DIMENSIONAL APPROXIMATIONS OF DIRICHLET KERNELS

Consider the infinite dimensional linear system defined by the convolution equation

$$y(t) = (T*u)(t)$$

where T is the function of the previous section

$$T(s) = (1-2^{1-s})Z(s) = \sum (-1)^{n+1}(n)^{-s}.$$

If we construct finite dimensional approximations based on the partial sums of the Dirichlet series

$$\sum (-1)^{n+1} \exp(-\ln(n)s)$$

they simply do not converge. However if we construct the typical means and in particular the expressions $w^{-k}C_{\mu}{}^{k}(w)$ we do have convergence to the kernel on the positive real numbers.

It is interesting to consider a particular case, for example k=4.

$$w^{-4}C_{-\ln(n)}{}^{4}(w) =$$

$$-1 + ((w-\ln 2)^4/w^4)\exp(-\ln 2)s - ((w-\ln 4)^4/w^4\exp(-\ln 4)s + \cdots\cdots$$

The effect of the typical means on the realization is not on the system operator but on the input and output functions. We can construct symmetric realizations of the approximations with the typical terms of

$$(w-\ln(n))^2/w^2,$$

in the input and output terms. Since the typical means converge in the right half plane Re(s)>-2 we have that the b_n and c_n converge in $L^2(S^1)$. Now we have no real evidence that the systems converge in any satisfactory sense but we do see that if in fact the multipliers $\{-\ln(n)\}$ define a useful operator on $L^2(S^1)$ we do have each set of coefficients converging. It is also worth noting that we have constructed in this case a one parameter family of finite dimensional controllable and observable system which are converging to a nonfinite dimensional system. This will be examined in the light of Krishnaprasad and Martin [8] and Hazewinkel [9] in a later paper.

The general case is of course more complicated than the above example might suggest. In general the typical means need not converge over any half plane at all and we may not be able to extend the convergence into the left half plane. Never the less the typical means do allow us to approximate a class of kernels which were outside of the usual methods. While there is a great deal known about convergence of summation methods there are still many functions for which there do not exist readily applicable theorems.

APPROXIMATION BY DIRICHLET POLYNOMIALS

A natural question to ask is to what class of functions can the above techniques be applied. This seems to be a very difficult question. There exists one fundamental paper which surveys most of the known results. The survey article of Leont'ev [10] covers much of what is known but the results of that paper are not directly applicable to the problems at hand. First Leont'ev allows complex coefficients which in itself is not bad but it destroys the property that convergence at a point implies convergence in a half plane. For his paper to be completely useful the theorems must be reinterpreted to exclude the case that there are complex exponentials. This would modify the nature of his results a great deal.

A second approach is to note that Dirichlet series are in fact Laplace transforms with respect to a measure that is concentrated at discrete points. There is a well developed theory of what functions can be Laplace transforms with respect to Lebesque measure and it is conceivable that these results could be extended to the case of point masses. However this would be a major research project in itself. On the other hand if a theory of finite dimensional approximation could be developed for single input single output linear systems this would in fact be a theory of when a function could be approximated by a Dirichlet polynomial. So as usual the results go both ways.

CONCLUSIONS

The theory of approximation of single input single output convolutional operators by convolutional operators that admit finite dimensional realizations is closely related to theory of approximation by Dirichlet polynomials. The technique of summation of divergent series can be used to construct approximations that converge in the sense of converging to the kernels. However in general the approximations are not good in the sense required by system theory. Even if there is uniform and absolute convergence of the kernels we need not have uniform convergence of the output functions.

BIBLOGRAPHY

[1] C. Byrnes, "Remarks on nonlinear planar control systems which are linearizable by feedback", System and Control Letters, Vol. 5, 1985, pp 363-368.

[2] R. Gardner and W. Shadwick, "The control problem for control systems with scalar control", Notices of the American Mathematics Society, Vol. 33, 1985, p. 105.

[3] R. Grossman and C. Martin, "Control of systems evolving on compact groups", Proceedings of the 1983 Berkeley-Ames Conference on Nonlinear Problems in Control and Fluid Dynamics.

[4] R. Grossman and C. Martin, "Approximation and control of symmetric systems on the circle," P. A. Fuhrmann (ed.), Mathematical Theory of Networks and Systems: Proceedings of the MTNS-83 International Symposium, Beer Sheva, Isarel, Lecture Notes in control and Information Sciences, vol 58, Springer-Verlag, Berlin, 1984.

[5] E. Stein, Topics in harmonic analysis: related to the Littlewood-Paley theory, Princeton University Press, 1970.

[6] M. Reisz, "Sur les sèries de Dirichlet et les sèries entières", Comptes Rendus, Vol. 22, 1909.

[7] G. Hardy and M. Reisz, The general theory of Dirichlet's series, Cambridge University Press, 1952.

[8] P. Krishnaprasad and C. Martin, "Families of systems and deformations," Int. J. Control, Vol. 38, pp. 1055-1079, 1983.

[9] M. Hazewinkel, "On families of of linear systems: degeneration phenomena", In Algebraic and Geometric methods in Linear Systems Theory, C. Byrnes and C. Martin, (eds.), American Mathematics Society, 1980.

[10] A. Leont'ev, "Representation of functions by generalized Dirichlet series", Russian Mathematics Surveys, Vol. 24, No. 2, 1969, pp 101-178.

Frequency Domain and State Space Methods for Linear Systems
C.I. Byrnes and A. Lindquist (Editors)

INTERCONNECTED SYSTEMS WITH DELAYS

L. PANDOLFI[+]
Dipartimento di Matematica
Politecnico di Torino
Torino - Italia

We define the zero structure of a class of distributed parameter systems. Applications tc the study of parallel, tandem and feedback connections of systems are presented.

1. INTRODUCTION

In the last few years, frequency domain conditions have been applied to study convolution systems (see for example [1, 2]). In the special case of delay systems, the connections between state space and frequency domain properties have been clarified ([4, 5, 6]), so that frequency domain analysis of state space concepts is possible. In fact, even in finite dimensions, some problems whose solution is involved in the state space setting have an easy solution in the frequency domain approach. In particular, this is the case of the study of interconnected systems which are both controllable and observable. It is quite natural to search for conditions under which the composite system retains these properties. A relatively easy solutions to this problem exists in the frequency domain ([8, 9]), while the corresponding conditions in the state space approach became soon involved ([10]). These observations justify our interest to study the cascade, parallel and feedback connections of a class of convolution systems which includes, in particular, systems with delays. Even more so, because interconnected systems are an important source of delays (since transportation delays may be large and not safely ignored). In fact, the cascade connection was studied in [7]. In this paper we abstract the foundamental ideas in [7] and we use them to treat parallel and feedback composition of systems from the class that we describe now.

2. DESCRIPTION OF THE SYSTEMS AND PRELIMINARIES

Let us consider the system

$$\begin{cases} \dot{x} = \int_{-h}^{0} dA'(s)x(t+s) + \int_{-h}^{0} dB'(s)u(t+s) \\ y = \int_{-h}^{0} dC'(s)x(t+s) + \int_{-h}^{0} dD'(s)u(t+s) \end{cases}$$

where u, x, y denote, respectively, m-, n-, p-vectors and the matrices have suitable dimensions. If the function u is Laplace transfor-

(+) PAPER SUPPORTED BY THE MINISTERO DELLA PUBBLICA ISTRUZIONE.

mable, then we can take Laplace transform of both members of the above relations and we find ($x(t) = 0$ for $t \leq 0$, $u(t) = 0$ for $t < 0$)

$$\begin{cases} A(z)x(z) = B(z)u(z) \\ y(z) = C(z)x(z) + D(z)u(z) \end{cases} \tag{1}$$

where the complex vector functions $u(z)$, $x(z)$, $y(z)$ are respectively the Laplace transforms of the functions $u(t)$, $x(t)$, $y(t)$ and

$$A(z) = zI - \int_{-h}^{0} e^{zs} dA'(s), \quad B(z) = \int_{-h}^{0} e^{zs} dB'(s)$$

and analogous definitions for $C(z)$, $D(z)$ (I always denotes the identity matrix of suitable dimensions). The elements of the matrices $A(z)$, $B(z)$, $C(z)$, $D(z)$ are entire functions and, moreover:

CONDITION 1: there exists a number z such that $\det(A(z)) \neq 0$.

In fact, examples of more general convolution systems suggest that we assume the less restrictive condition ([6]):

CONDITION 2: The elements of the matrix $A(z)$ are meromorphic functions in a region Ω of the complex plane.

In this paper the word "system" will indicate a relation like (1), which satisfies the above two conditions. We observe explicitly that this allows consideration of systems expressed by implicit differential equations, like the system

$$K\dot{x} - Ax = Bu,$$

provided that $\det(zK-A) \neq 0$.

Now we collect some preliminary material. It is known ([11]) that matrices of holomorphic functions admit a Smith normal form so that two matrices $E(z)$, $P(z)$ with the same number of rows are left coprime, i.e. have no (non invertible) common left divisor, if and only if the matrix $[E(z), P(z)]$ has full rank for every complex number z. If the matrices $E(z)$, $P(z)$ have the same number of columns, they are right coprime if and only if the matrices $[E^*(z), P^*(z)]$ are left coprime.

Since a Smith form exists for holomorphic matrices, a Mc Millan Smith form exists for meromorphic matrices. This implies that every matrix $A(z)$ of meromorphic functions admits left and right coprime fractional decompositions,

$$A(z) = E^{-1}(z)P(z) = \tilde{P}(z)\tilde{E}^{-1}(z).$$

We say that a point $z_o \in \mathbb{C}$ is a zero of the meromorphic matrix $A(z)$ when it is a zero of one of the non zero entries of its Mc Millan Smith form.

Now, let us go back to consider system (1). We say that the pair $(B(z), A(z))$ is <u>z_o-controllable</u> when z_o is not a zero of the matrix

$[A(z),B(z)]$. The pair $(A(z),C(z))$ is $\underline{z_o\text{-observable}}$ when the pair $(C^*(z),A^*(z))$ is z_o-controllable. "Controllable" and "observable" stand for z_o-controllable and z_o-observable for every $z_o \in \mathbb{C}$. "Canonical" means both controllable and observable.

REMARK. In the case of delay systems the above conditions are known as spectral controllability and spectral observability ([5, 12, 13]).
◊

The system matrix of system (1) is the matrix

$$S(z) = \begin{bmatrix} A(z) & -B(z) \\ C(z) & D(z) \end{bmatrix}$$

while its transfer function is the matrix $T(z) = C(z)A^{-1}(z)B(z)+D(z)$. The transfer function $T(z)$ is still a meromorphic function, because of CONDITION 1. We shall write left and right coprime decompositions for it:

$$T(z) = Q^{-1}(z)M(z) = \tilde{M}(z)\tilde{Q}^{-1}(z).$$

3. COMPOSITE SYSTEMS

From now on, we denote by (S) the system described by (1) and by (S') the system

$$\begin{cases} F(z)x'(z) = G(z)u'(z) \\ y'(z) = H(z)x'(z) + K(z)u'(z) \end{cases} \tag{2}$$

where u', x', y' are respectively m'-, n'-, p'-vectors. The matrices in (2) have suitable dimensions and satisfy CONDITIONS 1, 2. The system matrix of (S') is the matrix

$$S'(z) = \begin{bmatrix} F(z) & -G(z) \\ H(z) & K(z) \end{bmatrix}$$

and its transfer function is $T'(z) = H(z)F^{-1}(z)G(z) + K(z)$. We write the coprime fractional decompositions:

$$T'(z) = R^{-1}(z)N(z) = \tilde{N}(z)\tilde{R}^{-1}(z).$$

The parallel composition of the systems (S), (S') is the system (Sp) xhich is obtained when we impose the condition $u = u'$ and whose output is $(y+y')$. Of course, this is possible when $m=m'$, $p=p'$.

The system matrix of (Sp) is:

$$Sp(z) = \left[\begin{array}{cc|c} A & 0 & -B \\ 0 & F & -G \\ \hline C & H & D+K \end{array}\right]$$

(the argument z will no longer be indicated in the blocks of the system matrices).

It is obvious that (Sp) satisfies CONDITIONS 1, 2 and that its transfer function is $Tp(z) = T(z)+T'(z)$.

The tandem composition (St) of the systems (S) and (S') is obtained when $m'=p$ and we impose $y=u'$. The input of system (St) is u, while the output is y'. The system matrix is:

$$St(z) = \left[\begin{array}{cc|c} A & 0 & -B \\ -GC & F & -GD \\ \hline KC & H & KD \end{array}\right]$$

Again, CONDITIONS 1, 2 are satisfied by (St) and its transfer function is $Tt(z) = T'(z)T(z)$.

The feedback composition is possible when $p=m'$ and $m=p'$. In this case we can impose the conditions $u'=y$, $u=y'+v$. We obtain a system (Sf) whose input is v and whose output is y, provided that CONDITION 1 is verified. In this case, the system matrix is

$$Sf(z) = \left[\begin{array}{ccccc} A & 0 & 0 & -B & -B \\ 0 & F & -G & 0 & 0 \\ C & 0 & -I & D & D \\ 0 & H & K & -I & 0 \\ \hline C & 0 & 0 & D & D \end{array}\right]$$

CONDITION 1 is satisfied when the determinant of the upper left block in $Sf(z)$ is not identically zero. Now, rows and columns operations reduce it in the following way:

$$\begin{bmatrix} A & 0 & 0 & -B \\ 0 & F & -G & 0 \\ C & 0 & -I & D \\ 0 & H & K & -I \end{bmatrix} \longrightarrow \begin{bmatrix} A & 0 & 0 & -B \\ 0 & F & -G & 0 \\ C & DH & DK-I & 0 \\ 0 & H & K & -I \end{bmatrix} \longrightarrow \left[\begin{array}{ccc|c} A & -BH & -BK & 0 \\ 0 & F & -G & 0 \\ C & DH & DK-I & 0 \\ \hline 0 & H & K & -I \end{array}\right] \qquad (4)$$

so that we must give conditions in order that the left upper block of the last matrix be invertible.

Theorem 1. The following conditions are equivalent:

1. CONDITION 1 holds for (Sf).
2. $\det(T(z)T'(z)-I) \not\equiv 0$.
3. $\det[M\tilde{N} - Q\tilde{R}] \not\equiv 0$.
4. $\det[MR^{-1}N - Q] \not\equiv 0$.
5. $\det[\tilde{M}\tilde{Q}^{-1}\tilde{N} - \tilde{R}] \not\equiv 0$.

Proof. We can use Schur's formulas ([14], pag. 45) to calculate the determinant of the upper left block of the last matrix in (4). We find that it is zero if and only if

$$0 = \det\left\{[C,DH]\begin{bmatrix} A^{-1} & A^{-1}BHF^{-1} \\ 0 & F^{-1}\end{bmatrix}\begin{bmatrix} BK \\ G\end{bmatrix} + DK - I\right\} =$$

$$\det\left\{CA^{-1}BK + CA^{-1}BHF^{-1}G + DHF^{-1}G + DK - I\right\} =$$

$$\det\left\{(CA^{-1}B + D)K - (CA^{-1}B + D)HF^{-1}G - DHF^{-1}G + DHF^{-1}G - I\right\} =$$

$$\det\,(T(z)T'(z) - I).$$

Now,

$$\det\,(TT'-I) = \det\,(Q^{-1}MN\tilde{\tilde{R}}^{-1} - I) = \det\,Q^{-1}(M\tilde{N} - Q\tilde{R})R^{-1}.$$

This proves the equivalence between condition 1 and conditions 2, 3. Different factorizations for $T(z)$, $T'(z)$ give conditions 4, 5.

◇

When CONDITION 1 is satisfied by the feedback composition of (S) and (S'), it is easy to see that the corresponding transfer function is:

$$Tf(z) = (I-T(z)T'(z))^{-1}T(z),$$

which is well defined because of Theorem 3.1.

Now, we can describe our problem: we assume that both systems (S) and (S') are canonical. We want to find necessary and sufficient conditions under which the composite systems are canonical. The idea that works for lumped parameter systems is to associate to the system matrices $S(z)$, $S'(z)$ "system matrices" of a simpler structure. In the next section we extend this idea to the class of systems under study.

4. THE MAIN IDEA

In this section we abstract the main idea in [7]. We consider two meromorphic matrices, $L(z)$ and $\mathcal{L}(z)$, and, first of all, we give a method to investigate their zeros. After this, we give a set of conditions under which the two matrices have the same zeros. First of all we describe the method we shall use to study the zeros of $L(z)$.

Let $L(z)$ be a matrix which has m columns. We define the following subspaces of $\mathbb{C}^m$:

1. $G'_L(z_o)$ is the subspace of those vectors u' such that <u>for each</u> $r > 0$ we can find a holomorphic function $u(z)$, defined in a neighboorood of z_o, such that:

i. $u(z_o) = u'$

ii. $L(z)u(z)$ has a zero, for $z = z_o$, of order at least r.

2. $G''_L(z_o)$ is the subspace of those vectors u'' such that there <u>exists</u> a number $r > 0$ such that we can find a holomorphic function $u(z)$, defined in a neighboorhood of z_o, such that:

iii. $u(z_o) = u''$

iv. $L(z)u(z)$ has a zero, for $z=z_o$, of order r.

Clearly, $G''_L(z_o) \supseteq G'_L(z_o)$.

From now on, we intend that z_o is fixed (= 0 in the proofs, in order to simplify the notations) and it will not be indicated any more. Moreover, all the vector functions we meet are holomorphic near z_o unless explicitly stated (and with the exception of the transfer functions). Holomorphic functions will be called simply "functions".

It is clear that if $P(z)$ is a holomorphic matrix, $\det P(z_o)$ different from zero, then

$$G'_{LP} = P^{-1}(z_o)G'_L \ , \ G''_{LP} = P^{-1}(z_o)G''_L .$$

Hence, it is easy to see that $\dim (G'_L)$ is the number of the zero columns of the Smith form of $L(z)$. Moreover, let $u_1,\dots,u_k$ be a basis of $G''_L \ominus G'_L$ and s_j, $1 \le j \le k$ be the maximum of those numbers r such that a function $u(z)$, $u(z_o) = u_j$, exists, and $L(z)u(z)$ has a zero in z_o, of order r. Then,

$$\bar{m}_L(z_o) = \sum_{1}^{k} s_j$$

is the sum of the positive exponents in the Smith Mac Millan form of $L(z)$. We call it the order of the zero z_o. In fact, z_o is a zero of $L(z)$ if and only if $G''_L \setminus G'_L \neq \{0\}$.

The results above have been proved in [7]. From now on, in order to simplify the notations, we write G', G'' instead then G'_L, G''_L and $\mathcal{G}'$, $\mathcal{G}''$ instead then $G'_{\mathcal{L}}$, $G''_{\mathcal{L}}$.

Let us assume that $X(z)$ is any function such that $L(z)X(z)$ has a zero for $z = z_o$. Let us assume that we cangive a method, we call it METHOD 1, which associate a function $\mathcal{X}(z)$ to $X(z)$, may be not in a unique way. We assume that the following properties hold:

1. If $L(z)X(z)$ has a zero of order at least r at z_o, then $\mathcal{L}(z)\mathcal{X}(z)$ has a zero of order at least r at z_o.

2. Let $X_i(z)$ be functions such that $L(z)X_i(z)$ has a zero in z_o, and $\mathcal{X}_i(z)$ be one of the functions which correspond to $X_i(z)$. Then:

2i. $X_i(z_o) = X_j(z_o)$ implies that $\mathcal{X}_i(z_o) = \mathcal{X}_j(z_o)$

2ii. If the vectors $X_i(z_o)$ are linearly independent, the vectors $\mathcal{X}_i(z_o)$ are linearly independent.

2iii. The function $\sum\alpha_i\mathcal{X}_i(z_o)$ is one of the functions that METHOD 1 associate to the function $\sum\alpha_i X_i(z_o)$.

The above property are independent, since more then one function $\mathcal{X}(z)$ may be associated to the same function $X(z)$.

Moreover, we assume that we can find a second method, that we call METHOD 2, which associate functions $X(z)$ to functions $\mathcal{X}(z)$, such that $\mathcal{L}(z)\mathcal{X}(z)$ has a zero for $z = z_o$. The properties that METHOD 2 must haveare the same as above, with the roles of $L(z)$, $X(z)$, $\mathcal{L}(z)$, $\mathcal{X}(z)$ interchanged.

The key results of this paper are the following ones:

Lemma 1. Let us assume that METHOD 1 and METHOD 2 can be given. Let $X_i(z)$ be functions such that $\{X_i(z_o)\}$ is a basis of G'. Let $\mathcal{X}_i(z)$ be a function which corresponds to $X_i(z)$. Then, the vectors $\mathcal{X}_i(z_o)$ are a basis of $\mathcal{G}'$ and, in particular, $\dim G' = \dim\mathcal{G}'$.

Proof.Let $X_i = X_i(z_o)$ and r be a given number. We can find new functions, still denoted $X_i(z)$, such that:

i. $L(z)X_i(z)$ has a zero in z_o, of order larger then r.

ii. $X_i(z_o) = X_i$.

Now, METHOD 1 provides functions $X_i(z)$ such that

i'. $L(z)X_i(z)$ has a zero of order (at least) r in z_o

ii'. the vectors $X_i = X_i(z_o)$ are linearly independent (by 2") and do not depend on r.

These properties imply that $\{X_i\}$ is an independent set in G', so that dim $G' \geq$ dim G'. The converse inequality is obtained by METHOD 2. Hence, dim G' = dim G' and $\{X_i\}$ is a basis of G'.

◊

Lemma 2. Let us assume that METHOD 1 and METHOD 2 can be given. Let $X_i(z)$ be functions such that $\{X_i(z_o)\}$ is a basis for a complementary subspace of G' in G". Then, $\{\hat{X}_i(z_o)\}$ is a basis for a complementary subspace of G' in $G"$ and:

$\bar{m}_L(z_o) = \bar{m}_{L}(z_o)$.

Proof. It is not restrictive to assume that $L(z)X_i(z)$ has a zero of maximum order s_i, among those functions $\hat{X}(z)$ such that $\hat{X}(z_o) = X_i(z_o)$. In fact, a function $\hat{X}(z)$ will correspond to $\hat{X}(z)$ and, in general, $\hat{X}(z) \neq X_i(z)$; but, $\hat{X}(z_o) = X_i(z_o)$ because of property 2'. Hence, we get that $L(z)X_i(z)$ has a zero of order at least s_i and the order of z_o as a zero of $L(z)$ is not minor then the order of z_o as a zero of L(z) if we can prove the implication:

$$\sum \alpha_i X_i(z_o) \in G' \Rightarrow \alpha_i = 0 \quad \text{for each index } i.$$

Then, equality follows by reversing the previous arguments, *via* METHOD 2.

So, let us assume that $\sum \alpha_i X_i(z_o) \in G'$ and consider the function $X(z) = \sum \alpha_i X_i(z)$. Lemma 1 implies that we can find functions X'(z) such that:

1. $\{X'_i(z_o)\}$ is a basis of G'
2. $\{X'_i(z_o)\}$ is a basis of G'.

The vectors $X(z_o)$ and $X'_i(z_o)$ are linearly independent. But, property 2" asserts that $X(z)$ is one of the functions which correspond to $X(z) = \sum \alpha_i X_i(z)$ so that, by property 2", $\sum \alpha_i X_i(z_o) \in G'$. This implies that α_i is zero for each index i. As a consequence, $\bar{m}_L(z_o) \geq \bar{m}_L(z_o)$ and, moreover, dim $(G" \ominus G') \geq$ dim $(G" \ominus G')$. The proof is completed by interchanging the roles of L and L.

◊

The next lemmas ([5, 6]) will be used in subsequent applications, to derive METHOD 1 and METHOD 2.

Lemma 3 Let us assume that system (S) is z_o-controllable. Let $w(z) = (x^*(z), u^*(z))^*$ be a function such that $S(z)w(z)$ has a zero of order at least r for $z = z_o$. Then, it is possible to find a function $u'(z)$ such that $T(z)[u(z) + (z-z_o)^r u'(z)]$ has a zero of order at least r for $z = z_o$.

◊

Lemma 4 Let us assume that system (S) is z_o-observable. Then:

1. $T(z)u(z)$ bounded at z_o implies that $A^{-1}(z)B(z)u(z)$ is bounded at z_o.
2. If $S(z)w(z)$ has a zero for $z = z_o$, $w(z) = (x^*(z), u^*(z))^*$, and if $u(z_o) = 0$, then $x(z_o) = 0$ too.

◊

5. THE SYSTEMS AND THEIR ASSOCIATES

Let us consider system (S), whose transfer function is $T(z) = Q^{-1}(z)M(z) = \tilde{M}(z)\tilde{Q}^{-1}(z)$. It is easy to see that $T(z)$ is the transfer function which corresponds to the system matrices

$$S(z) = \begin{bmatrix} Q & -M \\ I & 0 \end{bmatrix}, \qquad \tilde{S}(z) = \begin{bmatrix} \tilde{Q} & -I \\ \tilde{M} & 0 \end{bmatrix}.$$

We denote by (S) and $(\tilde{S})$ the systems which formally can be associated to $S(z)$ and $\tilde{S}(z)$. These systems satisfy CONDITIONS 1, 2 and are canonical. But, we are not asserting that they "really" describe same "physical engine".

Systems (S), $(\tilde{S})$ will be said the associated systems to system (S). Analogously, we can define the associated systems to system (S'), whose matrices are

$$S'(z) = \begin{bmatrix} R & -N \\ I & 0 \end{bmatrix} \qquad \tilde{S}'(z) = \begin{bmatrix} \tilde{R} & -I \\ \tilde{N} & 0 \end{bmatrix}$$

Now, we start to study the observability property of the composite systems. This will be sufficient, as we shall see, because of the "duality" between the definitions of controllability and observability.

Let us recall that the output zeros of a system matrix (like S(z)) are the zeros of the matrix which is obtained by deleting the last block column.

The observability property can be studied in the simplest way if we can show that the output zeros of the matrices Sp(z), St(z), Sf(z) are the same as the output zeros of the following matrices:

$$Sp(z) = \left[\begin{array}{cc|c} Q & 0 & -M \\ 0 & R & -N \\ \hline I & I & 0 \end{array}\right] \qquad St(z) = \left[\begin{array}{cc|c} Q & 0 & -M \\ -N & R & 0 \\ \hline 0 & I & 0 \end{array}\right]$$

$$Sf(z) = \left[\begin{array}{cccc|c} Q & 0 & 0 & -M & -M \\ 0 & R & -N & 0 & 0 \\ I & 0 & -I & 0 & 0 \\ 0 & I & 0 & -I & 0 \\ \hline I & 0 & 0 & 0 & 0 \end{array}\right]$$

In the next sections we shall prove the above conjecture. Hence, the simple structure of the matrices $Sp(z)$, $St(z)$, $Sf(z)$ give the following result:

Theorem 1. The systems (Sp), (St), (Sf) are z_o-observable if and only if, respectively,

$$\operatorname{Ker}\left(\begin{bmatrix} Q(z_o) \\ R(z_o) \end{bmatrix}\right) = \{0\}; \quad \ker\left(\begin{bmatrix} Q(z_o) \\ N(z_o) \end{bmatrix}\right) = \{0\}; \quad \ker\left(\begin{bmatrix} M(z_o) \\ R(z_o) \end{bmatrix}\right) = \{0\}.$$

◇

6. THE PARALLEL AND TANDEM CONNECTIONS

In this section we study the parallel connection of systems (S) and (S') and we recall, from [7], the results that we already found for the tandem connection.

As we said in sect. 5, it is sufficient to investigate wether the matrices Sp(z) and $\mathcal{S}$p(z) have the same output zeros. This can be made by using the ideas of the previous sections, with $L = Sp$, $\mathcal{L} = \mathcal{S}p$. Of course, we assume that systems (S) and (S') are both z_o-canonical at a given point z_o. In order to simplify the notations, we assume (without any real restriction) that $z_o = 0$.

Now we describe:

METHOD 1. Let $X(z)$ have block components $x(z)$, $x'(z)$, $u(z)$ and satisfy

$$Sp(z)X(z) = z^r\begin{pmatrix}\varphi\\ \varphi'\\ \psi\end{pmatrix}$$

Rows 1, 3 and rows 2, 3 of this equality give:

$$\left\{\begin{array}{l}\left[\begin{array}{c|cc}A & 0 & -B\\ \hline C & H & D+K\end{array}\right]\begin{pmatrix}x\\ x'\\ u\end{pmatrix} = z^r\begin{pmatrix}\varphi\\ \psi\end{pmatrix}\\[2ex] \left[\begin{array}{c|cc}F & -G & 0\\ \hline H & D+K & C\end{array}\right]\begin{pmatrix}x'\\ u\\ x\end{pmatrix} = z^{\tau}\begin{pmatrix}\varphi'\\ \psi\end{pmatrix}\end{array}\right.$$

Lemma 4.3 applied to both the above systems give:

$$\left\{\begin{array}{l}\left\{CA^{-1}[0,B] + [H,D+K]\right\}\left\{\begin{pmatrix}x'\\ u\end{pmatrix} + z^r\begin{pmatrix}v'\\ v\end{pmatrix}\right\} = z^r\psi_1\\[2ex] \left\{HF^{-1}[G,0] + [D+K,C]\right\}\left\{\begin{pmatrix}u\\ x\end{pmatrix} \quad z^r\begin{pmatrix}\omega\\ \omega'\end{pmatrix}\right\} = z^r\psi_2\end{array}\right. \tag{1}$$

so that, by introducing the coprime factorizations of T(z), T'(z),

$$Q\{H(x'+z^r v') + K(u+z^r v) - z^r\psi_1\} + M(u+z^r v) = 0$$

$$N(u+z^r\omega) + R\{D(u+z^r\omega) + C(x+z^r\omega') - z^r\psi_2\} = 0.$$

Hence,

$$\left[\begin{array}{cc|c}Q & 0 & -M\\ 0 & R & -N\\ \hline I & I & 0\end{array}\right]\begin{pmatrix}-(Hx'+Ku)\\ -(Cx+Du)\\ u\end{pmatrix} = z^r\begin{pmatrix}Q(Hv'+Kv-\psi_1) + MV\\ R(D\omega+C\omega'-\psi_2) + N\omega\\ -\psi\end{pmatrix} \tag{2}$$

i.e. $\mathcal{X}(z)$, the vector $\mathcal{X}$ to which $\mathcal{S}p(z)$ is applied in (2), is a linear function of X(z). This at once shows property 2iii, while properties 1, 2 i are obvious from (2). To see that property 2ii holds, let us assume that $\mathcal{X}(0) = 0$. Then, $u(0) = 0$ and $H(0)x'(0) = 0$, $C(0)x(0) = 0$. The systems being observable, we get $x(0) = 0$, $x'(0) = 0$ (Lemma 4.4).

Now we describe METHOD 2. Let $x(z)$, $x'(z)$, $u(z)$ be the block componets of a vector $X(z)$ such that $Sp(z)X(z)$ has a zero of order r. Then, we have the following implications, still depending on the Lemma 4.3.

$$\left\{\begin{array}{l}\left[\begin{array}{cc|c} Q & 0 & -M \\ \hline I & I & 0\end{array}\right]\begin{pmatrix} x \\ x' \\ u\end{pmatrix} = z^r\begin{pmatrix}\varphi \\ \gamma\end{pmatrix} \\ \left[\begin{array}{cc|c} R & -N & 0 \\ \hline I & 0 & I\end{array}\right]\begin{pmatrix} x' \\ x \\ u\end{pmatrix} = z^r\begin{pmatrix}\varphi' \\ \gamma\end{pmatrix}\end{array}\right. \Rightarrow \left\{\begin{array}{l}\left\{Q^{-1}[0,M] + [I,0]\right\}\left\{\begin{pmatrix} x' \\ u\end{pmatrix} + z^r\begin{pmatrix} v' \\ v\end{pmatrix}\right\} = z^r\gamma_1 \\ \left\{R^{-1}[N,0] + [0,I]\right\}\left\{\begin{pmatrix} u \\ x\end{pmatrix} + z^r\begin{pmatrix}\omega \\ \omega'\end{pmatrix}\right\} = z^r\gamma_2\end{array}\right. \quad (3)$$

$$\Rightarrow \left\{\begin{array}{l}[I, CA^{-1}B+D]\left\{\begin{pmatrix} x' \\ u\end{pmatrix} + z^r\begin{pmatrix} v' \\ v\end{pmatrix}\right\} = z^r\gamma_1 \\ [HF^{-1}G+K, I]\left\{\begin{pmatrix} u \\ x\end{pmatrix} + z^r\begin{pmatrix}\omega \\ \omega'\end{pmatrix}\right\} = z^r\gamma_2\end{array}\right.$$

Now we note that $CA^{-1}B(u+z^r v)$, $HF^{-1}G(u+z^r\ )$ are bounded, so that - Lemma 4.4- $A^{-1}B(u+z^r v)$, $F^{-1}G(u+z^r\ )$ are bounded. Hence,

$$\begin{bmatrix} A & 0 & -B \\ 0 & F & -G \\ C & H & D+K\end{bmatrix}\begin{pmatrix} A^{-1}B(u+z^r v) \\ F^{-1}G(u+z^r\omega) \\ u\end{pmatrix} = z^r\begin{pmatrix} Bv \\ G\omega \\ \tilde{\gamma}\end{pmatrix} \quad (4)$$

for some function $\tilde{\varphi}$. Hence, METHOD 2 is the one which assigns to $X(z)$ the functione $X(z)$ which is multiplied by $Sp(z)$ in (4). Let us note explicitly that $X(z)$ is not uniquely determinede, since v and ω are note unique.

We show that properties 1 and 2 holds for METHOD 2. Property 1 is clear from (4). Let us show property 2iii. Let $X_i(z)$ be given, $X_i(z) = \mathrm{col}(x_i, x'_i, u_i)$ and let $X_i(z)$ be one of the associated functions,

$$X_i = \begin{pmatrix} A^{-1}B(u_i+z^r v_i) \\ F^{-1}G(u_i+z^r\ _i) \\ u_i\end{pmatrix}.$$

Let us consider the function $\hat{X}(z) = \sum\alpha_i X_i(z)$ whose block components are respectively: $A^{-1}B\sum\alpha_i(u_i+z^r v_i)$; $F^{-1}G\sum\alpha_i(u_i+z^r\omega_i)$; $\sum\alpha_i u_i$.

We can assert that $\hat{X}(z)$ is one of the functions that METHOD 2 associates to $\hat{X}(z) = \sum\alpha_i X_i(z)$ if we can show that relation (3) holds for $\hat{X}(z)$, with some functions $\hat{v}$, $\hat{v}'$, $\hat{\omega}$, $\hat{\omega}'$. But, let us write relation (3) for each index i. It is easily seen that

$$\left\{Q^{-1}[0,M] + [I,0]\right\}\left\{\sum\alpha_i\begin{pmatrix} x'_i \\ u_i\end{pmatrix} + z^r\sum\alpha_i\begin{pmatrix} v'_i \\ v_i\end{pmatrix}\right\} = z^r\sum\alpha_i\varphi_{1_i}$$

$$\left\{R^{-1}[N,0] + [0,I]\right\}\left\{\sum\alpha_i\begin{pmatrix} u^i \\ x_i\end{pmatrix} + z^r\sum\alpha_i\begin{pmatrix}\omega_i \\ \omega'_i\end{pmatrix}\right\} = z^r\sum\alpha_i\varphi_{2_i}$$

This proves property 2iii. Now, we show propertis 2i, 2ii. Let us assume that $X(0) = 0$ so that $\lim Q^{-1}M(u + z^r v) = 0$, $\lim R^{-1}N(u + z^r \omega) = 0$ and $u(0) = (0)$. Then, $A^{-1}B(u + z^r v)$, $F^{-1}G(u + z^r \omega)$ are bounded (see (3) and Lemma 4.4). Now, it is easy to see that

$$S(z)\begin{pmatrix} A^{-1}B(u + z^r v) \\ u + z^r v \end{pmatrix}$$

is zero for $z = 0$ so that, from Lemma 4.4., $\lim A^{-1}B(u + z^r v) = 0$ and $\lim F^{-1}G(u + z^r \omega) = 0$.

The above considerations and property 2iii imply property 2i.

Now, it is easy to see property 2ii. Let us assume that $X(0) = 0$, i.e. that $u(0) = 0$ and $\lim A^{-1}B(u(z) + z^r v(z)) = 0$, $\lim F^{-1}G(u(z) + z^r \omega(z)) = 0$ and, from (3), $x'(0) = 0$, $x(0) = 0$. This, and property 2iii proves 2i.

As a consequence, the matrices $Sp(z)$, $\mathcal{S}p(z)$ have the same zeros. Let us observe now that when the last block component of $X(z)$ (of $\mathcal{X}(z)$) is zero, then the corresponding component of $\mathcal{X}(z)$ (of $X(z)$) is zero. Hence, the matrices $Sp(z)$, $\mathcal{S}p(z)$ have the same output zeros, with equal multiplicities. Theorem 5.1 implies now that the parallel connection is 0-observable (observable) when

$$\operatorname{Ker}\begin{bmatrix} Q(0) \\ R(0) \end{bmatrix} = \{0\} \quad \left(\operatorname{Ker}\begin{bmatrix} Q(z) \\ R(z) \end{bmatrix} = \{0\} \quad \forall z \in \mathbb{C}\right).$$

Now, we observe that the parallel connection is spectrally controllable when the matrix $\tilde{S}p^*(z)$ has no output zeros, i.e. when the matrix

$$\begin{bmatrix} \tilde{Q}^* & 0 & -M^* \\ 0 & R^* & -N^* \\ I & I & 0 \end{bmatrix} = \tilde{\mathcal{S}}p^*(z)$$

has no output zero. Since the matrices $\mathcal{S}p(z)$, $\tilde{S}p^*(z)$ have the same structure, we have:

Theorem 1. The parallel composition of systems (S), (S') is spectrally controllable if and only if
$\operatorname{rank}[\tilde{Q}(z), \tilde{R}(z)] = n \quad \forall z \in \mathbb{C}$.

◇

An analogous method was used in [7] to study the tandem connection of systems (S), (S'). We quote the main result of that paper:

Theorem 2. The tandem connection of systems (S), (S') is:

observable iff: $\operatorname{Ker}\begin{bmatrix} Q(z) \\ N(z) \end{bmatrix} = \{0\} \quad \forall z \in \mathbb{C}$;

controllable iff: $\operatorname{rank}[R(z), M(z)] = n \quad \forall z \in \mathbb{C}$.

◇

7. THE FEEDBACK CONNECTION

In this section we study the properties of the feedback connection of systems (S), (S'). Of course, we assume the conditions of Theorem 3.1.

As in the previous section, we consider the observability property first and we show that the matrices $Sf(z)$ and $\mathcal{S}f(z)$ have the same output zeros. In order to do this, we must describe METHOD 1 and METHOD 2 with respect to the matrices $L(z) = Sf(z)$, $\mathcal{L}(z) = \mathcal{S}f(z)$. So, let us assume $z_o = 0$ (to simplify the notation) and

$$Sf(z)X(z) = \begin{bmatrix} A & 0 & 0 & -B & -B \\ 0 & F & -G & 0 & 0 \\ C & 0 & -I & D & D \\ 0 & H & K & -I & 0 \\ C & 0 & 0 & D & D \end{bmatrix} \begin{pmatrix} x \\ x' \\ u' \\ y' \\ v \end{pmatrix} = z^r \begin{pmatrix} \varphi \\ \varphi' \\ \nu \\ \nu' \\ \psi \end{pmatrix} \qquad (1)$$

Then, in particular, from rows 1, 5 and 2, 4 we get (using Lemma 4.3)

$$\begin{cases} \begin{bmatrix} A & -B & -B \\ C & D & D \end{bmatrix} \begin{pmatrix} x \\ y' \\ v \end{pmatrix} = z^r \begin{pmatrix} \varphi \\ \psi \end{pmatrix} \\ \begin{bmatrix} F & -G & 0 \\ H & K & -I \end{bmatrix} \begin{pmatrix} x' \\ u' \\ y' \end{pmatrix} = z^r \begin{pmatrix} \varphi' \\ \nu' \end{pmatrix} \end{cases} \Rightarrow$$

$$\begin{cases} \{ CA^{-1}[B,B] + [D,D] \} \begin{pmatrix} y' + z^r \hat{y}' \\ v + z^r \hat{v} \end{pmatrix} = z^r \tilde{\psi} \\ \{ HF^{-1}[G,0] + [K,-I] \} \begin{pmatrix} u' + z^r \tilde{u}' \\ y' + z^r \hat{y}' \end{pmatrix} = z^r \tilde{\nu}' \end{cases} \Rightarrow \qquad (2)$$

$$\begin{cases} Q^{-1}M[y'+v+z^r(\tilde{y}'+\tilde{v})] = z^r \tilde{\psi} \\ R^{-1}N[u'+z^r\tilde{u}'] - y' + z^r\hat{y}' = z^r \tilde{\nu}' \end{cases} \Rightarrow \begin{cases} M(v+y'+z^r(\tilde{y}'+\tilde{v})) = z^r Q \tilde{\psi} \\ N(u'+z^r\tilde{u}') - R(y'+z^r\hat{y}') = z^r R \tilde{\nu} \end{cases}$$

Now, from the third line of (1),
$z^r \nu = (Cx + Dy' + Dv) - u' = z^r \psi - u'$, i.e. $u' = -z^r(\nu - \psi)$.
The relations that we found can be written as

$$\mathcal{S}f(z) \begin{pmatrix} 0 \\ y' \\ u' \\ y' \\ v \end{pmatrix} = \begin{bmatrix} Q & 0 & 0 & -M & -M \\ 0 & R & -N & 0 & 0 \\ I & 0 & -I & 0 & 0 \\ 0 & I & 0 & -I & 0 \\ I & 0 & 0 & 0 & 0 \end{bmatrix} \begin{pmatrix} 0 \\ y' \\ u' \\ y' \\ v \end{pmatrix} = z^r \begin{pmatrix} \tilde{y}'+\tilde{v}-Q\tilde{\psi} \\ N\tilde{u}'-R\hat{y}'-R\tilde{\nu} \\ \nu - \psi \\ 0 \\ 0 \end{pmatrix}$$

This suggests that METHOD 1 should associate to the vector $X(z)$ the vector which is multiplied by $\mathcal{S}f(z)$ in the previous relation. Let us denote it $\mathcal{X}(z)$. Then, clearly, the transformation $X \to \mathcal{X}$ is linear and properties 1, 2', 2"' are self-evident. Let us prove 2". It is sufficient to see that if $\mathcal{X}(0) = 0$ then $X(0) = 0$ (since the transformation $X \to \mathcal{X}$ is linear!) If $\mathcal{X}(0) = 0$ then $y'(0) = 0$, $u'(0) = 0$, $v(0) = 0$ so that $x(0) = 0$ and $x'(0) = 0$ from Lemma 4.4 applied to the systems in (2).

Now we describe METHOD 2. Let us assume that

$$\mathcal{S}f(z)\mathcal{X}(z) = \begin{bmatrix} Q & 0 & 0 & -M & -M \\ 0 & R & -N & 0 & 0 \\ I & 0 & -I & 0 & 0 \\ 0 & I & 0 & -I & 0 \\ I & 0 & 0 & 0 & 0 \end{bmatrix} \begin{pmatrix} x \\ x' \\ u' \\ y' \\ v \end{pmatrix} = z^r \begin{pmatrix} \varphi \\ \varphi' \\ \nu \\ \nu' \\ \gamma \end{pmatrix} \quad (3)$$

Still from Lemma 4.3 we get

$$\left\{\begin{array}{l} \begin{bmatrix} Q & -M & -M \\ I & 0 & 0 \end{bmatrix} \begin{pmatrix} x \\ y' \\ v \end{pmatrix} = z^r \begin{pmatrix} \varphi \\ \gamma \end{pmatrix} \\ \begin{bmatrix} R & -N & 0 \\ I & 0 & -I \end{bmatrix} \begin{pmatrix} x' \\ u' \\ y' \end{pmatrix} = z^r \begin{pmatrix} \varphi' \\ \nu' \end{pmatrix} \end{array}\right. \Rightarrow \left\{\begin{array}{l} Q^{-1}[M,M]\left\{\begin{pmatrix} y' \\ v \end{pmatrix} + z^r \begin{pmatrix} \tilde{y}' \\ \tilde{v} \end{pmatrix}\right\} = z^r \tilde{\varphi} \\ R^{-1}[N,-I]\left\{\begin{pmatrix} u' \\ y' \end{pmatrix} + z^r \begin{pmatrix} \hat{u}' \\ \hat{y}' \end{pmatrix}\right\} = z^r \tilde{\nu}' \end{array}\right. \Rightarrow \quad (4)$$

$$\left\{\begin{array}{l} (CA^{-1}B+D)(y' + z^r\tilde{y}' + v + z^r\tilde{v}) = z^r\check{\varphi} \\ (HF^{-1}G+K)(u' + z^r\tilde{u}') - y' - z^r\hat{y}' = z^r\check{\nu}' \end{array}\right.$$

i.e.

$$\begin{bmatrix} A & 0 & 0 & -B & -B \\ 0 & F & -G & 0 & 0 \\ C & 0 & -I & D & D \\ 0 & H & K & -I & 0 \\ C & 0 & 0 & D & D \end{bmatrix} \begin{pmatrix} A^{-1}B(y' + z^r\tilde{y}' + v + z^r\tilde{v}) \\ F^{-1}G(u' + z^r\tilde{u}') \\ u' \\ y' \\ v \end{pmatrix} = z^r \begin{pmatrix} B\check{y}'+B\tilde{v} \\ G\tilde{u}' \\ \check{\varphi} - D\check{y}' - D\tilde{v} - \gamma + \nu \\ \check{\nu}' + \hat{y}' - K\tilde{u}' \\ \check{\varphi}' - D\tilde{v}' - D\tilde{y}' \end{pmatrix} \quad (5)$$

The third line because, from (3), $x - u' = z^r\nu$ and $x = z^r\gamma$.

The vectors $A^{-1}B(y'+z^r\tilde{y}' + v + z^r\tilde{v})$ and $F^{-1}G(u' + z^r\tilde{u}')$ are bounded, from Lemma 4.4. Now, METHOD 2 is the one which assigns to $\mathcal{X}(z)$ the vector $X(z)$ which is multiplied by $Sf(z)$ in the above relation.Of course, it is not uniquely identified, since y', y',v,u' are not unique.

The above relation shows directly property (1) for METHOD 2. In order to show property 2 we first of all show 2iii. Let $X_i(z)$ be functions, $X_i(z)$ associated to $\mathcal{X}_i(z)$. This means that if x_i, x'_i, u'_i, v_i are the block components of $\mathcal{X}_i(z)$, then we can find vectors $\tilde{y}'_i$, $\hat{y}'_i$, $\tilde{v}_i$, $\check{u}'_i$, $\tilde{\gamma}_i$, $\check{\nu}_i$ such that (4) holds. Since conditions (4) are linear, we see, as in the previous section, that $\sum \alpha_i X_i(z)$ corresponds to $\sum \alpha_i \mathcal{X}_i(z)$. Now, it is easy to show properties 2', 2". Since we know that property 2"' holds, we can confine ourselves to show that

i. $\mathcal{X}(0) = 0$ implies that $X(0) = 0$ (property 2');
ii. $\mathcal{X}(0) \neq 0$ implies that $X(0) \neq 0$ (property 2").

Let us assume that $\mathcal{X}(0) = 0$. Then, from (5) both the following vectors are zero for $z = 0$:

$$S(z)\begin{pmatrix} A^{-1}B(y' + z^r\tilde{y}' + v + z^r\tilde{v}) \\ y' + z^r\tilde{y}' + v + z^r\tilde{v} \end{pmatrix} \qquad S'(z)\begin{pmatrix} F^{-1}G(u' + z^r\tilde{u}') \\ u' + z^r\tilde{u}' \end{pmatrix}$$

Since $\mathcal{X}(0) = 0$ implies that $y'(0)$, $v(0)$, $u'(0)$ are the zero vectors, we have, from Lemma 4.4, that the functions $A^{-1}B(y' + z^r\tilde{y}' + v + z^r\tilde{v})$, $F^{-1}G(u' + z^r\tilde{u}')$ (which are bounded: see above) are in fact zero for $z = 0$. Hence, $X(0) = 0$ and this shows property 2i.

In order to see property ii, it is sufficient to observe that it cannot be $u'(0) = 0$, $y'(0) = 0$, $v(0) = 0$ since in this case, from (4), $x(0) = 0$ and $x'(0) = 0$, i.e. $X(0) = 0$.

Now, the arguments of Sect. 4 show that $z_o = 0$ is a zero of the matrix $Sf(z)$ if and only if it is a zero of the matrix $\mathcal{S}f(z)$. Even more, $z_o=0$ is an output zero of $Sf(z)$ if and only if it is an output zero of $\mathcal{S}f(z)$, since $X(z)$ and $\mathcal{X}(z)$ have the same u-components. Of course, this argument can be repeated for each z_o and Theorem 5.1 follows.

Now, we examine the spectral controllability property for the feedback connection of systems (S), (S').

The input zeros of the matrices $Sf(z)$ and $\mathcal{S}f(z)$,

$$\tilde{\mathcal{S}}f(z) = \left[\begin{array}{cccc|c} \tilde{Q} & 0 & 0 & -I & -I \\ 0 & \tilde{R} & -I & 0 & 0 \\ \tilde{M} & 0 & -I & 0 & 0 \\ 0 & \tilde{N} & 0 & -I & 0 \\ \hline \tilde{M} & 0 & 0 & 0 & 0 \end{array}\right]$$

are, respectively, the output zeros of the matrices $S^*f(z)$, $\mathcal{S}^*f(z)$. We can exchange columns 3, 4 and rows 3, 4 of the matrices $S^*f(z)$, $\mathcal{S}^*f(z)$ to get

$$\hat{S}f(z) = \left[\begin{array}{cccc|c} A^* & 0 & 0 & C & 0 \\ 0 & F^* & H^* & 0 & 0 \\ -B^* & 0 & -I & D^* & D^* \\ 0 & -G^* & K^* & -I & 0 \\ \hline -B^* & 0 & 0 & D^* & D^* \end{array}\right] \qquad \hat{\mathcal{S}}f(z) = \left[\begin{array}{cccc|c} \tilde{Q}^* & 0 & 0 & \tilde{M}^* & \tilde{M}^* \\ 0 & \tilde{R}^* & \tilde{N}^* & 0 & 0 \\ -I & 0 & -I & 0 & 0 \\ 0 & -I & 0 & -I & 0 \\ \hline -I & 0 & 0 & 0 & 0 \end{array}\right]$$

The matrices $\hat{S}f(z)$, $\hat{\mathcal{S}}f(z)$ have the same structure as the matrices $Sf(z)$, $\mathcal{S}f(z)$. Moreover, the above described operations on rows and columns do not affect the output zeros. Hence

Theorem 1. The feedback composition of the systems (S), (S') is:

i. spectrally observable when

$$\ker \begin{bmatrix} M(z) \\ R(z) \end{bmatrix} = \{0\} \qquad \forall z \in \mathbb{C}.$$

2. spectrally controllable when $\operatorname{rank}[\tilde{R}(z), \tilde{M}(z)] = n \quad \forall z \in \mathbb{C}$.

◇

REMARK. We observe explicitly that the conditions for the spectral controllability of the feedback and tandem connections coincide. This fact can easily be proved in the state space setting, if $D(z)$ and $K(z)$ are identically zero.

An analogous observation could be show to hold for the observability property.

◇

7. CONCLUSIONS

An approach to the study of the zero structure of convolution systems has been presented, which can be compared with the one in [15]. This approach was used to derive necessary and sufficient conditions for the spectral controllability and observability of composite systems which rests on coprime fractional decompositions of the transfer functions. Clearly, other equivalent conditions can be obtained if we make use of different decompositions.

REFERENCES

1. Callier, F.M., Desoer, C.A., *Stabilization, Tracking and Disturbance Rejection in Multivariable Convolution Systems*, Annales de la Societé Mathématiques de Bruxelles, T. 94, I, 7-51, 1980.

2. Desoer, C.A., Liu, R.W., Murray, J. Saeks, R., *Feedback Systems Design: the Fraction Representation Approach to Analysis and Synthesis*, I.E.E.E. Transaction on Automatic Control, AC-25, 399-412, 1980.

3. Anderson, B.D.O., Gevers, M.R., *On Multivariable Pole-Zero Cancellations and the Stability of Feedback Systems*, IEEE Transactions on Circuits and Systems, CAS-28, 830-833, 1981.

4. Pandolfi, L. *The Transmission Zeros of Systems with Delays*, International J. Control, 36, 959-976, 1982.

5. Pandolfi, L., *On the Zeros of Transfer Functions of Systems with Delays*, Systems and Control Letters, 1, 204-210, 1981.

6. Pandolfi, L., *The Pole-Zero Structure of Linear Systems*, in "Theory of Distributed Parameter Systems and Applications", F. Kappel, W. Kunish, W. Schappacher Ed.s, Lecture Notes in Control and Information Science n. 54, Springer-Verlag, Berlin, 1983.

7. Pandolfi, L., *Tandem Connections of Systems with Delays*, to appear in "Theory of Distributed Parameter Systems and Applications, II".

8. Callier, F. Nahum, C.D., *Necessary and Sufficient Conditions for the Complete Controllability and Observability of Systems in Series Using the Coprime Factorization of Rational Matrix*, IEEE Transaction on Circuit and Systems, CAS-22, 90-95, 1975.

9. Wolovich, W.A., Hwang, H.L., *Composite Systems Controllability and Observability*, Automatica, 10, 209-212, 1974.

10. Chen, C.T., Desoer, C.A., *Controllability and Observability of Composite Systems*, IEEE Transactions on Automatic Control, AC-12, 402-409, 1967.

11. Wedderburn, J.H.M., *On Matrices whose Coefficients are Functions of a Single Variable*, Transactions of the American Mathematical Society, 10, 328-333, 1915.

12. Pandolfi, L., *On Feedback Stabilization of Functional Differential Equations*, Boll. Un. Matamatica Italiana, (IV), 10, 114-123, 1974.

13. Pandolfi, L., *Canonical Realizations of Systems with Delays*, SIAM J. Control Opt., 21, 598-612, 1983.

14. Gantmacher, F.R., *The Theory of Matrices*, Vol. 1, Chelsea Publ. Co., New York, 1960

15 Wyman, B.F., Sain, M.K.S., *The Zero Module and Essential Inverse Systems*, IEEE Transaction on Circuit and Systems, CAS-28, 112-126, 1981.

Frequency Domain and State Space Methods for Linear Systems
C.I. Byrnes and A. Lindquist (Editors)

RKH SPACE METHODS FOR APPROXIMATING THE COVARIANCE KERNELS OF A CLASS OF STOCHASTIC LINEAR HEREDITARY SYSTEMS, II

James A. Reneke[1]
Clemson University
Clemson, South Carolina

Roland B. Minton
Virginia Polytechnic Institute and State University
Blacksburg, Virginia

Reproducing kernel Hilbert space methods have produced useful approximations for estimation and control problems of deterministic linear hereditary systems. By obtaining explicit Parzen type RKH space representations of stochastic processes governed by linear hereditary systems as spaces of Hellinger integrable functions these same approximation methods apply to the problem of finding the covariance kernel given the model. Some sample calculations will be presented.

INTRODUCTION

A convenient summary of the inner product space geometry of second order stochastic processes can be found in [3, Chapter 2]. Our concern is with the covariance kernel of such a process, the reproducing kernel of a Hilbert space representation of the process [5]. A recent collection of papers [9] covering the last two decades illustrates the utility of RKH space methods for computing various functionals of a process related to problems of estimation and detection.

The covariance kernel also plays a central role in the design of controls for linear finite dimensional state space systems [8]. In this work the diagonal of the kernel function, the variance of the system response to a general noise process, is most important and can be found by solving a matrix Riccati equation. However, for hereditary systems, systems modelled by delay differential equations for instance, the covariance kernel has not been so useful. One difficulty is that infinite dimensional state space methods lead to an operator Riccati equation for the variance of the states [2] rather than the variance of the process.

The approach of this paper to the problem of estimating the covariance kernel of the response of an hereditary system to a general noise process input is in the spirit of Parzen's original idea. We extend the method presented in [7] for scalar systems to obtain an explicit description of the reproducing kernel Hilbert space representing the process as a space of Hellinger integrable functions [4]. Convergence of a class of finite dimensional approximations to the covariance kernel is obtained using the approximation theory developed in [6].

SOME BACKGROUND NOTATION

We are concerned with second order processes $\{X(t),\ t \text{ in } S\}$ with values in R^d, i.e., the components of the d×d matrix $E(X(t)X'(t))$ are finite for each t. Let $\langle\cdot,\cdot\rangle$ denote the usual inner product for R^d and $|\cdot|$ the inner product norm.

Parzen's <u>covariance kernel</u> R is defined by $R(s,t) = E(X(s)X'(t))$, for each s and t in S. One has immediately that $\Sigma_{p,q=0}^{n}\langle c_p, R(s_p,s_q)c_q\rangle \geq 0$ for each sequence $\{s_p\}_{p=0}^{n}$ in S and sequence $\{c_p\}_{p=0}^{n}$ in R^d. Aronszajn's classic result [1] tells us that there is a complete Hilbert space G for which R is a reproducing kernel, i.e., denoting the inner product by $Q(\cdot,\cdot)$ we have

i) $R(\ ,t)c$ is in G for each t in S and c in R^d and

ii) $Q(f, R(\ ,t)c) = \langle f(t),c\rangle$ for each f in G, t in S, and c in R^d.

It is appropriate to think of $\{G,Q\}$ as a representation of $\{X(t),\ t \text{ in } S\}$ since, for each c in R^d, the function ψ_c from $L_2(X(t), t \text{ in } S)$ into $\{G,Q\}$ defined by $\psi_c(u) = E(Xu')c$ maps X(t) to $R(\ ,t)c$, for each t in S. However, obtaining $\{G,Q\}$ explicitly is basic to making use of this representation of the process.

Consider for instance the process given by the real valued stochastic differential equation $dX = \alpha X(t)dt + dW$ where W is the standard Wiener process. The covariance kernel for $\{X(t),\ t \geq 0\}$ is given by $R(s,t) = (1/\alpha)\exp(\alpha \max(t,s))\cdot \sinh(\alpha \min(t,s))$, for s,t in $S = [0,\infty)$. The corresponding RKH space G is the space of all functions f which are 0 at 0, absolutely continuous on compact subintervals of S, and such that $\int_0^\infty |f'(t) - \alpha f(t)|^2\,dt < \infty$. The inner product Q is given by

$Q(f,g) = \int_0^\infty (f'(t) - \alpha f(t)(g'(t) - \alpha g(t))dt$ for each f and g in G.

This elementary result can be generalized to include delay systems. However, Parzen's ideas apply to more complicated systems which we will attempt to encompass by introducing an abstract formulation. Assume $S = [-r,\infty)$ where $r \geq 0$ and k is an increasing function from S to the numbers with $k(-r)=0$. Let G_0 denote the space of continuous R^d valued functions on S and G_H the subset of G_0 to which f belongs only in case $f(-r) = 0$ and f is Hellinger integrable with respect to k on compact subintervals of S [4,6], i.e., for each compact subinterval [u,v] of S there is a number b such that $\Sigma_s |df|^2/dk = \Sigma_{p=1}^{m} |f(s_p)-f(s_{p-1})|^2/(k(s_p)-k(s_{p-1})) \leq b$ for each partition $\{s_p\}_{p=0}^{m}$ of [u,v]. The Hellinger integral $\int_u^v |df|^2/dk$ is the least such number b.

Note that elements of G_H are absolutely continuous on compact subintervals of of S. Furthermore, if each of f and g is in G_H and [u,v] is a compact subinterval then the integral $\int_u^v \langle df, dg\rangle / dk$ exists as a limit through refinement of partitions of $[u,v]$ of approximating sums of the form $\Sigma_s \langle df, dg\rangle/dk = \Sigma_{p=1}^{m} \langle f(s_p)-f(s_{p-1}), g(s_p)-g(s_{p-1})\rangle / \left(k(s_p)-k(s_{p-1})\right)$.

For each u in S and f in G_0 let $N_u(f) = \underset{x \le u}{\text{l.u.b.}} |f(x)|$ and

$$[P_u f](x) = \begin{cases} f(x) & x \le u \\ f(u) & u \le x. \end{cases}$$

Let $\mathcal{B}$ denote the class of linear transformations of G_0 to which B belongs only in case

i) $[Bf](u) = 0$ for each f in G_0 and $u < 0$ and

ii) for each compact subinterval S_0 of $\overline{S}$ there is a number c such that $|[Bf](v) - [Bf](u)| \le c \int_u^v N_x(f) dk(x)$, for each f in G_0 and subinterval $[u,v]$ of S_0.

Note that the formal integral operator

$$[Bf](u) = \begin{cases} 0 & u \le 0 \\ \alpha \int_0^u f(s) dk(s) + \beta \int_0^u f(s-r) dk(s) & 0 \le s, \end{cases}$$

where α and β are d×d matrices, restricted to G_0 is in $\mathcal{B}$.

Let $\mathcal{A}$ denote the space of linear transformations of G_0 to which A belongs only in case A - I is in $\mathcal{B}$, where I also denotes the identity transformation on G_0. If B is in $\mathcal{B}$ then I-B is a reversible function from G_0 onto G_0 and $(I-B)^{-1}$ is in $\mathcal{A}$. If A is in $\mathcal{A}$ then A is a reversible function from G_0 onto G_0 and $I-A^{-1}$ is in $\mathcal{B}$.

Consider the process

$$X(t) = \begin{cases} Z(t) & t \le 0 \\ Z(t) + [BX](t) & 0 \le t \end{cases}$$

$$= [AZ](t) ,$$

where B is in $\mathcal{B}$, $A = (I-B)^{-1}$, and Z is a mean-zero process in R^d with independent increments, and $E(dZ(dZ)') = (dk)I$, I the d×d identity. We assume that $Z(-r) = 0$. In the next section we want to obtain explicitly an RKH space which represents the process $\{X(t),\ t \text{ in } S\}$ in the sense of Parzen.

AN RKH SPACE PARAMETERIZATION OF R.

We begin with a generalization of the earlier elementary result for the process given by $X(t) = W(t) + \alpha\int_0^t X(s)ds$. Let G_∞ denote the subset of G_H to which f belongs only in case $\int_{-r}^{\infty} |df|^2/dk < \infty$ and Q_∞ the inner product for G_∞ given by $Q_\infty(f,g) = \int_{-r}^{\infty}\langle df,dg\rangle/dk$. Recall that elements of G_H are 0 at -r.

<u>Theorem</u>. If $R(\ ,t)c$ is in G_H for each t in S and c in R^d then the process $\{X(t), t \text{ in } S\}$ given above can be represented in the sense of Parzen by the RKH space $\{G,Q\}$ where

1) G is the subspace of G_H to which f belongs only in case $(I-B)f$ is in G_∞ and

2) Q is the inner product for G given by $Q(f,g) = Q_\infty\big((I-B)f, (I-B)g\big)$ for each (f,g) in $G\times G$.

<u>Indication of Proof</u>. As an indication that everything works, notice that there is a function M from S S to the d d matrices such that $X(t) = [AZ)(t) = \int_{-r}^{t} M(t,s)dZ(s)$ for each t in S and

$$E\big(dZ(u,v)([AZ](t))'\big) = \begin{cases} 0 & \text{if } t \le u \le v \\ \int_u^v M'(t,s)dk(s) & \text{if } u \le v \le t. \end{cases}$$

Also if $s \le t$ and c in R^d then

$$\begin{aligned} [(I-B)R(\ ,s)c](t) &= [(I-B)E(X(\cdot)\ X'(s))c](t) \\ &= [E(Z(\cdot)\ X'(s))c](t) \\ &= [E(Z(\cdot)\ X'(s))c](s) \\ &= [(I-B)R(\ ,s)c](s)\ , \end{aligned}$$

i.e., $(I-B)R(\ ,s)c$ is in G_∞ or $R(\ ,s)c$ is in G for all s in S and c in R^d.

Furthermore, if c_1 and c_2 are d-vectors,

$$\begin{aligned} \langle c_1, R(s,t)c_2\rangle &= \langle c_1, E(X(s)X'(t))c_2\rangle \\ &= \langle c_1, E\Big(\Big(\int_{-r}^{s} M(s,u)dZ(u)\Big)\Big(\int_{-r}^{t} M(t,v)dZ(v)\Big)'\Big)c_2\rangle \\ &= \langle c_1, \int_{-r}^{s} M(s,u)M'(t,u)dk(u)c_2\rangle \\ &= \int^{s} \langle M'(s,u)c\ ,M'(t,u)c\ \rangle dk(u) \end{aligned}$$

$$= \int_{-r}^{\infty} \langle dE\big(Z(\cdot)X'(s)\big)c_1, dE\big(Z(\cdot)X'(t)\big)c_2\rangle / dk$$

$$= Q_\infty\big((I-B)R(\ ,s)c_1,\ (I-B)R(\ ,t)c_2\big)$$

$$= Q\big(R(\ ,s)c_1,\ R(\ ,t)c_2\big).$$

Let K denote the function from $S \times S$ to the $d \times d$ matrices given by

$$K(s,t) = \begin{cases} k(s + r)I & s \le t \\ k(t + r)I & t \le s. \end{cases}$$

Note that K is the reproducing kernel for $\{G_\infty, Q_\infty\}$.

<u>Theorem</u>. For $s \le t$, $R(s,t) = [A(P_tA)^*K(\ ,t)](s)$.

<u>Proof</u>. Since P_tA is a continuous linear transformationof $\{G_\infty, Q_\infty\}$, we have the adjoint $(P_tA)^*$ exists. Also $AK(\ ,s)$ is in G for each s in S. Hence, for arbitrary d-vectors c_1 and c_2 and $s \le t$ in S,

$$\langle [AK(\ ,s)](t)c_1, c_2\rangle = Q\big(AK(\ ,s)c_1,\ R(\ ,t)c_2\big)$$

$$= \int_{-r}^{t} \langle dK(\ ,s)c_1, d(I-B)R(\ ,t)c_2\rangle / dk$$

$$= \langle c_1, [(I-B)R(\ ,t)](s)c_2\rangle\ .$$

On the other hand,

$$\langle [AK(\ ,s)](t)c_1, c_2\rangle = Q_\infty\big(AK(\ ,s)c_1,\ K(\ ,t)c_2\big)$$

$$= Q_\infty\big(K(\ ,s)c_1,\ (P_tA)^*\ K(\ ,t)c_2\big)$$

$$= \langle c_1, [(P_tA)^*\ K(\ ,t)](s)c_2\rangle\ .$$

Since c_1 and c_2 are arbitrary, $P_s(I-B)\ R(\ ,t) = P_s(P_tA)^*\ K(\ ,t)$. But then $P_sAP_s(I-B)R(\ ,t) = P_sR(\ ,t) = P_sA(P^tA)^*\ K(\ ,t)$ or $R(s,t) = [A(P_tA)^*\ K(\ ,t)](s)$.

We might note this can also be written $\langle R(s,t)c_1, c_2\rangle = Q_\infty\big(K(\ ,s)c_1,\ A(P_tA)^* \cdot K(\ ,t)c_2\big) = Q_\infty\big((P_sA)^*K(\ ,s)c_1,\ (P_tA)^*\ K(\ ,t)c_2\big)$. If we introduce the matrix L defined by $L(s,t) = \big[(P_tA)^*\ K(\ ,t)\big](s) = \big[P_t(P_tA)^*\ K(\ ,t)\big](s)$, then $L(s,t) = L(t,t)$ when $s \ge t$, $L(\ ,t)c_i$ is in G_∞ for $i=1,2$, and $\langle R(s,t)c_1, c_2\rangle = Q_\infty\big(L(\ ,s)c_1,\ L(\ ,t)c_2\big)$.

NUMERICAL METHODS.

For the purpose of illustration assume that everything is scalar valued, i.e., $d = 1$, and $k = I$ the identity on S. Computing [diag R](t), for a given t, splits into two parts. We are concerned first with function evaluations $L(s,t) = [AK(\ ,s)](t)$ and then with the integral $Q_\infty(L(\ ,t),L(\ ,t)) = \int_{-r}^{t} |dL(\ ,t)|^2/dI$. Working backward, observe that the approximating sums $\Sigma_{p=1}^{n} |L(s_p,t) - L(s_{p-1},t)|^2/(s_p-s_{p-1})$ for the intergral are nondecreasing with respect to refinement [4].

If $s_p-s_{p-1} = \delta > 0$ for $p = 1, 2, \ldots, n$, ε is a positive number, and $L(\ ,t)$ is an approximation for $L(\ ,t)$ such that $|L(s_p,t) - L(s_p,t)| < \varepsilon$ for $p = 1,2,\ldots,n$, then $|\Sigma_{p=1}^{n} |L(s_p,t) - L(s_{p-1},t)|^2/\delta - \Sigma_{p=1}^{n} |L(s_p,t) - L(s_{p-1},t)|^2/\delta| < 4\varepsilon[2N_t(L(\ ,t)+\varepsilon]/\delta$.

Thus, assuming the operator A is given formally by an integro-differential equation, we can compute for a fixed partition $\{s_p\}_{p=0}^{n}$ of $[-r,t]$ the approximating sum $\Sigma_s |dL(\ ,t)|^2/dI$ to any desired accuracy. By refining in turn the partitions s we obtain [diag R](t).

The method was implemented for the class of scalar systems of the form

$$h(t) = f(t) + \alpha\int_0^t h(u)du + \beta\int_0^t h(u-r_1)du + \gamma\int_0^t h(u-r_2)du.$$

Experiments with $t = 2$, $r_1 = 1/2$, $r_2 = 1$, and various values of α, β, and γ, $-2 \leq \alpha, \beta, \gamma \leq 2$, the algorithm always converged with err = 0.001, i.e., halving the step size for the function evaluation and the integral quadrature produced no change in in the first four decimals of [diag R](t).

CONCLUSIONS

RKH space methods lead to an explicit representation of the convariance kernel for a process governed by an integro-differential equation in terms of solutions that can be obtained by standard numerical methods. Although the presentation of these methods for the sake of clarity was in terms of scalar systems, the methods are readily extended to finite dimensional vector systems. The relative straightforward nature of these methods leads one to anticipate their incorporation into a general methodology for system identification and control design for hereditary systems.

BIBLIOGRAPHY

1. Aronszajn, N., Theory of reproducing kernels, Am. Math. Soc. Trans. 68 (1950), pp. 337-404.

2. Curtain, R. F., and Pritchard, A. J., Infinite dimensional linear systems theory, Springer-Verlag, New York, 1978.

3. Grenander, U., Abstract inference, Wiley-Interscience, New York, 1981.

4. Mac Nerney, J. S., Hellinger integrals in inner product spaces, J. Elisha Mitchell Sci. Soc. 76 (1960), pp. 252-273.

5. Parzen, E., An approach to time series analysis, Am. Math. Stat. 32 (1961), pp. 951-989.

6. Reneke, J. A., and Fennell, R. E., An identification problem for hereditary systems, Int. J. App. Anal. 11 (1981), pp. 167-183.

7. Reneke, J. A., and Minton, R. B., RKH space methods for approximating the covariance kernels of a class of stochastic linear hereditary systems, Estimation and Control of Distributed Systems, Proceedings of the University of Arkansas Annual Lecture Series in the Mathematical Sciences, Z. Jackiewicz, ed.

8. Russell, D. L., Mathematics of finite-dimensional control systems, Marcel Dekker Inc., New York, 1979.

9. Weinert, H. L., Reproducing kernel Hilbert spaces: Applications in statistical signal processing, Huchinson Ross Publishing Company, Stroudsburg, Pennsylvania, 1982.

[1] This research was supported in part by the Air Force Office of Scientific Research Under Grant AFOSR-84-0236

Frequency Domain and State Space Methods for Linear Systems
C.I. Byrnes and A. Lindquist (Editors)

STABILITY CRITERIA FOR TIME-VARYING DELAY SYSTEMS

Xu Daoyi

Mianyang Teachers' College
Mianyang, Sichuan
China

In this paper, we derive algebraic stability criteria of time-varying systems with time delays. First, some criteria for single linear delay systems are presented. Then these results are extended to the composite linear delay systems and nonlinear delay systems. The criteria are with simple forms, easily checked and applicable to systems with time-varying delays without any restriction.

INTRODUCTION

Stability testing or stabilization of control systems with time delays is not as easy, even if they are linear, as with systems without delays. For this reason, many people have been studying, and many papers have been published (Nazaroff,1973; Kwon et al,1977; Tokumaru et al,1974; Araki and Mori,1979), but they require such troublesome tasks as choosing many arbitrary parameters or solving nonlinear matrix equations. For constant linear delay systems Amemiya (1983) and Mori et al (1981b) gave the simple sufficient conditions for stability. For time-varying linear delay systems the stability criteria are few so far. In this paper we present some algebraic stability criteria for single, composite (Mori,1981b) and nonlinear time-varying delay systems with the aid of M-matrices and the norm of matrices. The criteria with simple forms are easily checked and applicable to systems with time-varying delays without any restriction.

NOTATION AND PRELIMINARIES

Throughout this paper $\lambda_m(A)$ and $\lambda_M(A)$ represent respectively the minimum and maximum real part of all eigenvalues of an $n\times n$ matrix A. $\|\bullet\|$ denotes Euclidean norm of a vector $X=(x_1,x_2,\dots,x_n)^T$ or a matrix $A=(a_{ij})_{n\times n}$, i.e.

$$\|X\|=(\sum_{i=1}^{n}x_i^2)^{\frac{1}{2}},\quad \|A\|=(\sum_{i=1}^{n}\sum_{j=1}^{n}a_{ij}^2)^{\frac{1}{2}}.$$

Consider first the general delay system

$$\dot{X}(t)=f(t,X(t),X(t-r(t))) \tag{1}$$

where f is smooth enough to guarantee the existence of solutions of (1) for $t\geq t_0$, $f(t,0,0)=0$, $X(t)$, $r(t)$ belong to R^n, and

$$X(t)=(x_1(t),\dots,x_n(t))^T,\ X(t-r(t))=(x_1(t-r_1(t)),\dots,x_n(t-r_n(t)))^T.$$

Definition. The equilibrium state of system (1) is exponentially stable in the large if there exist two positive constant π, $\varkappa$, which are independent of the initial conditions (t_0, X_0), such that

$$\|X\| \le \pi \|X_0\| \exp(-\varkappa(t-t_0)) \quad t \ge t_0 \quad X_0 = X(t_0). \tag{2}$$

Assume that all delays are sectionally smooth and bounded, i.e. there is a positive constant r_0 such that

$$0 \le r_i(t) \le r_0 \quad (i=1,2,\ldots,n).$$

Applying the comparison principle in Ladde (1976) and following the proof of Theorem in Xu's paper (1984), we have the following lemma.

Lemma 1. Let $V \in C(J \times S_\rho, R_+^m)$, in which $J=[\theta, +\infty)$, θ is a number or the symbol $-\infty$, and $S_\rho = \{X \in R^n: \|X\| < \rho\}$. And let $V(t,X)$ be locally Lipschitzian in X, $V(t,0)=0$ and there be positive numbers a, η_1, η_2 such that

$$\eta_1 \|X\|^a \le \sum_{i=1}^{m} d_i V_i(t,X) \le \eta_2 \|X\|^a \quad (d_i > 0,\ i=1,\ldots,m) \tag{3}$$

If the upper right derivative $D^+_{(1)}V(t,X)$ of $V(t,X)$ along the solution of the system (1) satisfies

$$D^+_{(1)}V(t,X) \le W(t,V,\bar{V})$$

where $\bar{V} = \sup_{\theta \in [-r_0, 0]} V(t, X(t+\theta))$, $W \in C(J \times R_+^m \times \mathfrak{Z}_+^m, R_+^m)$, in which $\mathfrak{Z}^n = C([-r_0, 0], R^n)$. And if $W(t,V,\bar{V})$ is quasimonotone nondecreasing in V and nondecreasing in $\bar{V}$ for $t \in R_+$ and $W(t,0,0)=0$. Then exponential stability in the large of the equilibrium state of the system

$$Y = W(t,Y,\bar{Y})$$

implies exponential stability in the large of the equilibrium state of the system (1).

Lemma 2 (Berman and Hershkowitz, 1983). If A is M-matrix, then there is a positive definite diagonal matrix D such that $AD+DA^T$ is positive definite.

SINGLE LINEAR DELAY SYSTEMS

Consider the linear time-varying delay system

$$\dot{X}(t) = A(t)X(t) + B(t)X(t-r(t)) \tag{4}$$

where $A(t) = (a_{ij}(t))$, $B(t) = (b_{ij}(t))$ are $n \times n$ continuous function matrices, $0 \le r_i(t) \le r_0$ (r_0 is constant).
Let matrices $A = (a_{ij})$, $B = (b_{ij})$ be defined as

$$a_{ii} = \sup_t a_{ii}(t) < 0, \qquad a_{ij} \ge \sup_t |a_{ij}(t)| \qquad (i \ne j)$$

$$b_{ij} \ge \sup_t |b_{ij}(t)| \qquad (i,j=1,2,\ldots,n).$$

We have

Theorem 1. The equilibrium state of the system (4) is exponentially stable in the large if the matrix $(-A-B)$ is an M-matrix, i.e. all of its leading principal minors are positive.

Proof. Let a vecter Lyapunov function for the system (1) be

$V=(V_1,\ldots,V_n)^T$, $V_i=|x_i(t)|$, and designate $|\bar{x}_i| = \max\limits_{r\in[0,r_0]} |x_i(t-r)|$. Calculate the function $D^+_{(4)}V_i$ as

$$D^+_{(4)}V_i = D^+_{(4)}|x_i(t)| = (\mathrm{sgn}\,x_i(t+0))\dot{x}_i(t)$$

$$= \mathrm{sgn}\,x_i(t+0)\Big(\sum_{j=1}^n a_{ij}(t)x_j(t) + \sum_{j=1}^n b_{ij}(t)x_j(t-r_j(t))\Big)$$

$$\le \sum_{j=1}^n a_{ij}|x_j| + \sum_{i=1}^n b_{ij}|\bar{x}_j| = \sum_{j=1}^n a_{ij}V_j + \sum_{j=1}^n b_{ij}\bar{V}_j .$$

That is,

$$D^+_{(4)}V \le AV + B\bar{V}.$$

From Theorem 2 in Tokumaru et al (1975), the equilibrium state of the system

$$\dot{U} = AU + B\bar{U}$$

is exponentially stable in the large. By Lemma 1, Theorem 1 goes true.

Substituting $x(t)=\exp(-st)w(t)$ ($s>0$ constant) into (4), we obtain easily the following corollary

<u>Corollary 1</u>. If the matrix $(-A-sI-\exp(sr_0)B)$ (I is a unit matrix) is an M-matrix for a certain positive constant $s\in R$, then every solution of (4) has decaying rate s.

<u>remark</u>. If $B(t)=0$ in (4), we get the main result in Mori et al (1981a). If $A(t)$, $B(t)$ are constant matrices, we get some results in Araki and Mori (1979) and Amemiya (1983).

<u>Theorem 2</u>. Suppose there exist positive numbers c, d such that

$$\lambda_M\big(\tfrac{1}{2}(A^T(t)+A(t))\big) \le -c, \quad \|B(t)\| \le d.$$

If $c>d$, the equilibrium state of the system (4) is exponentially stable in the large.

<u>Proof</u>. Let $V=X^TX=\sum_{i=1}^n x_i^2$, $\bar{V}=\sum_{i=1}^n |\bar{x}_i|^2$. Calculate the function $\dot{V}_{(4)}$ as

$$\dot{V}_{(4)} = \dot{X}^TX + X^T\dot{X}$$

$$= X^T(A^T(t)+A(t))X + X^T(t-r(t))B^T(t)X + X^TB(t)X(t-r(t))$$

$$\le -2cX^TX + \sum_{i=1}^n\sum_{j=1}^n 2|b_{ij}(t)|\,|\bar{x}_i|\,|\bar{x}_j|$$

Let $|\bar{X}| = (|\bar{x}_1|,|\bar{x}_2|,\ldots,|\bar{x}_n|)^T$, $|B(t)| = (|b_{ij}(t)|)$. And note that $\lambda_M(A) \le \|A\|$ for arbitrary matrx A, we have

$$\dot{V}_{(4)} \le -2cX^TX + |\bar{X}|^T(|B^T(t)| + |B(t)|)|\bar{X}|$$

$$\le -2cX^TX + \lambda_M(|B^T(t)| + |B(t)|)|\bar{X}|^T|\bar{X}|$$

$$\le -2cV + 2\|B(t)\|\bar{V} \le -2cV + 2d\bar{V}.$$

The rest can be proved as the proof of Theorem 1.

COMPOSITE LINEAR DELAY SYSTEMS

Now we consider the composite system made up of m subsystems described by

$$\dot{X}_i=A_i(t)X_i+\sum_{j=1}^{N}A_{ij}(t)X_j+\sum_{j=1}^{N}B_{ij}(t)X_j(t-r_j(t))$$
$$i=1,2,\ldots,N, \tag{5}$$

where $X_i\in R^{n_i}$, $\sum_{i=1}^{N}n_i=n$, and $A_{ij}(t)$, $B_{ij}(t)$ and $A_i(t)$ are $n_i\times n_j$ and $n_i\times n_i$ continuous function matrices respectively.

Theorem 3. If there are $n_i\times n_i$ differentiable function matrices $P_i(t)$ $(i=1,2,\ldots,N)$ such that

$$\alpha\leqslant\lambda_m(P_i(t))\leqslant\lambda_M(P_i(t))\leqslant\beta \quad (\forall t\geq t_0,\ i=1,2,\ldots,N) \tag{6}$$

and

$$\lambda_M[A_i^T(t)P_i(t)+P_i(t)A_i\ (t)+\dot{P}_i(t)]\leqslant-2\delta_i<0 \tag{7}$$

where α, β, δ_i $(i=1,\ldots,N)$ are positive constant numbers. And if

$$2\sum_{j=1}^{N}X_i^TP_i(t)A_{ij}(t)X_j+2\sum_{j=1}^{N}X_i^TP_i(t)B_{ij}(t)X_j(t-r_j(t))\leqslant$$
$$\leqslant\sum_{j=1}^{N}d_{ij}X_j^TX_j+\sum_{j=1}^{N}e_{ij}\overline{X}_j^T\overline{X}_j \tag{8}$$

where d_{ij}, e_{ij} are real nonnegative numbers. Let

$$\Delta=\mathrm{diag}(2\alpha\delta_i/\beta),\ D=(-d_{ij}),\ E=(-e_{ij}).$$

If $(\Delta+D+E)$ is an M-matrix, then the equilibrium state of system (5) is exponentially stable in the large.

Proof. Let $V=(V_1,V_2,\ldots,V_N)$, $V_i=X^TP_i(t)X$ be a vector Lyapunov function for system (5). It is well known that

$$\lambda_m[P_i(t)]X_i^TX_i\leqslant X_i^TP_i(t)X_i\leqslant\lambda_M[P_i(t)]X_i^TX_i$$

By the condition (6), the condition (3) in Lemma 1 is satisfied. Calculate the functuon $\dot{V}_i$ as

$$\begin{aligned}\dot{V}_i&=\dot{X}_i^TP_i(t)X_i+X_i^T\dot{P}_i(t)X_i+X_i^TP_i(t)\dot{X}_i\\
&=X_i^T[A_i^T(t)P_i(t)+P_i(t)A_i(t)+\dot{P}_i(t)]X_i+\\
&\quad+2\sum_{j=1}^{N}X_i^TP_i(t)A_{ij}(t)X_j+2\sum_{j=1}^{N}X_i^TP_i(t)B_{ij}(t)X_j(t-r_j(t))\\
&\leqslant-2\delta_iX_i^TX_i+\sum_{j=1}^{N}d_{ij}X_j^TX_j+\sum_{j=1}^{N}e_{ij}\overline{X}_j^T\overline{X}_j\\
&\leqslant(-2\delta_i/\beta)X_i^TP_i(t)X_i+\sum_{j=1}^{N}(d_{ij}/\alpha)X_j^TP_j(t)X_j+\sum_{j=1}^{N}(e_{ij}/\alpha)\overline{X}_j^TP_j(t)\overline{X}_j\\
&=-[(2\delta_i/\beta)V_i-\sum_{j=1}^{N}(d_{ij}/\alpha)V_j-\sum_{j=1}^{N}(e_{ij}/\alpha)\overline{V}_j]\end{aligned}$$

(where $\bar{V}_j=\bar{X}_j^T P_j(t)\bar{X}_j$), that is

$$\dot{V}\leq-(\Delta^*+D^*)V-E^*\bar{V}$$

where $\Delta^*=\mathrm{diag}(2\delta_i/\beta)$, $D^*=(-d_{ij}/\alpha)$, $E^*=(-e_{ij}/\alpha)$. It is evident that all leading principal minors of $(\Delta^*+D^*+E^*)$ and those of $(\Delta+D+E)$ have like signs. So $(\Delta^*+D^*+E^*)$ is an M-matrix and the equilibrium state of the system

$$U=-(\Delta^*+D^*)U-E^*\bar{U}$$

is exponentially stable in the large (Tokumaru et al, 1975). The rest is proved as Theorem 1.

In Theorem 3, taking $P_i(t)=I_i$ (unit matrix) and imitating the proof of Theorem 2, we obtain easily the following corollary

Corollary 2. Suppose there exist positive numbers c,d such that for all $t\in J$ and $i=1,\ldots,N$,

$$\lambda_M(\tfrac{1}{2}(A_i^T(t)+A_i(t)))\leq-c,\quad \sum_{j=1}^{N}\|B_{ij}(t)\|+\sum_{j=1}^{N}\|A_{ij}(t)\|\leq d.$$

If c>d, then the equilibrium state of the system (5) is exponentially stable in the large.

A CLASS OF NONLINEAR DELAY SYSTEM

Consider the nonlinear system with time delays

$$\dot{x}_i=\sum_{j=1}^{n}a_{ij}(t,x)f_j(x_j)+\sum_{j=1}^{n}b_{ij}(t,x)f_j(x_j(t-r_{ij})) \qquad (9)$$

$$i=1,2,\ldots,n,$$

where $a_{ij}(t,X)$, $b_{ij}(t,X)$, $f_j(x_j)$ are smooth enough to guarantee the existence of solutions of (9), r_{ij} are constant. Let there be constant numbers a_{ij}, N such that

$$a_{ii}(t,X)\leq a_{ii}<0,\quad |a_{ij}(t,X)|\leq a_{ij}\quad (i\neq j)$$
$$|b_{ij}(t,X)|\leq N\quad (i,j=1,2,\ldots,n). \qquad (10)$$

And let $A=(a_{ij})$, we have

Theorem 4. If the system (9) satisfies

i) $f_i(x_i)x_i>0$ $(x_i\neq0)$, $f_i(0)=0$, and $\int_0^{\pm\infty}f_i(x_i)dx_i=+\infty$,

ii) there is a positive definite diagonal matrix $D=\mathrm{diag}(d_i)$ $(i=1,2,\ldots,n)$ such that

$$\lambda_M(AD+DA^T)+2nd_iN<0 \qquad (11)$$

then the equilibrium state of (9) is globally asymptotically stable.

Proof. Let a Lyapunov function of the system (9) be

$$V=\sum_{i=1}^{n}d_i\int_0^{x_i}f_i(x_i)dx_i+k\sum_{i=1}^{n}\sum_{j=1}^{n}\int_{-r_{ij}}^{0}f_j^2(x_j(t+\theta))d\theta$$

where k>0 is undetermined constant. It is easily proved that V is a positive definite function with infinitesimal upper bound (Liao,1982). Calculate the function $\dot{V}_{(9)}$ as

$$\dot{V}_{(9)}=\sum_{i=1}^{n} d_i f_i(x_i)\dot{x}_i+\sum_{i=1}^{n}\sum_{j=1}^{n} k\left[f_j^2(x_j)-f_j^2(x_j(t-r_{ij}))\right]$$

$$\leq \sum_{i=1}^{n}\sum_{j=1}^{n} d_i a_{ij}|f_i(x_i)||f_j(x_j)|+\sum_{i=1}^{n}\sum_{j=1}^{n} d_i N|f_i(x_i)||f_j(x_j(t-r_{ij}))|$$

$$+k\sum_{i=1}^{n}\sum_{j=1}^{n}\left[f_i^2(x_i)-f_j^2(x_j(t-r_{ij}))\right]$$

Let $F=(|f_1|,|f_2|,\dots,|f_n|)^T$, $\lambda=\lambda_M(AD+DA^T)$, we have

$$\sum_{i=1}^{n}\sum_{j=1}^{n} d_i a_{ij}|f_i||f_j| =\tfrac{1}{2}F^T(AD+DA^T)F\leq \tfrac{1}{2}\lambda F^T F =\tfrac{1}{2}\lambda\sum_{i=1}^{n} f_i^2$$

Hence,

$$\dot{V}_{(9)}\leq\sum_{i=1}^{n}\sum_{j=1}^{n}\left[(\lambda/_{2n}+k)f_i^2(x_i)+d_i N|f_i(x_i)||f_j(x_j(t-r_{ij}))|-kf_j^2(x_j(t-r_{ij}))\right]$$

The expressions in the above summation sign are quadratics. These quadratics are negative definite if

$$\lambda/_{2n}+k<0,\quad \tfrac{1}{4}(d_iN)^2<-k(\lambda/_{2n}+k).$$

Taking $k=-\lambda/_{4n}$, we have $d_iN<-\lambda/_{2n}$. That is

$$\lambda_M(AD+DA^T)+2nd_iN<0,$$

the proof is completed.

When $d_i=1$ $(i=1,2,\dots,n)$, from the properties of dominant diagonal matrices, we obtain easily the following corollary

Corollary 3. The equilibrium state of (9) is globally asymptotically stable if conditions (11) are changed as

$$\sum_{j=1}^{n}(a_{ij}+a_{ji}+2N)<0 \quad \text{for } i=1,2,\dots,n. \tag{12}$$

In the system (9), let $b_{ij}(t,X)=0$ $(i,j=1,2,\dots,n)$, we get the following nonlinear system

$$\dot{x} =\sum_{j=1}^{n} a_{ij}(t,X)f_j(x_j) \quad i=1,2,\dots,n. \tag{13}$$

Applying Theorem 4, the system (13) is globally asymptotically stable if $\lambda_M(AD+DA^T)<0$, i.e. the matrix $(-AD-DA^T)$ is positive definite. From Lemma 2 and the properties of M-matrices (Poole and Boullion, 1974), we have

Corollary 4. The system (13) is globally asymptotically stable if condition i) in Theorem 4 is satisfied and

$$(-1)^k\begin{vmatrix} a_{11} & a_{12} & \cdots & a_{1k}\\ a_{21} & a_{22} & \cdots & a_{2k}\\ \cdots & \cdots & \cdots & \cdots\\ a_{k1} & a_{k2} & \cdots & a_{kk}\end{vmatrix}>0 \quad k=1,2,\dots,n. \tag{14}$$

Example.

$$\begin{cases} \dot{x}_1=-2x_1^3+x_2\cos(e^t x_1) \\ \dot{x}_2=x_1^3\operatorname{arctg}t-x_2 \end{cases} \tag{15}$$

Let $f_1(x_1)=x_1^3$, $f_2(x_2)=x_2$, then $a_{11}=-2$, $a_{22}=-1$, $|\cos(e^t x_1)|\leq 1 \triangleq a_{12}$, $|\operatorname{arctg}t|\leq \pi/2 \triangleq a_{21}$. So the matrix $A=(a_{ij})$ satisfies conditions (14). From Corollary 4, the system (15) is globally asymptotically stable.

REFERENCES

[1] Amemiya, T., Delay-independent stabilization of linear systems, Int. J. Control, 37,5 (1983) 1071-1079.

[2] Araki, M. and Mori, T., Trans. Soc. Instrum. Control Engrs., 19 (1979) 267.

[3] Berman, A. and Hershkowitz, D., Matrix diagonal stability and its implications, SIAM J. Algebraic Discrete Methods 4 (1983) 377-382.

[4] Kwon, W.H. and Pearson, A.E., IEEE Trans. Automat. Contr. AC-22 (1977) 468.

[5] Ladde, G.S., Pacific J. Math. 66 (1976) 161-171.

[6] Liao, X., On the stability of certain class of non-linear defferential-difference equations with separated variables, J. Math. (in Chinese), 2 (1982) 75-80.

[7] Mori, T., Fukuma, N. and Kuwahara, M.,Simple stability criteria for single and composite linear systems with time delays, Int. J. Control, 34, 6 (1981b) 1175-1184.

[8] Mori, T., Fukuma, N. and Kuwahara, M., A stability criterion for linear time-varying systems, Int. J. Control, 34, 3 (1981a) 585-591.

[9] Nazaroff, G.J., IEEE Trans. Automat. Contr. AC-18 (1973) 317.

[10] Poole, G. and Boullion, T., A survey on M-matrices, SIAM Review, 16, 4 (1974) 419-427.

[11] Tokumaru, H., Adachi, N. and Amemiya, T., Macroscopic stability of interconnected systems, Proc. of IFAC 6th world Congress, paper 44.4 (1975).

[12] Xu Daoyi, Comparison theorems and vector V-functions for stability of discrete systems, Proc. of IFAC 9th world Congress, paper 09.1/c (1984).

7. LINEAR GEOMETRIC CONTROL THEORY

Frequency Domain and State Space Methods for Linear Systems
C.I. Byrnes and A. Lindquist (Editors)

STABILITY WITHOUT EIGENSPACES IN THE GEOMETRIC APPROACH : SOME NEW RESULTS

G.Basile (1), G.Marro (2) and A.Piazzi (2)

(1) Department of Electrical Engineering, University of Florida, Gainesville, Florida.
(2) Department of Electronics, Computers and Systems, University of Bologna, Bologna, Italy.

A new solution to the disturbance localization problem through dynamic compensation with the stability requirement for the overall system, is given. The main advantage of this solution is that it relies only on the fundamental algorithms of the geometric approach (those for the computation of the maximum controlled invariant contained in a given subspace and its dual) and a simple "a posteriori" stabilizability check, so that numerically critical computations of eigenspaces are avoided. A further advantage of the proposed solution compared to some already known in the literature is a possible order reduction in the actual synthesis procedure.

INTRODUCTION

Disturbance localization is one of the earliest problems to which the geometric approach has been successfully applied. Disturbance localization whith state-to-input algebraic feedback was first inverstigated independently by Basile and Marro (1969) and Wonham and Morse (1970). In the former contribution some preliminary results were given even in the case where an output-to-input dynamic feedback (through an observer) is used. This case was subsequently investigated and solved by Akashi and Imai (1979) and Schumacher (1979).

The above contributions deal only with the "structural" solution of the problem, without taking into account any stability requirement for the overall system (controlled plant plus regulator). The disturbance localization problem with stability using state-to-input algebraic feedback is considered by Wonham (1971). The more complete solution with stability when the state is assumed to be inaccessible and an observer is included in the regulator, has been presented independently by Willems and Commault (1981) and Imai and Akashi (1981), who obtained the same results by using rather different procedures. These consist in elegant and simple necessary and sufficient conditions, which also embody a constructive synthesis method.

However, the mentioned solutions with stability rely on the derivation of some subspaces of the system state space, the maximum internally-stabilizable controlled invariant contained in a given subspace and its dual, the minimum externally-stabilizable conditioned invariant containing a given subspace, whose computations are certainly more difficult than those of the corresponding elements without any stabilizabilty constraints. In fact, computation of the maximum controlled invariant contained in a given subspace and its dual is based on recursive algorithms which surely converge in few steps (fewer than the state dimension), so that the corresponding results are quite reliable, while in the solutions with stability already known in the literature it is necessary to include computations of eigenspaces, which may be very critical in ill-conditioning situations. This drawback has been pointed out by Golub and Wilkinson (1976) in connection with the similar problem of deriving the Jordan canonical form of a square matrix and, more specifically, by Moore and Laub (1978).

The possibility of avoiding eigenspace computations in dealing with analysis and synthesis problems with stability in the framework of the geometric approach is being investigated by the authors of this paper, who have already introduced some basic tools and procedures and obtained significant preliminary results. The fundamental concepts on which the new approach is based, are *self-boundedness*, as a further characterization of controlled invariants, and *self-hiddenness*, as the dual property for conditioned invariants, introduced by two of the authors of this paper (1982). These characterizations allow the handling of lattices instead of semilattices and imply some nice stabilizability properties, one of which was conjectured by two of the authors (1982) and proved by Schumacher (1983).

A deeper analysis of the stabilizability features of self-bounded controlled and self-hidden conditioned invariants has been subsequently developed and applied to the solution without eigenspaces of some basic problems of control theory, such as disturbance localization with stability by state-to-input algebraic feedback in the two cases where disturbance is assumed inaccessible or accessible for measurement, and asymptotic estimation of a linear function of the state, in the two cases where a purely dynamic and a non-purely dynamic observer is used (1984).

In the present paper the authors move a step further and give a solution without eigenspaces to the disturbance localization problem by dynamic compensation with stability, based on a new set of necessary and sufficient conditions which are also constructive for synthesis purposes.

GENERAL BACKGROUND AND NOTATION

Consider the standard linear dynamic system

$$\dot{x}(t) = A\,x(t) + B\,u(t) + D\,d(t) \quad , \tag{1a}$$

$$y(t) = C\,x(t) \quad , \tag{1b}$$

$$e(t) = E\,x(t) \quad , \tag{1c}$$

where $x \in R^n$ denotes the state, $u \in R^p$ the control input, $d \in R^r$ the disturbance input, $y \in R^q$ the observed output and $e \in R^s$ the controlled output.

Let $\mathcal{B} := \text{im}\,B$, $\mathcal{C} := \ker C$. Denote by $\mathcal{S}$ a general $(A,\mathcal{C})$-conditioned invariant subspace, i.e. any subspace such that $A(\mathcal{S} \cap \mathcal{C}) \subseteq \mathcal{S}$, by $\mathcal{V}$ a general $(A,\mathcal{B})$-controlled invariant subspace, i.e. any subspace such that $A\mathcal{V} \subseteq \mathcal{V} + \mathcal{B}$.

It is well-known that the set of all $(A,\mathcal{C})$-conditioned invariants containing a given subspace $\mathcal{D}$ is a lower semilattice with respect to $\subseteq$, $\cap$, whose infimum will be denoted in the sequel by $\text{MN}.\mathcal{S}(A,\mathcal{C},\mathcal{D})$. Similarly, the set of all $(A,\mathcal{B})$-controlled invariants contained in a given subspace $\mathcal{E}$ is an upper semilattice with respect to $\subseteq$, $+$, whose supremum will be denoted by $\text{MX}.\mathcal{V}(A,\mathcal{B},\mathcal{E})$. The algorithms for computation of $\text{MN}.\mathcal{S}(A,\mathcal{C},\mathcal{D})$ and $\text{MX}.\mathcal{V}(A,\mathcal{B},\mathcal{E})$, which undoubtedly are the fundamental tools of the geometric approach, were first introduced by two of the authors of this paper (1969). Let $\mathcal{D} := \text{im}\,D$ and $\mathcal{E} := \ker E$, and, in order to simplify subsequent notations, introduce the symbols

$$\mathcal{K} := \text{MN}.\mathcal{S}(A,\mathcal{C},\mathcal{D}) \quad , \tag{2}$$

$$\mathcal{J} := \text{MX}.\mathcal{V}(A,\mathcal{B},\mathcal{E}) \quad . \tag{3}$$

Particular subsets of the above semilattices are

$$\Psi \ := \ \{ \mathcal{S} \ : \ A(\mathcal{S} \cap \mathcal{C}) \subseteq \mathcal{S}, \ \mathcal{S} \supseteq \mathcal{D}, \ \mathcal{S} \subseteq \mathcal{K} + \mathcal{C} \} \quad , \tag{4}$$

$$\Phi \ := \ \{ \mathcal{V} \ : \ A\,\mathcal{V} \subseteq \mathcal{V} + \mathcal{B}, \ \mathcal{V} \subseteq \mathcal{E}, \ \mathcal{V} \supseteq \mathcal{J} \cap \mathcal{B} \} \quad . \tag{5}$$

which have been proved to be lattices with respect to $\subseteq$, $+$, $\cap$ by Basile and Marro (1982). Let us refer to them as the set of all "(A,$\mathcal{C}$)-conditioned invariants self-hidden with respect to $\mathcal{D}$" and the set of all "(A,$\mathcal{B}$)-controlled invariants self-bounded with respect to $\mathcal{E}$".

The supremum of Ψ and the infimum of Φ have been proved by the authors of this paper (1984) to be respectively

$$\mathcal{K} + \text{MX}.\mathcal{V}(A,\mathcal{D},\mathcal{C}) \quad , \tag{6}$$

and

$$\mathcal{J} \cap \text{MN}.\mathcal{S}(A,\mathcal{E},\mathcal{B}) \quad , \tag{7}$$

while obviously the infimum of Ψ is $\mathcal{K}$ and the supremum of Φ is $\mathcal{J}$.

Subspace (7) is the reachable set on $\mathcal{E}$ from the origin acting on the control u, then internally pole-assignable, and (6) is its dual, externally pole-assignable.

Assume

$$\mathcal{K} \subseteq \mathcal{E} \quad , \tag{8}$$

and

$$\mathcal{J} \supseteq \mathcal{D} \quad . \tag{9}$$

Sublattices of (4) and (5) which will be used in the sequel are

$$\Psi_1 \ := \ \{ \mathcal{S} \ : \ \mathcal{S} \in \Psi, \ \mathcal{S} \subseteq \mathcal{E} \} \quad , \tag{10}$$

and

$$\Phi_1 \ := \ \{ \mathcal{V} \ : \ \mathcal{V} \in \Phi, \ \mathcal{V} \supseteq \mathcal{D} \} \quad . \tag{11}$$

Conditions (8),(9) guarantee that (10),(11) are not empty. The infimum of Ψ_1 and the supremum of Φ_1 are still $\mathcal{K}$ and $\mathcal{J}$ respectively, while the supremum of Ψ_1 and the infimum of Φ_1 have been proved by the authors (1984) to be in this case

$$\mathcal{S}_M \ := \ \mathcal{K} + \text{MX}.\mathcal{V}(A,\mathcal{D},\mathcal{C} \cap \mathcal{E}) \quad , \tag{12}$$

and

$$\mathcal{V}_m \ := \ \mathcal{J} \cap \text{MN}.\mathcal{S}(A,\mathcal{E},\mathcal{B}+\mathcal{D}) \quad . \tag{13}$$

$\mathcal{S}_M$ will be referred to as the maximum (A,$\mathcal{C}$)-conditioned invariant self-hidden with respect to $\mathcal{D}$ and contained in $\mathcal{E}$, $\mathcal{V}_m$ as the minimum (A,$\mathcal{B}$)-controlled invariant self-bounded with respect to $\mathcal{E}$ and containing $\mathcal{D}$. They have fine stabilizability properties, as the following lemma points out.

> *Lemma 1.* If $\mathcal{S}_M$ is not externally-stabilizable, no other externally-stabilizable (A,$\mathcal{C}$)-conditioned invariants containing $\mathcal{D}$ and contained in $\mathcal{E}$ exist; if $\mathcal{V}_m$ is not internally-stabilizable, no other internally-stabilizable (A,$\mathcal{B}$)-controlled invariants contained in $\mathcal{E}$ and containing $\mathcal{D}$ exist.

The proof of the above lemma is well-known in the geometric approach literature.

It has been presented by Schumacher (1983) and, in slightly different terms, by the authors of this paper (1984).

PROBLEM FORMULATION

Still referring to system (1), let us consider the problem of making the controlled output e insensitive to the disturbance input d by means of a dynamic feedback loop from the observed output y to the control input u (see Fig.1). This problem is known in the literature as the "disturbance localization problem by dynamic compensation".

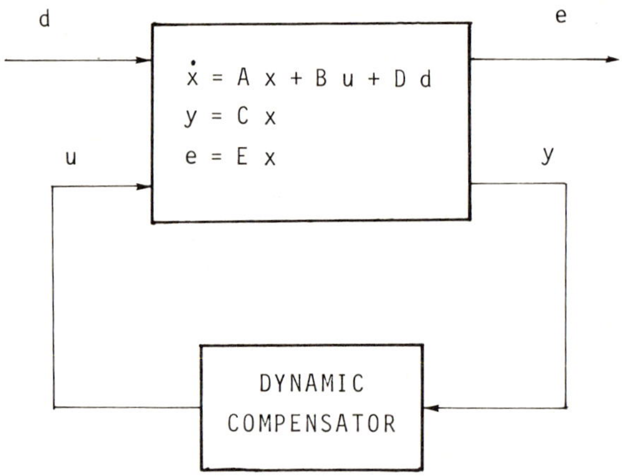

Figure 1
Controlled System and Dynamic Compensator

With the further requirement that the overall system (plant and compensator) must be stable, we have the "disturbance localization problem by dynamic compensation with stability".

For the problem with stability a customary assumption is that system (1) is stabilizable and detectable. This is understood here too. Necessary and sufficient conditions for a solution to the above problems are well-known and can be summarized as follows.

Proposition 1. The problem without stability has a solution if and only if there exist both an $(A,\mathcal{C})$-conditioned invariant $\mathcal{S}$ and an $(A,\mathcal{B})$-controlled invariant $\mathcal{V}$ such that

$$\mathcal{D} \subseteq \mathcal{S} \subseteq \mathcal{V} \subseteq \mathcal{E} \quad . \tag{14}$$

Another statement, equivalent to Proposition 1, is the following.

Proposition 2. The problem without stability has a solution if and only if

$$\mathcal{K} \subseteq \mathcal{J} \quad , \tag{15}$$

where $\mathcal{K}$ and $\mathcal{J}$ are defined as in (2),(3).

Condition (14) is non-constructive, while (15) is constructive, since the realization of an actual disturbance localization compensator is based on the availability of computational algorithms for the derivation of a conditioned invariant $\mathcal{S}$ and a controlled invariant $\mathcal{V}$ satisfying (14), as $\mathcal{K}$ and $\mathcal{J}$ are.

When the problem with stability is approached, the following propositions, which are very similar in their statements to the previous ones, must be considered.

Proposition 3. The problem with stability has a solution if and only if there exist both an externally-stabilizable (A,C)-conditioned invariant $\mathcal{S}^*$ and an internally-stabilizable (A,$\mathcal{B}$)-controlled invariant $\mathcal{V}^*$ such that

$$\mathcal{D} \subseteq \mathcal{S}^* \subseteq \mathcal{V}^* \subseteq \mathcal{E} \quad . \tag{16}$$

A statement equivalent to Proposition 3, but constructive, is the following.

Proposition 4. The problem with stability has a solution if and only if

$$\text{MN.ES.}\mathcal{S}(A,\mathcal{C},\mathcal{D}) \subseteq \text{MX.IS.}\mathcal{V}(A,\mathcal{B},\mathcal{E}) \quad , \tag{17}$$

where the notation in the left side member stands for "the minimum externally-stabilizable (A,C)-conditioned invariant containing $\mathcal{D}$" and that in the right side member for "the maximum internally-stabilizable (A,$\mathcal{B}$)-controlled invariant contained in $\mathcal{E}$".

The above propositions are directly or indirectly stated and proved in the existing literature. Algorithms for the computation of MN.ES.$\mathcal{S}(A,\mathcal{C},\mathcal{D})$ and MX.IS.$\mathcal{V}(A,\mathcal{B},\mathcal{E})$ have been presented by Marro (1973) and Wonham (1974).

THE MAIN RESULT

Note that "condition with stability" (17) implies "structural condition" (15). In the following theorem, which is the main result of this paper, two further conditions are joined to (15), so that a new set of necessary and sufficient conditions equivalent to (17) is obtained for the problem with stability. These conditions are constructive too, since the involved subspaces are completely algorithmically defined; moreover, they do not require any eigenspace computation.

Theorem 1. The problem with stability has a solution if and only if

i) $\quad \mathcal{K} \subseteq \mathcal{J} \quad ,$ (15)

ii) $\quad \mathcal{S}_m := \mathcal{S}_M \cap \mathcal{V}_m$ is externally-stabilizable , (18)

iii) $\quad \mathcal{V}_M := \mathcal{S}_M + \mathcal{V}_m$ is internally-stabilizable . (19)

In (18),(19) $\mathcal{S}_M$ and $\mathcal{V}_m$ are the same as defined in relations (12) and (13).

In the statement of Theorem 1 it is implicitly assumed that $\mathcal{S}_m$ is an (A,C)-conditioned invariant and $\mathcal{V}_M$ is an (A,$\mathcal{B}$)-controlled invariant. These properties are proved below in Lemma 2.

For a more compact notation, we introduce the new symbols

$$\mathcal{K}' := \text{MN.}\mathcal{S}(A,\mathcal{E},\mathcal{B}+\mathcal{D}) \quad , \tag{20}$$

$$\mathcal{J}' := \text{MX.}\mathcal{V}(A,\mathcal{D},\mathcal{C}\cap\mathcal{E}) \quad . \tag{21}$$

On the structural basic assumption (15), it can easily be shown that

$$K' \supseteq K \quad , \tag{22}$$

$$J' \subseteq J \quad , \tag{23}$$

$$S_m = (K + J') \cap K' = K + (J' \cap K') \quad , \tag{24}$$

$$V_M = (J \cap K') + J' = J \cap (K' + J') \quad . \tag{25}$$

Inclusion (22) is proved by the following steps:

$$K' \supseteq \text{MN}.S(A, C \cap E, B + D) \supseteq K \quad ,$$

where the first inclusion is a consequence of the algorithm for the computation of MN.$S(A,*,*)$, the second is due to K being contained in J, hence in E, so that it is an $(A, C \cap E)$-conditioned invariant. Inclusion (23) can be proved by a dual argument. Relations (24) derive from

$$S_m := (K + J') \cap (J \cap K') = (K \cap J + J') \cap K' = (K + J') \cap K' = K + J' \cap K' \quad .$$

where the first equality is obtained by virtue of the distributiveness of the intersection by J with respect to the sum $(K+J')$, due to (23), the second is due to (15) and the third to (22). Relations (25) are derived by a dual argument.

> *Lemma 2.* On the basic structural assumption (15) S_m, as defined in (18), is an (A,C)-conditioned invariant and V_M, as defined in (19), is an (A,B)-controlled invariant.

Proof. First let us consider S_m. From the first equality in (24) and (12) we obtain

$$S_m = S_M \cap K' \quad ,$$

where S_M has been defined as the maximum (A,C)-conditioned invariant self-hidden with respect to D contained in E. Both S_M and K' are $(A, C \cap E)$-conditioned invariants, hence S_m, as their intersection, is an $(A, C \cap E)$-conditioned invariant too, but, being contained in E, is also an (A,C)-conditioned invariant. The part of the statement concerning V_M can be proved by duality.

It is noted that S_m is an (A,C)-conditioned invariant self-hidden with respect to D and contained in E, hence it is an element of lattice Ψ_1 defined by (10) and, similarly, V_M is an (A,B)-controlled invariant self-bounded with respect to E and containing D, hence it is an element of lattice Φ_1 defined by (11).

It is interesting to note also that S_m is the maximum element of lattice Ψ_1 contained in V_m and V_M is the minimum element of lattice Φ_1 containing S_M.

PROOF OF THEOREM 1

Only if. Assume that the problem with stability has a solution, hence, by virtue of Proposition 3, there exist both an externally-stabilizable (A,C)-conditioned invariant S^* and an internally-stabilizable (A,B)-controlled invariant V^* which satisfy inclusions (16).

First we prove that in this case there also exists an externally-stabilizable (A,C)-conditioned invariant $\bar{S}^*$, self-hidden with respect to D, such that

$$S_m \subseteq \bar{S}^* \subseteq S_M \quad , \tag{26}$$

and an internally-stabilizable $(A,\mathcal{B})$-controlled invariant $\bar{\mathcal{V}}^*$, self-bounded with respect to $\mathcal{E}$, such that

$$\mathcal{V}_m \subseteq \bar{\mathcal{V}}^* \subseteq \mathcal{V}_M \quad , \tag{27}$$

which satisfy the similar inclusion conditions

$$\mathcal{D} \subseteq \bar{\mathcal{S}}^* \subseteq \bar{\mathcal{V}}^* \subseteq \mathcal{E} \quad . \tag{28}$$

To this end we define

$$\bar{\mathcal{S}}^* := (\mathcal{S}^* \cap \mathcal{S}_M) + \mathcal{S}_m \quad , \tag{29}$$

$$\bar{\mathcal{V}}^* := (\mathcal{V}^* + \mathcal{V}_m) \cap \mathcal{V}_M \quad . \tag{30}$$

Consider the right side member of (29); since, by virtue of Lemma 1, $\mathcal{S}_M$ is externally-stabilizable, $\mathcal{S}^* \cap \mathcal{S}_M$ is externally-stabilizable and self-hidden with respect to $\mathcal{D}$ (it belongs to Ψ_1), so that $\bar{\mathcal{S}}^*$, as the sum of two elements of lattice Ψ_1, one of which is externally-stabilizable, belongs to Ψ_1 and is externally-stabilizable. By a duality argument, it is easily proved that $\bar{\mathcal{V}}^*$, as defined by (30), is internally-stabilizable. Note that (26), (27) define sublattices of Ψ_1 and Φ_1 respectively. Inclusions (28) are easily checked as follows:

$$\mathcal{S}^* \cap \mathcal{S}_M \subseteq \mathcal{S}^* \subseteq \mathcal{V}^* \subseteq \mathcal{V}^* + \mathcal{V}_m \quad ,$$

$$\mathcal{S}^* \cap \mathcal{S}_M \subseteq \mathcal{S}_M \subseteq \mathcal{S}_M + \mathcal{V}_m = \mathcal{V}_M \quad ,$$

$$\mathcal{S}_m = \mathcal{S}_M \cap \mathcal{V}_m \subseteq \mathcal{V}_m \subseteq \mathcal{V}^* + \mathcal{V}_m \quad ,$$

$$\mathcal{S}_m = \mathcal{S}_M \cap \mathcal{V}_m \subseteq \mathcal{V}_m \subseteq \mathcal{V}_M \quad .$$

We now introduce a change of basis which will lead to the proof. This type of approach can be also used in order to point out connections between stabilizability features of self-bounded controlled invariants and related self-hidden conditioned invariants. Let $T = [T_1 T_2 T_3 T_4]$ be a nonsingular $n \times n$ matrix, with $\mathrm{im}\, T_1 = \mathcal{S}_m = \mathcal{S}_M \cap \mathcal{V}_m$, $\mathrm{im}\, [T_1 T_2] = \mathcal{V}_m$, $\mathrm{im}\, [T_1 T_3] = \mathcal{S}_M$, $\mathrm{im}\, [T_1 T_2 T_3] = \mathcal{V}_M = \mathcal{V}_m + \mathcal{S}_M$ and

$$\mathrm{im}\, [T_1 T_2 T_3] \supseteq \mathcal{K}' \quad . \tag{31}$$

The last condition is consistent with the previous ones because, as $\mathcal{V}_m = \mathcal{J} \cap \mathcal{K}'$, $\mathcal{V}_M = (\mathcal{J} \cap \mathcal{K}') + \mathcal{J}'$, we have

$$\mathrm{im}\, T_3 \subseteq \mathcal{J}' \subseteq \mathcal{J} \quad , \tag{32}$$

hence $\mathrm{im}\, T_3 \cap \mathcal{K}' = \mathcal{O}$.

In the new basis, matrices A, B and C are expressed as

$$A' = T^{-1} A T = \begin{bmatrix} A'_{11} & A'_{12} & A'_{13} & A'_{14} \\ A'_{21} & A'_{22} & 0 & A'_{24} \\ 0 & 0 & A'_{33} & A'_{34} \\ A'_{41} & A'_{42} & 0 & A'_{44} \end{bmatrix} \; ; \quad (33) \qquad B' = T^{-1} B = \begin{bmatrix} B'_1 \\ B'_2 \\ 0 \\ B'_4 \end{bmatrix} \; ; \tag{34}$$

$$C' = C T = \begin{bmatrix} C'_1 & C'_2 & 0 & C'_4 \end{bmatrix} \; ; \tag{35}$$

Condition (31) implies the particular structure of matrix B' (recall that $\mathcal{K}'\supseteq\mathcal{B}$) while (32) implies the particular structure of C' (recall that $\mathcal{J}'\subseteq\mathcal{C}$). As far as the structure of matrix A' is concerned, note that the zeros in the third row are due to the particular structure of B' and to $\mathcal{V}_m$ being an (A,$\mathcal{B}$)-controlled invariant, while the zeros in the third column are due to the particular structure of C' and to the fact that $\mathcal{S}_M$ is an (A,$\mathcal{C}$)-conditioned invariant.

If the structural zeros in A',B',C' are taken into account, from (26) and (27) it follows that basis matrices for all possible $\bar{\mathcal{S}}^*$ and $\bar{\mathcal{V}}^*$ can be expressed as

$$\bar{\mathcal{S}}^* = \text{im}\left(T\begin{bmatrix} I_1 & 0 \\ 0 & 0 \\ 0 & X_1 \\ 0 & 0 \end{bmatrix}\right) \quad ; \quad \bar{\mathcal{V}}^* = \text{im}\left(T\begin{bmatrix} I_1 & 0 & 0 \\ 0 & I_2 & 0 \\ 0 & 0 & X_2 \\ 0 & 0 & 0 \end{bmatrix}\right) ,$$

where I_1, I_2 are properly dimensioned identity matrices and X_1, X_2 are basis matrices of an externally stable and an internally stable $\underline{A}_{33}'$-invariant subspace respectively. These stability properties follow from $\bar{\mathcal{S}}^*$ being externally-stabilizable and $\bar{\mathcal{V}}^*$ internally-stabilizable.

Condition (28) clearly implies $\text{im}\, X_1 \subseteq \text{im}\, X_2$, so that A_{33}' must be a stable matrix, hence $\mathcal{S}_m$ is externally-stabilizable. In fact, the external stabilizability of $\mathcal{S}_m$ depends on A_{33}' being stable and $\mathcal{S}_M$ being externally-stabilizable (which is guaranteed by Lemma 1). Similarly, $\mathcal{V}_M$ is internally-stabilizable, A_{33}' being stable and $\mathcal{V}_m$ internally-stabilizable by virtue of Lemma 1.

If. Assume that (15), (18), (19) hold. Then there exist both an externally-stabilizable (A,$\mathcal{C}$)-conditioned invariant $\mathcal{S}^* := \mathcal{S}_m$ and an internally-stabilizable (A,$\mathcal{B}$)-controlled invariant $\mathcal{V}^* := \mathcal{V}_M$ such that (16) holds.

SOME FURTHER REMARKS AND CONCLUSIONS

In Theorem 1 the main result of this paper is stated in a concise and self-dual form. The following corollaries, reciprocally dual, also provide necessary and sufficient conditions without eigenspaces.

Corollary 1. The problem with stability has a solution if and only if

i) $\mathcal{K}\subseteq\mathcal{J}$, (15)

ii) $\mathcal{S}_m := \mathcal{S}_M \cap \mathcal{V}_m$ is externally-stabilizable , (18)

iii) $\mathcal{V}_m$ is internally-stabilizable . (36)

Corollary 2. The problem with stability has a solution if and only if

i) $\mathcal{K}\subseteq\mathcal{J}$, (15)

ii) $\mathcal{S}_M$ is externally-stabilizable , (37)

iii) $\mathcal{V}_M := \mathcal{S}_M + \mathcal{V}_m$ is internally-stabilizable . (19)

Proof.

Only if. The necessity of (15), (18) and (19) follows from Theorem 1, while the necessity of (36) and (37) follows from Lemma 1.

If. Use the same argument as in the proof of Theorem 1.

The results stated in this paper can be used in order to sketch a computational procedure which solves the disturbance localization problem with stability, i.e. derives a couple of subspaces S^*, V^* with the properties stated in Proposition 3:

a) check structural condition (15): if it is not satisfied, the problem has no solution;

b) check if S_M, as defined by (12), is externally-stabilizable and if V_m, as defined by (13), is internally-stabilizable (see ref. [13] for algorithms): even if one condition is not satisfied, the problem has no solution;

c) check if $S_M \subseteq V_m$: if so, it is possible to assume $S^* := S_M$ and $V^* := V_m$, if not, go to step e) if an increase of the dimensions of S^* is preferred to a reduction of the dimensions of V^*, to step d) in the opposite case;

d) check if S_m is externally-stabilizable: if not, the problem has no solution, if so, assume $S^* := S_m$ and $V^* := V_m$;

e) check if V_M is internally-stabilizable: if not, the problem has no solution, if so, assume $S^* := S_M$ and $V^* := V_M$.

In the above procedure there is an arbitrary choice which corresponds to different features of the compensator: step d), based on Corollary 1, in general allows the assignment of a greater number of poles in the controlled system, while step e), based on Corollary 2, may result in a reduction of the compensator state dimension. For instance it is possible to prove (but it is beyond the object of this paper) that with the latter synthesis method it is possible to reduce the state dimension of the compensator by $\dim S_M - \dim S_m$.

The research whose results have been in part reported above is being supported by M.P.I. - 40 % (Italian Education Ministry - Interuniversitary research projects).

REFERENCES

[1] Basile, G. and Marro, G., L'invarianza rispetto ai disturbi studiata nello spazio degli stati, Rendiconti della LXX Riunione Annuale AEI (1969), Rimini, Italy, Paper n. 1-4-01.

[2] Wonham, W.M., and Morse, A.S., Decoupling and pole assignment in linear multivariable systems: a geometric approach, SIAM J. Contr. 8 (1970), 1-18.

[3] Akashi, H. and Imai, H., Disturbance localization and output deadbeat control through an observer in discrete-time linear multivariable systems, IEEE Trans. Autom. Contr, AC-24 (1979), 621-627.

[4] Schumacher, J.M., (C,A)-invariant subspaces: some fact and uses, Vrije Universiteit, Amsterdam, Holland, Report No. 110 (july 1979).

[5] Wonham, W.M., Algebraic methods in linear multivariable control, in: Morse, A. (ed.), Sustem Structure (IEEE Cat. No. 71C61, New York, 1971).

[6] Willems, J.C. and Commault, C., Disturbance decoupling by measurement feedback with stability or pole placement, SIAM J. Contr. Optimiz. 19 (1981), 490-504.

[7] Imai, H. and Akashi, H., Disturbance localization and pole shifting by dynamic compensation, IEEE Trans. Aut. Contr. AC-26 (1981), 490-504.

[8] Golub, G.H. and Wilkinson, J.H., Ill-conditioned eigensystem and the computation of the Jordan canonical form, SIAM Review 18 (1976), 578-619.

[9] Moore, B.C. and Laub, A.J., Computation of supremal (A,B)-invariant and controllability subspaces, IEEE Trans. Aut. Contr. AC-23 (1978), 783-792.

[10] Basile, G. and Marro, G., Self-bounded controlled invariant subspaces: a straightforward approach to constrained controllability, J. Optimiz. Th. Applicat. 38 (1982), 71-81.

[11] Basile, G. and Marro, G., Dual lattice theorems in the geometric approach, J. Optimiz. Th. Applicat. 48 (1986).

[12] Schumacher, J.M., On a conjecture of Basile and Marro, J. Optimiz. Th. Applicat. 41 (1983), 371-377.

[13] Basile, G. and Marro, G., Self-bounded controlled invariants versus stabilizability, J. Optimiz. Th. Applicat. 48 (1986).

[14] Basile, G., Marro, G. and Piazzi, A., A new solution to the disturbance localization problem with stability and its dual, Proceedings of the 84 International AMSE Conference on Modelling and Simulation 1.2, Athens (1984), 19-27.

[15] Basile, G. and Marro, G., Controlled and conditioned invariant subspaces in linear system theory, J. Optimiz. Th. Applic. 3 (1969), 305-315.

[16] Marro, G., Controlled and conditioned invariance in the synthesis of unknown-input observers and inverse systems, Control and Cybernetics (Poland) 2 (1973). 81-98.

[17] Wonham, W.M., Linear multivariable control: a geometric approach (Springer-Verlag, New York, 1974).

Frequency Domain and State Space Methods for Linear Systems
C.I. Byrnes and A. Lindquist (Editors)

OBSERVABLE, UNOBSERVABLE SUBSPACES AND OBSERVABILITY OF AN UNKNOWN INPUT SYSTEM

Külmız Çevık İzzet Cem Göknar

Department of Circuits and Systems
Faculty of Electrical Engineering
Technical University of İstanbul
Ayazağa, Istanbul, Turkey

Though the relation between observability and the existence of Luenberger observers for linear, time invariant systems which are subject to only control inputs is well known; there exists no similar results for systems excited by unknown, unmeasurable disturbances. This work establishes the precise relation between extended observability and the existence of Luenberger observers for unknown input systems in terms of the newly introduced "known-initial state-observability". As a side result an easy method for computing (A,B)-invariant subspaces in kerC is developed.

I. INTRODUCTION

The problem of estimating the states of linear, time invariant, dynamical systems is considered in the presence of disturbances. After the original works in [1] and [2] this problem has been investigated by many researchers; see, for instance, [3], [4], [5] and references cited therein.

Instead of giving a new design algorithm, the purpose of the present paper is to introduce a new concept which provides the link between different types of estimation methods existing in literature.

The problems are treated in time domain using state space representation of systems and the geometric approach developed by Wonham [9]. Throughout the paper uppercase letters are used for linear mappings and their matrix representations and lowercase letters for vectors. Vector spaces and subspaces are denoted by bold face capitals. A' is used to denote the transpose of A. $\mathbf{S}^{\perp}$ stands for the orthogonal complement of **S**.

The symbols $\underline{I}(A;\ \mathbf{K})$, $\underline{V}(A,B;\mathbf{K})$, $\underline{R}(A,B;\mathbf{K})$ indicate respectively the classes of A-invariant, (A,B)-invariant and (A,B)-controllability subspaces contained in a given subspace **K**. Their supremal elements are denoted by $\mathbf{I}^*(A;\mathbf{K})$, $\mathbf{V}^*(A,B;\mathbf{K})$ and $\mathbf{Q}^*(A,B;\mathbf{K})$ respectively. Another class of subspaces that will be needed is the set of (C,A)-invariant subspaces containing a given subspace **K**, [2], which is denoted by $\underline{Q}(C,A;\mathbf{K})$. This set posseses the infimal element $\mathbf{Q}_*(C,A;\mathbf{K})$.

The model assumed for the plant is of the form

$$\frac{d}{dt}\,x(t) = Ax(t) + B\,u(t) + Eq(t) \tag{1.1a}$$

$$y(t) = Cx(t) \tag{1.1b}$$

$$z(t) = Dx(t) \tag{1.1c}$$

with $x\in\mathbf{X}$ representing the state, $u\in\mathbf{U}$ the control input, $q\in\mathbf{Q}$ the disturbance, $y\in\mathbf{Y}$ the measurements and $z\in\mathbf{Z}$ the outputs to be estimated. **X, Y, U, Q, Z** are finite dimensional inner product spaces having dimensions n, p, m, r and ℓ respectively.

A linear dynamical system which is driven by the known input u and the measurements y of (1.1) and giving the output $\hat{z}$ is said to be an observer for (1.1) if the estimation error defined by

$$e(t) = \hat{z}(t) - z(t) \tag{1.2}$$

satisfies

$$\lim_{t\to\infty} e(t) = 0 \text{ for all } u\in\mathbf{U},\ q\in\mathbf{Q},\ x(0)\in\mathbf{X}. \tag{1.3}$$

when (1.3) is satisfied $\hat{z}$ is called the estimate of z.

Basically there exists two different approaches to state estimation problems: the differentiator based observer of Basile and Marro [2] and the integrator based observer of Luenberger [1]. In the next section the precise relation between these two approaches, which are developed separately and independently in literature, will be established.

An observer based on differentiators has the general form :

$$\hat{z}(t) = \sum_{k=0}^{n-1} K_k \frac{d^k}{dt^k}\,y(t) + L_k \frac{d^k}{dt^k}\,u(t) \tag{1.4}$$

and an observer based on integrators can be described by

$$\frac{d}{dt} w(t) = N\, w(t) + M\, y(t) + G\, u(t) \tag{1.5a}$$

$$\hat{z}(t) = L\, w(t) + K\, y(t) \tag{1.5b}$$

where $w \in \mathbf{W}$ is the state of the observer.

An observer for which there is no feed through from the measurements y to the estimate $\hat{z}$ ($K=0$ in (1.5b)) is said to be a Kalman observer.

II. OBSERVABLE AND UNOBSERVABLE SUBSPACES OF THE STATE SPACE

Definition 2.1 : The largest subspace of the state space **X** on which the orthogonal projection of the state of (1.1) can be estimated is called the observable subspace and is denoted by **O**. The orthogonal complement $\mathbf{O}^{\perp}$ of **O** is defined to be the unobservable subspace.

Unlike systems whose inputs are all known functions of time, the observable subspace of (1.1) is different for observers based on integrators and for observers using differentiators.

In [2] and [4] necessary and sufficient conditions are given for the existence of observers of the forms (1.4) and (1.5) that estimate the function (1.1c). The following theorems about the observable subspace of (1.1) can be easily shown using these results.

Theorem 2.1 : The observable subspace of (1.1) for observers based on integrators is given by

$$\mathbf{O}_I = \mathrm{im}C' + \mathbf{R}^*(A',C';\ker E') \tag{2.1}$$ ■

Proof of theorem 2.1 is the same as that of theorem 2.4.

Theorem 2.2 : The observable subspace of (1.1) for observers based on differentiators is given by

$$\mathbf{O}_D = \mathbf{Q}_*(E',A';\mathrm{im}C')$$ ■

Proof : see [2].

We note that both $\mathbf{O}_I$ and $\mathbf{O}_D$ reduce to

$$\mathbf{O} = \mathbf{O}_I = \mathbf{O}_D = \langle A' | \mathrm{im}C' \rangle \overset{\Delta}{=} \sum_{k=0}^{n-1} (A')^k\, \mathrm{im}C' \tag{2.3}$$

the usual definition of observable subspace when $E=0$ ie., when there are no unknown inputs. The reason for this will be apparent when the precise relation between $\mathbf{O}_I$ and $\mathbf{O}_D$ is given.

In all observer design problems considered in the literature it is assumed that the initial state x(0) of (1.1) is unknown. In fact this is the only nontrivial situation when the inputs are known since, knowledge of x(0) is equivalent to knowing x(t) for all t>0 in this case. But when there are unknown inputs it makes sense to consider known-initial state-observer design or known-initial state-observability of (1.1). This observation motivates the theorem below which shows the possibility of perfect tracking of the to be estimated outputs.

<u>Theorem 2.3</u> : Assume that the initial state x(0) of (1.1) is given then there exists a dead-beat observer of the form (1.5) (i.e., an observer for which e(t)=0 for all t≥0) if and only if

$$\mathrm{im}D' \subset \mathrm{im}C' + \mathbf{V}^*(A',C';\ker E') \tag{2.4}$$

<u>Proof</u> : The theorem follows from the results in [3] or [4]. The desired dead-beat response can be obtained by starting the observer system with the initial condition w(0)=Vx(0), V being a matrix used in the construction of the observer system which satisfies $\mathrm{im}V' = \mathbf{V}^*(A',C';\ker E')$ ∎

<u>Theorem 2.4</u> : The observable subspace of (1.1) with x(0) known is given by

$$\mathbf{O}_K = \mathrm{im}C' + \mathbf{V}^*(A',C';\ker E') \tag{2.5a}$$

and the unobservable subspace

$$\mathbf{O}_K^{\perp} = \ker C \cap \mathbf{Q}_*(C,A;\mathrm{im}E) \tag{2.5b}$$

<u>Proof</u> : To see (2.5a), let D: $\mathbf{X}\rightarrow\mathbf{X}$ be the orthogonal projection operator on $\mathbf{S} = \mathrm{im}C' + \mathbf{V}^*(A',C';\ker E')$. Now, as imD'=imD=S the function Z=Dx can be estimated according to theorem 2.3. Furthermore $\mathbf{S}$ is the largest subspace having this property because $\mathbf{V}^*(A',C';\ker E')$ is the largest member of the family $\underline{V}(A',C';\ker E')$. (2.5b) follows by taking annihilators of subspaces and by making use of the duality between (A,B)-invariant and (C,A)-invariant subspaces [2]. ∎

The main result of this section which ties different types of observable subspaces together is given in the theorem below.

Theorem 2.5 : Let

$O_I \stackrel{\Delta}{=}$ Observable subspace of (1.1) for observers based on integrators.

$O_D \stackrel{\Delta}{=}$ Observable subspace of (1.1) for observers based on differentiators.

$O_K \stackrel{\Delta}{=}$ Observable subspace of (1.1) given the initial state x(0). and

$O_I^\perp$, $O_D^\perp$ and $O_K^\perp$ be the corresponding unobservable subspaces. Then

$$\text{(a) } O_I = O_D \cap O_K \qquad \text{(b) } O_I^\perp = O_D^\perp + O_K^\perp \qquad (2.6)$$

Proof : It suffices to prove only one of expressions (2.6), the other one follows by taking orthogonal complements of subspaces and recalling the identity $(S_1+S_2)^\perp = S_1^\perp \cap S_2^\perp$. From (2.1)

$$O_I = \text{im}C' + \mathbf{R}^*(A',C';\ker E')$$

Employing the basic relation

$$\mathbf{R}^*(A',C';\ker E') = \mathbf{V}^*(A',C';\ker E') \cap \mathbf{Q}_*(E',A';\text{im}C')$$

given in [7] and using the distributive rule we have

$$O_I = (\text{im}C' + \mathbf{V}^*(A',C';\ker E')) \cap \mathbf{Q}_*(E',A';\text{im}C')$$

which proves (2.6a) ■

Theorem 2.5 is also valid for Kalman observers. For a known input system $O_K=X$, $O_K^\perp=0$ so that O_I and O_D coincide as predicted by expression (2.6). The two types of estimation methods discussed in this section have been developed separately and independently in literature. The close relationship between the two methods, given by theorem 2.5, does not seem to be appreciated before. Of the two observer types, the differentiator based observer of Basile and Marro has the largest observable subspace but the use of differentiators has some obvious disadvantages. The integrator based observer of Luenberger on the other hand, has no such problem. It can provide an asymtotic, on-line identification of the to be estimated outputs. But these desirable properties are purchased at the price of increasing the unobservable subspace in fact, this price can be exactly seen in expression 2.6b.

Estimators that employ differentiators and integrators simultaneously could also be tried. In this case there might be a compromise between the number of differentiators used, and the dimension of the observable subspace of (1.1).

III. OBSERVABILITY OF AN UNKNOWN INPUT SYSTEM

The following trivial definition of observability is needed in the sequel.

Definition 3.1 : A system is said to be observable if its observable subspace is the whole state space or equivalently if its unobservable subspace is the zero subspace.

When the initial state x(0) is unknown, the discussion in the previous section naturally suggests two types of observability for an unknown input system. The system (1.1) may be called observable whenever

$$O_D^{\perp} = 0 \tag{3.1a}$$

or

$$O_I^{\perp} = 0 \tag{3.1b}$$

is satisfied.

The condition (3.1a) is known as "extended observability" in literature [6]. However the relation of this condition to the existence of Luenberger observers was not clearly understood; corollary 3.1 to theorem 2.5 immediately clarifies this relation. Moreover when (3.1b) holds the whole state vector can be estimated by an observer of the form (1.5) and the error poles can be arbitrarily assigned; such a spectral assignability property is not implied by (3.1a)

Corollary 3.1 : The system (1.1) is observable in the sense of (3.1b) if and only if it is

(i) known-initial state-observable
(ii) extended observable ■

The next two theorems give easily verifiable conditions for known-initial state-observability and observability in the sense of (3.1b)

Theorem 3.1 : The system (1.1) is known-initial state-observable ($\mathbf{O}_K=0$) if and only if

$$\ker C \cap \operatorname{im} E = 0 \qquad (3.2)$$

Proof : The condition $\ker C \cap \mathbf{Q}_*(C,A;\operatorname{im}E)=0$ is equivalent to (3.2) ■

The following lemma is needed to prove theorem 3.2

Lemma 3.1 : Assume $\ker C \cap \operatorname{im}E=0$. Let $\mathbf{S}$ be a subspace such that $\ker C \subset \mathbf{S}$ and $\mathbf{S} \oplus \operatorname{im}E=\mathbf{X}$. Let $P:\mathbf{X}\to\mathbf{X}$ be a projection on $\mathbf{S}$ along $\operatorname{im}E$. Then

$$\underline{V}(A,E;\ker C) = \underline{I}(PA;\ker C) \qquad (3.3)$$

and

$$\mathbf{V}^*(A,E;\ker C) = \bigcap_{k=1}^{n} \ker C(PA)^{k-1} \qquad (3.4)$$

Proof : Let $\mathbf{V}\in\underline{V}(A,E;\ker C)$. By definition $A\mathbf{V}\subset\mathbf{V}+\operatorname{im}E$, $\mathbf{V}\subset\ker C$ or $PA\mathbf{V}\subset P\mathbf{V}=\mathbf{V}\subset\ker C$ which means $\mathbf{V}\in\underline{I}(PA;\ker C)$ hence

$$\underline{V}(A,E;\ker C) \subset \underline{I}(PA;\ker C) \qquad (3.5)$$

For the reverse inclusion, let $\mathbf{V}\in\underline{I}(PA;\ker C)$ then

$$PA\mathbf{V}\subset\mathbf{V},\quad \mathbf{V}\subset\ker C$$

Operating with P^{-1} (the functional inverse of P)

$$P^{-1}(PA\mathbf{V}) = \operatorname{im}E+A\mathbf{V} \subset P^{-1}\mathbf{V} = \mathbf{V}+\operatorname{im}E \quad \text{or} \quad A\mathbf{V}\subset\mathbf{V}+\operatorname{im}E$$

Thus $\mathbf{V}\in\underline{V}(A,E;\ker C)$ and

$$\underline{I}(PA;\ker C) \subset \underline{V}(A,E;\ker C) \qquad (3.6)$$

From (3.5) and (3.6), (3.3) follows. (3.4) can also be proven by standard techniques ■

The result of lemma 3.1 is complementary to the work in [8] for the easy computation of maximal (A,E)-invariant subspace in kerC. In [8] a method has been given for the computation of $\mathbf{V}^*(A,E,\ker C)$ but this method can not be applied if

$$r = \text{No. of disturbances} < p = \text{No. of measurements}$$

whereas lemma 3.1 can be used exactly in this case since the assumption $\ker C\cap\operatorname{im}E=0$ generically holds when $r\leq p$.

Theorem 3.2 : The system (1.1) is observable in the sense of (3.1b) if and only if

(i) $\ker C \cap \operatorname{im} E = 0$

(ii) The pair (C,PA) is observable

where the projection P is defined as in lemma 3.1

Proof : As $O_D^{\perp} = V^*(A,E;\ker C)$, the theorem is an immediate consequence of corollary 3.1, theorem 3.1 and lemma 3.1 ∎

IV. CONCLUSION

A new type of observability is defined for unknown input systems which provides the relation between different state estimation methods proposed in literature. In view of the design procedures given in the references the results are completely constructive.

The work may be further extended to get similar results for detectable subspaces and detectability of an unknown input system.

In another direction, the problems could be investigated for a decomposed system which consists of separated algebraic and lossless parts. A complete set of results related to disturbance decoupling and unknown input observability of decomposed systems will be the subject of another publication.

REFERENCES

[1] D.G. Luenberger, Observers for multivariable systems, IEEE Trans. Automat. Contr. vol. AC.11, No.2 (1966) pp 190-197

[2] G. Basile and G. Marro, On the observability of linear time invariant systems with unknown inputs, J. Optimiz. Theory and Appl. vol. 3 No. 6 (1969) pp 410-415

[3] J.M. Schumacher, On the minimal stable observer problem, Int. J. Control vol. 32 No.1 (1980) pp 17-30

[4] J.C. Willems and C. Commault, Disturbance decoupling by measurement feedback with stability or pole placement, SIAM J. Contr. Optimiz. vol.19 No.4 (1981) pp 490-504

[5] J. Kurek, Observation of the state vector of linear multivariable systems with unknown inputs, Int. J. Control vol.36 No.3 (1982) pp 511-516

[6] B.P. Molinari, Extended controllability and observability for linear systems, IEEE Trans. Automat. Contr. vol.AC-21 (1976) pp 136-137

[7] A.S. Morse, Structural invariants of linear multivariable systems, SIAM J. Control vol.11 (1973) pp 446-465

[8] S.P. Bhattacharyya, On calculating maximal (A,B)-invariant subspaces, IEEE Trans. Automat. Contr. vol. AC-20 (1975) pp 264-265

[9] W.M. Wonham, Linear Multivariable Control: A Geometric Approach, 2nd ed., Springer-Verlag, New York, 1979

Frequency Domain and State Space Methods for Linear Systems
C.I. Byrnes and A. Lindquist (Editors)

DECOUPLING TECHNIQUES FOR DECENTRALIZED CONTROL SYSTEMS DESIGN

Sebastian Engell and Dieter Konik

Measurement & Control Group, Fachbereich 7
Universität - GH - Duisburg
Duisburg
Federal Republic of Germany

This paper deals with the design of decentralized controllers. We propose the use of certain decoupling techniques in order to allow a sequential design of local controllers of balanced order which assign the poles of the closed-loop system. The local controllers consist of observers which are decoupled from certain inputs to the overall system, and restricted static state feedback laws which assign closed-loop eigenvectors that are unobservable from certain outputs and correspond to acceptable eigenvalues. The local observers are generalized dynamical systems, i. e. they include differentiating terms, and also the case of approximate decoupling by "high gain" is included. From an analysis in terms of matrix pencils, we obtain detailed results on the order and structure of the local controllers. Finally, we discuss the problem in which order one should close the local feedback loops.

INTRODUCTION

The simplicity of the implementation of decentralized, purely local feedback laws for multivariable, especially "large" (i. e. spatially distributed and/or dynamically complex) systems favours decentralized control schemes against centralized, possibly optimal solutions. The emphasis of the work on decentralized control has so far been on the question whether a given system is stabilizable for a certain decentralized control structure, i. e. an ordering of the inputs and outputs. It has been shown [1], that a necessary and sufficient condition for this is the nonexistence of unstable decentralized fixed modes and a number of tests for fixed modes has been developped [2, 3]. Other contributions deal with the question, under which conditions local control laws which are determined for the isolated subsystems can stabilize the overall system [4, 5, 6].

The design problem however remains largely an open one. If a system is decentrally stabilizable, the local control laws can be chosen sequentially to produce a stable overall system. However, in each step the dimension of the problem increases by the order of the previously chosen controller, and if the design is based on the observer plus static state-feedback controller structure, the order of the last controller can be enormous. The use of static output feedback in all but one control loops [7] somewhat reduces this problem, but still the distribution of the controller dynamics seems very inappropriate. A balanced distribution of the controller dynamics results if the local control laws are calculated for the isolated subsystems, but then even the stability of the overall system is only guaranteed if all the eigenvalues tend to infinity (in modulus) which makes this concept difficult to apply.

An engineering approach to the design problem is to base the local controller design on an approximation of the controlled overall system where the couplings between the subsystem under consideration and the rest of the system (with all the other loops closed) are represented by roughly estimated coupling systems [8]. This solves the problem to achieve balanced orders of the local controllers but introduces a considerable amount of heuristics.

The design techniques which we present here are motivated by the following specifications:

- we want to place all closed-loop poles in a region Λ of the complex plane by decentralized control laws
- we want to make maximal use of the internal structure of the system
- the complexity of the local controllers should be balanced and the overall order of the controller should be as low as possible.

We assume that the system which is to be controlled is given as a state-space model

$$\underline{\dot{x}}(t) = \underline{A}\ \underline{x}(t) + \sum_{i=1}^{N} \underline{B}_i\ \underline{u}_i(t) \tag{1a}$$

$$\underline{y}_i(t) = \underline{C}_i\ \underline{x}(t)\ , \qquad i = 1 \dots N\ , \tag{1b}$$

where the $\underline{B}_i$ are $n \times r_i$-matrices of full column rank and the $\underline{C}_i$ are $m_i \times n$-matrices of full row rank. The N local control laws consist of local observers which use $\underline{u}_i$ and $\underline{y}_i$ as inputs, and local constant feedback of the estimated linear functional of the state $\underline{x}$. The local observers are constructed such that the error is decoupled from certain other inputs $\underline{u}_j$. Thus the order of the system does not increase when the local control loops are closed sequentially. The constant feedback gains are chosen such that invariant subspaces of the controlled system are produced which have associated eigenvalues in Λ and are in the kernel of certain of the other matrices $\underline{C}_j$. This is less restricted than to ask for decoupling of the controlled subsystems, and it allows sequential placement of all the closed-loop poles if, for example, decoupling can be achieved with respect to the inputs and outputs of the loops which have not yet been closed (because a triangular coupling structure is obtained).

As a theoretical basis for this approach to the calculation of a decentralized control scheme, we present a unified theory of disturbance decoupled state estimation with generalized linear systems, and of the assignment of unobservable eigenvalues by restricted constant state feedback. Methodically, our approach is based on matrix pencil characterizations of certain invariant subspaces. We believe that this provides the most detailed insight into the internal structure of linear systems because of the simultaneous characterization of invariant subspaces and their associated eigenvalues. This approach is also attractive from a computational point of view because the structural information in singular pencils of matrices can be obtained using numerically well behaving algorithms (e. g. [9]).

LOCAL OBSERVERS WITH DECOUPLED ERROR DYNAMICS

We consider the linear time-invariant system

$$\underline{\dot{x}}(t) = \underline{A}_i\ \underline{x}(t) + \underline{B}_i\ \underline{u}_i(t) + \underline{E}_i\ \underline{d}_i(t) \tag{2a}$$

$$\underline{y}_i(t) = \underline{C}_i\ \underline{x}(t)\ , \tag{2b}$$

$$\underline{x}(0) = \underline{x}_0\ , \tag{2c}$$

$\underline{x} \in \mathbb{R}^n$, $\underline{u}_i \in \mathbb{R}^{r_i}$, $\underline{y}_i \in \mathbb{R}^{m_i}$, $\underline{d}_i \in \mathbb{R}^{s_i}$, which represents the dynamics of the plant with certain local control loops closed at step i of the design procedure. $\underline{u}_i$ is the local input, $\underline{y}_i$ the local measurement vector. For (2a-c) we want to construct a local observer

$$\underline{Q}_i^o \, \dot{\underline{z}}_i(t) = \underline{F}_i^o \, \underline{z}_i(t) + \underline{G}_i \, \underline{y}_i(t) + \underline{J}_i \, \underline{u}_i(t) \tag{3a}$$

$$\underline{w}_i(t) = \underline{L}_i \, \underline{z}_i(t) + \underline{H}_i \, \underline{y}_i(t) \tag{3b}$$

$$\underline{z}(0) = \underline{z}_0 \quad , \tag{3c}$$

$\underline{z}_i \in \mathbb{R}^{n_i^o + \bar{n}_i^o}$, $\underline{w}_i \in \mathbb{R}^{o_i}$, for the linear functional $\underline{R}_i\underline{x}\,(t)$.

Without loss of generality, we can assume that $\underline{Q}_i^o$ and $\underline{F}_i^o$ have the structure

$$\underline{Q}_i^o = \text{diag}\,\{\underline{I}_{n_i^o} \, , \, \tilde{\underline{Q}}_i^o\} \tag{4a}$$

$$\underline{F}_i^o = \text{diag}\,\{\tilde{\underline{F}}_i^o \, , \, \underline{I}_{\bar{n}_i^o}\} \tag{4b}$$

where $\tilde{Q}_i^o$ is a Jordan matrix with $\bar{s}_i^o$ Jordan blocks of dimensions $\bar{n}_{1i}^o$ and eigenvalue zero.

Definition 1

The system (3a) - (3c) is called a local observer of the linear functional $\underline{R}_i\underline{x}(t)$ for the system (2a) - (2c) if

a) for some $\hat{\underline{T}}_i$, the dynamics of

$$\underline{\xi}_i = \underline{z}_i - \hat{\underline{T}}_i \, \underline{x} \tag{5}$$

are given by

$$\underline{Q}_i^o \, \dot{\underline{\xi}}_i = \underline{F}_i^o \, \underline{\xi}_i \tag{6}$$

b)
$$\underline{\xi}(0) = 0 \Rightarrow \underline{w}_i(t) = \underline{R}_i \, \underline{x}(t) \; \forall t \geq 0 \, . \tag{7}$$

One easily derives from these conditions that they are equivalent to the following set of equations:

$$\underline{Q}_i^o \, \hat{\underline{T}}_i \, \underline{E}_i = \underline{0} \tag{8a}$$

$$\underline{F}_i^o \, \hat{\underline{T}}_i + \underline{G}_i \, \underline{C}_i = \underline{Q}_i^o \, \hat{\underline{T}}_i \, \underline{A}_i \tag{8b}$$

$$\underline{L}_i \, \hat{\underline{T}}_i + \underline{H}_i \, \underline{C}_i = \underline{R}_i \tag{8c}$$

$$\underline{J}_i = \underline{Q}_i^o \, \hat{\underline{T}}_i \, \underline{B}_i \quad . \tag{8d}$$

(8a) and (8b) describe the possible matrices $\hat{\underline{T}}_i$ and dynamics $\underline{F}_i^o$, $\underline{Q}_i^o$, and (8c) restricts the possible linear functionals which can be estimated.

An elegant and numerically attractive way of exploring the possible solutions to (8a, b) is to use the theory of pencils of matrices [10 - 11]. We introduce matrices $\underline{M}^{Ci}$, $\underline{N}^{Ei}$ where $\underline{M}^{Ci}$ spans ker $\underline{C}_i$ and $(\underline{N}^{Ei})^T$ spans the complement of $\underline{E}_i$. Then

$$s\underline{N}^{Ei} \, \underline{M}^{Ci} - \underline{N}^{Ei} \, \underline{A}_i\underline{M}^{Ci} = \underline{P}_i^o(s) \tag{9}$$

is [12] the zero pencil of the system $\{\underline{A}_i, \underline{E}_i, \underline{C}_i\}$ and completely describes the possible triples ($\hat{\underline{T}}_i$, $\underline{F}_i^o$, $\underline{Q}_i^o$) and, therefore, the possible matrices $\underline{R}_i$ for various conditions on $\underline{F}_i^o$ and $\underline{Q}_i^o$.

Theorem 1

Choose $\underline{N}^{E_i}$ and $\underline{M}^{C_i}$ such that $\underline{P}_i^o(s)$ has canonical form [10, 13] (this is always possible):

$$\bar{\underline{P}}_i^o(s) = \text{block diag}\left\{\underline{P}_{\varepsilon^o_{1i}}(s), \ldots \underline{P}_{\varepsilon^o_{q^o_i i}}(s),\ s\underline{I}_{\bar{\beta}^o_i} - \underline{J}^{o\bar{\Lambda}}_{fi},\right.$$

$$\left. s\underline{I}_{\beta^o_i} - \underline{J}^{o\Lambda}_{fi},\ s\underline{J}^o_{oi} - \underline{I}_{\bar{\nu}^o_i},\ \underline{P}^T_{\mu^o_{1i}}(s) \ldots \underline{P}^T_{\mu_{p^o_i i}}(s)\right\} \tag{10}$$

$$\underline{P}_k(s) = \begin{bmatrix} s & -1 & 0 & \cdots & \\ 0 & s & -1 & & \\ \vdots & \ddots & \ddots & \ddots & 0 \\ \vdots & & & \ddots & \\ \cdot & & 0 & s & -1 \end{bmatrix} \quad ((k+1)\times k)\ , \tag{11}$$

where $\underline{J}^{o\Lambda}_{fi}$, $\underline{J}^{o\bar{\Lambda}}_{fi}$ are Jordan matrices with eigenvalues in Λ resp. $\mathbb{C} \setminus \Lambda$ and $\underline{J}^o_{oi}$ is a nilpotent Jordan matrix.

Partition $\bar{\underline{M}}^{C_i}$ (the special choice of $\underline{M}^{C_i}$ which gives the canonical pencil) corresponding to the blocks of $\bar{\underline{P}}_i^o(s)$:

$$\bar{\underline{M}}^{C_i} = [\bar{\underline{M}}^{C_i}_{\varepsilon}, \bar{\underline{M}}^{C_i}_{f\bar{\Lambda}}, \bar{\underline{M}}^{C_i}_{f\Lambda}, \bar{\underline{M}}^{C_i}_{\infty}, \bar{\underline{M}}^{C_i}_{\mu}]\ . \tag{12}$$

Then an observer (2a) - (2c) according to Definition 1 exists

(i) for arbitrary eigenvalues of $\tilde{\underline{F}}_i^o$ iff

$$\underline{R}_i\ \bar{\underline{M}}^{C_i}_{\varepsilon} = \underline{0}\ ; \tag{13a}$$

(ii) for $\tilde{\underline{F}}_i^o$ with eigenvalues in Λ iff

$$\underline{R}_i\ [\bar{\underline{M}}^{C_i}_{\varepsilon}, \bar{\underline{M}}^{C_i}_{f\bar{\Lambda}}] = \underline{0}\ ; \tag{13b}$$

(iii) as a regular system ($\underline{Q}_i^o = \underline{I}$) with eigenvalues in Λ iff

$$\underline{R}_i\ [\bar{\underline{M}}^{C_i}_{\varepsilon}, \bar{\underline{M}}^{C_i}_{f\bar{\Lambda}}, \bar{\underline{M}}^{C_i}_{\infty}] = \underline{0}\ ; \tag{13c}$$

(iv) for $\tilde{\underline{F}}_i^o$ with almost arbitrarily prescribed eigenvalues iff

$$\underline{R}_i\ [\bar{\underline{M}}^{C_i}_{\varepsilon}, \bar{\underline{M}}^{C_i}_{f\bar{\Lambda}}, \bar{\underline{M}}^{C_i}_{f\Lambda}] = \underline{0}\ ; \tag{13d}$$

(v) as a regular system with almost arbitrarily prescribed eigenvalues iff

$$\underline{R}_i\ [\bar{\underline{M}}^{C_i}_{\varepsilon}, \bar{\underline{M}}^{C_i}_{f\bar{\Lambda}}, \bar{\underline{M}}^{C_i}_{f\Lambda}, \bar{\underline{M}}^{C_i}_{\infty}] = \underline{0}\ . \tag{13e}$$

Proof:

Without loss of generality $\tilde{\underline{F}}_i^o$ can be assumed to have Jordan form. Then the row vectors $\hat{\underline{t}}_{ji}$ corresponding to a Jordan block of $\tilde{\underline{F}}_i^o$ with eigenvalue λ^o_{ki} and dimension n_{ki} must satisfy (l denotes the last row corresponding to the respective Jordan block)

$$\underline{t}_{ji} = \bar{\underline{t}}_{ji}\ \underline{N}^{E_i} \tag{14a}$$

$$\underline{\bar{t}}_{1i}(\lambda^o_{ki}\ \underline{N}^{Ei}\ \underline{M}^{Ci} - \underline{N}^{Ei}\ \underline{A}_i\ \underline{M}^{Ci}) = \underline{0} \tag{14b}$$

$$\underline{\bar{t}}_{ji}(\lambda^o_{ki}\ \underline{N}^{Ei}\ \underline{M}^{Ci} - \underline{N}^{Ei}\ \underline{A}_i\ \underline{M}^{Ci}) = \underline{\bar{t}}_{j+1,i}\ \underline{N}^{Ei}\ \underline{M}^{Ci} \tag{14c}$$

$$j = 1 - 1 \ldots 1 - n_{ki} \quad .$$

The row vectors corresponding to a Jordan block with eigenvalue zero of $\underline{Q}^o_i$ of dimension $\bar{n}_{ki}$ must satisfy

$$\underline{\hat{t}}_{1i}\ \underline{M}^{Ci} = \underline{0} \tag{15a}$$

$$\underline{\hat{t}}_{ji} = \underline{\bar{t}}_{ji}\ \underline{N}^{Ei} \quad , \quad j = 1 \ldots 1 - \bar{n}_{ki} + 1 \quad , \tag{15b}$$

$$\underline{\hat{t}}_{ji}\ \underline{M}^{Ci} = \hat{t}_{j+1,i}\ \underline{A}_i\ \underline{M}^{Ci} \quad , \quad j = 1 - 1 \ldots 1 - \bar{n}_{ki} \quad . \tag{15c}$$

Assume now, that $\underline{N}^{Ei}$, $\underline{M}^{Ci}$ are such that $\underline{P}^o_i(s)$ has canonical form. Then the solutions to (14b) consist of general solutions to free eigenvalues which are orthogonal to all submatrices of $\bar{\underline{M}}^{Ci}$ except $\bar{\underline{M}}^{Ci}_{\mu}$, and solutions to fixed eigenvalues λ^o_{fi} which are the eigenvalues of $\underline{J}^{o\Lambda}_{fi}$ or $\underline{J}^{o\Lambda}_{fi}$ (i. e. the finite zeros of $\{\underline{A}_i, \underline{E}_i, \underline{C}_i\}$ and these solutions are orthogonal to all the submatrices of $\bar{\underline{M}}^{Ci}$ except $\underline{\bar{M}}^{Ci}_{\mu}$ and $\bar{\underline{M}}^{Ci}_{f\Lambda}$ resp. $\bar{\underline{M}}^{Ci}_{f\Lambda}$. The stated orthogonality is maintained for the solutions to (14c). This gives (iii) and (v). Finally, the solutions to (15a) - (15c) are orthogonal to $\bar{\underline{M}}^{Ci}_{\varepsilon}$, $\underline{M}^{Ci}_{f\Lambda}$, $\bar{\underline{M}}^{Ci}_{f\Lambda}$, which gives, together with the previous result, (i), (ii) and (iv). The term "almost arbitrarily prescribed" refers to the fact that there are certain restrictions, e. g. if only simple eigenvalues are desired, not more than q^o_i of them can be equal□

Of course, Theorem 1 is related to the result [12, 14, 15] that $\bar{\underline{M}}^{Ci}_{\varepsilon}$ spans the maximal ($\underline{A}_i$, $\underline{E}_i$) - controllability subspace in ker $\underline{C}_i$, $[\bar{\underline{M}}^{Ci}_{\varepsilon}, \underline{M}^{Ci}_{f\Lambda}, \bar{\underline{M}}^{Ci}_{f\Lambda}]$ spans the maximal ($\underline{A}_i$, $\underline{E}_i$) - invariant subspace in ker $\underline{C}_i$ and $[\bar{\underline{M}}^{Ci}_{\varepsilon}, \underline{M}^{Ci}_{f\Lambda}, \bar{\underline{M}}^{Ci}_{f\Lambda}, \bar{\underline{M}}^{Ci}_{\infty}]$ spans the maximal almost ($\underline{A}_i$, $\underline{E}_i$) - invariant subspace.

The matrix pencil description also gives very usefull information about the necessary order of the observer.

Theorem 2

Suppose that in each case, $\underline{R}_i$ has the maximal possible rank. Let us introduce

$$\bar{\mu}^o_i = \sum_{j=1}^{p^o_i} \mu^o_{ji} \tag{16}$$

and let ν^o_{ji} , j = 1 ... $\bar{k}^o_i$ be the dimensions of the Jordan blocks of $\underline{J}^o_{oi}$.

Then

(i) under the restriction (13b),

$$n^o_i + \bar{n}^o_i \geq \bar{\mu}^o_i + \bar{\nu}^o_i + \bar{k}^o_i + \beta^o_i \tag{17}$$

is necessary and β^o_i eigenvalues of $\underline{\tilde{F}}^o_i$ are prescribed, moreover, $\underline{\tilde{Q}}^o_i$ must contain at least $\bar{k}^o_i$ Jordan blocks of dimensions ν^o_{ji}+1;

(ii) under the restriction (13c)

$$n^o_i \geq \bar{\mu}^o_i + \beta^o_i \tag{17b}$$

is necessary and β_i^o eigenvalues are prescribed;

(iii) under the restriction (13d),

$$n_i^o + \bar{n}_i^o \geq \bar{\mu}_i^o + \bar{\nu}_i^o + \bar{k}_i^o \tag{17c}$$

is necessary and the same condition as under (i) applies to $\tilde{\underline{Q}}_i^o$;

(iv) under the restriction (13e),

$$n_i^o \geq \bar{\mu}_i^o \tag{17d}$$

is necessary.

This follows also from the discussion of the solutions to (14a) - (14c) resp. (15a) - (15c) □

As the dimensions of the Jordan blocks of $\tilde{\underline{Q}}_i^o$ determine the order of the differentiation which is performed by the system (3a) - (3c),

$$n_{diff} = \max_{1 \leq j \leq \bar{s}_i^o} \bar{n}_{ji}^o - 1 = \max_{1 \leq j \leq \bar{k}_o^i} \nu_{ji}^o , \tag{18}$$

one is usually interested in keeping the $\bar{n}_{ji}^o$ as small as possible, and this bound is determined by those infinite zero orders ν_{ji}^o for which the corresponding columns of $\underline{\bar{M}}_\infty^{C_i}$ are not in ker $\underline{R}_i$.

Corollary 1

If

$$s_i \geq m_i \quad , \tag{19}$$

eigenvalues of $\tilde{\underline{F}}_i^o$ can only be prescribed if the system $\{\underline{A}_i, \underline{E}_i, \underline{C}_i\}$ is degenerate, i. e. the transfer matrix from $\underline{d}$ to $\underline{y}$ looses rank for all values of s.

Proof:

If (19) is satisfied, $\underline{P}_i^o(s)$ can only have column indices, if it also has row indices, this implies the degeneracy of $\{\underline{A}_i, \underline{E}_i, \underline{C}_i\}$ [13] □

If $\underline{u}_i$, $\underline{y}_i$ are scalar (the frequently studied case of coupled SISO-systems), there are only two possibilities:

a) the system is already dynamically decoupled, i. e. $\underline{A}_i$ has a block-triangular structure
b) only differentiation of the local measurements can produce estimates of non--measured states with decoupling from the input $\underline{d}_i(t)$ and arbitrary error dynamics.

This underlines the usefullness of the more general observer structure compared to e. g. [16]. Willems [17] calls the system (3a-c) a PID observer.

The possible fixed eigenvalues of $\tilde{\underline{F}}_i^o$ here have a slightly different role than in the conventional case without disturbance decoupling, because they are not necessarily eigenvalues of $\underline{A}_i$ (this is only the case if they are input-output decoupling zeros of $\{\underline{A}_i, \underline{E}_i, \underline{C}_i\}$). Therefore, in contrast to the conventional case where the inclusion of these dynamics in the observer is useless, the extension of $\underline{R}_i$ corresponding to the implementation of the prescribed eigenvalues in the observer may, depending on their location, be worthwhile.

ASSIGNMENT OF UNOBSERVABLE MODES BY LOCAL CONSTANT STATE FEEDBACK

Here we represent the system in the i-th step of the design by

$$\dot{\underline{x}}_i(t) = \underline{A}_i \, \underline{x}_i(t) + \underline{B}_i \, \underline{u}_i(t) \tag{20a}$$

$$\underline{v}_i = \underline{D}_i \, \underline{x}_i(t) \tag{20b}$$

$$\underline{x}_i(0) = \underline{x}_{io} \quad , \tag{20c}$$

where $\underline{v}_i(t)$ represents the outputs to all or some other local controllers. We want to use local state feedback

$$\underline{u}_i(t) = \underline{K}_i \, \underline{R}_i \, \underline{x}_i(t) \tag{21}$$

to obtain a closed-loop system of the form

$$\dot{\underline{x}}_i^*(t) = \underline{F}_i^C \, \underline{x}_i^*(t) + \underline{G}_i^* \, \underline{x}_{i+1}(t) \tag{22a}$$

$$\dot{\underline{x}}_{i+1}(t) = \bar{\underline{G}}_i \, \underline{x}_i^*(t) + \underline{A}_{i+1} \, \underline{x}_{i+1}(t) \tag{22b}$$

$$\underline{x}_i(t) = \underline{T}_i^* \, \underline{x}_i^*(t) + \underline{T}_{i+1} \, \underline{x}_{i+1}(t) \tag{22c}$$

where $\underline{T}_i^*$ and $\underline{T}_{i+1}$ are orthogonal, have full column rank, and

$$\underline{D}_i \, \underline{T}_i^* = \underline{0} \quad . \tag{23}$$

We prescribe the following structure for $\underline{F}_i^C$ and $\bar{\underline{G}}_i$:

$$\underline{F}_i^C = \begin{bmatrix} \underline{F}_{1i}^C & \tilde{\underline{G}}_i^C \\ \underline{0} & \underline{F}_{2i}^C \end{bmatrix} \tag{24}$$

where the submatrix $\underline{F}_{1i}^C$ has Jordan form and eigenvalues λ_{ki}^C of multiplicity s_{ki}^C and the submatrix $\underline{F}_{2i}^C$ has Jordan form and eigenvalues $\gamma_i^C \cdot \lambda_{ki}^C$ of multiplicity $\bar{s}_{ki}^C$ (all eigenvalues of $\underline{F}_i^C$ have to be in Λ).
For $\bar{\underline{G}}_i$, we require that the first columns corresponding to the Jordan blocks of $\underline{F}_{2i}^C$, $\bar{\underline{J}}_{ki}^C$, are bounded as $\gamma_i^C \to \infty$ and the remaining columns vanish.

<u>Theorem 3</u>

The following characterizations of subspaces τ_i^* are equivalent

(i) There exists a feedback

$$\underline{u}_i = \tilde{\underline{K}}_i \, \underline{x}_i \tag{25}$$

such that for some basis $\underline{T}_i^*$ which is polynomial in γ_i^C a closed-loop system (22a) - (22c) with the properties described above is obtained.

(ii) There exists a feedback (25) such that for any fixed initial condition $\underline{x}_{io} \in \tau_i^*$ and for any $\varepsilon > 0$, the maximal value of $\|\underline{v}_i(t)\|$ does not exceed ε and the eigenvalues of the restriction of the closed-loop system to τ_i^* are in Λ.

(iii) The pencil

$$\underline{P}_i^t(s) = s\underline{N}^{Bi} \, \underline{T}_i^* - \underline{N}^{Bi} \, \underline{A}_i \, \underline{T}_i^* \tag{26}$$

has no non-zero row indices and no finite singularities in $\bar{\Lambda}$.

<u>Proof:</u>

(ii) ⇔ (iii) : see [14], [15].
(iii) ⇔ (i) :
From (22a) - (22c), one obtains

$$(\underline{A}_i + \underline{B}_i \tilde{\underline{K}}_i)\, \underline{T}_i^* - \underline{T}_i^* \underline{F}_i^C = \underline{T}_{i+1} \bar{\underline{G}}_i \tag{27a}$$

or, as a condition for $\underline{T}_i^*$,

$$\underline{N}^{Bi} \underline{A}_i \underline{T}_i^* - \underline{N}^{Bi} \underline{T}_i^* \underline{F}_i^C = \underline{N}^{Bi} \underline{T}_{i+1} \bar{\underline{G}}_i \; . \tag{27b}$$

Write $\underline{T}_i^* = \bar{\underline{T}}_i^* \cdot \tilde{\underline{T}}_i^*$ where $\bar{\underline{T}}_i^*$ is a constant basis of τ_i^* such that the pencil $\bar{\underline{P}}_i^t(s) = \bar{\underline{N}}^{Bi} \underline{A}_i \bar{\underline{T}}_i^* - s\bar{\underline{N}}^{Bi} \bar{\underline{T}}_i^*$ is in canonical form for some $\bar{\underline{N}}^{Bi}$. From (24) and the structure of $\bar{\underline{G}}_i$, (27b) decomposes into

$$\bar{\underline{N}}^{Bi} \underline{A}_i \bar{\underline{T}}_i^* \tilde{\underline{T}}_{1i}^* - \bar{\underline{N}}^{Bi} \bar{\underline{T}}_i^* \tilde{\underline{T}}_{1i}^* \underline{F}_{1i}^C = \underline{0} \tag{27c}$$

and

$$\bar{\underline{N}}^{Bi} \underline{A}_i \bar{\underline{T}}_i^* \tilde{\underline{T}}_{2i}^* - \bar{\underline{N}}^{Bi} \bar{\underline{T}}_i^* \tilde{\underline{T}}_{2i}^* \underline{F}_{2i}^C = \bar{\underline{N}}^{Bi} (\underline{T}_{i+1} \bar{\underline{G}}_i + \bar{\underline{T}}_i^* \tilde{\underline{T}}_{1i}^* \tilde{\underline{G}}_i) \; . \tag{27d}$$

(27c) has solutions only corresponding to the column indices and to the finite singularities of $\underline{P}_i^t(s)$, whereas (27d) has solutions which are polynomial in γ_i^C only corresponding to the infinite singularities and to the column indices, even for an arbitrary constant right-hand side. Therefore, as a regular $\tilde{\underline{T}}_i^*$ is required, $\underline{P}_i^t(s)$ can only have finite singularities in Λ, column indices, infinite singularities and zero row indices. So we have shown (i) ⇒ (iii).

Assume now that $\underline{P}_i^t(s)$ has the structure described in (iii). Then (27c) has to be satisfied for the finite singularities as eigenvalues of $\underline{F}_{1i}^C$. The solutions to (27d) must satisfy

$$[\bar{\underline{N}}^{Bi} \underline{A}_i \bar{\underline{T}}_i^* - \gamma_i^C \lambda_{ki} \bar{\underline{N}}^{Bi} \bar{\underline{T}}_i^*]\, \tilde{\underline{t}}_{\bar{l}_k i}^* = \bar{\underline{N}}^{Bi} (\underline{T}_{i+1} \bar{\underline{g}}_{\bar{l}_k i} + \underline{T}_{1i}^* \tilde{\underline{g}}_{\bar{l}_k i}) \tag{28a}$$

$$[\bar{\underline{N}}^{Bi} \underline{A}_i \bar{\underline{T}}_i^* - \gamma_i^C \lambda_{ki} \bar{\underline{N}}^{Bi} \bar{\underline{T}}_i^*]\, \tilde{\underline{t}}_{ji}^* = \bar{\underline{N}}^{Bi} \bar{\underline{T}}_i^* \tilde{\underline{t}}_{j-1,i}^* + \bar{\underline{N}}^{Bi} (\underline{T}_{i+1} \bar{\underline{g}}_{ji} + \underline{T}_{1i}^* \tilde{\underline{g}}_{ji}),$$

$$j = \bar{l}_k + 1 \;\ldots\; \bar{l}_k + \bar{s}_{ki}^C \; . \tag{28b}$$

From the canonical form of $\underline{P}_i^t(s)$, one can see that (28b) can be solved for $\underline{t}_{ji}^*$ if $\bar{\underline{g}}_{ji}$, $\tilde{\underline{g}}_{ji}$ are zero. The solutions to (28a) for a general right-hand side have the form

$$(\tilde{\underline{t}}_{\bar{l}_k i})^T = (\alpha_{1k}, \alpha_{1k} s + c_{11}^{(k)}, \alpha_{1k} s^2 + c_{11}^{(k)} s + c_{21}^{(k)}, \ldots \alpha_{1k} s^{\varepsilon_{1i}^t} + c_{11}^{(k)} s^{\varepsilon_{1i}^t - 1} + \ldots + c_{\varepsilon_{1i}^t,1}^{(k)}$$

$$\alpha_{2k}, \;\ldots\ldots \qquad \alpha_{q_i^t,k}\, s^{\varepsilon_{q_i^t i}^t} + c_{1,q_i^t}^{(k)}\, s^{\varepsilon_{q_i^t i}^t - 1} + \ldots + c_{\varepsilon q_i^t i, q_i^t}^{(k)} \; ,$$

$$0, \ldots\; 0, \; \bar{\alpha}_{1k}\, s^{\nu_{1i}^t} + \bar{c}_{11}^{(k)} s^{\nu_{1i}^t - 1} + \ldots + \bar{c}_{\nu_{1i}^t,1}^{(k)}, \ldots, \bar{\alpha}_{1k} \; ,$$

$$\ldots\; \bar{\alpha}_{\bar{k}_i^t,k}\, s^{\nu_{\bar{k}_i^t i}^t} + \bar{c}_{1,\bar{k}_i^t}^{(k)}\, s^{\nu_{\bar{k}_i^t i}^t - 1} + \ldots + \bar{\alpha}_{\bar{k}_i^t,k}) \; , \; s = \gamma_i^C \cdot \bar{\lambda}_{ki} \; , \tag{29}$$

where $\varepsilon_{1i}^t \ldots \varepsilon_{q_i^t i}^t$ denote the column indices of $\underline{P}_i^t(s)$ and $\nu_{1i}^t \ldots \nu_{\bar{k}_i^t i}^t$ the orders of the infinite singularities of $\underline{P}_i^t(s)$, and the zeros correspond to $\bar{\underline{T}}_{f\Lambda_i}^*$. The parameters $\bar{\alpha}_{jk}$ are constants (not all zero), as well as the $c_{j\ell}^{(k)}$, $\bar{c}_{j\ell}^{(k)}$ (which may all vanish), the α_{jk} are arbitrary (except for rank conditions). The rhs of (28a) results as

$$(c_{11}^{(k)},\ldots c_{\varepsilon_{q_i^t,q_i^t}}^{(k)},0\ldots 0,\bar{c}_{\nu_{1i}^t 1}^{(k)},\ldots \bar{c}_{11}^{(k)},\bar{\alpha}_{1k},\bar{c}_{\nu_{2i}^t 2}^{(k)}\ldots \bar{c}_{12}^{(k)},\bar{\alpha}_{2k},\ldots \bar{\alpha}_{\bar{k}_i^t,k},0,\ldots 0)^T \ .$$

Now $\bar{\underline{N}}^{Bi}\ \bar{\underline{T}}_i^*$ has zero rows corresponding to the entries $\bar{\alpha}_{ik}$ on the rhs and to the last zeros, therefore $\bar{\underline{N}}^{Bi}\ \underline{T}_{i+1}$ has independent rows there. Also $\bar{\underline{N}}^{Bi}\cdot \underline{T}_{1i}^*$ has independent rows at least corresponding to the finite singularities. Therefore by choice of $\bar{\underline{g}}_{\bar{l}_k i}$, $\tilde{\underline{g}}_{\bar{l}_k i}$ and the free constants $c_{ij}^{(k)}$, $\bar{c}_{ij}^{(k)}$, (28a) can always be satisfied.

Remark: It is easy to see that (i) also implies (ii). It would be desirable to prove (i)$\Rightarrow$(iii) without using the fact that (ii)$\Rightarrow$(iii)$\square$

From the arguments in the proof of Theorem 3, we can draw some more conclusions.

Corollary

The maximal almost invariant subspace in ker $\underline{D}_i$ with associated eigenvalues in $\Lambda,\mathcal{S}_i^*$, is spanned by $[\bar{\underline{M}}_{\varepsilon}^{Di}\ \bar{\underline{M}}_{f\Lambda}^{Di}\ \bar{\underline{M}}_{\infty}^{Di}]$ where $\bar{\underline{M}}_{\varepsilon}^{Di}$ etc. are the partitions (compare (12)) of the matrix $\bar{\underline{M}}^{Di}$ which brings the pencil

$$\underline{P}_i^C(s) = s\ \underline{N}^{Bi}\ \underline{M}^{Di} - \underline{N}^{Bi}\ \underline{A}_i\ \underline{M}^{Di} \tag{30}$$

to canonical form. If $\underline{T}_i^*$ has maximal rank, $\underline{F}_{1i}^C$ has at most $\bar{\varepsilon}_i^C + \beta_i^C$ eigenvalues λ_{ji} of which β_i^C are prescribed and $\bar{\varepsilon}_i^C$ can be chosen almost arbitrarily (β_i^C, $\bar{\varepsilon}_i^o$ being defined as the dimensions of $\bar{\underline{M}}_{f\Lambda}^{Di}$, $\bar{\underline{M}}_{\varepsilon}^{Di}$ as in the previous section).

From the condition that $\bar{\underline{G}}_i$ has only a minimal number of nonzero columns, it follows that the maximal orders $\bar{s}_{ki}^C$ are given by the orders of the infinite zeros. It can also be derived that the eigenvalues of the system (22a,b) are given as $\{\lambda_{ki}\}$, $\{\gamma_i^C(\bar{\lambda}_{ki}+\bar{\delta}_{ki})\},\{\sigma_k(\underline{A}_{i+1})+\delta_{ki}\}$ where $\bar{\delta}_{ki}$, $\delta_{ki}\to 0$ as $\gamma_i^C\to\infty$. The eigenvalues of $\underline{F}_i^C$ therefore become (exactly resp. asymptotically) eigenvalues of the closed-loop system, independent of a feedback which uses $\underline{v}_i$ as input, if $\gamma_i^C\to\infty$ (but the external feedback remains fixed).

Now the feedback must be of the form (21) instead of (25). The feedback matrix has to satisfy

$$\underline{K}_i\ \underline{R}_i\ \underline{T}_i^* = \underline{B}_i^L\ (\underline{T}_i^*\ \underline{F}_i^C - \underline{A}_i\ \underline{T}_i^* + \underline{T}_{i+1}\ \bar{\underline{G}}_i) \ . \tag{31}$$

If im $\underline{T}_i^*$ and ker $\underline{R}_i$ do not intersect, (31) can always be solved for $\underline{K}_i$. As $\underline{R}_i$ arises from the problem of local state estimation with error decoupling which normally imposes quite restrictive conditions on $\underline{R}_i$, the prerequisite of the above corollary will in cases of practical interest only be met for more or less small subspaces of the possible (almost) invariant subspaces in ker $\underline{D}_i$. It is therefore necessary to ask whether it is possible to achieve (almost) invariant subspaces which intersect with ker $\underline{R}_i$ and have a spectrum in Λ.
We treat this problem separately for completely decoupled dynamics (invariant subspaces with finite spectrum) and asymptotically decoupled dynamics, because the results show some essential differences between the two cases.

To simplify the notation, we denote im $[\bar{\underline{M}}_{\varepsilon}^{Di}\ \bar{\underline{M}}_{f\Lambda}^{Di}]$ by $\mathcal{V}_i^*$ and im $[\bar{\underline{M}}_{\varepsilon}^{Di}\ \bar{\underline{M}}_{\infty}^{Di}]$ by $\bar{\mathcal{S}}_i^*$.

Theorem 4

Let $\mathcal{V}_{\ker R_i}^* = \mathcal{V}_i^* \cap \ker \underline{R}_i$ and $\underline{V}_i^*$, $\underline{V}_{\ker \underline{R}_i}^*$ be minimal bases of the resp. subspaces.

(i) The intersection of $\mathcal{V}^*_{\ker \underline{R}_i}$ with the unobservable subspace of $\{\underline{A}_i, \underline{D}_i\}$ is always an invariant subspace in ker $\underline{D}_i$ with n^C_{fi} associated fixed eigenvalues (which are eigenvalues of $\underline{A}_i$).

(ii) Let $\tilde{\underline{B}}_i$ be a matrix of maximal rank such that $\tilde{\underline{B}}_i^T\,[\bar{\underline{M}}^{Di}_\varepsilon\ \bar{M}^{Di}_{f\Lambda}] = \underline{0}$ and $\tilde{\underline{B}}_i^T\,\underline{B}_i$ has full row rank. Let $\tilde{\underline{M}}^{Di}$ be a minimal basis of $\mathcal{V}^*_{\ker \underline{R}_i} \cap \ker(\tilde{\underline{B}}_i^T\,\underline{A}_i)$. Then the maximal dimension of an invariant subspace in $\mathcal{V}^*_i$ which can be produced by restricted feedback is

$$n^C_{\max_i} = \bar{\varepsilon}^C_i + \beta^C_i - \text{rank}\,(\tilde{\underline{B}}_i^T\,\underline{A}_i) - \tilde{\mu}^C_i, \tag{32}$$

where $\tilde{\mu}^C_i$ is the sum of the first rank $(\tilde{\underline{B}}_i^T\,\underline{A}_i)$ row indices of the pencil $\underline{V}_i^{*T}\,\underline{A}_i\,\tilde{\underline{M}}^{Di} - s\underline{V}_i^{*T}\,\tilde{\underline{M}}^{Di}$.

(iii) If

$$q^C_i + n^C_{fi} > \dim \mathcal{V}^*_{\ker \underline{R}_i}, \tag{33}$$

an invariant subspace of maximal possible dimension can be produced such that all the associated eigenvalues can be placed in Λ except for the n^C_{fi} fixed eigenvalues, up to β^C_i of these eigenvalues may be fixed as finite zeros of $\{\underline{A}_i, \underline{B}_i, \underline{D}_i\}$ and the remaining eigenvalues can be placed almost arbitrarily.

Proof: See [18], the arguments are by and large the same as in the proof of the next theorem□

If rank $(\underline{R}_i\,\underline{V}^*_i) >$ rank $(\tilde{\underline{M}}^{Di})$ (i. e. the subspace which is available for feedback is larger than the remaining part of the possible invariant subspace), $\tilde{\mu}_i = 0$.

If we allow the eigenvalues to become large and admit the coupling term $\bar{\underline{G}}_i$ in (22b) (which becomes asymptotically negligible), we get a different result:

Theorem 5

Let $\underline{T}^*_{1i}$ be the basis of an invariant subspace in ker $\underline{D}_i$ which can be obtained by restricted feedback. Let $\mathcal{S}_{\ker \underline{R}_i}$ be the intersection of ker $\underline{R}_i$ with $\bar{\mathcal{S}}^*_i$. Then if

$$q^C_i > \dim \mathcal{S}_{\ker \underline{R}_i}, \tag{34}$$

a feedback $\underline{K}_i$ can be chosen such that for a matrix $\underline{S}$ which spans $\bar{\mathcal{S}}^*_i$

$$(\underline{A}_i + \underline{B}_i\underline{K}_i\underline{R}_i)\,\underline{S} = \underline{S}\,\underline{F}^{Cs}_i + \underline{T}_{i+1}\,\bar{\underline{G}}_i + \underline{T}^*_{1i}\,\tilde{\underline{G}}_i \tag{35}$$

is achieved where $\underline{F}^{Cs}_i$, $\bar{\underline{G}}_i$ have the structure prescribed above (the columns of $\bar{\underline{G}}_i$ vanish except for the first columns corresponding to Jordan blocks of $\underline{F}^{Cs}_i$ with eigenvalues $\gamma^C_i\,\bar{\lambda}_{ki}$, the nonvanishing columns are bounded as $\gamma^C_i \to \infty$), and the eigenvalues of $\underline{F}^{Cs}_i$ can be prescribed in a simple structure except for possible unobservable modes of $\{\underline{A}_i, \underline{D}_i\}$ with associated unobservable subspaces in ker $\underline{R}_i \cap \bar{\mathcal{S}}^*_i$ ($\underline{T}_{i+1}$ is orthogonal to $\underline{S}$ and $\underline{T}^*_{1i}$).

In other words, if (34) holds, $\bar{\mathcal{S}}^*_i \cup$ im $\underline{T}^*_{1i}$ can be made almost invariant by restricted feedback.

Proof:

We choose a basis $[\underline{S}_{1i}\ \underline{S}_{2i}\ \underline{S}_{3i}]$ for $\bar{\mathcal{S}}^*_i$ such that $\underline{S}_{3i}$ is in im $\underline{T}^*_{1i}$, $\mathcal{S}_{\ker \underline{R}_i} \subset$ im $[\underline{S}_{2i}, \underline{S}_{3i}]$ and $\underline{S}_{1i}$ has dimension $\dim \bar{\mathcal{S}}^*_i - \dim \mathcal{S}_{\ker \underline{R}_i}$. From (35) we get as sufficient conditions

$$[\underline{A}_i + \underline{B}_i \underline{K}_i \underline{R}_i] \underline{S}_{1i} = \underline{S}_{1i} \underline{F}_{1i}^{Cs} + \underline{T}_{i+1} \bar{\underline{G}}_{1i} + \underline{T}_{1i}^* \tilde{\underline{G}}_{1i} \tag{36a}$$

$$\underline{A}_i \underline{S}_{2i} = \underline{S}_{2i} \underline{F}_{2i}^{Cs} + \underline{S}_{1i} \bar{\underline{F}}_i^{Cs} + \underline{T}_{i+1} \bar{\underline{G}}_{1i} + \underline{T}_{1i}^* \tilde{\underline{G}}_{2i} \quad . \tag{36b}$$

($\underline{S}_{3i}$ spans a priori an invariant subspace). $\bar{\underline{G}}_{1i}$, $\bar{\underline{G}}_{2i}$ have nonvanishing columns where $\underline{F}_{1i}^{Cs}$, $\underline{F}_{2i}^{Cs}$ have eigenvalues γ_i^C $\bar{\lambda}_{ki}^{Cs}$.

Now form Theorem 3 there is always a solution $\{\underline{K}_i, \underline{S}_{1i}, \bar{\underline{G}}_{1i}, \tilde{\underline{G}}_{1i}\}$ for prescribed diagonal $\underline{F}_{1i}^{Cs}$, because $\underline{R}_i\underline{S}_{1i}$ has full column rank. There is a solution $\{\underline{S}_{2i}, \bar{\underline{F}}_i^{Cs}, \bar{\underline{G}}_{2i}, \tilde{\underline{G}}_{2i}\}$ to (36b) iff

$$\underline{N}^{S_{1i}} \underline{A}_i \underline{S}_{2i} - \underline{N}^{S_{1i}} \underline{S}_{2i} \underline{F}_{2i}^{Cs} = \underline{N}^{S_{1i}} (\underline{T}_{i+1} \bar{\underline{G}}_{2i} + \underline{T}_{1i}^* \tilde{\underline{G}}_{2i}) \tag{36c}$$

has a solution $\{\underline{S}_{2i}, \bar{\underline{G}}_{2i}, \tilde{\underline{G}}_{2i}\}$ for prescribed $\underline{F}_{2i}^{Cs}$. We can decompose $\underline{N}^{S_{1i}}$ as

$$\underline{N}^{S_{1i}} = [\bar{\underline{N}}_{\mu}^{Bi\,T} \; \bar{\underline{N}}_{f\Lambda}^{Bi\,T} \; \bar{\underline{N}}_{f\bar{\Lambda}}^{Bi\,T} \; \tilde{\underline{D}}_i^T \; \underline{U}_i]^T \tag{37}$$

where $\tilde{\underline{D}}_i^T$ spans $\{\text{im}\ \underline{D}_i^T \cap \text{im}\ [\underline{B}_i \ \underline{N}_{\infty}^{Bi}]\}$ and $\underline{U}_i$ is in $\mathcal{F}_i^*$. Using this, (36c) is equivalent to

$$\tilde{\underline{D}}_i^T \underline{A}_i \underline{S}_{2i} = \tilde{\underline{D}}_i^T \underline{T}_{i+1} \bar{\underline{G}}_{2i} \tag{38a}$$

$$\underline{U}_i^T \underline{A}_i \underline{S}_{2i} - \underline{U}_i^T \underline{S}_{2i} \tilde{\underline{F}}_i^C = \underline{U}_i^T \underline{T}_{1i}^* \tilde{\underline{G}}_{2i} \tag{38b}$$

$$\begin{bmatrix} \bar{\underline{N}}_{\mu}^{Bi} \\ \bar{\underline{N}}_{f\Lambda}^{Bi} \\ \bar{\underline{N}}_{f\bar{\Lambda}}^{Bi} \end{bmatrix} (\underline{T}_{i+1} \bar{\underline{G}}_{2i} + \underline{T}_{1i}^* \tilde{\underline{G}}_{2i}) = \underline{0} \tag{38c}$$

because $\tilde{\underline{D}}_i^T \underline{T}_{1i}^* = \tilde{\underline{U}}_i \underline{T}_{i+1} = \underline{0}$.

For $\underline{S}_{2i}$, we can choose a basis $[\underline{S}_{2i}^1 \; \underline{S}_{2i}^2]$ where $\underline{S}_{2i}^1$ is in the unobservable subspace of $\{\bar{\underline{A}}_i, \underline{D}_i\}$ such that

$$\underline{A}_i \underline{S}_{2i}^1 = \bar{\underline{F}}_{2i}^{Cs} \underline{S}_{2i}^1 \tag{39}$$

where $\bar{\underline{F}}_{2i}^{Cs}$ is a Jordan matrix, the corresponding columns of $\bar{\underline{G}}_{2i}$, $\tilde{\underline{G}}_{2i}$ vanish because (38a,b) are satisfied for vanishing right-hand side. $\underline{S}_{2i}^2$ is arbitrary. It is then possible to solve (38a) and (38c) for $\bar{\underline{G}}_{2i}$, $\tilde{\underline{G}}_{2i}$, and to determine a matrix $\tilde{\underline{U}}_i^T$ of rank $(\dim \mathcal{F}_i^* - \dim \mathcal{S}_{\ker \underline{R}_i})$ such that

$$\tilde{\underline{U}}_i^T [\bar{\underline{M}}_{\varepsilon}^{Di} \; \bar{\underline{M}}_{\infty}^{Di}]^T (\underline{A}_i \underline{S}_{2i} - \underline{S}_{2i} \underline{F}_{2i}^{Cs} - \underline{T}_{1i}^* \tilde{\underline{G}}_{21}) = \underline{0} \tag{40}$$

for any $\underline{F}_{2i}^{Cs}$ of the form diag $\{\bar{\underline{F}}_{2i}^{Cs}, \gamma_i^C \; \bar{\lambda}_{ki}^{Cs}\}$ for precribed $\bar{\lambda}_{ki}^{Cs}$. Now if (34) holds, each column of $\underline{S}_{1i}$ contains at least q_i^C free parameters, namely the values α_{ik} in (29), which can be chosen such that

$$\tilde{\underline{U}}_i^T [\bar{\underline{M}}_{\varepsilon}^{Di} \; \bar{\underline{M}}_{\infty}^{Di}] \underline{S}_{1i} = \underline{0} \tag{41}$$

So for this choice of $\underline{S}_{1i}$ (which also determines $\underline{K}_i$), (38a) - (38c) and hence (36c) are satisfied, therefore for a suitable $\bar{\underline{F}}_i^{Cs}$, (36a,b) hold. It remains to show that there is a transformation which diagonalizes $\underline{F}_i^{Cs}$ and maintains the prescribed structure of $\bar{\underline{G}}_i$. This will be the case if

$$\begin{bmatrix} \underline{I} & -\underline{X}_i \\ \underline{0} & \underline{I} \end{bmatrix} \begin{bmatrix} \underline{F}_{1i}^{Cs} & \bar{\underline{F}}_i^{Cs} \\ \underline{0} & \underline{F}_{2i}^{Cs} \end{bmatrix} \begin{bmatrix} \underline{I} & \underline{X}_i \\ \underline{0} & \underline{I} \end{bmatrix} = \begin{bmatrix} \underline{F}_{1i}^{Cs} & \underline{0} \\ \underline{0} & \underline{F}_{2i}^{Cs} \end{bmatrix} \tag{42}$$

has a solution $\underline{X}_i$ such that $\underline{G}_{1i}\ \underline{X}_i$ remains bounded as $\gamma_i^C \to \infty$. (42) is satisfied if

$$\underline{F}_{1i}^{Cs}\ \underline{X}_i - \underline{X}_i\ \underline{F}_{2i}^{Cs} = \bar{\underline{F}}_i^{Cs} = (\underline{S}_{1i})^L \Big[\underline{A}_i\ \underline{S}_{2i} - \underline{S}_{2i}\ \underline{F}_{2i}^{Cs} - \underline{T}_{i+1}\ \bar{\underline{G}}_{2i} - \underline{T}_{1i}^*\ \tilde{\underline{G}}_{2i} \Big]. \tag{43}$$

As $\underline{F}_{1i}^{Cs}$ and $\underline{F}_{2i}^{Cs}$ can always have different eigenvalues, there is always such a solution because $\underline{S}_{1i}$ can be chosen such that the relevant elements of $(\underline{S}_{1i})^L$ remain bounded as $\gamma_i^C \to \infty$ simply by choosing either all $\bar{\alpha}_{ki}$ nonzero or all zero. The basis of the almost invariant subspace is then given as $[\underline{T}_{1i}^*, \underline{S}_{1i}, \underline{S}_{2i} + \underline{S}_{1i}\ \underline{X}_i]$ □

The restriction of the feedback to the form (21) is thus less restrictive for the generation of almost invariant subspaces than for the generation of invariant subspaces as long as the conditions (33) (34) are satisfied. Note that if $\underline{P}_i^C(s)$ has infinite zeros, $\underline{F}_{1i}^C$ must also have at least some large eigenvalues even if $\bar{\underline{M}}_\infty^{Di}$ is in ker $\underline{R}_i$, if an almost invariant subspace of maximal dimension is to be generated.

It is of course possible that the restriction of the feedback causes fixed invariant subspaces in ker $\underline{D}_i$ which are not in $\mathscr{S}_i^*$ if $\bar{\underline{M}}_{f\bar{\Lambda}}^{Di}$ does not vanish. Moreover, if $\{\underline{A}_i, \underline{B}_i\ \underline{D}_i\}$ has r_i+1 zeros with the same value in $\bar{\Lambda}$, there is always, even for unrestricted feedback, an eigenvalue in $\bar{\Lambda}$ (the system $\{\underline{A}_i, \underline{B}_i, \underline{E}_i, \underline{C}_i, \underline{D}_i\}$ has a decentralized fixed mode).

DESIGN PROCEDURES

1. Independent design / completely decoupled dynamics

If we choose

$$\underline{A}_i = \underline{A} \tag{44a}$$

$$\underline{E}_i = [\underline{B}_1 \dots \underline{B}_{i-1}, \underline{B}_{i+1} \dots \underline{B}_N] \tag{44b}$$

$$\underline{D}_i^T = [\underline{C}_1^T \dots C_{i-1}^T, C_{i+1}^T \dots \underline{C}_N^T] \quad , \tag{44c}$$

we require that each local observer is decoupled from all other inputs and the poles which are assigned in step i are unobservable from all other outputs. The local controllers can be designed independently and the closed-loop system has completely decoupled dynamics, maximal robustness against failures and changes in other loops is obtained.

The following lemma shows that this is however only possible under very strong restrictions on the structure of the system.

Lemma 1

Let $\mathscr{S}_i$ be a subspace which can be decoupled from $\underline{v}_i$ by local state feedback exactly or asymptotically with associated eigenvalues in Λ, and let ker $\underline{R}_i$ be the unobservable subspace (with error decoupling) for a local observer with nice dynamics, possibly with differentiation. Then

$$\bigcup_{j=1}^{N} \mathscr{S}_j = \mathbb{R}^n \tag{45}$$

and

$$\mathscr{S}_i \cap \ker \underline{R}_i = \{0\},\ i = 1 \dots N, \tag{46}$$

imply

$$\mathcal{S}_i \cap \bigcup_{\substack{j=1\\j\neq i}}^{N} \mathcal{S}_j = \{0\} \tag{47a}$$

$$\bigcap_{j=1}^{N} \ker \underline{R}_j = \{0\} \tag{47b}$$

$$\ker \underline{R}_i \cup \bigcap_{\substack{j=1\\j\neq i}}^{N} \ker \underline{R}_i = \mathbb{R}^n \ . \tag{47c}$$

Proof:

This results from the two basic inclusions

$$\mathcal{S}_i \subset \bigcap_{\substack{j=1\\j\neq i}}^{N} \ker \underline{R}_j \tag{48}$$

and

$$\ker \underline{R}_i \supset \bigcup_{\substack{j=1\\j\neq i}}^{N} \mathcal{S}_j \quad . \tag{49}$$

(48) means that the subspace which can be decoupled from $\underline{v}_i$ using $\underline{u}_i$ must be included in all unobservable subspaces from the components of $\underline{v}_i$ for arbitrary inputs, (49) means that the unobservable subspace for output $\underline{y}_i$ must contain all the subspaces which can be decoupled from all the outputs including $\underline{y}_i$ using local feedback.

From (46) and (49), (47a) is immediate. From (45) and (48), (49)

$$\mathbb{R}^n = \mathcal{S}_i \cup \bigcup_{\substack{j=1\\j\neq i}}^{N} \mathcal{S}_j \subset \bigcap_{\substack{j=1\\j\neq i}}^{N} \ker \underline{R}_j \cup \ker \underline{R}_i \subset \mathbb{R}^n$$

yields (47c) and, in fact, equality in (48) and (49).

Finally, if (47b) were not true, $\bigcap_{j=1}^{N} \ker \underline{R}_i = \mathcal{R}^*$, no $\mathcal{S}_i$ could contain $\mathcal{R}^*$ which would contradict (45) □

Remark: Theorems 1, 3 leave a certain amount of ambiguity in the choices of $\mathcal{S}_i$, $\underline{R}_i$ because the additional possibilities of differentiation resp. high gain are present in both problems, however $\mathcal{S}_i$ and $\underline{R}_i$ must be chosen such that (48), (49) hold, i. e. decoupling by high gain is not neutralized by differentiation.

(45) and (47a) imply that $\mathbb{R}^n$ can be decomposed uniquely into the subspaces $\mathcal{S}_i$. (47b) states that the system must be observable by completely decoupled local observers.

If the subspaces $\mathcal{S}_i$ can be decoupled exactly with free dynamics or if the local observers do not include differentiating terms and have free dynamics, (45) and (46) imply that the system can be diagonalized simply by using only certain combinations of the inputs resp. of the outputs.

If the coupling systems are not completely degenerate (and there is therefore no design problem at all), the condition (46) can only be relaxed if a subsystem has at least three inputs. Thus fully decoupled control practically requires that the system satisfies very strong structural conditions. From the point of view of applicability it is necessary to relax the decoupling conditions.

2. Sequential design / triangular coupling structure

If we choose

$$\underline{A}_i = \underline{A}_{i-1} + \underline{B}_{i-1}\,\underline{\tilde{K}}_{i-1} \tag{50a}$$

$$\underline{E}_i = [\underline{B}_{i+1} \;\cdots\; \underline{B}_N] \tag{50b}$$

$$\underline{D}_i^T = [\underline{C}_{i+1}^T \;\cdots\; \underline{C}_N^T] \quad , \tag{50c}$$

we only require that each local observer is decoupled from the inputs which are used in the following design steps, and the poles which are assigned in step i are unobservable from the outputs which are used by the not yet fixed local controllers. This allows a sequential design which leads to a triangular (backwards) coupling structure.

Obviously, the conditions for decoupling are easier to meet in this case because the dimensions of $\underline{E}_i$, $\underline{D}_i$ decrease in each step. Also, only the restriction of the dynamics to the subspace spanned by $\underline{T}_{i+1}$ as described by $\underline{A}_{i+1}$ (compare (22a-c)) has to be considered in the next step.

Under two conditions, it is possible to assign all the poles in Λ by such a procedure.

(i) the dynamics in the output decoupled (almost or exactly) subspace can be prescribed in Λ for each step i,

(ii) the system $(\underline{A}_{i+1}, \underline{B}_{i+1})$ is controllable or has only uncontrollable modes in Λ, i = 1 ... N-1.

Let us discuss these conditions in detail, assuming that the overall system is controllable and observable from the inputs $\underline{u}_1 \ldots \underline{u}_N$ resp. outputs $\underline{y}_1 \ldots \underline{y}_N$.

(i): there are two possibilities which can lead to a violation of (i)
- the existence of "bad" modes which are decoupled from $\underline{v}_i$ is due to the restriction $\underline{\tilde{K}}_i = \underline{K}_i\,\underline{R}_i$
- there are eigenvalues which are unobservable from $\underline{v}_i$ for all $\underline{K}_i$ but also uncontrollable from $\underline{u}_i$. If this is independent of feedback from $\underline{y}_1 \ldots \underline{y}_{i-1}$ to $\underline{u}_1 \ldots \underline{u}_{i-1}$, these eigenvalues are fixed modes and the system cannot be controlled satisfactorily by a decentralized control scheme.

If the violation of (i) is due to the restriction of the local feedback, the decoupling condition for the local observer must be relaxed such that these modes can be assigned or made observable. This leads to a certain increase in the order of $\underline{A}_{i+1}$ (by the order of an additional, not decoupled observer). If the second case occurs, and the resp. modes are not fixed modes, the design in steps 1...i-1 must be modified such that this situation is avoided, this can always be done because we have some freedom to choose $\underline{K}_i\,\underline{R}_i\,\underline{T}_{i+1}$ for given $\underline{K}_i\,\underline{R}_i\,\underline{T}_i^*$; in the worst case, $\underline{R}_i$ must be modified as above.

(ii): Assuming that this condition was met in the previous step, the resp. modes are controllable from $\underline{u}_i$. So if they are observable from $\underline{y}_i$, they can be placed in Λ using a non-decoupled observer. If not, there are three exclusive possibilities:
- these modes are fixed modes under the given decentralized control scheme
- these modes are also unobservable from $\underline{y}_1 \ldots \underline{y}_{i-1}$ and the chosen sequence is inappropriate
- the situation can be avoided by a modification of the previous loops.

A sequence of loop closures is only suitable for the sequential design process if the subspace which is unobservable from outputs $\underline{y}_1 \ldots \underline{y}_i$ is contained in the subspace which is controllable from $\underline{u}_{i+1} \ldots \underline{u}_N$, i=1...N. This condition may be used to test the possible sequences.

Also, the feedback $\underline{K}_i$ at each step should be used to modify the dynamics of the remaining system by shifting poles into Λ and, for the uncontrollable modes, by improving the observability and controllability in order to obtain as large as possible subspaces which are both controllable from $\underline{u}_k$ and not unobservable from $\underline{y}_k$, k=i+1...N.

Apart from the above condition, the design sequence should be chosen such that it starts with subproblems which have many available inputs and outputs and where the observability and controllability of the system from the rest of the inputs and outputs is bad.

EXAMPLE

A 13th order system is given as

$$\dot{\underline{x}} = \begin{bmatrix} * & 0 & * & * & * & * & * & * & * & * & \circledast & * & * \\ 0 & b & * & * & * & * & 0 & & & & & * & \\ 0 & 0 & * & \circledast & * & * & 0 & & & & & * & \\ 0 & 0 & * & * & \circledast & * & 0 & & & & & * & \\ 0 & 0 & * & * & * & \circledast & 0 & & & & & * & \\ 0 & 0 & * & * & * & * & \circledast & & & & & * & \\ 0 & 0 & * & * & * & * & 0 & & & & & \circledast & \\ & & & & & & & * & * & * & 0 & 0 & \circledast \\ & & & & & & & \circledast & 0 & 0 & 0 & 0 & 0 \\ a & -a & 0 & 0 & 0 & 0 & 0 & * & * & * & 0 & 0 & * \\ * & * & \circledast & * & * & * & 0 & * & * & * & * & * & * \\ \circledast & 0 & * & * & * & * & 0 & * & * & * & 0 & * & * \\ c & -c & 0 & 0 & 0 & 0 & 0 & * & \circledast & * & 0 & 0 & * \end{bmatrix} \cdot \underline{x} + \begin{bmatrix} 1 & 0 & & & & \\ 0 & 1 & & & & \\ 0 & 0 & & & & \\ 0 & 0 & & & & \\ 0 & 0 & & & & \\ 0 & 0 & & & & \\ 0 & 0 & & & & \\ & & 0 & 0 & 0 & 0 \\ & & 0 & 0 & 0 & 0 \\ & & 1 & 0 & 0 & 0 \\ & & 0 & 1 & 0 & 0 \\ & & 0 & 0 & 1 & 0 \\ & & 0 & 0 & 0 & 1 \end{bmatrix} \begin{bmatrix} \underline{u}_1 \\ \underline{u}_2 \end{bmatrix}$$

$$\underline{y}_1 = \begin{bmatrix} x_3 \\ x_5 \\ x_{12} \end{bmatrix} \qquad \underline{y}_2 = \begin{bmatrix} x_{10} \\ x_{13} \end{bmatrix},$$

$*$: arbitrary entries
$\circledast$: nonzero entries with arbitrary values
a, c $\neq$ 0.

The system is controllable and observable both from the pairs ($\underline{u}_1$, $\underline{y}_1$) and ($\underline{u}_2$, $\underline{y}_2$) alone and does not have decoupled dynamics.
From the pencils $\underline{P}_1^o = \underline{P}_2^c$, $\underline{P}_1^c = \underline{P}_2^o$, one finds

$\underline{P}_1^o = \underline{P}_2^c$: $\mu_{11}^o = 1$ $\qquad\qquad$ $\varepsilon_{12}^c = \underline{0}$

$\mu_{21}^o = 2$ $\qquad\qquad$ $\varepsilon_{22}^c = 1$

finite singularity at s = b $\qquad\qquad$ $\varepsilon_{32}^c = 2$

$\underline{P}_2^o = \underline{P}_1^c$ $\quad$ $\mu_{12}^o = 2$ $\qquad\qquad$ $\varepsilon_{11}^c = 7$

$\nu_{12}^o = 1$

The system is not decentrally observable, but all the poles can be prescribed by local feedback. For a sequential design, we start with the second controller. We have dim $\mathcal{S}_2^*$ = dim $\mathcal{V}_2^*$ = 6, dim $\mathcal{V}^*_{\ker \underline{R}_i}$ = 1, q_2^c = 3, and an unobservable subspace of dimension 6 can be obtained with prescribed eigenvalues. The second observer has 2nd order dynamics and includes one differentiation (which in practice must be approximated). The rest of the eigenvalues can be assigned with a conventional controller for the first loop. In fact this is possible using only a suitable combination of the inputs and with output decoupling with respect to $\underline{y}_2$.

ACKNOWLEDGEMENTS

The work of the first author was made possible by grants of the Fritz Thyssen Stiftung and the Deutsche Forschungsgemeinschaft (DFG En 152/1-1). The work of the second author was performed as part of a research project on hierarchical control under the direction of Prof. Dr. H. Schwarz, funded by the Deutsche Forschungsgemeinschaft (grants Schw 120/34 and Schw 120/39). Part of the work was done while the first author was with the Department of EE, McGill University, Montréal, Canada. The support of all persons and institutions mentioned is gratefully acknowledged.

REFERENCES

[1] Wang, S.H.; Davison, E.J., On the Stabilization of Decentralized Control Systems, IEEE-Tr.-AC 18 (1973), 473-479.

[2] Anderson, B.D.O.; Clements, D.J., Algebraic Characterization of Fixed Modes in Decentralized Control, Automatica 17 (1981), 703-712.

[3] Seraji, H., On Fixed Modes in Decentralized Control Systems, International Journal of Control 35 (1982), 775-784.

[4] Siljak, D.D., On Decentralized Control of Large Scale Systems, IFAC World Congress 1978, Preprints p. 1849-1856.

[5] Bachmann, W., On Decentralized Output Feedback Stabilizability, Large Scale Systems 4 (1983), 165-177.

[6] Bachmann, W.; Konik, D., On Stabilization of Decentralized Dynamic Output Feedback Systems, Systems and Control Letters 5 (1984), 89-95.

[7] Corfmat, J.P.; Morse, A.S., Decentralized Control of Linear Multivariable Systems, Automatica 12 (1976), 479-495.

[8] Litz, L., Dezentrale Regelung (München und Wien: Oldenbourg Verlag, 1983).

[9] Van Dooren, P.; Dewilde, P., The Eigenstructure of an Arbitrary Polynomial Matrix: Computational Aspects, Lin. Algebr. and Appl. 50 (1983), 545-579.

[10] Gantmacher, F.R., Matrizenrechnung II (Berlin, 1959).

[11] Wilkinson, J.H., Linear Differential Equations and Kronecker's Canonical Form, in: de Boor, C. and Golub, G.H. (eds.), Recent Advances in Numerical Analysis (New York, 1978).

[12] Karcanias, N.; Kouvaritakis, B., The Output Zeroing Problem and its Relationship to the Invariant Zero Structure: A Matrix Pencil Approach, Int. J. Control 30 (1979), 395-415.

[13] Kailath, T., Linear Systems (Prentice-Hall, Englewood Cliffs, 1980).

[14] Jaffe, S.; Karcanias, N., Matrix Pencil Characterization of Almost ($\underline{A}$, $\underline{B}$)-invariant Subspaces: A Classification of Geometric Concepts, Int. J. Control 33 (1981), 51-93.

[15] Schumacher, J.M., Algebraic Characterisations of Almost Invariance, Int. J. Control 38 (1983), 107-124.

[16] Viswanadham, N.; Ramakrishna, A., Decentralized Estimation and Control for Interconnected Systems, Large Scale Systems 3 (1983), 255-266.

[17] Willems, J.C., Almost Invariant Subspaces: An Approach to High Gain Feedback Design - Part I: Almost Controlled Invariant Subspaces, IEEE Tr. on AC 26 (1981), 235-252, and Part II: Almost Conditionally Invariant Subspaces, IEEE Tr. on AC 27 (1982), 1071-1085.

[18] Engell, S., Assignment of Unobservable Eigenvalues by Restricted Constant State Feedback, submitted for publication.

Frequency Domain and State Space Methods for Linear Systems
C.I. Byrnes and A. Lindquist (Editors)

ALMOST INVARIANT SUBSPACES AND INFINITE ZERO STRUCTURE

M. MALABRE

Laboratoire d'Automatique de Nantes, Unité Associée au C.N.R.S.
E.N.S.M., 1 rue de la Noë, 44072 NANTES Cédex, FRANCE

Since their introduction in 1980, the Almost Invariant Subspaces (A.I.S) have been described through many facets but we only enhance here some of their algebraic properties in connection with the concept of infinite Zero Structure (I.Z.S.). We first recall ("lattice aspect of A.I.S.") how the A.I.S. have contributed to the important geometric characterization of the I.Z.S. which lead to a multitude of structural solvability conditions for some control problems (namely the decoupling problem with static state feedback and regular input transformations and the dynamic model matching problem).
We secondly show ("feedback-chain aspect of the A.I.S.") how the introduction of the maximal input chain and associated feedbacks gave insight into the manner the I.Z.S. can be modified via non regular static state feedback laws, that is laws of the type u = Fx + Gv with G non invertible. This has been used in the solution of the general static state feedback decoupling problem for systems with 2 outputs.

These last five years, the study of linear systems has known two important contributions :

⁖ the introduction of Almost Invariant Subspaces (A.I.S.) [5],[8]
⁖ the structural solution of control problems in terms of Infinite Zero Structures (I.Z.S.) [3],[11],[12],[13],[14],[15],[19],[21],[25]...

Since their introduction in 1980, the A.I.S. have been described through various facets [6],[8],[9],[16],[26], but we shall only enhance here some of their algebraic properties in connection with the concept of I.Z.S.. In particular, though this is at the origin of their introduction, we shall not pay interest to their topological properties : this very important aspect plays a key role in the solution of "almost problems" and in the use of high gain feedbacks [8].

We shall present here some recent results on structural properties of linear systems and on the solvability of classical control problems such as model matching or decoupling with emphasis on the role played by A.I.S. ; more precisely, after having recalled some geometric basic concepts, we shall collect these results through two particular facets of A.I.S., namely their "lattice" aspect (§ 2) and their "feedback-chain" descriptions (§ 3).

1. BRIEF BACKGROUND

We shall deal with linear (C,A,B) systems described by :

$$\begin{cases} \dot{x} = Ax + Bu & x \in \mathcal{X} = R^n,\ u \in \mathcal{U} = R^m \\ y = Cx & y \in \mathcal{Y} = R^p \end{cases}$$

Let $\mathcal{K}$ =: Ker C and $\mathcal{B}$ =: Im B.

The basic algorithms of the geometric approach are the following [4],[8] :

(I.S.A.)
Invariant Subspace
Algorithm

$$\begin{cases} \mathcal{V}_0 = \mathcal{X} \\ \mathcal{V}_{i+1} = \mathcal{K} \cap A^{-1}(\mathcal{B} + \mathcal{V}_i) \end{cases}$$

which is non increasing and converges towards $\mathcal{V}^*$, the supremal (A,B) or controlled invariant subspace contained in $\mathcal{K}$, and :

(C.I.S.A.)
Conditionally Invariant
Subspace Algorithm

$$\begin{cases} \mathcal{S}_0 = 0 \\ \mathcal{S}_{i+1} = \mathcal{B} + A\,(\mathcal{K} \cap \mathcal{S}_i) \end{cases}$$

which is non decreasing and converges towards $\mathcal{S}^*$, the infimal (C,A) or conditionally invariant subspace containing $\mathcal{B}$.

Let us recall, for instance, that $\mathcal{R}_i =: \mathcal{V}^* \cap \mathcal{S}_i$ is the ith step of the well known Controllability Subspace Algorithm (C.S.A.) [4],[8], converging towards $\mathcal{R}^*$, the supremal controllability subspace contained in $\mathcal{K}$:

(C.S.A.)
Controllability Subspace
Algorithm

$$\begin{cases} \mathcal{R}_0 = 0 \\ \mathcal{R}_{i+1} = \mathcal{V}^* \cap (A\,\mathcal{R}_i + \mathcal{B}) \end{cases}$$

In the same way, $\mathcal{R}_{ai} =: \mathcal{K} \cap \mathcal{S}_i$ is the ith step of the well known Almost Controllability Subspace Algorithm (A.C.S.A.) converging towards $\mathcal{R}_a^*$, the supremal almost controllability subspace contained in $\mathcal{K}$:

(A.C.S.A.)
Almost Controllability
Subspace Algorithm

$$\begin{cases} \mathcal{R}_{ao} = 0 \\ \mathcal{R}_{ai+1} = \mathcal{K} \cap (A\,\mathcal{R}_{ai} + \mathcal{B}) \end{cases}$$

We shall also use some particular lists of integers, namely the MORSE's lists I_2 and I_4 of (C,A,B) [1]. They are defined as follows :

The MORSE's list I_2 : ([1])

Let
$$\pi_i =: d(\mathcal{R}_i/\mathcal{R}_{i-1}), \quad i \geq 1 \tag{1}$$

then : $I_2(i)$ =: number of π_j's which are greater than or equal to i.

The Infinite Zero Structure, otherwise said the MORSE's list I_4 : ([1],[10])

Let
$$p_i =: d\{(\mathcal{S}_i + \mathcal{V}^*)/(\mathcal{S}_{i-1} + \mathcal{V}^*)\} \quad i \geq 1$$

then $I_4(i)$ = number of p_j's which are greater than or equal to i.

2. THE LATTICE ASPECT OF A.I.S.

The word lattice is used in the spirit of [9] ; we shall in fact consider dimensional relationships between subspaces constructed from the $\mathcal{S}_i$'s and $\mathcal{V}_i$'s with the operations + and $\cap$. The main important result is that, for any $i \geq 0$ and any $j \geq 0$, we have [24] :

$$\begin{aligned} d\{(S^* + V_j)/(S_i + V_j)\} &= d\{(S_j \cap V_i)/(S_j \cap V^*)\} \\ &= d\{(S^* \cap V_i)/(S^* \cap V_{i+j})\} \\ &= d\{(V^* + S_{i+j})/(V^* + S_i)\} \,. \end{aligned} \tag{2}$$

These are very powerful relationships (this holds for any i and any j), a first particular use of which has lead to new descriptions for the I.Z.S.

Indeed, setting j = 1 in (2) leads to the following new expressions for the list p_i defined in (1) [13] :

$$\begin{aligned} p_{i+1} =: d\{(S_{i+1} + V^*)/(S_i + V^*)\} &= d\{(S^* \cap V_i)/(S^* \cap V_{i+1})\} \\ &= d\{(K + S^*)/(K + S_i)\} \\ &= d\{(B \cap V_i)/(B \cap V^*)\} \end{aligned} \tag{3}$$

This is this last description of the p_i's (in terms of B, V_i, and V^*) which played a key role in the geometric solution of the regular decoupling problem [19] and of the dynamic model matching problem [13],[21] in terms of I.Z.S., indeed :

<u>The regular block-decoupling problem</u> :

The regular block-decoupling problem is solvable if and only if [19] the I.Z.S. of $(C,A,B_{ })$ equals the union of the I.Z.S. of each block subsystem (C_i,A,B) with $C = [C_1^T | C_2^T | \ldots | C_k^T]^T$ and where k denotes the number of blocks and T stands for the transpose :

<u>The dynamic model matching problem</u> :

A linear system $\Sigma_1 = (C_1,A_1,B_1)$ can dynamically follow a given linear model $\Sigma_2 = (C_2,A_2,B_2)$ if and only if [13],[21] the I.Z.S. of Σ_1 equals the I.Z.S. of the composite system $\Sigma = (C,A,B)$ where :

$$A = \begin{bmatrix} A_1 & 0 \\ 0 & A_2 \end{bmatrix} \qquad B = \begin{bmatrix} B_1 & 0 \\ 0 & B_2 \end{bmatrix} \qquad C = [C_1 \;\vdots\; -C_2]$$

These results have also been obtained in the transfer matrix approach ([2],[3],[7] for the model matching problem and [11],[12],[14] for the decoupling one) this is one of the main interests of these structural approach independent solvability conditions.

Each approach has its own advantages and drawbacks : the transfer matrix approach usually leads to shorter proofs and directly works with proper systems ; the geometric one is often a bit heavier (in particular when dealing with non strictly proper systems ((A,B,C,D) quadruples), see for instance [21]). However, the main advantage of the latter is that, with adequate definitions and tools, the given proofs have been extended, sometimes almost up to the word, to the affine nonlinear case : by NIJMEIJER and SCHUMACHER for the regular decoupling problem [18], and by DI BENEDETTO and ISIDORI for the model matching one [22].

To make a link with the next section, let us remark that setting i = 1 in relations (2) leads to :

$$d(R_a^i \,/\, R_a^{i-1}) = p_{i+1} + \pi_i \,, \tag{4}$$

this precisely shows how the steps of A.C.S.A. contain the information about (C,A,B) MORSE's lists I_2 and I_4.

3. FEEDBACK-CHAIN ASPECT OF A.I.S.

Another important facet of A.I.S. is the following feedback-chain description of Almost Controllability Subspaces (A.C.S.) [5],[8] :

$\mathcal{R}_a$ is an A.C.S. if and only if there exist a state feedback F and a chain $\{\mathcal{B}_i\}$ of $\mathcal{B}$ (i.e. a sequence of subspaces $\mathcal{B}_i$ satisfying : $\mathcal{B} \supset \mathcal{B}_1 \supset \mathcal{B}_2 \dots \supset \mathcal{B}_n$), such that, with $A_F =: A + BF$:

$$\mathcal{R}_a = \mathcal{B}_1 + A_F\mathcal{B}_2 + A_F^2 \mathcal{B}_3 + \dots + A_F^{n-1} \mathcal{B}_n \qquad (5)$$

Each A.C.S. has many possible feedback-chain descriptions like (5) (non unicity of the chain and, for each possible chain, non unicity of the feedback). The following result [17] establishes the existence of both a maximal input chain and associated feedbacks (for the sake of shortness, this is written for $\mathcal{R}_a^*$ but obviously holds for any A.C.S. $\mathcal{R}_a$ (chose $K = \mathcal{R}_a$)).

3.1. The maximal input chain :

There always exist (non unique) ϕ satisfying $A_\phi \mathcal{V}^* \subset \mathcal{V}^*$, such that :

$$\mathcal{R}_a^* = \mathfrak{B}_1 + A_\phi\mathfrak{B}_2 + A_\phi^2\mathfrak{B}_3 + \dots + A_\phi^{n-1}\mathfrak{B}_n$$

with : (6)

$$\mathfrak{B}_i =: \mathcal{B} \cap \mathcal{V}_i \ ;$$

Moreover, any feedback-chain description of $\mathcal{R}_a^*$ like (5) satisfies :

$$\mathcal{B}_i \subset \mathfrak{B}_i \quad , \quad \forall\ i.$$

Hence $\{\mathfrak{B}_i\}$ is called <u>the maximal input chain</u> of $\mathcal{R}_a^*$.

ϕ can be selected by a procedure the dual of which has been given by MORSE [1] or, more easily, by a "triangular procedure" [23] since these feedbacks ϕ satisfy :

$$A_\phi \ \mathcal{V}_i \subset \mathcal{V}_{i-1} \quad , \forall\ i \geq 1.$$

This maximal input chain and these associated feedbacks ϕ have many interesting properties :

∷∷the previous description and maximality property hold not only for $\mathcal{R}_a^*$, but also for each step of A.C.S.A. :

$$\mathcal{R}_{a\ i} = \mathfrak{B}_1 + A_\phi\mathfrak{B}_2 + \dots + A_\phi^{i-1}\mathfrak{B}_i \qquad , \forall\ i \geq 1$$

∷∷even more generally, this also holds for any subspace like $\mathcal{S}_i \cap \mathcal{V}_j$ (let us recall that $\mathcal{R}_{a\ i} = K \cap \mathcal{S}_i = \mathcal{V}_1 \cap \mathcal{S}_i$) :

$$\mathcal{S}_i \cap \mathcal{V}_j = \mathfrak{B}_j + A_\phi\mathfrak{B}_{j+1} + \dots + A_\phi^{i-1}\mathfrak{B}_{i+j-1} \qquad \forall\ i \geq 1, \forall\ j \geq 1$$

∷∷the maximal input chain is directly related to the I.Z.S. of (C,A,B) (list I_4 of (C,A,B)), indeed, since $\mathfrak{B}_i =: \mathcal{B} \cap \mathcal{V}_i$, and because of (3) :

$$p_{i+1} = d(\mathfrak{B}_i / \mathfrak{B}_n) \qquad \forall i \geq 0$$

∷∷when the system if left invertible (B monic and $\mathcal{R}^* = 0$), the maximal input chain is in fact <u>the only possible one</u> to write $\mathcal{S}_i \cap \mathcal{V}_j$ in a feedback chain form.

∷∷they have contributed to new results concerning the row-by-row decoupling problem with no a priori constraints on the feedback law u = Fx + Gv (such as regularity of G). This is recalled below.

3.2. Applications to the row-by-row decoupling problem (Morgan's Problem) :

The maximal input chain and its associated feedbacks (see (6)) when combined with the general relationships (2) have been useful in the study of the general decoupling problem, general being understood in the sense that no artificial constraint (such as regularity of G or Restricted Static State Feedback (R.S.S.F.) [20]) is added. This is a very important problem since there obviously exist systems which are decouplable and for which no regular [19] or even R.S.S.F. [20] solution exists. This is also a difficult problem mainly because the transformation performed on the system do not belong to a group.

3.2.1. A new sufficient condition for the p output case [23]

Let us first recall that an obvious necessary condition for a linear (C,A,B) system to be decouplable is its right invertibility. Let us then assume that rank (C,A,B) = p where p is the number of outputs ; this also means that (C,A,B) has p infinite zeros.

We shall assume that the order of the infinite zero of each row subsystem equals 1. As far as we are concerned with the row by row decoupling problem, this assumption is unrestrictive at all since it can easily be shown [23] that :

(C,A,B) is decouplable with a control law (F,G) if and only if (Γ,A,B) is so, where (Γ,A,B) is the shifted system associated with (C,A,B) in the following way :

$$\Gamma =: \begin{vmatrix} c_1 A^{n_1-1} \\ \vdots \\ c_p A^{n_p-1} \end{vmatrix} \qquad \text{with } C =: \begin{vmatrix} c_1 \\ \vdots \\ c_p \end{vmatrix}$$

and where n_i is the I.Z.S. of (c_i,A,B).

This shifted system has the required property.

Let β denote the supremal term in the I.Z.S. of (C,A,B).

Then, decoupling with pure static state feedback is possible if :

$$d(\mathcal{R}_{a\,\beta-1}/\,\mathcal{R}_{a\,\beta-2}) \geqslant p$$

where $\mathcal{R}_{ai}$ is the ith step of (A.C.S.A.) (see section 1).

3.2.2. Necessary and sufficient condition when p = 2 [25]

Let $\{n_1', n_2'\}$ be the I.Z.S. of $([-\frac{c_1}{c_2}], A,B)$

Let n_i be the I.Z.S. of (c_i,A,B), i = 1,2.

Let $\{I_2(1), I_2(2), \ldots, I_2(m-2)\}$ be the Morse's list I_2 of $([-\frac{c_1}{c_2}], A,B)$.

Then, decoupling with pure static state feedback is possible if and only if :

$$(n_1' + n_2') - (n_1 + n_2) \leqslant \sum_{i=1}^{m-2} I_2(i) \qquad (7)$$

The existence of regular solutions to the decoupling problem is equivalent to the nullity of the left hand side of (3) (see section 2).

When regular solutions do not exist, the basic problem for decoupling is to find a control law (F_o,G_o) which conveniently increases the initial infinite zero structures (global and row by row) until obtaining equality. The fundamental fact is that this increasing comes from terms of the list I_2 which are transformed into I_4 terms !

When p = 2, the condition (7) means that R^*, the supremal controllability subspace contained in the Kernel of C, must be big enough with respect to the required increasing $(n_1' + n_2') - (n_1 + n_2)$.

When p > 2, the basic idea is the same : the list I_2 must be well behaved with respect to the required increasings, however the précise conditions as well as a good procedure for the construction of (F_o,G_o) have to be found !

Let us mention that recent results from J.J. LOISEAU [27] completely characterize the set of the I.Z.S. which can be obtained from (C,A,B) by using non regular state feedback laws (F,G) : this is only a part of the answer ...

REFERENCES

[1] A.S. MORSE : Structural invariants of linear multivariable systems, SIAM J. Control, 11, 3, pp. 446-465, 1973.

[2] A.S. MORSE : System invariants under feedback and cascade control, Proc. Int. Symp., Udine, Springer Verlag, pp. 61-74, 1975.

[3] G.C. VERGHESE : Infinite frequency behaviour in generalized dynamical systems, Ph. D. Thesis, Elect. Eng. Dpt., Stanford University, 1978.

[4] W.M. WONHAM : Linear multivariable control : a geometric approach, 2nd Edition, Applications of Mathematics, Vol. 10, Springer Verlag, 1979.

[5] J.C. WILLEMS : Almost A (mod B) invariant subspaces, Astérisque, Vol. 75-76, pp. 239-248, 1980.

[6] S. JAFFE, N. KARCANIAS : Matrix pencil characterization of almost (A,B) invariant subspaces : a classification of geometric concepts, INT. J. Control, 33, 1, pp. 51-93, 1981.

[7] L. PERNEBO : An algebraic theory for the desing of controllers for linear multivariable systems, IEEE Trans. on Automat. Contr., AC-26, 1, pp. 171-194, 1981.

[8] J.C. WILLEMS : Almost invariant subspaces : an approach to high gain feedback design, IEEE Trans. on Automat. Contr., Part 1, AC-26, 1, pp. 235-252, 1981 ; Part 2, AC-27, 5, pp. 1071-1095, 1982.

[9] M. MALABRE : Almost invariant subspaces, transmission and infinite zeros : a lattice interpretation, Systems & Control Letters, Vol. 1, 6, pp. 347-355, 1982.

[10] C. COMMAULT, J.M. DION : On the structure at infinity of linear multivariable systems : a geometric approach, IEEE Trans. on Automat. Contr., AC-27, 3, pp. 693-696, 1982.

[11] J. DESCUSSE, J.M. DION : On the structure at infinity of linear square decouplable systems, IEEE Trans. on Automat. Contr., AC-27, 4, pp. 971-974, 1982.

[12] J.M. DION : Découplage par blocs des systèmes linéaires multivariables, Analysis and Optimization of Systems, Lecture Notes in Control and Information Sciences, Vol. 44, Springer Verlag, 1982.

[13] M. MALABRE : Structure à l'infini des triplets invariants, Application à la Poursuite de Modèle, Analysis and Optimization of Systems, Lecture Notes in Control and Information Sciences, Vol. 44, Springer Verlag, 1982.

[14] J.M. DION : Feedback block decoupling and infinite structure of linear systems, INT. J. Control, Vol. 37, 3, pp. 521-533, 1983.

[15] J. DESCUSSE : Sur la structure à l'infini des systèmes linéaires découplables : le cas des systèmes inversibles à droite, Outils et Modèles Mathématiques pour l'Automatique, l'Analyse de Systèmes et le Traitement du Signal, Vol. 3, Editions du CNRS, Paris, 1983.

[16] J.M. SCHUMACHER : Algebraic characterization of almost invariance, INT. J. Control, Vol. 38, 1, pp. 107-124, 1983.

[17] M. MALABRE : A complement about almost controllability subspaces, Systems & Control Letters, Vol. 3, 3, pp. 119-122, 1983.

[18] H. NIJMEIJER, J.M. SCHUMACHER : Zeros at infinity for affine nonlinear control systems, Memorandum NR 441, Technische Hogeschool Twente, The Netherlands.

[19] J. DESCUSSE, J.F. LAFAY, M. MALABRE : On the structure at infinity of linear block decouplable systems : the general case, IEEE Trans. on Automat. Contr., AC-28, 12, pp. 1115-1118, 1983.

[20] J. DESCUSSE, J.F. LAFAY, V. KUČERA : Decoupling by restricted static state feedback : the general case, IEEE Trans. on Automat. Contr., AC-29, 1, pp. 79-81, 1984.

[21] M. MALABRE, V. KUČERA : Infinite structure and exact model matching problem : a geometric approach, IEEE Trans. on Automat. Contr., AC-29, 3, pp. 266-268, 1984.

[22] M.D. DI BENEDETTO, A. ISIDORI : The matching of nonlinear models via dynamic state feedback, Rap. 04-84, Universita Degli Studi di Roma "La Sapienza", Italy.

[23] J. DESCUSSE, J.F. LAFAY, M. MALABRE : Further results on Morgan's problem, Systems & Control Letters, Vol. 4, 4, pp. 203-208, 1984.

[24] M. MALABRE : A complement about the structure of linear systems, Systems & Control Letters, Vol. 4, 6, pp. 339-342, 1984.

[25] J. DESCUSSE, J.F. LAFAY, M. MALABRE : Solution of the static state feedback decoupling problem for linear systems with 2 outputs, IEEE Trans. on Automat. Contr., AC-30,7, 1985.

[26] J.J. LOISEAU : Some geometric considerations about the Kronecker normal form, INT. J. Control., to appear.

[27] J.J. LOISEAU : Structural modifications of linear systems : a matrix pencil approach, 7th International Symposium on the Mathematical Theory of Networks and Systems, June 1985, Stockholm.

Frequency Domain and State Space Methods for Linear Systems
C.I. Byrnes and A. Lindquist (Editors)

DESCRIPTION OF PROPERTIES OF DECENTRALIZED SYSTEMS BY (A,B)-CHARACTERISTIC SUBSPACE

Zheng, Yu Fan

Mathematics Department, East-China Normal University
Shanghai, China

Han, Zheng Zhi, Shanghai Xhui District part-time College

The decentralized systems with ν stations, denoted as $\{c_i, A, B_i; i \in \underline{\nu}\}$, are discussed. The sufficient and necessary conditions for D-controllability and D-observability and for non-existence of decentralized fixed-mode are described in (A,B)-characteristic subspace. We also define the decentralized input-fixed-mode and output-fixed-mode of a given station, as well as their multiplicities. In addition, we give a formula to calculate these multiplicities. At the same time, same geometric characterizations of the multiplicity are advanced in a simple case.

INTRODUCTION

The decentralized controller configuration is one of the most important approaches to control a large scale system. In such configuration certain inputs and outputs of systems are naturally (according to locations, for example) associated with a definite control station and each controller is only provided with incomplete information of system´s outputs and can exert an effect upon the system only through certain local input channels.

Let us describe a decentralized system with ν stations in following:

$$\dot{x}(t) = Ax(t) + \sum_{i=1}^{\nu} B_i u_i(t) = Ax(t) + (B_1 B_2 \ldots B_\nu) \begin{vmatrix} u_1(t) \\ u_2(t) \\ \cdot \\ \cdot \\ \cdot \\ u_\nu(t) \end{vmatrix} \tag{1.1}$$

$$y(t) = \begin{vmatrix} y_1(t) \\ y_2(t) \\ \cdot \\ \cdot \\ \cdot \\ y_\nu(t) \end{vmatrix} = \begin{vmatrix} C_1 \\ C_2 \\ \cdot \\ \cdot \\ \cdot \\ C_\nu \end{vmatrix} x(t) \qquad (1.2)$$

or simply by $\{C_i, A, B_i;\ i \in \underline{\nu}\}$
where x, u_i, y_i are n, m_i, r_i dimesional vectors, respectively. A, B_i, C_i are real constant matrices of compatible orders, write $A \in \mathbb{R}^{n\times n}$, $B_i \in \mathbb{R}^{n\times m_i}$, $C_i \in \mathbb{R}^{r_i\times n}$; $i \in \underline{\nu} := \{1,2,\ldots,\nu\}$. Let $m = \sum_{i=1}^{\nu} m_i$, $r = \sum_{i=1}^{\nu} r_i$. Matrices B_i and C_1, $i \in \underline{\nu}$, are assumed, without loss of generality, to be of full rank.

There are different approaches to deal with the control problems concerning decentralized systems. In this paper the concept called (A,B)-characteristic subspace or/and (C,A)-characteristic subspace ((A,B)-ch.s. and (C,A)-ch.s. for short) are adapted to describe the properties of them under local output feedback transformations. The fundamental properties and structure of (A,B)-ch.s. have been studied intensively by authors [1],[2],[3]. Here the definition of it and some properties relevant are introduced briefly in what to follow.

Write $\mathbb{C}$ and $\mathbb{R}$ are complex and real fields, respectively. Given the compatible matrices $\{A,B\}$ and $\{C,A\}$, for any $\lambda \in \mathbb{C}$, we have

<u>Definition 1-1</u>

If $\{A,B\}$ is given, then

$$S(\lambda) := \{x \in \mathbb{C}^n / (A-\lambda I)x \in \operatorname{Im} B\}$$

is called the (A,B)-characteristic subspace of λ.
If $\{C,A\}$ is given, then

$$T(\lambda) := \{x \in \mathbb{C}^n / (A' - \lambda I)x \in \operatorname{Im} C\}$$

is called the (C,A)-characteristic subspace of λ.

Proposition 1-2:

$\{A,B\}$ is a controllable pair, if and only if for each $\lambda \in C$

$$\dim S(\lambda) = m = \text{Rank } B \tag{1.3}$$

$\{A,B\}$ is a compatible matrix pair.

Notations:

$R := \langle A/\text{Im}\,B \rangle$, the controllable subspace of $\{A,B\}$;

$\overline{X} := X/R$, where X is the state space;

$p : X \to \overline{X}$ be canonical projection;

$X(\lambda) := \ker(A-\lambda I)$, $\lambda \in C$;

$\overline{X(\lambda)} := pX(\lambda)$;

$Q : \overline{X} \to X$ be a monomorphism;

$S_c(\lambda) := S(\lambda) \cap R$.

Proposition 1-3:

(1) $\dim S(\lambda) \geq m$ (1.4)

(2) $S(\lambda) = S_c(\lambda) \oplus \overline{QX(\lambda)}$ (1.5)

Corollary 1-4:

$$\dim S_c(\lambda) = m \quad \text{for each } \lambda \in C \tag{1.6}$$

By above propositions and corollary, it becomes clear that the dimension of $\overline{X(\lambda)}$ gives a measurement of uncontollability of the systems $\{A,B\}$. If $\dim \overline{X(\lambda)} = p$, then, in order to reform the original systems completely controllable, we have to add at least p extra input channels, write $\overline{B} = (b_1 b_2 \ldots b_p)$, such that $(A,(B\overline{B}))$ is completely controllable.

Let $\overline{A}$ be the induced map of A on $\overline{X}(=X/\langle A/\text{Im}B\rangle)$.

Lemma 1-5;

$\overline{A}$ is cyclic matrix if and only if for each $\lambda \in C$, $\dim S(\lambda) \leq m+1$.

(2) On D-controllability and D-observability

Notations:

$R^i : = \langle A/ImB_i\rangle = ImB_i + A\ ImB_i + \ldots + A^{n-1}ImB_i,$

$A_{k_ik_j\ldots k_f} : = A + B_iK_iC_i + B_jK_jC_j + \ldots + B_fK_fC_f,$

$S^i(\lambda) : = (A,B_i)$-ch.s. of λ ,

$T^i(\lambda) : = (C_i,A)$-ch.s. of λ ,

$S^i_c(\lambda) : = S^i(\lambda) \cap R^i = S^i(\lambda)/R^i,$

$\bar{A}^i(\bar{A}^i_k)$ is the induced map on $X/R^i = \bar{X}^i$ by $A(A_k)$,

$S^i_{k_j}(\lambda) : = (A_{k_j},B_i)$-ch.s. of λ.

We start discussing from the simplest case when $\nu = 2$. Assume that $(A, (B_1B_2))$ is controllable and $R^1 \neq X$. Without loss of generality, let $ImB_2 \cap R^1 = 0$. If it is not so, write $ImB_2 = Im\bar{B}_2 \oplus \{ImB_2 \cap R^1\}$. Thus, $(A, (B_1\bar{B}_2))$ is controllable, and $Im\bar{B}_2 \cap R^1 = 0$. Furthermore, when $\lambda \in \sigma(\bar{A})$ (notice: when $\nu = 2$, write $\bar{A} : = \bar{A}^1$), $Im(A - \lambda I) + ImB_1 + ImB_2 \neq Im(A-\lambda I)+ImB_1$, which implies that $ImB_2 \cap \{Im(A - \lambda I) + ImB_1\}$, $\lambda \in \sigma(\bar{A})$, is a proper subspace in ImB_2, and $\bigcup_{\lambda\in\sigma(\bar{A})} \{[Im(A - \lambda I) + ImB_1] \cap ImB_2\}$ is a proper variety in ImB_2. Therefore, for almost all vector b_2 in ImB_2, it is true that

$$b_2 \bar{\in} \bigcup_{\lambda\in\sigma(\bar{A})} \{[Im(A - \lambda I) + ImB_1] \cap ImB_2\} \tag{2.1}$$

Theorem 2-1

There is $K_2 : Y_2 \to U_2$ which makes $\bar{A}_{k_2}$ $(: = \overline{A + B_2K_2C_2})$ cyclic if and only if

$$\dim \{KerC_2 \cap S^1(\lambda)\} \leq m_1 + 1 \tag{2.2}$$

for each $\lambda \in C$.

Proof: <only if> If there exists $\lambda_0 \in C$ such that $\dim\{\mathrm{Ker}C_2 \cap \cap S^1(\lambda_0)\} = m_1+p$, $p \geq 2$, then write $\mathrm{Ker}C_2 \cap S^1(\lambda_0) = = \mathrm{span}\{x_1, x_2, \ldots, x_{m_1+p}\}$. Furthermore, let $\{x_i;\ i \in \underline{n}\}$ be a basis for X. For any $K : Y_2 \to U_2$, $(A + B_2KC_2 - \lambda_0 I)x_i = (A - \lambda_0 I)x_i \in \mathrm{Im}B$, $i \in \underline{m_1+p}$, which implies that $x_1 \in \mathrm{Ker}\ C_2 \cap S^1(\lambda_0)$.

Thus, $\dim S_k^1(\lambda_0) \geq m_1 + p$. By proposition 1-3, lemma 1-5, $\bar{A}_k$ is not cyclic.

<If> If there exists $\lambda_0 \in \sigma(\bar{A})$ such that $\dim \bar{X}_{\lambda_0} = m_1 + p$, $p \geq 2$, then write $S^1(\lambda_0) = \mathrm{span}\{x_1, x_2, \ldots, x_{m_1+p}\}$. Thus there exists at least one vector, e.g., $x_1 \in S^1(\lambda_0)$ and $x_1 \in \mathrm{Ker}\ C_2$. Define $K(\in R^{m_2 \times r_2})$ in following

$$B_2KC_2 : \quad \begin{array}{l} x_1 \to b_2, \\ x_i \to 0, \text{ when } x_1 \text{ is real vector}, \\ (i \geq 2); \end{array}$$

$$\text{or} \quad \begin{array}{l} x_1 \to b_2 \\ x_2 = \bar{x}_1 \to \bar{b}_2 = b_2, \text{ when } x_1 \text{ is complex vector}, \\ x_i \to 0, \\ (i \geq 3), \end{array}$$

where $b_2(\in \mathrm{Im}B_2)$ satisfies (2.1).

When $\lambda \in \sigma(\bar{A})$, for each $X \in S_k^1(\lambda) := (A + B_2KC_2, B_1)$-ch.s. of λ, write $X = \sum_{i=1}^{n} \alpha_i X_i$. $(A + B_2KC_2 - \lambda I)x = (A - \lambda I)x + \alpha_1 b_2 = b_1 \in \mathrm{Im}B_1$.

By (2.1), $\alpha_1 b_2 = 0$. Thus, $x \in S^1(\lambda)$, which implies that $\dim S_k^1(\lambda) \leq \dim S^1(\lambda)$, when $\lambda \in \sigma(\bar{A})$,

If $\lambda = \lambda_0$, then $x_1 \in S^1(\lambda)$. But $(A + B_2KC_2 - \lambda I)x_1 = = (A - \lambda I)x + b_2 = b_1 + b_2 \bar{\in} \mathrm{Im}B_1$, which implies that

$$\dim S_k^1(\lambda_0) < \dim S^1(\lambda_0).$$

When $\lambda \bar{\in} \sigma(\bar{A})$, for each $x \in S_k^1(\lambda)$, $x = \sum_{i=1}^{n} \alpha_i x_i$. If $x \bar{\in} S_{kc}^1(\lambda)$:

$= S_k^1(\lambda) \cap R^1$, write $P: X \to \bar{X} = X/R^1$, the canonical projection, $(A + B_2KC_2 - \lambda I)x = (A - \lambda I)x + \alpha_1 b_2 = b_1$. Therefore, $(\bar{A} - \lambda I)\bar{x} + \alpha_1 \bar{b}_2 = 0$, or say, $\bar{x} \in (\bar{A} - \lambda I)^{-1} \mathrm{Im}\bar{b}_2$, which implies that

$$\dim \overline{S_k^1(\lambda)} = \dim PS_k^1(\lambda) = 1.$$

On the other hand, if $x \in S_k^1(\lambda) \cap R^1$, then it is valid that $\alpha_1 b_2 = 0$ and $x \in S_{kc}^1(\lambda)$. Therefore, when $\lambda \in \sigma(\bar{A})$, $\dim S_k^1(\lambda) \leq m_1 + 1$ as $\dim S_{kc}^1(\lambda) = m_1$.

Repeating the forgoing procedure in finite steps, there exists $K_2 : Y_2 \to U_2$ such that $\dim S_{k_2}^1(\lambda) \leq m_1 + 1$, which implies that $\bar{A}_{k_2}$ is cyclic by lemma 1-5.

Corollary 2-2:

There exists $\lambda \in \sigma(A)$ such that $\dim\{\mathrm{Ker}C_2 \cap S^1(\lambda)\} = m_1 + p$ $(p \geq 1)$, then we have to add at least new input channels in order to make $(A_{k_2}, (B_1\ b_1\ b_2 \ldots b_p))$ controllable provided that $K_2 : Y_2 \to U_2$ is chosen properly.

Lemma 2-3:

$\{C, A, B\}$ is given, then

(1) $\dim\{\mathrm{Ker}C \cap S(\lambda)\} = n + m - \mathrm{rank} \begin{vmatrix} \lambda I - A & -B \\ C & 0 \end{vmatrix}$

(2) $\dim\{\mathrm{Ker}B' \cap T(\lambda)\} = n + r - \mathrm{rank} \begin{vmatrix} \lambda I - A' & -C' \\ B' & 0 \end{vmatrix}$

Proof: The two results are dual to each other. We can varify them by the dimension theorem of linear map.

Lemma 2-4:

provided that $\dim\{\mathrm{Ker}C \cap S(\lambda)\} \leq m$, the following statements are equivalent:

(1) $$\text{rank}\begin{vmatrix} \lambda I-A & B \\ C & 0 \end{vmatrix} \equiv n \tag{2.5}$$

(2) $$\dim\{\text{Ker}C \cap S(\lambda)\} \equiv m \tag{2.6}$$

(3) $$C < A \,/\, \text{Im}> = 0 \tag{2.7}$$

Proof: We varify that (1) and (3) are equivalent to each other only. If (3) holds, then $C\,\text{Im}(B\ AB \ldots A^{n-1}B \ldots) = 0$, i.e. $C(\lambda I-A)^{-1}B \equiv 0$. Write (A,B) in Kalman controllability form

$$\begin{vmatrix} A_{11} & A_{12} & \cdot & B_1 \\ 0 & A_{22} & \cdot & 0 \\ \cdots & \cdots & \cdots & \cdots \\ C_1 & C_2 & \cdot & 0 \end{vmatrix} \tag{2.8}$$

Thus, $0 \equiv C(\lambda I-A)^{-1}B = C_1(\lambda I-A_{11})^{-1}B_1$. As (A_{11},B_1) is controllable, $C_1 = 0$. By lemma 2-3,

$$\text{Rank}\begin{vmatrix} \lambda I-A_{11} & -A_{12} & B_1 \\ 0 & \lambda I-A_{22} & 0 \\ 0 & C_2 & 0 \end{vmatrix} \geq n$$

But $\text{Rank}\begin{vmatrix} \lambda I-A_{22} \\ C_2 \end{vmatrix} \leq \dim A_{22}$, which implies that

$\text{Rank}\begin{vmatrix} \lambda I-A & B \\ C & 0 \end{vmatrix} \equiv 0$. Therefore, (1) holds.

On the other hand, if (1) holds, then transform (2.8) into following

$$\begin{vmatrix} A_{11}^1 & A_{11}^2 & A_{12}^3 & B_{11} \\ 0 & A_{11}^3 & A_{12}^2 & B_{12} \\ 0 & 0 & A_{22} & 0 \\ 0 & C_{12} & C_2 & 0 \end{vmatrix} \tag{2.9}$$

In (2.9) (C_{12},A_{11}^3,B_{12}) is complete controllable and observable. If $\lambda \overline{\in} \sigma(A_{11}^3) \cup \sigma(A_{11}^3) \cup \sigma(A_{22}) = \Sigma$, then, by (2.9) $n \equiv$

$$\text{Rank}\begin{vmatrix} \lambda I-A & B \\ C & 0 \end{vmatrix} = n_1+n_2+\text{Rank}\begin{vmatrix} \lambda I-A_{11}^3 & B_{12} \\ C_{12} & 0 \end{vmatrix} =$$

$$= n_1 + n_2 + \text{Rank}\begin{vmatrix} (\lambda I-A_{11}^3) & B_{12} \\ 0 & C_{12}(\lambda I-A_{11}^3)^{-1}B_{12} \end{vmatrix},$$

which implies that $C_{12}(\lambda I-A_{11}^3)^{-1}B_{12} \equiv 0$, as Σ is a finite subset of C. As (A_{11}^3,B_{12}) is controllable, $C_{12} = 0$. Therefore, (3) holds.

Theorem 2-5

There exists $K_2 : Y_2 \to U_2$ such that (A_{k_2},B_1) is controllable if and only if

(1) $\dim\{\text{Ker}C_2 \cap S^1(\lambda)\} \leq m_1$ (2.10)

(2) $\dim\{\text{Ker}C_2 \cap S^1(\lambda)\} \not\equiv m_1$ (2.11)

(2.11) implies that there exists $\lambda_0 \in C$ such that

$$\dim\{\text{Ker}C_2 \cap S^1(\lambda_0)\} < m_1 \tag{2.11'}$$

(2.10) and (2.11) can be written in

$$m_1 \not\equiv \dim\{\text{Ker}C_2 \cap S^1(\lambda)\} \leq m_1 \tag{2.12}$$

By means of Lemma 2-4, using a way similar to the proof of Theorem 2-1, it is not difficult to varify it. The proof is omitted to save space.

Remark 1: (2.10) is equivalent to that

$$r_2 \not\equiv \dim\{\text{Ker}B_1' \cap T^2(\lambda)\} \leq r_2 \tag{2.13}$$

Remark 2: (2.12) is the connectiveness condition. If at the same time, the inequality that

$$\dim\{\text{Ker}B_2' \cap T^1(\lambda)\} < r_2 \quad \text{for some } \lambda \in C \tag{2.14}$$

holds, then the system is strong connective. (2.14) is equivalent to

$$\dim\{\mathrm{Ker}C_1 \cap S^2(\lambda)\} < m_2 \tag{2.15}$$

Remark 3: The property that (A_{k_2}, B_1) be controllable is generic, $K_2 \in R^{m_2 \times r_2}$.

Corollary 2-6

There exists $K_2 : Y_2 \to U_2$ such that (C_1, A_{k_2}) is observable if and only if

$$r_1 \not\equiv \dim\{\mathrm{Ker}B_2' \cap T^1(\lambda)\} \leq r_1 \tag{2.16}$$

or

$$m_2 \not\equiv \dim\{\mathrm{Ker}C_1 \cap S^2(\lambda)\} \leq m_2 \tag{2.17}$$

The following lemma is an important midstep to generalized Theorem 2-8 into the case when $\nu > 2$.

Lemma 2-7

When $\nu = 3$ in (1.1), there exists local feedbacks $K_2 : Y_2 \to U_2$ and $K_3 : Y_3 \to U_3$ such that (A_{k_2,k_3}, B_1) is controllable if and only if

1) $m_1 + m_3 \not\equiv \dim\{\mathrm{Ker}C_2 \cap S^{1,3}(\lambda)\} \leq m_1 + m_3$ (2.18)

2) $m_1 + m_2 \not\equiv \dim\{\mathrm{Ker}C_3 \cap S^{1,2}(\lambda)\} \leq m_1 + m_2$ (2.19)

3) $m_1 \not\equiv \dim\{\mathrm{Ker}C_{2,3} \cap S^1(\lambda)\} \leq m_1$ (2.20)

where $C_{2,3} := \begin{vmatrix} C_2 \\ C_3 \end{vmatrix}$, $S^{i,j}(\lambda) := (A, (B_i B_j))$-ch.s. of λ

Proof: (only if) If K_2, K_3 are given to make $(A_{k_2,k_3} B_1)$ controllable, then

$$(A + (B_2 B_3) \begin{vmatrix} K_2 & 0 \\ 0 & K_3 \end{vmatrix} \begin{vmatrix} C_2 \\ C_3 \end{vmatrix}, B_1)$$

is controllable, which implies that (2.20) holds. Furthermore, both

$(A_{k_2},(B_1B_3))$ and $(A_{k_3},B_1B_2))$ are controllable, then (2.18) and (2.19) hold.

(If) By (2.18), there exists $K_2 \in R^{m_2 \times r_2}$ such that $(A+B_2K_2C_2,(B_1B_3))$ is controllable. As $\dim\{\mathrm{Ker}C_{2,3} \cap S^1(\lambda)\} = \dim\{\mathrm{Ker}C_{2,3} \cap S^1_{k_2}(\lambda)\}$, where $S^1_{k_2}(\lambda)$ is (A_{k_2},B_1)-ch.s.. There exists $(\bar{K}_2\bar{K}_3)$ such that (A^*,B_1) is controllable, where $A^* := A + B_2K_2C_2 + B_3(\bar{K}_2\bar{K}_3)\begin{vmatrix} C_2 \\ C_3 \end{vmatrix} =$

$= A + B_3\bar{K}_3C_3 + (B_2B_3)\begin{vmatrix} K_2 \\ \bar{K}_2 \end{vmatrix} C_2.$

By Theorem 2-5, we have $m_1 \neq \dim\{\mathrm{Ker}C_2 \cap S^1_{k_3}(\lambda)\} \leq m_1$.

According to the Remark 3 of Theorem 2-5, it is possible to choose $K_3 \in R^{m_3 \times r_3}$, which satisfies that (1) $(A + B_3K_3C_3, (B_1B_2))$ is controllable; (2) $m_1 \neq \dim\{\mathrm{Ker}C_2 \cap S^1_{k_3}(\lambda)\} \leq m_1$ simulteneously.

Using Theorem 2-5 again, we find out K_2, such that $(A_{k_2k_3},B_1)$ is controllable.

Up to now the results we give are concerning about D-controllability of decentralized systems. The corresponding results about D-observability can be obtained by means of dual principle of controllability and observability.

By T_1 we denote the set of proper subsets of $\underline{\nu}:=\{1,2,\ldots,\nu\}$, which contain $\{1\}$.e.g., $\nu=3$, $T_1 = \{\{1\}, \{1,2\}, \{1,3\}\}$. The condition of Lemma 2-7 can be described as $\sum_{i\in\tau} m_i \neq \dim\{\mathrm{Ker}C_{\bar{\tau}} \cap S^{\tau}(\lambda)\} \leq \sum_{i\in\tau} m_i$ for any $\tau \in T_1$, $\tau \cap \bar{\tau} = \phi$ and $\tau \cup \bar{\tau} = \underline{\nu}$.

Now turn to study the case that $\nu > 3$.

<u>Theorem 2-8</u>

There exists $K_2,K_3,\ldots,K_\nu,K_i \in R_i^{m_i \times r_i}$; $i \in \{2,3,\ldots,\nu\}$, such that $(A_{k_2k_3\ldots k_\nu},B_1)$ is controllable if and only if for any $\tau \in T_1$

$$\sum_{i\in\tau} m_i \neq \dim\{\mathrm{Ker}C_{\bar{\tau}} \cap S^{\tau}(\lambda)\} \leq \sum_{i\in\tau} m_i \qquad (2.21)$$

Remark:

$\tau = \{i_1, i_2, \ldots, i_j\}$, $S^{\tau}(\lambda) := (A, (B_{i_1} B_{i_2} \ldots B_{i_j}))$-ch.s. of λ and $\bar{\tau} = \{i_{j+1}, i_{j+2}, \ldots, i_{\nu}\}$, $C' := (C'_{i_{j+1}} C'_{i_{j+2}} \ldots C'_{i_{\nu}})$. $\bar{\tau}$ is the complementary set of τ relative to $\underline{\nu}$.

Proof: Let τ_1 be a proper subset of $\underline{\nu-1}$ containing $\{1\}$. By (2.21), we have

$$\sum_{i \in \tau_1} m_i + m_{\nu} \neq \dim\{\mathrm{Ker} C_{\bar{\tau}_1} \cap S^{(\tau_1, \nu)}(\lambda)\} \leq \sum_{i \in \tau} m_i \; m_{\nu} ,$$

$$\sum_{i=1}^{\nu-1} m_i \neq \dim\{\mathrm{Ker} C_{\nu} \cap S^{\underline{\nu-1}}(\lambda)\} \leq \sum_{i=1}^{\nu-1} m_i ,$$

$$\sum_{i \in \tau_1} m_i \neq \dim\{\mathrm{Ker} C_{\bar{\tau}_1, \nu} \cap S^{\tau_1}(\lambda)\} \leq \sum_{i \in \tau} m_i .$$

In a way similar to the proof of Lemma 2-7, there exists $K_{\nu}: Y_{\nu} \to U_{\nu}$ such that both $(A + B_{\nu} K_{\nu} C_{\nu}, (B_{\tau_1} B_{\bar{\tau}_1}))$ and $(A_{k_{\nu}} + B_{\bar{\tau}_1} K_{\bar{\tau}_1} C_{\bar{\tau}_1}, B_{\bar{\tau}_1})$ are controllable. ($K_{\bar{\tau}_1}$ is not block diagonal matrix). The later is equivalent to

$$\sum_{i \in \tau} m_i \neq \dim\{\mathrm{Ker} C_{\bar{\tau}_1} \cap S_{k_{\nu}}^{\tau_1}(\lambda)\} \leq \sum_{i \in \tau} m_i .$$

By induction the theorem is varified.

Corollary 2-9

For any $j \in \underline{\nu}$ there exists $K_i, K_i \in R^{m_i \times r_i}$, $i \in \underline{\nu}$, such that $(C_j, A_{k_1 k_2 \ldots k_{\nu}}, B_j)$ is complete controllable and observable if and only if for any two proper subsets of $\underline{\nu}$ that $\tau_i \neq \underline{\nu}$ $(i = 1,2)$ $\tau_1 \cap \tau_2 = \phi$

$$\sum_{i \in \tau_1} m_i \neq \dim\{\mathrm{Ker} C_{\tau_2} \cap S^{\tau_1}(\lambda) \leq \sum_{i \in \tau} m_i \tag{2.22}$$

By duality, it is easy for us to have all corresponding results about D-observability.

(3) On Decentralized Fixed Modes

Decentralized fixed mode is a key concept for decentralized systems. The first definition was given by Wong and Davidson [4], (1973). In this paper we go further to classify them into input fixed modes and output fixed modes, and to define the multiplicity of them. As the space limitation we don´t give the proof in detail.

system we deal with is still (1.1). Define

$$\kappa := \{K;\ K = \text{block diag}(K_1 K_2 \ldots K_\nu);\ K_i \in R^{m_i \times r_i}\}$$

Definition 3-1

(1) $$\Lambda_i^{\bar{c}} := \bigcap_{\substack{K \in \kappa \\ F \in R^{m_i \times n}}} \sigma(A + BKC + B_i F)$$

$\Lambda_i^{\bar{c}}$ is named the input fixed mode set of i-th channel of (1.1).

(2) $$\Lambda_j^{\bar{o}} := \bigcap_{\substack{K \in \kappa \\ G \in R^{n \times r_j}}} \sigma(A + BKC + GC_j);$$

$\Lambda_j^{\bar{o}}$ is named the output fixed mode set of j-th channel of (1.1).

By comparison with the definition of fixed mode set denoted by Λ, it is obvious that for any $i,j \in \underline{\nu}$

$$\Lambda_i^{\bar{c}} \subset \Lambda \quad \text{and} \quad \Lambda_j^{\bar{o}} \subset \Lambda$$

We define that $\Lambda_{\bar{c}} = \bigcup_{i \in \underline{\nu}} \Lambda_i^{\bar{c}}$ and $\Lambda_{\bar{o}} = \bigcup_{j \in \underline{\nu}} \Lambda_j^{\bar{o}}$.

Thus, the following proposition can be obtained by the results of Section (2).

Proposition 3-2:

1) $\lambda \in \Lambda_{\bar{c}}$ if and only if there exists $j \in \underline{\nu}$ such that for any $K \in \kappa$

$$\dim S_k(\lambda) > m_j \tag{3.1}$$

where $S_k^j(\lambda) := (A + \sum_{i=1}^{\nu} B_i K_i C_i, B_j)$-ch.s. of λ.

2) $\lambda \in \Lambda_{\bar{o}}$ if and only if there exists $j \in \underline{\nu}$ such that for any $K \in \kappa$

$$\dim T_k^j(\lambda) > r_j \tag{3.2}$$

where $T_k^j(\lambda) = (C_j, A + \sum_{i=1}^{\nu} B_i K_i C_i)$-ch.s. of λ.

Corollary 3-3:

1) $\Lambda_j^{\bar{o}} = \phi$ if and only if for almost all K in κ, (A_k, B_i) is controllable pair.

2) $\Lambda_j^{\bar{o}} = \phi$ if and only if for almost all K in κ, (C_j, A_k) is observable pair.

Theorem 3-4:

$$\Lambda = \Lambda_i^{\bar{c}} \cup \Lambda_i^{\bar{o}} \quad \text{for each } i \in \underline{\nu}$$

Corollary 3-5:

$\Lambda = \phi$ if and only if for all $\lambda \in C$, and any proper subset τ of $\underline{\nu}$

$$\dim\{\mathrm{Ker}C_{\bar{\tau}} \in S^{\tau}(\lambda)\} \leq \sum_{i \in \tau} m_i \tag{3.3}$$

Corollary 3-6:

$$\Lambda = \Lambda_{\bar{c}} = \Lambda_{\bar{o}} \tag{3.4}$$

By Theorem 3-4 and its corollary 3-6, in order to study the multiplicity of fixed mode it is enough to study the multiplicity of input-fixed mode or output-fixed mode. Now we turn to study input-fixed mode and start our discussing from the case that $\nu = 2$, i.e.

the system $\{C_i, A, B_i,\ i \in \{1,2\}\}$. Let $C' = (C_1' \ C_2')'$ and $B = (B_1 B_2)$, write $R_k^1 = \langle A_k/\mathrm{Im}B_1\rangle$, $\bar{A}_k^1$ is the induced map of A_k on $X/R_k^1 = \bar{X}_k^1$. The characteristic subspace of $\bar{A}_k^1$ is denoted by $\overline{X_k^1(\lambda)}(:=\mathrm{Ker}(\bar{A}_k^1-\lambda I))$.

Definition 3-7:

$\lambda(\in \Lambda_i^{\bar{c}})$ is called a l_λ^i multi-input fixed mode of (A,B_i) if

$l^i = \min_{k\in\kappa} \dim \overline{X_k^i(\lambda)}$.

Under the assumption that $\{C_i, A, B_i;\ i \in \underline{2}\}$ be strong connective system, if $k \in \kappa$ is found to make that

$$l_\lambda^i = \dim \overline{X_k^i(\lambda)}.$$

Then we say, (A_k, B_i) has the maximal controllability structure (m.c.s. for short) of B_i. Obviously, even if (A_k, B_i) has m.c.s., this λ is still an uncontrollable mode of (A_k, B_i). Without loss of generality, let $i=1$.

Theorem 3-8:

For each $\lambda \in \Lambda_i^{\bar{c}}$

$$l_\lambda^1 = \dim\{\mathrm{Ker}C_2 \cap S^1(\lambda)\} - m_1 \tag{3.5}$$

Theorem 3-9:

(1) For almost all $K \in \kappa$, (A_k, B_i) has m.c.s. of B_i

(2) If K is found that (A_k, B_i) has m.c.s., then there exists exactly l_λ^i independent eigenvectors of λ, which are unobservable by C_2 and uncontrollable by B_1 simultaneously.

Remark: In general, if K^* is given to make (A_{K^*}, B_1) have m.c.s., then it is not always deduced that $\dim \langle A_{k^*}/\mathrm{Im}B_1\rangle = \max_{k\in K} \dim\langle A_k/\mathrm{Im}B_1\rangle$. But we conclude that the cyclic index of $\bar{A}_{k^*}^1$ is minimal. When $\nu=2$, the calculations of l_λ^i can be done by Theorem 3-8. In the general case, the formula is quite complicate. We will discuss it in the other paper.

References

[1] Zheng, Yu Fan and Han, Zheng Zhi, "What can be done for linear systems by feedback?", Proceedings to IFAC'84 1984.

[2] Zheng, Yu Fan and Han, Zheng Zhi "On (A,B)-characteristic subspace of linear systems: Fundamental properties and its structure". CONTROL THEORY AND ITS APPLICATIONS Vol.1.No.3 1984. (In Chinese)

[3] Zheng, Yu Fan and Han, Zheng Zhi, "On th local feedback in decentralized control systems", Science Report of Mathematics Department, East-China Normal University, China. 1983. (In Chinese)

[4] Wong, S.H. and Davison, E.J., "On the stability of decentralized control systems" IEEE AC-18 No.5 pp. 437-478, 1973.

8. RICCATI AND LYAPUNOV EQUATIONS

Frequency Domain and State Space Methods for Linear Systems
C.I. Byrnes and A. Lindquist (Editors)

Singular Riccati Equations for Singular Systems

Douglas J. Bender

Dept. of Electrical and Computer Engineering
University of California
Santa Barbara, CA 93106

ABSTRACT

In this paper we extend Silverman's structure algorithm to descriptor systems. Our work is in the spirit of [5] in that we concentrate on the relationships of the structure algorithm to the discrete-time Riccati equation and to the geometric subspace recursions introduced by Wonham [10] and others. The key to this extension is the use of a special coordinate system derived by performing a singular value decomposition of the leading matrix E. Using this coordinate system we define a system much like a state-space system whose state vector has dimension r, the rank of E. It is easy to derive a structure algorithm for this system. We prove that this algorithm can also be used to compute $\mathbf{V}^*$ for the original descriptor system. We derive a singular Riccati equation whose solution has dimension equal to the rank of E.

I. Introduction

In this paper we extend some well-known results from linear system theory to the discrete-time descriptor system

$$Ex_{k+1} = Ax_k + Bu_k, \tag{1a}$$

$$y_k = Cx_k + Du_k, \tag{1b}$$

where $x_k \in \mathbf{R}^n$, $u_k \in \mathbf{R}^m$, $y_k \in \mathbf{R}^p$, E, A, B, C, and D are conformably dimensioned, real matrices, and $\text{rank}(E) = r \leq n$. E and A are assumed to be square and the pencil $(zE-A)$ is assumed to be regular (i.e., $|zE-A| \not\equiv 0$). We define an iterative geometric subspace algorithm for computing $\mathbf{V}^*$, the supremal output-nulling controlled invariant subspace (or unknown-input unobservable subspace) for the system (1). This is called the $\mathbf{V}^*$ algorithm. We then present the descriptor system version of Silverman's "left structure algorithm," which is a matrix realization of the $\mathbf{V}^*$ algorithm. We also derive the associated singular discrete-time control Riccati difference equation. The solution P_i of this Riccati equation is only an $r\times r$ matrix. From the solution at the i'th iteration of this equation an $n\times n$ matrix can be formulated whose kernel is $\mathbf{V}^{(i)}$, the subspace computed at the i'th iteration of the $\mathbf{V}^*$ algorithm. We shall also show that the associated feedback gains ("friends of $\mathbf{V}^{(i)}$") can be computed from this Riccati solution. There are up to $m(n-r)$ more degrees of freedom in the choice of this feedback gain than in the state space version. In the case of the nonsingular problem ($D^TD > 0$) for state-space systems $\mathbf{V}^*$ is n-dimensional and there is one unique friend of $\mathbf{V}^*$. However, in descriptor systems even in the nonsingular case there are up to $m(n-r)$ degrees of freedom in the choice of friends.

The basic theory of descriptor systems is fairly well worked out; for example, controllability and other abilities are treated in, e.g., [3] and [4]. This paper extends some more advanced topics in linear system theory to descriptor systems. Verghese [9] defined the concept of controlled invariant subspaces for descriptor systems; he gave the same definition which we give here. Van der Weiden & Bosgra [8] derived rather exhaustive results on the structural properties (poles, zeros, controllability,

*This research was supported by a Hughes Aircraft Company Doctoral Fellowship.

observability, etc) of the alternate system, which we will define later. However, they did not investigate the subspace recursions which we examine here.

In Section II we present a coordinate system which can be derived by using the singular value decomposition of E. Naturally, we call this an "SVD coordinate system." We then use this coordinate system to define a state-space-like system whose state vector has dimension r, which we call the "alternate system."

In Section III we briefly state some results related to $\mathbf{V}^*$ and the $\mathbf{V}^*$ algorithm for descriptor systems.

In Section IV we present two versions (nested and "stacked") of the structure algorithm and show that they compute $\mathbf{V}^*$. We then show how a certain discrete-time optimal control Riccati equation for the alternate system can be derived from the structure algorithm. We also give some simple results which relate this Riccati equation to the optimal regulator problem for the system (1).

First we discuss the notation to be used.

Notation

Vectors are represented by lower-case italic letters (e.g., x), matrices are represented by italic capital letters (e.g., A), and linear subspaces are represented by boldface capital letters (e.g., $\mathbf{Q}$). The range space of matrix is represented by the corresponding boldface capital letter; thus, the range space of A is represented by $\mathbf{A}$. $\mathbf{R}^n$ represents Euclidean n-space. If $x \in \mathbf{R}^n$ then $\mathbf{X} = \mathbf{R}^n$. The zero subspace is represented by 0; i.e., $\mathbf{0} = \{ x \in \mathbf{X} \mid x = 0 \}$. The orthogonal complement of the subspace $\mathbf{Q}$ is denoted by $\mathbf{Q}^{\perp}$. If A is a linear map with $A : \mathbf{X} \to \mathbf{Y}$, and $\mathbf{S} \subset \mathbf{X}$ and $\mathbf{Q} \subset \mathbf{Y}$ are linear subspaces, then the image function $A\mathbf{S}$ and the inverse image function $A^{-1}\mathbf{Q}$ are defined by

$$A\mathbf{S} = \{y \in \mathbf{Y} \mid y = As \text{ for some } s \in \mathbf{S}\},$$

and $A^{-1}\mathbf{Q} = \{x \in \mathbf{X} \mid Ax \in \mathbf{Q}\}$.

II. SVD Coordinate System for Descriptor Systems and the Alternate System

The key to deriving the results in this paper is the singular value decomposition of E:

$$UEV = \begin{bmatrix} \Sigma^2 & 0 \\ 0 & 0 \end{bmatrix}, \tag{2}$$

where $\Sigma^2 = diag(\sigma_1, \sigma_2, \ldots, \sigma_r, 0, \ldots, 0)$ (the σ_i's are the nonzero singular values of E) and U and V are square orthogonal matrices. If (1) is transformed to the corresponding coordinate system, it becomes:

$$\begin{bmatrix} \Sigma^2 & 0 \\ 0 & 0 \end{bmatrix} \begin{bmatrix} x_{1_{k+1}} \\ x_{2_{k+1}} \end{bmatrix} = \begin{bmatrix} A_{11} & A_{21} \\ A_{21} & A_{22} \end{bmatrix} \begin{bmatrix} x_{1_k} \\ x_{2_k} \end{bmatrix} + \begin{bmatrix} B_1 \\ B_2 \end{bmatrix} u_k, \tag{3a}$$

$$y_k = \begin{bmatrix} C_1 & C_2 \end{bmatrix} \begin{bmatrix} x_{1_k} \\ x_{2_k} \end{bmatrix} + Du_k \tag{3b}$$

where $\begin{bmatrix} x_{1_i} \\ x_{2_i} \end{bmatrix} = V^T x_i$.

We say that the descriptor system (3) is in an SVD coordinate system. The coordinate system (3) is not unique (specifically we do not constrain the order of the singular values of E in Σ^2); therefore, we say a system is in *an* SVD coordinate system, instead of *the* SVD coordinate system.

In the sequel we shall assume without loss of generality that the system (1) is in an SVD coordinate system (3). It is not absolutely necessary to use the singular value decomposition to derive the coordinate system (3). The essential idea is that a row and column compression of E is done; i.e., that the matrix Σ^2 be square and invertible.

Alternate System

In this paper we shall be interested in finding results pertinent to the descriptor system (1) or (3). However, much of the theory will be motivated by considering an r-dimensional state-space-like system.

Definition 1 Alternate System

Assume that the system (1) is in an SVD coordinate system. Then the system

$$\Sigma^2 x_{1_{k+1}} = A_{11} x_{1_k} + \begin{bmatrix} A_{12} & B_1 \end{bmatrix} \begin{bmatrix} x_{2_k} \\ u_k \end{bmatrix}, \tag{4a}$$

$$\begin{bmatrix} y_{1_k} \\ y_{2_k} \end{bmatrix} = \begin{bmatrix} A_{21} \\ C_1 \end{bmatrix} x_{1_k} + \begin{bmatrix} A_{22} & B_2 \\ C_2 & D \end{bmatrix} \begin{bmatrix} x_{2_k} \\ u_k \end{bmatrix}. \tag{4b}$$

where $x_{1_k} \in \mathbf{R}^r$, $\begin{bmatrix} x_{2_k} \\ u_k \end{bmatrix} \in \mathbf{R}^{\bar{m}}$, and $\begin{bmatrix} y_{1_k} \\ y_{2_k} \end{bmatrix} \in \mathbf{R}^{\bar{p}}$, where $\bar{m} \triangleq m+(n-r)$ and $\bar{p} \triangleq p+(n-r)$ is called the *alternate system*.

Further, $\begin{bmatrix} x_{2_k} \\ u_k \end{bmatrix}$, is called the *extended control vector*, and $\begin{bmatrix} y_{1_k} \\ y_{2_k} \end{bmatrix}$ is the *extended output vector*. ■

Remarks 1

1) If $\Sigma^2 = I$ or if the first equation in (4) is premultiplied by Σ^{-2} (4) becomes a state-space system. The approach taken in this paper will be to use slightly modified state-space techniques on (4) and relate the results back to (3).
2) The systems (3) and (4) are not equivalent because y_{1_k} is assumed always to be equal to zero in (3) and it is not necessarily so in (4).
3) The number of finite and infinite dynamic modes of the system (1) is equal to the rank of E; thus, the dimension of the descriptor vector of (4) is equal to the number of dynamic modes of (1) as well as (4). However, none of the dynamic modes of (4) are infinite while they can be in (1).
4) As stated in the introduction, Van der Weiden & Bosgra [8] derived many results directly for the system (4). ■

III. $\mathbf{V}^*$ and the Structure Algorithm

In this Section we motivate the theory to follow by making some basic statements on geometric control theory for descriptor systems. We define output-nulling controlled invariant subspaces and state the subspace recursion for computing $\mathbf{V}^*$, the supremal such subspace.

Definition 2a. Output-Nulling Controlled Invariant Subspace for the Descriptor System

We call any linear subspace $\mathbf{V}$ of the descriptor space $\mathbf{R}^n$ of (1) an output-nulling controlled invariant subspace for (1) if it satisfies

$$\begin{bmatrix} A \\ C \end{bmatrix} \mathbf{V} \subset \begin{bmatrix} E \\ 0 \end{bmatrix} \mathbf{V} + \begin{bmatrix} \mathbf{B} \\ \mathbf{D} \end{bmatrix}. \quad ■ \tag{5}$$

Definition 2b. Output-Nulling Controlled Invariant Subspace for the Alternate System

We call any linear subspace $\mathbf{V}_1$ of the descriptor space $\mathbf{R}^r$ of (4) an output-nulling controlled invariant subspace for (4) if it satisfies

$$\begin{pmatrix} A_{11} \\ A_{21} \\ C_1 \end{pmatrix} \mathbf{V}_1 \subset \begin{pmatrix} \Sigma^2 \\ 0 \\ 0 \end{pmatrix} \mathbf{V}_1 + \begin{pmatrix} \mathbf{A}_{12} & \mathbf{B}_1 \\ \mathbf{A}_{22} & \mathbf{B}_2 \\ \mathbf{C}_2 & \mathbf{D} \end{pmatrix}. \quad ■ \tag{6}$$

The usual feedback characterization of an output-nulling controlled invariant subspace extends to the descriptor system (1) and the alternate system (4).

Lemma 1a

The subspace $\mathbf{V}$ is an output-nulling controlled invariant subspace for (1) if and only if there exists a matrix K such that

$$\left[\begin{pmatrix} A \\ C \end{pmatrix} + \begin{pmatrix} B \\ D \end{pmatrix} K \right] \mathbf{V} \subset \begin{pmatrix} E \\ 0 \end{pmatrix} \mathbf{V}. \tag{7}$$

Lemma 1b

The subspace $\mathbf{V}_1$ is an output-nulling controlled invariant subspace for (4) if and only if there exists a matrix K' such that

$$\left[\begin{pmatrix} A_{11} \\ A_{21} \\ C_1 \end{pmatrix} + \begin{pmatrix} A_{12} & B_1 \\ A_{22} & B_2 \\ C_2 & D \end{pmatrix} K' \right] \mathbf{V}_1 \subset \begin{pmatrix} \Sigma^2 \\ 0 \\ 0 \end{pmatrix} \mathbf{V}_1. \tag{8}$$

Proof

The proofs of both parts of this Lemma are much like the one for state-space systems [10, Lemma 4.2]. ■

Definition 2a.

If $\mathbf{V}$ is an output-nulling controlled invariant subspace for the descriptor system (1) we call any descriptor variable feedback matrix K satisfying (7). a ***friend*** of $\mathbf{V}$. Notationally, we denote the set of friends of $\mathbf{V}$ by $\mathbf{F}(\mathbf{V})$. ■

Definition 2b.

If $\mathbf{V}_1$ is an output-nulling controlled invariant subspace for the alternate system (4) we call any feedback matrix K' satisfying (8). a ***friend*** of $\mathbf{V}_1$. Notationally, we denote the set of friends of $\mathbf{V}_1$ by $\mathbf{F}'(\mathbf{V}_1)$. ■

It is well-known that for state-space systems there exists a supremal output-nulling controlled invariant subspace; this property extends to descriptor systems:

Lemma 2a

For the descriptor system (1) there exists a unique supremal output-nulling controlled invariant subspace satisfying (7) which contains all other output-nulling controlled invariant subspaces.

Lemma 2b

For the alternate system (4) there exists a unique supremal output-nulling controlled invariant subspace satisfying (8) which contains all other output-nulling controlled invariant subspaces.

Proof

These statments are trivial extensions of Theorem 4.1 in [10].

Definition 4a.

$\mathbf{V}^*$ will denote the supremal output-nulling controlled invariant subspace for (1). ■

Definition 4b.

$\mathbf{V}_1^*$ will denote the supremal output-nulling controlled invariant subspace for (4). ■

It is possible to compute $\mathbf{V}^*$ and $\mathbf{V}_1^*$ using some geometric subspace algorithms much like the $\mathbf{V}^*$ algorithm for state-space systems in [10]:

Lemma 3a

Define the sequence $\mathbf{V}^{(i)}$ according to:

$$\mathbf{V}^{(0)} = \mathbf{R}^n, \tag{9a}$$

$$\mathbf{V}^{(i+1)} = \begin{bmatrix} A \\ C \end{bmatrix}^{-1} \left[\begin{bmatrix} E \\ 0 \end{bmatrix} \mathbf{V}^{(i)} + \begin{bmatrix} \mathbf{B} \\ \mathbf{D} \end{bmatrix} \right], i = 1, 2, \cdots \tag{9b}$$

Then $\mathbf{V}^{(i+1)} \subset \mathbf{V}^{(i)}$, and for some $0 \le \mu \le r$, $\mathbf{V}^{(\mu+1)} = \mathbf{V}^{(\mu)} \triangleq \mathbf{V}^*$.

Lemma 3b

Define the sequence $\mathbf{V}_1^{(i)}$ according to:

$$\mathbf{V}_1^{(0)} = \mathbf{R}^r, \tag{10a}$$

$$\mathbf{V}_1^{(i+1)} = \begin{bmatrix} A_{11} \\ A_{21} \\ C_1 \end{bmatrix}^{-1} \left[\begin{bmatrix} \Sigma^2 \\ 0 \\ 0 \end{bmatrix} \mathbf{V}_1^{(i)} + \begin{bmatrix} \mathbf{A}_{12} & \mathbf{B}_1 \\ \mathbf{A}_{22} & \mathbf{B}_2 \\ \mathbf{C}_2 & \mathbf{D} \end{bmatrix} \right], i = 1, 2, \cdots \tag{10b}$$

Then $\mathbf{V}_1^{(i+1)} \subset \mathbf{V}_1^{(i)}$, and for some $0 \le \mu \le r$, $\mathbf{V}_1^{(\mu+1)} = \mathbf{V}_1^{(\mu)} \triangleq \mathbf{V}_1^*$.

Proof

These are trivial extensions of Theorem 4.3 of [10]. ■

Definition 5a

In algorithm (9) we call $\mathbf{V}^{(i)}$ the *i'th output-nulling controlled invariant subspace* for the system (1). ■

Definition 5b

In algorithm (10) we call $\mathbf{V}_1^{(i)}$ the *i'th output-nulling controlled invariant subspace* for the system (4). ■

Remarks 2

1) Since the $\mathbf{V}^*$ algorithm for state space-systems is usually written:

$$\mathbf{V}^{(0)} = \mathbf{R}^n, \quad \mathbf{V}^{(i+1)} = C^{-1}\mathbf{D} \cap A^{-1}\left\{\mathbf{V}^{(i)} + \mathbf{B}\right\}, \tag{11}$$

it might be expected that algorithm (11) would be equivalent to:

$$\mathbf{V}^{(0)} = \mathbf{R}^n, \quad \mathbf{V}^{(i+1)} = C^{-1}\mathbf{D} \cap A^{-1}\left\{E\mathbf{V}^{(i)} + \mathbf{B}\right\}. \tag{12}$$

However, this is not true. If $\mathbf{V}^{(i+1)}$ satisfies (11) then it satisfies (12) but the reverse implication does not hold. This can be seen by a simple example:

Take

$$E = \begin{bmatrix} 1 & 0 \\ 0 & 0 \end{bmatrix}, \quad A = \begin{bmatrix} 0 & 1 \\ 1 & 0 \end{bmatrix}, \quad B = \begin{bmatrix} 0 \\ 1 \end{bmatrix}, \quad C = \begin{bmatrix} 0 & 1 \end{bmatrix}, \quad D = 1. \tag{13}$$

Then $\mathbf{V}^{(1)}$ from (11) is spanned by $\begin{bmatrix} 1 \\ 1 \end{bmatrix}$ but $\mathbf{V}^{(1)}$ from algorithm (12) is $\mathbf{R}^2$.

It is easy to define a system with the same transfer function as (1) which has no D matrix by appending u_k or y_k to the descriptor vector. This system will have a descriptor vector of dimension $\min\left[(n+m), (n+p)\right]$. In this case algorithms (11) and (12) are equivalent.

2) Because $\mathbf{V}^{(i)}$ lies in $\mathbf{R}^n$ and $\mathbf{V}_1^{(i)}$ lies in $\mathbf{R}^r$, $\mathbf{V}^{(i)}$ can be up to an $(n-r)$ higher dimensional space than $\mathbf{V}_1^{(i)}$. This means that $\mathbf{V}^{(i)}$ can span a higher-dimensional space than the number of dynamic modes being placed by a friend of $\mathbf{V}^{(i)}$, because this number of dynamic modes is equal to the dimension of $\mathbf{V}_1^{(i)}$. In [2] it is shown how to compute from $\mathbf{V}_1^*$ a subspace $\mathbf{V}_d^*$ for the descriptor system (1) whose dimension is equal to the number of dynamic modes being placed. ■

IV. Structure Algorithms and Riccati Equations

We now consider actual matrix algorithms for computing $\mathbf{V}^*$; specifically, we present two versions of Silverman's structure algorithm. This Section extends the results in Molinari [5] to descriptor systems and follows his exposition fairly closely.

Recall that the descriptor system (1) is assumed to be in an SVD coordinate system.

Most of our results will use Algorithm 1:

Algorithm 1. "Stacked Version" of the Structure Algorithm

This algorithm is initialized at $i = 0$ with $M_0 = 0$. At the i'th step F_{i+1}, G_{i+1}, and M_{i+1} are computed from M_i from the following algorithm:

$$T_i \begin{bmatrix} M_i A_{11} & M_i A_{12} & M_i B_1 \\ A_{21} & A_{22} & B_2 \\ C_1 & C_2 & D \end{bmatrix} = \begin{bmatrix} G_{11_{i+1}} & G_{12_{i+1}} & F_{i+1} \\ G_{21_{i+1}} & G_{22_{i+1}} & 0 \\ M_{i+1}\Sigma^2 & 0 & 0 \end{bmatrix}. \tag{14}$$

where T_i is invertible $\forall i \geq 0$, and $G_{22_{i+1}}$ and F_{i+1} both have full row rank $\forall i \geq 0$. ■

There exists another version of the structure algorithm which minimizes storage by computing at each step only the new information in the matrix M_i and not the entire matrix.

Algorithm 2 "Nested Version" of the Structure Algorithm

This algorithm is initialized at $i = 0$ with $\Delta_0 = 0$ and $(G_{11_0}\ G_{12_0}\ F_0) = \begin{bmatrix} A_{21} & A_{22} & B_2 \\ C_1 & C_2 & D \end{bmatrix}$. The matrix $(G_{21_0}\ G_{22_0})$ is vacuous. The algorithm is then

$$S_i \begin{bmatrix} \Delta_i A_{11} & \Delta_i A_{12} & \Delta_i B_1 \\ G_{11_i} & G_{12_i} & F_i \\ G_{21_i} & G_{22_i} & 0 \end{bmatrix} = \begin{bmatrix} G_{11_{i+1}} & G_{12_{i+1}} & F_{i+1} \\ G_{21_{i+1}} & G_{22_{i+1}} & 0 \\ \Delta_{i+1}\Sigma^2 & 0 & 0 \end{bmatrix} \tag{15}$$

where S_i is invertible $\forall i \geq 0$, and $G_{22_{i+1}}$ and F_{i+1} both have full row rank $\forall i \geq 0$. ■

Remark 3

S_i and T_i could be selected orthogonal. ■

Algorithm 2 is the version of the structure algorithm presented in Silverman's work (e.g., [6], [7]). However, we will see in the sequel that it is often easier to use Algorithm 1 to prove theoretical results.

Algorithms 1 and 2 are intimately related; Algorithm 2 is the "differencing" version of Algorithm 1. We show this in the following Theorem:

Theorem 1: Relationship Between the Stacked and Nested Versions

If the sequences $\{\Delta_i\}$, $\{F_i\}$, and $\{G_i\}$ are generated by Algorithm 2 for some sequence $\{S_i\}$, then there exists a sequence $\{T_i\}$ so that Algorithm 1 generates $\{M_i\}$, $\{F_i\}$, and $\{G_i\}$, where

$$M_i = \begin{bmatrix}\Delta_i\\ \Delta_{(i-1)}\\ \cdot\\ \cdot\\ \cdot\\ \Delta_1\\ \Delta_0\end{bmatrix} \quad \forall i \tag{16}$$

Proof

This can be proved as in [5] or [7]. At the i'th step,

$$\begin{bmatrix} M_iA_{11} & M_iA_{12} & M_iB_1\\ A_{21} & A_{22} & B_2\\ C_1 & C_2 & D\end{bmatrix} \triangleq \begin{bmatrix} \begin{bmatrix}\Delta_i\\ \Delta_{(i-1)}\\ \cdot\\ \cdot\\ \Delta_1\\ \Delta_0\end{bmatrix}A_{11} & \begin{bmatrix}\Delta_i\\ \Delta_{(i-1)}\\ \cdot\\ \cdot\\ \Delta_1\\ \Delta_0\end{bmatrix}A_{12} & \begin{bmatrix}\Delta_i\\ \Delta_{(i-1)}\\ \cdot\\ \cdot\\ \Delta_1\\ \Delta_0\end{bmatrix}B_1\\ G_{11_0} & G_{12_0} & F_0\\ G_{21_0} & G_{22_0} & 0\end{bmatrix} \sim$$

$$\begin{bmatrix} \begin{bmatrix}\Delta_i\\ \Delta_{(i-1)}\\ \cdot\\ \cdot\\ \Delta_1\end{bmatrix}A_{11} & \begin{bmatrix}\Delta_i\\ \Delta_{(i-1)}\\ \cdot\\ \cdot\\ \Delta_1\end{bmatrix}A_{12} & \begin{bmatrix}\Delta_i\\ \Delta_{(i-1)}\\ \cdot\\ \cdot\\ \Delta_1\end{bmatrix}B_1\\ \Delta_1\Sigma^2 & 0 & 0\\ G_{11_1} & G_{12_1} & F_1\\ G_{21_1} & G_{22_1} & 0\end{bmatrix} \sim \cdots \sim \begin{bmatrix} \begin{bmatrix}\Delta_{i+1}\\ \Delta_i\\ \cdot\\ \cdot\\ \cdot\\ \Delta_0\end{bmatrix}\Sigma^2 & \begin{bmatrix}0\\ 0\\ \cdot\\ \cdot\\ \cdot\\ 0\end{bmatrix} & \begin{bmatrix}0\\ 0\\ \cdot\\ \cdot\\ \cdot\\ 0\end{bmatrix}\\ G_{11_{i+1}} & G_{12_{i+1}} & F_{i+1}\\ G_{21_{i+1}} & G_{22_{i+1}} & 0\end{bmatrix} \sim$$

$$\sim \begin{bmatrix} G_{11_{i+1}} & G_{12_{i+1}} & F_{i+1}\\ G_{21_{i+1}} & G_{22_{i+1}} & 0\\ M_{i+1}\Sigma^2 & 0 & 0\end{bmatrix}. \tag{17}$$

where we recall that $\Delta_0 = 0$. (Note that $\sim$ means that two matrices have the same row space and are related by a row transformation.) ■

Remark 4

Theorem 1 and, in fact, Algorithm 2 only hold if $\Delta_0 = 0$. Silverman [7] treats the case where $\Delta_0 \neq 0$. ■

We now examine the relationship between the $\mathbf{V}^*$ algorithm and the structure algorithm.

Theorem 2: Relationship Between the $\mathbf{V}^*$ and the Structure Algorithm

a) If $\mathbf{V}^{(i)} = ker \begin{bmatrix} G21_i & G22_i \\ M_i\Sigma^2 & 0 \end{bmatrix}$ then $\mathbf{V}^{(i)}$ is the i'th output-nulling controlled invariant subspace for (1).

b) If $\mathbf{V}_1^{(i)} = ker(M_i\Sigma^2)$ then $\mathbf{V}_1^{(i)}$ is the i'th output-nulling controlled invariant subspace for (4).

Proof

From (14),

$$T_i \begin{bmatrix} M_iA_{11} & M_iA_{12} & M_iB_1 \\ A_{21} & A_{21} & B_2 \\ C_1 & C_2 & D \end{bmatrix} \begin{bmatrix} x_1 \\ x_2 \\ u \end{bmatrix} = \begin{bmatrix} G11_{i+1} & G12_{i+1} & F_{i+1} \\ G21_{i+1} & G22_{i+1} & 0 \\ M_{i+1}\Sigma^2 & 0 & 0 \end{bmatrix} \begin{bmatrix} x_1 \\ x_2 \\ u \end{bmatrix}. \tag{18}$$

Since $G22_{i+1}$ and F_{i+1} both have full row rank, the following three conditions are all equivalent:

1) $M_{i+1}\Sigma^2 x_1 = 0.$ (19)

2) There exists a vector $\begin{bmatrix} x_2 \\ u \end{bmatrix}$ such that the entire right hand side of (18) is zero and thus

$$\begin{bmatrix} M_iA_{11} & M_iA_{12} & M_iB_1 \\ A_{21} & A_{21} & B_2 \\ C_1 & C_2 & D \end{bmatrix} \begin{bmatrix} x_1 \\ x_2 \\ u \end{bmatrix} = 0. \tag{20}$$

This can be rewritten as

$$\begin{bmatrix} A_{11} \\ A_{21} \\ C_1 \end{bmatrix} x_1 = \begin{bmatrix} \Sigma^2 \\ 0 \\ 0 \end{bmatrix} z_1 - \begin{bmatrix} A_{12} & B_1 \\ A_{22} & B_2 \\ C_2 & D \end{bmatrix} \begin{bmatrix} x_2 \\ u \end{bmatrix} \tag{21}$$

for some $\begin{bmatrix} x_2 \\ u \end{bmatrix}$ and for some z_1 satisfying $M_i\Sigma^2 z_1 = 0$.

3) There exists x_2 such that

$$\begin{bmatrix} G21_{i+1} & G22_{i+1} \\ M_{i+1}\Sigma^2 & 0 \end{bmatrix} \begin{bmatrix} x_1 \\ x_2 \end{bmatrix} = 0, \tag{22}$$

and u such that the entire right hand side of (18) is zero and thus

$$\begin{bmatrix} M_iA_{11} & M_iA_{12} & M_iB_1 \\ A_{21} & A_{21} & B_2 \\ C_1 & C_2 & D \end{bmatrix} \begin{bmatrix} x_1 \\ x_2 \\ u \end{bmatrix} = \begin{bmatrix} M_i & 0 & 0 \\ 0 & I_{(n-r)} & 0 \\ 0 & 0 & I \end{bmatrix} \begin{bmatrix} A & B \\ C & D \end{bmatrix} \begin{bmatrix} x \\ u \end{bmatrix} = 0 \tag{23}$$

where $x = \begin{bmatrix} x_1 \\ x_2 \end{bmatrix}$. Equation (23) can be rewritten as

$$\begin{bmatrix} A \\ C \end{bmatrix} x = \begin{bmatrix} E \\ 0 \end{bmatrix} z - \begin{bmatrix} B \\ D \end{bmatrix} u \tag{24}$$

for some u and some z satisfying

$$\begin{bmatrix} M_i & 0 \\ 0 & I \end{bmatrix} Ez = \begin{bmatrix} M_i & 0 \\ 0 & I \end{bmatrix}\begin{bmatrix} \Sigma^2 & 0 \\ 0 & 0 \end{bmatrix} z = \begin{bmatrix} M_i\Sigma^2 & 0 \\ 0 & 0 \end{bmatrix} z \tag{25}$$

The latter condition is equivalent to $M_i\Sigma^2 z_1 = 0$ if $z \triangleq \begin{bmatrix} z_1 \\ z_2 \end{bmatrix}$.

The equivalence of (24) and (22) proves assertion (a), and the equivalence of (21) and (19) proves assertion (b). ■

Remark 5

Both output-nulling controlled invariant subspaces ($\mathbf{V}^*$ for (1) and $\mathbf{V}_1^*$ for (4)) can be computed from the same structure algorithm. ■

We can derive a discrete-time Riccati equation from Algorithm 1. We state this result as:

Theorem 3 Riccati Equation

Select T_i orthogonal in (14). Then, $P_i \triangleq M_i^T M_i$ satisfies the singular discrete-time Riccati equation

$$\Sigma^2 P_{i+1}\Sigma^2 = A_{11}^T P_i A_{11} + (A_{21}^T \; C_1^T)\begin{bmatrix} A_{21} \\ C_1 \end{bmatrix} - \left[A_{11}^T P_i (A_{12} \; B_1) + (A_{21}^T \; C_1^T)\begin{bmatrix} A_{22} & B_2 \\ C_2 & D \end{bmatrix}\right] \cdot$$

$$\cdot \left[\begin{bmatrix} A_{22}^T & C_2^T \\ B_2^T & D^T \end{bmatrix}\begin{bmatrix} A_{22} & B_2 \\ C_2 & D \end{bmatrix} + \begin{bmatrix} A_{12}^T \\ B_1^T \end{bmatrix} P_i (A_{12} \; B_1)\right]^{\dagger} \left[\begin{bmatrix} A_{12}^T \\ B_1^T \end{bmatrix} P_i A_{11} + \begin{bmatrix} A_{22}^T & C_2^T \\ B_2^T & D^T \end{bmatrix}\begin{bmatrix} A_{21} \\ C_1 \end{bmatrix}\right]. \tag{26}$$

Proof

The technique of proof is well-known [5]. Briefly, if (14) is premultiplied by its transpose, we get (using $T_i^T T_i = I$):

$$\begin{bmatrix} \left[A_{11}^T P_i A_{11} + (A_{21}^T \; C_1^T)\begin{bmatrix} A_{21} \\ C_1 \end{bmatrix}\right] & \left[A_{11}^T P_i (A_{12} \; B_1) + (A_{21}^T \; C_1^T)\begin{bmatrix} A_{22} & B_2 \\ C_2 & D \end{bmatrix}\right] \\ \left[\begin{bmatrix} A_{12}^T \\ B_1^T \end{bmatrix} P_i A_{11} + \begin{bmatrix} A_{22}^T & C_2^T \\ B_2^T & D^T \end{bmatrix}\begin{bmatrix} A_{21} \\ C_1 \end{bmatrix}\right] & \left[\begin{bmatrix} A_{22}^T & C_2^T \\ B_2^T & D^T \end{bmatrix}\begin{bmatrix} A_{22} & B_2 \\ C_2 & D \end{bmatrix} + \begin{bmatrix} A_{12}^T \\ B_1^T \end{bmatrix} P_i (A_{12} \; B_1)\right] \end{bmatrix}$$

$$= \begin{bmatrix} \left[(G_{11_{i+1}}^T \; G_{21_{i+1}}^T)\begin{bmatrix} G_{11_{i+1}} \\ G_{21_{i+1}} \end{bmatrix} + \Sigma^2 P_{i+1}\Sigma^2\right] & \left[(G_{11_{i+1}}^T \; G_{21_{i+1}}^T)\begin{bmatrix} G_{12_{i+1}} & F_{i+1} \\ G_{22_{i+1}} & 0 \end{bmatrix}\right] \\ \left[\begin{bmatrix} G_{12_{i+1}}^T & G_{22_{i+1}}^T \\ F_{i+1}^T & 0 \end{bmatrix}\begin{bmatrix} G_{11_{i+1}} \\ G_{21_{i+1}} \end{bmatrix}\right] & \left[\begin{bmatrix} G_{12_{i+1}}^T & G_{22_{i+1}}^T \\ F_{i+1}^T & 0 \end{bmatrix}\begin{bmatrix} G_{12_{i+1}} & F_{i+1} \\ G_{22_{i+1}} & 0 \end{bmatrix}\right] \end{bmatrix} \tag{27}$$

The upper left corner of this equation is

$$\Sigma^2 P_{i+1}\Sigma^2 = A_{11}^T P_{11} A_{11} + (A_{21}^T \; C_1^T)\begin{bmatrix} A_{21} \\ C_1 \end{bmatrix} - (G_{11_{i+1}}^T \; G_{21_{i+1}}^T)\begin{bmatrix} G_{11_{i+1}} \\ G_{21_{i+1}} \end{bmatrix}. \tag{28}$$

Because $\begin{bmatrix} G_{12_{i+1}} & F_{i+1} \\ G_{22_{i+1}} & 0 \end{bmatrix}$ has full row rank, we can write

$$(G_{11_{i+1}}^T \; G_{21_{i+1}}^T)\begin{bmatrix} G_{11_{i+1}} \\ G_{21_{i+1}} \end{bmatrix} = (G_{11_{i+1}}^T \; G_{21_{i+1}}^T)\begin{bmatrix} G_{12_{i+1}} & F_{i+1} \\ G_{22_{i+1}} & 0 \end{bmatrix} \cdot$$

$$\cdot \left[\begin{bmatrix} G_{12_{i+1}}^T & G_{22_{i+1}}^T \\ F_{i+1}^T & 0 \end{bmatrix}\begin{bmatrix} G_{12_{i+1}} & F_{i+1} \\ G_{22_{i+1}} & 0 \end{bmatrix}\right]^{\dagger}\begin{bmatrix} G_{12_{i+1}}^T & G_{22_{i+1}}^T \\ F_{i+1}^T & 0 \end{bmatrix}\begin{bmatrix} G_{11_{i+1}} \\ G_{21_{i+1}} \end{bmatrix}. \tag{29}$$

The terms of the last expression are known from (27) and can be substituted into **(28)** to derive the Riccati equation (26). ■

This Riccati equation can be used directly to compute $\mathbf{V}_1^{(i)}$:

Corollary 1 (to Theorem 3)

$$\mathbf{V}_1^{(i)} = ker\left[\Sigma^2 P_i \Sigma^2\right] \tag{30}$$

Proof

Obvious from Theorem 2b and $P_i = M_i^T M_i$. ■

Algorithm 2 is a "differencing" version of the structure algorithm; the solution Δ_i relates directly to the incremental cost in the Riccati equation. We state this formally in:

Theorem 4

$P_i = P_{i-1} + \Delta_i^T \Delta_i$.

Proof

Follows directly from (16) and $P_i = M_i^T M_i$. ■

Remarks 6

1) The Riccati equation (26) is the optimal control Riccati equation for the alternate system (4). The solution of (26) is an $r \times r$ matrix, and $x_{1_i}^T \Sigma^2 P_i \Sigma^2 x_{1_i}$ is the optimal cost-to-go for the corresponding linear-quadratic optimal control problem:

$$J_N = x_{1_{N-i}}^T \Sigma^2 P_{N-i} \Sigma^2 x_{1_{N-i}} + \sum_{k=0}^{N-(i+1)} \left(y_{1_k}^T y_{2_k}^T\right)\begin{bmatrix} y_{1_k} \\ y_{2_k} \end{bmatrix}$$

$$= x_{1_{N-i}}^T \Sigma^2 P_{N-i} \Sigma^2 x_{1_{N-i}} + \sum_{k=0}^{N-(i+1)} \left(x_{1_k}^T \; x_{2_k}^T \; u_k^T\right)\begin{bmatrix} A_{21}^T & C_1^T \\ A_{22}^T & C_2^T \\ B_2^T & D^T \end{bmatrix}\begin{bmatrix} A_{21} & A_{22} & B_2 \\ C_1 & C_2 & D \end{bmatrix}\begin{bmatrix} x_{1_k} \\ x_{2_k} \\ u_k \end{bmatrix} \tag{31}$$

subject to (4a) and the initial condition x_{1_0}.

2) If the initial condition x_{1_0} is in the kernel of $\Sigma^2 P_0 \Sigma^2$, there exists a feedback K' for the system (4) so that the cost (31) is equal to zero. Thus, with this initial condition and feedback $\begin{bmatrix} y_{1_k} \\ y_{2_k} \end{bmatrix} = \begin{bmatrix} 0 \\ 0 \end{bmatrix} \forall 0 \le k \le (N-i)$. ■

3) It is also possible to prove Corollary 1 directly using the technique of [1].

4) The alternate system (4) is not equivalent to the descriptor system because the variable y_{1_k} is not constrained to be zero. However, the output-nulling controlled

invariant subspace for the two problems can be computed with the same algorithm because in system (4) "output nulling" means that both y_{1_k} and y_{2_k} are constrained to be zero. If in (4) $y_{1_k} = 0 \,\forall k$, system (4) is equivalent to system (1). Then, if $y_{1_k} = 0$, y_{2_k} in (4) is equal to y_k in (1), and x_{1_k} being in an output-nulling subspace of (4) is equivalent to x_k being in an output-nulling subspace of (1). ■

Feedback; The Friends of $\mathbf{V}^*$

We can use Algorithm 1 to compute the friends of $\mathbf{V}^*$ or $\mathbf{V}_1^*$. That is, we can compute for the descriptor system (1) or the system (4) all possible feedback gains which relate the control and descriptor vector as in Theorem 2. The expressions which these gains must satisfy are:

Descriptor system (1):

$$\left[\begin{pmatrix} A \\ C \end{pmatrix} + \begin{pmatrix} B \\ D \end{pmatrix} K_i\right] \mathbf{V}^{(i)} \subset \begin{pmatrix} E \\ 0 \end{pmatrix} \mathbf{V}^{(i)}, \tag{32}$$

Alternate system (4):

$$\left[\begin{pmatrix} A_{11} \\ A_{21} \\ C_1 \end{pmatrix} + \begin{pmatrix} A_{12} & B_1 \\ A_{22} & B_2 \\ C_2 & D \end{pmatrix} K'_i\right] \mathbf{V}_1^{(i)} \subset \begin{pmatrix} \Sigma^2 \\ 0 \\ 0 \end{pmatrix} \mathbf{V}_1^{(i)}. \tag{33}$$

We now characterize all possible feedback gains which satisfy (32) and (33).

Theorem 5a

All feedback gains satisfying (32) (i.e., $\left\{K_i \ni K_i \in \mathbf{F}\left[\mathbf{V}^{(i)}\right]\right\}$) can be computed from

$$K_i = -F_i^{\dagger}(G_{11_i}\; G_{12_i}) + (I - F_i^{\dagger}F_i)W_i + \left(Y_{2_i}^T\, Y_{3_i}^T\right)\begin{pmatrix} G_{21_i} & G_{22_i} \\ M_i\Sigma^2 & 0 \end{pmatrix} \tag{34}$$

where W_i and $\begin{pmatrix} Y_{2_i} \\ Y_{3_i} \end{pmatrix}$ are properly dimensioned arbitrary real matrices.

Proof

From Theorem 2a, if $x = \begin{pmatrix} x_1 \\ x_2 \end{pmatrix} \in \mathbf{V}^{(i)}$, there exists u such that

$$\begin{pmatrix} G_{11_i} & G_{12_i} & F_i \\ G_{21_i} & G_{22_i} & 0 \\ M_i\Sigma^2 & 0 & 0 \end{pmatrix}\begin{pmatrix} x_1 \\ x_2 \\ u \end{pmatrix} = 0. \tag{35}$$

Using $u = K_i x$ gives

$$\left[\begin{pmatrix} G_{11_i} & G_{12_i} \\ G_{21_i} & G_{22_i} \\ M_i\Sigma^2 & 0 \end{pmatrix} + \begin{pmatrix} F_i \\ 0 \\ 0 \end{pmatrix} K_i\right] x = 0. \tag{36}$$

The general solution to this problem for K_i can be written (Rao & Mitra 1971, p. 26)

$$K_i = -\begin{pmatrix} F_i \\ 0 \\ 0 \end{pmatrix}^{\dagger}\begin{pmatrix} G_{11_i} & G_{12_i} \\ G_{21_i} & G_{22_i} \\ M_i\Sigma^2 & 0 \end{pmatrix} + \left(Y_{1_i}^T\;\; Y_{2_i}^T\;\; Y_{3_i}^T\right)\left[I - \begin{pmatrix} F_i \\ 0 \\ 0 \end{pmatrix}\begin{pmatrix} F_i \\ 0 \\ 0 \end{pmatrix}^{\dagger}\right]\begin{pmatrix} G_{11_i} & G_{12_i} \\ G_{21_i} & G_{22_i} \\ M_i\Sigma^2 & 0 \end{pmatrix} + \left[I - \begin{pmatrix} F_i \\ 0 \\ 0 \end{pmatrix}^{\dagger}\begin{pmatrix} F_i \\ 0 \\ 0 \end{pmatrix}\right] W_i \tag{37}$$

where W_i and $Y_i = \begin{bmatrix} Y_{1_i} \\ Y_{2_i} \\ Y_{3_i} \end{bmatrix}$ are appropriately dimensioned arbitrary real matrices.

Simplifying (37) gives (34). The following facts are helpful:

i) $$\begin{bmatrix} F_i \\ 0 \\ 0 \end{bmatrix}^{\dagger} = \begin{bmatrix} F_i^{\dagger} & 0 & 0 \end{bmatrix} \tag{38}$$

ii) $$F_i F_i^{\dagger} = I, \tag{39}$$

iii) $$\left[I - \begin{bmatrix} F_i \\ 0 \\ 0 \end{bmatrix} \begin{bmatrix} F_i \\ 0 \\ 0 \end{bmatrix}^{\dagger} \right] = \begin{bmatrix} 0 & 0 & 0 \\ 0 & I & 0 \\ 0 & 0 & I \end{bmatrix}. \tag{40}$$

Fact ii) follows from the full row rank of F_i, and fact iii) follows from i) and ii).

Theorem 5b

For the alternate system (4), all feedback gain matrices satisfying (33) (i.e., $\left\{K'_i \ni K'_i \in \mathbf{F}\left[\mathbf{V}_1^{(i)}\right]\right\}$) can be computed from

$$K'_i = -\begin{bmatrix} G_{12_i} & F_i \\ G_{22_i} & 0 \end{bmatrix}^{\dagger} \begin{bmatrix} G_{11_i} \\ G_{21_i} \end{bmatrix} + \left[I - \begin{bmatrix} G_{12_i} & F_i \\ G_{22_i} & 0 \end{bmatrix}^{\dagger} \begin{bmatrix} G_{12_i} & F_i \\ G_{22_i} & 0 \end{bmatrix} \right] \begin{bmatrix} W_{1_i} \\ W_{2_i} \end{bmatrix} + Y_{3_i}^T M_i \Sigma^2 \tag{41}$$

where $\begin{bmatrix} W_{1_i} \\ W_{2_i} \end{bmatrix}$ and Y_{3_i} are appropriately dimensioned arbitrary real matrices.

Proof

K'_i can be any solution of:

$$\begin{bmatrix} G_{12_i} & F_i \\ G_{22_i} & 0 \\ 0 & 0 \end{bmatrix} K'_i = -\begin{bmatrix} G_{11_i} \\ G_{21_i} \\ M_i \Sigma^2 \end{bmatrix} \tag{42}$$

Again, from Rao & Mitra (1971, p.26), the solutions of (42) are characterized by (41). ■

Remarks 7

1) There is some freedom in the choice of K_i, the friends of $\mathbf{V}^{(i)}$, which does not exist for K'_i, the friends of $\mathbf{V}_1^{(i)}$. In particular, $Y_{2_i}^T$ is arbitrary. The number of columns of this matrix is the same as the number of rows of G_{22_i}; G_{22_i} has less than or equal to $(n-r)$ rows because G_{22_i} has full row rank and has $(n-r)$ columns. Thus there are up to $m(n-r)$ additional degrees of freedom in the choice of K_i which stem from the fact that what is required is a relationship between u_i and x_i and not a relationship between $\begin{bmatrix} x_{2_i} \\ u_i \end{bmatrix}$ and x_{1_i}. By comparing equations (34) and (41) it is easy to see that these extra degrees of freedom are characterized by the range of $\begin{bmatrix} G_{21_i} & G_{22_i} \end{bmatrix}$. We show in [2] that $\mathbf{V}_d^* = \mathbf{V}^*$ (see Remark 2.2) then G_{22_i} has row rank $(n-r)$ and K_i in this case has $m(n-r)$ extra degrees of freedom.

2) In state-space systems a control input in the kernel of F_i is exactly the control input which can be used to place some of the eigenvalues of the closed-loop transition matrix $(A+BK)$ restricted to $\mathbf{V}^*$ if a controllability subspace exists (see [5] and [10], Chapter 5). In descriptor systems this would seem to be true as well. The freedom in choosing $Y_{2_i}^T$ in (34) gives up to $m(n-r)$ extra degrees of freedom in choosing the feedback gain but using this freedom will not cause the poles to move. ■

The minimum-norm output-nulling feedback gain (41) can be computed in terms of the solution of the Riccati equation (26). From Theorem 5b, the minimum-norm feedback gain K'_i with $\begin{bmatrix} W_{1_i} \\ W_{2_i} \end{bmatrix}$ and Y_{3_i} both set to zero is

$$K'_i = -\begin{bmatrix} G_{12_i} & F_i \\ G_{22_i} & 0 \end{bmatrix}^{\dagger} \begin{bmatrix} G_{11_i} \\ G_{21_i} \end{bmatrix}. \tag{43}$$

Using the same technique as in the proof of Theorem 3,

$$K'_i = = -\left[\begin{bmatrix} A_{22}^T & C_2^T \\ B_2^T & D \end{bmatrix}\begin{bmatrix} A_{22} & B_2 \\ C_2 & D \end{bmatrix} + \begin{bmatrix} A_{12}^T \\ B_1^T \end{bmatrix} P_i (A_{12}\ B_1)\right]^{\dagger} \left[\begin{bmatrix} A_{12}^T \\ B_1^T \end{bmatrix} P_i A_{11} + \begin{bmatrix} A_{22}^T & C_2^T \\ B_2^T & D \end{bmatrix}\begin{bmatrix} A_{21} \\ C_1 \end{bmatrix}\right] \tag{44}$$

It is also well-known how to derive a different (feedback) form of the Riccati equation by substituting this feedback gain into (26).

Theorem 6

If K'_i is computed from (43), then the Riccati equation (26) can be reformulated as

$$\Sigma^2 P_{i+1}\Sigma^2 = \left[A_{11} + (A_{12}\ B_1)K_i\right]^T P_i \left[A_{11} + (A_{12}\ B_1)K_i\right] + \left[\begin{bmatrix} A_{21} \\ C_1 \end{bmatrix} + \begin{bmatrix} A_{22} & B_2 \\ C_2 & D \end{bmatrix} K_i\right]^T \left[\begin{bmatrix} A_{21} \\ C_1 \end{bmatrix} + \begin{bmatrix} A_{22} & B_2 \\ C_2 & D \end{bmatrix} K_i\right]. \tag{45}$$

Proof

Substituting $\begin{bmatrix} x_2 \\ u \end{bmatrix} = K'_i x_1$ into (18) and using (43), (18) becomes

$$T_i \begin{bmatrix} M_i[A_{11} + (A_{12}\ B_1)K'_i] \\ \left[\begin{bmatrix} A_{21} \\ C_1 \end{bmatrix} + \begin{bmatrix} A_{22} & B_2 \\ C_2 & D \end{bmatrix} K'_i\right] \end{bmatrix} x_1 = \begin{bmatrix} 0 \\ 0 \\ M_{i+1}\Sigma^2 \end{bmatrix} x_1 \tag{46}$$

Equation (45) can now be derived simply by pre-multiplying (46) by its transpose. ■

The Riccati equation solution P_i yields another matrix whose kernel is $\mathbf{V}^{(i)}$ for the descriptor system (1).

Theorem 7

$$\mathbf{V}^{(i+1)} = \ker\left\{A^T \begin{bmatrix} P_i & 0 \\ 0 & I \end{bmatrix} A + C^T C - \left[A^T \begin{bmatrix} P_i & 0 \\ 0 & I \end{bmatrix} B + C^T D\right]\left[B^T \begin{bmatrix} P_i & 0 \\ 0 & I \end{bmatrix} B + D^T D\right]^{\dagger} \left[B^T \begin{bmatrix} P_i & 0 \\ 0 & I \end{bmatrix} A + D^T C\right]\right\} \tag{47}$$

Proof

From Theorem 2a,

$$\mathcal{V}^{(i+1)} = ker\left\{\begin{pmatrix} G21_{i+1} & G22_{i+1} \\ M_{i+1}\Sigma^2 & 0 \end{pmatrix}\right\} = ker\left\{\begin{pmatrix} G21^T_{i+1} & \Sigma^2 M^T_{i+1} \\ G22^T_{i+1} & 0 \end{pmatrix}\begin{pmatrix} G21_{i+1} & G22_{i+1} \\ M_{i+1}\Sigma^2 & 0 \end{pmatrix}\right\}. \tag{48}$$

In order to compute the entity on the right side of (47) it is helpful to use (27) with a different partitioning:

$$\begin{bmatrix} \left[A^T\begin{pmatrix} P_i & 0 \\ 0 & I \end{pmatrix}A + C^TC\right] & \left[A^T\begin{pmatrix} P_i & 0 \\ 0 & I \end{pmatrix}B + C^TD\right] \\ \left[B^T\begin{pmatrix} P_i & 0 \\ 0 & I \end{pmatrix}A + D^TC\right] & \left[B^T\begin{pmatrix} P_i & 0 \\ 0 & I \end{pmatrix}B + D^TD\right] \end{bmatrix}$$

$$= \begin{bmatrix} \left[(A^T\ C^T)\begin{pmatrix} P_i & 0 & 0 \\ 0 & I & 0 \\ 0 & 0 & I \end{pmatrix}\begin{pmatrix} A \\ C \end{pmatrix}\right] & \left[(A^T\ C^T)\begin{pmatrix} P_i & 0 & 0 \\ 0 & I & 0 \\ 0 & 0 & I \end{pmatrix}\begin{pmatrix} B \\ D \end{pmatrix}\right] \\ \left[(B^T\ D^T)\begin{pmatrix} P_i & 0 & 0 \\ 0 & I & 0 \\ 0 & 0 & I \end{pmatrix}\begin{pmatrix} A \\ C \end{pmatrix}\right] & \left[(B^T\ D^T)\begin{pmatrix} P_i & 0 & 0 \\ 0 & I & 0 \\ 0 & 0 & I \end{pmatrix}\begin{pmatrix} B \\ D \end{pmatrix}\right] \end{bmatrix}$$

$$= \begin{bmatrix} \left[\begin{pmatrix} G11^T_{i+1} & G21^T_{i+1} & \Sigma^2 M^T_{i+1} \\ G12^T_{i+1} & G22^T_{i+1} & 0 \end{pmatrix}\begin{pmatrix} G11_{i+1} & G12_{i+1} \\ G21_{i+1} & G22_{i+1} \\ M_{i+1}\Sigma^2 & 0 \end{pmatrix}\right] & \left[\begin{pmatrix} G11^T_{i+1} \\ G12^T_{i+1} \end{pmatrix}F_{i+1}\right] \\ \left[F^T_{i+1}(G11_{i+1}\ G12_{i+1})\right] & \left[F^T_{i+1}F_{i+1}\right] \end{bmatrix}. \tag{49}$$

Then, using the technique of the proof of Theorem 3, we can show that

$$\begin{pmatrix} G21^T_{i+1} & \Sigma^2 M^T_{i+1} \\ G22^T_{i+1} & 0 \end{pmatrix}\begin{pmatrix} G21_{i+1} & G22_{i+1} \\ M_{i+1}\Sigma^2 & 0 \end{pmatrix} = A^T\begin{pmatrix} P_i & 0 \\ 0 & I \end{pmatrix}A + C^TC -$$

$$\left[A^T\begin{pmatrix} P_i & 0 \\ 0 & I \end{pmatrix}B + C^TD\right]\left[B^T\begin{pmatrix} P_i & 0 \\ 0 & I \end{pmatrix}B + D^TD\right]^{\dagger}\left[B^T\begin{pmatrix} P_i & 0 \\ 0 & I \end{pmatrix}A + D^TC\right] \quad \blacksquare \tag{50}$$

Remarks 8

1) Again this Theorem can be proved directly without resort to Theorem 3 by using the technique of [1].
2) If a gain K_i is computed as

$$K_i = -\left[B^T\begin{pmatrix} P_i & 0 \\ 0 & I \end{pmatrix}B + D^TD\right]^{\dagger}\left[B^T\begin{pmatrix} P_i & 0 \\ 0 & I \end{pmatrix}A + D^TC\right]. \tag{51}$$

(47) can be rewritten as

$$\begin{bmatrix} (G21^T_{i+1}G21_{i+1} + \Sigma^2 P_{i+1}\Sigma^2) & G21^T_{i+1}G22_{i+1} \\ G22^T_{i+1}G21_{i+1} & G22^T_{i+1}G22 \end{bmatrix} = (A+K_i)^T\begin{pmatrix} P_i & 0 \\ 0 & I \end{pmatrix}(A+K_i) + (C+DK_i)^T(C+DK_i). \blacksquare \tag{52}$$

References

[1] Aoki, M. and M. T. Li, *Partial Reconstruction of State Vectors in Decentralized Dynamic Systems*, IEEE Trans. Aut. Contr., vol. AC-18, no. 3, pp. 289-292, 1973.

[2] Bender, D. J., *Descriptor Systems and Geometric Control Theory*, Ph.D. Dissertation, Dept. of Electrical and Computer Engineering, University of California at Santa Barbara, September 1985.

[3] Cobb, D. J., *Controllability, Observability and Duality in Singular Systems*, IEEE Trans. Aut. Contr., vol. AC-29, no. 12, pp. 1076-1082, 1984.

[4] Lewis, F. L. and K. Ozcaldiran, *Reachability and Controllability for Descriptor Systems*, Proc. 27th Midwest Symp. on Circ. & Syst., Morgantown, WV, 1984.

[5] Molinari, B. P., *A Strong Controllability and Observability in Linear Multivariable Control*, IEEE Trans. Auto. Contr., vol. 21, no. 5, pp. 761-764, 1976.

[6] Silverman, L. M., *Inversion of Multivariable Linear Systems*, IEEE Trans. Aut. Contr., vol. AC-14, no. 3, pp. 270-276, 1969.

[7] Silverman, L. M., *Discrete Riccate Equations: Alternative Algorithms, Asymptotic Properties, and System Theory Interpretations*, in: Leondes, C. T. (ed.), *Control and Dynamic Systems: Advances in Theory and Applications, vol. 12*, (New York: Academic, 1976).

[8] Van Der Weiden, A. J. J. and O. H. Bosgra, *The Determination of Structural Properties of a Linear System by Operations of System Similarity; 2. Non-Proper Systems in Generalized State-Space Form*, Int. J. Ctrl., vol. 32, no. 3, pp. 489-537, 1980.

[9] Verghese, G. C., *Further Notes on Singular Descriptions*, Proc. Joint Aut. Contr. Conf., Paper TA-4., 1981.

[10] Wonham, W. M., *Linear Multivariable Control: A Geometric Approach*, Third Edition (Berlin: Springer-Verlag, 1985).

Frequency Domain and State Space Methods for Linear Systems
C.I. Byrnes and A. Lindquist (Editors)

EXTENDED INTERTIA THEOREMS FOR LYAPUNOV AND RICCATI EQUATIONS

Sergio Bittanti
Politecnico di Milano
Dipartimento di Elettronica
Piazza Leonardo da Vinci, n. 32
20133 MILANO - ITALY

Patrizio Colaneri
Centro di Teoria dei Sistemi CNR
c/o Politecnico di Milano
Dipartimento di Elettronica
Piazza Leonardo da Vinci, n. 32
20133 MILANO - ITALY

In this paper, the discrete-time algebraic Lyapunov equation (DALE)

$$0 = A'PA - P + C'C$$

is firstly considered. An extended inertia theorem (EIT) is derived under the assumption of detectability of the underlying system. It extends a previous result based on the osservability assumption.

On the basis of the DALE EIT, it is possible to derive an extended inertia theorem for the periodic Lyapunov differential equation (PLDE)

$$-\dot{P}(t) = A(t)'P(t)+P(t)A(t) + C(t)'C(t),$$

where A(.) and C(.) are periodic matrices of period T. In turn, from the PLDE EIT, an extended inertia theorem for the periodic Riccati differential equation (PRDE)

$$-\dot{P}(t) = A(t)'P(t) + P(t)A(t)-P(t)B(t)B(t)'P(t) + C(t)'C(t)$$

can be worked out (B(.) is T-periodic too).

From the PLDE EIT and PRDE EIT, two extended inertia theorems for the continuous-time algebraic Lyapunov equations (CALE) and continuous-time algebraic Riccati equation (CARE) obviously follow.

Summarizing, in this paper, five inertia theorems are proved, i.e. the DALE EIT, CALE EIT, CARE EIT, PLDE EIT and PRDE EIT.

In all the derivations, a key role is played by a modal characterization of detectability.

1. INTRODUCTION

1.1 *Purpose of the paper*

The purpose of this paper is to provide inertia theorems for the differential Lyapunov and differential Riccati equation with periodic coefficients. By focusing on the time-invariant equations as a particular case, new inertia theorems for the algebraic Lyapunov equation and the algebraic Riccati equation are obtained. The results of this paper have the feature of being stated under the detectability assumptions of the underlying system. As such, they extend previously known results, which hold under the more restrictive observability assumption.

To put the new results into perspective, we are well advised to firstly acquire a comprehensive overview of the existing literature on the subject. To this purpose, some notational convenctions are in order.

1.2 *The inertia of a matrix*

Characterizing the number of stable and unstable eigenvalues of a matrix leads to the notion of inertia. In the sequel, both continuous-time and discrete-time systems will be dealt with. Correspondingly, two notions of inertia, called c-inertia and d-inertia, are introduced in this subsection.

For any square matrix M, the c-inertia,cin (M) is the triple of integers

$$cin(M) = (\Pi_c(M),\ \delta_c(M),\ \nu_c(M)),$$

where

$\Pi_c(M)$ = number of eigenvalues of M with positive real part

$\delta_c(M)$ = number of eigenvalues of M with vanishing real part

$\nu_c(M)$ = number of eigenvalues of M with negative real part.

The d-inertia of M, din(M), is defined as follows

$$din(M) = (\Pi_d(M),\ \delta_d(M),\ \nu_d(M)),$$

where

Π_d (M) = number of eigenvalues of M with modulus greater than 1

δ_d (M) = number of eigenvalues of M with modulus equal to 1

ν_d (M) = number of eigenvalues of M with modulus less than 1.

1.3 *Inertia theorems for the algebraic Lyapunov equation and the algebraic Riccati equation*

In basic linear algebra, the invariance of the signature of quadratic forms under certain matrix transformations is known as inertia law. For the Lyapunov equation, the inertia theorems relate the inertia of any solutions to the inertia of the dinamic matrix of the underlying system. Precisely, consider the system

$$\dot{x}(t) = A\,x\,(t) \qquad (1.a)$$

$$y(t) = C\,x\,(t), \qquad (1.b)$$

where $u(t) \varepsilon R^m$, $x(t) \varepsilon R^n$, $y(t) \varepsilon R^p$, and suppose that the system is observable. The associated continuous-time algebraic Lyapunov equation (CALE) is defined as follows :

$$0 = A'P + PA + C'C \qquad (2)$$

The classical inertia theorem (IT) relates the inertia of a symmetric solution of (2) to the inertia of matrix A , (Ostrowski and Schneider (1962), Carlson and Schneider (1963), Chen (1973), Tausski (1961), Wimmer and Ziebur (1975)).

Precisely it can be stated as follows.

CALE IT

If system (1) is observable and equation (2) has a symmetric solution P, then $\delta_c(A) = 0$ and cin (-P) = cin (A). □

Thus, the number of positive /negative/ eigenvalues of any solution coincides with the number of eigenvalues of A in the left /right/ half-plane.

In the discrete-time case, i.e. for the system

$$x(t+1) = A\,x(t) \tag{3.a}$$

$$y(t) = C\,x\,(t), \tag{3.b}$$

the Lyapunov equation (DALE) is given by

$$0 = A'PA + C'C. \tag{4}$$

In fact, equation (4) is also referred to as the Stein equation. In Wimmer and Ziebur (1975), the discrete-time counterpart of the previous inertia theorem is proved.

DALE IT

If system (3) is observable and equation (4) has a symmetric solution P, then $\delta_d(A) = 0$ and $\text{cin}(-P) = \text{din}\,(A)$. □

Thus, the number of positive (negative) eigenvalues of P coincides with the number of eigenvalues of A with modulus less (greater) then 1.

An important corollary of the CALE /DALE/ inertia theorems is that, if there exists a positive definite solution of (2) /(4)/, then (1) /(3)/ is stable. Such a conclusion is part of the celebrated *Lyapunov lemma*, Kalman (1963), which states that, if the system is observable, the algebraic Lyapunov equation has a positive definite solution if and only if the underlying system is stable.

On the basis of CALE IT, an inertia theorem can be derived for the continuous-time Riccati equation (CARE):

$$0 = A'P + PA - PBB'P + C'C \tag{5}$$

where $B \in R^{nxm}$. The inertia of a solution P is then related to the inertia of the dynamic matrix of the so-called closed-loop system

$$\dot{x}(t) = F\,x\,(t)$$

where

$$F = A - BB'P. \tag{6}$$

CARE IT

If system (1) is observable and equation (5) has a symmetric solution P, then $\delta_c(F) = 0$ and $\text{cin}(-P) = \text{cin}(F)$. □

For a proof of this theorem, see Wimmer (1976). Notice in passing that the CALE IT, DALE IT and CARE IT obviously imply that, under the observability assumption of the underlying system, equations (2), (4) and (5) cannot admit singular symmetric solutions.

1.4 *Inertia theorems for the periodic Lyapunov equation and periodic Riccati equation*

Consider now the time-varying system

$$\dot{x}(t) = A(t)\, x(t) \tag{7.a}$$

$$y(t) = C(t)\, x(t) \tag{7.b}$$

where $A(.) : R \longrightarrow R^{nxn}$ and $C(.) : R \longrightarrow R^{pxn}$ are continuous and *periodic* of period T.

As is well known, see e.g. Yacubovich and Starzhinskii (1972), the system stability is determined by the eigenvalues of the *monodromy matrix* of A(.). Denoting by $\Phi_A(t,\tau)$ the transition matrix of A(.), $\Phi_A(T,0)$ is the monodromy matrix of A(.). Its eigenvalues are called the characteristic multipliers of A(.). The periodic system (7) is stable if and only if its characteristic multipliers belong to the open unit disk. We will also say that (7) is anti-stable if its characteristics multipliers do not belong to the closed unit disk.

As far as the structural properties of periodic system is concerned, the interested reader is referred to Kalman, Falb and Arbib (1969, Prop. 2.26), Hewer (1975), Kano and Nishimura (1979), Bittanti, Guardabassi, Maffezzoni and Silverman (1978), Bittanti, Colaneri and Guardabassi (1984), Bittanti and Bolzern (1984), Bittanti and Bolzern (1985a, b, c, d). Here, we recall only that, if a periodic system is observable at a given time point, it is observable at any time point. Hence, it makes sense to speak of *system observability* without any further specification.

The periodic Lyapunov differential equation (PLDE) associated with (7) is the equation

$$-\dot{P}(t) = A(t)'P(t)+P(t)A(t)+C(t)'C(t). \tag{8}$$

It can be easily shown that finding a periodic solution of (8) is equivalent to finding a solution of a suitably defined DALE, see e.g. Hewer (1975) ,Shayman(1984) and Bittanti, Bolzern and Colaneri (1984). The invariant system underlying this DALE is observable if and only if the periodic system (8) is observable. Then, from the DALE IT, the following inertia theorem was proved by Shayman (1984).

PLDE IT

If system (7) is observable and equation (8) has a T-periodic symmetric solution P(.), then $\delta_d(\Phi_A(T,0)) = 0$ and $\text{cin}(-P(t)) = \text{din}\, \Phi_A(T,0)$, $\forall t$.

The PLDE IT can be applied to the case when (8) admits a positive definite /negative definite/ periodic solution, so leading to the conclusion that system (7) is stable /anti-stable/. In the positive definite case, this is part of the *periodic Lyapunov Lemma*, which claims that, assuming (7) observable, the PLDE has a positive definite T-periodic solution if and only if the system is stable. For the periodic Lyapunov lemma, see Bittanti, Bolzern and Colaneri (1984) and Bittanti, Bolzern and Colaneri (1985).

Finally, letting $B(.) : R \longrightarrow R^{nxm}$ be a T-periodic matrix, the periodic Riccati equation (PRDE) is given by

$$-\dot{P}(t) = A(t)'P(t)+P(t)A(t) - P(t)B(t)B(t)'P(t)+C(t)'C(t) . \tag{9}$$

For any T-periodic solution P(.) of (8), we define

$$F(t) = A(t) - B(t)B(t)'P(t). \tag{10}$$

From the PLDE IT, and inertia theorem for the Riccati equation can be worked out, (Shayman (1984)).

PRDE IT

If system (7) is observable and equation (9) has a T-periodic symmetric solution P(.), then, $\delta_d(\Phi_F(T,0)) = 0$ and $\text{cin}(-P(.)) = \text{din}(\Phi_F(T,0))$, $\forall t$. □

Interestingly enough, observe that the PLDE IT and PRDE IT have the following consequences. (i) Any T-periodic symmetric solution of (8) or (9) is nonsingular (at each time point). (ii) The spectrum of any T-periodic solution of (8) or (9) is periodically time-varying. However, no eigenvalue can vanish and the inertia is constant.

Obviously, the CALE IT and CARE IT can be seen as corollaries of the PLDE IT and PRDE IT, respectively.

1.5 *Extended inertia theorems*

In this paper, we prove the following extended inertia theorems (EIT), see also Bittanti, Colaneri (1984).

PLDE EIT

If system (7) is detectable and equation (8) has a T-periodic symmetric solution P(.), then $\delta_d(\Phi_A(T,0)) = 0$ and $\Pi_c(P(t)) + \delta_c(P(t)) = \nu_d(\Phi_A(T,0))$, $\forall t$.

PRDE EIT

If system (7) is detectable and equation (9) has a T-periodic symmetric solution P(.), then $\delta_c(\phi_F(T,0)) = 0$ and $\Pi_c(P(t)) + \delta_c(P(t)) = \nu_d(\phi_F(T,0))$, $\forall t$. □

Thus, if system (7) is detectable only, P(.) may be singular.

From the PLDE EIT, it follows that if system (7) is detectable and (8) is a positive semidefinite /negative definite/ periodic solution, then the system is stable /anti-stable/. The conclusion relative to the positive semidefinite case is part of the *extended periodic Lyapunov lemma*, (Bittanti, Bolzern and Colaneri (1984)). An analogous result for the Riccati equation can be easily worked out from the PRDE EIT.

The PRDE EIT can be proved on the basis of the PLDE EIT. In turn, to derive the PLDE EIT, a key role is played by the following extension of the DALE IT.

DALE EIT

If system (3) is detectable and equation (4) has a symmetric solution P, then $\delta_d(A) = 0$ and $\Pi_c(P) + \delta_c(P) = \gamma_d(A)$.

Precisely, the PLDE EIT can be derived from the DALE EIT by the same rationale used to derive the PLDE IT from the DALE IT, previously described in Section 1.4.

The DALE EIT, PLDE EIT and PRDE EIT are the main results of this paper.

The following extended inertia theorems for the algebraic Lyapunov equation and the algebraic Riccati equation are obvious corollaries of the PLDE EIT and PRDE EIT. In the PRDE EIT, F is defined as in (6).

CALE EIT

If system (1) is detectable and equation (2) has a symmetric solution P, then $\delta_c(A) = 0$ and $\pi_c(P) + \delta_c(P) = \nu_c(A)$.

CARE EIT

If system (1) is detectable and equation (5) has a symmetric solution P, then $\delta_c(F) = 0$ and $\pi_c(P) + \delta_c(P) = \nu_c(F)$.

The whole picture of the inertia theorems can be sketched as in Fig. 1, where the arrows point out the derivation routes followed by the various authors.

It goes without saying that analogous results hold for the dual Lyapunov and Riccati equations, by considering stabilizability in place of detectability.

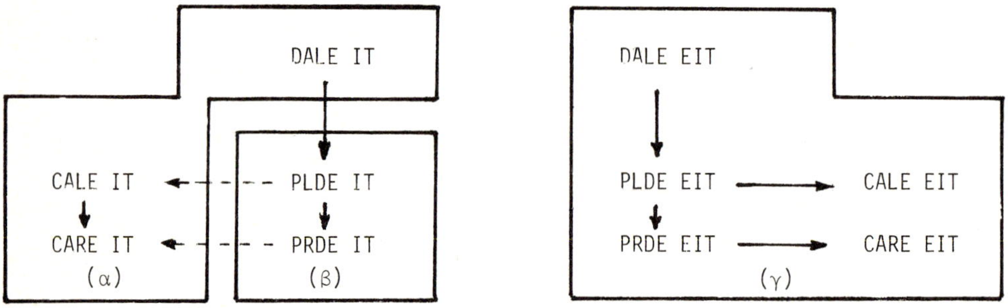

(α) : Tausky (1961); Ostrowski, Schneider (1962); Carlson, Schneider (1963); Chen (1973); Wimmer, Ziebur (1975), Wimmer (1976).

(β) : Shayman (1984

(γ) : This paper; Bittanti, Colaneri (1984).

Figure 1

Inertia theorems and extended inertia theorems-main references

1.7 *Paper organization*

In Section 2, some modal characterizations of observability and detectability for time-invariant and periodic systems are preliminarly introduced. These characterizations play a major role in the subsequent inertia analysis.

The key result of this paper is the DALE EIT, which will be proved in Section 3 by resorting to a decomposition technique. In Section 4, the PLDE EIT and PRDE EIT are dealt with. Precisely, the PLDE EIT is derived on the basis of the DALE EIT. In turn, the proof of the PRDE EIT relies on the PLDE EIT.

2. MODAL NOTIONS OF OBSERVABILITY AND DETECTABILITY

2.1 *Time-invariant systems*

A well-known test of observability for time-invariant systems is the following Popov-Belevitch-Hautus test, (Popov (1966), Belevitch (1968), Hautus (1969)).

PBH test of observability

A pair (A,C) is observable if and only if, for each eigenvalue λ of A, $Ax=\lambda x$ and $Cx=0$ imply $x=0$. □

This modal characterization of observability applies to the continuous-time system (1) as well as to the discrete-time system (2). On the contrary, when dealing with the notion of detectability, two different tests have to be introduced.

PBH test of c-detectability

A pair (A,C) is c-detectable if and only if, for each eigenvalues λ of A with nonnegative real part, $Ax=\lambda x$ and $Cx=0$ imply $x=0$.

PBH test of d-detectability

A pair (A,C) is d-detectable if and only if, for each eigenvalue λ of A not belonging to the open unit, disk, $Ax=\lambda x$ and $Cx=0$ imply $x=0$.

2.2 *Periodic systems*

In the periodic case, only continuous-time systems are considered. The generalization of the PBH-tests goes back to a paper by Hewer (1975). A slightly different, yet equivalent, modal condition, (Kano and Nishimura (1979)), can be stated as follows.

H test of observability

A periodic pair (A(.),C(.)) is observable if and only if, for each characteristics multiplier λ of A(.) and for some time paint t, $\phi_A(t+T,t)x = \lambda x$ and $c(\sigma)\ \phi_A(\sigma,t)x = 0,\ \forall\ \sigma \in [t,t+T]$, imply $x = 0$. □

The proof of the equivalence between this characterization and the usual Kalman definition of observability is given in Bittanti, Calaneri and Guardabassi (1984). The extension to detectability is discussed in Bittanti, Bolzern (1984) and in Bittanti, Bolzern (1984).

H test of detectability

A periodic pair (A(.), C(.)) is detectable if and only if, for each characteristic multiplier λ of A(.) not belonging to the open unit disk and for some time point t, $\Phi_A(t+T,t)x = \lambda x$ and $C(\sigma)\ \Phi_A(\sigma,t)x = 0,\ \forall\sigma\in [t,t+T]$, imply $x=0$.

3. THE DALE EXTENDED INERTIA THEOREM

In this section, the proof of the DALE EIT is outlined. For more details, the reader is referred to Bittanti, Colaneri (1984).

Proof of the DALE EIT (outline)

First of all, it is shown that $\delta_d(A) = 0$.

Indeed, suppose that A has an eigenvalue λ with $|\lambda| = 1$ and let x, $x \neq 0$, and associated eigenvector. Then, from (4), it follows that

$$0 = (|\lambda|^2 - 1)\, x^* Px + x^*C'Cx = x^*C'Cx.$$

Here, * denotes conjugate transpose. Therefore $Cx = 0$, so that the PBH test is violated.

Since $\delta_d(A) = 0$, there exists a suitable state transformation S such that $\tilde{A}=SAS^{-1}$ has the form

$$\tilde{A} = \begin{bmatrix} A_o & 0 \\ 0 & A_i \end{bmatrix}, \tag{11}$$

where the eigenvalues of A_o and A_i are outside or inside the unit disk, respectively Notice that $\text{din}\,(\tilde{A})^o = \text{din}(A)$.

Let $\tilde{C} = CS^{-1}$ be accordingly partitioned :

$$\tilde{C} = \begin{bmatrix} C_o & C_i \end{bmatrix} \tag{12}$$

Consider a solution $\tilde{P}$ of the DALE equation associated with $(\tilde{A}, \tilde{C})$,

$$0 = \tilde{A}'\tilde{P}\tilde{A} - \tilde{P} + \tilde{C}'\tilde{C} \tag{13}$$

and let

$$\tilde{P} = \begin{bmatrix} P_o & P_{oi} \\ P_{oi}' & P_i \end{bmatrix}, \tag{14}$$

where P_o and P_i have the same dimensions of A_o and A_i, respectively.

It is easy to see that : (i) If P is a solution of (4), $\tilde{P} = S\,P\,S^{-1}$ is a solution of (13). (ii) $\text{cin}\,(\tilde{P}) = \text{cin}(P)$. Therefore, the DALE inertia analysis can be performed with reference to (13).

From (11) - (14), it follows that

$$0 = A_o'\, P_o\, A_o - P_o + C_o'C_o. \tag{15}$$

Equation (15) is a DALE in the pair (A_o, C_o). We now prove that (A_o, C_o) is observable.

By contradiction, suppose that A_o admits an eigenvalue λ, with an associated non-zero eigenvector x_o, such that $C_o x_o = 0$. From (11), it follows that λ is an eigenvalue of $\tilde{A}$ and $\tilde{x} = |x_o' \;\; 0'|'$ is an associated eigenvector of $\tilde{A}$. Moreover $\tilde{C}\tilde{x} = 0$. Since $|\lambda| > 1$, this is contrary to the PBH test.

(A_o, C_o) being observable, the Lyapunov lemma implies that P_o is negative definite.

From this conclusion and (14), it follows that

$$\nu_c(P) = \nu_c(\tilde{P}) \geq \dim P_o = \dim A_o = \Pi_d(\tilde{A}) = \Pi_d(A).$$

The converse inequality, $\nu_c(P) \leq \Pi_d(A)$, follows from the discrete-time version of the inequality derived in Wimmer and Ziebur (1975, Theorem 2). This version can be worked out by applying to the DALE the same rationale used in Wimmer and Ziebur (1975) for the CALE.

In conclusion, $\delta_d(A) = o$ and $\nu_c(P) = \Pi_d(A)$, so that the proof is completed.

4. THE PLDE AND PRDE EXTENDED INERTIA THEOREM

Let D(t) be any factor of the observability gramian over (t,t+T), i.e.

$$D(t)'D(t) = \int_t^{t+T} \Phi_A(\sigma,t)'C'(\sigma)'C(\sigma)\,\Phi_A(\sigma,t)\,d\sigma .$$

It is easy to see that, if P(.) is any T-periodic solution of the PLDE equation (8), then, for each t, P(t) is a solution of the following DALE

$$0 = \Phi_A(t+T,t)'P(t)\,\Phi_A(t+T,t)-P(t)+D(t)'D(t) .$$

On the other hand, it can be shown (Bittanti and Colaneri (1984)) that, if the periodic pair (A(.), C(.)) is detectable, the constant pair $(\phi_A(t+T,t), D(t))$ is d-detectable at each time point t , see also Bittanti and Bolzern (1985c).

Then, the PLDE EIT straightforwardly follows from the DALE EIT.

Finally, consider the PRDE equation (9) and for any T-periodic symmetric solution P(.), let H(t) be any matrix such that

$$H(t)'H(t) = P(t)\,B(t)\,B(t)'P(t) + C(t)'C(t).$$

The solution P(.) satisfies then the PLDE:

$$-\dot{P}(t) = F(t)'P(t) + P(t)\,F(t) + H(t)'\,H(t).$$

where F(.) is the closed-loop dynamic matrix defined in (10). In Bittanti and Colaneri (1984), it is demonstrated that the periodic pair (F(.), H(.)) is detectable provided that (A(.), C(.)) is detectable. Therefore, from the PLDE EIT, the PRDE EIT is proved.

5. CONCLUSIONS

To the best knowledge of the authors, the inertia theorems for the Lyapunov equation and Riccati equation have always been stated under the assumption of observability (or reachability) of the underlying system. In this paper, a number of inertia theorems are derived by assuming detectability (or stabilizability) only. The algebraic Lyapunov equations and algebraic Riccati equation, as well as the periodic Lyapunov equation and periodic Riccati equation are considered. Thus, a comprehensive set of extended inertia theorems is provided. For an analogous discussion relative to discrete-time and periodic systems, see Bolzern and Colaneri (1985).

The inertia theorems were used in the analysis of the stability of linear multivariable systems (Datta, 1976 and 1978) and in the study of Lyapunov and Riccati equations, see e.g. Coppel (1974) and Wimmer (1976). In particular, they were used to analyze the existence of periodic solutions, Shayman (1984). It is expected that these results can be generalized by means of the extended inertia theorems given in this paper. Some generalizations are reported in Bittanti, Colaneri (1985).

ACKNOWLEDGEMENT

This paper was partially supported by the Centro di Teoria dei Sistemi CNR and by MPI.

REFERENCES

Belevitch, V. (1968). *Classical Network Theory*, Holden-Day, San Francisco.

Bittanti, S., and P. Bolzern (1984). Four equivalent notions of stabilizability of periodic system. *American Control Conference*, San Diego, USA, 1321-1323

Bittanti, S. and P. Bolzern (1985-a). Reachability and controllability of discrete-time linear periodic systems. *IEEE Trans-AC* 30, 399-401.

Bittanti, S. and P. Bolzern (1985-b). Discrete-time linear periodic systems: Gramian and modal criteria for reachability and controllability. *Int. J. of Control*, 41, 909-928.

Bittanti, S. and P. Bolzern (1985-c). Stabilizability and detectability of linear periodic systems. *Systems and Control Letters* (to appear).

Bittanti, S. and P. Bolzern, (1985-d). On the structure theory of discrete-time linear systems. *Int. J. of Systems Science* (to appear).

Bittanti, S., P. Bolzern and P. Colaneri (1984). Stability analysis of linear periodic systems via the Lyapunov equation. *IFAC World Congress*, Budapest, Vol. 8, 169-172.

Bittanti, S., P. Bolzern and P. Colaneri (1985). The extended periodic Lyapunov Lemma, *Automatica*(to appear)

Bittanti, S. and P. Colaneri (1984). Lyapunov and Riccati equations: periodic inertia theorems. *Int. Rep. LCA* 84-13, Dipartimento di Elettronica, Politecnico di Milano

Bittanti, S. and P. Colaneri (1985). Lyapunov and Riccati equations:periodic inertia theorems - an addendum. *Int. Rep. LCA* -85-7, Dipartimento di Elettronica, Politecnico di Milano

Bittanti, S., P. Colaneri and G. Guardabassi (1984). H-controllability and observability of linear periodic systems. *SIAM J. Contr. and Opt.* 22, 889-893.

Bittanti, S., G. Guardabassi, C. Maffezzoni and L. Silverman, Periodic systems: controllability and the matrix Riccati equation, *SIAM J. Cont. and Opt.* 16, 37-40.

Bolzern, P. and Colaneri, P. (1985). Inertia theorems for the periodic Lyapunov difference equation and periodic Riccati difference equation. *Int. Rep. LCA 85-6*, Dipartimento di Elettronica, Politecnico di Milano.

Carlson, D. and Schneider, H. (1963). Inertia theorems for matrices: the semidefinite case. *J. Math. Anal. Appl. 6*, 430-446.

Chen. C.T. (1973). A generalization of the inertia theorem. *SIAM J. Appl. Math.* 25, 158-161.

Coppel, W.A. (1974). Matrix quadratic equations, *Bull. Austral. Math. Soc.* 10, 377-401.

Datta, B.N. (1976). Application of Hankel matrix to the root location problem, *IEEE Trans.-AC* , 21, 610-612.

Datta, B.N. (1978). Two inertia theorems for Hessenberg matrices and their applications to stability analysis of linear control systems, *Matrix and tensor Quart.* 29, 55-62.

Hautus, M.L.J. (1969). Controllability and observability conditions of linear autonomous systems. *Proc. J. Nederl. Akad. Wetendr. Ser. A.* 72, 443-448.

Hewer, G.A. (1975). Periodicity, detectability and the matrix Riccati Equation. *Siam J. Control*, 13, 1235-1251.

Kalman, R.R. (1963). Lyapunov functions for the problem for Lur'e in automatic control. *Nat. Acad. Sci.* 49, 201-205.

Kalman, R.E., P.L. Falb, and M.A. Arbib, (1969). *Topics in Mathematical System Theory*, Mc Graw-Hill, New York, 1969.

Kano, H. and T. Nishimura (1979). Periodic solutions of matrix Riccati equations with detectability and stabilizability, *Int. J. Control 29*, 471-487.

Ostrowki A. and H. Schneider (1962). Some theorems on the inertia of general matrices. *J.Math. Anal. Appl. 4*, 72-84.

Popov, V.M. (1966) *Hyperstability of Control Systems*, Springer Berlin, 1973.

Shayman,M. (1984). Inertia theorems for the periodic Lypunov equation and periodic Riccati equation. *System and Control Letters*, 4, 27-32.

Tausski, O. (1961). A generalization of a theorem of Lyapunov, *SIAM J. Appl. Math.* 9, 640-643.

Wimmer, H.K. (1976). On the algebraic Riccati equation. *Bull. Austral. Math. Soc.* 14, 457-461.

Wimmer, H.K. and A.D. Ziebur (1975). Remarks on inertia theorems for matrices. *Czech. Math. J.* 25, 556-561.

Yakubovich, V.A. and V.M. Starzhiskii (1972). *Linear differential equations with periodic coefficients*, J. Wiley and Sons, New York.

Frequency Domain and State Space Methods for Linear Systems
C.I. Byrnes and A. Lindquist (Editors)

WHEN IS THE SOLUTION OF THE RICCATI EQUATION STABILIZING AT EVERY INSTANT?

Michel R.Gevers, Robert R. Bitmead, Ian R. Petersen
Department of Systems Engineering
Australian National University
Canberra ACT
Australia

and

R. John Kaye*
School of Electrical Engineering
University of New South Wales
Kensington NSW
Australia

We consider the Kalman filter associated with a continuous-time or discrete-time state-space model with constant parameters. This Kalman filter has a time-varying closed-loop state transition matrix, which is directly obtained from the time-varying solution of a Riccati differential or difference equation (RDE). We consider the problem of selecting an initial covariance matrix P_0 for the RDE to ensure that the closed loop filter is exponentially asymptotically stable at every instant as a time-invariant filter. The results have application in observer design.

INTRODUCTION

This paper is concerned with the stabilizing properties of the solutions P(t) of the Riccati difference or differential equation (RDE) of optimal filtering. We consider either the discrete-time Kalman filter

$$\hat{x}(t+1) = [F-K(t)H]\hat{x}(t) + K(t)y(t) \tag{1.1a}$$

$$K(t) = FP(T)H^T[HP(t)H^T+I]^{-1} \tag{1.1b}$$

and the associated discrete-time RDE:

$$P(t+1) = FP(t)F^T - FP(t)H^T[HP(t)H^T+I]^{-1}HP(t)F^T + LL^T \tag{1.1c}$$

or the continuous-time filter

$$\dot{\hat{x}}(t) = [F-K(t)H]\hat{x}(t) + K(t)y(t) \tag{1.2a}$$

$$K(t) = P(t)H^T \tag{1.2b}$$

and the associated continuous-time RDE:

$$\dot{P}(t) = P(t)F^T + FP(t) - P(t)H^THP(t) + LL^T \tag{1.2c}$$

These Kalman filters are associated with the following signal models.

*The order of the authors' names was decided by a random walk.

In discrete-time:

$$x(t+1) = Fx(t) + w(t) \tag{1.3a}$$

$$y(t) = Hx(t) + v(t)$$

where x and y have dimension n and p respectively, and where w(t) and v(s) are uncorrelated with

$$E(w(t)w^T(s)] = LL^T\delta_{ts} \tag{1.4a}$$

$$E[v(t)v^T(s)] = I\delta_{ts} \tag{1.4b}$$

In continuous-time:

$$\dot{x}(t) = Fx(t) + w(t) \tag{1.5a}$$

$$y(t) = Hx(t) + v(t) \tag{1.5b}$$

with the same conditions on x,y,w and v where δ_{ts} is replaced by $\delta(t-s)$.

We then ask the question: Under what conditions on the underlying signal model (i.e. H,F,L) and for what initial conditions P(∘) of the RDE are the closed-loop filtering matrices {F-K(t)H} exponentially asymptotically stable (e.a.s.) matrices for all t≧0? We say that a matrix A is e.a.s. if

- $|\lambda_i(A)| < 1$, i=1,...,n, in discrete-time (1.6a)

- $\text{Re}\lambda_i(A) < 0$, i=1,...,n, in continuous-time (1.6b)

Surprisingly, this problem does not seem to have been studied before. Besides its theoretical interest, the major practical motivation for studying this problem is as follows. The conditions under which the Kalman filter is asymptotically stable as a time-varying filter are well established: see e.g. [1]. However, here we want to study the exponential stability of the time-invariant filter

$$\hat{x}(s+1) = [F-K(t)H]\hat{x}(s) + K(t)y(s) \tag{1.7}$$

in discrete-time, or

$$\dot{\hat{x}}(s) = [F-K(t)H]\hat{x}(s) + K(t)y(s) \tag{1.8}$$

in continuous-time, where K(t) is given, respectively, by (1.1b,c) and (1.2b,c), for some arbitrary but <u>fixed</u> t.

There are conditions under which the Kalman filter is asymptotically stable as a time-varying filter and converges, but the limiting steady-state filter need not be asymptotically stable. This is the case when the pair [F,L] has uncontrollable modes on the unit circle (in discrete-time) or on the jω-axis (in continuous-time): see [2], [3]. The results of this paper then allow the use of the Kalman filter as an optimal time-varying filter from an initial start-up time until a switching-off or freezing time, from which the Kalman gain update is stopped and the filter is run as a time-invariant exponentially asymptotically stable observer. This particular problem arises naturally in short-time Fourier analysis [4], where L=0 and F has eigenvalues equally spaced around the unit circle.

In this paper, sets of sufficient conditions on H,F,L and P(∘) will be presented under which the matrices {F-K(t)H} are e.a.s. for all t≧0. Our results will be obtained by proving the monotonic non-increasing behaviour of the solution P(t) of the RDEs (1.1c) and (1.2c) under those sets of sufficient conditions. The major results of this paper were presented for the discrete-time problem in [5]. In section 2 we shall briefly recall the major discrete-time results without proof. The continuous-time results will be presented and proved in section 3: the proofs of the continuous-time results closely parallel those of [5] although some simplifications occur in continuous-time. The stability results on $\{F-K(t)H\}$ are obtained by transforming the RDE into a Fake Algebraic Riccati Equation (FARE) and by using existing results on the Algebraic Riccati Equation (ARE) established in [2]-[3].

For the purposes of this paper, the most relevant previous contributions on the RDE and the ARE are those of Willems [6], Kucera [7], Martensson [8], Caines and Mayne [9], Payne and Silverman [10], Chan, Goodwin and Sin [2] and de Souza, Gevers and Goodwin [3].

2. THE DISCRETE-TIME RESULTS

We consider the discrete-time Kalman filter (1.1). The closed-loop state transition matrix of the Kalman filter is

$$\bar{F}(t) = F - K(t)H = F - FP(t)H^T[HP(t)H^T+I]^{-1}H \tag{2.1}$$

To the RDE (1.1c) there is associated an algebraic Riccati equation (ARE):

$$P = FPF^T - FPH^T(HPH^T+I)^{-1}HPF^T + LL^T \tag{2.2}$$

We shall assume throughout that H has full rank. For each P(t), solution of the RDE (1.1c), we define a symmetric matrix Q(t) as follows:

$$Q(t) \triangleq P(t) - FP(t)F^T + FP(t)H^T[HP(t)H^T+I]^{-1}HP(t)F^T \tag{2.3}$$

Note that Q(t) can also be written as:

$$Q(t) = LL^T + P(t) - P(t+1) \tag{2.4}$$

For fixed t, equation (2.3) can be viewed as an ARE, provided Q(t) can be shown to be nonnegative definite: (2.3) will therefore be called a Fake Algebraic Riccati Equation (FARE). The main trick in our proofs is the introduction of the FARE (2.3) and the use of stability results for the ARE recently obtained in [2] and [3]. The major task then is to obtain conditions on H,F,L and P(o) which guarantee that Q(t) is nonnegative definite. As is clear from (2.4), this property of Q(t) is closely linked to the monotonically nonincreasing behaviour of P(t). The introduction of the FARE has given rise to a whole set of new techniques called Fake Algebraic Riccati Techniques (FART): see [11].

It is well-known (see e.g. [9]) that if $P(\circ)=0$, then the sequence P(t) increases monotonically, in the sense that $P(t+1) \geq P(t)$ (i.e. $P(t+1)-P(t)$ is nonnegative definite). However, this monotonic behaviour breaks down if $P(\circ)$ is an arbitrary nonnegative definite matrix. In deriving conditions that ensured the monotonic decreasing behaviour of the discrete-time RDE, we were led to prove the following lemma, which we believe is of independent interest; it is inspired by a result of Nishimura [12].

Lemma 1. Consider two RDEs (1.1c) with the same F and H matrices but possibly different $Q=LL^T$ matrices, $\hat{Q}$ and $\bar{Q}$, and possibly different initial conditions. Let the solutions to these RDEs be written

$$\hat{P}(t+1) = f[\hat{P}(t),\hat{Q}], \quad \bar{P}(t+1) = f[\bar{P}(t),\bar{Q}] \tag{2.5}$$

where the functional form of f is identical and is given by (1.1c) with LL^T

replaced by Q. Then

$$\hat{P}(t+1) \geqq \bar{P}(t+1) \text{ if } \hat{P}(t) \geqq P(t) \text{ and } \hat{Q} \geqq Q \tag{2.6}$$

The proof relies on the following expression which can be obtained from [12] after lengthy manipulations (for simplicity P(t) will be written P_t):

$$\hat{P}_{t+1} - \bar{P}_{t+1} = F[I-\bar{P}_tH^T(H\bar{P}_tH^T+I)^{-1}H]g(\hat{P}_t-\bar{P}_t)$$
$$. [I-\bar{P}_tH^T(H\bar{P}_tH^T+I)^{-1}H]^TF^T + \hat{Q} - \bar{Q} \tag{2.7a}$$

where

$$g(\hat{P}_t-\bar{P}_t) = (\hat{P}_t-\bar{P}_t)H^T[H(\hat{P}_t-\bar{P}_t)H^T + H(\hat{P}_t-\bar{P}_t)H^T(H\bar{P}_tH^T+I)^{-1}H(\hat{P}_t-\bar{P}_t)H^T]^{-1}H(\hat{P}_t-\bar{P}_t)$$
$$+ \{I - (\hat{P}_t-\bar{P}_t)H^T[H(\hat{P}_t-\bar{P}_t)H^T]^{-1}H\}(\hat{P}_t-\bar{P}_t)$$
$$. \{I - (\hat{P}_t-\bar{P}_t)H^T[H(\hat{P}_t-\bar{P}_t)H^T]^{-1}H\}^T \tag{2.7b}$$

No similar continuous-time result is needed to prove the continuous-time results, as will appear from section 3.

Our main results on the discrete-time Riccati equation are as follows.

Theorem 1 [5]. Consider the RDE (1.1c) with initial condition P(∘). Define Q(∘) from (2.3). If

(1) [H,F] is detectable
(2) [F,L] is stabilizable
(3) $P(\circ) \geqq 0$ is such that $Q(\circ) \geqq LL^T$

then the solution sequence P(t) of (1.1c) is stabilizing for all $t \geqq 0$, i.e. $|\lambda_k[\bar{F}(t)]| < 1$ for all $t \geqq 0$ and for k=1,2,...,n with $\bar{F}(t)$ defined by (2.1) and λ_k denoting the eigenvalues.

Two comments are in order

(1) First note that the set of P(∘) satisfying condition (3) is clearly non-empty. Indeed, take any $Q(\circ) \geqq LL^T$; then it is easy to show that $[F,Q^{\frac{1}{2}}(\circ)]$ is a stabilizable pair because [F,L] is stabilizable. Conditions (1) and (2) then ensure that the ARE (2.3), with t=0, has a unique nonnegative definite solution P(∘) (see e.g. [3]).

(2) The conditions of Theorem 1 guarantee that the sequence {P(t)} is monotonically nonincreasing and convergent to a steady-state P^+. By condition (2), P^+ is actually stabilizing. As we argued earlier, we envisage the stability results on P(t) to be of particular use when the steady-state P^+ is not stabilizing, due to the pair [F,L]} not being stabilizable. We would therefore want to replace condition (2) of Theorem 1 by a weaker condition which would only involve the initial condition. Our aim is to replace the stabilizability condition on [F,L] by a condition on $[F,Q(\circ)^{\frac{1}{2}}]$. this is achieved in the next Theorem.

Theorem 2 [5]. Consider the RDE (1.1c). Define Q(∘) as in (2.3) and assume that

(1) [H,F] is detectable
(2) $[F,Q(\circ)^{\frac{1}{2}}]$ is stabilizable
(3) $P(\circ) \geqq 0$ is such that $Q(\circ) \geqq LL^T$.

Then the solution sequence {P(t)} with initial condition P(∘) is stabilizing for each $t \geqq 0$.

3. CONTINUOUS TIME RESULTS

We now consider the continuous-time Kalman filter (1.2). The results presented are directly analogous to the discrete-time results presented in the previous section. Furthermore, in this section, we include the proofs of these continuous-time results. Most of the proofs are directly analogous to the proofs of the discrete-time results given in [5].

Associated with the RDE (1.2c) is an ARE

$$PF^T + FP - PH^THP + LL^T = 0 \tag{3.1}$$

A solution P^+ to this ARE is said to be a strong solution if all eigenvalues λ of $F - P^+H^TH$ satisfy $Re(\lambda) \leq 0$. This definition was first introduced in [2], in a discrete-time context.

Proposition 1. If the RDE (1.2c) is such that $[H,F]$ is detectable then, for any initial condition $P(o)\geq 0$, there exists a solution to (1.2c) defined on $[0,\infty)$. Furthermore, if the initial condition $P(o)$ is such that $P(o)\geq P^+$ then $\lim_{t\to\infty} P(t) = P^+$ where P^+ is the unique strong solution to the ARE (3.1).

Proof. The existence of solutions to the RDE (1.2c) follows from remark M in [13]. Furthermore, the fact that the ARE (3.1) has a unique strong solution P^+ and that $\lim_{t\to\infty} P(t) = P^+$ follows from Poubelle et al. [14]. A discrete-time version of this result has been proved in [3].

We now introduce the Fake Algebraic Riccati Equation for the continuous-time case. If $P(t)$ is the solution to RDE (1.2c), then for each $P(t)$, we define an associated symmetric matrix $Q(t)$ according to

$$Q(t) \triangleq P(t)H^THP(t) - P(t)F^T - FP(t). \tag{3.2}$$

This is the Fake Algebraic Riccati Equation corresponding to the RDE (1.2c). Comparing RDE (1.2c) to ARE (3.2), we obtain

$$Q(t) = LL^T - \dot{P}(t). \tag{3.3}$$

The following lemma concerns the monotonicity of the solutions to the RDE (1.2c).

Lemma 2. Consider the RDE (1.2c) with $[H,F]$ detectable and let $Q(t)$ be defined as in (3.2). If the initial condition $P(o)\geq 0$ is such that $Q(o)\geq LL^T$ then $P(t_2)\leq P(t_1)$ for all times $t_2\geq t_1\geq 0$ and hence $\dot{P}(t)\leq 0$ for all $t\geq 0$.

Proof. By Proposition 1 and the assumptions of the lemma, the RDE (1.2c) has a solution $P(t)$ for all $t\geq 0$. We now introduce the notation

$$\tilde{F} = F - P(o)H^TH;$$

$$\tilde{P}(t) = P(t) - P(o).$$

It follows that $\tilde{P}(t)$ satisfies the RDE

$$\dot{\tilde{P}}(t) = \tilde{P}(t)\tilde{F}^T + \tilde{F}\tilde{P}(t) - \tilde{P}(t)H^TH\tilde{P}(t) + LL^T - Q(o) \tag{3.4}$$

with initial condition $\tilde{P}(o)=0$. We will use a duality argument to interpret $\tilde{P}(t)$ in terms of the solution to a certain optimal control problem.

Indeed, using Theorem 21.1 of [15], it follows that given any vector x_1 and $t_1>0$ then

$$x_1^T\tilde{P}(t_1)x_1 = \min_{u(\cdot)} \int_0^{t_1} [x(t)^T(LL^T-Q(\circ))x(t) + u(t)^Tu(t)]dt$$

subject to

$$\dot{x}(t) = -\tilde{F}^Tx(t) + H^Tu(t); \quad x(t_1) = x_1.$$

As in [15], this optimal control problem has a solution $\tilde{x}(\cdot)$, $\tilde{u}(\cdot)$. We let $t_2 \geq t_1$ be given and let $\Delta = t_2 - t_1$. A control law $\hat{u}(\cdot)$ is now defined according to

$$\hat{u}(t) \triangleq 0 \text{ for } t \in [o,\Delta]$$

$$\hat{u}(t) \triangleq \tilde{u}(t-\Delta) \text{ for } t \in [\Delta,t_2]$$

Let $\hat{x}(t)$ be the corresponding solution to the state equation

$$\dot{x}(t) = -\tilde{F}^Tx(t) + H^Tu(t); \quad x(t_2) = x_1.$$

Using the fact that this system is time-invariant, it follows that $\hat{x}(t) = \tilde{x}(t-\Delta)$ for all t. Therefore

$$x_1^T\tilde{P}(t_1)x_1 = \int_0^{t_1} [\tilde{x}(t)^T(LL^T-Q(\circ))\tilde{x}(t) + \tilde{u}(t)^T\tilde{u}(t)]dt$$

$$= \int_\Delta^{t_2} [\tilde{x}(t-\Delta)^T(LL^T-Q(\circ))\tilde{x}(t-\Delta) + \tilde{u}(t-\Delta)^T\tilde{u}(t-\Delta)]dt$$

$$= \int_\Delta^{t_2} [\hat{x}(t)^T(LL^T-Q(\circ))\hat{x}(t) + \hat{u}(t)^T\hat{u}(t)]dt.$$

However, $LL^T-Q(\circ) \leqq 0$ and $\hat{u}(t) = 0$ for $t \in [0,\Delta]$. Hence

$$\int_\Delta^{t_2} [\hat{x}(t)^T(LL^T-Q(\circ))\hat{x}(t) + \hat{u}(t)^T\hat{u}(t)]dt$$

$$\geq \int_0^{t_2} [\hat{x}(t)^T(LL^T-Q(\circ))\hat{x}(t) + \hat{u}(t)^T\hat{u}(t)]dt$$

$$\geqq x_1^T\tilde{P}(t_2)x_1$$

using the optimal control interpretation of $\tilde{P}(t_2)$. Thus, we have established that for arbitrary x_1

$$x_1^T\tilde{P}(t_1)x_1 \geq x_1^T\tilde{P}(t_2)x_1.$$

That is $\tilde{P}(t_1) \geq \tilde{P}(t_2)$ and hence $\dot{P}(t_1) \geqq P(t_2)$. The fact that $\dot{P}(t) \leqq 0$ now follows directly from the definition of the $\dot{P}(t)$. □

Using this lemma, we obtain the following theorem, which is the continuous-time version of Theorem 1 in [5].

<u>Theorem 3</u>. Consider the RDE (1.2c). Define $Q(\circ)$ as in (3.2) and assume that

(1) $[H,F]$ is detectable
(2) $[F,L]$ is stabilizable
(3) $P(\circ) \geqq 0$ is such that $Q(\circ) \geqq LL^T$,

then the solution $P(t)$ of the RDE (1.2c) is stabilizing for all $t \geqq 0$, i.e. all eigenvalues λ of the matrix $F - P(t)H^TH$ satisfy $Re(\lambda) < 0$.

<u>Proof</u>: Since $P(\circ)$ and $Q(\circ)$ are related by the FARE (3.2), it follows from Theorem 1 of [16] that $Q(\circ) \geqq LL^T$ implies $P(\circ) \geqq P^+$, where P^+ is the unique

strong solution of (3.1). Using Proposition 1, it follows that the RDE (1.2c) has a solution $P(t) \to P^+$. Futhermore, Lemma 2 implies that $P(t) \geq 0$ for all $t \geq 0$. Thus, equation (3.3) implies that given any $t \geq 0$

$$Q(t) \geq LL^T \tag{3.5}$$

Using this fact and Condition (2) of the theorem statement, it is straightforward to establish that $[F, Q(t)^{\frac{1}{2}}]$ is stabilizable. However, P(t) satisfies the FARE

$$FP(t) + P(t)F^T - P(t)H^THP(t) + Q(t) = 0$$

where [H,F] is detectable and $[F, Q(t)^{\frac{1}{2}}]$ is stabilizable. Hence using Theorem 4.11 of [17], it follows that the matrix $F - P(t)H^TH$ is a stability matrix. □

As in the discrete-time case, it would be desirable to weaken the requirement that [F,L] is stabilizable. This is achieved in the following theorem.

Theorem 4. Consider the RDE (1.2c). Define $Q(\circ)$ as in (3.2) and assume that

(1) [H,F] is detectable
(2) $[F, Q(\circ)^{\frac{1}{2}}]$ is stabilizable
(3) $P(\circ) \geq 0$ is such that $Q(\circ) \geq LL^T$.

Then the solution P(t) with initial condition $P(\circ)$ is stabilizing for each $t \geq 0$.

In order to prove this theorem, we first establish a number of preliminary lemmas.

Lemma 3. Let P(t) be the solution to the RDE (1.2c) with initial condition $P(\circ) \geq 0$, then $P(t) \geq 0$ for all $t \geq 0$. Furthermore, if x_1 is an eigenvector of F^T such that $x_1^*P(t_1)x_1 = 0$ for some $t_1 \geq 0$ then $x_1^*P(\circ)x_1 = 0$ (where x_1^* denotes the complex conjugate transpose of the vector x_1).

Proof. This proof relies on a duality argument in which P(t) is related to a certain optimal control problem. We observe that it is straightforward to extend Theorem 21.1 of [15] to include the case where both x(t) and u(t) are complex vectors. Hence, given any $t_1 \geq 0$ and any vector x_1,

$$x_1^*P(t_1)x_1 = \min_{u(\cdot)} \int_0^{t_1} [x(t)^*Qx(t) + u(t)^*u(t)]dt + x(\circ)^*P(\circ)x(\circ) \tag{3.6}$$

subject to

$$\dot{x}(t) = -F^Tx(t) + H^Tu(t); \quad x(t_1) = x_1.$$

Since all terms on the right hand side of (3.6) are nonnegative, it follows that $x_1^*P(t_1)x_1 \geq 0$. Furthermore, if x_1 is an eigenvector of F^T such that $x_1^*P(t_1)x_1 = 0$, we obtain

$$\tilde{x}(\circ)^*P(\circ)\tilde{x}(\circ) = 0, \tag{3.7}$$

$$\tilde{x}(t)^*Q\tilde{x}(t) \equiv 0 \text{ on } [0,t_1]$$

and

$$\tilde{u}(t)^*\tilde{u}(t) \equiv 0 \quad \text{on } [0,t_1],$$

where $\tilde{x}(\cdot)$, $\tilde{u}(\cdot)$ solves the given optimal control problem. Therefore

$$\dot{\tilde{x}}(t) = -F^T\tilde{x}(t); \quad \tilde{x}(t_1) = x_1.$$

Using the fact that x_1 is an eigenvector of F^T with eigenvalue λ, it follows that

$$\tilde{x}(0) = e^{+\lambda t_1} x_1.$$

Substituting this into (3.7) we obtain

$$x_1^* P(0) x_1 = 0. \quad \square$$

Lemma 4. Let $P(t) \geq 0$ be a solution to RDE (1.2c) such that $\dot{P}(t) \leq 0$ for all t and $HP(t_1)x = 0$ for some $t_1 \geq 0$ and some vector x. Then $HP(t)x = 0$ for all $t \geq t_1$.

Proof. For each t, we can write

$$P(t) = V_1(t)^T V_1(t) + V_2(t)^T V_2(t)$$

where $V_1(t)^T$ is in the range space of H^T and $V_2(t)^T$ is in the null space of H. Similarly, write $x = x_1 + x_2$. Then $HP(t_1)x = 0$ implies that $V_1(t_1)x_1 = 0$. Furthermore, given any $t \geq t_1$, the fact that $P(t) \leq P(t_1)$ implies that $V_1(t_1)^T V_1(t_1) \geq V_1(t)^T V_1(t)$. Therefore $V_1(t)x_1 = 0$ and hence $HP(t)x = 0$. $\square$

Proof of Theorem 4. Consider the RDE (1.2c) with initial condition $P(0)$. Using Proposition 1 and Lemma 3, it follows that there exists a solution $P(t)$ such that $P(t) \geq 0$ for all $t \geq 0$. Furthermore, using Lemma 2, $\dot{P}(t) \leq 0$ for all $t \geq 0$. Now let $t_1 \geq 0$ be given and let x be an eigenvector of $\bar{F}^T \triangleq F^T - H^T HP(t_1)$ with eigenvalue λ. We must show that $Re(\lambda) < 0$.

The matrix $P(t_1)$ satisfies the ARE (3.2) and hence

$$\bar{F}P(t_1) + P(t_1)\bar{F}^T + P(t_1)H^T HP(t_1) + Q(t_1) = 0.$$

Therefore

$$(\lambda+\lambda^*)x^*P(t_1)x + x^*P(t_1)H^T HP(t_1)x + x^*Q(t_1)x = 0. \quad (3.8)$$

We note that $\dot{P}(t_1) \leq 0$ and hence (3.3) implies $Q(t_1) \geq 0$.

We now consider two cases:

Case 1. $x^*P(t_1)x \neq 0$.

In this case, (3.8) implies $\lambda+\lambda^* \leq 0$, i.e. $Re(\lambda) \leq 0$. We will show $Re(\lambda) \neq 0$ by contradiction. Suppose $\lambda+\lambda^* = 0$. Using (3.8) we conclude that $x^*Q(t_1)x = 0$ and $HP(t_1)x = 0$. Hence, using Lemma 4, it follows that $HP(t)x = 0$ for all $t \geq t_1$. Now using [16] and Proposition 1, it follows that $P(t) \to P^+$ and hence $HP^+x = 0$. We also observe that $(F^T - H^T HP(t_1))x = \lambda x$ and hence $F^T x = \lambda x$. Furthermore, substituting $\dot{P}(t_1) \leq 0$ and $x^*Q(t_1)x = 0$ into (3.3) it follows that $x^*LL^T x = 0$. Now, returning to RDE (1.2c), it follows that for all $t \geq t_1$,

$$\begin{aligned} x^*\dot{P}(t)x &= x^*P(t)F^T x + x^*FP(t)x - x^*P(t)H^T HP(t)x + x^*LL^T x \\ &= (\lambda+\lambda^*)x^*P(t)x \\ &= 0. \end{aligned}$$

Hence, $x^*P(t)x = x^*P(t_1)x$ for all $t \geq t_1$ and therefore $x^*P^+x = x^*P(t_1)x \neq 0$.

Now

$$FP^+ + P^+F^T - P^+HH^T P^+ + LL^T = 0.$$

Therefore $FP^+x + P^+F^T x = 0$ and hence, if we let $z = P^+x$ then

$$Fz = -\lambda P^{+}x = -\lambda z.$$

Furthermore, $z = P^{+}x \neq 0$ and therefore z is an eigenvector of F with eigenvalue $-\lambda$. Also, note that $Hz = 0$. The fact that $Re(\lambda) = 0$ now contradicts the detectability of the pair [H,F].

<u>Case 2</u>. $x^{*}P(t_1)x = 0$.

In this case, it follows from (3.8) that $HP(t_1)x = 0$ and $x^{*}Q(t_1)x = 0$. Now, we have $(F^T - HH^TP(t_1))x = \lambda x$ and hence $F^Tx = \lambda x$. That is, x is an eigenvector of F^T. Furthermore, it follows from Lemma 3 that $P(\circ)x = 0$. We now conclude from ARE (3.2) that

$$\begin{aligned} x^{*}Q(\circ)x &= x^{*}P(\circ)H^THP(\circ)x - x^{*}P(\circ)F^Tx - x^{*}FP(\circ)x \\ &= 0 \end{aligned}$$

That is, x is an eigenvector of F^T (with eigenvalue λ) such that $x^{*}Q(\circ)x = 0$. The detectability of $[Q(\circ)^{\frac{1}{2}}, F^T]$ now implies that $Re(\lambda) < 0$. □

4. CONCLUSION

Our results provide a rather easy procedure to initialize a Riccati equation in such a way that the RDE can be stopped at any time and the corresponding Kalman gain used as the constant gain of an observer, while guaranteeing the exponential stability of that observer. While our conditions are only sufficient, it has proved remarkably difficult to obtain necessary and sufficient conditions. In [5] we have given a number of counterexamples in discrete-time, which show just how badly the Riccati equation solution can behave if one of the conditions of our main theorems is violated.

REFERENCES

[1] Anderson, B.D.O. and Moore, J.B., Optimal Filtering, (Prentice-Hall, Englewood Cliffs, NJ, 1979).

[2] Chan, S.W., Goodwin, G.C. and Sin, K.S., Convergence properties of the Riccati difference equation in optimal filtering of nonstabilizable systems, IEEE Trans. Automat. Control, AC-29, No. 2 (Feb. 1984) 110-118.

[3] de Souza, C.E., Gevers, M. and G.C. Goodwin, Riccati equations in optimal filtering of nonstabilizable systems having singular state transition matrices, Tech. Rept. EE 8522, University of Newcastle, submitted for publication.

[4] Bitmead, R.R., Tsoi, A.-C. and Parker, P.J., A Kalman-filtering approach to short-time Fourier analysis, submitted for publication.

[5] Bitmead, R.R., Gevers, M.R., Petersen, I.R. and Kaye, R.J., Monotonicity and stabilizability properties of solutions of the Riccati difference equation: Propositions, lemmas, theorems, fallacious conjectures and counterexamples, Systems and Control Letters, 5, No. 5 (1985) 309-315.

[6] Willems, J.C., Least squares stationary optimal control and the algebraic Riccati equation, IEEE Trans. Automat. Control, AC-16, No. 6 (Dec. 1971) 621-634.

[7] Kucera, V., The discrete Riccati equation of optimal control, Kybernetika 8 (1972) 430-447.

[8] Martensson, K., On the matrix Riccati equation, Inform. Sci., 3, (1971) 17-49.

[9] Caines, P.E. and Mayne, D.Q., On the discrete-time matrix Riccati equation of optimal control, Int. J. Control, 12 (Nov. 1970) 785-794.

[10] Payne, H.J. and Silverman, L.M., On the discrete-time algebraic Riccati equation, IEEE Trans. Automat. Control, AC-18, No. 3 (June 1973) 226-234.

[11] Poubelle, M.A., What can FARTs be used for? submitted for publication.

[12] Nishimura, T., On the a priori information in sequential estimation problems, IEEE Trans. Automat. Control, AC-11, No. 2 (April 1966) 147-204; See also AC-12, No. 1 (Feb. 1967) 123.

[13] Callier, F.M and Willems, J.L., Criterion for the convergence of the solution of the Riccati Differential Equation, IEEE Trans. Autom. Control, AC-26 No. 6 (Dec. 1981) 1232-1242.

[14] Poubelle, M.A., Bitmead, R.R., Gevers, M.R. and Petersen, I.R., A miscellany of results on an equation of Count J.F. Riccati, in preparation.

[15] Brockett, R.W., Finite dimensional linear systems (J. Wiley, 1970).

[16] Wimmer, H.K., Monotonicity of maximal solutions of algebraic Riccati equations, Systems and Control Letters, Vol. 5, No. 5 (1985) 317-319.

[17] Kwakernaak, H. and Sivan, R., Linear Optimal Control Systems (J. Wiley, 1972).

Frequency Domain and State Space Methods for Linear Systems
C.I. Byrnes and A. Lindquist (Editors)

RICCATI AND LYAPUNOV SYSTEMS

Vicente Hernández Lucas Jódar

Department of Mathematics
Polytechnical University of Valencia
P.O. Box 22.012- Valencia
SPAIN

By application of the annihilating polynomial method, the resolution problem of a Riccati system, $A_i+B_iX+XC_i+XD_iX=0$, $i=1,2$, is reduced to a linear system of the type $A+XB=0$, $C+DX=0$.

INTRODUCTION

The resolution problem of an algebraic quadratic equation of Riccati type

$$A_1 + B_1 X + X C_1 + X D_1 X = 0 \tag{1.1}$$

or the resolution problem of a linear equation of Lyapunov type

$$A_1 + B_1 X + X C_1 = 0 \tag{1.2}$$

appears, in the context of system theory, in linear-quadratic optimal control problems, filtering problems, stability, ... , [1].

In this paper we study the more general problem of Riccati systems, that is, systems of the type

$$\left.\begin{aligned} A_1 + B_1 X + X C_1 + X D_1 X = 0 \\ A_2 + B_2 X + X C_2 + X D_2 X = 0 \end{aligned}\right\} \tag{1.3}$$

as well as the Lyapunov systems

$$\left.\begin{aligned} A_1 + B_1 X + X C_1 = 0 \\ A_2 + B_2 X + X C_2 = 0 \end{aligned}\right\} \tag{1.4}$$

A motivation for the study of such systems appears in the analysis of perturbation problems. So, for instance, notice that a solution of (1.3) is a solution of (1.1) and of the perturbed Riccati equation

$$(A_1 + A_2) + (B_1 + B_2)X + X(C_1 + C_2) + X(D_1 + D_2)X = 0$$

In the systems and matrices bibliography a great deal of information on the resolution methods of matrix algebraic equations is shown, [2]. The annihilating polynomial method applied by Jameson [3], and Jones [4], in the resolution problems of equations (1.2) and (1.1), respectively, is extended to the resolution problems of systems (1.3) and (1.4).

RESULTS

The first result permits one to eliminate the bilateral character of the quadratic system (1.3), as well as to reduce the degree of the problem.

Theorem 1.

Let

$$\left.\begin{aligned} A_1 + B_1 X + X C_1 + X D_1 X = 0 \\ A_2 + B_2 X + X C_2 + X D_2 X = 0 \end{aligned}\right\} \tag{2.1}$$

be a Riccati quadratic system where $A_i \in \mathbb{C}_{mxn}$, $B_i \in \mathbb{C}_{mxm}$, $C_i \in \mathbb{C}_{nxn}$, $D_i \in \mathbb{C}_{nxm}$, i=1,2. If X is a solution of (2.1) and

$$p(z)=\sum_{j=o}^{r} a_j z^j \ , \qquad q(z)=\sum_{k=o}^{s} b_k z^k$$

are annihilating polynomials of the matrices B_1+XD_1, C_2+D_2X, respectively; then it is verified that X is a solution of the linear system

$$\left.\begin{aligned} A + X B = 0 \\ C + D X = 0 \end{aligned}\right\} \tag{2.2}$$

where

$$\begin{aligned} A&=\Big(\sum_{j=o}^{r} a_j U_j\Big)C_2 \ , \qquad & B&=\Big(\sum_{j=o}^{r} a_j V_j\Big)C_2 \\ C&=B_1\Big(\sum_{k=o}^{s} b_k Y_k\Big) \ , \qquad & D&=B_1\Big(\sum_{k=o}^{s} b_k Z_k\Big) \end{aligned} \tag{2.3}$$

with the matrices U_j, V_j, Y_k, Z_k, defined by the expressions

$$\begin{bmatrix} U_j \\ V_j \end{bmatrix} = \begin{bmatrix} B_1 & -A_1 \\ D_1 & -C_1 \end{bmatrix} \begin{bmatrix} U_{j-1} \\ V_{j-1} \end{bmatrix} \quad j=1,\dots,r \qquad \begin{bmatrix} U_o \\ V_o \end{bmatrix} = \begin{bmatrix} 0 \\ I \end{bmatrix} \tag{2.4}$$

$$\begin{bmatrix} Y_k & Z_k \end{bmatrix} = \begin{bmatrix} Y_{k-1} & Z_{k-1} \end{bmatrix} \begin{bmatrix} C_2 & D_2 \\ -A_2 & -B_2 \end{bmatrix} \quad k=1,\dots,s; \quad \begin{bmatrix} Y_o & Z_o \end{bmatrix} = \begin{bmatrix} 0 & I \end{bmatrix} \tag{2.5}$$

Proof: If X is a solution of (2.1), then X satisfies the system

$$\left.\begin{aligned} A_1 + E_1 X + X C_1 = 0 \\ A_2 + E_2 X + X C_2 = 0 \end{aligned}\right\} \tag{2.6}$$

where $E_1=B_1+XD_1$, $E_2=B_2+XD_2$. From the second equation of (2.6) it follows that

$$\begin{aligned} A_2 + E_2 X + X C_2 &= 0 \\ E_1 A_2 + E_1 E_2 X + E_1 X C_2 &= 0 \\ &\cdots \\ E_1^r A_2 + E_1^r E_2 X + E_1^r X C_2 &= 0 \end{aligned} \tag{2.7}$$

From the first equation of (2.6) it follows

$$E_1X = -A_1 - XC_1 = U_1 + XV_1, \quad U_1 = -A_1, \; V_1 = -C_1$$

and by premultiplying E_1 successively

$$E_1^jX = U_j + XV_j, \qquad j=o,1,\ldots,r \tag{2.8}$$

the coefficient matrices U_j, V_j, can be recurrently calculated. Therefore,

$$U_j + XV_j = E_1^j X = E_1E_1^{j-1}X = E_1(U_{j-1}+XV_{j-1}) =$$

$$= E_1U_{j-1} + E_1XV_{j-1} = (B_1 + XD_1)U_{j-1} + (-A_1 - XC_1)V_{j-1} =$$

$$= (B_1U_{j-1} - A_1V_{j-1}) + X(D_1U_{j-1} - C_1V_{j-1})$$

thus U_j, V_j satisfy the recurrent equations (2.4).

By multiplying by a_j the j-th equation of (2.7), and taking into account substitutions (2.8), one gets

$$a_jE_1^jA_2 + a_jE_1^jE_2X + a_j(U_j + XV_j)C_2 = 0 \;, \; j=o,1,\ldots,r$$

and by adding these equations

$$p(E_1)A_2 + p(E_1)E_2X + \left(\sum_{j=o}^{r} a_jU_j\right)C_2 + X\left(\sum_{j=o}^{r} a_jV_j\right)C_2 = 0$$

whereby $p(E_1)=0$, we conclude that X is a solution of the equation A+XB=0, where matrices A and B are given by (2.3).

On the other hand, if X is a solution of (2.1), then X also satisfies the system

$$A_1 + B_1X + XF_1 = 0$$

$$A_2 + B_2X + XF_2 = 0$$

where $F_1=C_1+D_1X$, $F_2=C_2+D_2X$. In a similar way it is proved that if q(z) is an annihilating polynomial of F_2, then X is a solution of C+DX=0, where C and D are given by (2.3).

The following result is a reciprocal one of theorem 1.

<u>Theorem 2.</u>

Let $A_i \in \mathbb{C}_{mxn}$, $B_i \in \mathbb{C}_{mxn}$, $C_i \in \mathbb{C}_{nxn}$, $D_i \in \mathbb{C}_{nxm}$, i=1,2. If the polynomials $p(z)=\sum_{j=o}^{r} a_jz^j$, $q(z)=\sum_{k=o}^{s} b_kz^k$, are such that the matrices

$$p\left(\begin{bmatrix} B_1 & -A_1 \\ D_1 & -C_1 \end{bmatrix}\right) = \begin{bmatrix} M & A \\ N & B \end{bmatrix} \;, \quad q\left(\begin{bmatrix} C_2 & D_2 \\ -A_2 & -B_2 \end{bmatrix}\right) = \begin{bmatrix} P & Q \\ C & D \end{bmatrix} \tag{2.9}$$

satisfy the conditions: (i) B invertible, $M=AB^{-1}N$, (ii) D invertible, $P=QD^{-1}C$, (iii) $DA=CB$, then it follows that the only solution of the linear system

$$\left.\begin{aligned} A + X B &= 0 \\ C + D X &= 0 \end{aligned}\right\} \tag{2.10}$$

is a solution of the Riccati system (2.1).

Proof: Under the invertibility of matrices B and D, and (iii), it follows that (2.10) admits only one solution, $X=-AB^{-1}= -D^{-1}C$. We are going to prove that said solution satisfies the Riccati system (2.1). Postmultiplying the first equation of (2.10) by $B^{-1}N$, and by applying the hypothesis $M=AB^{-1}N$, one gets

$$M + X N = 0 \tag{2.11}$$

By the conmutativity of matrices $\begin{bmatrix} B_1 & -A_1 \\ D_1 & -C_1 \end{bmatrix}$, $p\left(\begin{bmatrix} B_1 & -A_1 \\ D_1 & -C_1 \end{bmatrix}\right)$, it follows

$$B_1A - A_1B = -M A_1 - A C_1 \quad , \quad D_1A - C_1B = -N A_1 - B C_1$$

thus

$$A C_1 = -M A_1 - B_1A + A_1B \quad , \quad B C_1 = -N A_1 - D_1A + C_1B \tag{2.12}$$

If we take into account that the only solution of the equation $A+XB=0$, satisfies equation (2.11), it follows that

$$A = - X B \quad , \quad M = - X N \tag{2.13}$$

Postmultiplying $A+XB=0$ by C_1 and sustituting the expressions (2.12) and (2.13), one gets

$$A C_1 + X B C_1 = 0$$

$$(-M A_1 - B_1A + A_1B) + X(-N A_1 - D_1A + C_1B) = 0$$

$$X N A_1 + B_1 X B + A_1 B - X N A_1 + X D_1 X B + X C_1 B = 0$$

$$(A_1 + B_1X + X C_1 + X D_1X) B = 0$$

If we postmultiply the last equation by B^{-1}, then $A_1+B_1X + X C_1 + X D_1X = 0$.

On the other hand, premultiplying the second equation of (2.10) by QD^{-1}, and by applying the hypothesis $P=QD^{-1}C$, we obtain $P+QX=0$. Because the matrices

$$\begin{bmatrix} C_2 & D_2 \\ -A_2 & -B_2 \end{bmatrix} \quad , \quad q\left(\begin{bmatrix} C_2 & D_2 \\ -A_2 & -B_2 \end{bmatrix}\right)$$

are conmutative, we can similarly prove that $A_2 + B_2X + X C_2 + X D_2X = 0$.

Remark.

Matrices A,B,C and D of theorem 2 can be obtained by means of expressions which are similar to (2.3)-(2.5) of theorem 1. The following result proves that the class of polynomials which verify the hypothesis of theorem 2 is contained in the class of the polynomials which satisfies the hypothesis of theorem 1.

Proposition 1.

If $p(z)=\sum_{o\leq j\leq r} a_j z^j$, $q(z)=\sum_{o\leq k\leq s} b_k z^k$, are polynomials that verify the hypothesis of theorem 2, then $p(z)$, $q(z)$ are annihilating polynomials of matrices B_1+XD_1, C_2+D_2X, respectively, with X being a solution of the Riccati system (2.1).

Proof: From theorem 2 we see that the solutions of linear system (2.10) satisfies the Riccati system (2.1). This solution given by $X=-AB^{-1}$, satisfies

$$\begin{bmatrix} I & X \\ 0 & I \end{bmatrix} \begin{bmatrix} B_1 & -A_1 \\ D_1 & -C_1 \end{bmatrix} \begin{bmatrix} I & X \\ 0 & I \end{bmatrix}^{-1} = \begin{bmatrix} B_1+XD_1 & 0 \\ D_1 & -C_1-D_1X \end{bmatrix} \tag{2.14}$$

By applying $p(z)$ to two members of (2.14) one gets

$$\begin{bmatrix} I & X \\ 0 & I \end{bmatrix} \begin{bmatrix} M & A \\ N & B \end{bmatrix} \begin{bmatrix} I & X \\ 0 & I \end{bmatrix}^{-1} = \begin{bmatrix} p(B_1+XD_1) & 0 \\ * & p(-C_1-D_1X) \end{bmatrix} \tag{2.15}$$

Because X also satisfies $M+XN=0$, as we have already seen in the proof of theorem 2 it therefore follows that from (2.15)

$$\begin{bmatrix} 0 & 0 \\ * & * \end{bmatrix} = \begin{bmatrix} p(B_1+XD_1) & 0 \\ * & p(-C_1-D_1X) \end{bmatrix}$$

Thus we conclude that $p(B_1+XD_1)=0$. In a similar way it is proved that $q(C_2+D_2X)=$ $=0$, from the equality

$$\begin{bmatrix} I & 0 \\ X & I \end{bmatrix}^{-1} \begin{bmatrix} C_2 & D_2 \\ -A_2 & -B_2 \end{bmatrix} \begin{bmatrix} I & 0 \\ X & I \end{bmatrix} = \begin{bmatrix} C_2+D_2X & D_2 \\ 0 & -B_2-XD_2 \end{bmatrix}$$

When we consider the case $D_1=D_2=0$, we get

Corollary 1.

Let

$$\left.\begin{aligned} A_1+B_1X+XC_1&=0 \\ A_2+B_2X+XC_2&=0 \end{aligned}\right\} \tag{2.16}$$

be a Lyapunov system, where $A_i\in\mathbb{C}_{mxn}$, $B_i\in\mathbb{C}_{mxm}$, $C_i\in\mathbb{C}_{nxn}$, $i=1,2$. If X is a solution of (2.16) and

$$p(z)=\sum_{j=o}^{r} a_j z^j \quad , \quad q(z)=\sum_{k=o}^{s} b_k z^k$$

are annihilating polynomials of B_1 and C_2, respectively, then it is verified that

X is a solution of the linear system

$$\left.\begin{aligned} A + X B &= 0 \\ C + D X &= 0 \end{aligned}\right\} \tag{2.17}$$

where

$$A=\sum_{j=1}^{r}\sum_{h=1}^{j}(-1)^h a_j B_1^{j-h} A_1 C_1^{h-1} C_2\ , \qquad B= p(-C_1)C_2$$

$$C=\sum_{k=1}^{s}\sum_{f=1}^{k}(-1)^f b_k\ B_1 B_2^{f-1} A_2 C_2^{k-f}\ , \qquad D= B_1 q(-B_2) \tag{2.18}$$

Proof: Let X be a solution of (2.16). From theorem 1, X satisfies the linear system A+XB=0, C+DX=0, where the coefficient matrices are given by (2.3)-(2.5). In this case, expression (2.4) takes the form

$$\begin{bmatrix} U_j \\ V_j \end{bmatrix} = \begin{bmatrix} B_1 & -A_1 \\ 0 & -C_1 \end{bmatrix} \begin{bmatrix} U_{j-1} \\ V_{j-1} \end{bmatrix} \quad 1\le j\le r, \qquad \begin{bmatrix} U_o \\ V_o \end{bmatrix} = \begin{bmatrix} 0 \\ 0 \end{bmatrix}$$

and from where it follows that

$$\begin{bmatrix} U_j \\ V_j \end{bmatrix} = \begin{bmatrix} B_1 & -A_1 \\ 0 & -C_1 \end{bmatrix}^j \begin{bmatrix} 0 \\ I \end{bmatrix} \quad j=o,1,\dots,r \tag{2.19}$$

where

$$\begin{bmatrix} B_1 & -A_1 \\ 0 & -C_1 \end{bmatrix}^j = \begin{bmatrix} B_1^j & \sum_{h=1}^{j}(-1)^h B_1^{j-h} A_1 C_1^{h-1} \\ 0 & (-C_1)^j \end{bmatrix}, \quad 1\le j\le r \tag{2.20}$$

By sustituting (2.19)-(2.20) into (2.3), it follows that matrices A and B are given by (2.18). Similarly it is proved that C and D are given by (2.18).

The following result is a necessary condition for the compatibility of the Lyapunov system (2.16).

Corollary 2.

Under the hypothesis of corollary 1 it is verified that if (2.16) is compatible then DA=CB, with A, B, C and D given by (2.18).

Proof: It is deduced inmediatly from corollary 1.

Corollary 3.

Let $A_i \in \mathbb{C}_{mxn}$, $B_i \in \mathbb{C}_{mxm}$, $C_i \in \mathbb{C}_{nxn}$, i=1,2, and $p(z)=\sum_{j=o}^{r} a_j z^j$, $q(z)=\sum_{k=o}^{s} b_k z^k$, annihilating polynomials of the matrices B_1, C_2, respectively. If the matrices

$$p\left(\begin{bmatrix} B_1 & -A_1 \\ 0 & -C_1 \end{bmatrix}\right)= \begin{bmatrix} 0 & A \\ 0 & B \end{bmatrix}, \quad q\left(\begin{bmatrix} C_2 & 0 \\ -A_2 & -B_2 \end{bmatrix}\right)= \begin{bmatrix} 0 & 0 \\ C & D \end{bmatrix}$$

are such that: (i) B and D are invertibles, (ii) DA=CB, then it is verified that the only solution of

$$\left.\begin{array}{l} A + X B = 0 \\ C + D X = 0 \end{array}\right\}$$

is a solution of (2.16).

Proof: It is easily deduced from theorem 2.

Remark.

Matrices A, B, C and D of corollary 3 can be obtained by means of expressions which are similar to (2.18).

ACKNOWLEDGEMENTS

This work has been supported by the Acción Integrada Hispano-Portuguesa,1985, nº 19/2 and by a Grant from the Consellería de Cultura Educació i Ciencia de la Generalitat Valenciana,11-12-84.

REFERENCES

1. J.L. Casti,"Dynamical systems and their applications:Linear Theory",Academic Press, New York, 1977.
2. S. Barnett,"Matrices in Control Theory", Van Nostrand Reinhold Company,London, 1971.
3. A. Jameson,"Solution of the equation AX+XB=C by inversion of an MxM or NxN matrix", SIAM J. Appl. Math.,16(1968),1020-1023.
4. J. Jones, Jr.,"Solution of certain matrix equation", Proc. Amer. Math. Soc., 31(1972),333-339.

Frequency Domain and State Space Methods for Linear Systems
C.I. Byrnes and A. Lindquist (Editors)

RICCATI EQUATIONS, LINEAR FLOWS ON THE FLAG MANIFOLD, AND THE BRUHAT DECOMPOSITION*

Mark A. Shayman
Department of Systems Science and Mathematics
Washington University
St. Louis, Missouri 63130

Abstract: The linear flow ϕ_B on the flag manifold corresponding to the natural action of a 1-parameter subgroup e^{Bt} of the general linear group can be viewed as a generalization of the Riccati flow. We characterize the nonwandering set and the stable and unstable manifolds of ϕ_B, and their relationship to the Bruhat decomposition of the flag manifold. Although ϕ_B is generally not a Morse-Smale flow, it satisfies Morse-type equalities.

1. Introduction

The matrix Riccati equation refers to the quadratic differential equation

$$\dot{K} = B_{21} + B_{22}K - KB_{11} - KB_{12}K$$

defined on the space $\mathbb{R}^{m\times n}$ of real m x n matrices. In this equation, B_{21}, B_{22}, B_{11}, and B_{12} are constant real matrices of dimensions m x n, m x m, n x n, and n x m respectively. The Riccati equation is crucial to a diversity of applications including among others transmission line phenomena, optimal control and filtering, diffusion problems, and invariant imbedding [1].

Let $G^n(\mathbb{R}^{n+m})$ denote the Grassmann manifold of n-dimensional subspaces of $\mathbb{R}^{n+m}$. Define a mapping ψ: $\mathbb{R}^{m\times n} \to G^n(\mathbb{R}^{n+m})$ by $\psi(K) = \mathrm{Sp}\begin{bmatrix} I \\ K \end{bmatrix}$, the column space of the (n+m) x n matrix $\begin{bmatrix} I \\ K \end{bmatrix}$. ψ imbeds $\mathbb{R}^{m\times n}$ as an open and dense subset of $G^n(\mathbb{R}^{n+m})$. Consequently, we may view the Grassmann manifold as a compactification of the space of m x n matrices.

Define a flow θ_B on $G^n(\mathbb{R}^{n+m})$ by $\theta_B(t, S_0) = e^{Bt}(S_0)$, where B denotes the (n + m) x (n + m) matrix $\begin{bmatrix} B_{11} & B_{12} \\ B_{21} & B_{22} \end{bmatrix}$. The flow θ_B on $G^n(\mathbb{R}^{n+m})$ is related to the flow $K_B(t, K_0)$ of the Riccati equation on $\mathbb{R}^{m+n}$ by the equation

$$\psi(K_B(t, K_0)) = \theta_B(t, \psi(K_0))$$

*Research partially supported by the National Science Foundation under Grants ECS-8301015 and ECS-8451486.

which is valid as long as the solution $K_B(t, K_0)$ continues to exist. (See e.g. [2], [3], [4].) Consequently, the flow θ_B may be regarded as the extension of the Riccati flow from $\mathbb{R}^{m\times n}$ to its compactification $G^n(\mathbb{R}^{n+m})$.

A natural generalization of the Grassmann manifold is the so-called (partial) flag manifold. Let $k_1, \ldots, k_d$ and n be nonnegative integers satisfying $k_1 \leq k_2 \leq \ldots \leq k_d \leq n$, and let $k = (k_1, \ldots, k_d)$. Let $G^{k_i}(\mathbb{R}^n)$ denote the Grassmann manifold of k_i-dimensional subspaces of $\mathbb{R}^n$, and let Flag(k, $\mathbb{R}^n$) denote the flag manifold defined by

$$\text{Flag}(k, \mathbb{R}^n) = \{(S_1, \ldots, S_d) \in \prod_{i=1}^{d} G^{k_i}(\mathbb{R}^n):\ S_1 \subset \cdots \subset S_d\}.$$

Flag(k, $\mathbb{R}^n$) is a nonsingular subvariety of the indicated product of Grassmannians.

We now let B denote an n x n real matrix. B determines a flow ϕ_B on Flag(k, $\mathbb{R}^n$) given by

$$\phi_B(t, (S_1, \ldots, S_d)) = (e^{Bt}(S_1), \ldots, e^{Bt}(S_d)),$$

where $e^{Bt}(S_i)$ is the image of the k_i-dimensional subspace S_i under the nonsingular linear transformation e^{Bt}. We refer to ϕ_B as the linear flow on Flag(k, $\mathbb{R}^n$) induced by B.

In the special case where d = 1, Flag(k, $\mathbb{R}^n$) is a Grassmann manifold. The flow ϕ_B then coincides with the extended Riccati flow θ_B. Thus, the flow ϕ_B on the flag manifold can be regarded as a generalization of the (extended) Riccati flow.

A second motivation for studying ϕ_B comes from numerical linear algebra. Recently, it has been shown [5] [6] that the QR-algorithm can be interpreted as a discrete dynamical system on Flag(k, $\mathbb{R}^n$) defined by a diffeomorphism

$$(S_1, \ldots, S_d) \to (C(S_1), \ldots, C(S_d)),$$

where C is a nonsingular n x n matrix. Some properties of this discrete system are discussed in these references. The flow ϕ_B is clearly a continuous analogue of this discrete system.

In the sections which follow, we study the properties of the flow ϕ_B. We characterize the nonwandering set and the stable and unstable manifolds under the assumption that B is semisimple (i.e. diagonalizable). We also describe the relationship of the phase portrait of ϕ_B to so-called Bruhat decompositions of the flag manifold. This relationship can be exploited to show that ϕ_B satisfies

Morse-type equalities even though it is generally not a Morse-Smale flow. Each of the results has an obvious analogue for the discrete dynamical system referred to above.

2. Nonwandering Set

For the remainder of this paper we make the simplifying assumption that B is semisimple. Let $\mu_1 < \mu_2 < \dots < \mu_r$ denote the distinct real parts of the eigenvalues of B. Let E_j denote the direct sum of the primary components of B corresponding to all those eigenvalues of B with real part μ_j, and let n_j denote the dimension of E_j.

We associate with B two flags (i.e. increasing sequences of subspaces). By the stable flag determined by B, we mean the sequence $0 = M_0 \subset M_1 \subset \dots \subset M_r = \mathbb{R}^n$ where $M_j = \bigoplus_{p=1}^{j} E_p$. By the unstable flag determined by B, we mean the sequence $0 = N_0 \subset N_1 \subset \dots \subset N_r = \mathbb{R}^n$ where $N_j = \bigoplus_{p=1}^{j} E_{r-p+1}$. As we will see, the stable (unstable) flag plays a crucial role in describing the asymptotic behavior of solutions in forward (backward) time.

It is useful to note that B also induces a flow on $\prod_{i=1}^{d} G^{k_i}(\mathbb{R}^n)$, namely

$$\hat{\phi}_B(t, (S_1, \dots, S_d)) = (e^{Bt}(S_1), \dots, e^{Bt}(S_d)).$$

The submanifold Flag(k, $\mathbb{R}^n$) is invariant under the flow $\hat{\phi}_B$, and ϕ_B is the restriction of $\hat{\phi}_B$ to Flag(k, $\mathbb{R}^n$). In previous work [4], we have given a rather complete description, both set-theoretically and topologically, of the phase portrait of the linear flow $e^{Bt}(S_i)$ on the Grassmann manifold $G^{k_i}(\mathbb{R}^n)$, i.e. of the extended Riccati flow. The phase portrait for the flow $\hat{\phi}_B$ is obtained immediately from these results. Since ϕ_B is the restriction to Flag(k, $\mathbb{R}^n$) of $\hat{\phi}_B$, the nonwandering set, stable manifolds and unstable manifolds for ϕ_B are the intersections with the submanifold Flag(k, $\mathbb{R}^n$) of the nonwandering set, stable and unstable manifolds for $\hat{\phi}_B$.

From the definition of the flow ϕ_B, it is clear that the equilibrium set consists of those flags which are invariant under the linear transformation B - i.e. $\{(S_1, \dots, S_d) \in \text{Flag}(k, \mathbb{R}^n): B(S_i) \subseteq S_i, i = 1, \dots, d\}$. Using the fact that the primary components of $B|S_i$ are the intersections of S_i with the primary components of B, together with the assumption that B is semisimple, it follows easily that the equilibrium set is topologically a disjoint union of products of real and complex flag manifolds.

For a flow ϕ_t on a manifold M, $x \in M$ is called a wandering point [7] [8] if there exists a neighborhood U of x in M and some $t_0 > 0$ such that $\phi_t(U) \cap U$ is empty whenever $|t| > t_0$. The subset of M consisting of all points which are not wandering points is called the nonwandering set.

Let Ω and $\hat{\Omega}$ denote the nonwandering sets of ϕ_B and $\hat{\phi}_B$ respectively, and let $\hat{\Omega}_i$ denote the nonwandering set of the flow $e^{Bt}(S_i)$ on $G^{k_i}(\mathbb{R}^n)$. Then it follows trivially that

$$(1) \qquad \Omega = \hat{\Omega} \cap \mathrm{Flag}(k, \mathbb{R}^n) = [\prod_{i=1}^{d} \hat{\Omega}_i] \cap \mathrm{Flag}(k, \mathbb{R}^n).$$

Let $\ell_i = (\ell_{i1}, \ldots, \ell_{ir})$ be an r-tuple of integers such that $0 \leq \ell_{ij} \leq n_j$ and $\ell_{i1} + \ldots + \ell_{ir} = k_i$. Let

$$T_i(\ell_i) = \{S_{i1} \oplus \ldots \oplus S_{ir}:\ S_{ij} \in G^{\ell_{ij}}(E_j),\ j = 1, \ldots, r\},$$

where $G^{\ell_{ij}}(E_j)$ is the Grassmann manifold of ℓ_{ij}-dimensional subspaces of E_j. $T_i(\ell_i)$ is a submanifold of $G^{k_i}(\mathbb{R}^n)$ homeomorphic to the product of Grassmannians $\prod_{j=1}^{r} G^{\ell_{ij}}(\mathbb{R}^{n_j})$ and is both positively and negatively invariant under the flow $e^{Bt}(S_i)$ on $G^{k_i}(\mathbb{R}^n)$. It is shown in [4] that the nonwandering set for the flow on $G^{k_i}(\mathbb{R}^n)$ is

$$(2) \qquad \hat{\Omega}_i = \coprod_{\ell_i} T_i(\ell_i).$$

Furthermore, every motion on $\hat{\Omega}_i$ is either constant, periodic, or almost periodic. Combining (1) and (2) gives

$$(3) \qquad \Omega = \coprod_{\ell_1} \ldots \coprod_{\ell_d} [(\prod_{i=1}^{d} T_i(\ell_i)) \cap \mathrm{Flag}(k, \mathbb{R}^n)].$$

It is easy to see that $(\prod_{i=1}^{d} T_i(\ell_i)) \cap \mathrm{Flag}(k, \mathbb{R}^n)$ is nonempty if and only if $\ell_{ij} \leq \ell_{i+1,j}$ for all i and j, in which case it is homeomorphic to $\prod_{j=1}^{r} \mathrm{Flag}((\ell_{1j}, \ldots, \ell_{dj}), \mathbb{R}^{n_j})$.

Let I(B) denote the set of all d x r matrices L of nonnegative integers ℓ_{ij} with the following properties:

(a) $\ell_{1j} \leq \ldots \leq \ell_{dj} \leq n_j$, $j = 1, \ldots, r$.

(b) $\ell_{i1} + \ldots + \ell_{ir} = k_i$, $i = 1, \ldots, d$.

Then the preceding analysis yields the following description of the nonwandering set of the flow ϕ_B on Flag(k, $\mathbb{R}^n$) induced by B:

Theorem 1: Ω is homeomorphic to

$$\coprod_{L\varepsilon I(B)} \prod_{j=1}^{r} \text{Flag}((\ell_{1j}, \ldots, \ell_{dj}), \mathbb{R}^{n_j}).$$

Each motion on Ω is either constant, periodic, or almost periodic.

Remark 1: Suppose that B satisfies the generic conditions that its n eigenvalues be distinct, and if λ_i and λ_j are a pair of eigenvalues with the same real part, then $\lambda_i = \bar{\lambda}_j$. (Overbar denotes complex conjugation.) In this case, n_j is either 1 or 2. Then Flag((ℓ_{1j}, ..., ℓ_{dj}), $\mathbb{R}^{n_j}$) is a point unless $n_j = 2$ and at least one of the numbers ℓ_{1j}, ..., ℓ_{dj} is equal to 1. In the latter case, Flag((ℓ_{1j}, ..., ℓ_{dj}), $\mathbb{R}^{n_j}$) is homeomorphic to the projective line $G^1(\mathbb{R}^2)$ which is topologically a circle. It follows from Theorem 1 that the nonwandering set Ω is a union of tori of various dimensions.

Remark 2: Suppose that B satisfies the non-generic condition that its n eigenvalues have distinct real parts. In particular, this implies that every eigenvalue is real. Then $n_j = 1$ for all j, and Flag((ℓ_{1j}, ..., ℓ_{dj}), $\mathbb{R}^{n_j}$) is a point. Consequently, Ω is finite, and every element of Ω is an equilibrium point.

3. Stable and Unstable Manifolds

Let L ε I(B), and let T(L) denote the connected component of Ω associated with L. By Theorem 1, T(L) is topologically a product of flag manifolds. Let $W^s(T(L))$ and $W^u(T(L))$ be the stable manifold and unstable manifold respectively of T(L). In other words

$$W^s(T(L)) = \{(S_1, \ldots, S_d) \in \text{Flag}(k, \mathbb{R}^n): \phi_B(t, (S_1, \ldots, S_d)) \to T(L) \text{ as } t \to \infty\}$$

$$W^u(T(L)) = \{(S_1, \ldots, S_d) \in Flag(k, \mathbb{R}^n): \phi_B(t, (S_1, \ldots, S_d)) \to T(L) \text{ as } t \to -\infty\}.$$

Let $W^s(T_i(\ell_i))$ and $W^u(T_i(\ell_i))$ be the stable manifold and unstable manifold of $T_i(\ell_i)$ for the flow on $G^{k_i}(\mathbb{R}^n)$. Then

$$W^s(T(L)) = [\prod_{i=1}^{d} W^s(T_i(\ell_i))] \cap Flag(k, \mathbb{R}^n) \tag{4}$$

$$W^u(T(L)) = [\prod_{i=1}^{d} W^u(T_i(\ell_i))] \cap Flag(k, \mathbb{R}^n). \tag{5}$$

It is shown in [4] that

$$W^s(T_i(\ell_i)) = \{S_i \in G^{k_i}(\mathbb{R}^n): \dim S_i \cap M_j = \sum_{p=1}^{j} \ell_{ip}, \ j = 1, \ldots, r\} \tag{6}$$

$$W^u(T_i(\ell_i)) = \{S_i \in G^{k_i}(\mathbb{R}^n): \dim S_i \cap N_j = \sum_{p=1}^{j} \ell_{i,r-p+1}, \ j = 1, \ldots, r\}. \tag{7}$$

From (4)-(7), we obtain

<u>Theorem 2</u>:

(a) $W^s(T(L)) = \{(S_1, \ldots, S_d) \in Flag(k, \mathbb{R}^n):$

$$\dim S_i \cap M_j = \sum_{p=1}^{j} \ell_{ip}, \ \forall i, j\}.$$

(b) $W^u(T(L)) = \{(S_1, \ldots, S_d) \in Flag(k, \mathbb{R}^n):$

$$\dim S_i \cap N_j = \sum_{p=1}^{j} \ell_{i,r-p+1}, \ \forall i, j\}.$$

Theorem 2 shows that an arbitrary motion $\phi_B(t, (S_1, \ldots, S_d))$ converges to an invariant manifold $T(L)$ as $t \to \infty$ and an invariant manifold $T(L')$ as $t \to -\infty$. Furthermore, L and L' are determined by the dimensions in which $S_1, \ldots, S_d$ intersect the stable flag $\{M_j\}$ and the unstable flag $\{N_j\}$ respectively.

Next, we will determine exactly which motions on T(L) and T(L') $\phi_B(t, (S_1, \ldots, S_d))$ approaches as $t \to \infty$ and $t \to -\infty$ respectively. Let η_j denote the projection onto the subspace E_j along the subspace $\bigoplus_{\substack{p=1 \\ p\neq j}}^{r} E_p$. Define mappings

$\Pi_i^+ : G^{k_i}(\mathbb{R}^n) \to G^{k_i}(\mathbb{R}^n)$ and $\Pi_i^- : G^{k_i}(\mathbb{R}^n) \to G^{k_i}(\mathbb{R}^n)$ by

$$\Pi_i^+(S_i) = \bigoplus_{j=1}^{r} \eta_j(S_i \cap M_j)$$

$$\Pi_i^-(S_i) = \bigoplus_{j=1}^{r} \eta_{r-j+1}(S_i \cap N_j).$$

It is shown in [4] that Π_i^+ and Π_i^- map $G^{k_i}(\mathbb{R}^n)$ onto the nonwandering set $\hat{\Omega}_i$.

Let $(\bar{S}_1, \ldots, \bar{S}_d) \in T(L)$. Let

$$W^s(\bar{S}_1, \ldots, \bar{S}_d) = \{(S_1, \ldots, S_d) \in W^s(T(L)): \phi_B(t, (S_1, \ldots, S_d)) \to \phi_B(t, (\bar{S}_1, \ldots, \bar{S}_d)) \text{ as } t \to \infty\}$$

$$W^u(\bar{S}_1, \ldots, \bar{S}_d) = \{(S_1, \ldots, S_d) \in W^u(T(L)): \phi_B(t, (S_1, \ldots, S_d)) \to \phi_B(t, (\bar{S}_1, \ldots, \bar{S}_d)) \text{ as } t \to -\infty\}.$$

Similarly, for $\bar{S}_i \in T_i(\ell_i)$, define

$$W^s(\bar{S}_i) = \{S_i \in W^s(T_i(\ell_i)): e^{Bt}(S_i) \to e^{Bt}(\bar{S}_i) \text{ as } t \to \infty\}$$

$$W^u(\bar{S}_i) = \{S_i \in W^u(T_i(\ell_i)): e^{Bt}(S_i) \to e^{Bt}(\bar{S}_i) \text{ as } t \to -\infty\}.$$

It is shown in [4] that $W^s(\bar{S}_i) = (\Pi_i^+)^{-1}(\bar{S}_i)$ and $W^u(\bar{S}_i) = (\Pi_i^-)^{-1}(\bar{S}_i)$. The next result is an immediate consequence.

<u>Theorem 3</u>: Let $(\bar{S}_1, \ldots, \bar{S}_d) \in T(L)$.

(a) $W^s(\bar{S}_1, \ldots, \bar{S}_d) = \{(S_1, \ldots, S_d) \in W^s(T(L)): \Pi_i^+(S_i) = \bar{S}_i, \forall i\}$.

(b) $W^u(\bar{S}_1, \ldots, \bar{S}_d) = \{(S_1, \ldots, S_d) \in W^u(T(L)): \Pi_i^-(S_i) = \bar{S}_i, \forall i\}$.

In other words, given $(S_1, \ldots, S_d) \in \mathrm{Flag}(k, \mathbb{R}^n)$, the motion $\phi_B(t, (S_1, \ldots, S_d))$ approaches the motion $\phi_B(t, (\Pi_1^+(S_1), \ldots, \Pi_d^+(S_d)))$ as $t \to \infty$ and approaches the motion $\phi_B(t, (\Pi_1^-(S_1), \ldots, \Pi_d^-(S_d)))$ as $t \to -\infty$. From Theorem 1, each limiting motion is either constant, periodic, or almost periodic.

4. Bruhat Decomposition

In this section, we show how the phase portrait of the flow ϕ_B is related to the Bruhat cell decomposition of $\mathrm{Flag}(k, \mathbb{R}^n)$.

We begin by describing the Schubert cell decomposition of the Grassmann manifold $G^{k_i}(\mathbb{R}^n)$. It is constructed by first choosing a flag of subspaces $0 = V_0 \subset V_1 \subset \ldots \subset V_n = \mathbb{R}^n$ such that $\dim V_j = j$. To emphasize that this flag contains a subspace for each dimension between 0 and n, we refer to it as a complete flag.

(Except in the special case where every eigenvalue of B is real and distinct - i.e. where $\dim E_j = 1$ for all j, the stable flag $\{M_j\}$ and unstable flag $\{N_j\}$ are not complete flags.) For each n-tuple $a_i = (a_{i1}, \ldots, a_{in})$ such that $a_{ip} = 0$ or $a_{ip} = 1$ and $\sum_{p=1}^{n} a_{ip} = k_i$, let

$$U_i(a_i) = \{S_i \in G^{k_i}(\mathbb{R}^n): \dim S_i \cap V_j = \sum_{p=1}^{j} a_{ip}, \forall j\}.$$

The $\binom{n}{k_i}$ sets $\{U_i(a_i)\}$ partition $G^{k_i}(\mathbb{R}^n)$, and each $U_i(a_i)$ is homeomorphic to a Euclidean space. This partition is the Schubert cell decomposition of $G^{k_i}(\mathbb{R}^n)$ induced by the complete flag $\{V_j\}$. (For more details see e.g. [9].)

Since $\{V_j\}$ induces a decomposition of $G^{k_i}(\mathbb{R}^n)$ for each $i = 1, \ldots, d$, we have the decomposition

$$\prod_{i=1}^{d} G^{k_i}(\mathbb{R}^n) = \coprod_{a_1} \ldots \coprod_{a_d} \prod_{i=1}^{d} U_i(a_i).$$

Consequently,

$$\mathrm{Flag}(k, \mathbb{R}^n) = \coprod_{a_1} \ldots \coprod_{a_d} [\mathrm{Flag}(k, \mathbb{R}^n) \cap \prod_{i=1}^{d} U_i(a_i)]. \tag{8}$$

It is not hard to show that $\mathrm{Flag}(k, \mathbb{R}^n) \cap \prod_{i=1}^{d} U_i(a_i)$ is nonempty if and only if $a_{1p} \leq a_{2p} \leq \ldots \leq a_{dp}$ for all p. Let $J(k, n)$ denote the set of all $d \times n$ matrices A such that

(a) $a_{ip} = 0$ or $a_{ip} = 1$

(b) $\sum_{p=1}^{n} a_{ip} = k_i$

(c) $a_{1p} \leqq a_{2p} \leqq \dots \leqq a_{dp}$.

For each $A \in J(k, n)$, let

$$U(A) = \{(S_1, \dots, S_d) \in \text{Flag}(k, \mathbb{R}^n): \dim S_i \cap V_j = \sum_{p=1}^{j} a_{ip}, \forall i, j\}.$$

Then (8) can be rewritten as

$$\text{Flag}(k, \mathbb{R}^n) = \coprod_{A \in J(k,n)} U(A). \tag{9}$$

This is a partition of $\text{Flag}(k, \mathbb{R}^n)$ into $\binom{n}{k_1}\binom{n-k_1}{k_2-k_1}\dots\binom{n-k_{d-1}}{k_d-k_{d-1}}$ subsets, each of which is homeomorphic to a Euclidean space. The decomposition (9) is called the Bruhat cell decomposition of $\text{Flag}(k, \mathbb{R}^n)$ induced by the complete flag $\{V_j\}$. Each subset $U(A)$ is called a Bruhat cell. (For more details, see e.g. [10].)

Remark 3: If the eigenvalues of B are all real and distinct, then $\{M_j\}$ and $\{N_j\}$ are complete flags. By Theorem 2, the partition of $\text{Flag}(k, \mathbb{R}^n)$ into stable manifolds is precisely the Bruhat decomposition of $\text{Flag}(k, \mathbb{R}^n)$ induced by $\{M_j\}$. Similarly, the partition into unstable manifolds coincides with the Bruhat decomposition induced by $\{N_j\}$.

In the general case where $\{M_j\}$ and $\{N_j\}$ are not complete flags, we proceed by refining them so as to obtain complete flags. For each j, we arbitrarily choose a direct sum decomposition

$$E_j = E_{j1} \oplus \dots \oplus E_{jn_j}$$

into 1-dimensional subspaces. Then define complete flags $\{M_{jp}: p = 1, \dots, n_j; j = 1, \dots, r\}$ and $\{N_{jp}: p = 1, \dots, n_{r-j+1}; j = 1, \dots, r\}$ by

$$M_{jp} = M_{j-1} \oplus E_{j1} \oplus \dots \oplus E_{jp}$$

$$N_{jp} = N_{j-1} \oplus E_{r-j+1,n_{r-j+1}} \oplus \dots \oplus E_{r-j+1,n_{r-j+1}-p+1}.$$

In particular, $M_{jn_j} = M_j$ and $N_{jn_{r-j+1}} = N_j$.

The complete flags $\{M_{jp}\}$ and $\{N_{jp}\}$ each induce Bruhat decompositions of Flag(k, $\mathbb{R}^n$), say

$$\text{Flag}(k, \mathbb{R}^n) = \coprod_{A \varepsilon J(k,n)} U_M(A) \tag{10}$$

$$\text{Flag}(k, \mathbb{R}^n) = \coprod_{A \varepsilon J(k,n)} U_N(A). \tag{11}$$

It follows from Theorem 2 and the definitions of $\{M_{jp}\}$ and $\{N_{jp}\}$ that the Bruhat decomposition (10) decomposes each stable manifold into a union of Bruhat cells $U_M(A)$, while the Bruhat decomposition (11) decomposes each unstable manifold into a union of Bruhat cells $U_N(A)$. Let $m_j = \dim M_j = n_1 + \ldots + n_j$. Then we have

Theorem 4: Let $L \varepsilon I(B)$.

(a) $W^s(T(L))$ is the union of those Bruhat cells $U_M(A)$ for which

$$\sum_{p=m_{j-1}+1}^{m_j} a_{ip} = \ell_{ij}, \ \forall i, j.$$

(b) $W^u(T(L))$ is the union of those Bruhat cells $U_N(A)$ for which

$$\sum_{p=n-m_j+1}^{n-m_{j-1}} a_{ip} = \ell_{ij}, \ \forall i, j.$$

5. Morse Equalities

It is apparent from Theorem 1 that ϕ_B is generally not a Morse-Smale flow. However, we will see that the phase portrait of ϕ_B is related to the topology of the flag manifold by a generalized form of the Morse inequalities for a Morse-Smale dynamical system. In fact, for ϕ_B, the inequalities are actually satisfied as equalities. The key tool for establishing this result is the Bruhat decomposition, which is related on one hand to the homology of Flag(k, $\mathbb{R}^n$) and on the other hand, via Theorem 4, to the phase portrait of ϕ_B.

Let $b_q(N, K)$ denote the q-th Betti number of a compact manifold N for the coefficient field K. Let $P_K(N; t)$ denote the Poincaré polynomial of N for the coefficient field K. In other words,

$$P_K(N; t) = \sum_{q=0}^{\nu} b_q(N, K)t^q,$$

where ν is the dimension of N.

By analogy with the case of the gradient flow associated with a Morse-Bott function, we define the <u>Morse series</u> for the flow ϕ_B on Flag(k, $\mathbb{R}^n$) as follows: For each $L \in I(B)$, define the <u>index</u> of T(L) to be

$$(12) \qquad i(T(L)) = \dim W^s(T(L)) - \dim T(L).$$

Then the Morse series for ϕ_B is defined to be the polynomial

$$(13) \qquad M_B(t) = \sum_{L \in I(B)} P_{Z_2}(T(L); t)\, t^{i(T(L))}.$$

Using Theorem 1, (13) can be rewritten as

$$(14) \quad M_B(t) = \sum_{L \in I(B)} t^{i(T(L))} \prod_{j=1}^{r} P_{Z_2}(\mathrm{Flag}((\ell_{1j}, \ldots, \ell_{dj}), \mathbb{R}^{n_j}); t).$$

We have the following result:

<u>Theorem 5</u>: $M_B(t) = P_{Z_2}(\mathrm{Flag}(k, \mathbb{R}^n); t)$.

Theorem 5 gives Morse-type <u>equalities</u> for the flow ϕ_B since it asserts that the coefficient of t^q in the Morse series for ϕ_B is equal to the q-th Betti number of the underlying manifold Flag(k, $\mathbb{R}^n$) for the coefficient field Z_2. Note that the Morse series $M_B(t)$ is defined in terms of data from the phase portrait of ϕ_B - namely the dimensions of the stable manifolds together with the dimensions and topological structure of the connected components {T(L)} of the nonwandering set of ϕ_B. On the other hand, the Betti numbers of Flag(k, $\mathbb{R}^n$) are topological invariants which are completely independent of B. Thus, Theorem 5 may be regarded as describing constraints on the flow ϕ_B imposed by the topological structure of the underlying manifold Flag(k, $\mathbb{R}^n$).

The proof of Theorem 5 is based on the characterization of the stable manifolds as unions of Bruhat cells given by Theorem 4 and on the well-known fact that the number of q-dimensional cells in a Bruhat decomposition of Flag(k, $\mathbb{R}^n$) is equal to the q-th mod 2 Betti number, $b_q(\mathrm{Flag}(k, \mathbb{R}^n), Z_2)$. The details of this proof will be published elsewhere [11].

<u>Remark 4</u>: Theorem 5 remains true if the definition of the index of T(L) given in (12) is replaced by

$$i(T(L)) = \dim W^u(T(L)) - \dim T(L).$$

References

1. W. T. Reid, Riccati Differential Equations, Academic Press, New York, 1972.

2. R. Herman, Cartanian Geometry, Nonlinear Waves, and Control Theory, Part A, Interdisciplinary Mathematics, Vol. XX, Math. Sci. Press, Brookline, MA, 1979.

3. C. F. Martin, "Grassmannian manifolds, Riccati equations and feedback invariants of linear systems," in Geometrical Methods for the Theory of Linear Systems (ed. C. I. Byrnes and C. F. Martin), Reidel, Dordrecht, 1980.

4. M. A. Shayman, "Phase portrait of the matrix Riccati equation," SIAM J. Contr., 24 (Jan. 1986).

5. G. S. Ammar, "Riccati equations and the numerical matrix eigenvalue methods," Ph.D. thesis, Case Western Reserve University, Cleveland, Ohio, 1983.

6. G. S. Ammar and C. F. Martin, "The geometry of the matrix eigenvalue methods," preprint.

7. G. D. Birkhoff, Dynamical Systems, American Mathematical Society Colloquium Publication, Vol. 9, American Mathematical Society, Providence, RI, 1927.

8. S. Smale, "Differentiable dynamical systems," Bull. Amer. Math. Soc., 73 (1967), 747-817.

9. P. Griffiths and J. Harris, Principles of Algebraic Geometry, John Wiley, New York, 1978.

10. H. Hiller, Geometry of Coxeter Groups, Pitman, London, 1982.

11. M. A. Shayman, "Morse equalities for linear flows on the flag manifold," Report No. SSM 8501, Department of Systems Science and Mathematics, Washington University, St. Louis, MO, September, 1985.